CONVERSION OF U.S. CUSTOMARY UNITS TO SI UNITS (*Continued*)

Multiply (U.S. Customary)	By	To Obtain (SI)
$lb \cdot in. \cdot s^2$ (pound · inch · second^2)	$1.1298\ (10)^{-1}$	$kg \cdot m^2$ (kilogram · meter^2)
Spring Constant		
lb/in. (pound/inch)	$1.7513\ (10)^2$	N/m (newton/meter)
Stress, Pressure		
psi (pound/inch^2)	$6.8947\ (10)^3$	N/m^2 (newton/meter^2 or pascal)
Velocity		
ft/s (foot/second)	$3.048\ (10)^{-1}$	m/s (meter/second)
in./s (inch/second)	$2.54\ (10)^{-2}$	m/s (meter/second)
Volume		
ft^3 (foot^3)	$2.8317\ (10)^{-2}$	m^3 (meter^3)
in.^{3} (inch^3)	$1.6387\ (10)^{-5}$	m^3 (meter^3)
Work, Energy		
ft-lb (foot-pound)	1.3558	J (joule)
in.-lb (inch-pound)	$1.1298\ (10)^{-1}$	J (joule)

PREFIXES FOR SI UNITS

Prefix	Numerical Value	SI Designation
nano	10^{-9}	n
micro	10^{-6}	μ
milli	10^{-3}	m
kilo	10^3	k
mega	10^6	M
giga	10^9	G

Vibration of Mechanical and Structural Systems

VIBRATION OF MECHANICAL AND STRUCTURAL SYSTEMS: With Microcomputer Applications

M. L. James
G. M. Smith
J. C. Wolford
P. W. Whaley
University of Nebraska—Lincoln

HARPER & ROW, PUBLISHERS, New York
Cambridge, Philadelphia, San Francisco, London,
Mexico City, São Paulo, Singapore, Sydney

Sponsoring Editor: James Cook
Project Editor: Ellen MacElree
Cover Design: Wanda Lubelska Design
Text Art: RDL Artset Ltd
Production Manager: Jeanie Berke
Production Assistant: Beth Maglione
Compositor: TAPSCO, Inc.
Printer and Binder: R. R. Donnelley & Sons Company
Cover Printer: Phoenix Color Corp

Vibration of Mechanical and Structural Systems, with Microcomputer Applications

Library of Congress Cataloging-in-Publication Data

Vibration of mechanical and structural systems: with microcomputer
applications / by M. L. James . . . [et al.].
p. cm.
Includes index.
ISBN 0-06-043261-6
1. Vibration. 2. Structural dynamics. 3. Vibration—Data
processing. I. James, M. L. (Merlin L.)
TA355.V523 1989
620.3—dc19 88-4014
CIP

90 91 92 9 8 7 6 5 4 3

Contents

Preface ix

Chapter 1
Basic Concepts/Mathematics 1

1-1. Introduction 1
1-2. Sinusoidal Waves/Harmonic Motion 4
1-3. Complex Notation 8
1-4. Orthogonal Relations 10
1-5. Periodic Motion/Fourier Series 11
1-6. Least-Squares Error Fit by Truncated Fourier Series 21
1-7. Fourier Integral/Fourier Transforms 22
1-8. Basic Instrumentation for Vibration Analysis 32
1-9. Decibel/Mean-Square Value/Root-Mean-Square Value 34
Problems 37

Chapter 2
Free Vibration of Single-Degree-of-Freedom Systems 43

2-1. Introduction 43
2-2. Differential Equations of Damped Free Vibration 43
2-3. Solving Differential Equations of Motion 55
2-4. Experimental Measurement of System Damping 66
2-5. Coulomb Damping 70
2-6. Unstable Systems 72
2-7. Equivalent Springs 74
2-8. Determining Natural Frequencies by Rayleigh's Energy Method 77
2-9. Rayleigh's Energy Method Applied to Beams 89
Problems 112

Chapter 3
Harmonic Excitation of Single-Degree-of-Freedom Systems 134

3-1. Introduction 134
3-2. Excitation due to an Unbalanced Rotating Mass 136
3-3. Critical Speed of Rotating Shafts 147
3-4. Vortex Shedding (Flow-Induced Vibration) 155
3-5. Beating 160
3-6. Support Excitation 162
3-7. General Periodic Excitation/Steady-State Response 170

3-8. Vibration Isolation 177
3-9. Transducers (Force and Vibratory Motion) 186
3-10. Steady-State Vibration Measurements 194
3-11. Equivalent Viscous Damping 201
Problems 211

Chapter 4
Transient Vibration from Nonharmonic Excitation 227

4-1. Introduction 227
4-2. Nonharmonic Force Excitation 229
4-3. Nonharmonic Support Excitation 241
4-4. Rotational Transient Response 245
4-5. Response Spectrums 246
4-6. Numerical Integration for Obtaining Response Spectrums 247
Problems 268

Chapter 5
Free Vibration of Multiple-Degree-of-Freedom Systems (Discrete and Lumped-Mass Systems) 277

5-1. Introduction 277
5-2. Differential Equations of Motion 277
5-3. Influence Coefficients (Stiffness, Flexibility, Damping) 283
5-4. Coordinate Coupling (Dynamic and Static) 308
5-5. Natural Frequencies and Mode Shapes (Eigenvalues and Eigenvectors) 311
5-6. Orthogonality Properties of the Normal Modes 327
5-7. Modal Analysis—Principal Coordinates 333
5-8. Methods for Solution of Eigenvalue Problems—General 342
5-9. Power Method for Determining Eigenvalues and Eigenvectors 343
5-10. Iteration for Intermediate Eigenvalues and Eigenvectors—Hotelling's Deflation Method 349
5-11. Jacobi's Method for Determining Eigenvalues and Eigenvectors 365
5-12. Lumped-Mass Modeling of Distributed Masses (Rods and Beams) 379
Problems 390

Chapter 6
Forced Vibration of Discrete and Lumped-Mass Systems/Modal Analysis 408

6-1. Introduction 408
6-2. Equations of Motion—Forced Vibration 409
6-3. General Discussion of Determining Forced-Vibration Response 414
6-4. Vibration Absorber 415
6-5. Computer Program for the Simultaneous Solution of Coupled Differential Equations of Motion 423
6-6. Modal Analysis (Modal Superposition Method) 433
6-7. Determining Structural Response to Earthquakes by Modal Analysis 450

6-8. Comments Concerning Computer Solutions 456
Problems 464

Chapter 7
Lagrange's Equations and the Finite-Element Method in Vibration Analysis 482

7-1. Introduction 482
7-2. Lagrange's Equations 483
7-3. Mass and Stiffness Matrices for Rod and Beam Elements 491
7-4. Coordinate Transformations—Use of Transformation Matrix 507
7-5. Equations of Motion for Systems Composed of Finite Elements 512
7-6. Outline for Use of Finite-Element Method in Vibration Analysis 518
7-7. Computer Program for the Vibration Analysis of Plane Frames Using Beam Elements (FINITEL.FOR) 541
7-8. Computer Program for the Vibration Analysis of Plane Trusses (TRUSS.FOR) 562
Problems 573

Chapter 8
Continuous Systems 584

8-1. Introduction 584
8-2. One-Dimensional Wave Equation 585
8-3. Transcendental Frequency Equations (Wave Equation) 595
8-4. Surge in Helical Springs 598
8-5. Beam Equations 600
8-6. Properties of Normal-Mode Functions 609
8-7. Analyzing the Vibration of Continuous Systems by the Modal-Superposition Method 612
Problems 624

Appendix A Laplace Transforms 631

A-1 Laplace Transformation 631
A-2 Laplace Transformation of Derivatives 631

Appendix B Matrix Algebra 635

B-1 Multiplication 635
B-2 Matrix Inversion 636
B-3 Transpose of a Matrix 637
B-4 Orthogonality Principle of Symmetric Matrices 637
B-5 Orthogonality Principle of the Form $AX = \lambda BX$ 638

Index 639

Preface

The purpose of this text is to present comprehensive coverage of the fundamental principles of vibration theory, with emphasis on the application of these principles to practical engineering problems. While it is written primarily for use as a text for engineering students, practicing engineers should find it convenient for reference and self-study, as well as a source of some useful computer programs.

Although it is intended as an introductory-level text, sufficient material is included for its use in a two-quarter or two-semester sequence of courses, and the material is organized so that considerable flexibility is allowed in arranging both course level and content. For example, the basic experimental techniques presented for making vibration measurements would be desirable in a course accompanied by a laboratory but could be deleted in a course that was not.

The student studying this text should have a basic knowledge of dynamics, mechanics of materials, and differential equations, and some knowledge of matrix algebra (a discussion of essential matrix operations required is presented in Appendix B). Although not prerequisite, some background in numerical methods would give the reader added understanding of the computer programs presented.

To keep the text to a reasonable length, it was necessary to omit some important topics, and we elected to omit nonlinear vibration theory and random vibrations on the premise that these topics are generally presented at the graduate level and thus are beyond the intended scope of this text.

Following our feeling that texts written for students should present material with sufficient detail to be easily followed, we included many more details of topics in the explanations and applications of principles. The computer programs included are also presented in some detail, not just as "canned programs." The numerical methods used are discussed, and the variable and array names are carefully defined to aid the reader in using and understanding the programs.

The FORTRAN version used for the programs is compatible with IBM personal computers. In using some versions of FORTRAN it may be necessary to modify some of the statements slightly. All the programs can be run on a microcomputer having as little as 64k of random-access memory. Readers having computers with more memory who wish to use the programs to analyze systems larger than those discussed in the examples in the text will generally find that they can accomplish this by simply changing the DIMENSION statements to allow for larger arrays. Comment statements appearing on the console indicate the data necessary for input to the programs.

A good selection of problems is given at the end of each chapter, grouped together by section. Since engineers must be familiar with both the SI and English systems of units during the current transition period, which it seems may last for some time, both systems are used in the examples and problems.

Chapter 1 covers many of the basic concepts of vibration theory, some mathematical concepts pertinent to vibration analysis, and a brief introduction to the instrumentation used in experimental vibration analysis. An instructor may find it desirable to cover this chapter in its entirety at the beginning of the course or to present some of the topics at that time and postpone others until they are needed in a particular portion of the text.

Chapter 2 examines the vibration of damped and undamped single-degree-of-freedom systems. The direct application of Newton's second law and the concept of dynamic equilibrium are used to derive the differential equations of motion of these systems. Rayleigh's energy method for determining the fundamental natural frequencies of various types of systems is presented, with emphasis on its use with beams.

Chapter 3 begins with the study of single-degree-of-freedom systems subjected to a harmonic force or support excitation. This is followed by discussions concerned with the critical speed of shafts, the phenomena of beating, flow-induced vibration, vibration isolation, and transducers and vibration measurements. The latter should be particularly useful in courses accompanied by laboratory investigations.

Chapter 4 is concerned with the response of single-degree-of-freedom systems subjected to nonharmonic force or support excitations. Considerable emphasis is placed here on the numerical integration of transient-type problems by the fourth-order Runge-Kutta method to obtain data for plotting response spectrums. Such spectrums are used later in Chapter 6 to obtain the maximum response of multiple-degree-of-freedom systems subjected to nonharmonic excitations.

Chapter 5 is involved with the free vibration of multiple-degree-of-freedom systems (consideration was given to a separate chapter for two-degree-of-freedom systems, but it was decided to treat them along with other multiple-degree-of-freedom systems). A thorough discussion of influence coefficients is included, and the power method used with Hotelling's deflation technique, and Jacobi's method, are the numerical methods presented for determining eigenvalues and eigenvectors on the computer. Modal analysis is also introduced at this point, and further attention is given to the lumped-mass modeling of distributed-mass of rods and beams.

In Chapter 6 multiple-degree-of-freedom systems subjected to either force or support excitations are analyzed. Two general approaches are employed: the simultaneous solution of coupled differential equations of motion and modal superposition. Special attention is given here to the use of modal analysis to obtain the response of systems subjected to nonharmonic force or support excitation.

Chapter 7 consists essentially of an introduction to the use of the finite-element method in vibration analysis. However, it opens with a discussion of the application of Lagrange's equations in deriving the differential equations of motion of systems. Although we considered introducing this topic earlier in the book, it was delayed to this point because of the importance of these equations in deriving the mass and stiffness matrices of different types of finite elements obtained in this chapter, which we felt should be fresh in the reader's mind. An instructor wishing to introduce the use of Lagrange's equations earlier can easily do so since the discussion of them in this chapter is separated from the subsequent discussion of the finite-element method. The development of the finite-element method is presented in some detail, so that the reader can become familiar with its applications before the computer programs for utilizing the method are presented.

In Chapter 8 continuous systems are studied. The solutions found in this chapter provide a reference for comparing some of the approximate results obtained earlier using lumped-mass models of continuous systems with exact results. The modal-superposition method receives special emphasis in this chapter.

We wish to thank the reviewers of the manuscript for their constructive comments. They include Professors Daniel J. Inman, University at Buffalo; Jeffrey C. Huston, Iowa State University; David W. Nicholson, Stevens Institute of Technology; William B. Bickford, Arizona State University; and Alan Haddow, Michigan State University.

We are also grateful to our wives for their patience and understanding during the preparation of this text, and to the staff of Harper & Row for their cooperation, expertise, and planning in the production of the text.

M. L. James
G. M. Smith
J. C. Wolford
P. W. Whaley

Vibration of Mechanical and Structural Systems

Chapter 1

Basic Concepts/Mathematics

1-1 INTRODUCTION

When designing and analyzing mechanical and structural systems, it is essential that the designer have a thorough understanding of the fundamental principles of vibration theory and the attendant mathematical concepts.

The prevention or control of the vibration of machines and structures is an important design consideration. Under certain conditions moving machine parts can cause vibration, which annoys people or even makes them ill, adversely affects the operation of instruments on or near the machine, or can even damage the machine itself by causing excessive levels of stress. Fluctuating stresses caused by vibration cause fatigue failures, and earthquake motions and wind forces can excite structures to such large amplitudes of vibration that they are severely damaged.

In the instances just mentioned, vibration was undesirable and needed to be reduced or eliminated. In other cases vibration is deliberately introduced into systems such as paint mixers, concrete vibrators, air hammers, massaging devices, and musical instruments. In these systems the designer must be able to obtain desired amplitudes and frequencies of vibration.

The design of devices such as gun recoil mechanisms, pressure-measuring transducers, seismic instruments, and suspension systems requires vibration theory in their design so that they adequately perform their intended functions.

There are two general types of vibration, *free* and *forced.* When a system is displaced from its static equilibrium position and then released, it vibrates freely with a frequency that depends upon the mass and stiffness of the system. Such vibration diminishes with time because of energy losses from the system, which are referred to as damping losses and occur to some extent in all real systems.

Forced vibration occurs when a system is subjected to some type of external excitation that adds energy to the system. In general, the amplitude of such vibration depends upon the natural frequencies of the system and the damping inherent in it, as well as upon the

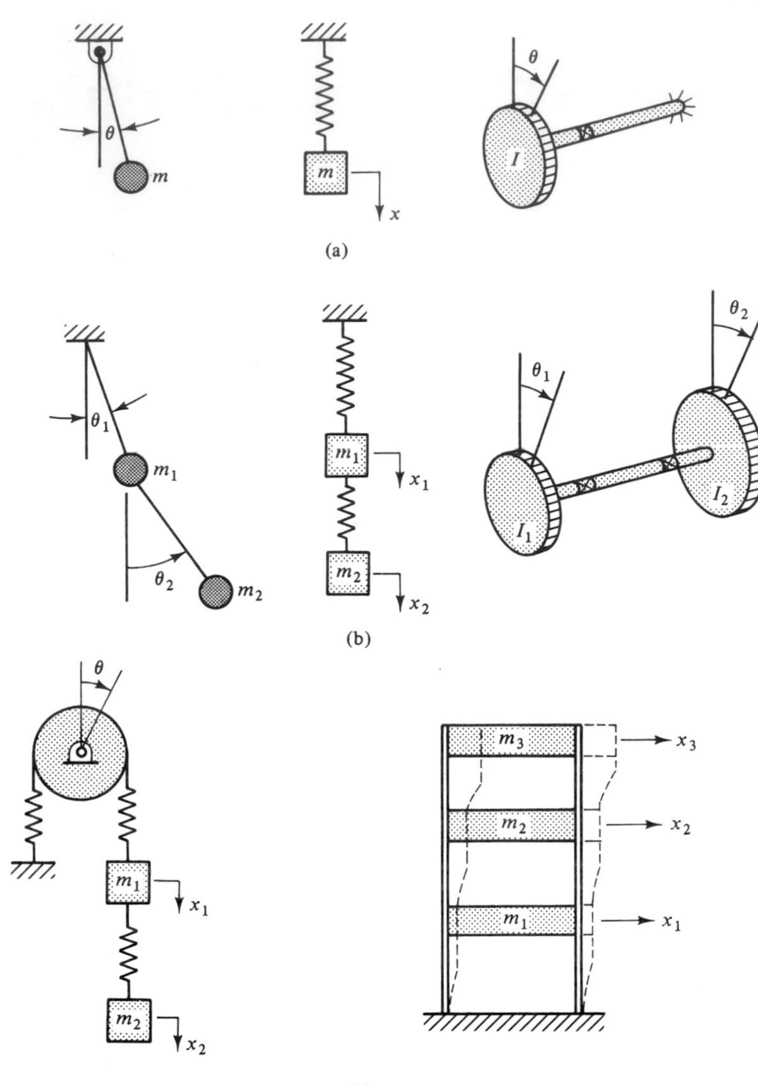

Figure 1-1 One-, two-, and three-degree-of-freedom systems. (a) One degree of freedom. (b) Two degrees of freedom. (c) Three degrees of freedom.

frequency components present in the exciting force. The amplitude of a forced vibration can become very large when a frequency component of the excitation source approaches one of the natural frequencies of the system, particularly the fundamental one. Such a condition is referred to as *resonance,* and the attendant stresses and strains have the potential of causing failures in both machines and structures. For this reason it is very important that a designer have a means of determining the natural frequencies of various types of systems both experimentally and by the use of mathematical models of the systems.

The number of independent coordinates required to describe the configurations of a system during its vibration depends upon the number of degrees of freedom the system

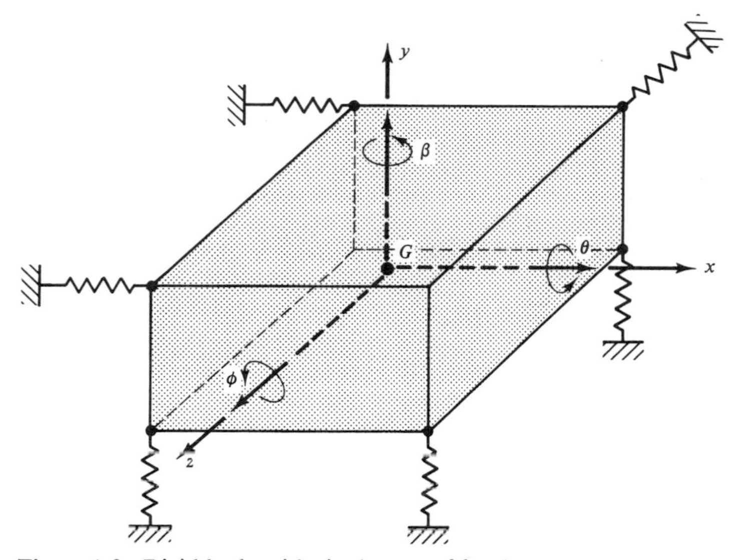

Figure 1-2 Rigid body with six degrees of freedom.

has, or more precisely upon the number of degrees of freedom of the model used to represent the system. Examples of one-, two-, and three-degree-of-freedom systems described by one, two, and three independent coordinates are shown in Fig. 1-1. A rigid body is shown in Fig. 1-2, which can have as many as six degrees of freedom, requiring six independent coordinates to describe its configurations during vibration; three translational coordinates x, y, and z, and three angles of rotation θ, β, and ϕ.

Many systems such as rods, beams, and plates, which have *distributed* parameters such as mass, stiffness, and damping, theoretically have an infinite number of degrees of freedom and are referred to as continuous systems.[1] As we shall see later, a rigorous analysis of such systems involves the solution of partial differential equations. However, as we shall also see later, the analysis of these systems is facilitated by modeling them as discrete lumped-mass systems having finite numbers of degrees of freedom. An example of a four-lumped-mass model of an overhanging beam is shown in Fig. 1-3.

Second-order linear vibration systems can be mathematically modeled quite satisfactorily by ordinary linear differential equations of motion having the form of

$$\ddot{x} + 2\zeta\omega_n\dot{x} + \omega_n^2 x = f(t)$$

in which $\ddot{x}$ and $\dot{x}$ are the notations commonly used for the time derivatives d^2x/dt^2 and dx/dt, respectively. The solution of these equations is basic to the study of vibrations.

The vibration characteristics of machines and structures can be described in either the *time domain* or the *frequency domain.* As we shall see in Sec. 1-7, these domains can be related mathematically so that the vibration characteristics described in the frequency domain (frequency spectrum) can be obtained from the time domain, and vice versa.

Other mathematical concepts pertinent to the study of vibration include complex

[1] Continuous systems are also referred to as distributed-parameter systems. The use of the word "continuous" to describe systems having distributed parameters should not be confused with its use to describe systems for which *time* is the continuous variable.

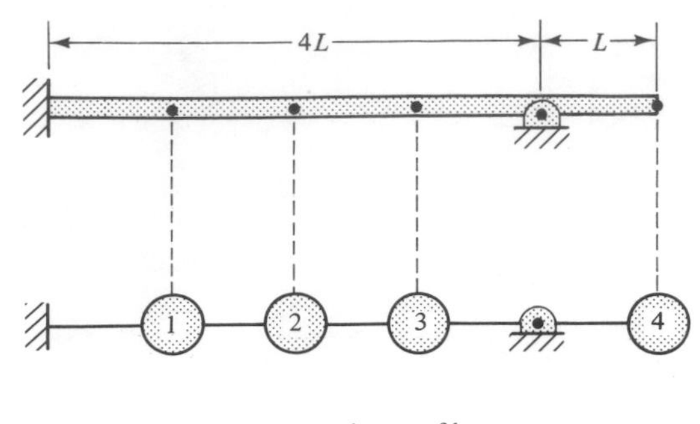

m_b = total mass of beam
$m_1 = m_2 = m_3 = m_b/5$
$m_4 = m_b/10$

Figure 1-3 Typical lumped-mass model of a beam.

algebra, matrix algebra, Fourier series, and Fourier transforms. The latter two are essential to understanding the use of digital real-time spectrum analyzers that are in wide use in experimental studies. The basic concepts of matrix algebra are presented in Appendix B, and the other concepts mentioned will be discussed later in this chapter.

1-2 SINUSOIDAL WAVES/HARMONIC MOTION

The basic parameters of a sinusoidal wave are its amplitude, frequency, and initial phase. A sinusoidal wave

$$y = A \sin \theta = A \sin \omega t$$

can be generated by a point P moving in a circular path as shown in Fig. 1-4. The amplitude of the sine wave is the radius A of the circle. The position vector of length A rotates counterclockwise with an angular velocity of ω rad/s, which is known as the *circular* frequency (rad/s) of the wave. The time it takes for the vector to complete 1 cycle is the period τ, usually measured in seconds. Thus, the frequency in cycles per second (Hertz) is given by

$$f = \frac{1}{\tau} = \frac{\omega}{2\pi} \tag{1-1}$$

It can also be seen that the horizonal projection of the vector A generates the function

$$x = A \cos \omega t$$

Using the relationship in Eq. 1-1, the functions $A \sin \omega t$ and $A \cos \omega t$ can be written in terms of the frequency f as

$$\left.\begin{aligned} y &= A \sin 2\pi f t \\ x &= A \cos 2\pi f t \end{aligned}\right\} \tag{1-2}$$

It is important to note in Fig. 1-4 that $\cos \omega t$ leads $\sin \omega t$ by 90° since $\cos \omega t$ is at its peak 90° ahead of $\sin \omega t$. The frequency spectrum for a single sine wave, with a frequency f and amplitude A, is shown in Fig. 1-5.

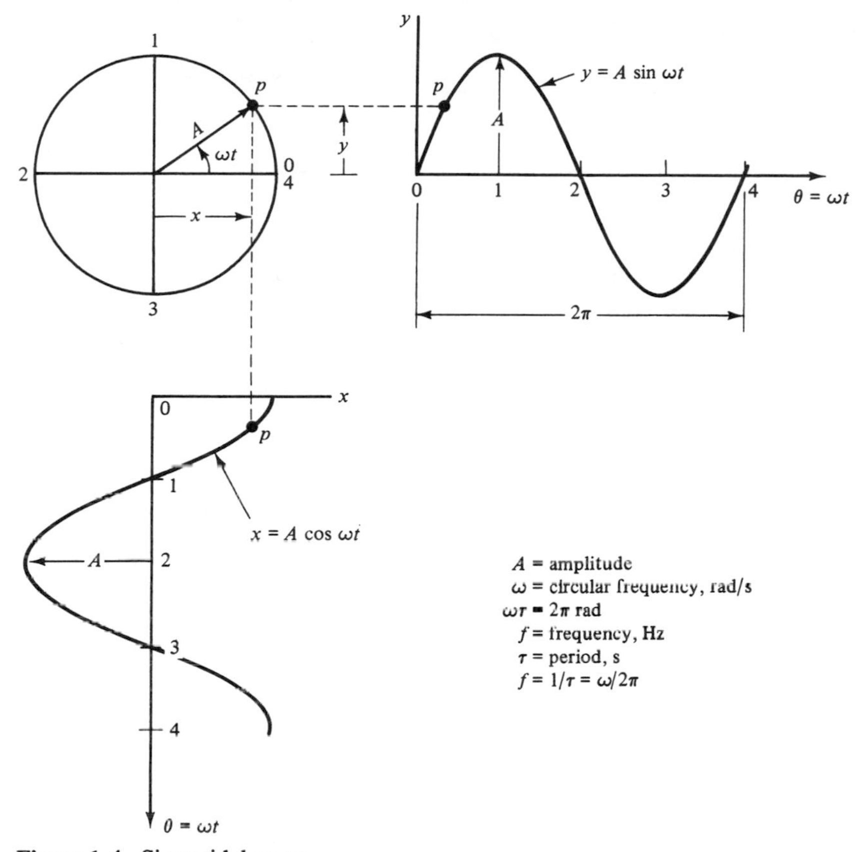

Figure 1-4 Sinusoidal wave.

If the sine wave is not zero at the instant that we start measuring time, as shown in Fig. 1-6, then

$$y = A \sin(\omega t + \phi) \tag{1-3}$$

where $(\omega t + \phi)$ is the *phase* of the motion and ϕ is the *phase angle*, or *initial phase*, as shown in Fig. 1-6. Note that the positive ϕ indicates that the sine wave of Fig. 1-6 leads $y = A \sin \omega t$ by the phase angle ϕ.

Equation 1-3 can also be written as

$$y = C \cos \omega t + D \sin \omega t \tag{1-4}$$

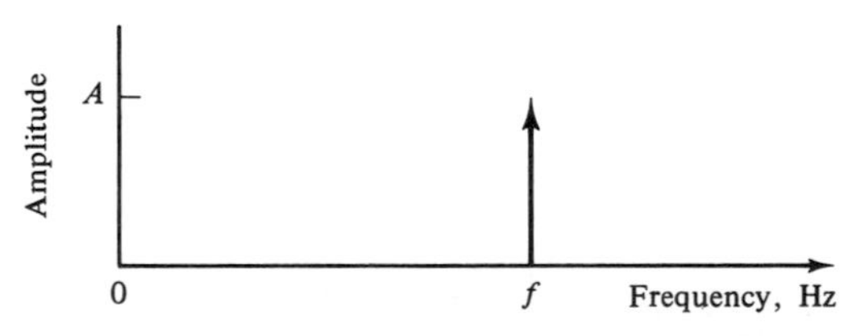

Figure 1-5 Frequency spectrum for a single sine wave.

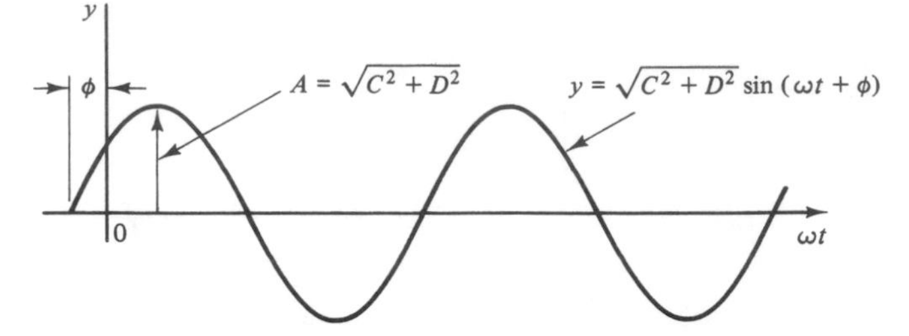

Figure 1-6 Sine wave leading sin ωt by phase angle ϕ.

To see that Eqs. 1-3 and 1-4 are equivalent, let us refer to the rotating vectors shown in Fig. 1-7, and note that $C \cos \omega t$ leads $D \sin \omega t$ by 90°, and that $A = \sqrt{C^2 + D^2}$. If the phase angle ϕ is as shown in Fig. 1-8, then

$$y = A \sin (\omega t - \phi) \tag{1-5}$$

and the negative sign preceding ϕ indicates that the sine wave of Eq. 1-5 lags $y = A \sin \omega t$ by the phase angle ϕ.

Harmonic Motion

If a mass m is attached to either a light elastic spring of stiffness k or a light cantilever beam as shown in Fig. 1-9a and b, either system will oscillate vertically when displaced from its static equilibrium position and then released. We shall see later that, with damping neglected, the motion of m for each system can be described in the time domain by the equation

$$y = A \sin \omega t \tag{1-6}$$

in which the displacement y is sinusoidal, and the natural circular frequency is $\omega = \sqrt{k/m}$.

The velocity and acceleration of m can be obtained by successive differentiations of Eq. 1-6 as

$$\dot{y} = A\omega \cos \omega t \qquad \text{(velocity)}$$
$$\ddot{y} = -A\omega^2 \sin \omega t \qquad \text{(acceleration)}$$

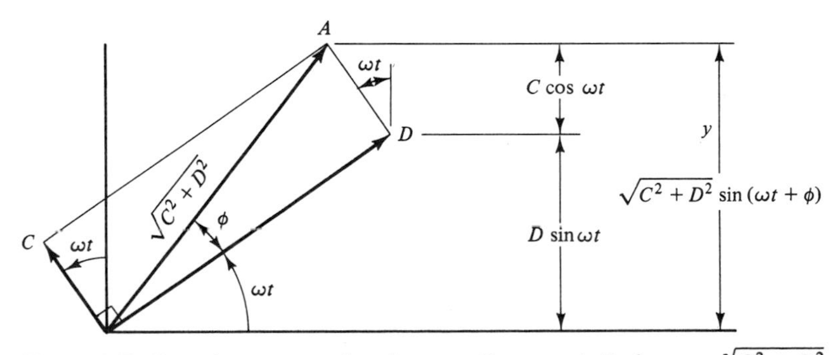

Figure 1-7 Rotating vectors showing $y = C \cos \omega t + D \sin \omega t = \sqrt{C^2 + D^2} \times \sin(\omega t + \phi)$.

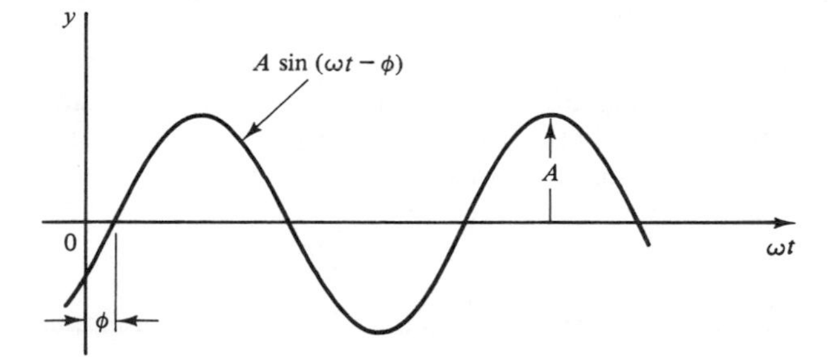

Figure 1-8 Sine wave that lags sin ωt by phase angle ϕ.

The latter term shows that the acceleration of the mass m is proportional to the displacement, but opposite in sense to it. Such motion is defined as *simple harmonic motion.* It is also important to note from the above that the maximum velocity and acceleration of m are the amplitudes $A\omega$ and $A\omega^2$, respectively.

Electrical signals from vibration-measuring transducers are usually observed on an oscilloscope, oscillograph, or spectrum analyzer. It should be emphasized that such signals appear only as functions of time t as shown in Fig. 1-9, and not as functions of ωt as shown in Figs. 1-4, 1-6, and 1-8.

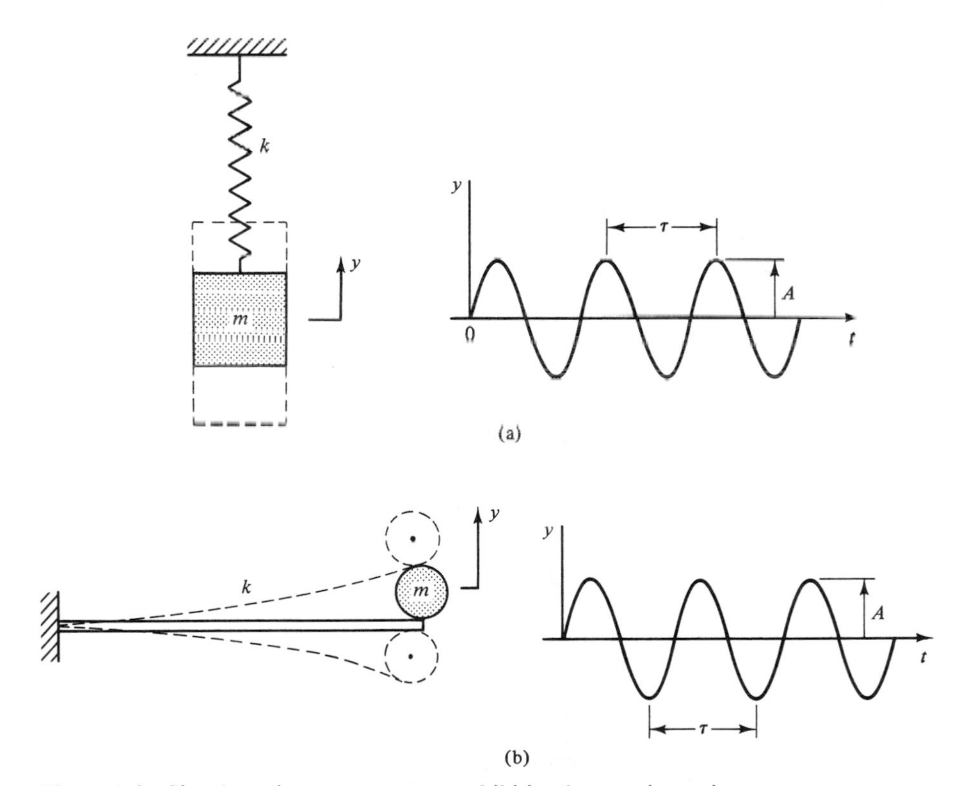

Figure 1-9 Simple spring-mass systems exhibiting harmonic motion.

1-3 COMPLEX NOTATION

The use of complex algebra facilitates the mathematical analysis of many physical phenomena and is particularly useful in the study of vibrations. Therefore, let us briefly review some of the basic concepts of complex algebra.

The complex quantity

$$z = a + jb \qquad (j = \sqrt{-1}) \tag{1-7}$$

is shown in Fig. 1-10. The *real* part of the vector r is a, and the *imaginary* part is b. It should be noted from the figure that

$$\begin{aligned} a &= r \cos \theta \\ b &= r \sin \theta \\ |z| &= r = \sqrt{a^2 + b^2} = \text{modulus (absolute value of } z) \\ \tan \theta &= \frac{b}{a} \end{aligned}$$

It follows then that Eq. 1-7 can be written as

$$z = r(\cos \theta + j \sin \theta) \tag{1-8}$$

Then, considering that $j^2 = -1$, $j^3 = -j$, $j^4 = 1$, and so forth, we can write the following two series:

$$\cos \theta = 1 - \frac{\theta^2}{2!} + \frac{\theta^4}{4!} - \cdots = 1 + \frac{(j\theta)^2}{2!} + \frac{(j\theta)^4}{4!} + \cdots$$

and

$$j \sin \theta = j\left(\theta - \frac{\theta^3}{3!} + \frac{\theta^5}{5!} - \cdots\right) = j\theta + \frac{(j\theta)^3}{3!} + \frac{(j\theta)^5}{5!} + \cdots$$

Adding these series, we obtain

$$\cos \theta + j \sin \theta = 1 + j\theta + \frac{(j\theta)^2}{2!} + \frac{(j\theta)^3}{3!} + \frac{(j\theta)^4}{4!} + \cdots \tag{1-9}$$

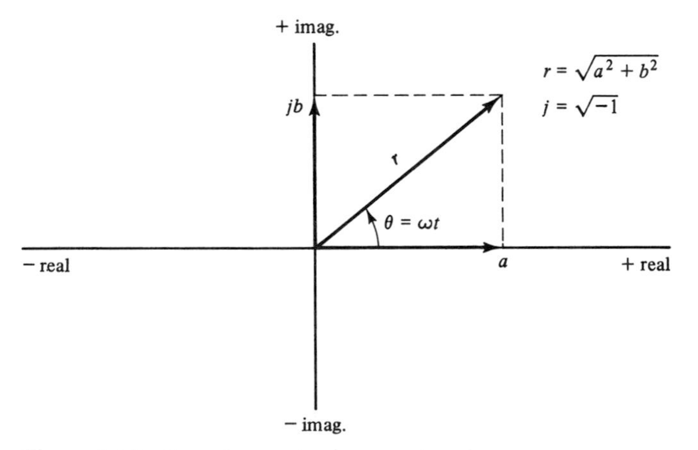

Figure 1-10 Complex vector in complex plane.

Since the right-hand side of Eq. 1-9 is the series for $e^{j\theta}$, we can write that

$$e^{j\theta} = \cos\theta + j\sin\theta \qquad (1\text{-}10)$$

This equation is known as Euler's equation. Letting $\theta = -\theta$ in Eq. 1-10 shows that

$$e^{-j\theta} = \cos\theta - j\sin\theta \qquad (1\text{-}11)$$

The word *imaginary* arises from the term $j = \sqrt{-1}$ since, as is easily proved, no negative number has a *real* square root. It does not mean nonexistent but is simply a mathematical expression associated with complex notation.

If $\theta = \omega t$, the vector **r** shown in Fig. 1-10 is rotating *counterclockwise* with a circular frequency of ω rad/s so that

$$re^{j\omega t} = r(\cos\omega t + j\sin\omega t) \qquad (1\text{-}12)$$

If $\theta = -\omega t$, then **r** is rotating *clockwise*, and

$$re^{-j\omega t} = r(\cos\omega t - j\sin\omega t) \qquad (1\text{-}13)$$

The term $e^{-j\omega t}$ does not imply a negative circular frequency, since we are concerned only with positive frequencies in practical systems. The negative sign arises only from the mathematical notation used, and simply means, as mentioned above, that the rotating vector **r** has a clockwise sense.

The two vectors given by Eqs. 1-12 and 1-13, which are shown in Fig. 1-11, can be added to obtain

$$r\cos\omega t = r\left(\frac{e^{j\omega t} + e^{-j\omega t}}{2}\right) \qquad (1\text{-}14)$$

If Eq. 1-13 is then subtracted from Eq. 1-12, we find that

$$r\sin\omega t = r\left(\frac{e^{j\omega t} - e^{-j\omega t}}{2j}\right) \qquad (1\text{-}15)$$

Equations 1-14 and 1-15 will be used in Sec. 1-7 to derive the complex form of the Fourier series.

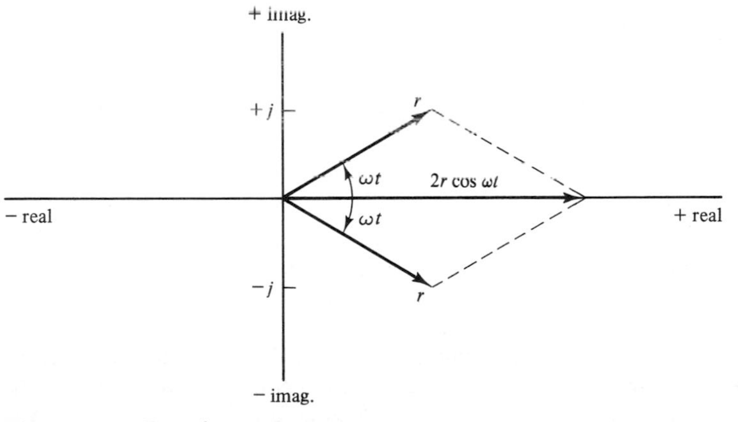

Figure 1-11 Rotating vectors.

The addition, subtraction, and multiplication of complex numbers follow the ordinary rules of algebra. Examples of the sum and product of the complex numbers

$$z_1 = a + jb = r_1 e^{j\theta} \quad \text{and} \quad z_2 = c + jd = r_2 e^{j\phi}$$

are as follows:

$$z_1 + z_2 = a + jb + c + jd = (a + c) + j(b + d)$$
$$z_1 z_2 = (a + jb)(c + jd) = (ac - bd) + j(ad + bc)$$

or in *exponential* form as

$$z_1 + z_2 = r_1 e^{j\theta} + r_2 e^{j\phi}$$
$$z_1 z_2 = r_1 r_2 e^{j(\theta+\phi)}$$

The quotient of z_1/z_2 is

$$\frac{z_1}{z_2} = \frac{a + jb}{c + jd} \qquad (z_2 \neq 0)$$

This equation can be simplified by multiplying the numerator and denominator by $c - jd$, which is the *conjugate* of $c + jd$. Thus,

$$\frac{z_1}{z_2} = \frac{(a + jb)(c - jd)}{(c + jd)(c - jd)} = A + jB$$

where

$$A = \frac{ac + bd}{c^2 + d^2}$$
$$B = \frac{bc - ad}{c^2 + d^2}$$

The quotient z_1/z_2 in exponential form is simply

$$\frac{z_1}{z_2} = \frac{r_1 e^{j\theta}}{r_2 e^{j\phi}} = \frac{r_1 e^{j(\theta-\phi)}}{r_2}$$

1-4 ORTHOGONAL RELATIONS

If a set of functions $f_1(x), f_2(x), \ldots, f_m(x), \ldots, f_n(x), \ldots$ is such that

$$\int_a^b f_n(x) f_m(x)\, dx = \begin{Bmatrix} 0(m \neq n) \\ \lambda(m = n) \end{Bmatrix} \tag{1-16}$$

the integral is referred to as an *orthogonal* relation. This means that the area under a curve generated by the product of any two of these functions is zero over the interval from a to b, while the area under a curve generated by multiplying any one of these functions by itself is equal to some value that we shall call λ.

In the discussion of periodic motion and Fourier series in the following section, it will be important to recognize that certain integrals of sine and cosine functions are orthogonal relations. It can be shown that for the functions $\sin n\theta$ and $\cos n\theta$

$$\int_{-\pi}^{\pi} \sin n\theta \sin m\theta \, d\theta = \begin{Bmatrix} 0 & n \neq m \\ \pi & n = m \end{Bmatrix} \tag{1-17}$$

$$\int_{-\pi}^{\pi} \cos n\theta \cos m\theta \, d\theta = \begin{Bmatrix} 0 & n \neq m \\ \pi & n = m \end{Bmatrix} \tag{1-18}$$

$$\int_{-\pi}^{\pi} \sin n\theta \cos m\theta \, d\theta = \begin{Bmatrix} 0 & n \neq m \\ 0 & n = m \end{Bmatrix} \tag{1-19}$$

where $n = 1, 2, 3, \ldots$
$m = 1, 2, 3, \ldots$

1-5 PERIODIC MOTION/FOURIER SERIES

When machines or structures are subjected to an excitation involving more than one frequency, the resulting vibrations are usually characterized by the presence of several frequencies, and such vibrations that show a repetitive pattern in the time domain are said to be periodic. In general, periodic motion is described by waveforms containing a combination of sines and/or cosines having frequencies present in the wave. A simple example of periodic motion described by

$$y = A_1 \sin(2\pi f_1 t) + A_2 \sin(2\pi f_2 t)$$

is shown in Fig. 1-12, in which $f_2 = 3f_1$ and the fundamental period τ corresponds to the fundamental frequency f_1. That is,

$$\tau = \frac{1}{f_1} \tag{1-20}$$

The use of Fourier series as a means of describing periodic motion and/or periodic excitation is essential to the study of vibration. Familiarity with Fourier series also aids in understanding the significance of experimentally determined frequency spectrums.

Any single-valued function $f(\theta)$ that is continuous, except possibly for a finite number of finite discontinuities in an interval of length 2π, and which has only a finite number of maxima and minima in the interval, can be represented by a convergent Fourier series. For $f(\theta)$ defined in the interval of $(-\pi, \pi)$, the Fourier series can be written as

$$f(\theta) = \frac{a_0}{2} + (a_1 \cos\theta + b_1 \sin\theta) + (a_2 \cos 2\theta + b_2 \sin 2\theta) + (a_3 \cos 3\theta + b_3 \sin 3\theta) + \cdots$$

or more simply as

$$f(\theta) = \frac{a_0}{2} + \sum_{n=1}^{\infty} a_n \cos n\theta + \sum_{n=1}^{\infty} b_n \sin n\theta \tag{1-21}$$

in which a_n and b_n are the *Fourier coefficients*. For example, the function

$$f(\theta) = 0 \qquad 0 \le \theta < \pi$$
$$f(\theta) = -\theta \qquad -\pi < \theta \le 0$$

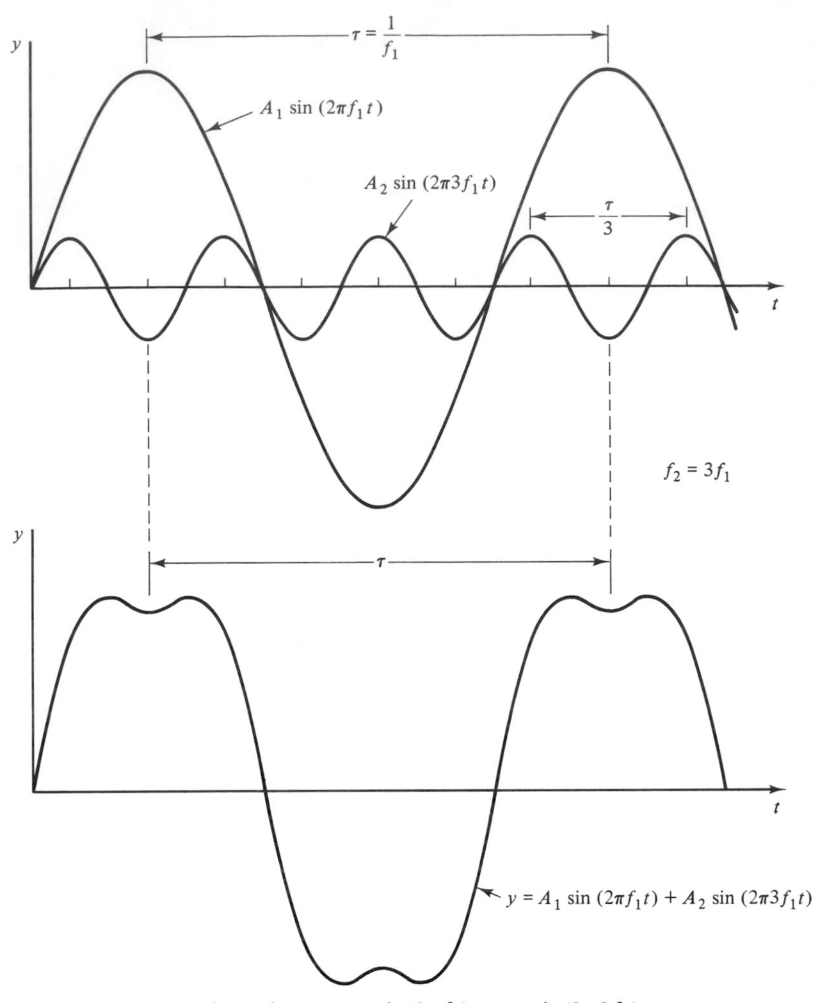

Figure 1-12 Waveform for $y = A_1 \sin(2\pi f_1 t) + A_2 \sin(2\pi 3 f_1 t)$.

as represented by the series of Eq. 1-21 is periodic and repeats itself as shown in Fig. 1-13. It should be noted that the period is 2π, and that $f(\theta + 2\pi) = f(\theta)$.

A periodic function in the time domain in the interval $(-\tau/2, \tau/2)$ has a period τ as shown in Fig. 1-14. To transform an interval of length 2π such as that shown in Fig. 1-13 into an interval of length τ such as that shown in Fig. 1-14 requires a transformation of the independent variable t. Comparing Figs. 1-13 and 1-14, it can be seen that the relationship between the variables θ and t is given by

$$\frac{\theta}{2\pi} = \frac{t}{\tau}$$

from which

$$\theta = \frac{2\pi t}{\tau} \tag{1-22}$$

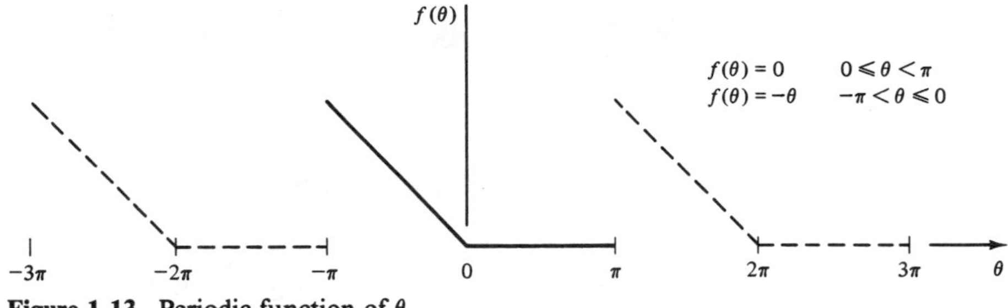

Figure 1-13 Periodic function of θ.

From Eq. 1-1,

$$\omega = \frac{2\pi}{\tau} \tag{1-23}$$

Then, since $f = 1/\tau$,

$$\omega = 2\pi f \tag{1-24}$$

Utilizing these relationships, the Fourier series of Eq. 1-21 can be written in the following forms:

$$f(t) = \frac{a_0}{2} + \sum_{n=1}^{\infty} a_n \cos \frac{n2\pi t}{\tau} + \sum_{n=1}^{\infty} b_n \sin \frac{n2\pi t}{\tau} \tag{1-25}$$

$$f(t) = \frac{a_0}{2} + \sum_{n=1}^{\infty} a_n \cos n\omega t + \sum_{n=1}^{\infty} b_n \sin n\omega t \tag{1-26}$$

$$f(t) = \frac{a_0}{2} + \sum_{n=1}^{\infty} a_n \cos n2\pi f t + \sum_{n=1}^{\infty} b_n \sin n2\pi f t \tag{1-27}$$

In Eq. 1-27, f is the fundamental frequency. When $n = 1$, the arguments of the terms correspond to the fundamental frequency. When $n > 1$, the arguments of the terms correspond to the higher harmonic frequencies. The fundamental circular frequency is $\omega = 2\pi f$, and the circular frequency of the $(n-1)$th harmonic is given by $n\omega = n(2\pi f)$.

Fourier Coefficients

With the series in the form of Eq. 1-26, the equations for determining the *Fourier coefficients* a_n and b_n for a full-range expansion from $-\pi/\omega$ to π/ω are as follows:

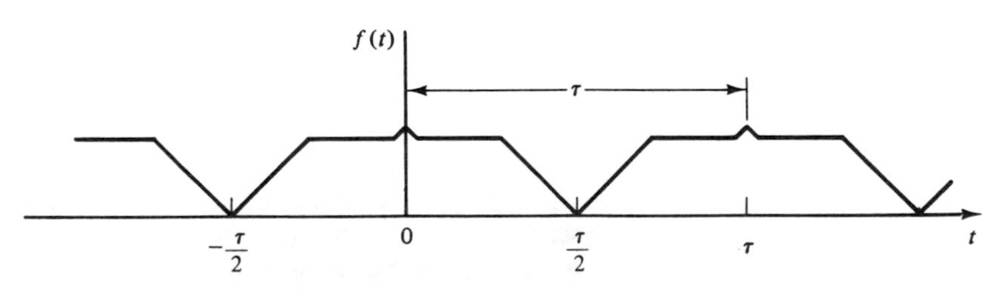

Figure 1-14 Periodic function in the time domain.

$$a_n = \frac{\omega}{\pi}\int_{-\pi/\omega=-\tau/2}^{\pi/\omega=\tau/2} f(t)\cos n\omega t\,dt \qquad n = 0, 1, 2, 3, \ldots \tag{1-28}$$

$$b_n = \frac{\omega}{\pi}\int_{-\pi/\omega=-\tau/2}^{\pi/\omega=\tau/2} f(t)\sin n\omega t\,dt \qquad n = 1, 2, 3, \ldots \tag{1-29}$$

It is instructive to see how Eqs. 1-28 and 1-29 were determined. To obtain Eq. 1-28, we multiply Eq. 1-26 by cos $m\omega t$ and integrate from $-\pi/\omega$ to π/ω to obtain

$$\int_{-\pi/\omega}^{\pi/\omega} f(t)\cos m\omega t\,dt = \frac{a_0}{2}\int_{-\pi/\omega}^{\pi/\omega}\cos m\omega t\,dt + \int_{-\pi/\omega}^{\pi/\omega}\left(\sum_{n=1}^{\infty} a_n\cos n\omega t\right)\cos m\omega t\,dt$$

$$+ \int_{-\pi/\omega}^{\pi/\omega}\left(\sum_{n=1}^{\infty} b_n\sin n\omega t\right)\cos m\omega t\,dt \tag{1-30}$$

We now consider Eq. 1-30, first with $m = 0$ and then with $m = 1$, $m = 2$, and so forth. Before doing this, however, we should note the following:

$$\frac{a_0}{2}\int_{-\pi/\omega}^{\pi/\omega}\cos m\omega t\,dt = \begin{Bmatrix} 0 & m \neq 0 \\ \dfrac{a_0\pi}{\omega} & m = 0 \end{Bmatrix}$$

$$\int_{-\pi/\omega}^{\pi/\omega}(a_n\cos n\omega t)\cos m\omega t\,dt = \begin{Bmatrix} 0 & m \neq n \\ \dfrac{a_n\pi}{\omega} & m = n,\ n \neq 0 \end{Bmatrix}$$

$$\int_{-\pi/\omega}^{\pi/\omega}(b_n\sin n\omega t)\cos m\omega t\,dt = \begin{Bmatrix} 0 & m \neq n \\ 0 & m = n \end{Bmatrix}$$

The second and third integrals above can be verified by utilizing the respective trigonometric relationships

$$\cos n\omega t\cos m\omega t = \tfrac{1}{2}[\cos(n+m)\omega t + \cos(n-m)\omega t]$$
$$\sin n\omega t\cos m\omega t = \tfrac{1}{2}[\sin(n+m)\omega t + \sin(n-m)\omega t]$$

If we now let m take on successive values of 0, 1, 2, . . . , in Eq. 1-30, and use the integrals shown above,

$$a_n = \frac{\omega}{\pi}\int_{-\pi/\omega}^{\pi/\omega} f(t)\cos n\omega t\,dt \qquad n = 0, 1, 2, 3, \ldots$$

which is Eq. 1-28. Equation 1-29 can be determined in a similar manner by multiplying Eq. 1-26 by sin $m\omega t$, and integrating over the interval $(-\pi/\omega, \pi/\omega)$.

Half-Range Expansions

If the functions represented by a Fourier series are either *odd* or *even,* it is usually more convenient to use a half-range series over the interval $(0, \pi/\omega)$. If $f(-t) = -f(t)$, the function is an odd function, and the cosine terms of Eq. 1-26 will vanish because each is an even function. Conversely, if $f(-t) = f(t)$, the function is an even function, and the sine terms of Eq. 1-26 vanish because each is an odd function. That is,

$$f(t) = \sum_{n=1}^{\infty} b_n \sin n\omega t \qquad \text{(odd function)}$$

$$f(t) = \frac{a_0}{2} + \sum_{n=1}^{\infty} a_n \cos n\omega t \qquad \text{(even function)}$$

Consider the function $f(t) = -t + 1$, defined in the interval $(0, \frac{1}{2})$, as shown in Fig. 1-15. The Fourier cosine series represents the even function shown in Fig. 1-15a, and the Fourier sine series represents the odd function shown in Fig. 1-15b. In either case, the series will converge to the same value in the interval $(0, \frac{1}{2})$, except at $t = 0$ and $t = \frac{1}{2}$. For the even function, the cosine series converges to 1 at $t = 0$ and to $\frac{1}{2}$ at $t = \frac{1}{2}$. However, the odd function is discontinuous at $t = 0$, and the sine series converges to the *mean value* of the function at the points of discontinuity. That is, the sine series converges to zero at $t = 0$ and $t = \frac{1}{2}$.

To determine the equations for calculating the Fourier coefficients a_n and b_n for half-range expansions, let us refer to the odd and even functions as

$$f_o = \text{any odd function}$$
$$f_e = \text{any even function}$$

and note that the products of these functions are odd or even as follows:

$$f_o \cdot f_e = g_o \qquad \text{(odd function)}$$
$$f_o \cdot f_o = g_e \qquad \text{(even function)}$$
$$f_e \cdot f_e = g_e \qquad \text{(even function)}$$

By inspection of the odd and even functions shown in Fig. 1-16, it is apparent from the areas shown that

$$\int_{-\pi/\omega}^{\pi/\omega} (f_o \cdot f_e)\, dt = \int_{-\pi/\omega}^{\pi/\omega} g_o\, dt = 0 \tag{1-31}$$

and that

$$\int_{-\pi/\omega}^{\pi/\omega} (f_o \cdot f_o)\, dt = 2 \int_{0}^{\pi/\omega} g_e\, dt \tag{1-32}$$

To illustrate the use of Eqs. 1-31 and 1-32, let us assume that the function $f(t)$ we wish to represent is an even function. Remembering that $\cos n\omega t$ is an even function and that $\sin n\omega t$ is an odd function,

$$f(t)\cos n\omega t = (f_e)(f_e) = g_e$$

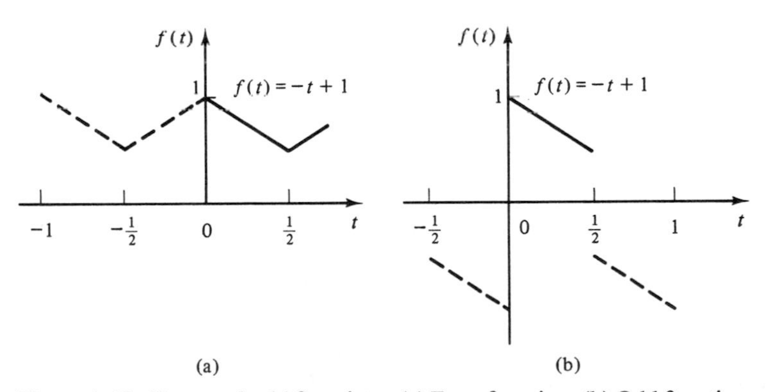

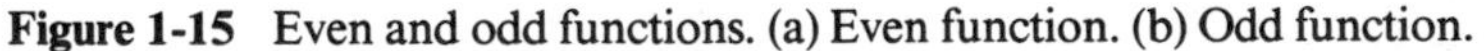

Figure 1-15 Even and odd functions. (a) Even function. (b) Odd function.

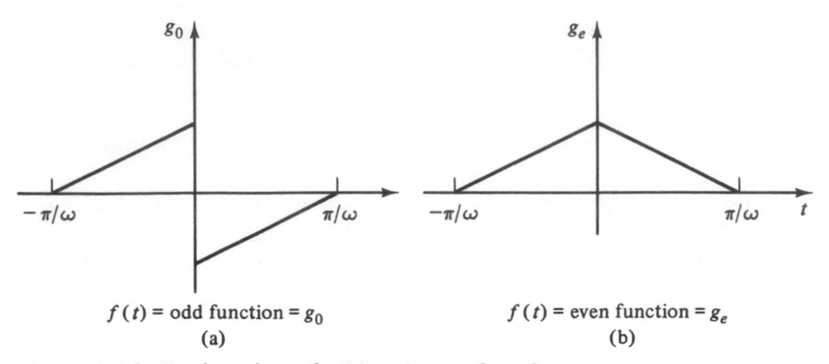

Figure 1-16 Designation of odd and even functions.

$$f(t)\sin n\omega t = (f_e)(f_o) = g_o$$

Correlating the above with Eqs. 1-31 and 1-32, we can write Eqs. 1-28 and 1-29 as

$$\left.\begin{aligned} a_n &= \frac{\omega}{\pi}\int_{-\pi/\omega}^{\pi/\omega} f(t)\cos n\omega t\, dt = \frac{2\omega}{\pi}\int_0^{\pi/\omega} f(t)\cos n\omega t\, dt \\ &\qquad\qquad n = 0, 1, 2, 3, \ldots \\ \text{and} \\ b_n &= \frac{\omega}{\pi}\int_{-\pi/\omega}^{\pi/\omega} f(t)\sin n\omega t\, dt = 0 \qquad n = 1, 2, 3, \ldots \end{aligned}\right\} f(t) \text{ even} \qquad (1\text{-}33)$$

Similarly, if $f(t)$ is an odd function, the results from Eqs. 1-31 and 1-32 enable us to write Eqs. 1-28 and 1-29 as

$$\left.\begin{aligned} b_n &= \frac{2\omega}{\pi}\int_0^{\pi/\omega} f(t)\sin n\omega t\, dt \qquad n = 1, 2, 3, \ldots \\ a_n &= 0 \qquad n = 0, 1, 2, 3, \ldots \end{aligned}\right\} f(t) \text{ odd} \qquad (1\text{-}34)$$

Let us now consider several examples to illustrate the use of Fourier series.

EXAMPLE 1-1

It is desired to determine the Fourier series for the square wave shown in Fig. 1-17 which has a period of $\tau = 0.2$ s, giving it a circular frequency of $\omega = 10\pi$ rad/s.

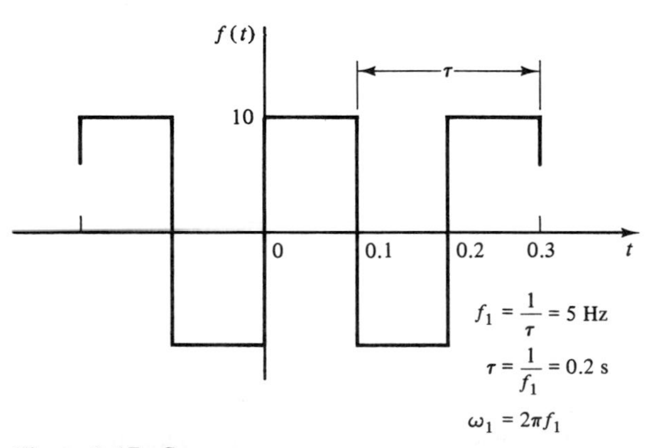

Figure 1-17 Square wave.

Solution. With the origin selected as shown in Fig. 1-17 the square wave is an odd function, with $f(t) = 10$ over the interval $(0, \pi/\omega)$, which is 0.1 s. Since it is an odd function, we can use the half-range expansion given by Eq. 1-34 to find that

$$a_n = 0$$

$$b_n = 2(10)\int_0^{0.1} 10 \sin(n10\pi t)\, dt = \frac{-200}{n10\pi}\cos(n10\pi t)\Big|_0^{0.1}$$

$$b_n = \frac{40}{n\pi} = \frac{b_1}{n} \qquad n = 1, 3, 5, \ldots$$

$$b_n = 0 \qquad n = 2, 4, 6, \ldots$$

Then from Eq. 1-26,

$$f(t) = \frac{40}{\pi}\sum_{n=1,3,5,\ldots}^{\infty} \frac{1}{n}\sin n\omega t$$

or

$$f(t) = \frac{40}{\pi}\left(\sin \omega_1 t + \frac{1}{3}\sin 3\omega_1 t + \frac{1}{5}\sin 5\omega_1 t + \cdots\right)$$

in which ω_1 is the fundamental circular frequency. Since $\omega_1 = 2\pi f_1$ and $f_1 = 5$ Hz, the frequency spectrum for a frequency bandwidth of 0 to 25 Hz is shown in Fig. 1-18. It should be apparent that the *frequency spectrum is simply a graphical plot of the Fourier coefficients,* which are in turn simply the *amplitudes of the frequency components* present in the time-domain wave.

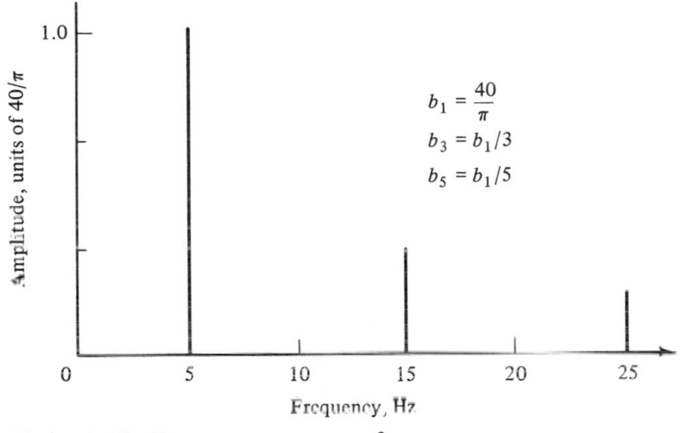

Figure 1-18 Frequency spectrum for square wave.

EXAMPLE 1-2

The cam shown in Fig. 1-19a, which is rotating at N rpm, subjects the follower system to the sawtooth displacement function shown in Fig. 1-19b. This function can be expressed as

$$\left.\begin{aligned} f(t) &= y_0\frac{t}{\tau} \qquad && 0 \le t \le \tau \\ f(t+\tau) &= f(t) \qquad && \text{(for all values of time } t) \end{aligned}\right\} \tag{1-35}$$

in which τ is the period of the function.

We wish to determine the Fourier series corresponding to the sawtooth function given by Eq. 1-35.

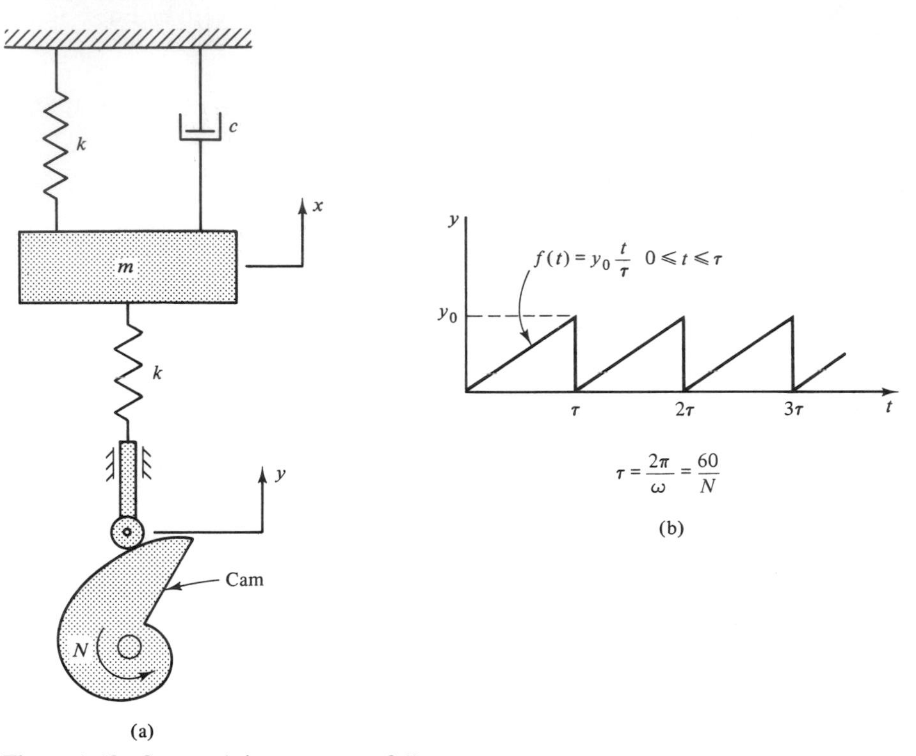

Figure 1-19 Sawtooth input to cam follower system.

Solution. Since an origin for $t = 0$ that would make $f(t)$ an odd or even function cannot be found, the Fourier coefficients must be determined from a full-range expansion using Eqs. 1-25, 1-28, and 1-29. From the latter two of these we can write that

$$a_n = \frac{2}{\tau} \int_0^\tau y_0 \frac{t}{\tau} \cos \frac{n2\pi t}{\tau} \, dt$$

$$b_n = \frac{2}{\tau} \int_0^\tau y_0 \frac{t}{\tau} \sin \frac{n2\pi t}{\tau} \, dt$$

integrating these equations by parts or by the use of integral tables, we find that

$$a_0 = y_0 \qquad n = 0$$

$$a_n = 0 \qquad n = 1, 2, 3, \ldots$$

$$b_n = -\frac{y_0}{n\pi} \qquad n = 1, 2, 3, \ldots$$

Using these coefficients in Eq. 1-25, the Fourier series for the displacement of the follower is

$$y = f(t) = y_0\left(\frac{1}{2} - \frac{1}{\pi}\sum \frac{1}{n}\sin\frac{n2\pi t}{\tau}\right) \qquad n = 1, 2, 3, \ldots$$

or

$$y = y_0\left(\frac{1}{2} - \frac{1}{\pi}\sum \frac{1}{n}\sin(n2\pi f_1 t)\right) \qquad n = 1, 2, 3, \ldots$$

in which f_1 is the fundamental frequency corresponding to the sawtooth period τ. From the series, it can be seen that the frequency spectrum for a bandwidth of 0 to $5f_1$ is as shown in Fig. 1-20. The term $y_0/2$ is analogous to a dc voltage with zero frequency.

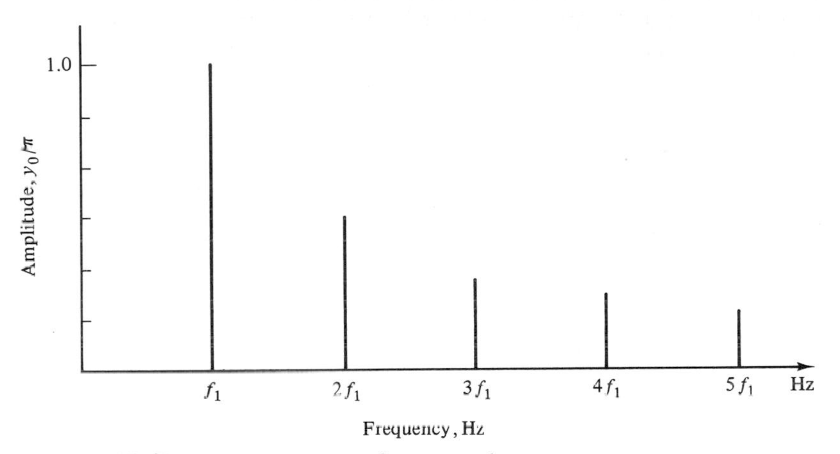

Figure 1-20 Frequency spectrum for sawtooth wave.

Frequency Analyzers (Spectrum Analyzers)

Digital frequency analyzers, often referred to as spectrum analyzers, are widely used at present to experimentally determine the Fourier coefficients of complicated vibrations of machines and structures. Basically, such analyzers have a *fast Fourier transform* (FFT) processor which transforms digitally sampled time-domain signals into a *finite* number of frequency components that are actually the Fourier coefficients.

The term *fast Fourier transform* refers to a digital computer algorithm, which is a fast and efficient scheme for computing a "finite" number of Fourier coefficients. One commonly used FFT algorithm is the Cooley-Tukey fast Fourier transform algorithm.[2]

As one might expect from the preceding discussion of Fourier series, digital frequency analyzers can also synthesize the original time-domain signal from the frequency components. Other mathematical concepts, such as the least-squares error fit by truncated Fourier series and Fourier transforms, which are pertinent to frequency analysis and frequency analyzers, are discussed in Secs. 1-6 and 1-7.

Figure 1-21 shows a triangular wave that has a period of $\frac{1}{150}$ s, and Fig. 1-22 shows the corresponding frequency spectrum of the wave as observed on a high-resolution frequency analyzer. The amplitudes of the Fourier coefficients can be read either in decibels or as root-mean-square (rms) values of the amplitudes (see Sec. 1-9).

[2] J. W. Cooley and J. W. Tukey, "An Algorithm for the Machine Calculation of Complex Fourier Series," *Mathematics of Computations,* vol. 19, pp. 297–301, 1965.

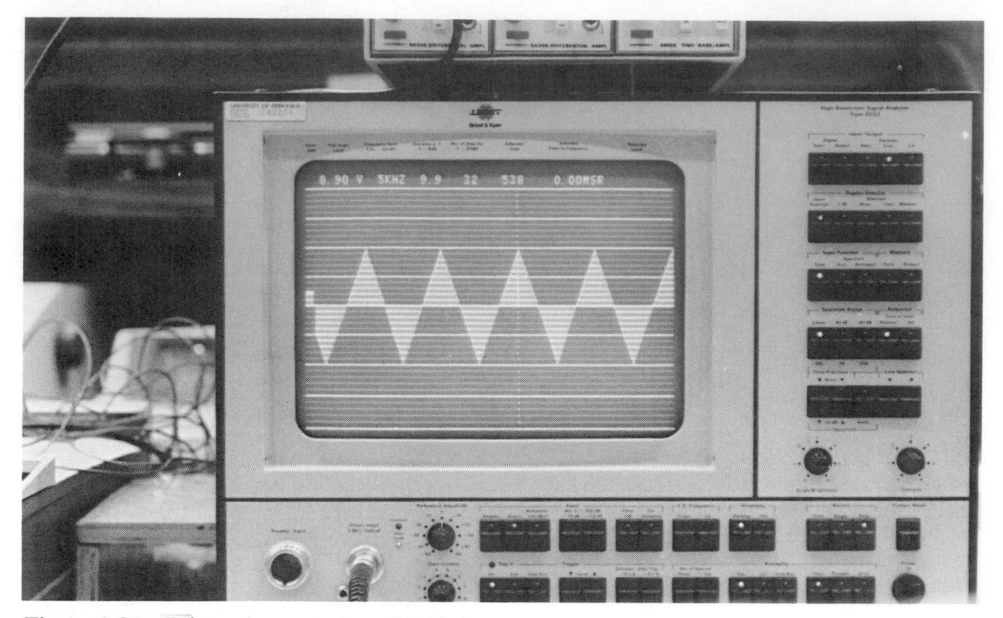

Figure 1-21 Triangular wave (τ = 1/150 s).

It was shown earlier that it is not difficult to determine the frequency components or Fourier coefficients of easily defined waves such as square waves and sawtooth waves using straightforward mathematics. However, the vibrations of machines and structures

Figure 1-22 Frequency spectrum of triangular wave shown on frequency analyzer (0 to 5000 Hz).

usually involve complicated motions that require the use of electronic equipment such as frequency analyzers to study and analyze them experimentally.

Frequency analyzers are also widely used at present as part of preventive-maintenance procedures for machines. For example, such a procedure might involve the periodic check of a machine's vibration characteristics to determine if any significant changes have occurred in them because of bearing or gear wear, loose fasteners, fractures, and so on.

1-6 LEAST-SQUARES ERROR FIT BY TRUNCATED FOURIER SERIES

In practical applications, we must deal with a finite number of terms of a series rather than an infinite number of terms. A series consisting of a finite number of terms is referred to as a *truncated series.* A truncated series of m terms used to represent $f(t)$ will be only approximately equal to $f(t)$. That is,

$$f(t) \cong \frac{a_0}{2} + \sum_{n=1}^{m} a_n \cos n\omega t + \sum_{n=1}^{m} b_n \sin n\omega t \tag{1-36}$$

We shall now show that a truncated series of m terms gives the "best fit" to $f(t)$ according to the *least-squares* criteria, in which a_n and b_n are the same Fourier coefficients that appear in the infinite series. The least-squares residue R, or error, in the interval $(-\pi/\omega, \pi/\omega)$ is

$$R = \int_{-\pi/\omega}^{\pi/\omega} \left[f(t) - \sum_{n=0}^{m} \right]^2 dt \tag{1-37}$$

in which $\Sigma_{n=0}^{m}$ represents the truncated series of Eq. 1-36 with the a_n's and b_n's unknown. Since R is a function of the unknown a_n's and b_n's, it will be a minimum if

$$\frac{\partial R}{\partial a_k} = 0 = 2 \int_{-\pi/\omega}^{\pi/\omega} \left[f(t) - \sum_{n=0}^{m} \right] (-\cos k\omega t)\, dt \tag{1-38}$$

and

$$\frac{\partial R}{\partial b_k} = 0 = 2 \int_{-\pi/\omega}^{\pi/\omega} \left[f(t) - \sum_{n=0}^{m} \right] (-\sin k\omega t)\, dt \tag{1-39}$$

In which k denotes any *particular* value of n in the series. Rearranging Eq. 1-38, we see that

$$\int_{-\pi/\omega}^{\pi/\omega} f(t) \cos k\omega t\, dt = \int_{-\pi/\omega}^{\pi/\omega} \left(\sum_{n=0}^{m} \right) \cos k\omega t\, dt \tag{1-40}$$

which is identical to Eq. 1-30, with the exception that Eq. 1-30 has an infinite number of terms.

By virtue of the *orthogonality properties* of the terms on the right-hand side of Eq. 1-40, it can be seen that

$$\int_{-\pi/\omega}^{\pi/\omega} \left(\sum_{n=0}^{m} \right) \cos k\omega t\, dt = \begin{matrix} 0 & n \neq k \\ \dfrac{a_n \pi}{\omega} & n = k \end{matrix} \Bigg\}$$

Thus,

$$\int_{-\pi/\omega}^{\pi/\omega} f(t)\cos n\omega t\, dt = \frac{a_n \pi}{\omega}$$

and

$$a_n = \frac{\omega}{\pi} \int_{-\pi/\omega}^{\pi/\omega} f(t)\cos n\omega t\, dt \qquad (n = 0, 1, 2, \ldots, m)$$

This is the same expression for a_n as that given in Eq. 1-28 for an infinite series.
Similarly, it can be shown that

$$b_n = \frac{\omega}{\pi} \int_{-\pi/\omega}^{\pi/\omega} f(t)\sin n\omega t\, dt \qquad n = 1, 2, 3, \ldots, m$$

is identical to the Fourier coefficient b_n of the infinite series given by Eq. 1-29 for an infinite series.

1-7 FOURIER INTEGRAL/FOURIER TRANSFORMS

As the period τ of a periodic function approaches infinity, the frequency spectrum approaches being continuous, rather than discrete as shown in Figs. 1-18 and 1-20. At the same time, the Fourier series representing the periodic function approaches an integral that is referred to as the *Fourier integral.* Thus, the Fourier integral can be thought of as the limiting case of the Fourier series as the period τ approaches infinity.

The *Fourier transform pair* consists of two integrals, one that transforms a time function to the frequency domain and one that transforms a frequency function to the time domain.

To obtain some background for a discussion of the Fourier integral and the Fourier transform pair that follows, let us consider the series of pulses shown in Fig. 1-23, often referred to as a *rectangular pulse train.* Recognizing that $f(t)$ is an even function for the pulse train, the Fourier series is

$$f(t) = \frac{a_0}{2} + \sum_{n=1}^{\infty} a_n \cos \frac{n2\pi t}{\tau} \tag{1-41}$$

From Eq. 1-33,

$$a_n = \frac{4}{\tau} \int_0^{t_0/2} A \cos \frac{n2\pi t}{\tau}\, dt$$

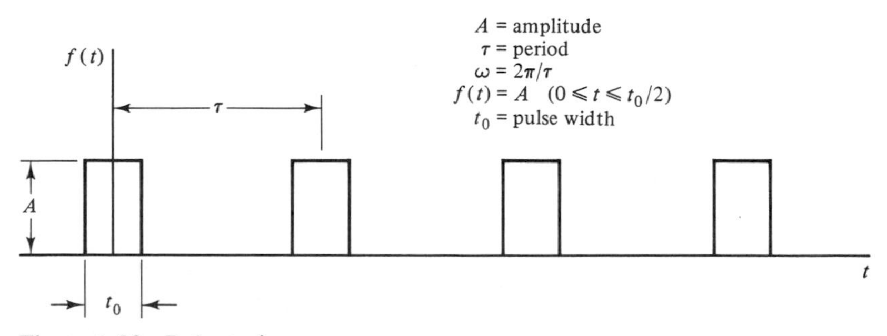

Figure 1-23 Pulse train.

the integration of which yields

$$\left.\begin{aligned} a_0 &= \frac{2t_0A}{\tau} && n = 0 \\ a_n &= \frac{2At_0}{\tau}\left[\frac{\sin(n\pi t_0/\tau)}{n\pi t_0/\tau}\right] && n = 1, 2, \ldots \end{aligned}\right\} \qquad (1\text{-}41a)$$

Let us suppose that the frequency spectrum shown in Fig. 1-24a is that of the pulse train of Fig. 1-23 with a period of τ_1 and a value of $t_0/\tau_1 = \frac{1}{10}$. This spectrum is obtained by plotting the absolute values of the a_n's in Eq. 1-41a as a function of the harmonic number n over a frequency bandwidth corresponding to $n = 30$. The Fourier series for this pulse train is given by

$$f(t) = \frac{0.1(2A)}{2} + \left[\sum_{n=1}^{30} \frac{2A}{n\pi}\sin\left(\frac{n\pi}{10}\right)\cos\left(\frac{n2\pi}{\tau_1}t\right)\right] + \sum_{n=31}^{\infty} \frac{2A}{n\pi}\sin\left(\frac{n\pi}{10}\right)\cos\left(\frac{n2\pi}{\tau_1}t\right)$$

in which the a_n terms within the brackets are those represented in the bandwidth shown in Fig. 1-24a.

Now let us think of another pulse train with a period τ_2 that is four times that of the first pulse train discussed, so that $t_0/\tau_2 = \frac{1}{40}$ with t_0 remaining unchanged. The Fourier series for this pulse train is

$$f(t) = \frac{0.025(2A)}{2} + \left[\sum_{n=1}^{120} \frac{2A}{n\pi}\sin\left(\frac{n\pi}{40}\right)\cos\left(\frac{n2\pi}{4\tau_1}t\right)\right] + \sum_{n=121}^{\infty} \frac{2A}{n\pi}\sin\left(\frac{n\pi}{40}\right)\cos\left(\frac{n2\pi}{4\tau_1}t\right)$$

in which the a_n terms within the brackets are those represented in the bandwidth shown in Fig. 1-24b.

Note from the arguments of the cosine functions in the two series that for a given frequency the n values of the second series must be four times as large as those of the first series. This means that in covering the same range of frequencies with the two series, the second series will have four times as many a_n values as the first series. This is shown in Fig. 1-24, and demonstrates that discrete frequency spectrums become more dense as t_0/τ decreases because of increasing τ.

As $\tau \rightarrow \infty$, we would expect the frequency spectrum to become continuous, and as we shall see later, the Fourier transform of a single pulse does result in a continuous spectrum.

The quantity in Eq. 1-41a,

$$\frac{\sin(n\pi t_0/\tau)}{n\pi t_0/\tau}$$

will have zero values (nulls) and maximum values (peaks) at identical frequency values in the arguments of the cosine terms for all different values of t_0/τ as τ increases. However, as can be seen in comparing the Fourier series written for the two pulse trains, the amplitudes decrease as τ increases, as shown in Fig. 1-24.

It is somewhat perplexing to note that $2At_0/\tau$ in Eq. 1-41a goes to zero as $\tau \rightarrow \infty$, which would seem to indicate the disappearance of the entire spectrum when a continuous spectrum is reached. However, Eq. 1-41a applies to individual harmonics of a periodic function for which τ is finite, although perhaps very large, and individual harmonics cannot be considered in a continuous spectrum. As mentioned before, the existence of a continuous frequency spectrum is demonstrated later utilizing the Fourier transform.

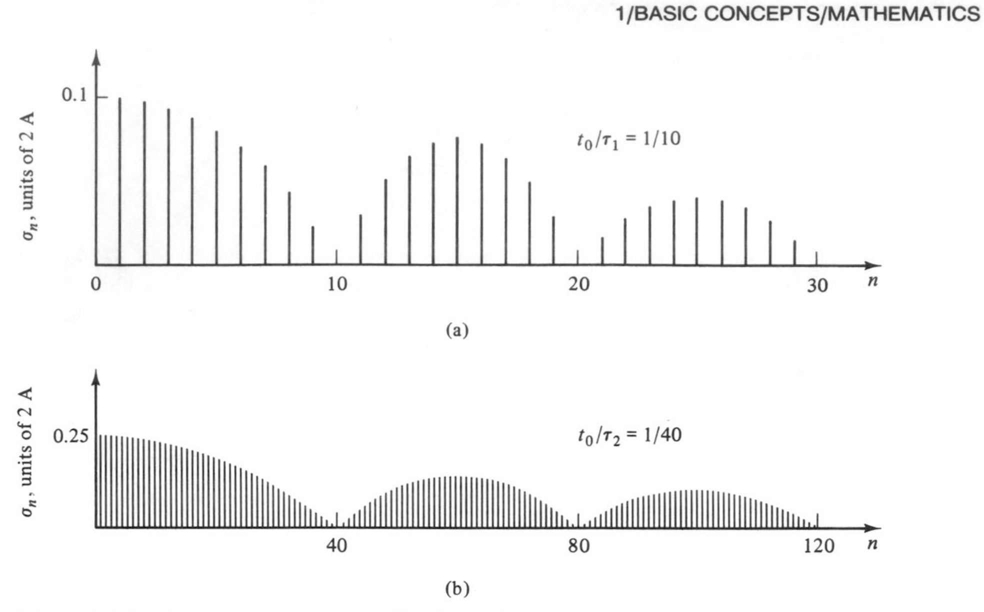

Figure 1-24 Frequency spectrums of pulse train.

Complex Form of Fourier Series

To develop the *Fourier integral* and the *Fourier transform pair,* it is advantageous to express the Fourier series in the complex form

$$f(t) = \sum_{n=-\infty}^{n=\infty} C_n e^{jn2\pi t/\tau} \tag{1-42}$$

To obtain this complex form, we note from Eqs. 1-14 and 1-15 that

$$\cos\frac{n2\pi t}{\tau} = \frac{e^{jn2\pi t/\tau} + e^{-jn2\pi t/\tau}}{2}$$

and

$$\sin\frac{n2\pi t}{\tau} = j\,\frac{e^{-jn2\pi t/\tau} - e^{jn2\pi t/\tau}}{2}$$

Substituting the right-hand side of these expressions into Eq. 1-25 gives

$$f(t) = \tfrac{1}{2}\sum_{n=0,1,2,\ldots}^{\infty} (a_n - jb_n)e^{jn2\pi t/\tau} + \tfrac{1}{2}\sum_{n=1,2,\ldots}^{\infty} (a_n + jb_n)e^{-jn2\pi t/\tau} \tag{1-43}$$

or

$$f(t) = \sum_{n=0,1,2,\ldots}^{\infty} C_n e^{jn2\pi t/\tau} + \sum_{n=1,2,\ldots}^{\infty} C_n^* e^{-jn2\pi t/\tau} \tag{1-44}$$

in which

$$C_n = \frac{a_n - jb_n}{2} \tag{1-45}$$

$$C_n^* = \frac{a_n + jb_n}{2} \qquad \text{(complex conjugate of } C_n\text{)} \tag{1-46}$$

(We shall see later that C_n for negative n equals C_n^* for positive n.) Thus, Eq. 1-44 can be written as

$$f(t) = \sum_{n=-\infty}^{n=\infty} C_n e^{jn2\pi t/\tau}$$

which is Eq. 1-42. The summation from $-\infty$ to $+\infty$ is necessary because Eq. 1-44 has terms containing both $e^{jn2\pi t/\tau}$ and $e^{-jn2\pi t/\tau}$. We can now write from Eqs. 1-28 and 1-29 that

$$\frac{a_n}{2} = \frac{1}{\tau}\int_{-\tau/2}^{\tau/2} f(t)\cos\frac{n2\pi t}{\tau}\,dt \tag{1-47}$$

and

$$\frac{jb_n}{2} = \frac{j}{\tau}\int_{-\tau/2}^{\tau/2} f(t)\sin\frac{n2\pi t}{\tau}\,dt \tag{1-48}$$

Subtracting Eq. 1-48 from Eq. 1-47 shows that

$$C_n = \frac{a_n - jb_n}{2} = \frac{1}{\tau}\int_{-\tau/2}^{\tau/2} f(t)\left[\cos\frac{n2\pi t}{\tau} - j\sin\frac{n2\pi t}{\tau}\right]dt \tag{1-49a}$$

or

$$C_n = \frac{1}{\tau}\int_{-\tau/2}^{\tau/2} f(t)e^{-jn2\pi t/\tau}\,dt \tag{1-49b}$$

Examination of the bracketed term in Eq. 1-49a reveals that C_n for a negative n equals C_n^* for a positive n, a relationship that was used earlier to write Eq. 1-44 in the form of Eq. 1-42.

Substituting Eq. 1-49b into Eq. 1-42 yields the Fourier series in complex form as

$$f(t) = \sum_{n=-\infty}^{n=\infty} \frac{1}{\tau}\left[\int_{-\tau/2}^{\tau/2} f(t)e^{-jn2\pi t/\tau}\,dt\right]e^{jn2\pi t/\tau} \tag{1-50}$$

Fourier Integral

The Fourier integral is obtained from Eq. 1-50 by considering that $\tau \to \infty$. When τ is very large, the circular frequency is conversely very, very small. Therefore, let us designate this very, very small circular frequency as $\Delta\omega$. If $\Delta\omega$ is then considered as the fundamental circular frequency of a time-varying function, and ω represents the circular frequency components of such a function, then in general

$$\omega = 0,\ \Delta\omega,\ 2\Delta\omega,\ \text{etc.}$$

or

$$\omega = n\,\Delta\omega$$

Using this notation, the period τ can be expressed as

$$\frac{1}{\tau} = \frac{\Delta\omega}{2\pi}$$

and Eq. 1-50 becomes

$$f(t) = \sum_{n=-\infty}^{n=\infty} \frac{\Delta\omega}{2\pi} \left[\int_{-\tau/2}^{\tau/2} f(t) e^{-jn\Delta\omega t} \, dt \right] e^{jn\Delta\omega t} \tag{1-51}$$

It now follows that as $\tau \to \infty$, $\Delta\omega \to d\omega$, and the terms of the series of Eq. 1-51 become infinitesimally close together in frequency (continuous frequency spectrum). The series of Eq. 1-51 becomes an integral having the form of

$$f(t) = \int_{-\infty}^{\infty} \frac{1}{2\pi} \underbrace{\left[\int_{-\infty}^{\infty} f(t) e^{-j\omega t} \, dt \right]}_{F(\omega)} e^{j\omega t} \, d\omega \tag{1-52}$$

which is the *Fourier integral.*

Fourier Transform Pair

The Fourier integral shown in Eq. 1-52 can be separated into the following two integral equations:

$$f(t) = \frac{1}{2\pi} \int_{-\infty}^{\infty} F(\omega) e^{j\omega t} \, d\omega \tag{1-53}$$

and

$$F(\omega) = \int_{-\infty}^{\infty} f(t) e^{-j\omega t} \, dt \tag{1-54}$$

Equation 1-54 defines the *Fourier transform* of $f(t)$, and the two together are known as the *Fourier transform pair.* Noting that $\omega = 2\pi f$, the Fourier transform pair can also be written as

$$f(t) = \int_{-\infty}^{\infty} F(f) e^{j2\pi ft} \, df \tag{1-55}$$

and

$$F(f) = \int_{-\infty}^{\infty} f(t) e^{-j2\pi ft} \, dt \tag{1-56}$$

The Fourier transform pair is the mathematical basis for signal analysis, and Eqs. 1-53 and 1-54, or Eqs. 1-55 and 1-56, are of considerable importance in analyzing the response of mechanical or structural systems subjected to shocks, nonperiodic excitations, and so forth. It is important to note that a function $f(t)$ in the time domain is transformed into a function $F(f)$ in the frequency domain, in which $F(f)$ is the frequency spectrum of $f(t)$. It should also be emphasized that $F(f)$ contains all of the frequency components present in $f(t)$. Conversely, the frequency function $F(f)$ is transformed into the time-domain function $f(t)$ by Eq. 1-55.

The Laplace Transform from the Fourier Transform

If we let $j\omega = s$ in Eq. 1-54, we obtain

$$F(s) = \int_{-\infty}^{\infty} f(t) e^{-st} \, dt$$

This integral does not converge for certain functions $f(t)$. However, if s is permitted to be a complex quantity, $s = \sigma + j\omega$, instead of a pure imaginary one, then σ can be made large enough to force convergence of the integral for a wide class of functions $f(t)$. With $s = \sigma + j\omega$, the above integral is called the *two-sided Laplace transform* of $f(t)$. In many instances, we are not concerned with what happened prior to $t = 0$, so the integration is performed from $t = 0$ to $t = \infty$, which gives the more commonly used *one-sided Laplace transform,*

$$F(s) = \int_0^{\infty} f(t)e^{-st}\,dt$$

This integral has the desirable property of being able to transform a *differential equation* in the independent variable, time t, to an algebraic equation in the variable s. After some algebraic manipulation of the equation in s, an *inverse transformation* can be used to obtain the solution of the differential equation. There is an inverse Laplace transform integral analogous to the one in Eq. 1-53, which theoretically could be used for the inverse transformation but which is not used in practice. Instead, tables of Laplace transform pairs, $f(t)$ and $F(s)$, are formed (see Appendix A), and they are used both for transforming the differential equation and for performing the inverse transformation to obtain the solution of the differential equation. The use of a table of Laplace transform pairs leads to a great simplification of many engineering problems.

EXAMPLE 1-3

It is desired to transform the rectangular pulse shown in Fig. 1-25 from the time domain to the frequency domain using a Fourier transform.

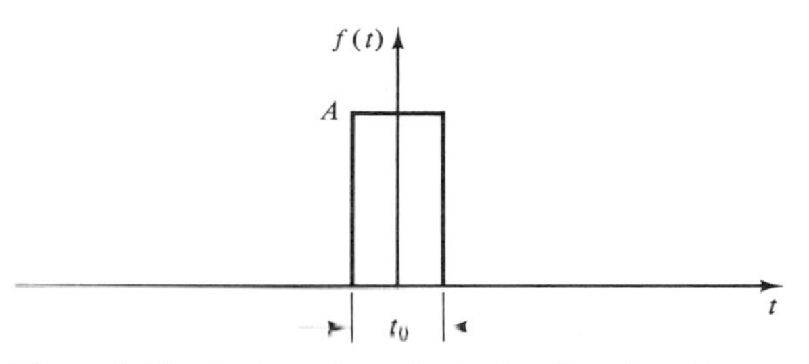

Figure 1-25 Rectangular pulse in the time domain.

Solution. In the time domain

$$f(t) = \begin{Bmatrix} A & -t_0/2 < t < t_0/2 \\ 0 & -t_0/2 > t > t_0/2 \end{Bmatrix}$$

Thus, from Eq. 1-56,

$$F(f) = \int_{-t_0/2}^{t_0/2} Ae^{-j2\pi ft}\,dt$$

or

$$F(f) = \frac{A}{2\pi f}\left(\frac{e^{j\pi ft_0} - e^{-j\pi ft_0}}{j}\right)$$

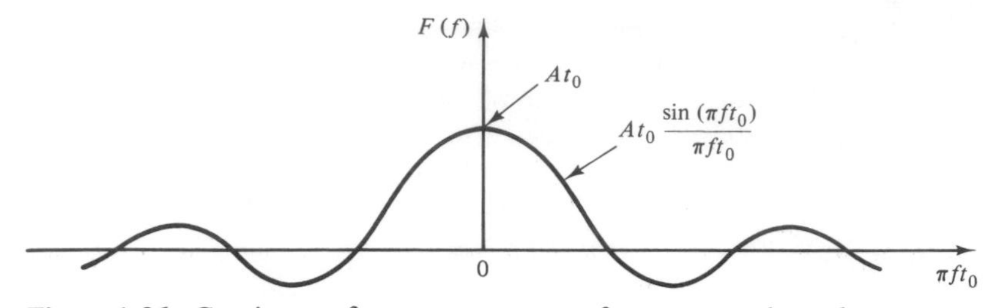

Figure 1-26 Continuous frequency spectrum for a rectangular pulse.

Referring to Eq. 1-15, it can be seen that $F(f)$ can be written more simply as

$$F(f) = At_0 \frac{\sin(\pi f t_0)}{\pi f t_0} \tag{1-57}$$

Figure 1-26 shows that Eq. 1-57 plots as a continuous function in the frequency domain. This means that all the possible frequencies are theoretically present in the rectangular pulse. Therefore, a single pulse is ideally suited for the excitation of the various modes of vibration of a system.

Since negative frequencies have no physical significance, and spectrum analyzers recognize only positive amplitudes, the frequency spectrum as observed on a spectrum analyzer for a single pulse would appear with *side lobes* as shown in Fig. 1-27. For a very small pulse width t_0, Eq. 1-57 shows that the initial amplitudes of the frequency spectrum are essentially constant over a fairly broad frequency range. This means that the frequency spectrum of a single pulse is somewhat like band-limited *white noise* in this region. In fact, as we see in Example 1-4, a *unit impulse* (one of theoretically zero width and infinite height) gives a constant amplitude over the frequency range of $-\infty \leq f \leq \infty$, which is defined as white noise.

Before determining the frequency spectrum of such a unit impulse, let us define the *unit impulse function.* Suppose that the pulse of height A and duration t_0 shown in Fig. 1-25 has a magnitude of unity (an area of unity in a time plot). If t_0 is now decreased and the force is increased (with the area remaining unity) until the duration of the pulse becomes very short in comparison with the undamped natural period τ_n of the system upon which it acts, the force is referred to as an impulsive one, and the rectangular pulse is called a unit impulse. Letting the process continue further, the impulse approaches being of infinitesimal width and infinite height, while still retaining its unit area. The limit of this process

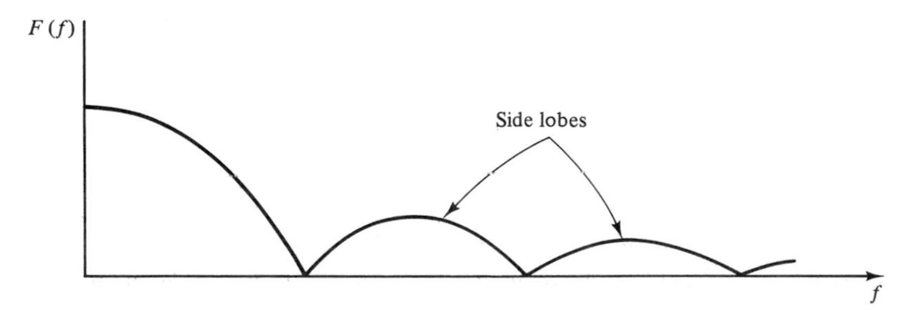

Figure 1-27 Frequency spectrum showing positive frequencies and amplitudes.

is referred to as a *unit impulse* function or the *Dirac delta function* (often shortened to *delta function*). The delta function is denoted by $\delta(t)$ if it occurs at $t = 0$ and by $\delta(t - a)$ if it occurs at some general time $t = a$ as shown in Fig. 1-28. Mathematically, the delta function is sometimes defined as

$$\left.\begin{aligned} \delta(t - a) &= 0 \qquad (t \neq a) \\ \int_{-\infty}^{+\infty} \delta(t - a)\, dt &= 1 \end{aligned}\right\} \qquad \begin{aligned} &(1\text{-}58a) \\ &(1\text{-}58b) \end{aligned}$$

in which Eq. 1-58b indicates that the unit impulse has unit area. If we multiply a unit impulse $\delta(t - a)$ by some constant C that is different from 1, we obtain an impulse with a different "magnitude" (of different area, but still having infinite height). Such an impulse is expressed by $C\delta(t - a)$.

The limiting process described above for obtaining the delta function from a unit rectangular pulse can be applied to a unit pulse of different shape (triangular, for example) to obtain a delta function as defined by Eq. 1-58.

An impulse as defined by Eq. 1-58 cannot occur in a physical system and is thus sometimes referred to as an "ideal" impulse. However, if the magnitude of a force input to a real system is very large and of very short duration as compared with the natural period of the system, the impulse input can be approximated by the ideal impulse function $C\delta(t - a)$. In this instance the effect on the system depends primarily upon the magnitude of the impulse and not upon its exact shape (recall that a unit pulse of *any shape* becomes the delta function in the limit).

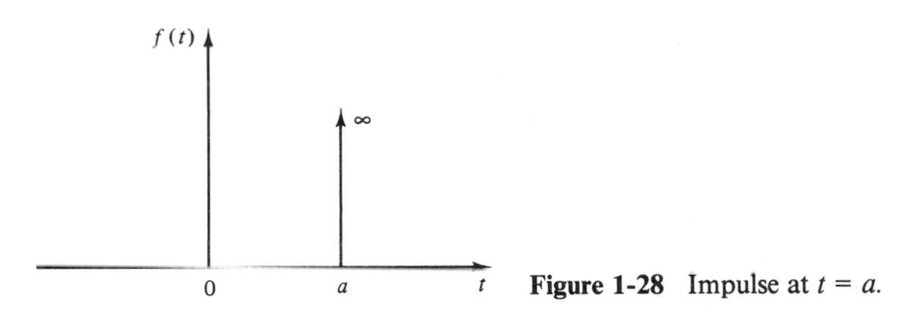

Figure 1-28 Impulse at $t = a$.

EXAMPLE 1-4

We are interested in determining the frequency spectrum, or Fourier transform, of the function

$$f(t) = C\delta(t) \qquad (1\text{-}59)$$

in which $\delta(t)$ is the delta function shown in Fig. 1-29.

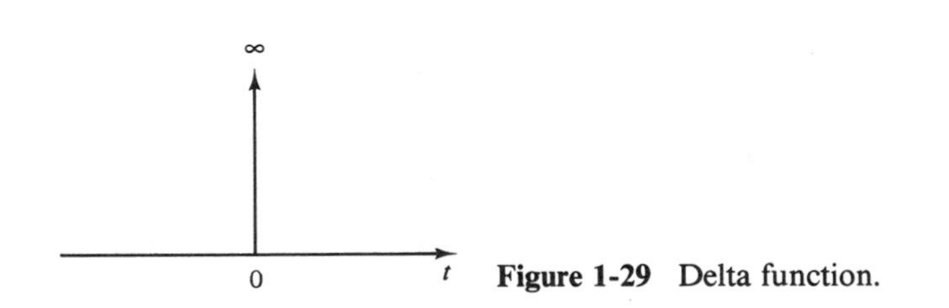

Figure 1-29 Delta function.

Solution. To obtain the frequency spectrum of the function given in Eq. 1-59, we write from Eq. 1-56 that

$$F(f) = C \int_{-\infty}^{\infty} \delta(t) e^{-j2\pi ft} \, dt$$

Noting that $e^{-j2\pi ft}$ is equal to 1 at $t = 0$, we obtain

$$F(f) = C$$

Thus, the frequency spectrum of the function $C\delta(t)$ is a constant as shown in Fig. 1-30.

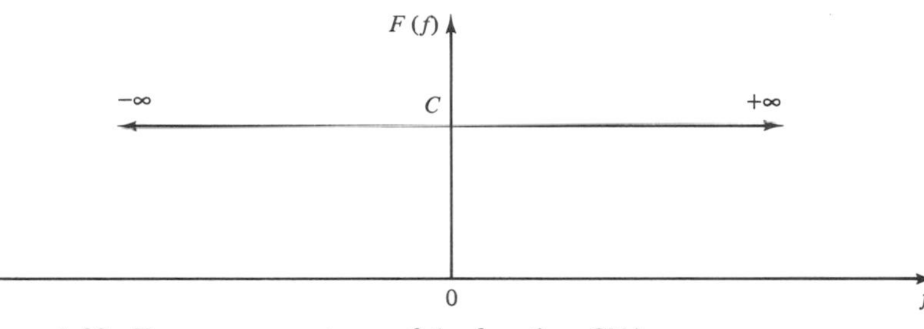

Figure 1-30 Frequency spectrum of the function $C\delta(t)$.

Impact Hammers

Small impact hammers such as the one shown in Fig. 1-31 are used to excite small structures and machines with an impulse. The width of the impulse, and hence the frequency range

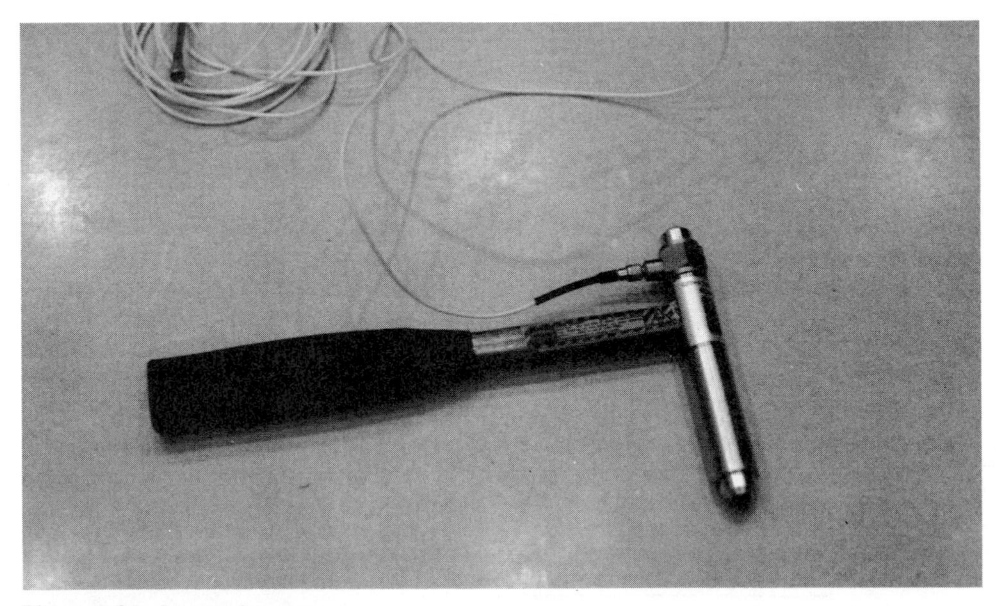

Figure 1-31 Impact hammer.

(a)

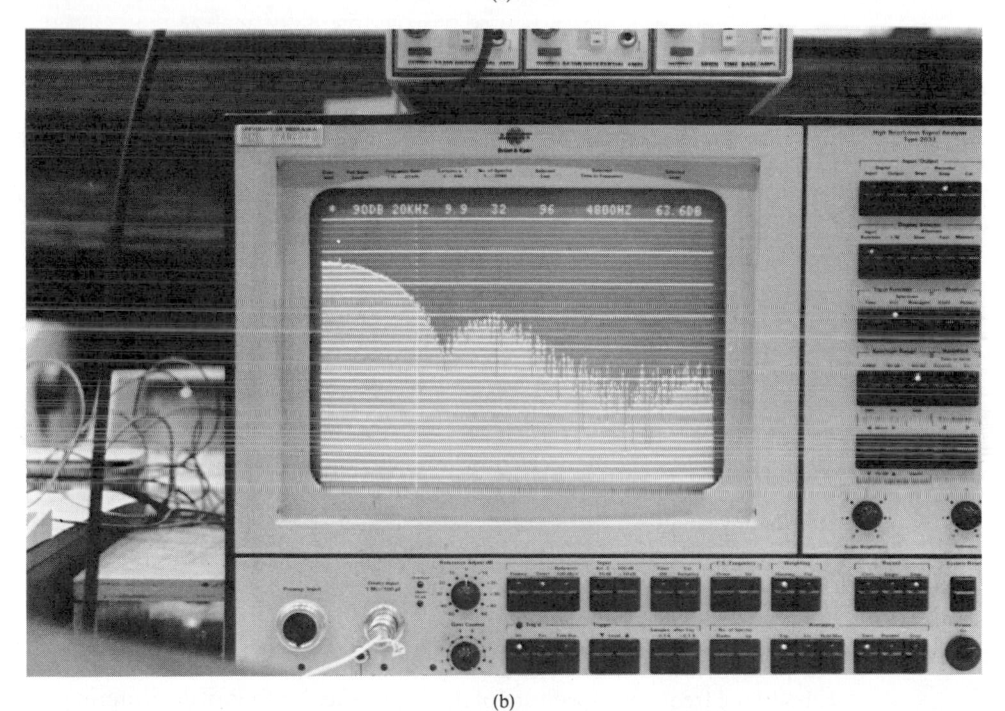

(b)

Figure 1-32 Recorded impulse and corresponding frequency spectrum. (a) Recorded impulse $f(t)$ from impact hammer (steel tip) striking concrete floor. (b) Frequency spectrum $F(f)$ of impulse for (0 to 20 kHz).

over which the amplitudes are essentially constant, depends upon the hardness of the hammer striker tip and upon the material and stiffness of the system to which the hammer is applied. The striker tips are interchangeable and are usually made of steel, aluminum, plastic, or rubber. The impulse $f(t)$ in the time domain is obtained from the output of a force transducer in the head of the impact hammer.

Figure 1-32a shows a recorded impulse on a spectrum analyzer that was obtained from the output of a steel-tipped impact hammer striking a concrete floor. The width of the impulse at its base was measured as 0.23 millisecond (ms). The frequency spectrum of 0 to 20 kHz generated by the spectrum analyzer's processor for this impulse is shown in Fig. 1-32b. It should be noted that the frequency spectrum is essentially continuous over the 0- to 20-kHz range. The reader should also note that the amplitude of the spectrum is practically constant up to approximately 2000 Hz, which means that the energy of each frequency component is also essentially constant in this region. The frequency spectrum exhibits two side lobes.

1-8 BASIC INSTRUMENTATION FOR VIBRATION ANALYSIS

It is hoped that the preceding discussions of Fourier series, Fourier transforms, frequency spectrums, impact hammers, and fast-Fourier-transform frequency analyzers have enhanced the reader's understanding not only of theoretical vibration analysis but also of the types of instrumentation presently used to experimentally analyze the vibration of mechanical and structural systems.

The introduction of fast-Fourier-transform frequency analyzers and impact hammers, along with other sophisticated means of electronic instrumentation, has resulted in significant and important testing techniques for measuring and analyzing the dynamic behavior of structural systems.

Figure 1-33 shows a schematic diagram of a simple instrumentation setup for determining the natural frequencies of a structural system. It includes a piezoelectric accelerometer, a fast-Fourier-transform frequency analyzer, and an X-Y plotter for making a permanent record of the experimental results.

Beeswax is commonly used to attach accelerometers to a machine or structure if the acceleration does not exceed 10 g's. The use of beeswax permits the accelerometers to be moved easily to various locations. When the acceleration to be measured is greater than 10 g's, the accelerometers should be secured by other means such as epoxy cement, threaded studs, or some other type of light, secure fastener.

The output of the accelerometer is the input to a charge amplifier (signal conditioner), which in turn provides the input to the frequency analyzer. As previously mentioned, the frequency analyzer can display the vibration response from the accelerometer in either the time or the frequency domain. An oscilloscope is also very helpful in monitoring the vibration response in the time domain.

The plate structure shown in Fig. 1-33 can be excited by a simple tap of an impact hammer. The impulse excitation can also be provided by simply using a small metal rod or piece of wood. However, an impact hammer should be used when it is important to obtain the frequency spectrum of the impulse as a means of determining which frequencies of the structure will be excited. For example, the magnitude of the frequency spectrum for the impulse shown in Fig. 1-32 is essentially flat up to approximately 2000 Hz, which

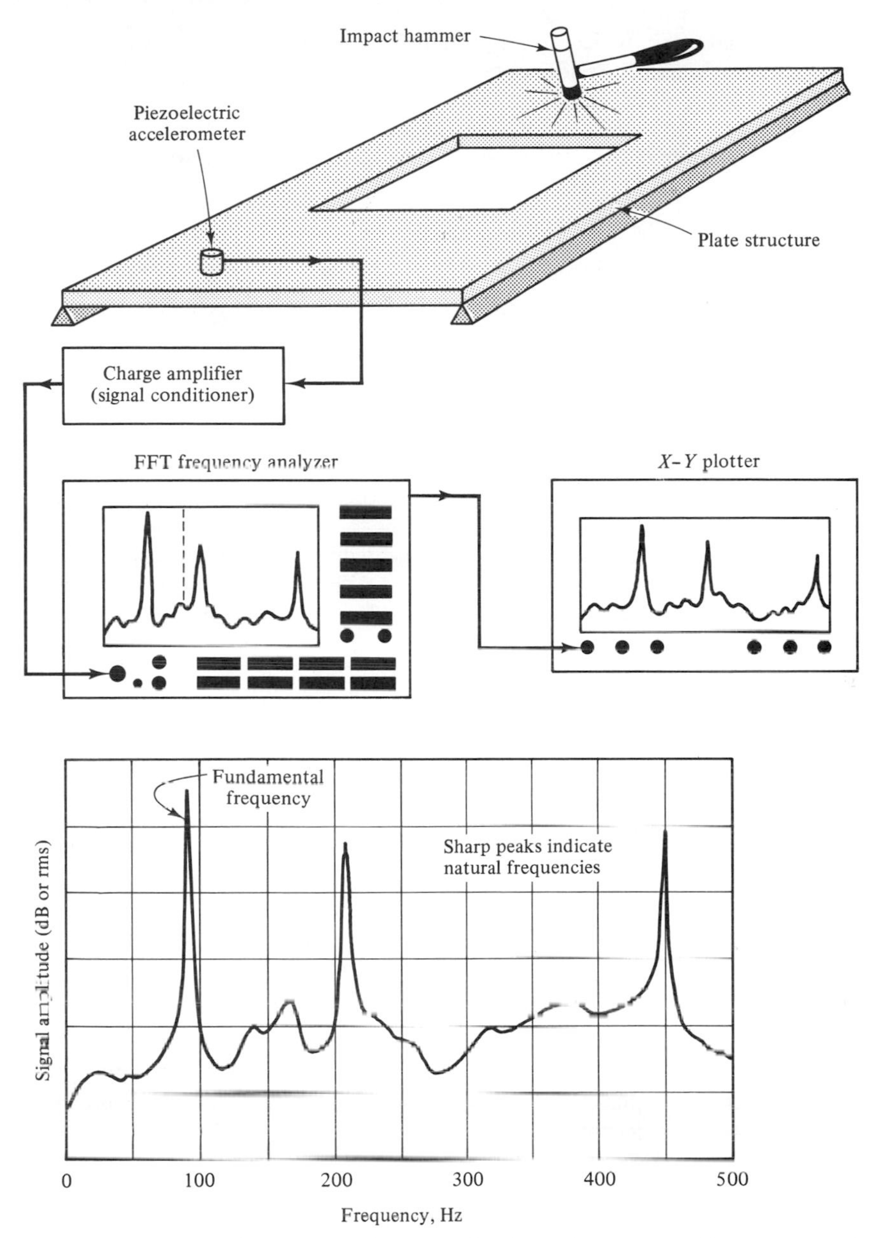

Figure 1-33 Simple instrumentation setup for determining natural frequencies of a structural system.

means that such an impulse has the necessary energy and frequency content to excite all the natural frequencies of a structure within the range of 0 to 2000 Hz.

Even though this is a rather simple instrumentation system compared with that for an automated modal-analysis system, it is capable of determining modal frequencies and modal damping, which are of primary importance in evaluating the dynamic behavior of structural systems. Besides obtaining frequency spectrums, most FFT frequency analyzers can perform other operations that are useful in vibration studies. Some of these operations include the measurement of time, division to obtain transfer functions, scanning to obtain modal damping, and the storage of spectrums for comparative purposes.

1-9 DECIBEL/MEAN-SQUARE VALUE/ROOT-MEAN-SQUARE VALUE

Let us now discuss some commonly used terms and concepts associated with the use of various types of instruments to obtain vibration measurements.

Decibel The advantages derived from representing dynamic quantities over a wide range of up to several decades have resulted in the use of the decibel as a popular measure in sound and vibration measurements. The decibel (dB) scale evolved from the field of transmission-line theory, in which the original definition of a decibel was based upon the *power ratio* P_2/P_1 as

$$\text{dB} = 10 \log \frac{P_2}{P_1} \tag{1-60}$$

in which P_1 is the *reference power.* Since power is proportional to the square of the voltage, the decibel can also be expressed as

$$\text{dB} = 10 \log\left(\frac{X_2}{X_1}\right)^2$$

or as

$$\text{dB} = 20 \log \frac{X_2}{X_1} \tag{1-61}$$

In these expressions the ratio X_2/X_1 can be the ratio of a variety of quantities such as displacements, velocities, accelerations, and volts, as long as X_2 is the same *type* of quantity as the *reference* quantity X_1. For example, an instrument employing an appropriate fixed internal reference voltage could give a readout in decibels corresponding to the voltage input from some type of transducer such as an accelerometer.

Manufacturers of vibration equipment such as transducers, amplifiers, and recorders will normally specify the frequency range over which the response of the equipment is essentially independent of the frequency of the phenomena being measured. The manufacturers will also frequently specify a frequency for which the response of the equipment is 3 dB down, which indicates the point at which the equipment deviates considerably from the flat-frequency response region.

To understand this better, let us determine the ratio X_2/X_1 from Eq. 1-61, which yields

$$-3\ \text{dB} = 20 \log \frac{X_2}{X_1}$$

and

$$\frac{X_2}{X_1} = 0.71$$

where X_2 = instrument output at a frequency outside the flat-frequency response range
X_1 = instrument output if it were operating within the flat-frequency response range

The ratio above shows that the output of the instrument would be approximately 30 percent too low for an accurate measurement. The -3 dB also corresponds to the half-power level, that is, $(X_2/X_1)^2 = 0.5$.

Mean-Square Value The mean-square value is defined mathematically as

$$\overline{x^2} = \frac{1}{T}\int_0^T [f(t)]^2\, dt \tag{1-62}$$

which reveals that the mean-square value is the average of the squared values of the time function $f(t)$ over a time period T. The time T necessary to obtain an accurate mean-square value depends upon the frequencies present in $f(t)$. In general, the presence of low frequencies requires a long time T. Ideally, T should be infinite, but for practical purposes it must of course be finite.

Root-Mean-Square Value The root-mean-square value (rms) is the square root of the mean-square value. That is,

$$x_{\text{rms}} = \sqrt{\frac{1}{T}\int_0^T [f(t)]^2\, dt} \tag{1-63}$$

As an example, the mean-square value of the sine wave, $A \sin \omega t$, or $A \sin 2\pi t/\tau$, averaged over a very long interval of time T is

$$\begin{aligned}\overline{x^2} &= \lim_{n\to\infty} \frac{n}{n\tau/2}\int_0^{\tau/2} A^2 \sin^2\frac{2\pi t}{\tau}\, dt \\ &= \frac{2A^2}{\tau}\int_0^{\tau/2} \frac{1}{2}\left(1 - \cos\frac{4\pi t}{\tau}\right) dt \\ &= \frac{A^2}{2}\end{aligned} \tag{1-64}$$

in which n denotes a large number of half sine waves as shown in Fig. 1-34, so that $T = n\tau/2$.

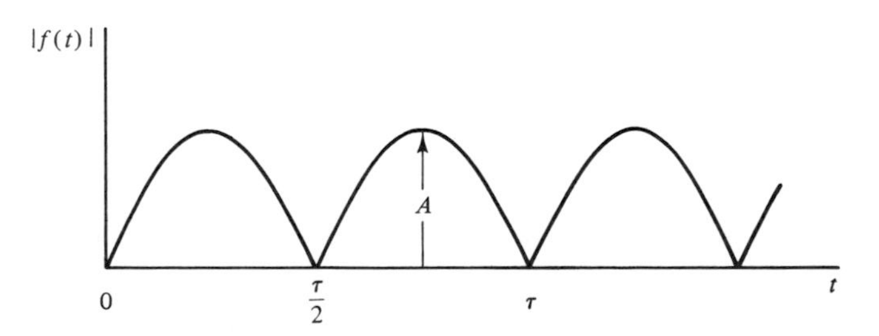

Figure 1-34 Absolute value of $A \sin 2\pi t/\tau$.

Then, from Eq. 1-64, the root-mean-square value of the sine wave is

$$x_{\text{rms}} = \frac{A}{\sqrt{2}} = 0.707A$$

In this instance, the same value is obtained for a sine wave averaged over a short period of time, $T = \tau/2$.

EXAMPLE 1-5

It is desired to determine the mean-square value and the root-mean-square value of the triangular wave shown in Fig. 1-35.

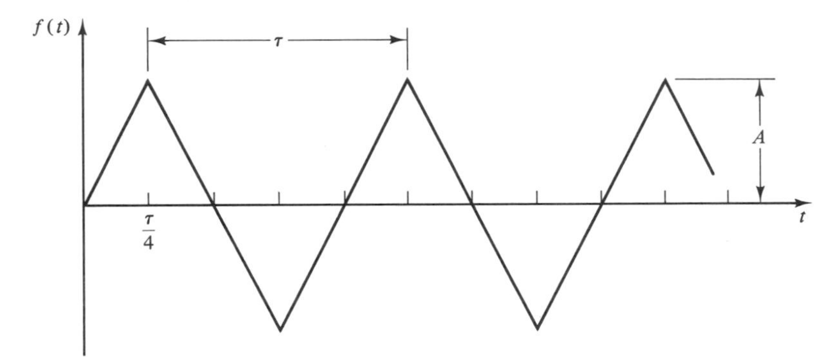

Figure 1-35 Triangular wave.

Solution. Considering that

$$f(t) = \frac{4At}{\tau} \qquad 0 \leq t \leq \frac{\tau}{4}$$

and the repetitive nature of the function, the mean-square value is

$$\overline{x^2} = \frac{4}{\tau}\int_0^{\tau/4}\left(\frac{4At}{\tau}\right)^2 dt = \frac{A^2}{3}$$

The root-mean-square value of the triangular wave is then

$$x_{\text{rms}} = \frac{A}{\sqrt{3}} = 0.577A$$

Absolute Maximum Value Quite often the response $x(t)$ of a system will consist of several, or numerous, sine waves with different amplitudes X_i and phase angles ϕ_i, such as

$$x(t) = X_1 \sin(\omega_1 t - \phi_1) + X_2 \sin(\omega_2 t - \phi_2) + \cdots + X_n \sin(\omega_n t - \phi_n) \qquad (1\text{-}65)$$

In some instances it may not be feasible or practical to determine the maximum value of $x(t)$. In such instances the *absolute maximum value* of the response could be determined by simply summing the amplitudes X_i. That is,

$$x(t)\Big|_{\text{abs. max.}} = X_1 + X_2 + X_3 + \cdots + X_n \tag{1-66}$$

This value is of course too high an estimate for the maximum response, but it would serve to establish an upper bound, which would be on the "safe side" for design.

PROBLEMS

Problems 1-1 through 1-10 (Sections 1-1 through 1-3)

1-1. The vibration of a mass is characterized by a sinusoidal wave having an amplitude of $A = 2.5$ mm and a circular frequency of $\omega = 75$ rad/s. **(a)** Write an equation defining the displacement of the mass at any time t, and **(b)** determine the period, maximum velocity, and maximum acceleration of the mass.

Partial ans: $\tau = 0.084$ s
$\dot{x}_{max} = 0.188$ m/s
$\ddot{x}_{max} = 14.06$ m/s^2

1-2. An accelerometer is used to measure the sinusoidal vibration of a structure. The acceleration is monitored on an oscilloscope which reveals that the amplitude of the acceleration is 3.2 g's (g = 9.81 m/s^2) and the period of the oscillation is 45 ms. Determine the amplitude of the displacement.

Ans: $A = 1.61$ mm

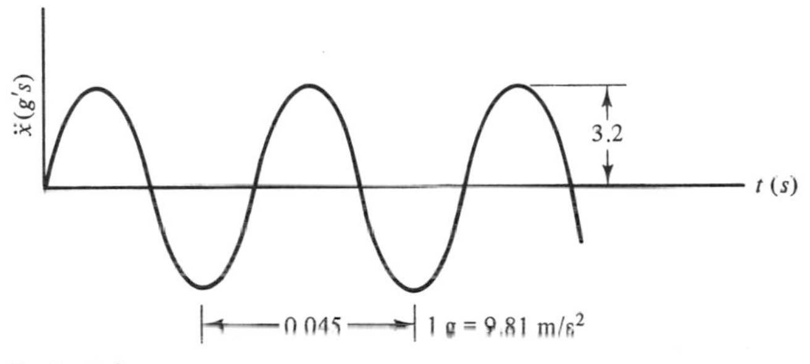

Prob. 1-2

1-3. The period and maximum displacement of the sinusoidal motion of a point on a structure are 0.125 s and 0.25 in., respectively. For that point determine **(a)** the frequency in Hz, **(b)** the maximum velocity, and **(c)** the maximum acceleration in g's (g = 386 in./s^2).

Partial ans: $\ddot{x}_{max} = 1.64$ g's

1-4. A vibration-measuring system that is capable of measuring the displacement, velocity, and acceleration of a system is used to measure the amplitude of both the displacement and the velocity of a vibrating floor of a building. The measurements indicate a maximum sinusoidal displacement of 0.005 in. and a maximum velocity of 0.5 in./s. Determine **(a)**

the frequency of vibration, **(b)** the period of vibration, and **(c)** the maximum acceleration of the floor at the point of measurements.

Partial ans: $\ddot{x}_{max} = 0.13\ g$'s

1-5. Determine the magnitude and direction of the resultant complex vector of the complex vectors $4e^{j\pi}$ and $3e^{j\pi/2}$.

1-6. Show that $je^{j\theta}$ is a complex vector that is 90° ahead of the complex vector $e^{j\theta}$.

1-7. If $z_1 = 3 + j4$ and $z_2 = 4 + j2$, determine the magnitude and direction of z_1/z_2.

Ans: $z_1/z_2 = 1.118e^{j(0.464)}$

1-8. Consider that $x = 8e^{jn\pi}$ with $n = 0, 2, 4, \ldots$, and show that the cube roots of x are $x_1 = 2$, $x_2 = 2[-0.5 + j(0.866)]$, and $x_3 = 2[-0.5 - j(0.866)]$.

1-9. Recalling the hyperbolic functions

$$\cosh u = \frac{e^u + e^{-u}}{2}$$

$$\sinh u = \frac{e^u - e^{-u}}{2}$$

use complex algebra to prove that

$$\cos ju = \cosh u$$

$$\sin ju = \frac{-\sinh u}{j}$$

in which u is real and $j = \sqrt{-1}$. Using these results, show also that the magnitude of $\sin(a + jb)$ is equal to

$$\sqrt{(\sin a \cosh b)^2 + (\cos a \sinh b)^2}$$

in which a and b are real.

1-10. Use complex algebra and prove that

$$\cosh(a + jb) = \cosh a \cos b + j \sinh a \sin b$$

in which a and b are real. The reader should recall that

$$\cosh u = \frac{e^u + e^{-u}}{2}$$

and

$$\sinh u = \frac{e^u - e^{-u}}{2}$$

Problems 1-11 through 1-22 (Sections 1-4 through 1-9)

1-11. Determine the Fourier coefficients for $f(t)$, which is the absolute value of the sine wave $A \sin(\pi t/T)$, as shown in the accompanying figure.

Ans: $a_0 = 4A/\pi$
$a_n = 0 \qquad n = 1, 3, 5, 7, \ldots$
$a_n = -4A/(n^2 - 1)\pi \qquad n = 2, 4, 6, \ldots$
$b_n = 0 \qquad n = 1, 2, 3, \ldots$

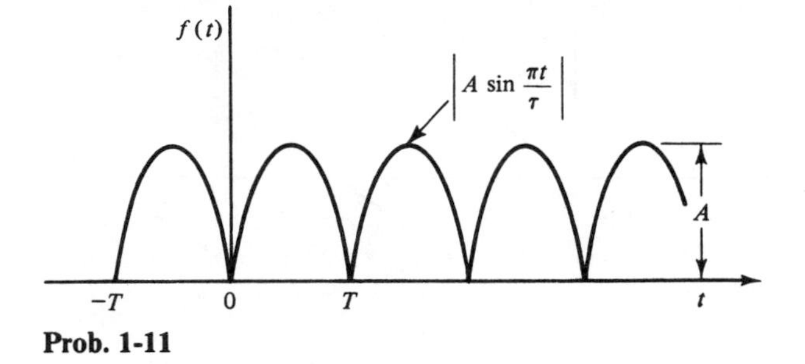

Prob. 1-11

1-12. Determine the Fourier series for the triangular wave having the origin shown in the accompanying figure. Sketch the frequency spectrum for $T = 0.1$ s.

Partial ans: $f(t) = F_0\left[\frac{1}{2} + \frac{4}{\pi^2}\left(\cos\frac{\pi t}{T} + \frac{1}{(3)^2}\cos\frac{3\pi t}{T} + \cdots\right)\right]$

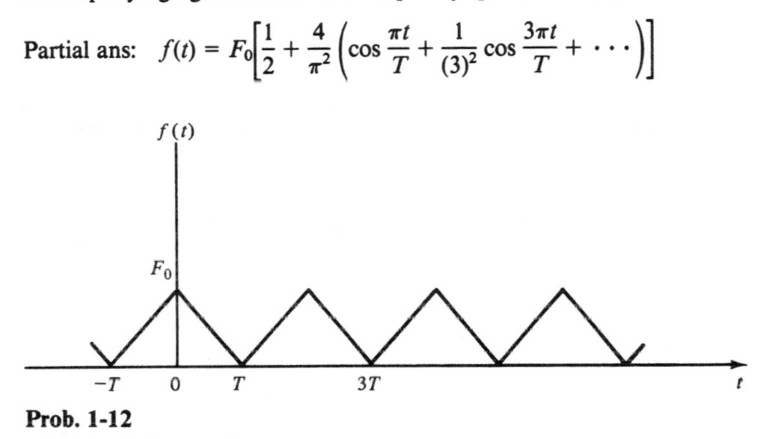

Prob. 1-12

1-13. Determine the Fourier series for the triangular wave having the origin shown in the accompanying figure. Sketch the frequency spectrum for $T = 0.1$ s. If you have worked Prob. 1-12 also, which series (and corresponding origin) gives the better fit for $f(t)$ at $t = 0$ and $t = T$?

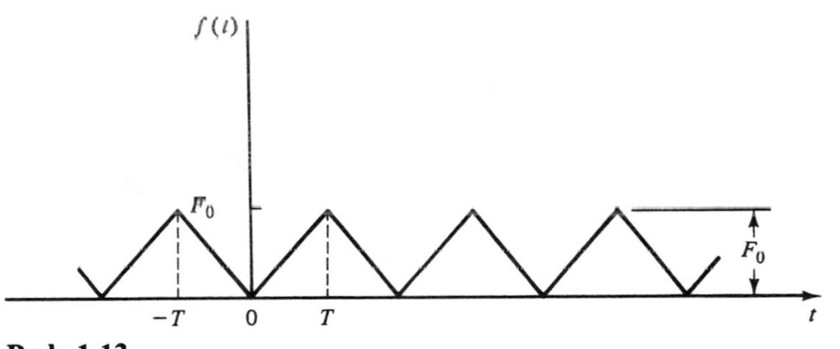

Prob. 1-13

1-14. Determine the Fourier series for the $f(t)$ shown in the accompanying figure. Sketch the frequency spectrum for $T = 0.05$ s.

Partial ans: $f(t) = \frac{2A}{\pi}\sum_{n=1}^{\infty}\frac{1}{n}(-\cos n\pi)\sin\frac{n\pi t}{T}$

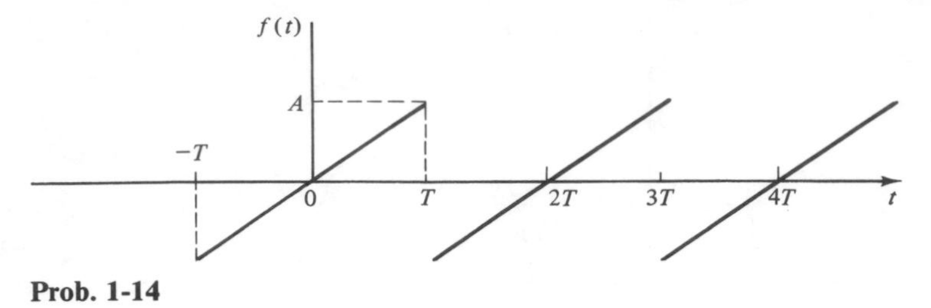

Prob. 1-14

1-15. A fixed-fixed beam is subjected to a uniformly distributed force w_0 over the length $2c$ of the beam as shown in the accompanying figure. The magnitude of the distributed force also varies with time such that

$$w(x, t) = f(x)g(t)$$

in which $f(x)$ is the spatial function of the uniformly distributed force w_0, which varies with time according to $g(t)$. Express the spatial distribution $f(x)$ as a Fourier series. (Hint: Comparing Fig. 1-16 and the figure for this problem, note that x corresponds to t and l to π/ω.)

Ans: $$f(x) = \frac{w_0 c}{l} + \frac{2w_0}{\pi} \sum_{n=1,2,3,\ldots}^{\infty} \frac{1}{n} \sin \frac{n\pi c}{l} \cos \frac{n\pi x}{l}$$

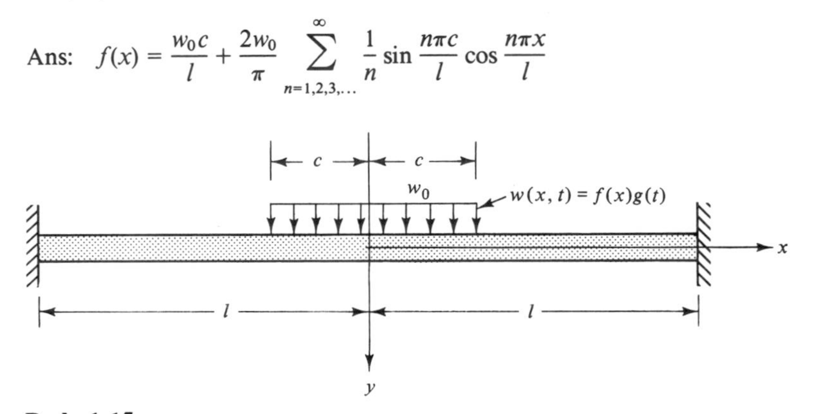

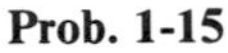

Prob. 1-15

1-16. It was shown in Example 1-1 that $f(t)$ for the square wave shown in Fig. 1-17 is represented by the Fourier series

$$f(t) = \frac{40}{\pi}\left(\sin \omega_1 t + \frac{1}{3} \sin 3\omega_1 t + \frac{1}{5} \sin 5\omega_1 t + \cdots\right)$$

Noting that $\omega_1 = 2\pi/\tau_1$ and that $f(t) = 10$ at $t = \tau_1/4$, show that the above series becomes

$$1 - \frac{1}{3} + \frac{1}{5} - \frac{1}{7} + \cdots = \frac{\pi}{4}$$

1-17. The frequency spectrum shown in the accompanying figure is defined by

$$F(f) = A[\delta(f - f_n) + \delta(f + f_n)]$$

in which $\delta(f - f_n)$ and $\delta(f + f_n)$ are delta functions in the frequency domain. Transform $F(f)$ to the time domain to determine $f(t)$.

Ans: $f(t) = A(e^{j2\pi f_n t} + e^{-j2\pi f_n t})$

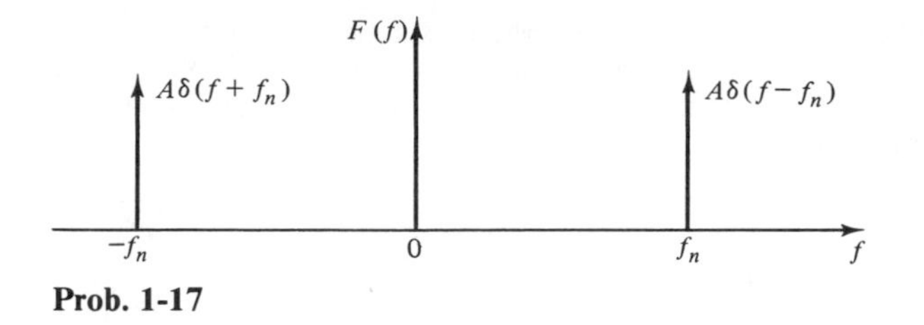

Prob. 1-17

1-18. Referring to the solution of Prob. 1-17, transform

$$f(t) = A \cos 2\pi f_n t$$

to the frequency domain to obtain $F(f)$.

Ans: $F(f) = \frac{A}{2}[\delta(f - f_n) + \delta(f + f_n)]$

1-19. The frequency spectrum observed on a spectrum analyzer for a vibration measured with an accelerometer indicates two sinusoidal waves for which the frequencies are $f_1 = 10$ Hz and $f_2 = 30$ Hz. The magnitudes of the frequency components are shown in volts (rms) on the accompanying figure. If the calibration of the accelerometer is 10 V/g, determine the amplitudes A_1 and A_2 of the sinusoidal acceleration components in in./s^2 ($g = 386$ in./s^2).

Ans: $A_1 = 43.68$ in./s^2
$A_2 = 27.30$ in./s^2

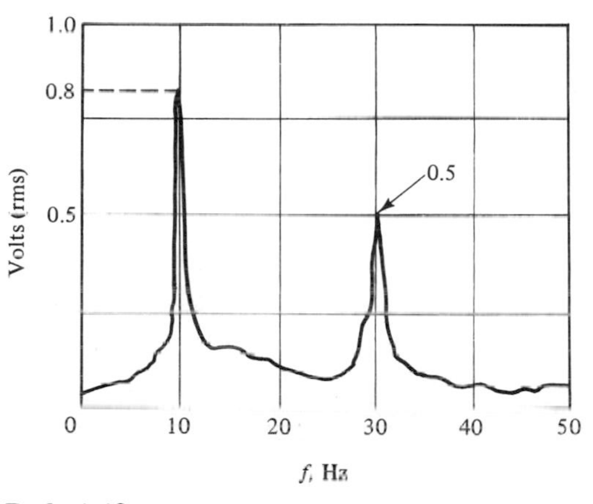

Prob. 1-19

1-20. Referring to Prob. 1-19 and considering that the reference voltage e_r of the spectrum analyzer is 0.001 V (rms), what would the amplitude readings be in decibels for the acceleration amplitude components shown?

Ans: $A_1 = 58.06$ dB
$A_2 = 53.98$ dB

1-21. The scale of the spectrum analyzer giving the amplitudes of the frequency spectrum shown in part a of the accompanying figure is in volts (rms). The amplitudes of the peaks of the spectrum at f_1 and f_2 are 10 and 1.5 V, respectively. Since the amplitude at f_2 is small compared with that at f_1, better resolution can be obtained by switching the spectrum analyzer to a scale giving the amplitudes in decibels. Noting that the amplitude at f_1 is 80 dB as shown in part b of the figure, determine the reference voltage e_r (rms) of the spectrum analyzer and the amplitude A_2 at f_2 in decibels.

Partial ans: $A_2 = 63.52$ dB

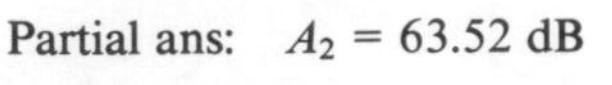

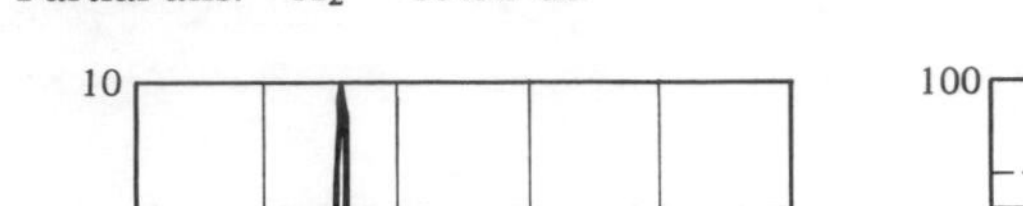

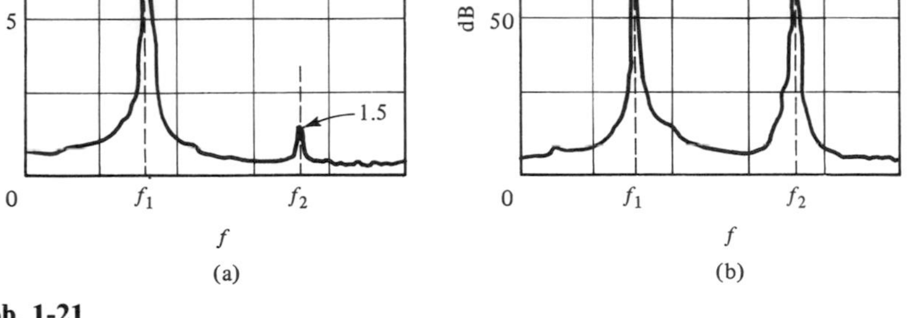

Prob. 1-21

1-22. Determine the mean-square value and the rms value of the sawtooth wave shown in the accompanying figure.

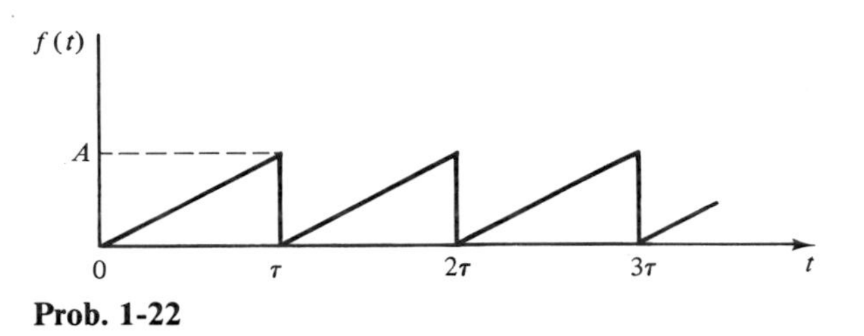

Prob. 1-22

Chapter 2

Free Vibration of Single-Degree-of-Freedom Systems

2-1 INTRODUCTION

Machines and structural systems vibrate freely about their static-equilibrium positions when displaced from those positions and then released. The frequencies at which they vibrate, known as their natural frequencies, depend primarily upon the mass and elasticity (stiffness) of the systems. This unforced motion is sometimes also referred to as natural motion.

All physical systems contain some inherent type of mechanism that dissipates energy, and this energy dissipation is referred to as *damping*. While such mechanisms cause the amplitude of a free vibration to diminish with time, the magnitude of damping in most real systems has little effect on their natural frequencies. The concepts in this chapter are fundamental to the understanding of the vibration of machines and structures, and should be carefully studied.

2-2 DIFFERENTIAL EQUATIONS OF DAMPED FREE VIBRATION

In analyzing the free-vibration characteristics of machines and structures, one of the initial steps is to model the system by some schematic arrangement of masses, elastic elements, and damping mechanisms. The differential equations of motion of this physical model are then derived to form the mathematical model of the system. The solution of these differential equations then enables us to analyze the characteristics of the system and predict its behavior under various conditions.

In this section we look at some schematic models of systems and at various ways of deriving their differential equations of motion. We limit our discussions to linear, single-degree-of-freedom systems with damping that can be represented by a viscous-damping mechanism. As we shall learn by the end of this section, a general form of the differential equations of motion for such systems can be written as

$$\ddot{x} + 2\zeta\omega_n\dot{x} + \omega_n^2 x = 0 \tag{2-1}$$

where x = a coordinate giving the displacement of the mass element from a reference position (usually the static-equilibrium position)
$\dot{x} = dx/dt$ (velocity of the mass)
$\ddot{x} = d^2x/dt^2$ (acceleration of the mass)
ζ = damping factor, which depends upon the damping mechanism(s) and system parameters such as mass and geometry
ω_n = undamped natural circular frequency, which depends upon system parameters such as mass, stiffness, and geometry

When damping can be represented by a term proportional to the velocity such as the middle term in Eq. 2-1, it is referred to as *viscous damping.* Although damping is always present to some degree in all real systems, valuable insights into systems can often be obtained by analyzing them as being theoretically undamped. In making such analyses, they are in effect treated as *conservative* systems,[1] and Eq. 2-1 reduces to

$$\ddot{x} + \omega_n^2 x = 0 \tag{2-2}$$

Elements of Physical Models

One of the simpler models of a physical system that is used in vibration analysis is the damped spring-and-mass system shown in Fig. 2-1. Several typical physical systems that this model might represent are shown in Fig. 2-2. The mass m shown in Fig. 2-1 is a measure of the inertia (the resistance to a change in the motion) of the vibrating body. The spring constant k defines the stiffness of the elastic elements of the system, and the spring force is equal to the product of the spring constant and the change in length of the spring from its free length. The spring in this model might represent the stiffness of the piping in a pumping system, the stiffness of a beam element, or the stiffness of the cable system of an elevator. Complex systems can often involve multiple elastic elements, in which case they are often combined into an equivalent spring of constant k_e.

Various damping models have been used to describe the energy-dissipating characteristics of different types of vibrating systems. One of the most widely used is the viscous-damping model, which assumes that the damping force is proportional to the velocity, and is generally represented schematically by a dashpot such as the one shown in Fig. 2-1. Since the viscous-damping force is proportional to the velocity and always has a sense opposite to that of the velocity, it can be expressed vectorially as

$$F_d = -c\dot{x} \tag{2-3}$$

with c being a *damping coefficient.* The value of the damping coefficient cannot be obtained directly from experimental measurements on most real systems but can be combined with other system parameters into a *damping factor* ζ (see Eq. 2-1) that can be determined experimentally from a free-vibration record of a system.

There are several reasons for the frequent use of viscous-damping models. One is that many real systems have damping that can be approximated quite satisfactorily by such models. Another is that closed-form mathematical solutions can be obtained for the differential equations of motion using this model. Viscous damping will be discussed in some detail in the next section.

[1] Conservative mechanical or structural systems are systems that are assumed to dissipate no energy, and on which no work is done by external excitation forces or moments.

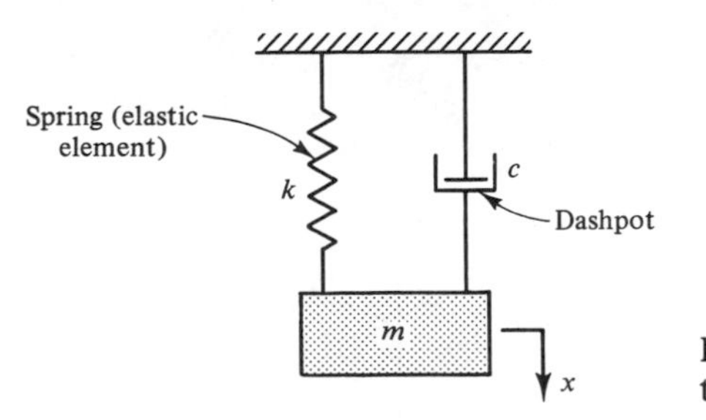

Figure 2-1 Damped spring-and-mass system.

The Derivation of Differential Equations of Motion (Mathematical Models)

One fundamental approach to deriving the differential equations of vibrating systems is to sketch a free body of the system and apply Newton's second law. However, there are those who prefer to add the inertia effects to the free body and consider D'Alembert's principle (dynamic equilibrium), and others who prefer the use of energy methods. Each approach has its merits for a particular analysis. Quite often the addition of the inertia effects to the free body helps to visualize the physical effects an accelerating system is experiencing, particularly for systems in general plane motion. Energy methods are convenient for use with systems that consist of several component parts, since the differential equations can be obtained without having to consider the forces acting between the component parts.

Of the methods mentioned above, let us first consider the direct application of Newton's second law. Errors are often made in using this approach when a sign convention is used that does not recognize that the displacements, velocities, and accelerations of a vibrating body have both positive and negative states depending upon the time t. The following paragraphs outline a procedure for use in applying Newton's second law to derive the differential equations of motion of single-degree-of-freedom systems that are vibrating with translation, or rotation about a fixed axis.

Systems Vibrating with Translation

1. Sketch a free body of the system in its static-equilibrium position, and determine any spring forces acting in this position.
2. Select a *common* positive sense for the displacement, velocity, and acceleration.
3. Sketch a free body of the system during a positive displacement *from the static-equilibrium position.* Show the forces acting, with the sense of each consistent with the force it exerts on a member during the displacement, remembering that the velocity is also assumed positive during this time.
4. Apply Newton's second law

$$\sum F = ma$$

All forces having the same sense as the assumed positive sense referred to in step 2 are summed positive. All forces with opposite sense are summed negative.

To illustrate the steps above, consider the spring-and-mass system shown in Fig. 2-3. A free body of the mass in the static-equilibrium position is shown in Fig. 2-3b with the spring stretched x_s from its free-length position. Summing forces vertically, it can be seen that

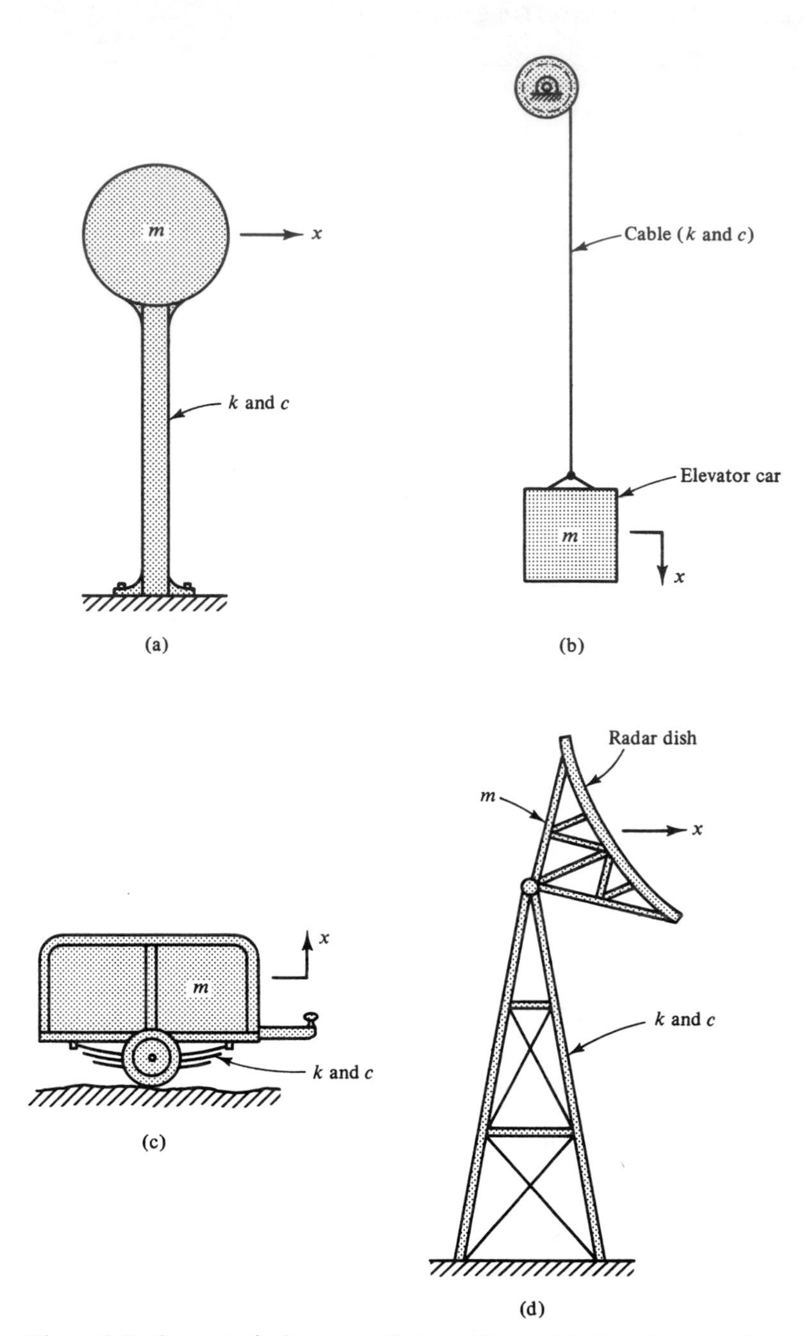

Figure 2-2 Some typical systems that may be modeled as simple spring-mass systems. (a) Water tank. (b) Elevator. (c) Trailer. (d) Radar dish and tower.

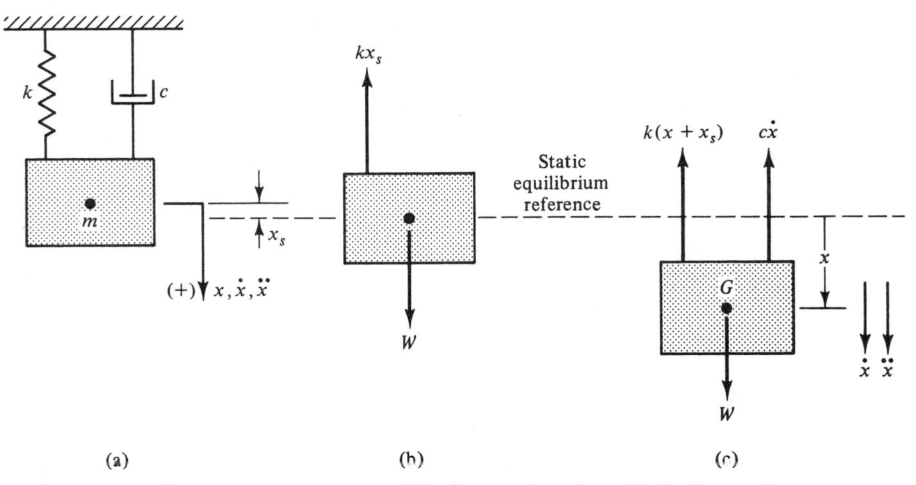

Figure 2-3 Spring-and-mass system with viscous damping. (a) Spring-and-mass system. (b) Static-equilibrium position. (c) Positive displacement of mass.

$$W = kx_s \tag{2-4}$$

Figure 2-3c shows the free-body diagram of the mass when it has a positive displacement x from the static-equilibrium position with the positive downward velocity and acceleration shown. The forces acting on the mass are the spring force $k(x + x_s)$ and the viscous-damping force $c\dot{x}$. The spring force acts upward because the spring is in tension during the positive displacement of the mass. The damping force is shown with a sense opposite to that of the assumed positive velocity during the displacement since damping always opposes motion. The sense of W is the same for all time t.

Applying Newton's second law,

$$W - k(x + x_s) - c\dot{x} = m\ddot{x}$$

Referring to Eq. 2-4, and dividing through by m, the differential equation assumes the form of

$$\ddot{x} + \frac{c}{m}\dot{x} + \frac{k}{m}x = 0 \tag{2-5}$$

Comparing Eqs. 2-5 and 2-1, we see that

$$\frac{k}{m} = \omega_n^2 \qquad \frac{c}{m} = 2\zeta\omega_n$$

It will be shown in Sec. 2-3 that ω_n^2 is the square of the undamped natural circular frequency.

If elastic forces support a system in the static-equilibrium position, the gravity forces and/or their moments acting on the system will be canceled by the forces and/or moments produced by the elastic elements in that position. After becoming experienced in writing equations of motion, the analyst can often save time by omitting both the gravity forces and the static-equilibrium-position elastic support forces from the free-body diagram.

Systems Vibrating with Rotation about Fixed Axes

1. Sketch a free body of the system in its static-equilibrium position, and determine any spring forces acting in this position.
2. Select a common positive sense for the angular displacement, velocity, and acceleration of the system. The common positive sense for the displacement, velocity, and acceleration of any *point* of the system will be obvious, as it must be consistent with the positive sense assumed for the angular variables.
3. Sketch a free body of the system during a positive angular displacement *from its static-equilibrium position.* Show the forces and couples acting on the system, with the sense of each consistent with the force or moment it exerts on the system during this time, and with the angular velocity and the velocity of the mass center also assumed positive during this time.
4. Apply Newton's second law,

$$\sum M_0 = I_0\ddot{\theta} \tag{2-6}$$

in which $\sum M_0$ is the summation of the moments of the couples and forces about the fixed axis O, and I_0 is the mass moment of inertia of the system about the fixed axis. All moments having the same sense as the sense assumed positive in step 2 are summed as positive. All moments with opposite sense are summed as negative. When the fixed axis is also a centroidal axis, Eq. 2-6 becomes

$$\sum M_G = \bar{I}\ddot{\theta}$$

in which $\bar{I}$ is the centroidal mass moment of inertia.

To illustrate the preceding steps, let us consider the slender rod of length l and weight W shown in Fig. 2-4. The rod has plane motion about a fixed noncentroidal axis through O as shown in the figure. In addition to the pin at O, the rod is supported by a spring of stiffness k at its right end. Viscous damping is supplied by a dashpot with a damping coefficient c.

A free-body diagram of the system in static equilibrium is shown in Fig. 2-4b. Summing moments about an axis through O, we find that the spring force required for static equilibrium is

$$F = \frac{W}{2}$$

Assuming θ, $\dot{\theta}$, and $\ddot{\theta}$ as having a common positive sense clockwise, we give the rod a positive angular displacement from its static-equilibrium position, and sketch the free body shown in Fig. 2-4c. The spring force has the sense shown since the spring is in tension during the positive displacement. The damping force has a sense opposite to that of the assumed positive velocity $b\dot{\theta}$ of point D on the rod since damping always opposes motion. The sense of W is the same for all time t, and the senses shown for the components of the pin reaction at O are arbitrary since they do not appear in the moment summation. Note that the spring force F appearing as part of the spring force is the spring force that supports the rod in its static-equilibrium position.

Summing clockwise moments as positive, the application of Eq. 2-6 yields

$$W\frac{l}{2}\cos\theta - cb^2\dot{\theta}\cos\theta - (F + kl\sin\theta)l\cos\theta = I_0\ddot{\theta}$$

This equation is nonlinear but can be linearized by considering only small oscillations of

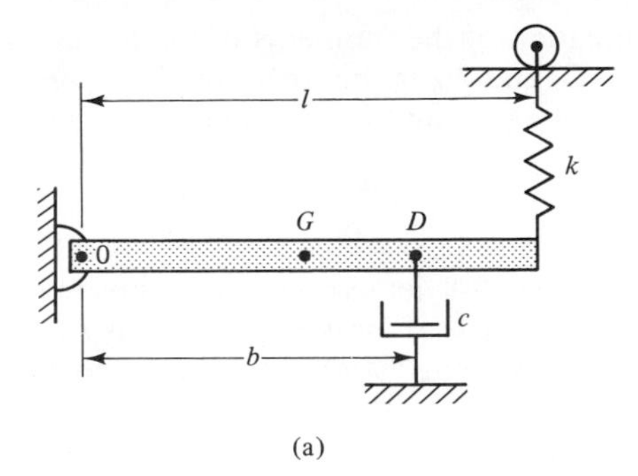

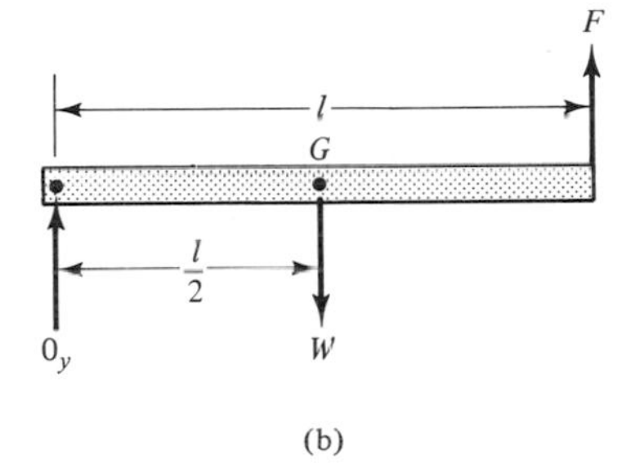

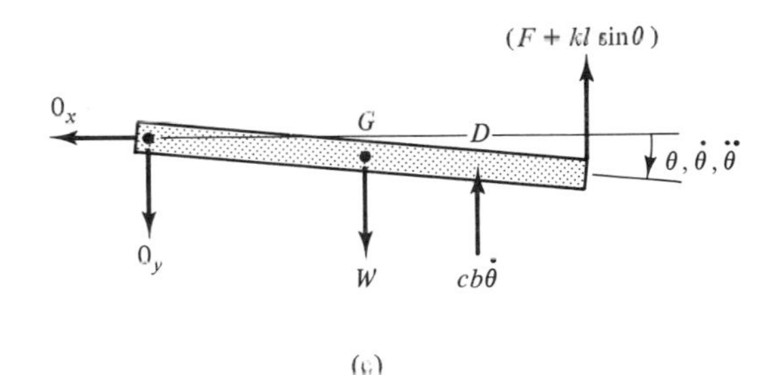

Figure 2-4 Vibration of a slender rod about a fixed axis.

the rod. Using the approximations that $\cos\theta \simeq 1$ and $\sin\theta \simeq \theta$, and remembering that $F = W/2$, the differential equation of motion of the rod can be written as

$$\ddot{\theta} + \underbrace{c\frac{b^2}{I_0}}_{2\zeta\omega_n}\dot{\theta} + \underbrace{\frac{kl^2}{I_0}}_{\omega_n^2}\theta = 0 \tag{2-7}$$

The differential equation of motion of a body vibrating about a fixed axis can also be determined by considering Newton's second law in the form of

$$\sum M_0 = \frac{dH_0}{dt} \tag{2-8}$$

in which ΣM_0 is again the summation of the moments of the forces and couples acting on the body about the fixed axis O, and H_0 is the total *angular momentum* of the body about the fixed axis. When the fixed axis is also a centroidal axis, Eq. 2-8 becomes

$$\sum M_G = \frac{dH_G}{dt}$$

To illustrate the application of Eq. 2-8, let us consider the inverted pendulum shown in Fig. 2-5. This system consists of a sphere of mass m_1 and radius r attached to a slender rod of length l_2 and mass m_2. The rod is pinned at O and a spring and dashpot are attached to the sphere as shown in the figure.

Clockwise is chosen as the common positive sense for θ, $\dot{\theta}$, and $\ddot{\theta}$. Giving the pendulum a positive displacement from its static-equilibrium position, and remembering that the angular velocity is also assumed positive during this displacement, the forces are as shown on the free-body diagram in Fig. 2-5b (assuming small oscillations of the pendulum).

The angular momentum of the system about the fixed axis O is

$$H_0 = I_{02}\dot{\theta} + I_{01}\dot{\theta} = \underbrace{I_{02}\dot{\theta}}_{\text{Rod}} + \underbrace{(\tfrac{2}{5}m_1 r^2 + m_1 l_1^2)\dot{\theta}}_{\text{Sphere}}$$

where
I_{02} = mass moment of inertia of rod about O
I_{01} = mass moment of inertia of sphere about O
$\frac{2}{5}m_1 r^2\dot{\theta}$ = angular momentum of sphere about its centroidal axis parallel to fixed axis O
$m_1 l_1^2\dot{\theta}$ = moment of linear momentum of sphere about fixed axis O

Noting in Fig. 2-5 that $r \ll l_1$, the sphere can be considered as a particle, and the angular momentum about the fixed axis O becomes

$$H_0 = (I_{02} + m_1 l_1^2)\dot{\theta}$$

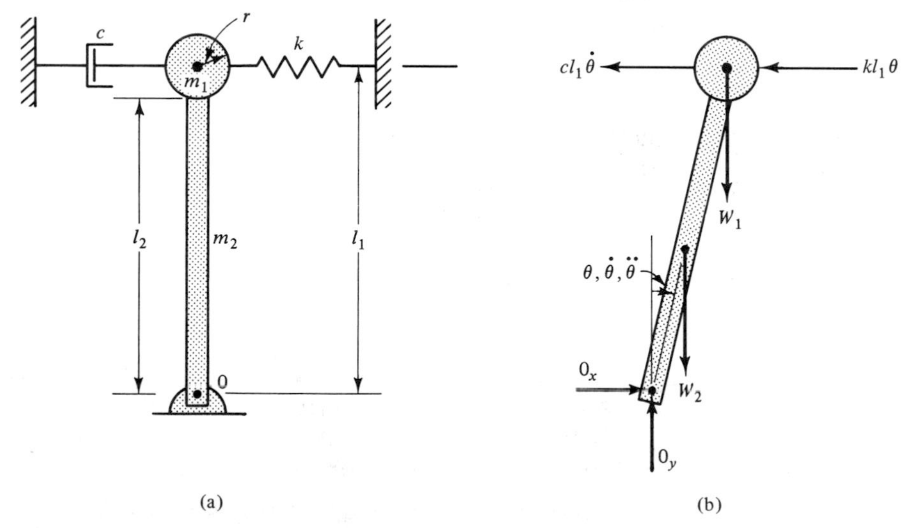

Figure 2-5 Inverted pendulum system.

Referring to the free-body diagram in Fig. 2-5b, and applying Eq. 2-8,

$$W_2 \frac{l_2}{2} \theta + W_1 l_1 \theta - k l_1 \theta l_1 - c l_1 \dot{\theta} l_1 = (I_{02} + m_1 l_1^2)\ddot{\theta}$$

or

$$\ddot{\theta} + \underbrace{\frac{c l_1^2}{I_0}}_{2\zeta\omega_n} \dot{\theta} + \underbrace{\left[\frac{k l_1^2 - (W_2 l_2/2) - W_1 l_1}{I_0}\right]}_{\omega_n^2} \theta = 0 \qquad (2\text{-}9)$$

in which

$$I_0 = I_{02} + m_1 l_1^2$$

is the mass moment of inertia of the rod-and-sphere system about the fixed axis O.

The reader should note that for the inverted pendulum ω_n^2 can be positive, zero, or negative, depending upon the magnitude of $k l_1^2$ relative to the magnitude of the quantity $(W_2 l_2/2 + W_1 l_1)$. The significance of $\omega_n^2 \leq 0$ will be discussed in Sec. 2-6.

D'Alembert's Principle (Dynamic Equilibrium)

As mentioned earlier, the concept of dynamic equilibrium is found convenient by some, particularly in analyzing systems having fixed-axis rotation or general plane motion. We shall limit our discussion of the use of this concept at this point to single-degree-of-freedom systems vibrating with fixed-axis rotation, and a special case of vibration in general plane motion that involves only one degree of freedom.

Systems Vibrating about Fixed Axes

1. Sketch a free body of the system in its static-equilibrium position, and determine any spring forces acting in this position.
2. Select a common positive sense for the angular displacement, velocity, and acceleration of the system. The common positive sense for the displacement, velocity, and acceleration of any point of the system will be obvious as it must be consistent with the positive sense assumed for the angular variables.
3. Sketch a free body of the system during a positive angular displacement *from its static-equilibrium position*. Show the forces and couples acting on the system, with the sense of each consistent with the force or moment it exerts on the free body, with the angular and mass-center velocities also assumed positive during this time.
4. Add the angular inertia effect $\bar{I}\ddot{\theta}$ (the inertia couple) to the free body by means of a dashed curl having a sense opposite the assumed positive acceleration of the system. Next add the inertia effect $m\bar{a}$ associated with the mass center of the system. For fixed-axis rotation this inertia effect is most conveniently expressed in terms of the normal and tangential components of acceleration of the mass center of the system. Therefore, a tangential effect and a normal effect are added to the free body as dashed vectors acting through the mass center. The tangential effect has a sense opposite the assumed positive tangential component of acceleration of the mass center. The normal effect has a sense away from the axis of rotation, which gives it a sense opposite to the normal component of acceleration of the mass center.

5. Determine the differential equation of motion of the system by summing to zero about the fixed axis:
   a. The moments of the forces and couples acting on the system
   b. The moment of the tangential inertia effect
   c. The moment of the inertia couple

To illustrate the use of the preceding steps, let us consider the slender rod of length l and weight W shown in Fig. 2-6. The rod has plane motion about a fixed axis through O. In addition to the pin at O, the rod is supported by a spring of stiffness k at its right end. Viscous damping is supplied by the dashpot, which has a damping coefficient c.

A free-body diagram of the system in its static-equilibrium position is shown in Fig. 2-6b. Summing moments about an axis through O of this free body, we find that the spring force required for static equilibrium of the system is

$$F = \frac{W}{2}$$

Assuming θ, $\dot{\theta}$, and $\ddot{\theta}$ as positive clockwise, we give the rod a positive displacement from its static-equilibrium position, and sketch the free body shown in Fig. 2-6c. The spring and damping forces shown are for small amplitudes of vibration such that $\sin \theta \cong \theta$ and $\cos \theta \cong 1$. The spring force has the sense shown since the spring is in tension during the positive displacement, while the damping force is shown with a sense opposite to the assumed positive sense of the velocity $b\dot{\theta}$ of point D on the rod since damping forces always oppose motion. The sense of W is the same for all time t, and the senses shown for the components of the pin reaction at O are arbitrary as they act through the fixed axis and thus do not appear in the moment summation. Note that the spring force F appearing as part of the spring force in the displaced position of the rod is the spring force supporting the rod in its static-equilibrium position.

The inertia couple $\bar{I}\ddot{\theta}$ is shown by the dashed curl with a sense opposite the assumed positive angular acceleration of the system, with $\bar{I}$ being the mass moment of inertia of the rod about its mass center. The inertia effects associated with the normal and tangential components of acceleration are shown by the dashed vectors acting through G, having magnitudes of $ml\dot{\theta}^2/2$ and $ml\ddot{\theta}/2$, respectively. The normal inertia effect has a sense away from the axis of rotation, while the tangential effect has a sense opposite the assumed positive tangential component of acceleration of the mass center.

Referring to Fig. 2-6c, and remembering that only small oscillations are being considered, we sum the moments of the forces and inertia effects about an axis through O to obtain

$$\sum M_0 = 0 = \bar{I}\ddot{\theta} + (F + kl\theta)l + (cb\dot{\theta})b - \frac{Wl}{2} + \left(m\frac{l}{2}\ddot{\theta}\right)\frac{l}{2}$$

Recalling that the spring force in the static-equilibrium position is $F = W/2$, and that the mass moment of inertia of the rod about an axis through O is given by the parallel-axis theorem as

$$I_0 = \bar{I} + m\left(\frac{l}{2}\right)^2$$

the above equation can be written as

$$\ddot{\theta} + \frac{cb^2}{I_0}\dot{\theta} + \frac{kl^2}{I_0}\theta = 0$$

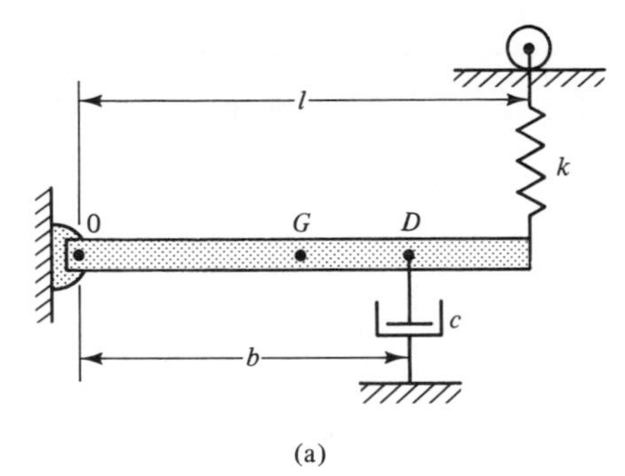

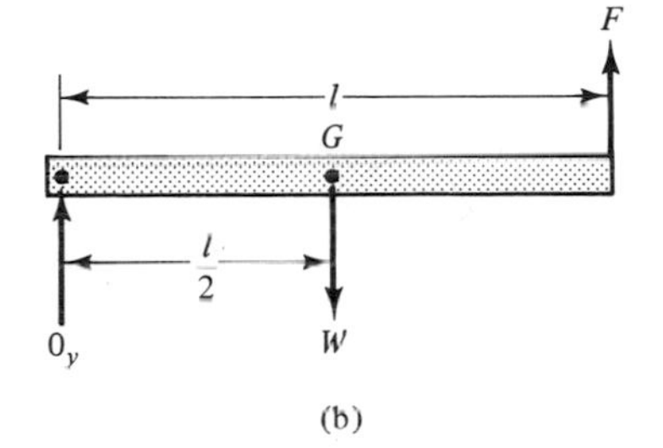

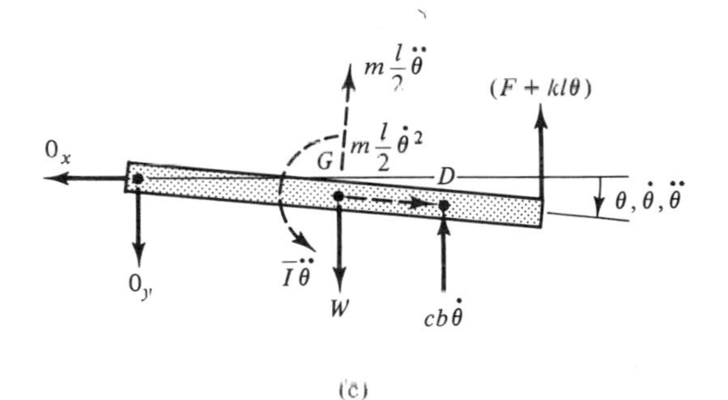

Figure 2-6 Vibration of a slender rod about a fixed axis.

which is the same as Eq. 2-7, which was obtained by the direct application of Newton's second law.

Wheel Rolling without Slipping

A wheel rolling without slipping is a special case of a body vibrating with general plane motion whose differential equation of motion can be determined by considering it as vibrating about a fixed axis through its point of rolling contact, since that point has zero velocity. It is also a special case of a body vibrating with general plane motion which has but a single degree of freedom. We examine this special case in Example 2-1.

EXAMPLE 2-1

Let us determine the differential equation of motion of the solid cylinder shown in Fig. 2-7 which is of radius r and mass m. It is attached to a spring of stiffness k and a dashpot of damping coefficient c which provides viscous damping in the system. Assume that there is sufficient friction for the wheel to roll without slipping.

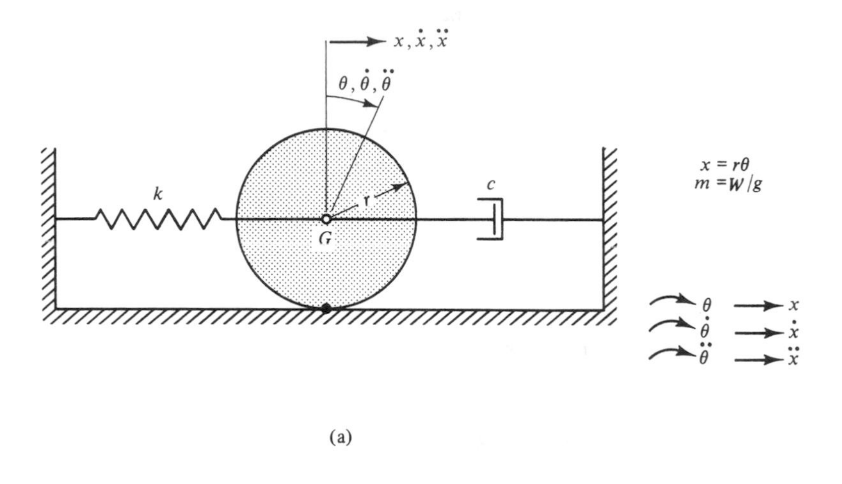

(a)

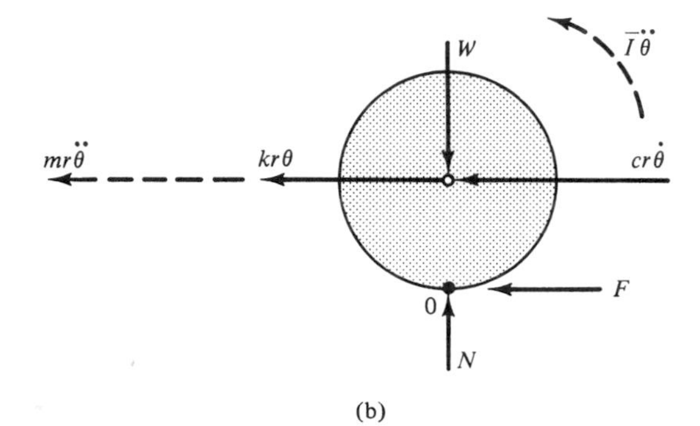

(b)

Figure 2-7 Cylinder of Example 2-1. (a) Oscillating cylinder. (b) Cylinder in dynamic equilibrium.

Solution. It is not necessary to sketch a free body of the system in its static-equilibrium position since inspection shows that there are no spring forces acting in that position, only the weight W and the normal force N.

We shall select clockwise as the positive sense for θ, $\dot{\theta}$, and $\ddot{\theta}$. Since a positive angular displacement θ is accompanied by a displacement x of the mass center to the right, we must select to the right as the positive sense of x, $\dot{x}$, and $\ddot{x}$ to be physically consistent.

A free body of the cylinder in dynamic equilibrium during a positive displacement is shown in Fig. 2-7b. The spring force $kr\theta$ acts to the left since the spring is in tension during the positive displacement. The damping force $cr\dot{\theta}$ acts to the left since that is opposite the assumed positive sense of the velocity of the mass center during the displacement. The senses of F, W, and N are not pertinent since they are eliminated from the differential equations of motion of the system.

The inertia couple $\bar{I}\ddot{\theta}$, in which $\bar{I}$ is the mass moment of inertia of the cylinder about its mass center G, is shown by the dashed curl with a sense opposite the positively assumed $\ddot{\theta}$ of the cylinder. The inertia effect associated with the translation of the mass center is shown by the dashed vector $mr\ddot{\theta}$ since the coordinates x and θ are related by the constraint equation[2]

$$x = r\theta$$

The moments of the forces and inertia effects can be summed to zero about any axis using the free body in dynamic equilibrium. However, we shall perform the summation about an axis through O since this eliminates N and F, and

$$\sum M_0 = 0 = kr^2\theta + cr^2\dot{\theta} + \bar{I}\ddot{\theta} + mr^2\ddot{\theta}$$

Recalling the parallel-axis theorem, we can write that

$$I_0 = \bar{I} + mr^2$$

Substituting this into the differential equation above, we obtain

$$kr^2\theta + cr^2\dot{\theta} + I_0\ddot{\theta} = 0$$

which when divided through by I_0 takes the form of

$$\ddot{\theta} + \underbrace{\frac{cr^2}{I_0}}_{2\zeta\omega_n}\dot{\theta} + \underbrace{\frac{kr^2}{I_0}}_{\omega_n^2}\theta = 0 \tag{2-10}$$

The relationship $x = r\theta$ and its derivatives can be substituted into Eq. 2-10 to obtain the differential equation of motion in terms of the independent coordinate x as

$$\ddot{x} + \frac{cr^2}{I_0}\dot{x} + \frac{kr^2}{I_0}x = 0$$

Either of the last two differential equations could also have been obtained by summing the moments of the forces and inertia effects about an axis through the mass center G, summing the forces and translational inertia effect along a horizontal axis, and then combining the two equations.

Equation 2-10 can also be obtained by considering the cylinder as vibrating about a fixed axis through its point of rolling contact (a point of zero velocity) and applying

$$\sum M_0 = I_0\ddot{\theta}$$

2-3 SOLVING DIFFERENTIAL EQUATIONS OF MOTION

At this point it should be apparent from the discussion in Sec. 2-2 that the differential equation of motion of a viscously damped single-degree-of-freedom system can be conveniently written as

[2] This constraint equation indicates that the system has but a single degree of freedom.

$$\ddot{x} + 2\zeta\omega_n\dot{x} + \omega_n^2 x = 0 \qquad \text{(translation)} \tag{2-11}$$

or

$$\ddot{\theta} + 2\zeta\omega_n\dot{\theta} + \omega_n^2\theta = 0 \qquad \text{(rotation)} \tag{2-12}$$

in which $2\zeta\omega_n$ and ω_n^2 are functions of the system parameters. As we shall see later, the damping factor ζ and the undamped natural circular frequency ω_n play significant roles in describing both the free and forced-vibration characteristics of systems.

Since Eqs. 2-11 and 2-12 are ordinary linear differential equations with constant coefficients, we can assume solutions for them having the form of

$$\left.\begin{aligned} x &= e^{st} \\ \theta &= e^{st} \end{aligned}\right\} \tag{2-13}$$

and determine an expression for s that satisfies the differential equation.

Selecting Eq. 2-11, we can substitute $x = e^{st}$, $\dot{x} = se^{st}$, and $\ddot{x} = s^2e^{st}$ into it to obtain

$$(s^2 + 2\zeta\omega_n s + \omega_n^2)e^{st} = 0$$

For the assumed solution to satisfy the differential equation, the expression in parentheses above must equal zero since $e^{st} \neq 0$ for finite values of t. This expression

$$s^2 + 2\zeta\omega_n s + \omega_n^2 = 0 \tag{2-14}$$

is referred to as the *characteristic equation,* and its solution yields the characteristic roots (eigenvalues)[3]

$$s_{1,2} = \omega_n(-\zeta \pm \sqrt{\zeta^2 - 1})$$

Since two arbitrary constants are required in the solution of a second-order ordinary differential equation, the general solution of Eq. 2-11 is

$$x = C_1e^{s_1t} + C_2e^{s_2t} \tag{2-15}$$

which can be written as

$$x = e^{-\zeta\omega_n t}(C_1e^{\omega_n\sqrt{\zeta^2-1}\,t} + C_2e^{-\omega_n\sqrt{\zeta^2-1}\,t}) \tag{2-16}$$

Examination of Eq. 2-16 reveals that the physical behavior of a system modeled by Eq. 2-11 depends primarily upon the magnitude of the damping factor ζ. There are three distinct damping cases: (a) $\zeta > 1$, (b) $\zeta = 1$, and (c) $\zeta < 1$. For these cases, the numerical values within the radicals of Eq. 2-16 are positive, zero, and negative, respectively. We shall consider each of these cases in turn.

Case a When $\zeta > 1$, the exponents of Eq. 2-15 are *real* and *negative,* and the system will not oscillate when displaced from its static-equilibrium position. Such systems are said to be *overdamped,* and a typical displacement-time curve for one is shown in Fig. 2-8. For the particular curve shown the mass was given an initial displacement x_0 from its static-equilibrium position and then released with zero initial velocity. It is apparent that an overdamped system takes an extremely long time (theoretically infinite) to return to its static-equilibrium position.

[3] The term eigenvalue comes from the German word "Eigenwert" in which Eigen means "characteristic" and Wert means "value."

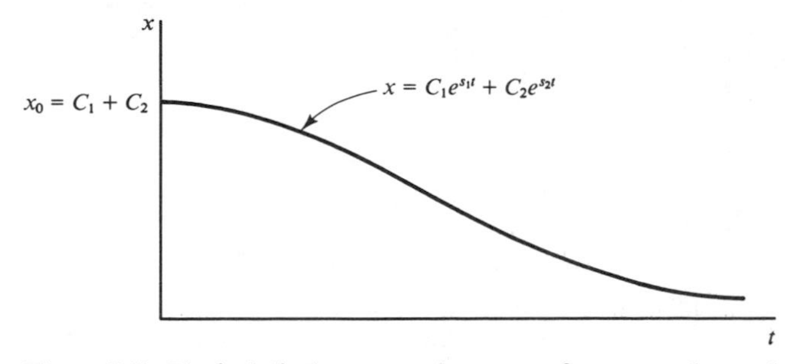

Figure 2-8 Typical displacement-time curve for an overdamped system ($\zeta > 1$).

Case b When $\zeta = 1 = \zeta_c$, a system is said to be *critically damped*. For this case Eq. 2-16 no longer provides a solution since there is only one arbitrary constant; that is,

$$x = e^{-\omega_n t}(C_1 + C_2)$$

It is easy to verify by substitution that

$$x = (A + Bt)e^{-\omega_n t} \tag{2-17}$$

is a general solution of Eq. 2-11 when $\zeta = 1$.[4]

A typical displacement-time curve for a critically damped system is shown in Fig. 2-9a. It reveals that such a system does not oscillate when displaced from its static-equilibrium position but returns to that position in a manner similar to that of an overdamped system. Like the overdamped system, it theoretically takes an infinite time to return to its static-equilibrium position. However, for all practical purposes such systems do return to their original position in a finite time, and the critically damped system will return in the *minimum* time possible without oscillating.

The recoil mechanism of an artillery piece is an example of a critically damped system. A simplified system would consist of a spring to store energy during recoil, and a dashpot to provide the critical damping. When fired, the barrel of the piece suddenly acquires an initial velocity $\dot{x}_0$ while still in its initial position of $x = 0$. Substituting these initial conditions into Eq. 2-17 and its first derivative, we find that

$$A = 0$$
$$B = \dot{x}_0$$

The displacement of the gun is then given by

$$x = \dot{x}_0 te^{-\omega_n t}$$

which is shown graphically in Fig. 2-9b. It is left as an exercise for the reader to determine that the maximum displacement occurs when $t = 1/\omega_n$, and is equal in magnitude to $(\dot{x}_0/\omega_n)e^{-1}$ (see Prob. 2-22).

Case c When $\zeta < 1$, the term $(\zeta^2 - 1)$ is negative, and, as we shall soon see, complex exponents lead to trigonometric functions associated with vibratory motion. Since the

[4] Refer to any elementary differential equations text for a discussion of solutions involving repeated roots.

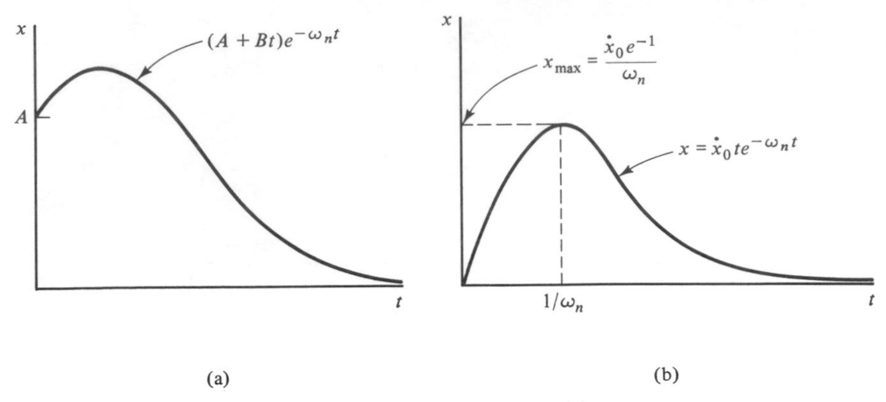

Figure 2-9 Typical displacement-time curves for a critically damped system ($\zeta = 1$).

damping inherent in many machines and structural systems is described by $\zeta < 1$, this underdamped case is a very important one in analyzing vibrating systems.

With $\zeta < 1$, Eq. 2-16 may be written as

$$x = e^{-\zeta\omega_n t}(C_1 e^{j\omega_n\sqrt{1-\zeta^2}t} + C_2 e^{-j\omega_n\sqrt{1-\zeta^2}t}) \tag{2-18}$$

in which $j = \sqrt{-1}$. From Euler's equation (Eq. 1-10), we can write that

$$e^{j\omega_n\sqrt{1-\zeta^2}t} = \cos \omega_n\sqrt{1-\zeta^2}t + j \sin \omega_n\sqrt{1-\zeta^2}t$$

and

$$e^{-j\omega_n\sqrt{1-\zeta^2}t} = \cos \omega_n\sqrt{1-\zeta^2}t - j \sin \omega_n\sqrt{1-\zeta^2}t$$

Substituting these relationships into Eq. 2-18, we obtain

$$x = e^{-\zeta\omega_n t}[(C_1 + C_2)\cos \omega_n\sqrt{1-\zeta^2}t + j(C_1 - C_2)\sin \omega_n\sqrt{1-\zeta^2}t] \tag{2-19}$$

Since x is *real*, the coefficients $(C_1 + C_2)$ and $j(C_1 - C_2)$ must be real. For the latter coefficient to be real, the constants C_1 and C_2 must be *complex conjugates*. Let us assume that

$$C_1 = \frac{A - jB}{2}$$

and

$$C_2 = \frac{A + jB}{2}$$

in which A and B are *real* constants. Adding and subtracting these equations in turn yields

$$C_1 + C_2 = A$$

and

$$j(C_1 - C_2) = B$$

If we then substitute the real constants A and B into Eq. 2-19, we obtain

$$x = e^{-\zeta\omega_n t}(A \cos \omega_n\sqrt{1-\zeta^2}t + B \sin \omega_n\sqrt{1-\zeta^2}t) \tag{2-20}$$

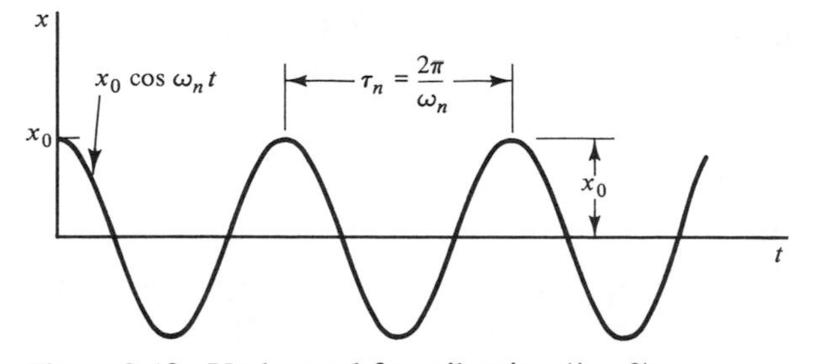

Figure 2-10 Undamped free vibration ($\zeta = 0$).

which is the general solution of the differential equation for damped free vibration when $\zeta < 1$. The constants A and B depend upon the initial conditions of the system.

Let us now examine Eq. 2-20 and relate ζ, ω_n, and $\omega_n\sqrt{1 - \zeta^2}$ to the vibrational characteristics of mechanical and structural systems. A special case of $\zeta < 1$ exists when $\zeta = 0$. This would indicate a system with no damping, but this cannot be a real case since damping is present to some degree in all physical systems. However, many machines and structures have relatively small magnitudes of damping, and we shall find that analyzing them as theoretically undamped systems provides valuable insights into their damped free-vibration characteristics, including a convenient method of determining very close approximations of their actual (damped) frequencies.

In considering a theoretically undamped system with $\zeta = 0$, Eq. 2-20 reduces to

$$x = A \cos \omega_n t + B \sin \omega_n t \tag{2-21}$$

which is the general solution of the differential equation for undamped free vibration.[5] Using the initial conditions

$$t = 0 \left| \begin{array}{l} x = x_0 \\ \dot{x} = 0 \end{array} \right.$$

with Eq. 2-21, we find that

$$x = x_0 \cos \omega_n t$$

which is shown graphically in Fig. 2-10. This shows that $\cos \omega_n t$ is periodic and repeats itself every 2π rad. Therefore, the *undamped* natural period (the time required for one cycle) is

$$\tau_n = \frac{2\pi}{\omega_n} \tag{2-22}$$

while the *undamped* natural frequency is

$$f_n = \frac{\omega_n}{2\pi} \tag{2-23}$$

which, as can be seen, is the reciprocal of τ_n. In these expressions τ_n is expressed in seconds, ω_n in radians per second, and f_n in Hz (cycles per second). The amplitude of the vibration

[5] In Sec. 2-2 it was noted that ω_n^2, which was given as the equivalent of k/m in the differential equation of motion of the spring-and-mass system of Fig. 2-3, would be shown to be the square of the undamped natural circular frequency of the system. It should now be evident from Eq. 2-21 that ω_n is indeed that entity.

is x_0, as shown in Fig. 2-10. Equations 2-22 and 2-23 express relationships that are fundamental to the study of vibrations and should be thoroughly understood by the reader.

Damped Free-Vibration Characteristics of Viscously Damped Systems

As mentioned earlier, some damping is present in all real systems. If this were not true, perpetual motion would be possible. The damping in real systems includes air resistance, fluid friction, Coulomb (dry) friction, solid (material) friction, and other mechanisms not yet well understood. Although each such mechanism may have some unique characteristics, the vibrational behavior of many machines and structures can be analyzed quite satisfactorily by assuming that their damping characteristics can be approximated by a viscous-damping model. Thus, Eqs. 2-11 and 2-12 are often used as the mathematical models of real systems, and a procedure for experimentally determining the viscous-damping factors ζ of such real systems is discussed in Sec. 2-4.

In the discussion that follows, it will be found convenient to rewrite Eq. 2-20 in the form of a single trigonometric function and a phase angle ϕ (see Sec. 1-2) as follows:

$$\left.\begin{aligned} x &= e^{-\zeta\omega_n t}[\sqrt{A^2 + B^2}\sin(\omega_n\sqrt{1-\zeta^2}t + \phi)] \\ &\text{or} \\ x &= X_0 e^{-\zeta\omega_n t}\sin(\omega_n\sqrt{1-\zeta^2}t + \phi) \end{aligned}\right\} \qquad (2\text{-}24)$$

in which $X_0 = \sqrt{A^2 + B^2}$. Equation 2-24 is shown graphically in Fig. 2-11. Note that the successive amplitudes $x_1, x_2, \ldots$ decrease with time according to the term $e^{-\zeta\omega_n t}$. Such a decrease in the amplitudes is referred to as *exponential decay,* and if one were to measure $x_1, x_2, \ldots$ and determine the successive ratios $x_1/x_2, x_2/x_3, \ldots$, it would be found that these ratios were constant, which indicates that the *rate of decay* is independent of the amplitude of vibration. The natural log of the constant is referred to as the *logarithmic decrement* of decay, and is often used as a measure of system damping. While the rate of

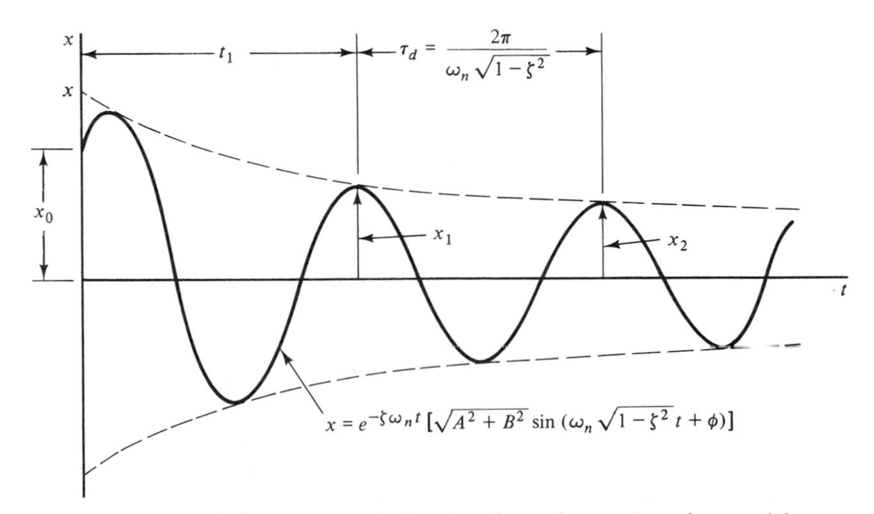

Figure 2-11 Typical free-damped vibration for a viscous-damping model.

decay is independent of the amplitude of vibration, Eq. 2-24 reveals that it does vary with both the magnitude of the damping (ζ) and the frequency of the vibration.

Since the sine wave repeats itself every 2π rad, the damped natural period τ_d and frequency f_d are given by

$$\tau_d = \frac{2\pi}{\omega_n\sqrt{1-\zeta^2}} \tag{2-25}$$

and

$$f_d = \frac{1}{\tau_d} = \frac{\omega_n\sqrt{1-\zeta^2}}{2\pi} \tag{2-26}$$

Recalling that the theoretically undamped natural period and frequency are given, respectively, by

$$\tau_n = \frac{2\pi}{\omega_n}$$

and

$$f_n = \frac{\omega_n}{2\pi}$$

it is apparent that $\tau_d > \tau_n$ and that $f_d < f_n$. However, as mentioned earlier, the damping in many real systems is relatively small, often less than 20 percent ($\zeta < 0.2$). In such instances the quantity $\sqrt{1-\zeta^2} \cong 1$, so that $\tau_d \simeq \tau_n$ and $f_d \cong f_n$. Therefore, it is common practice to calculate the theoretically undamped quantities ω_n, τ_n, and f_n for such systems, and use these as actual system values, since the damping factors of real systems are generally unknown unless they have been determined experimentally.

It follows from the preceding discussion that when the natural frequency of a real system is determined experimentally, it is the actual (damped) frequency f_d that is obtained. However, since $f_d \cong f_n$ as discussed above, f_n is generally used to refer to either frequency unless f_d is the subject under discussion as it is here, and we shall follow this practice throughout the text.

At this point the reader might wonder about the importance of the damping factor ζ since it has so little effect on the natural frequencies of real systems. However, as we saw earlier, it does affect the rate of decay of a damped free vibration, and as we shall see later, it has a significant effect on the vibration response of a system subjected to external disturbances such as exciting forces or support excitations.

EXAMPLE 2-2

Let us consider once again the slender rod shown in Fig. 2-4, with the following parameter values:

$W = 10$ lb (weight of rod)
$l = 24$ in. (length of rod)
$b = 20$ in. (location of dashpot from O)
$k = 4$ lb/in. (spring constant)
$c = 0.15$ lb · s/in. (damping coefficient)

We wish to determine (a) the undamped natural frequency f_n of the system, (b) the damping factor ζ, and (c) the damped natural frequency f_d.

Solution. **a.** Referring to Eq. 2-7, we see that

$$\omega_n^2 = \frac{kl^2}{I_0}$$

and

$$\frac{cb^2}{I_0} = 2\zeta\omega_n$$

If an in. · lb · s system of units is used, the acceleration of gravity is $g = 386$ in./s^2, and the mass moment of inertia about O is

$$I_0 = \frac{Wl^2}{3g} = \frac{10(24)^2}{3(386)} = 4.97 \text{ lb} \cdot \text{in.} \cdot \text{s}^2$$

The theoretical undamped natural circular frequency is

$$\omega_n = \sqrt{\frac{4(24)^2}{4.97}} = 21.53 \text{ rad/s}$$

from which

$$f_n = \frac{\omega_n}{2\pi} = 3.43 \text{ Hz}$$

b. The damping factor ζ can be determined from the second term of Eq. 2-7 as

$$\zeta = \frac{cb^2}{2I_0\omega_n} = \frac{0.15(20)^2}{2(4.97)(21.53)} = 0.28$$

c. Substituting the dimensionless value determined for ζ into Eq. 2-26, the damped natural frequency is found to be

$$f_d = \frac{21.53\sqrt{1 - (0.28)^2}}{2\pi} = 3.29 \text{ Hz}$$

The reader might pause at this point to observe the relationships of the system parameters to f_n, f_d, and ζ.

As mentioned before, the damping in many machines and structures will be identified by damping factors considerably less than the value of $\zeta = 0.28$ found in the preceding example. However, there are systems in which large amounts of damping are introduced to suppress their vibration. An example is the use of shock absorbers in the suspension systems of automobiles. The reader should note that even with the relatively large damping factor found in the example, the undamped and damped frequencies calculated for the rod varied by only a bit over 4 percent.

The Use of Initial Conditions

When it is desired to obtain the motion of a system as a function of time, the initial conditions associated with the motion can be used to determine the constants A and B in Eq. 2-20 or X_0 and ϕ in Eq. 2-24.

In determining either A and B or X_0 and ϕ, the signs assigned to the initial conditions must be consistent with the sign convention assumed for the coordinate system used. The use of initial conditions is illustrated in Example 2-3.

EXAMPLE 2-3

The simply supported beam in Fig. 2-12 has a static deflection of $\Delta_s = 0.1$ in. when it supports a weight W at its center. The weight W is then dropped from a height of $h = 10\Delta_s$ onto the middle of the beam. The weight remains in contact with the beam after striking it, so a plastic impact can be assumed. Neglecting the energy lost during the impact, determine (a) the ratio of the maximum beam deflection y_M to the static deflection Δ_s with damping neglected, and (b) the same ratio if the damping factor is $\zeta = 0.1$.

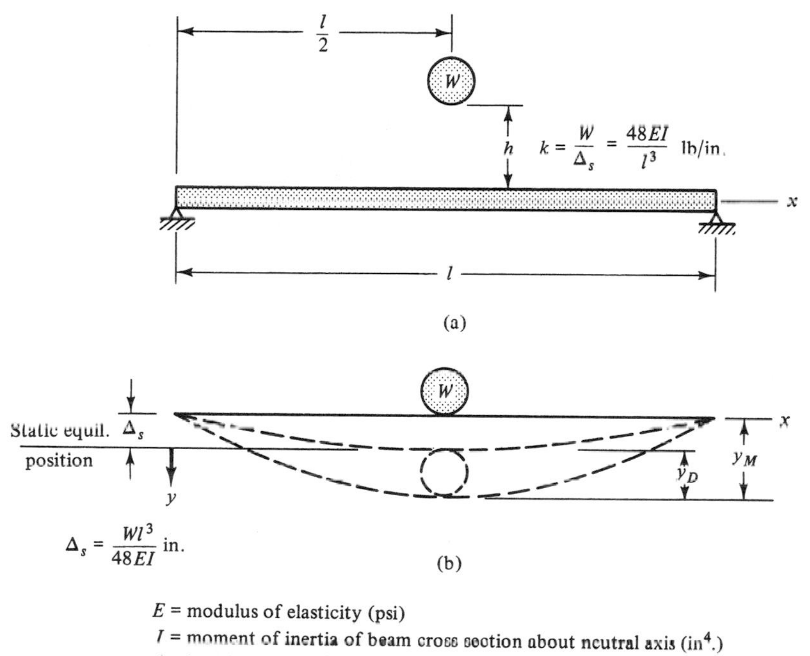

E = modulus of elasticity (psi)
I = moment of inertia of beam cross section about neutral axis (in^4.)
l = length of beam (in.)
y = displacement of beam from static equilibrium position (in.)
y_D = maximum displacement of beam from static equilibrium position (in.)
y_M = maximum deflection of beam from x axis (in.)
Δ_s = static deflection of beam (in.)

Figure 2-12 Beam of Example 2-3. (a) Simply supported beam. (b) Beam deflections.

Solution. Assuming that the beam acts as a weightless spring with a spring constant of $k = 48EI/l^3$ lb/in., the undamped and damped natural circular frequencies ω_n and ω_d can be calculated as follows:

$$\Delta_s = \frac{W}{k} = \frac{mg}{k}$$

$$\omega_n = \sqrt{\frac{k}{m}} = \sqrt{\frac{g}{\Delta_s}} = \sqrt{\frac{386}{0.1}} = 62.13 \text{ rad/s}$$

$$\omega_d = \omega_n\sqrt{1-\zeta^2} = 62.13\sqrt{1-(0.1)^2} = 61.82 \text{ rad/s}$$

Noting that the displacement of the beam is measured from its static-equilibrium position, and considering a displacement below that position and a downward velocity both as positive, the initial conditions are the displacement and velocity of W the instant after impact,

$$y_0 = -\Delta_s = -0.1 \text{ in.}$$

and

$$\dot{y}_0 = \sqrt{2gh} = \sqrt{2(386)(10\Delta_s)} = 27.78 \text{ in./s}$$

a. The maximum deflection of the center of the beam from the x axis is

$$y_M = \Delta_s + y_D$$

as shown in Fig. 2-12. Therefore, we must determine the maximum displacement y_D of the center of the beam from its static-equilibrium position. Substituting the initial conditions into Eq. 2-20 with $t = 0$ and $\zeta = 0$, we obtain

$$A = -\Delta_s$$

$$B = \frac{\sqrt{2gh}}{\omega_n}$$

Thus,

$$y = -\Delta_s \cos \omega_n t + \frac{\sqrt{2gh}}{\omega_n} \sin \omega_n t$$

or

$$y = \sqrt{\Delta_s^2 + \frac{2gh}{\omega_n^2}} \sin(\omega_n t + \phi)$$

When $\omega_n t + \phi = \pi/2$, the displacement y of the beam from the static-equilibrium position is a maximum so that

$$y_D = \sqrt{\Delta_s^2 + \frac{2gh}{\omega_n^2}}$$

and the maximum deflection of the center of the beam (and W) from the unloaded position (the x axis) is

$$y_M = \Delta_s + \sqrt{\Delta_s^2 + \frac{2gh}{\omega_n^2}}$$

Substituting values for Δ_s, h, and ω_n into the above equation yields

$$y_M = 0.1 + \sqrt{(0.1)^2 + \left(\frac{27.78}{62.13}\right)^2} = 0.56 \text{ in.}$$

and

$$\frac{y_M}{\Delta_s} = 5.6$$

We should expect this ratio to be higher than the actual ratio since damping was neglected as well as the energy lost during the impact. However, even with these idealizations, it is apparent that when a falling body impacts a structural system, there is a significant increase in the magnitude of the resulting displacements and accompanying stresses over those caused by statically loading the system with a body of the same weight.

In fact if we refer to the next-to-last equation above, it can be seen that if the weight W is suddenly released when $h = 0$,

$$y_M = 2\Delta_s$$

with damping neglected. This indicates that if a weight were barely in contact with a theoretically undamped structure, and it were then suddenly released, the resulting deflection would be twice the deflection the structure would experience if the same weight were slowly lowered onto it.

b. To determine the maximum displacement of the center of the beam with damping such that $\zeta = 0.1$, let us use an approach involving Eq. 2-24. Recalling that

$$\omega_d = \omega_n\sqrt{1 - \zeta^2}$$

we can put Eq. 2-24 in terms of the damped natural circular frequency as

$$y = Y_0 e^{-\zeta\omega_n t} \sin(\omega_d t + \phi)$$

from which

$$\dot{y} = Y_0 e^{-\zeta\omega_n t}[-\zeta\omega_n \sin(\omega_d t + \phi) + \omega_d \cos(\omega_d t + \phi)]$$

Using the initial condition $y_0 = -\Delta_s$, we obtain from the first of the above equations that

$$-\Delta_s = Y_0 \sin \phi$$

or

$$Y_0 = \frac{-\Delta_s}{\sin \phi}$$

Using the initial condition $\dot{y}_0$ in the equation for $\dot{y}$ $(t = 0)$ above yields

$$\dot{y}_0 = Y_0(-\zeta\omega_n \sin \phi + \omega_d \cos \phi)$$

or

$$\dot{y}_0 = \zeta\omega_n \Delta_s - \omega_d \Delta_s \frac{\cos \phi}{\sin \phi}$$

Rearranging this last equation algebraically shows that

$$\tan \phi = \frac{\Delta_s \omega_d}{\zeta\omega_n \Delta_s - \dot{y}_0} = \frac{0.1(61.82)}{0.1(62.13)(0.1) - 27.78}$$

and

$$\phi = -12.84° = -0.224 \text{ rad}$$

It can now be determined that

$$Y_0 = \frac{-\Delta_s}{\sin \phi} = \frac{-0.1}{\sin(-0.224)} = 0.45 \text{ in.}$$

We must now determine the time t_m at which the maximum displacement of the beam occurs (when $\dot{y} = 0$). Referring to the equation for $\dot{y}$ above, we can write that

$$0 = Y_0 e^{-\zeta\omega_n t_m}[-\zeta\omega_n \sin(\omega_d t_m + \phi) + \omega_d \cos(\omega_d t_m + \phi)]$$

from which

$$\tan(\omega_d t_m + \phi) = \frac{\omega_d}{\zeta\omega_n} = \frac{\sqrt{1 - \zeta^2}}{\zeta}$$

Substituting $\omega_d = 61.82$, $\phi = -0.224$, and $\zeta = 0.1$ into the above equation, we find that

$$t_m = 0.027 \text{ s}$$

Thus,

$$y_D = 0.45e^{-0.1(62.13)(0.027)}[\sin(61.82 \cdot 0.027 - 0.224)]$$
$$= 0.378 \text{ in.}$$

and

$$y_M = 0.1 + 0.378 = 0.478 \text{ in.}$$

We can now calculate the ratio

$$\frac{y_M}{\Delta_s} = 4.78$$

which is around 15 percent lower than the value of 5.6 obtained in part a with damping neglected.

2-4 EXPERIMENTAL MEASUREMENT OF SYSTEM DAMPING

The damping factor ζ can be determined experimentally for a system by obtaining a free-vibration record of the system showing how its amplitude of vibration varies with time. As mentioned earlier, a quantity that is frequently used as a measure of the magnitude of damping in a system is the *logarithmic decrement* δ, which is the natural logarithm of the ratio of two successive amplitudes. For a viscously damped system the value of the logarithmic decrement is the same for any two successive amplitudes, and this characteristic is used as a means of experimentally recognizing damping that is viscous in nature from a free-vibration record.

To relate the logarithmic decrement to the damping factor, let us observe two successive amplitudes such as x_1 and x_2 shown in Fig. 2-11, occurring at times t_1 and $(t_1 + \tau_d)$, respectively. Substituting these values of x and t into Eq. 2-24, we can write the logarithm of the ratio of the two amplitudes as

$$\ln\left(\frac{x_1}{x_2}\right) = \ln\left\{\frac{X_0 e^{-\zeta\omega_n t_1} \sin(\omega_n\sqrt{1 - \zeta^2}t_1 + \phi)}{X_0 e^{-\zeta\omega_n(t_1+\tau_d)} \sin[\omega_n\sqrt{1 - \zeta^2}(t_1 + \tau_d) + \phi]}\right\} \tag{2-27}$$

in which the damped period is $\tau_d = 2\pi/(\omega_n\sqrt{1 - \zeta^2})$.

Since the sine terms in the numerator and denominator are equal in magnitude at t_1 and $(t_1 + \tau_d)$, Eq. 2-27 can be written as

$$\ln\left(\frac{x_1}{x_2}\right) = \ln\left(\frac{X_0 e^{-\zeta\omega_n t_1}}{X_0 e^{-\zeta\omega_n(t_1+\tau_d)}}\right)$$

or

$$\ln\left(\frac{x_1}{x_2}\right) = \ln e^{\zeta\omega_n\tau_d}$$

The logarithmic decrement is then

$$\delta = \ln\left(\frac{x_1}{x_2}\right) = \zeta\omega_n\tau_d \tag{2-28}$$

If we next substitute the expression for τ_d from Eq. 2-25 into Eq. 2-28, the logarithmic decrement can be expressed as

$$\delta = \frac{\zeta 2\pi}{\sqrt{1-\zeta^2}} \tag{2-29}$$

Since the damping factor ζ is often less than 0.1 as observed before, $\sqrt{1-\zeta^2} \cong 1$, and Eq. 2-29 reduces to the approximation

$$\delta \cong \zeta 2\pi \tag{2-30}$$

As a practical matter, this relationship is valid for systems having a damping factor ζ as large as 0.2 (20 percent damping).

For such magnitudes of damping the differences between successive amplitudes are small, and measuring these differences accurately from oscillograph records or storage oscilloscopes is very difficult. One method of increasing accuracy in using such data is to plot a number of amplitudes from an experimental record on *semilog* paper as a function of the cycle number. Such a procedure provides a means of averaging the data, and has the effect of minimizing the errors made in reading the amplitudes. Such a plot will be a straight line for viscously damped systems.

To see that this plot is a straight line, we consider two amplitudes s periods apart instead of two consecutive amplitudes as was done in writing Eq. 2-27. In considering these two amplitudes, we can rewrite Eq. 2-28 as

$$\ln\left(\frac{x_1}{x_{1+s}}\right) = \zeta\omega_n(s\tau_d) = \frac{\zeta(2\pi)s}{\sqrt{1-\zeta^2}} \tag{2-31}$$

in which s is the number of elapsed cycles, x_{1+s} is the amplitude after that number of cycles, and $s\tau_d$ is the time elapsed for s cycles. Equation 2-31 can then be written as

$$\underbrace{\ln x_{1+s}}_{y} = \underbrace{-\frac{\zeta(2\pi)}{\sqrt{1-\zeta^2}}\,s}_{mx} + \underbrace{\ln x_1}_{b} \tag{2-32}$$

which is the slope-intercept form of the equation of a straight line ($y = mx + b$) on semi-log paper.

For the usual magnitudes of damping inherent in actual systems, we can approximate Eq. 2-32 with

$$\ln x_{1+s} = \ln x_1 - \zeta 2\pi s \tag{2-33}$$

From the preceding discussion it should be apparent that a semilog plot of the experimental data obtained from a free-vibration record will be essentially a straight line if the system damping is viscous in nature. If the data appear to deviate appreciably from a straight-line plot or show some scatter, an *average* damping factor ζ that is suitable for many engineering purposes can be determined by curve-fitting techniques providing the best fit of the data to a straight-line plot on semilog paper.[6] While curve fitting will always give the best fit of a specified function to the experimental data it is representing, a reasonably good straight-line fit can usually be obtained by simply drawing a straight line that *appears* to be an average of the data plotted on semilog paper.

Although the damping factor ζ can be calculated using Eq. 2-31 with two amplitude values from a free-vibration record that are s cycles apart, a better procedure is to use values obtained from a straight-line plot determined as the average of all the data. This avoids the possibility of inadvertently selecting one or more bad data values. The following example illustrates the procedure for determining values for the damping factor ζ and logarithmic decrement δ from an oscillograph record of the free vibration of a cantilever beam.

EXAMPLE 2-4

A photograph of the oscillograph record of the free vibration of a small plywood cantilever beam is shown in Fig. 2-13. This record was obtained by recording the output of an accelerometer attached to the free end of the beam as shown in Fig. 2-14. The beam was put in motion by simply displacing the free end from its static-equilibrium position, and then releasing it. We are interested in determining (a) if the free-vibration record indicates that the beam's damping is viscous in nature, and (b) the value of the damping factor ζ if viscous damping is indicated, or an average value for ζ if viscous damping is not indicated.

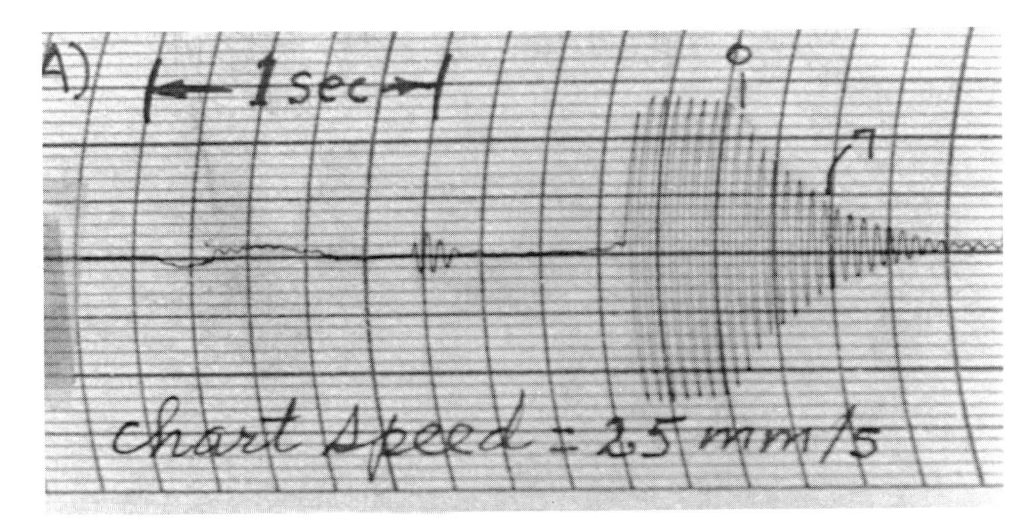

Figure 2-13 Photograph of oscillograph record.

[6] See M. L. James, G. M. Smith, and J. C. Wolford, *Applied Numerical Methods for Digital Computations,* 3d ed., pp. 315–324, Harper & Row, New York, 1985.

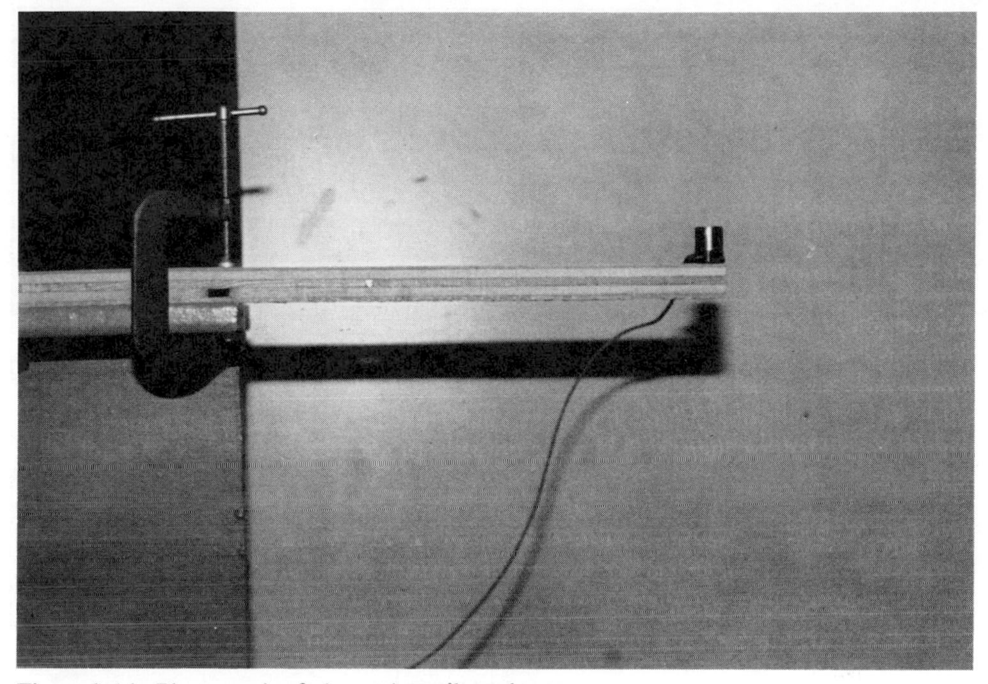

Figure 2-14 Photograph of plywood cantilever beam.

Solution. **a.** To determine the nature of the damping in the beam, the amplitudes obtained from the free-vibration record are plotted as a function of the number of elapsed cycles s on semilog paper as shown in Fig. 2-15. Since the first few cycles may depend upon just how the system was set in motion and may thus not be representative of the free vibration of the system, we start measuring the amplitudes anywhere on the record *after* the system has assumed its free-vibration mode. The initial and final amplitudes plotted in Fig. 2-15 correspond to $s = 0$ and $s = 7$, respectively (see 0 and 7 on the oscillograph record). Since we are only concerned with the amplitude *ratios,* the units in which the amplitudes are measured is unimportant, and the values used for plotting are obtained by merely counting the number of small divisions on the record. As can be seen from the scale of the record, care must be used in interpolating between divisions.

It should be apparent from Fig. 2-15 that the plot of the data points is essentially a straight line. This indicates that the logarithmic decrement is essentially a constant and that the damping factor ζ is independent of amplitude for practical purposes. Therefore, the damping of the plywood beam is viscous in nature and can be described by a viscous-damping model.

To average the errors made in reading the amplitudes from the oscillograph record, we can merely draw a straight line that *appears* to be an average of the data. As mentioned previously, a curve-fitting procedure could be used to provide a curve representing the data, but such a curve would be found to vary little from the one shown in Fig. 2-15 that was "eyeballed."

b. To determine an appropriate value for the damping factor ζ, we select two points on the straight line plotted in part a, and use Eq. 2-31 to calculate a value. We use the following points:

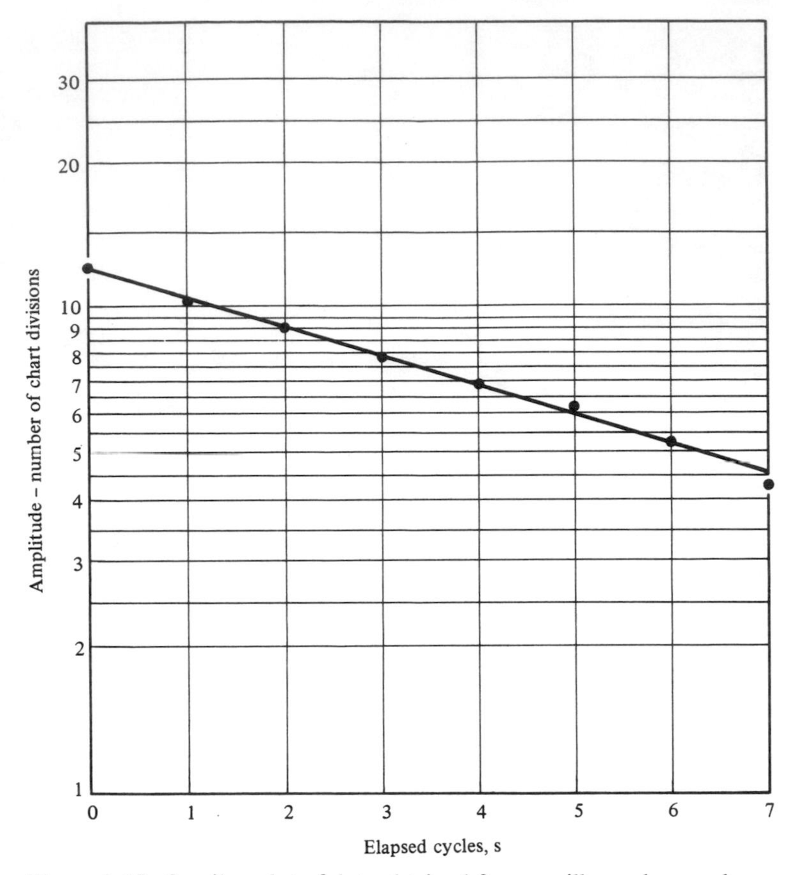

Figure 2-15 Semilog plot of data obtained from oscillograph record.

$$x|_{s=0} = 12$$
$$x|_{s=5} = 6$$

Then using Eq. 2-31, and making the assumption that $\sqrt{1 - \zeta^2} \cong 1$, we obtain

$$\ln(\tfrac{12}{6}) = \zeta 2\pi(5)$$

from which

$$\zeta = 0.022 = 2.2 \text{ percent}$$

The value obtained for ζ shows that the assumption that $\sqrt{1 - \zeta^2} \cong 1$ was justified, so that including this factor in Eq. 2-31 would have had but negligible effect in computing the value of ζ.

2-5 COULOMB DAMPING

Up to now we have considered only viscous damping and the use of viscous-damping models to represent system damping. Let us now turn our attention to damping that results from dry friction, commonly known as *Coulomb damping.*

Consider the system shown in Fig. 2-16, in which the kinetic coefficient of friction is μ. Since friction forces always oppose motion, the friction force F is shown with a sense opposite to the direction of motion indicated. Since the friction force is a discontinuous function (a square wave 180° out of phase with the velocity), we cannot use a single free body to determine the differential equation of motion, as was done for viscously damped systems. Instead, we use the two free-body diagrams shown in Fig. 2-16b, one for motion to the right and the other for motion to the left. Considering each in turn, we find that

$$\ddot{x} + \omega_n^2 x = -\frac{F}{m} \qquad \text{(motion to the right)}$$

and

$$\ddot{x} + \omega_n^2 x = \frac{F}{m} \qquad \text{(motion to the left)}$$

in which $\omega_n^2 = k/m$. Each equation is valid for only the half cycle of motion indicated.

Let us now consider that the mass m is released from position B with the initial displacement x_0. The solution for motion to the left is readily found to be

$$x = A \cos \omega_n t + B \sin \omega_n t + \frac{F}{k} \tag{2-34}$$

in which the trigonometric terms form the *complementary*, or *homogeneous*, solution, and F/k is the *particular* solution. The particular solution can be easily determined by inspection of the differential equation of motion, since F/m divided by ω_n^2 will satisfy the differential equation. That is, the second derivative of a constant is zero, and $F/m\omega_n^2 = F/k$.

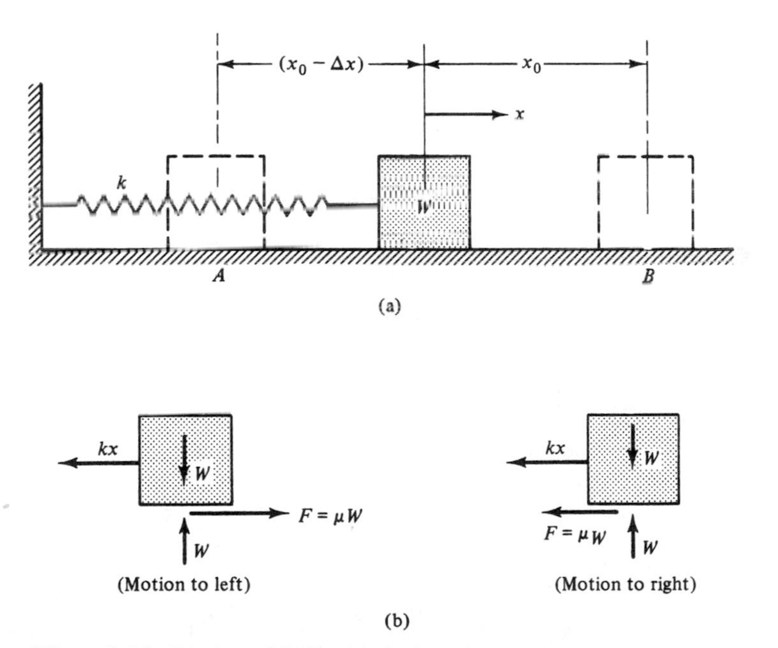

Figure 2-16 System with Coulomb damping.

Using the initial conditions $x = x_0$ and $\dot{x} = 0$ with Eq. 2-34, we find that

$$A = x_0 - \frac{F}{k}$$
$$B = 0$$

The motion to the left for the first half cycle is

$$x = \left(x_0 - \frac{F}{k}\right)\cos \omega_n t + \frac{F}{k} \tag{2-35}$$

The decrease Δx for a half cycle can be determined from Eq. 2-35. Observing that the displacement at the extreme left position is $-(x_0 - \Delta x)$ when $\omega_n t = \pi$, we find from Eq. 2-35 that

$$-(x_0 - \Delta x) = -x_0 + \frac{F}{k} + \frac{F}{k}$$

which in turn shows that

$$\Delta x = \frac{2F}{k} \tag{2-36}$$

Since the amplitude decay for a half cycle is independent of the displacement, we can conclude that the decay is the same for each half cycle. Therefore, the decay for a complete cycle is $4F/k$, as shown in Fig. 2-17.

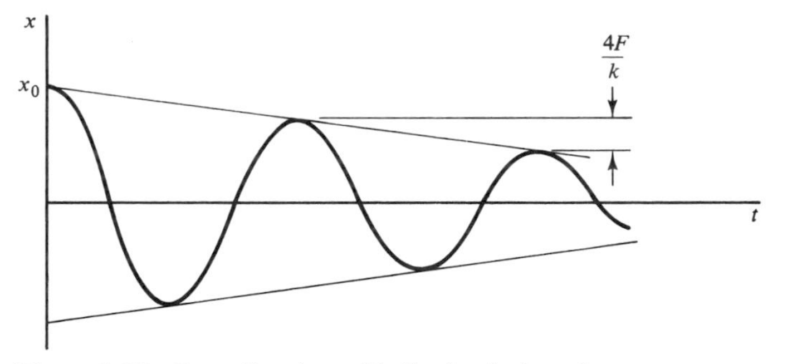

Figure 2-17 Free vibration with Coulomb damping.

2-6 UNSTABLE SYSTEMS

In Sec. 2-3 we discussed the solution of the differential equation of motion of viscously damped systems, in which the system parameters were such that ω_n^2 was always a positive quantity. However, as has been previously noted, the inverted pendulum shown in Fig. 2-5 has a differential equation of motion in which ω_n^2 may be positive, zero, or negative depending upon the relative magnitudes of the system parameters. The differential equation of motion is repeated here and is

$$\ddot{\theta} + \frac{cl_1^2}{I_0}\dot{\theta} + \left[\frac{kl_1^2 - (W_2 l_2/2) - W_1 l_1}{I_0}\right]\theta = 0 \tag{2-9}$$

If

$$kl_1^2 > \frac{W_2 l_2}{2} + W_1 l_1$$

then ω_n^2 is positive, and the solutions discussed in Sec. 2-3 are applicable to the solution of Eq. 2-9. However, if

$$kl_1^2 = \frac{W_2 l_2}{2} + W_1 l_1$$

then $\omega_n^2 = 0$, and the system will not oscillate because it is in a state referred to as *neutral equilibrium*. If the system is displaced some small angle θ_0 and then released, it will remain in this position.

Let us next consider the case in which

$$kl_1^2 < \frac{W_2 l_2}{2} + W_1 l_1$$

Equation 2-8 can then be written as

$$\ddot{\theta} + 2b\dot{\theta} - \omega_n^2\theta = 0 \tag{2-37}$$

in which

$$2b = \frac{cl_1^2}{I_0}$$

and

$$\omega_n^2 = \left|\frac{kl_1^2 - (W_2 l_2/2) - W_1 l_1}{I_0}\right| \quad \text{(absolute value)} \tag{2-38}$$

To see that Eq. 2-37 is the mathematical model of an *unstable* system requires that we determine a solution for it. Assuming a solution having the form of

$$\theta = e^{st} \tag{2-39}$$

and substituting it and its appropriate derivatives into Eq. 2-37, we obtain

$$(s^2 + 2bs - \omega_n^2)e^{st} = 0$$

which yields the real roots

$$s_{1,2} = -b \pm \sqrt{b^2 + \omega_n^2} \tag{2-40}$$

In these root expressions ω_n^2 is the absolute value of the quantity shown in Eq. 2-38. Thus, the solution of Eq. 2-37 is

$$\theta = e^{-bt}(C_1 e^{\sqrt{b^2+\omega_n^2}t} + C_2 e^{-\sqrt{b^2+\omega_n^2}t}) \tag{2-41}$$

To interpret the behavior of the system from Eq. 2-41, we first note that $\sqrt{b^2 + \omega_n^2} > b$. If the inverted pendulum shown in Fig. 2-5 is disturbed from its vertical equilibrium position by initial conditions, it will respond according to Eq. 2-41, and

$$C_2(e^{-bt})e^{-\sqrt{b^2+\omega_n^2}t}$$

approaches zero. However, the term

$$C_1(e^{-bt})e^{\sqrt{b^2+\omega_n^2}t} = C_1 e^{(-b+\sqrt{b^2+\omega_n^2})t}$$

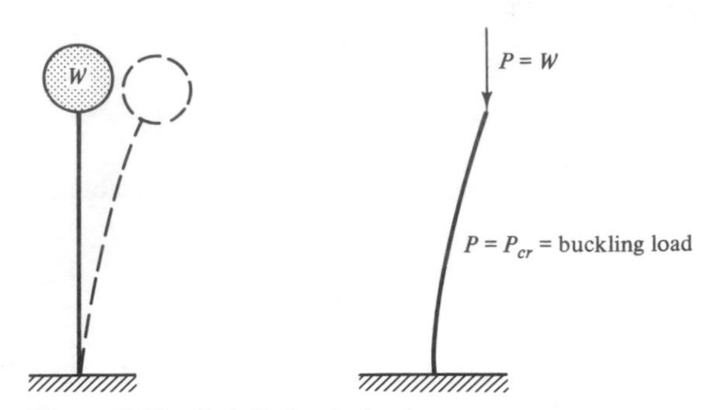

Figure 2-18 Axially loaded column.

continues to *increase* with time since the exponent is *real* and *positive.* Thus, the system is *unstable* when the term of Eq. 2-9,

$$\frac{kl_1^2 - (W_2 l_2/2) - W_1 l_1}{I_0}$$

is negative.

Columns subjected to axial loads such as the one shown in Fig. 2-18 exhibit characteristics similar to those just discussed for the inverted pendulum, in that as the axial load P increases, the natural frequency of the lateral vibration of the column decreases and approaches zero as P approaches the critical (the buckling) load P_{cr}. This will be illustrated in Example 2-11.

2-7 EQUIVALENT SPRINGS

As mentioned before, examples of elastic elements in systems might include cables, beams, plates, columns, and helical springs. All can be considered as springs when they are employed in a system that has a mass considerably greater than that of the elastic element. In some instances the vibration of a system containing a number of such springs in various *series* and *parallel* combinations can be analyzed more conveniently by replacing the combinations with a single *equivalent* spring.

The springs in a system can be in parallel, in series, or in combinations of these as shown in Fig. 2-19. When springs are in parallel as shown in Fig. 2-19a, the *deformation* of each spring is the same owing to the force F. Thus, the forces exerted by the three springs are

$$F_1 = k_1 x$$
$$F_2 = k_2 x$$
$$F_3 = k_3 x$$

The sum of these forces must be equal to the applied force F, and the *equivalent* spring constant k_e can be found by dividing F by the deformation x. Therefore,

$$F = k_1 x + k_2 x + k_3 x = x(k_1 + k_2 + k_3)$$

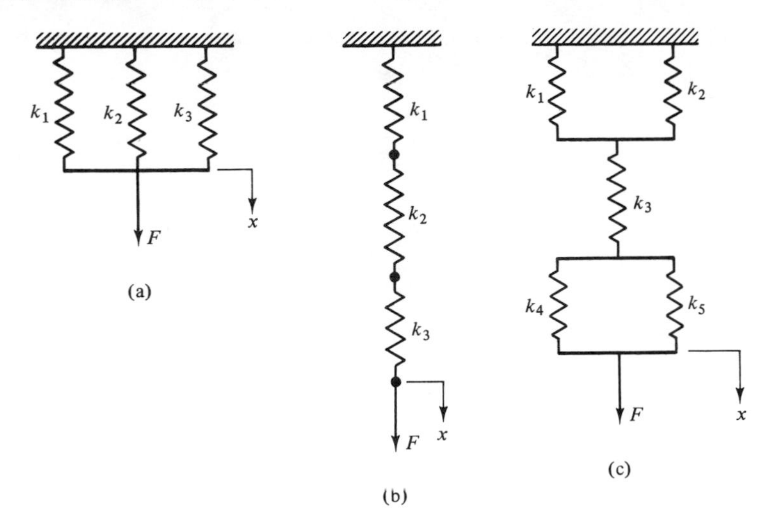

Figure 2-19 (a) Springs in parallel. (b) Springs in series. (c) Parallel and series combination.

and the equivalent spring constant is

$$k_e = \frac{F}{x} = k_1 + k_2 + k_3$$

We can conclude from the above that the equivalent spring constant for n springs in parallel is

$$k_e = \sum_{i=1}^{n} k_i \qquad \text{(parallel)} \tag{2-42}$$

When the springs are in series as shown in Fig. 2-19b, the force in each spring is the same as the applied force F, so that the total deformation x of the springs is the sum of their individual deformations. Thus, with

$$F = k_1x_1 = k_2x_2 = k_3x_3$$

and

$$x = x_1 + x_2 + x_3$$

we find that

$$x = F\left(\frac{1}{k_1} + \frac{1}{k_2} + \frac{1}{k_3}\right)$$

The equivalent spring constant for the three springs in series is then

$$k_e = \frac{F}{x} = \frac{1}{1/k_1 + 1/k_2 + 1/k_3} \qquad \text{(series)}$$

It follows that the equivalent spring constant for n springs in series is given by

$$k_e = \frac{1}{\sum_{l=1}^{n} \frac{1}{k_i}} \qquad \text{(series)}$$

or expressed inversely

$$\frac{1}{k_e} = \sum_{i=1}^{n} \frac{1}{k_i} \qquad \text{(series)} \tag{2-43}$$

With equations developed for determining the equivalent spring constants of parallel and series combinations, let us turn our attention to determining equivalent spring constants for combinations of these two. A general procedure for doing this is to determine the equivalent spring constants of parallel combinations in the system, and then combine them with the series elements to obtain yet another equivalent spring constant. This process continues until all parallel combinations have been reduced and an overall equivalent spring constant k_e determined. For example, the equivalent spring constant for the system shown in Fig. 2-19c would be obtained as

$$\frac{1}{k_e} = \frac{1}{k_1 + k_2} + \frac{1}{k_4 + k_5} + \frac{1}{k_3}$$

in which k_1 and k_2, and k_4 and k_5, are the parallel elements that are in series with k_3.

EXAMPLE 2-5

A weight W is attached to the end of a small cantilever beam of length l by a cable of stiffness k as shown in Fig. 2-20. If

$$W = 100 \text{ lb}$$
$$EI = 200{,}000 \text{ lb} \cdot \text{in.}^2 \text{ (for beam)}$$
$$k = 100 \text{ lb/in.}$$

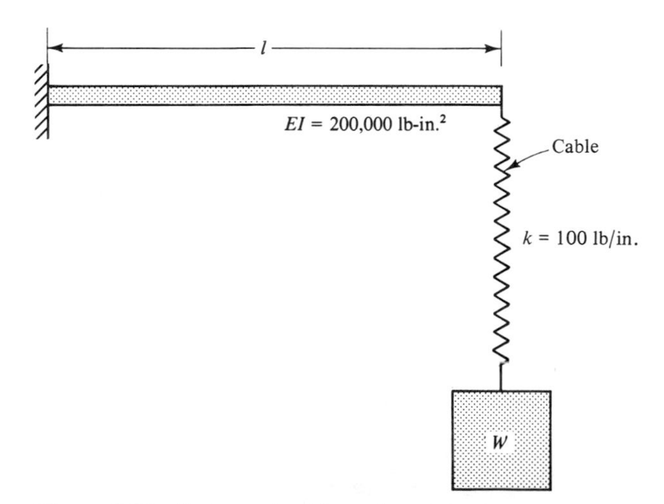

Figure 2-20 Beam and cable spring system.

determine the length l of the beam that will make the system's undamped natural frequency $f_n = 2$ Hz. Consider the mass of both cable and beam as negligible compared with the mass of the weight W.

Solution. The beam and the cable act as springs in series. With the undamped natural frequency of the system given by

$$f_n = \frac{\omega_n}{2\pi} = \frac{1}{2\pi}\sqrt{\frac{k_e}{m}}$$

the equivalent spring constant is

$$k_e = (2\pi f_n)^2 m = \frac{4(6.28)^2 100}{386} = 40.87 \text{ lb/in.}$$

Designating the spring constant of the beam as k_b, we can use Eq. 2-43 to write

$$\frac{1}{k_e} = \frac{1}{40.87} = \frac{1}{100} + \frac{1}{k_b}$$

from which $k_b = 69.12$ lb/in. Since the spring constant of a cantilever beam is $3EI/l^3$,

$$k_b = \frac{3EI}{l^3} = \frac{3(200{,}000)}{l^3} = 69.12$$

and

$$l = 20.6 \text{ in.}$$

Table 2-1 (see page 78) gives the spring constants and deflection equations of some commonly used elastic elements. The elastic curves shown for some typical beams will be used later as shape functions in Rayleigh's energy method for beams.

2-8 DETERMINING NATURAL FREQUENCIES BY RAYLEIGH'S ENERGY METHOD

Rayleigh's energy method is a convenient method of determining the fundamental natural frequencies of systems in which distributed masses must be considered.[7] In this section we use this method with several types of systems other than beams, and apply it particularly to beams in the next section.

Rayleigh's method does not require the determination of the differential equation of motion of a system to determine its fundamental natural frequency. It is particularly advantageous for use with systems consisting of several bodies with relative motion, and for systems in which the distributed masses of elastic elements must be considered.

Rayleigh's method is based upon the assumption that the energy of a system is conserved, which means that damping must be neglected. However, such an idealization

[7] J. W. S. Rayleigh, *The Theory of Sound,* 2d ed., vol. 1, Dover Publications, New York, 1945.

TABLE 2-1 SPRING CONSTANTS AND DEFLECTION EQUATIONS OF ELASTIC ELEMENTS

A = area of cross section
E = modulus of elasticity
I = area moment of inertia about neutral axis
G = modulus of rigidity
J = polar moment of inertia

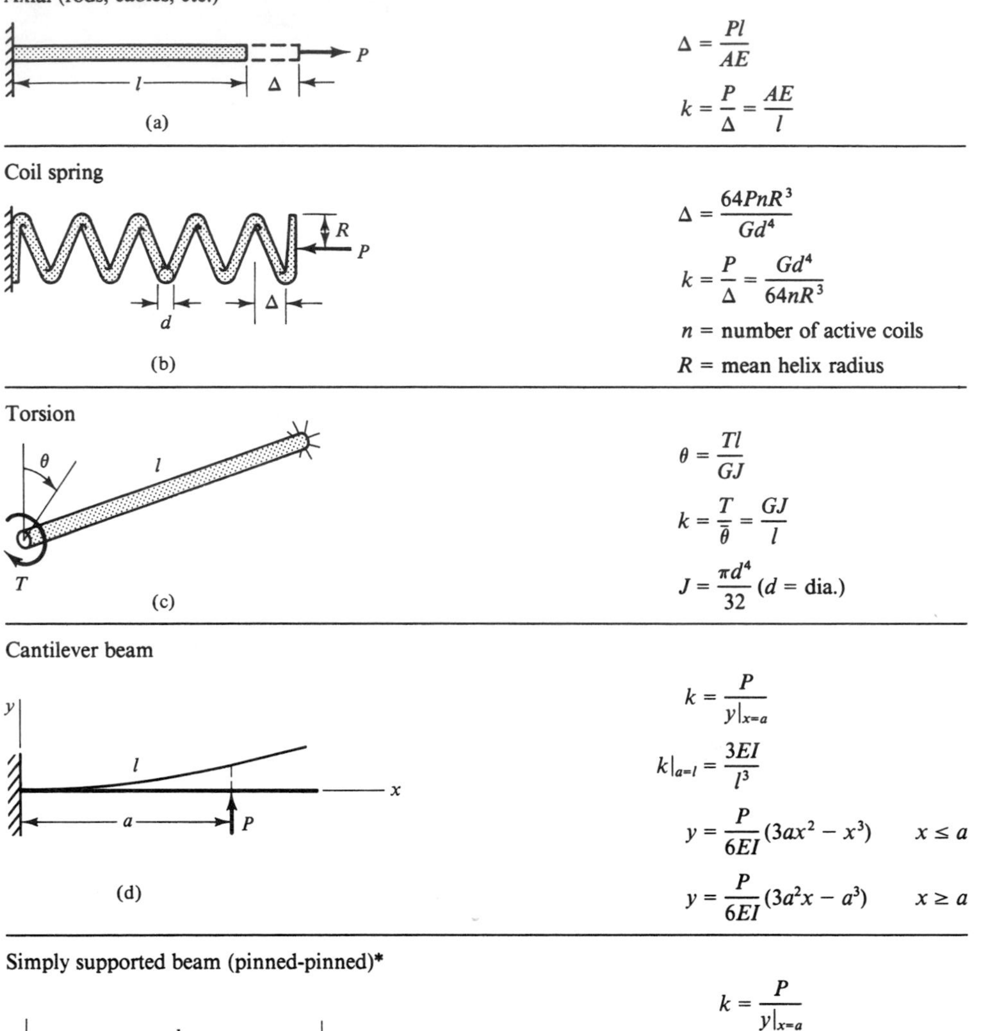

Axial (rods, cables, etc.)

(a)

$$\Delta = \frac{Pl}{AE}$$

$$k = \frac{P}{\Delta} = \frac{AE}{l}$$

Coil spring

(b)

$$\Delta = \frac{64PnR^3}{Gd^4}$$

$$k = \frac{P}{\Delta} = \frac{Gd^4}{64nR^3}$$

n = number of active coils

R = mean helix radius

Torsion

(c)

$$\theta = \frac{Tl}{GJ}$$

$$k = \frac{T}{\theta} = \frac{GJ}{l}$$

$$J = \frac{\pi d^4}{32} \ (d = \text{dia.})$$

Cantilever beam

(d)

$$k = \frac{P}{y|_{x=a}}$$

$$k|_{a=l} = \frac{3EI}{l^3}$$

$$y = \frac{P}{6EI}(3ax^2 - x^3) \qquad x \le a$$

$$y = \frac{P}{6EI}(3a^2x - a^3) \qquad x \ge a$$

Simply supported beam (pinned-pinned)*

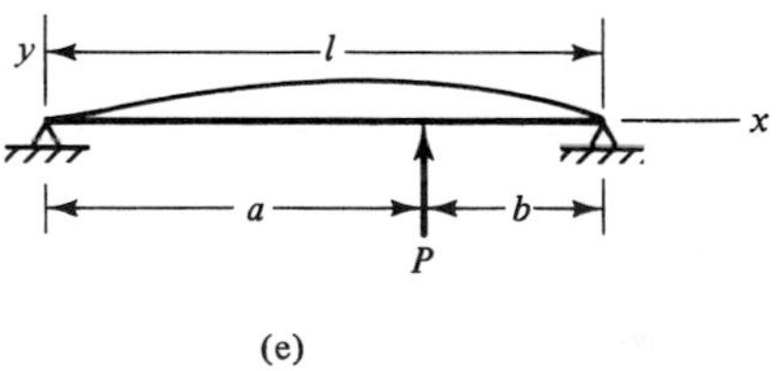

(e)

$$k = \frac{P}{y|_{x=a}}$$

$$k|_{a=l/2} = \frac{48EI}{l^3}$$

$$y = \frac{Pbx}{6EIl}(l^2 - x^2 - b^2) \qquad x \le a$$

$$y = \frac{Pb}{6EIl}\left[(l^2 - b^2)x - x^3 + \frac{l}{b}(x - a)^3\right] \qquad x \ge a$$

TABLE 2-1 (*Continued*)
A = area of cross section
E = modulus of elasticity
I = area moment of inertia about neutral axis
G = modulus of rigidity
J = polar moment of inertia

Pinned-pinned beam with overhang*

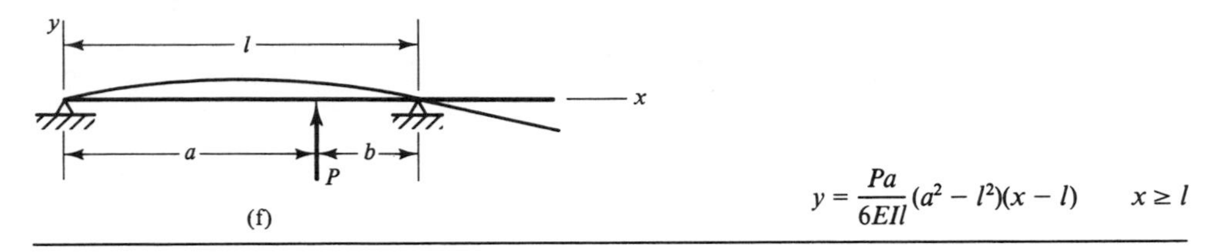

(f)

$$y = \frac{Pa}{6EIl}(a^2 - l^2)(x - l) \qquad x \geq l$$

Pinned-pinned beam with overhang (P at $x = l + a$)*

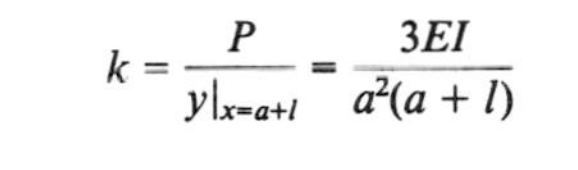

$$k = \frac{P}{y|_{x=a+l}} = \frac{3EI}{a^2(a + l)}$$

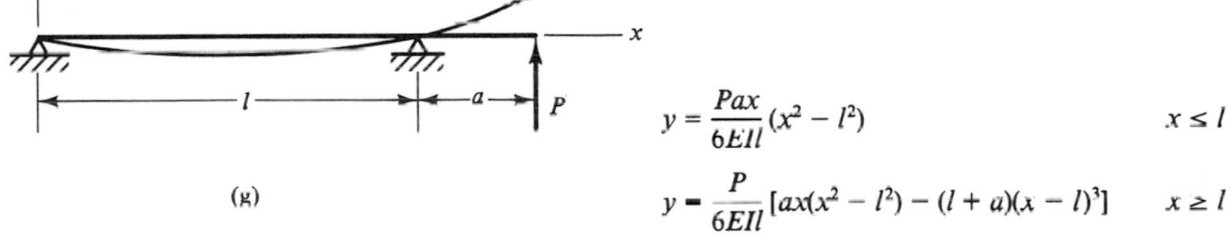

$$y = \frac{Pax}{6EIl}(x^2 - l^2) \qquad x \leq l$$

(g)

$$y = \frac{P}{6EIl}[ax(x^2 - l^2) - (l + a)(x - l)^3] \qquad x \geq l$$

Fixed-fixed beam with lateral displacement

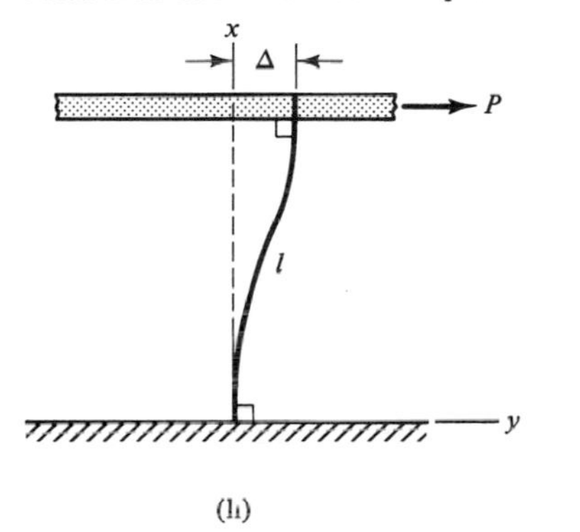

(h)

$$\Delta = \frac{Pl^3}{12EI}$$

$$k = \frac{12EI}{l^3}$$

$$y = \frac{P}{12EI}(3lx^2 - 2x^3)$$

TABLE 2-1 (*Continued*)
A = area of cross section
E = modulus of elasticity
I = area moment of inertia about neutral axis
G = modulus of rigidity
J = polar moment of inertia

Fixed-fixed beam*

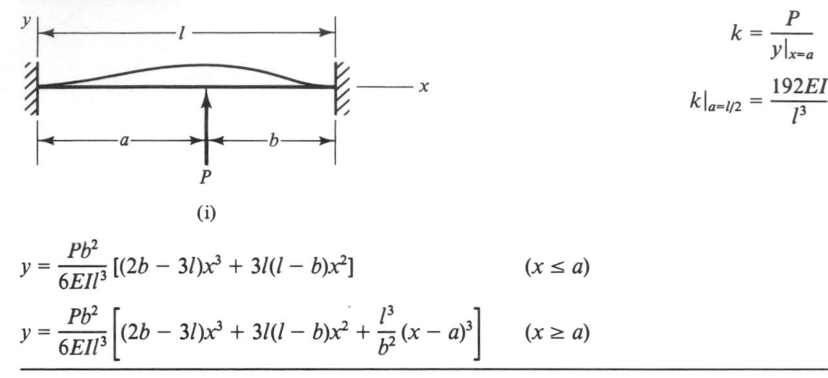

$$k = \frac{P}{y|_{x=a}}$$

$$k|_{a=l/2} = \frac{192EI}{l^3}$$

(i)

$$y = \frac{Pb^2}{6EIl^3}\left[(2b - 3l)x^3 + 3l(l - b)x^2\right] \qquad (x \leq a)$$

$$y = \frac{Pb^2}{6EIl^3}\left[(2b - 3l)x^3 + 3l(l - b)x^2 + \frac{l^3}{b^2}(x - a)^3\right] \qquad (x \geq a)$$

Fixed-pinned beam with overhang*

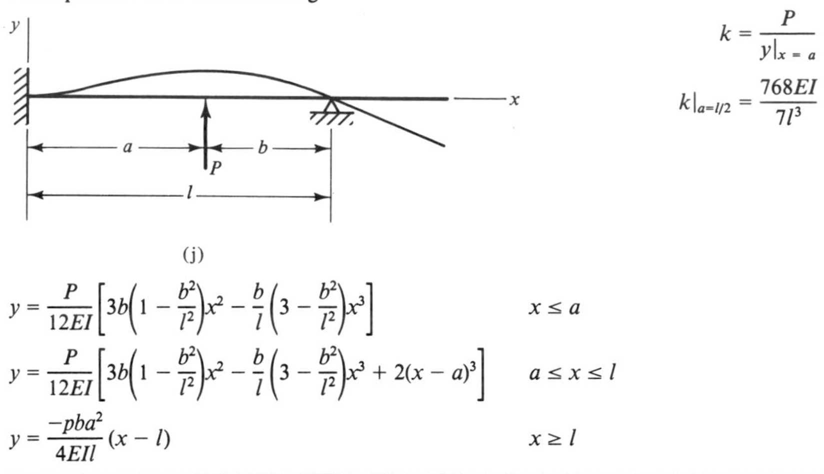

$$k = \frac{P}{y|_{x=a}}$$

$$k|_{a=l/2} = \frac{768EI}{7l^3}$$

(j)

$$y = \frac{P}{12EI}\left[3b\left(1 - \frac{b^2}{l^2}\right)x^2 - \frac{b}{l}\left(3 - \frac{b^2}{l^2}\right)x^3\right] \qquad x \leq a$$

$$y = \frac{P}{12EI}\left[3b\left(1 - \frac{b^2}{l^2}\right)x^2 - \frac{b}{l}\left(3 - \frac{b^2}{l^2}\right)x^3 + 2(x - a)^3\right] \qquad a \leq x \leq l$$

$$y = \frac{-pba^2}{4EIl}(x - l) \qquad x \geq l$$

Fixed-pinned beam with overhang (P at $x = l + a$)*

$$k = \frac{P}{y|_{x=l+a}} = \frac{12EI}{a^2(3l + 4a)}$$

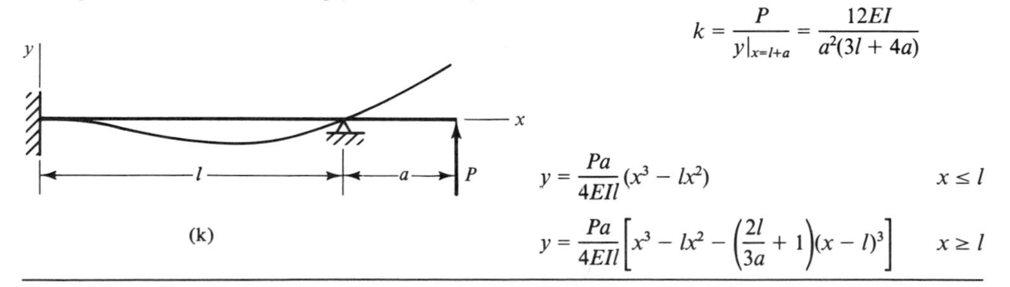

(k)

$$y = \frac{Pa}{4EIl}(x^3 - lx^2) \qquad x \leq l$$

$$y = \frac{Pa}{4EIl}\left[x^3 - lx^2 - \left(\frac{2l}{3a} + 1\right)(x - l)^3\right] \qquad x \geq l$$

* Axial extensions due to axial end constraints considered negligible.

does not introduce any serious error for systems in which the damping is relatively small, as discussed earlier in this chapter. Assuming a conservative system, we can write that

$$T + U = \text{constant} \tag{2-44}$$

where T = kinetic energy of the system
U = *change* in potential energy of the system from its potential energy in the *static-equilibrium position*

The kinetic energy of a system is a function of the velocities of the system masses. The potential energy of a system consists of the strain energy U_e stored in elastic elements, and the energy U_g which is a function of the vertical distances between the system masses and some arbitrary datum. Rayleigh's method utilizes a *displacement* function (often referred to as a *shape* function) to determine the potential and kinetic energy of a distributed mass. This function expresses the displacement of each particle of the distributed mass, and it is then assumed that the *shape* of this function is also the *shape* of the *amplitude* curve of the distributed mass as it vibrates. Rayleigh found that the displacement function selected need not have the exact shape of the amplitude curve to obtain a good approximation of the exact fundamental natural frequency of the distributed mass.

Before discussing the application of Rayleigh's energy method to systems containing distributed masses, let us review briefly several basic concepts of kinetic and potential energy by considering the simple undamped spring-and-mass system shown in Fig. 2-21. If we consider the mass of the spring as negligible, and treat the mass m as a particle since its vibration is in translation, this system will contain no distributed masses that require the use of a displacement function.

If we measure x from the static-equilibrium position, and start measuring time as the mass passes through the static-equilibrium position,

$$x = x_0 \sin \omega_n t \tag{2-45}$$

The velocity of the mass at any time t is then

$$\dot{x} = x_0\omega_n \cos \omega_n t$$

and the maximum velocity is

$$\dot{x}_{max} = x_0\omega_n$$

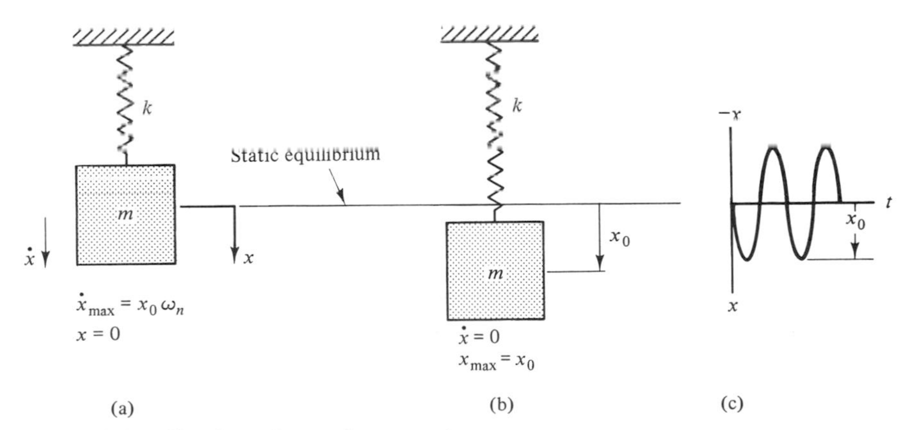

Figure 2-21 Simple spring-and-mass system.

when $\omega t = 0$ or multiples of 2π. That is, the maximum velocity is the product of the amplitude x_0 and the natural circular frequency ω_n, and occurs as the mass passes through the static-equilibrium position. Since U is the *change* in the potential energy of the system from its potential energy in the static-equilibrium position, $U = 0$ as the mass passes through this position with its maximum velocity. It follows then that

$$\left.\begin{aligned} U &= 0 \\ T &= T_{max} = \tfrac{1}{2}m(x_0\omega_n)^2 \end{aligned}\right\} \quad (x = 0) \tag{2-46}$$

To determine the maximum *change* in the potential energy of the system, we note that when the displacement is a maximum (x_0), the velocity is zero. Therefore,

$$\left.\begin{aligned} T &= 0 \\ U &= U_{max} = \tfrac{1}{2}kx_0^2 \end{aligned}\right\} \quad (x = x_0) \tag{2-47}$$

Since Eqs. 2-46 and 2-47 show that each energy function is a maximum when the other is zero, it follows from Eq. 2-44 that

$$T_{max} = U_{max} = \text{constant} \tag{2-48}$$

At this point the reader might question why the maximum change in the potential energy is just the term shown in Eq. 2-47, since there is a change in the potential energy of $U_g = -mgx_0$ when the mass is at a position x_0 below the static-equilibrium position. This term is canceled by a strain-energy term $U_e = mgx_0$ that results from the fact that the strain energy stored in an elastic element is a function of how much it is deformed from its free length, rather than from the static-equilibrium position from which the amplitude x_0 is measured. This will be illustrated in Example 2-6.

If the differential equation of motion is desired as well as the fundamental natural frequency, the derivative of Eq. 2-44

$$\frac{d}{dt}(T + U) = 0$$

will provide this equation, which contains ω_n^2 in terms of the system parameters. However, if only the natural frequency is desired, ω_n can be determined without obtaining the differential equation of motion. Substituting Eqs. 2-46 and 2-47 into Eq. 2-48

$$\tfrac{1}{2}m(x_0\omega_n)^2 = \tfrac{1}{2}kx_0^2$$

which yields the natural circular frequency as

$$\omega_n = \sqrt{\frac{k}{m}}$$

Noting that x_0^2 appears in both T_{max} and U_{max}, and consequently cancels, we confirm the fact that the frequency is independent of the amplitude of the vibration. This will occur for any linear system.[8]

EXAMPLE 2-6

Let us consider the rack-and-gear system shown in Fig. 2-22, in which the mass of the spring is significant in comparison with the rack and the gears, and thus requires the use

[8] A linear system is one whose equations of motion are linear differential equations.

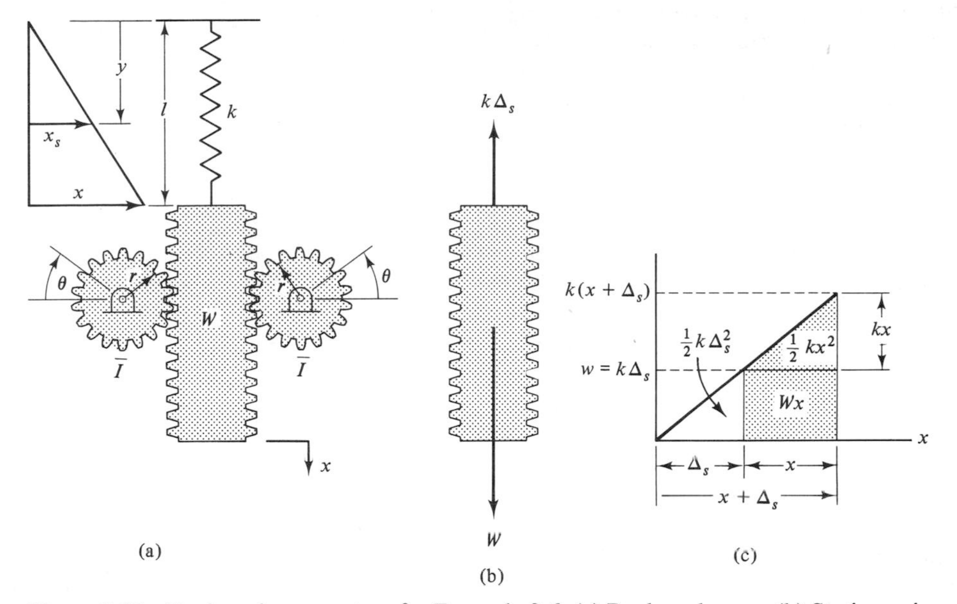

Figure 2-22 Rack-and-gear system for Example 2-6. (a) Rack and gears. (b) Static-equilibrium position ($W = k\ \Delta_s$). (c) Spring-force diagram.

of a displacement function. The system consists of two identical gears of pitch radius r and centroidal mass moment of inertia $\bar{I}$, a rack of weight W, and a linear spring of stiffness k, length l, and a mass of γ per unit length. We wish to determine

a. The differential equation of motion of the system using

$$\frac{d}{dt}(T + U) = 0$$

b. The undamped natural circular frequency ω_n of the system using

$$T_{max} = U_{max}$$

Solution. **a.** Since the static displacement x_s of any point on the spring is proportional to its distance y from the spring support, we can express these static deflections as

$$x_s = \frac{yx}{l}$$

which has the plot shown in Fig. 2-22a. Assuming that the amplitudes of vibration of points on the spring vary in the same manner as the static displacements, the static-deflection equation above can be used as our *displacement function.* That is, the amplitude curve is assumed to have the same *shape* as the static-deflection curve in Fig. 2-22a.

When the rack is displaced a distance x below the static-equilibrium position, the gears have the angular displacements θ shown in Fig. 2-22a, which are related to the displacement of the rack by

$$\theta = \frac{x}{r}$$

The angular velocity of each gear is then related to the velocity of the rack by

$$\dot{\theta} = \frac{\dot{x}}{r}$$

Since $x = r\theta$, we can write the displacement function as

$$x_s = \frac{yr\theta}{l}$$

The velocity of any point on the spring can now be found from the time derivative of the above as

$$\dot{x}_s = \frac{yr\dot{\theta}}{l}$$

The kinetic energy of a differential element of length dy of the spring is

$$dT = \tfrac{1}{2}\gamma(\dot{x}_s)^2\,dy$$

The total kinetic energy of the system in terms of θ is

$$T = \underbrace{\bar{I}\dot{\theta}^2}_{\text{Gears}} + \underbrace{\frac{1}{2}\frac{W}{g}(r\dot{\theta})^2}_{\text{Rack}} + \underbrace{\frac{\gamma}{2}\int_0^l \left(\frac{yr\dot{\theta}}{l}\right)^2 dy}_{\text{Spring}}$$

or

$$T = \left[\bar{I} + \frac{1}{2}\frac{W}{g}r^2 + \frac{m_s r^2}{2(3)}\right]\dot{\theta}^2 \tag{2-49}$$

in which $m_s = \gamma l$ = total mass of the spring.

The spring-force diagram in Fig. 2-22c shows that the *change* in the strain energy U_e for a *positive* downward displacement x of the rack from the static-equilibrium position (where the spring is already displaced Δ_s from its free length) is the shaded area in the figure given by

$$U_e = \tfrac{1}{2}kx^2 + Wx$$

The accompanying *change* in potential energy U_g as the rack moves a distance x *below* the static-equilibrium position is

$$U_g = -Wx$$

and the total *change* in potential energy is

$$U = U_e + U_g = \tfrac{1}{2}kx^2$$

or (since $x = r\theta$)

$$U = \tfrac{1}{2}k(r\theta)^2 \tag{2-50}$$

Substituting Eqs. 2-49 and 2-50 into

$$T + U = \text{constant}$$

gives

$$\left[\bar{I} + \frac{1}{2}\frac{W}{g}r^2 + \frac{m_s r^2}{2(3)}\right]\dot{\theta}^2 + \frac{1}{2}kr^2\theta^2 = \text{constant} \tag{2-51}$$

Differentiating Eq. 2-51 with respect to time, we obtain

$$\ddot{\theta} + \left[\frac{kr^2}{2\bar{I} + (W/g)r^2 + (m_s r^2/3)}\right]\theta = 0 \tag{2-52}$$

which is the differential equation of motion of the system. Since ω_n^2 is the coefficient of θ in the above equation, the natural circular frequency of the system is given by

$$\omega_n = r\sqrt{\frac{k}{2\bar{I} + (W/g)r^2 + (m_s r^2/3)}}$$

It is apparent from the above equation that if the mass m_s of the spring is small compared with either the mass W/g of the rack and/or the mass moment of inertia $\bar{I}$ of the gears, its effect on the magnitude of ω_n is negligible.

b. Referring to Eq. 2-45, the displacement of the rack during vibration can be expressed as

$$x = x_0 \sin \omega_n t$$

Analogously, the angular displacement of the gears can be expressed as

$$\theta = \theta_0 \sin \omega_n t$$

The derivative of the last expression gives us the maximum angular velocity as

$$\dot{\theta}_{max} = \theta_0 \omega_n$$

Substituting this relationship into Eq. 2-49, we find that

$$T_{max} = \left[\bar{I} + \frac{1}{2}\frac{W}{g}r^2 + \frac{m_s r^2}{2(3)}\right](\theta_0\omega_n)^2$$

Since the maximum angular displacement of each gear is its amplitude θ_0, the maximum potential energy is obtained from Eq. 2-50 as

$$U_{max} = \tfrac{1}{2}kr^2\theta_0^2$$

Then using the concept of $T_{max} = U_{max}$, we equate the above two equations to obtain

$$\omega_n = r\sqrt{\frac{k}{2\bar{I} + (W/g)r^2 + (m_s r^2/3)}}$$

which is the same result obtained in part a.

The reader is referred at this point to Fig. 2-22c as a reminder that the loss in potential energy $-Wx$ was canceled by a portion of the strain energy Wx indicated in the figure by the rectangular portion of the shaded area. In general, the only forces that the elastic elements must provide to maintain the system in its static-equilibrium position are those necessary to overcome gravity forces. In such instances, when the system is displaced from its static-equilibrium position, the *maximum change* in the potential energy of the system from its potential energy in the static-equilibrium position will be a term having the general form of

$$U_{max} = \tfrac{1}{2}(\text{stiffness factor})(\text{displacement from static-equilibrium position})^2$$

There are systems in which the change in the potential energy of position U_g is not canceled by a portion of the strain-energy change. This occurs when no spring force is required to maintain the system in its static-equilibrium position. An example is the inverted pendulum shown later in Example 2-8.

EXAMPLE 2-7

Let us use the concept of $T_{max} = U_{max}$ to determine the undamped natural circular frequency ω_n of the system shown in Fig. 2-23 which consists of a weight W attached to a slender rod. The rod has a length l, a mass per unit length γ, and a stiffness factor $k = AE/l$ (see Table 2-1 on page 78).

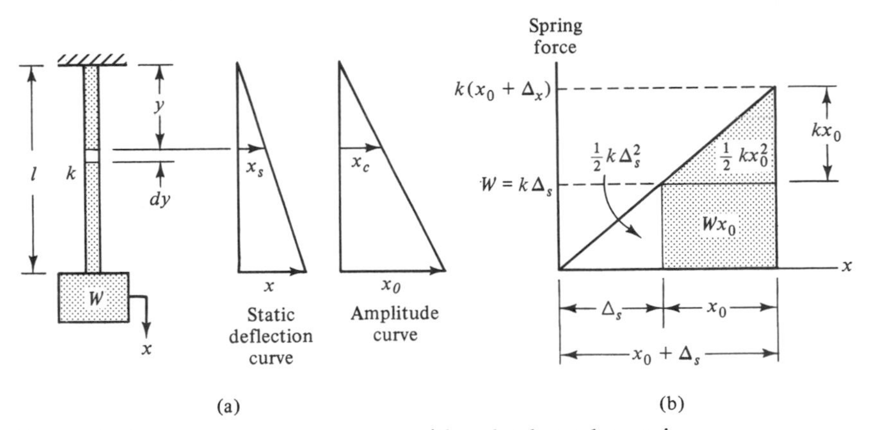

Figure 2-23 Spring-and-mass system with a slender rod as spring.

Solution. Our first step is to select a displacement function to use in determining the kinetic and potential energy of the system. We know from elementary mechanics of materials that the static displacement x_s of any cross section of the rod is proportional to its distance y from the fixed end. Thus, the static-deflection curve is as shown in Fig. 2-23a. If we now make the reasonable assumption that the *amplitude curve* of the rod has the same *shape* as the static-deflection curve as indicated in Fig. 2-23a, we can use as our shape function

$$x_c = \frac{x_0 y}{l}$$

in which x_c is the *amplitude* of any cross section of the rod and x_0 is the *amplitude* of the lower end of the rod and the weight W, as shown in Fig. 2-23a.

Since the natural circular frequency ω_n of each cross section of the rod is the same, the *maximum* velocity of any cross section along the rod is simply

$$x_c\omega_n = \frac{x_0 y}{l}\,\omega_n$$

The maximum kinetic energy of a differential element of length dy of the rod is

$$dT = \frac{1}{2}\gamma\left(\frac{x_0 y\omega_n}{l}\right)^2 dy$$

in which $\gamma\, dy$ is the mass of the differential element.

The maximum kinetic energy *of the system* is

$$T_{\max} = \frac{1}{2}\frac{W}{g}(x_0\omega_n)^2 + \frac{1}{2}\gamma\int_0^l \left(\frac{x_0 y}{l}\right)^2 \omega_n^2\, dy$$

Integrating and simplifying,

$$T_{\max} = \left[\frac{M}{2} + \frac{m_s}{2(3)}\right] x_0^2\omega_n^2 \tag{2-53}$$

where $M = W/g$ = mass of weight W
$m_s = \gamma l$ = mass of rod

Referring to the spring-force diagram in Fig. 2-23b we see that the strain energy in the static-equilibrium position is $k\Delta_s^2/2$. Therefore, the *maximum change* in the strain energy U_e for a positive amplitude x_0 measured from the static-equilibrium position is the shaded area in the figure

$$U_e = \tfrac{1}{2}kx_0^2 + Wx_0$$

The accompanying *maximum change* in the potential energy of position U_g as the weight W moves x_0 *below* the static-equilibrium position is

$$U_g = -Wx_0$$

The maximum change in the potential energy U of the system from its potential energy in the static-equilibrium position is thus

$$U_{\max} = U_e + U_g = \tfrac{1}{2}kx_0^2 \tag{2-54}$$

Since $T_{\max} = U_{\max}$, we set Eq. 2-53 equal to Eq. 2-54 to find that

$$\omega_n = \sqrt{\frac{k}{M + m_s/3}}$$

or

$$\omega_n = \sqrt{\frac{k}{M(1 + \alpha/3)}} \tag{2-55}$$

in which the mass ratio $(m_s/M) = \alpha$.

It should be apparent from Eq. 2-55 that if the mass m_s of the rod is small in comparison with the mass M of the rigid body, it has little effect on the natural circular frequency ω_n of the system. The mass term $(M + m_s/3)$ is frequently referred to as the *effective mass* M_e of a system. The approximate solution given by Eq. 2-55 is within 1 percent of the exact solution of the wave equation obtained in Sec. 8-3 when $m_s/M = 1.0$. However, as m_s/M increases above this value, the approximate solution begins to deviate quite rapidly from the exact solution. When $m_s/M = 1.5$, for example, the error is on the order of 32 percent.

EXAMPLE 2-8

The spring-restored inverted pendulum shown in Fig. 2-24 is a slender rod of length l, mass m, and mass moment of inertia of $I_0 = ml^2/3$ about the pivot axis o. Each of the springs attached to the end of the rod has a stiffness k. Neglect the mass of the springs, and determine the natural circular frequency of the pendulum using $T_{max} = U_{max}$. Assume that the amplitude θ_0 of the oscillations is small.

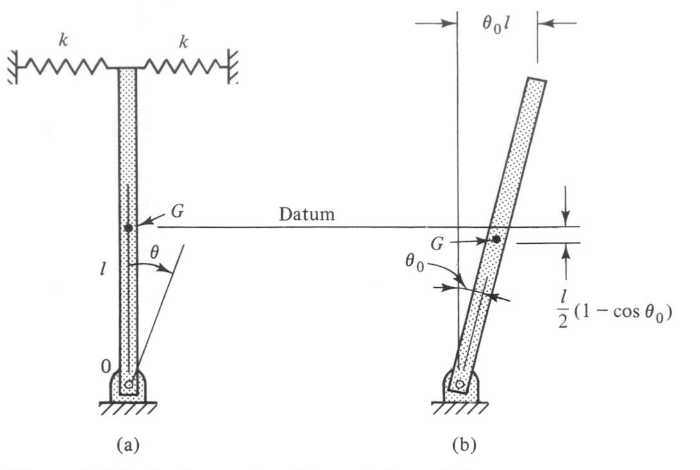

Figure 2-24 Spring-restored inverted pendulum.

Solution. The maximum angular velocity of the rod is the amplitude θ_0 times ω_n. Thus,

$$\dot{\theta}_{max} = \theta_0 \omega_n$$

and, since we are neglecting the masses of the elastic elements, the maximum kinetic energy of the system is the kinetic energy of the rod

$$T_{max} = \tfrac{1}{2} I_0 (\theta_0 \omega_n)^2 \tag{2-56}$$

When the pendulum is in its vertical position, there is no strain energy in either spring due to any gravity force, so the maximum *change* in the strain energy U_e stored in the two springs when the pendulum is at its maximum displacement θ_0 is

$$U_e = k(\theta_0 l)^2 \tag{2-57}$$

for small θ_0. The two springs may be either strained or unstrained when the pendulum is in its vertical static-equilibrium position, without altering the solution, since U_e depends only upon the *change* in strain energy between $\theta = 0$ and $\theta = \theta_0$.

The vertical displacement of the mass center G from the datum shown in the figure is $l(1 - \cos\theta_0)/2$. Thus, the *change* in potential energy U_g is

$$U_g = -\frac{mgl}{2}(1 - \cos\theta_0) \tag{2-58}$$

in which mg is the weight W of the rod. Since θ_0 is small, the first inclination might be to let $\cos\theta_0 = 1$. However, as Eq. 2-58 reveals, this would make the change in U_g zero, which we know is not true. To include the change in U_g, let us write $\cos\theta_0$ in terms of the series

$$\cos\theta_0 = 1 - \frac{\theta_0^2}{2!} + \frac{\theta_0^4}{4!} - \cdots$$

Then, for small values of θ_0, we can let

$$\cos\theta_0 \cong 1 - \frac{\theta_0^2}{2}$$

Substituting $1 - \theta_0^2/2$ for $\cos\theta_0$ in Eq. 2-58 gives

$$U_g = -\frac{mgl}{2}\left(\frac{\theta_0^2}{2}\right) \tag{2-59}$$

Summing Eqs. 2-57 and 2-59, we obtain the maximum *change* in potential energy as

$$U_{\max} = U_e + U_g = k(\theta_0 l)^2 - \frac{mgl\theta_0^2}{4} \tag{2-60}$$

Substituting Eqs. 2-56 and 2-60 into $T_{\max} = U_{\max}$, we can solve for the natural circular frequency of the pendulum as

$$\omega_n = \sqrt{\frac{2kl^2 - mgl/2}{I_0}} \tag{2-61}$$

in which $I_0 = ml^2/3$.

It should be noted in Eq. 2-61 that if $mgl/2 = 2kl^2$, then $\omega_n = 0$; and if $mgl/2 > 2kl^2$, the system is *unstable,* as discussed in Sec. 2-6. It might also be noted that if the rod in Fig. 2-24 were suspended from o at the *top* of the rod, with the springs attached at the bottom, the potential-energy terms would both be positive so that

$$U_{\max} = k(\theta_0 l)^2 + \frac{mgl\theta_0^2}{4}$$

which would make

$$\omega_n = \sqrt{\frac{2kl^2 + mgl/2}{I_0}}$$

which indicates a stable system.

2-9 RAYLEIGH'S ENERGY METHOD APPLIED TO BEAMS

Beams are among the most commonly used structural elements. Examples are floor supports, rafters, and truck and car chassis members. The fuselages and wings of aircraft are essentially large complex beams composed of other structural elements including small beams, trusses, and skin. Small beams are used as elements in various types of instruments and control devices. Very small beams have even been used as transducers for measuring tongue and cheek forces acting on people's teeth.[9]

[9] D. C. Haack et al., "On an Equilibrium Theory of Tooth Position," *Angle Orthodontist,* vol. 33, no. 1, pp. 1–26, 1963.

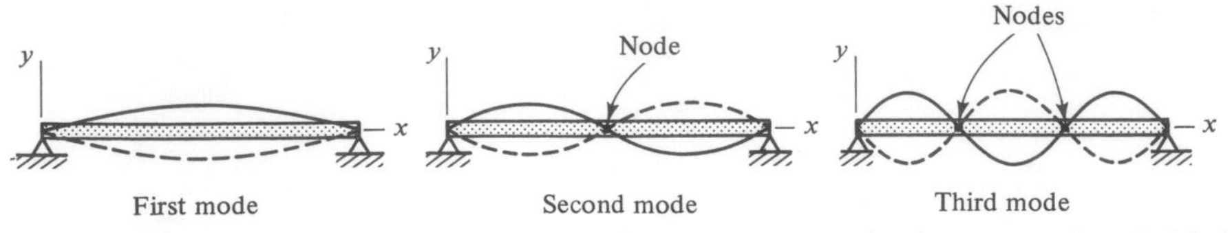

Figure 2-25 Simply supported beam showing mode shapes for first, second, and third modes.

A vibrating beam is an example of an elastic body of distributed mass that theoretically has an infinite number of degrees of freedom and thus a similar number of modes of vibration and natural frequencies. In many practical applications, however, only a few of the lower natural frequencies are of interest, with the lowest (the fundamental) natural frequency usually being the most important of these. Configurations for the first, second, and third modes of vibration of a simply supported beam are shown in Fig. 2-25. The *nodal points* shown for the second and third modes are points of zero deflection for the beam when vibrating in those modes.

Rayleigh's method is generally limited to determining the fundamental natural frequency, since a different shape function must be used for each higher mode as evidenced by Fig. 2-25. When several of the higher natural frequencies are desired, a lumped-mass model of the beam can be used, and these frequencies determined as discussed in Chap. 5.

If concentrated masses are attached to a beam, they can have an appreciable effect on the natural frequencies of the beam. If these masses are large in comparison with the mass of the beam, the mass of the beam can be neglected, and the beam treated as a "weightless spring."

Beam Shape Functions

In Sec. 2-8 we discussed the use of Rayleigh's energy method in determining the fundamental natural frequencies of various systems other than beams using

$$T_{\max} = U_{\max} \tag{2-62}$$

In this section we use the same method for determining the fundamental natural frequencies of beams. The displacement, or shape, functions for beams are usually static-deflection curves such as those shown in Table 2-1 on page 78 or trigonometric functions. If the shape function is a close approximation of the amplitude curve (satisfies the deflection, slope, and moment conditions at each end of the beam), Eq. 2-62 will give a close approximation of the exact ω_n value (usually within a few percent).[10] The use of an approximate shape function acts as an artificial constraint on the beam, which acts as a stiffening effect and results in an ω_n value higher than the exact value.

The basic *beam equations* from elementary mechanics of materials are shown here for convenience; they are

[10] See Chap. 8 for exact beam solutions.

$$\left.\begin{aligned} y &= \text{deflection} \\ \frac{dy}{dx} &= \text{slope} \\ EI\frac{d^2y}{dx^2} &= M \text{ (moment)} \\ EI\frac{d^3y}{dx^3} &= V \text{ (shear)} \\ EI\frac{d^4y}{dx^4} &= w \text{ (load intensity)} \end{aligned}\right\} \tag{2-63}$$

where E = modulus of elasticity of beam material
I = area moment of inertia about neutral axis

These equations relate to the coordinate axes shown in Fig. 2-26a, with a positive slope and deflection being as shown in the figure. Figure 2-26b shows the positive moment, shear, and load-intensity convention.

If a beam has a nonuniform cross section, the first three expressions of Eq. 2-63 are still valid, but the shear V and load intensity w become

$$\left.\begin{aligned} E\frac{d}{dx}\left[I(x)\frac{d^2y}{dx^2}\right] &= V \\ E\frac{d^2}{dx^2}\left[I(x)\frac{d^2y}{dx^2}\right] &= w \end{aligned}\right\} \tag{2-64}$$

in which $I(x)$ indicates that I is a function of x.

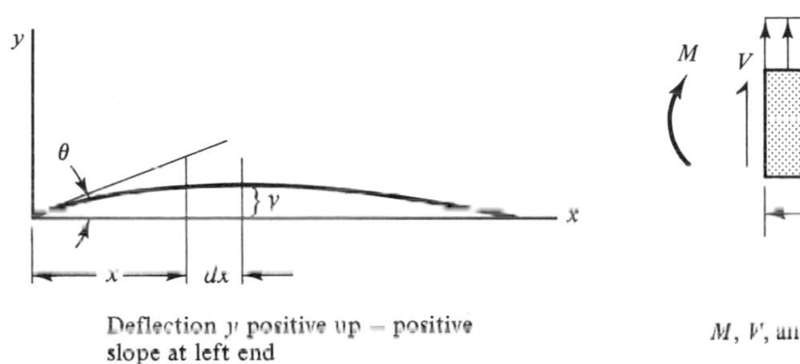

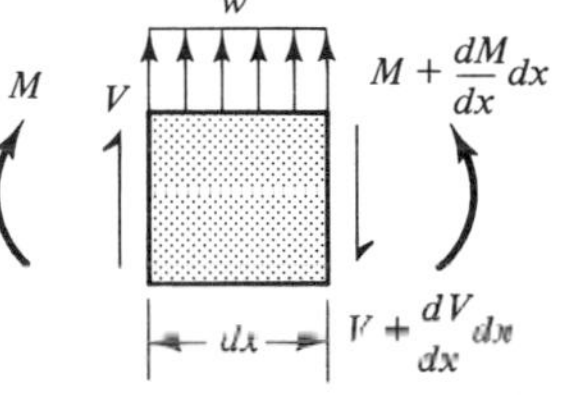

Figure 2-26 Coordinate system and sign convention for beam equations.

Strain Energy Due to Bending

To determine the strain energy accompanying the deformation of elastic bodies due to bending, let us consider a segment of a beam that is dx long as shown in Fig. 2-27a. At a section of a beam acted upon by a moment M the normal stress σ of a fiber a distance z from the neutral axis is

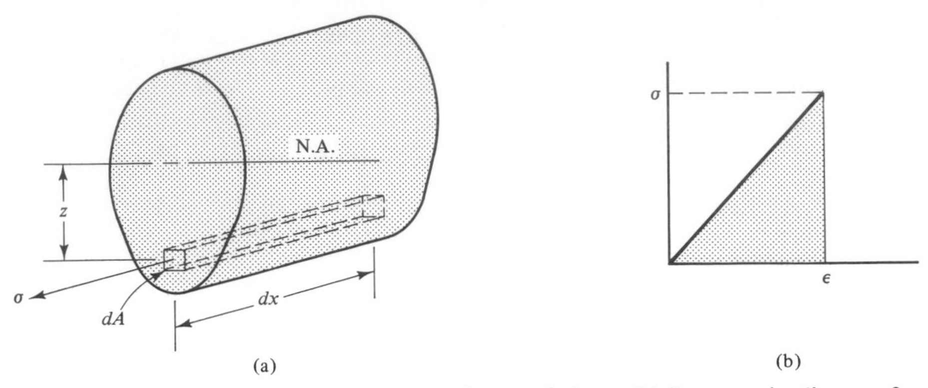

Figure 2-27 Beam segment. (a) Segment of beam dx long. (b) Stress-strain diagram for fiber shown.

$$\sigma = \frac{Mz}{I} \tag{2-65}$$

Referring to the stress-strain diagram in Fig. 2-27b, the strain ϵ corresponding to σ is

$$\epsilon = \frac{\sigma}{E} = \frac{Mz}{EI} \tag{2-66}$$

The strain energy stored in the material *per unit volume* (the shaded area in Fig. 2-27b) is

$$dU = \frac{\sigma\epsilon}{2} = \frac{\sigma^2}{2E}$$

or utilizing Eq. 2-65,

$$dU = \frac{1}{2E}\left(\frac{Mz}{I}\right)^2 \tag{2-67}$$

Thus, integrating over the *volume* of a beam of length l yields

$$U = \frac{1}{2E}\int_0^l \int_A \left(\frac{Mz}{I}\right)^2 dA\, dx \tag{2-68}$$

in which $\int_A$ indicates an integration over the cross-sectional area A of the beam. Since

$$\int_A z^2\, dA = I$$

and $M = EI(d^2y/dx^2)$, Eq. 2-68 can be written as

$$U = \frac{EI}{2}\int_0^l \left(\frac{d^2y}{dx^2}\right)^2 dx \tag{2-69}$$

which expresses the strain energy stored in a beam of uniform cross section that has a deflection equation given by $y = f(x)$.

We are now interested in obtaining the *maximum change* in the potential energy of the system. A beam vibrates about its static-equilibrium position the same as does the mass of a simple spring-and-mass system. If a function $y = f(x)$ is now used as a shape function, and we assume that y is measured from the *static-equilibrium position,* then any change in y involves a change in both the strain energy U_e and the energy of position U_g analogous to the changes discussed for the slender rod in Example 2-7 (see Fig. 2-23b). It follows then that the *maximum change* in the potential energy U of the beam from its potential energy at the static-equilibrium position is given by Eq. 2-69, so that

$$U_{\max} = \frac{EI}{2}\int_0^l \left(\frac{d^2y}{dx^2}\right)^2 dx \tag{2-70}$$

Kinetic Energy of Vibration

The time-varying displacement $Y(x, t)$ of any cross section along a beam that is vibrating with a natural circular frequency ω_n can be expressed as

$$Y(x, t) = f(x)\sin \omega_n t \tag{2-71}$$

as shown in Fig. 2-28. The displacement function $y = f(x)$ is the function used to define the *shape* of the *amplitude* curve during vibration.

From the time derivative of Eq. 2-71, the velocity of any cross section along the beam is given by

$$v = \dot{Y}(x, t) = f(x)\omega_n \cos \omega_n t$$

The maximum velocity of any cross section is

$$v_m = f(x)\omega_n = y\omega_n$$

in which y is the *amplitude* of any cross section.

It follows that the maximum kinetic energy of a uniform beam of mass γ per unit length is

$$T_{\max} = \frac{\gamma}{2}\int_0^l (y\omega_n)^2\, dx \tag{2-72}$$

If a rigid body of mass m is attached to the beam at a position $x = a$, as shown in Fig. 2-28, the maximum kinetic energy of the system is

$$T_{\max} = \frac{\gamma}{2}\int_0^l (y\omega_n)^2\, dx + \frac{m}{2}(y|_{x=a}\omega_n)^2 \tag{2-73}$$

in which $y|_{x=a}$ is the value of $f(x)$ at $x = a$.

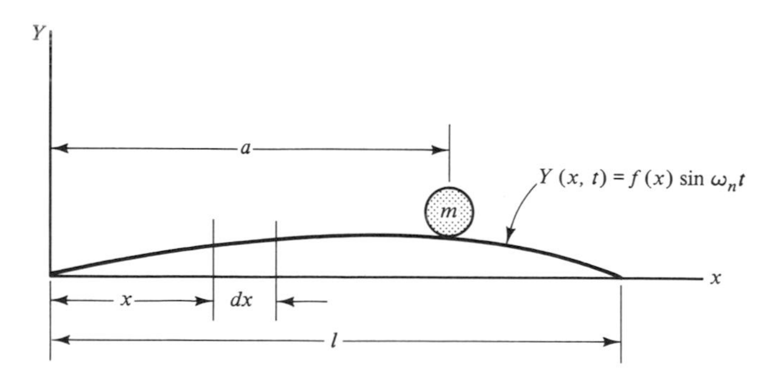

Figure 2-28 Typical configuration of a beam vibrating with frequency ω_n.

Equating T_{max} and U_{max} as obtained from Eqs. 2-70 and 2-72, the square of the fundamental natural circular frequency of a beam of uniform cross section is given by

$$\omega_n^2 = \frac{EI \int_0^l (d^2y/dx^2)^2 \, dx}{\gamma \int_0^l y^2 \, dx} \tag{2-74}$$

If the cross section of a beam is not uniform, then

$$\omega_n^2 = \frac{E \int_0^l I(x)(d^2y/dx^2)^2 \, dx}{\gamma' \int_0^l A(x)y^2 \, dx} \tag{2-75}$$

in which the area moment of inertia and the cross-sectional area are functions of x as indicated by $I(x)$ and $A(x)$, respectively, and γ' is the mass per unit *volume*.

EXAMPLE 2-9

A uniform beam of mass γ per unit length and length l is fixed at both ends as shown in Fig. 2-29a.[11] A rigid mass m is attached at $x = l/2$. We wish to determine the natural circular frequency of both the beam-and-mass system and the beam alone by utilizing $T_{max} = U_{max}$ with each of the following shape functions:

a. The static-deflection curve $y = f(x)$ of a beam with a force P applied at $x = l/2$ as shown in Fig. 2-29b.

b. The trigonometric function

$$y = A\left(1 - \cos \frac{2\pi x}{l}\right)$$

Solution. **a.** From Table 2-1 on page 80, we find the static-deflection curve of a fixed-fixed beam with a load applied at its center to be

$$y = \frac{P}{48EI}(3lx^2 - 4x^3) \qquad 0 \le x \le \frac{l}{2} \tag{2-76}$$

This equation could also be determined using the beam equations in Eq. 2-63, statics, and symmetry, with the following conditions: (a) $y = 0$ at $x = 0$; (b) $dy/dx = 0$ at $x = 0$ and $x = l/2$. Therefore, we can expect that the use of this function as a shape function will yield an ω_n value fairly close to the exact value. We also note that since y is symmetrical about $x = l/2$, the integrations involving T_{max} and U_{max} can be determined over the interval $x = 0$ to $x = l/2$, and then doubled.

The kinetic energy of the beam-and-mass system obviously includes that of both the beam and the mass m. With the mass m considered as concentrated at the center of the beam, its kinetic energy involves the velocity of the center of the beam. To allow us to also express the kinetic energy of the beam in terms of the velocity of its center, let us first express Eq. 2-76 in terms of the deflection Δ of the center of the beam. We can do this by letting $x = l/2$ in Eq. 2-76, from which

$$\Delta = \frac{Pl^3}{192EI} = \left(\frac{P}{48EI}\right)\left(\frac{l^3}{4}\right)$$

[11] The axial extensions in the beam fibers due to the axial constraints are usually negligible (see Example 2-13).

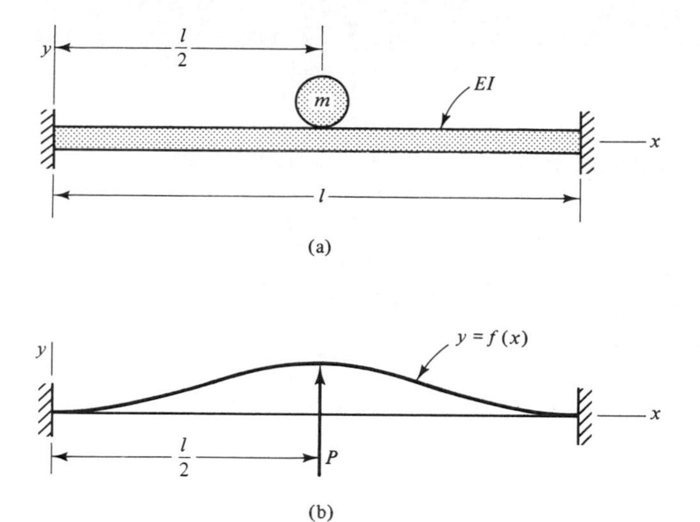

Figure 2-29 Uniform fixed-fixed beam with attached rigid mass m.

so that

$$\frac{P}{48EI} = \frac{4\Delta}{l^3}$$

Using this relationship, Eq. 2-76 can now be expressed as

$$y = \frac{4\Delta}{l^3}(3lx^2 - 4x^3) \tag{2-77}$$

The maximum kinetic energy of the system, which includes that of the beam and the rigid mass, is

$$T_{max} = 2\frac{\gamma}{2}\int_0^{l/2}(y\omega_n)^2\,dx + \frac{m}{2}(\Delta\omega_n)^2 \tag{2-78}$$

Substituting Eq. 2-77 into Eq. 2-78, we have the maximum kinetic energy in terms of the velocity of the center of the beam $\Delta\omega_n$ as

$$T_{max} = \frac{\gamma(\Delta\omega_n)^2}{2l^6}(32)\int_0^{l/2}(3lx^2 - 4x^3)^2\,dx + \frac{m}{2}(\Delta\omega_n)^2$$

Integrating the above and substituting limits gives the maximum kinetic energy as

$$T_{max} = \frac{(\Delta\omega_n)^2}{2}(0.3714\gamma l + m) \tag{2-79}$$

Since y is measured from the static-equilibrium position, the maximum change in the potential energy of the system from its potential energy in the static-equilibrium position is given by Eq. 2-70 as

$$U_{max} = 2\frac{EI}{2}\int_0^{l/2}\left(\frac{d^2y}{dx^2}\right)^2 dx$$

Substituting the second derivative of Eq. 2-77 into the above expression gives

$$U_{\max} = \frac{16\Delta^2 EI}{l^6} \int_0^{l/2} (6l - 24x)^2 \, dx$$

Integrating this expression, and substituting limits, yields

$$U_{\max} = \frac{96\Delta^2 EI}{l^3} \tag{2-80}$$

Referring to part i of Table 2-1 on page 80, it can be seen that the spring constant $k = 192EI/l^3$ for a fixed-fixed beam is associated with the deflection of the center of the beam due to a force P acting at the center. Since the shape function we chose to determine $T_{\max}$ and $U_{\max}$ in the preceding discussion is the deflection curve associated with the same force P, the maximum change in potential energy can also be determined from

$$U_{\max} = \frac{1}{2} k \, \Delta^2 = \frac{1}{2} \frac{192EI}{l^3} \Delta^2$$

which is the same as Eq. 2-80.

Substituting Eqs. 2-79 and 2-80 into $T_{\max} = U_{\max}$, we obtain

$$\omega_n = \sqrt{\frac{192EI}{l^3(0.3714\gamma l + m)}} \tag{2-81}$$

which is the approximate natural circular frequency of a fixed-fixed beam with a rigid mass m attached at its center. If we let $m = 0$ in Eq. 2-81, we obtain the approximate fundamental natural circular frequency of a fixed-fixed beam *alone* as

$$\omega_n = \frac{22.74}{l^2} \sqrt{\frac{EI}{\gamma}} \qquad (m = 0)$$

which is only 1.6 percent higher than the exact value of

$$\omega_1 = \frac{22.373}{l^2} \sqrt{\frac{EI}{\gamma}}$$

determined from the solution of Euler's beam equation (see Sec. 8-5). Obtaining such a close approximation with the shape function used illustrates Rayleigh's finding that the shape function used need only be a good approximation of the exact shape of the amplitude curve during vibration.

If we neglect the mass γl of the beam, Eq. 2-81 reduces to

$$\omega_n = \sqrt{\frac{192EI}{ml^3}} = \sqrt{\frac{k}{m}}$$

which in effect has the beam acting as a *massless spring,* as in a simple spring-and-mass system.

A more accurate value of the natural circular frequency of a fixed-fixed beam could be determined by considering only the beam, and choosing the deflection curve of a fixed-fixed beam with a uniform load over its length (see Prob. 2-42).

b. Let us now consider the trigonometric function

$$y = A\left(1 - \cos\frac{2\pi x}{l}\right)$$

as a shape function for the same beam and mass considered in part a. As stated earlier, the shape function should have approximately the shape of the amplitude curve of the beam while it is vibrating, which is accomplished by satisfying the deflection, slope, and moment conditions at the beam ends. Checking this function for these requirements, we find that

$$\left.\begin{aligned}
y &= A\left(1 - \cos\frac{2\pi x}{l}\right) &\quad \therefore \quad & y = 0 \text{ at } x = 0 \text{ and } x = l \\
& & & y = y_{\max} = 2A \text{ at } x = l/2 \\
\frac{dy}{dx} &= \frac{A2\pi}{l}\sin\frac{2\pi x}{l} &\quad \therefore \quad & y' = 0 \text{ at } x = 0 \text{ and } x = l \\
\frac{d^2y}{dx^2} = \frac{M}{EI} &= \frac{A4\pi^2}{l^2}\cos\frac{2\pi x}{l} &\quad \therefore \quad & \frac{M}{EI} = \frac{A4\pi^2}{l^2} \text{ at } x = 0 \text{ and } x = l
\end{aligned}\right\} \tag{2-82}$$

Since the function and its derivatives satisfy the deflections y, the slopes y', and the equal moments M at the ends of the beam, we should obtain a fairly accurate value for ω_n.

Since the mass m is at the center of the beam, its *amplitude* is $2A$, and the maximum kinetic energy of the system is

$$T_{\max} = \frac{\gamma}{2}\int_0^l (y\omega_n)^2\,dx + \frac{m}{2}(2A\omega_n)^2$$

which, with the shape function substituted, becomes

$$T_{\max} = \frac{\gamma}{2}\omega_n^2 A^2 \int_0^l \left(1 - \cos\frac{2\pi x}{l}\right)^2 dx + \frac{m}{2}(2A\omega_n)^2$$

Integrating this expression and substituting limits yields[12]

$$T_{\max} = \frac{A^2\omega_n^2}{2}(1.5\gamma l + 4m) \tag{2-83}$$

Since y is measured from the static-equilibrium position, the maximum change in potential energy is the strain energy stored in the beam as given by Eq. 2-70,

$$U_{\max} = \frac{EI}{2}\int_0^l \left(\frac{d^2y}{dx^2}\right)^2 dx = \frac{EI}{2}\frac{A^2 16\pi^4}{l^4}\int_0^l \cos^2\frac{2\pi x}{l}\,dx$$

from which

$$U_{\max} = \frac{EI}{2}\frac{A^2 16\pi^4}{l^4}\frac{l}{2} = \frac{EI}{2}\frac{A^2 779.3}{l^3} \tag{2-84}$$

[12] The trigonometric identities $\sin^2 x = (1 - \cos 2x)/2$ and $\cos^2 x = (1 + \cos 2x)/2$ facilitate the evaluation of integrals involving $\sin^2 x$ and $\cos^2 x$.

Then, substituting Eqs. 2-83 and 2-84 into $T_{\max} = U_{\max}$, we find that

$$\omega_n = \sqrt{\frac{779.3EI}{l^3(1.5\gamma l + 4m)}} \tag{2-85}$$

If we neglect the rigid mass (let $m = 0$ in Eq. 2-85), we obtain the natural circular frequency of the beam *alone* as

$$\omega_n = \frac{22.79}{l^2}\sqrt{\frac{EI}{\gamma}}$$

which yields a value 1.9 percent higher than the exact value (see the end of part a). If the mass γl of the beam is neglected, Eq. 2-85 reduces to

$$\omega_n = \sqrt{\frac{194.8EI}{ml^3}} \cong \sqrt{\frac{k}{m}}$$

since $k = 192EI/l^3$. We are in effect then considering the beam and mass m as a simple spring-mass system.

It should be pointed out that if the mass were not attached at the center of the beam, the functions used in this example would be poor ones, since the boundary conditions shown in Eq. 2-82, which these functions satisfied, would not all be applicable (the shape curve would not be symmetrical, and the end moments would not be equal) with a different mass location. In general, when a beam has a rigid mass attached to it, a good shape function to use is the static-deflection curve of the beam with a concentrated load applied at the point where the rigid mass is attached. Such functions can be determined from tables of deflection equations for various types of beams and beam loadings (see Table 2-1).

EXAMPLE 2-10

The steel cantilever beam shown in Fig. 2-30 is 24 ft long and consists of two segments of equal length, one with a diameter of 4 in., the other with a diameter of 3 in. The modulus of elasticity of the material is $E = 30(10)^6$ psi, and its weight density is $w = 0.283$ lb/in.^{3}.

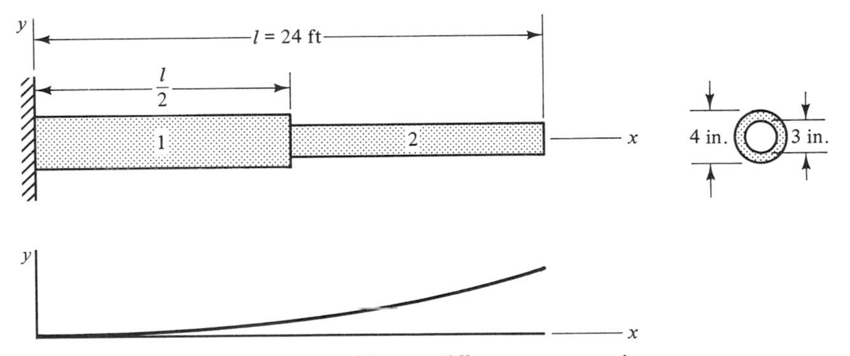

Figure 2-30 Cantilever beam with two different cross sections.

We wish to determine the fundamental natural frequency f_n of the beam using the concept of $T_{\max} = U_{\max}$. Two trigonometric functions have been suggested for use as a shape function,

$$y = A\left(1 - \cos\frac{\pi x}{l}\right) \tag{a}$$

and

$$y = A\left(1 - \cos\frac{\pi x}{2l}\right) \tag{b}$$

so the first step in our procedure will be to determine which of these functions will give us the best shape function for the beam under consideration.

Solution. As criteria for determining which of the two trigonometric functions to use as the shape function, let us examine the general shape of the amplitude curve we should expect when the beam is vibrating, and the end conditions associated with a cantilever beam. From these considerations we can determine which of the proposed functions is the better one. The end conditions are

$$x = 0\begin{Bmatrix} y = 0 \\ y' = 0 \\ y'' = \dfrac{M}{EI} \neq 0 \end{Bmatrix} \qquad x = l\begin{Bmatrix} y = \text{maximum} \\ y' = \text{maximum} \\ y'' = \dfrac{M}{EI} = 0 \end{Bmatrix}.$$

and we might expect the beam to vibrate with the amplitudes along the beam having the shape shown in Fig. 2-30. A comparison of the two functions on the above basis shows that

(a)	$y = A\left(1 - \cos\dfrac{\pi x}{l}\right)$	$y = 0$ at $x = 0$	(OK)
		$y \neq 0$ at $x = l$	(OK)
	$\dfrac{dy}{dx} = \dfrac{A\pi}{l}\sin\dfrac{\pi x}{l}$	$y' = 0$ at $x = 0$	(OK)
		$y' = 0$ at $x = l$	(Not OK)
	$\dfrac{d^2y}{dx^2} = \dfrac{M}{EI} = \dfrac{A\pi^2}{l^2}\cos\dfrac{\pi x}{l}$	$y'' \neq 0$ at $x = 0$	(OK)
		$y'' \neq 0$ at $x = l$	(Not OK)
(b)	$y = A\left(1 - \cos\dfrac{\pi x}{2l}\right)$	$y = 0$ at $x = 0$	(OK)
		$y \neq 0$ at $x = l$	(OK)
	$\dfrac{dy}{dx} = \dfrac{A\pi}{2l}\sin\dfrac{\pi x}{2l}$	$y' = 0$ at $x = 0$	(OK)
		$y' \neq 0$ at $x = l$	(OK)
	$\dfrac{d^2y}{dx^2} = \dfrac{M}{EI} = \dfrac{A\pi^2}{4l^2}\cos\dfrac{\pi x}{2l}$	$y'' \neq 0$ at $x = 0$	(OK)
		$y'' = 0$ at $x = l$	(OK)

It is apparent from this analysis that the first function is not a good shape function for this beam. However, the second function does generally satisfy the deflection, slope, and moment conditions associated with the beam constraints, and its use should yield a fairly accurate value of ω_n.

Using the function selected, the maximum kinetic energy can be expressed as

$$T_{\max} = \frac{\omega_n^2 \gamma_1}{2} \int_0^{144} A^2\left(1 - \cos\frac{\pi x}{2l}\right)^2 dx + \frac{\omega_n^2 \gamma_2}{2} \int_{144}^{288} A^2\left(1 - \cos\frac{\pi x}{2l}\right)^2 dx \tag{2-86}$$

Integrating Eq. 2-86, and substituting the following values for γ_1, γ_2, and l,

$$\gamma_1 = \frac{\pi}{4} d_1^2 \frac{0.283}{386} = 9.2(10)^{-3} \text{ lb} \cdot \text{s}^2/\text{in.}^2 \qquad (0 \le x \le 144)$$

$$\gamma_2 = \frac{\pi}{4} d_2^2 \frac{0.283}{386} = 5.2(10)^{-3} \text{ lb} \cdot \text{s}^2/\text{in.}^2 \qquad (144 \le x \le 288)$$

$$l = 288 \text{ in.}$$

yields

$$T_{\max} = \frac{A^2\omega_n^2}{2}(0.35) \tag{2-87}$$

Since y is measured from the static-equilibrium position, the maximum change in potential energy is the strain energy stored in the beam, which is given by

$$U_{\max} = \frac{EI_1}{2} \int_0^{144} \left(\frac{d^2y}{dx^2}\right)^2 dx + \frac{EI_2}{2} \int_{144}^{288} \left(\frac{d^2y}{dx^2}\right)^2 dx$$

Substituting

$$\frac{d^2y}{dx^2} = \frac{A\pi^2}{4l^2} \cos\frac{\pi x}{2l}$$

we obtain

$$U_{\max} = \frac{A^2\pi^4}{2(16)l^4}\left(EI_1 \int_0^{144} \cos^2\frac{\pi x}{2l}\, dx + EI_2 \int_{144}^{288} \cos^2\frac{\pi x}{2l}\, dx\right) \tag{2-88}$$

Integrating Eq. 2-88, and substituting the following values for l, EI_1, and EI_2:

$$l = 288 \text{ in.}$$

$$EI_1 = E\left(\frac{\pi d_1^4}{64}\right) = 37.70(10)^7 \text{ lb} \cdot \text{in.}^2$$

$$EI_2 = E\left(\frac{\pi d_2^4}{64}\right) = 11.93(10)^7 \text{ lb} \cdot \text{in.}^2$$

we find that

$$U_{\max} = \frac{A^2}{2}(42.09) \tag{2-89}$$

Substituting Eqs. 2-87 and 2-89 into $T_{\max} = U_{\max}$ gives

$$\frac{A^2\omega_n^2}{2}(0.35) = \frac{A^2}{2}(42.09)$$

from which

$$\omega_n = 10.97 \text{ rad/s}$$

and

$$f_n = \frac{\omega_n}{2\pi} = 1.75 \text{ Hz}$$

Since the exact value of ω_n is unknown, we cannot determine how close the value of ω_n obtained above is to being exact. We do know from prior observation that it is larger than the exact value since an inexact shape function acts as a constraint that has the effect of stiffening the beam.

EXAMPLE 2-11

The slender column in Fig. 2-31 has a length l and a stiffness factor EI, and supports the weight W as shown. As a part of the column design, it is desired to determine (a) the approximate natural frequency of the system and (b) the magnitude of the critical buckling load P_{cr}.

Solution. **a.** Let us use the trigonometric function

$$y = A\left(1 - \cos\frac{\pi x}{2l}\right)$$

as the shape function for the column. This function was used in Example 2-10 as the shape function of a cantilever beam that has the same deflection, slope, and moment conditions as the column we are considering.

In using the approach that $T_{max} = U_{max}$, we treat the column as a massless spring since columns generally support structures having masses much larger than their own. Thus, the kinetic energy of the system will involve only the weight W, and the maximum kinetic energy is

$$T_{max} = \frac{W}{2g}(y|_{x=l}\,\omega_n)^2 = \frac{WA^2\omega_n^2}{2g} \tag{2-90}$$

In determining the maximum change in the potential energy U, we note that with the column in its vertical static-equilibrium position, and with the datum shown

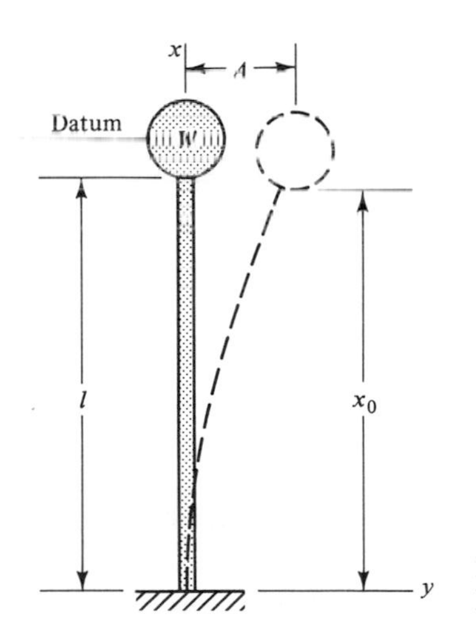

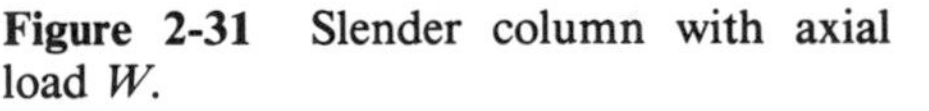
Figure 2-31 Slender column with axial load W.

in the figure, there is no potential energy in this initial position. With no initial strain energy, the change in the potential energy of position $U_g = -W(l - x_0)$ when the column is displaced from its static-equilibrium position will not be canceled by a portion of the change in the strain energy, as has been the case in the preceding examples. Therefore, the maximum change in potential energy from the zero potential energy in the static-equilibrium position will be the sum of the maximum changes in the strain energy U_e and the potential energy of position U_g.

Before we can determine the maximum value of U_g, we must relate the vertical change in position $(l - x_0)$ of W to the shape function we are using. To accomplish this, we first note from Fig. 2-32 that the total vertical displacement of W is the sum of the differences between ds and dx over the length of the column, so that

$$l - x_0 = \int_0^l (ds - dx) \tag{2-91}$$

From elementary calculus

$$ds = \sqrt{(dx)^2 + (dy)^2}$$

or

$$ds = \left[1 + \left(\frac{dy}{dx}\right)^2\right]^{1/2} dx \tag{2-92}$$

If we take the binomial expansion of the radical term above, and note that the slope dy/dx of the column is very small, we can take the approximation that

$$\left[1 + \left(\frac{dy}{dx}\right)^2\right]^{1/2} \cong 1 + \frac{1}{2}\left(\frac{dy}{dx}\right)^2$$

Substituting the approximation into Eq. 2-92, we find that

$$ds - dx = \frac{1}{2}\left(\frac{dy}{dx}\right)^2 dx$$

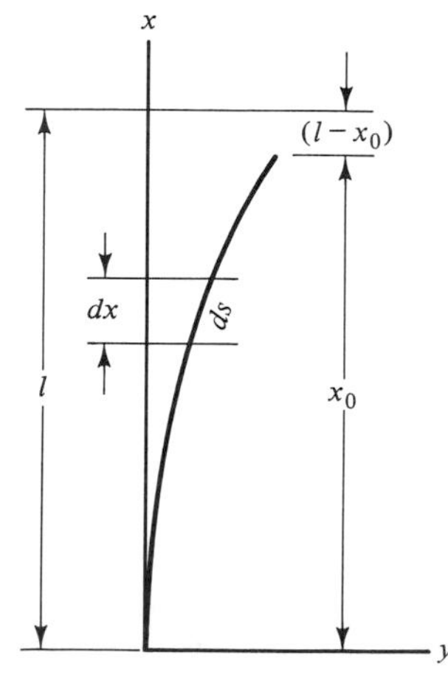

Figure 2-32 Elastic curve of column showing difference in ds and dx.

Equation 2-91 then becomes

$$l - x_0 = \frac{1}{2}\int_0^l \left(\frac{dy}{dx}\right)^2 dx \tag{2-93}$$

We can now write the *maximum change* in the potential energy as

$$U_{\max} = \frac{EI}{2}\int_0^l \left(\frac{d^2y}{dx^2}\right)^2 dx - \frac{W}{2}\int_0^l \left(\frac{dy}{dx}\right)^2 dx \tag{2-94}$$

with

$$\frac{dy}{dx} = \frac{A\pi}{2l}\sin\frac{\pi x}{2l}$$

and

$$\frac{d^2y}{dx^2} = \frac{A\pi^2}{4l^2}\cos\frac{\pi x}{2l}$$

The first integral in Eq. 2-94, which gives the maximum strain energy stored in the column, was derived earlier in this section (see Eq. 2-70). Integrating Eq. 2-94 with the first and second derivatives substituted yields

$$U_{\max} = \frac{A^2}{2}\left(\frac{EI\pi^4}{32l^3} - \frac{W\pi^2}{8l}\right) \tag{2-95}$$

Substituting Eqs. 2-90 and 2-95 into $T_{\max} = U_{\max}$, we obtain

$$f_n = \frac{\omega_n}{2\pi} = \frac{1}{2}\sqrt{\frac{g}{W}\left(\frac{EI\pi^2}{32l^3} - \frac{W}{8l}\right)} \tag{2-96}$$

b. Equation 2-96 shows that the frequency is positive, which indicates a stable system, when

$$\frac{EI\pi^2}{32l^3} > \frac{W}{8l}$$

It also reveals that the frequency is zero when

$$W = \frac{EI\pi^2}{4l^2}$$

The magnitude of W at which a column does not vibrate when displaced laterally from its static-equilibrium position is the *critical buckling load* for the column, and the value of W just determined agrees with the expression

$$P_{cr} = \frac{EI\pi^2}{4l^2}$$

which can be found in any elementary mechanics of materials text.

Beams of Two Materials

Beams of two materials are often used in structural systems to take advantage of the different properties of the two materials. For example, in reinforced-concrete beams, the concrete

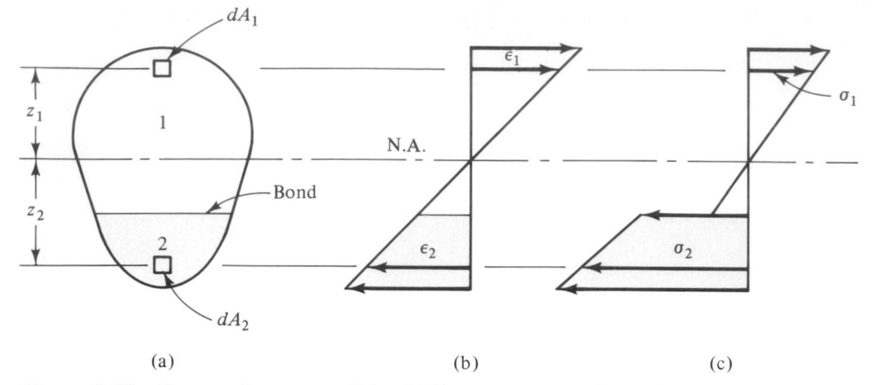

Figure 2-33 Beam of two materials. (a) Beam cross section. (b) Strain distribution. (c) Stress distribution.

is strong in compression but very weak in tension; so the beam is reinforced with steel to enable it to resist tensile stresses caused by bending.

To determine the fundamental natural circular frequency of a beam of two materials using $T_{max} = U_{max}$, it is first necessary to derive the equations for determining those two energy expressions. Figure 2-33 shows the cross section of a beam consisting of materials designated by 1 and 2. The assumption that a plane section before bending remains plane after bending is valid for a beam of two materials if the bond between the materials is capable of developing the longitudinal shearing stresses necessary to keep them from slipping with respect to each other. With this assumption, the strain distribution across a section is linear as shown in Fig. 2-33b. However, because of the difference in the elastic moduli of the two materials, E_1 and E_2, the normal *stress* does change abruptly where the materials join as shown in Fig. 2-33c.

It can be shown from elementary mechanics of materials that the stresses σ_1 and σ_2 and strains ϵ_1 and ϵ_2 in the respective materials can be expressed as follows:[13]

$$\left.\begin{aligned} \epsilon_1 &= z_1 \frac{d^2y}{dx^2} \\ \sigma_1 &= E_1\epsilon_1 = E_1 z_1 \frac{d^2y}{dx^2} \\ \epsilon_2 &= z_2 \frac{d^2y}{dx^2} \\ \sigma_2 &= E_2\epsilon_2 = E_2 z_2 \frac{d^2y}{dx^2} \end{aligned}\right\} \qquad (2\text{-}97)$$

where z_1 and z_2 = respective distances from neutral axis to differential areas dA_1 and dA_2

E_1 and E_2 = respective elastic moduli for materials 1 and 2

d^2y/dx^2 = curvature of beam's elastic curve

[13] A. Higdon, E. H. Ohlsen, W. B. Stiles, J. A. Weese, and W. F. Riley, *Mechanics of Materials,* 4th ed., pp. 356–359, John Wiley & Sons, Inc., New York, 1985.

The strain energies per unit volume stored in the materials are

$$\left.\begin{aligned} dU_1 &= \frac{\sigma_1\epsilon_1}{2} = \frac{E_1\epsilon_1^2}{2} = \frac{E_1}{2} z_1^2\left(\frac{d^2y}{dx^2}\right)^2 \\ dU_2 &= \frac{\sigma_2\epsilon_2}{2} = \frac{E_2\epsilon_2^2}{2} = \frac{E_2}{2} z_2^2\left(\frac{d^2y}{dx^2}\right)^2 \end{aligned}\right\} \tag{2-98}$$

Integrating these expressions over the *volume* of a beam of length l,

$$U_{\max} = \frac{E_1}{2}\int_0^l \int_{A_1} z_1^2\left(\frac{d^2y}{dx^2}\right)^2 dA_1\,dx + \frac{E_2}{2}\int_0^l \int_{A_2} z_2^2\left(\frac{d^2y}{dx^2}\right)^2 dA_2\,dx \tag{2-99}$$

By definition

$$I_1 = \int_{A_1} z_1^2\,dA_1$$

and

$$I_2 = \int_{A_2} z_2^2\,dA_2 \tag{2-100}$$

are the area moments of inertia of the areas A_1 and A_2 with respect to the *neutral* axis of the *composite* cross section. Substituting these terms into Eq. 2-99, it can be seen that the maximum strain energy stored in the beam is given by

$$U_{\max} = \frac{E_1I_1 + E_2I_2}{2}\int_0^l \left(\frac{d^2y}{dx^2}\right)^2 dx \tag{2-101}$$

Before I_1 and I_2 can be calculated, we must determine the location of the neutral axis of the *composite* cross section. From elementary mechanics of materials we know that this can be determined by using an *equivalent,* or *transformed,* cross section theoretically consisting of just one of the materials. The equivalent cross section is obtained by replacing the area of one of the materials by an equivalent area of the other, with the magnitude of the equivalent area depending upon the ratio of the elastic moduli of the materials.[14]

As an example, consider the composite cross section shown in Fig. 2-34. Assuming that $E_1 < E_2$ and that $E_2/E_1 = n$, both of the possible transformed cross sections are as shown in the figure. Note that in each case *only dimensions parallel to the neutral axis* have been changed to obtain the respective equivalent areas.

The mass per unit length of the composite beam of two materials shown in Fig. 2-34 is

$$\gamma = \rho_1A_1 + \rho_2A_2 \tag{2-102}$$

in which ρ_1 and ρ_2 are the mass densities of the materials. Since the maximum velocity of any cross section of the beam is $y\omega_n$, the maximum kinetic energy of the beam is

$$T_{\max} = \frac{\rho_1A_1 + \rho_2A_2}{2}\int_0^l (y\omega_n)^2\,dx \tag{2-103}$$

in which $y = f(x)$ is the shape function used.

[14] It can be shown that $E_1I_1 + E_2I_2 = E_eI_e$ in which E_e is the elastic modulus of the transformed cross-section material, and I_e is the area moment of inertia of the transformed cross section about the neutral axis.

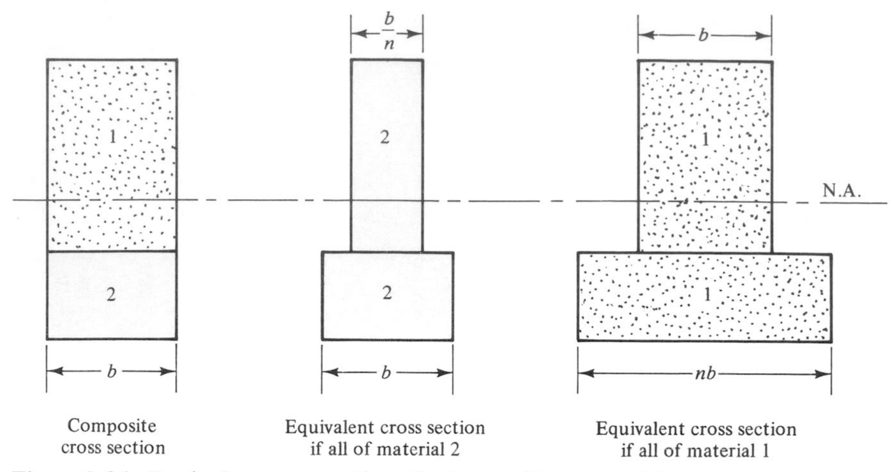

Figure 2-34 Equivalent cross sections for beam of two materials.

It should be apparent from the preceding discussion that the expressions determined for $T_{\max}$ and $U_{\max}$ can be easily extended to cover beams of more than two materials.

EXAMPLE 2-12

As a means of illustrating the procedure just discussed, let us consider a simply supported beam 20 ft in length and fabricated from wood (Douglas fir) and steel as shown in the cross-sectional view in Fig. 2-35. The $\frac{1}{2}$- by 6-in. steel plate is securely attached to the bottom of the 6- by 10-in. wooden beam. The properties of the materials are as follows:

	Specific weight, lb/in.^{3}	Modulus of elasticity, psi
Wood	$w_w = 0.022$	$E_w = 1.5(10)^6$
Steel	$w_s = 0.284$	$E_s = 30(10)^6$

We wish to determine the fundamental natural frequency of the beam using Rayleigh's energy method and the shape function

$$y = A \sin \frac{\pi x}{l}$$

Solution. Referring to Eq. 2-103, we see that the maximum kinetic energy, in terms of the shape function we are using, is given by

$$T_{\max} = \frac{\rho_w A_w + \rho_s A_s}{2} \int_0^l \omega_n^2 A^2 \sin^2 \frac{\pi x}{l}\, dx$$

in which $\rho_w A_w$ and $\rho_s A_s$ are the masses per unit length of the wood and steel portions of the beam, respectively. After integrating and substituting limits,

$$T_{\max} = \left(\frac{\rho_w A_w + \rho_s A_s}{2}\right) \frac{A^2 l}{2}\, \omega_n^2 \tag{2-104}$$

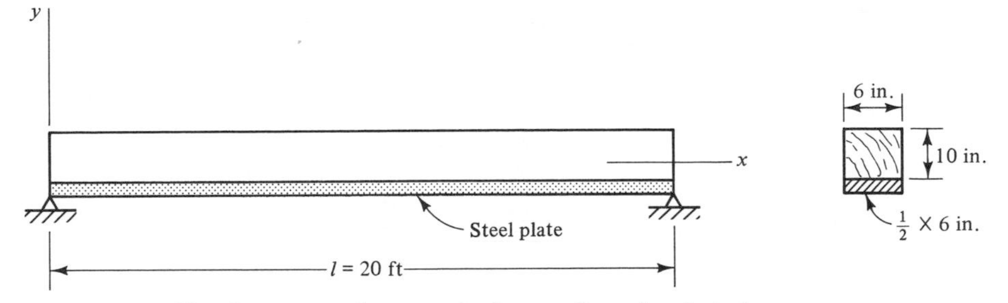

Figure 2-35 Simply supported composite beam of wood and steel.

Since the amplitude y is measured from the static-equilibrium position for the vibrating beam, the maximum *change* in the potential energy of the system from its potential energy in the static-equilibrium position is given by Eq. 2-101, which was derived with y measured from the unstrained position of the beam. Referring to Eq. 2-101, and using the second derivative of the shape function, the maximum change in the potential energy of the system is

$$U_{\max} = \frac{E_w I_w + E_s I_s}{2} \int_0^l \left(\frac{A\pi^2}{l^2}\right)^2 \sin^2 \frac{\pi x}{l}\, dx$$

After integrating and substituting limits,

$$U_{\max} = \left(\frac{E_w I_w + E_s I_s}{2}\right) \frac{A^2 \pi^4}{2l^3} \tag{2-105}$$

Substituting Eqs. 2-104 and 2-105 into $T_{\max} = U_{\max}$ gives

$$\omega_n = \frac{\pi^2}{l^2} \sqrt{\frac{E_w I_w + E_s I_s}{\rho_w A_w + \rho_s A_s}} \tag{2-106}$$

To this point the procedure followed has varied little from that followed in determining the fundamental natural frequency of a homogeneous beam. However, Eq. 2-106 shows that we must know I_w and I_s before ω_n can be determined. Referring to Eq. 2-100 for a definition of these quantities, we note that the neutral axis of the composite cross section must be known before we can evaluate them. Therefore, our next step is to determine an equivalent cross section and its centroidal axis about which bending occurs, as this axis will be the neutral axis of the composite cross section.

Using the ratio of the elastic moduli, $E_s/E_w = 20$, we can determine either an equivalent steel cross section or an equivalent wood cross section. Both are shown in Fig. 2-36 for purposes of illustration. The neutral axis can be found using either of the transformed sections. Using the equivalent steel cross section in this instance, the distance $\bar{y}$ from the bottom of the steel plate to the neutral axis is found to be

$$\bar{y} = \frac{(\frac{1}{2})6(\frac{1}{4}) + (\frac{6}{20})10(5.5)}{6(\frac{1}{2}) + (\frac{6}{20})10} = 2.88 \text{ in.}$$

With the neutral axis determined, we now refer to the composite cross section shown in Fig. 2-36a, and compute the area moment of inertia of each of these areas about the neutral axis just determined as

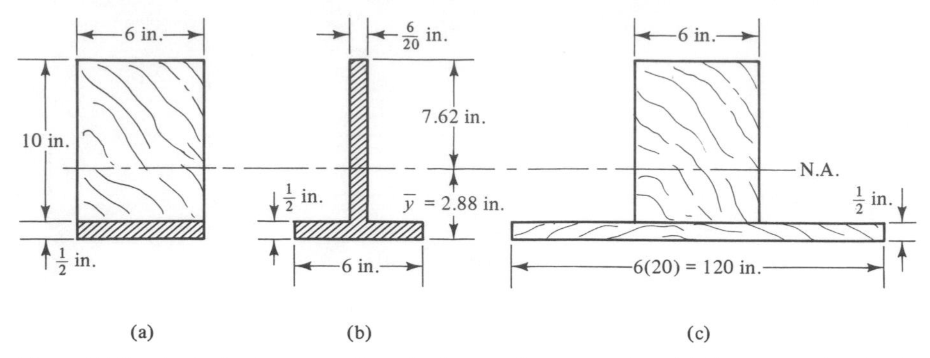

Figure 2-36 Equivalent cross sections for beam of two materials. (a) Composite (original). (b) Equivalent steel. (c) Equivalent wood.

$$I_w = \frac{6(10)^3}{12} + 60(7.62 - 5)^2 = 911.86 \text{ in.}^4$$

$$I_s = \frac{6(\frac{1}{2})^3}{12} + 3(2.88 - 0.25)^2 = 20.81 \text{ in.}^4$$

Using the given data and the values just computed above, we can now determine the following values for substitution into Eq. 2-106:

$$E_w I_w = 1.5(10)^6 911.86 = 13.68(10)^8 \text{ lb}\cdot\text{in.}^2$$
$$E_s I_s = 30(10)^6 20.81 = 6.24(10)^8 \text{ lb}\cdot\text{in.}^2$$

and

$$\rho_w A_w = \frac{0.022}{386} 60 = 3.42(10)^{-3} \text{ lb}\cdot\text{s}^2/\text{in.}^2$$

$$\rho_s A_s = \frac{0.284}{386} 3 = 2.21(10)^{-3} \text{ lb}\cdot\text{s}^2/\text{in.}^2$$

Substituting these values into Eq. 2-106, we find that

$$\omega_n = \frac{\pi^2}{l^2}\sqrt{\frac{(13.68 + 6.24)10^8}{(3.42 + 2.21)10^{-3}}} = 101.92 \text{ rad/s}$$

so that

$$f_n = \frac{\omega_n}{2\pi} = 16.22 \text{ Hz}$$

As will be seen in Chap. 8, the shape function used in this example gives the exact shape of the amplitude curve of a simply supported beam vibrating in its fundamental mode. Thus, the frequency value just obtained is an exact value.

The Effect of Axial Constraints on Natural Frequencies of Beams

When the ends of a beam are constrained so that their axial motion (motion along the x axis in Fig. 2-37) is restricted, lateral deflections of the beam are accompanied by length changes in the beam fibers that are in addition to the changes caused by bending. The tensile strains resulting from these *axial extensions* add to the strains caused by bending to produce higher strain and accompanying stress levels in the beam. This means that more strain energy is stored in an axially constrained beam than in a beam with ends free to move axially. Thinking in terms of $T_{max} = U_{max}$, it should be apparent that axial constraints act to increase the natural frequencies of beams.

A shape function for a beam with axially constrained ends is selected to meet the same deflection, slope, and moment conditions as if the beam were not axially constrained. Using a shape function $y = f(x)$ such as the one shown in Fig. 2-37, we see from the elastic curve that the tensile strain in the beam fibers due to axial extension can be expressed as

$$\epsilon_e = \frac{ds - dx}{dx} \tag{2-107}$$

It was shown in Example 2-11 that

$$(ds - dx) - \frac{1}{2}\left(\frac{dy}{dx}\right)^2 dx$$

so that Eq. 2-107 can also be written as

$$\epsilon_e = \frac{1}{2}\left(\frac{dy}{dx}\right)^2 \tag{2-108}$$

Since the strain distribution across a section is linear as shown in Fig. 2-33b, the strain due to bending of a beam fiber a distance z from the neutral axis is

$$\epsilon_b = z\,\frac{d^2y}{dx^2}$$

The total strain of a beam fiber a distance z from the neutral axis is thus

$$\epsilon = \epsilon_b + \epsilon_e = z\,\frac{d^2y}{dx^2} + \frac{1}{2}\left(\frac{dy}{dx}\right)^2 \tag{2-109}$$

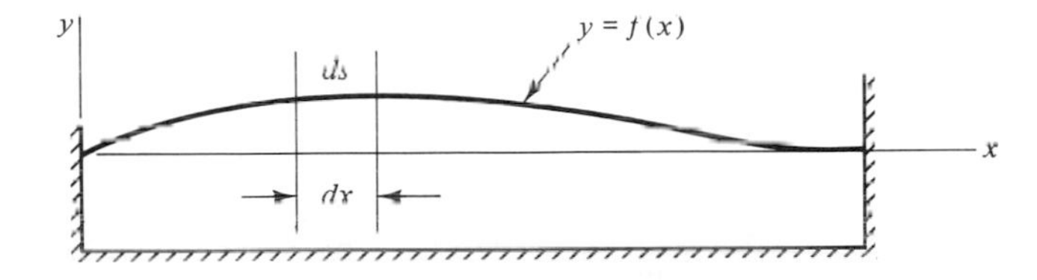

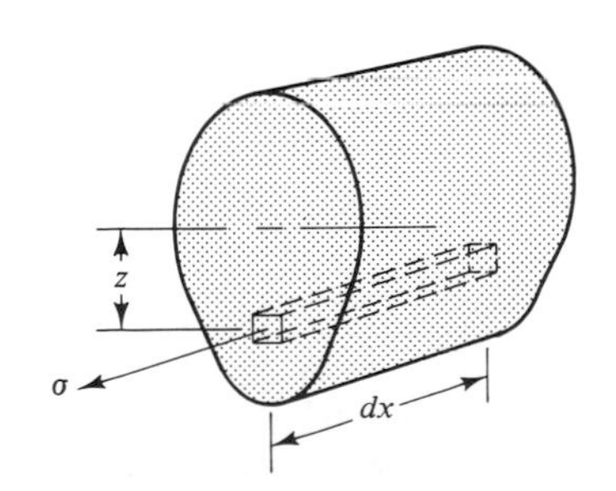

Figure 2-37 Beam with axial constraint.

The strain energy stored *per unit volume* is

$$dU = \frac{1}{2}\sigma\epsilon = \frac{E}{2}\epsilon^2$$

Substituting Eq. 2-109 into this expression,

$$dU = \frac{E}{2}\left[z^2\left(\frac{d^2y}{dx^2}\right)^2 + z\frac{d^2y}{dx^2}\left(\frac{dy}{dx}\right)^2 + \frac{1}{4}\left(\frac{dy}{dx}\right)^4\right] \tag{2-110}$$

The maximum strain energy stored in the beam is found by integrating Eq. 2-110 over the *volume* of the beam to obtain

$$U_{\max} = \frac{E}{2}\int_0^l \int_A \left[z^2\left(\frac{d^2y}{dx^2}\right)^2 + z\frac{d^2y}{dx^2}\left(\frac{dy}{dx}\right)^2 + \frac{1}{4}\left(\frac{dy}{dx}\right)^4\right] dA\, dx \tag{2-111}$$

Noting that

$\int_A z^2\, dA = I$ (area moment of inertia with respect to neutral axis)

$\int_A z\, dA = A\bar{z} = 0$ (statical moment of area with z measured from neutral axis)

$\int_A dA = A$ (cross-sectional area)

in Eq. 2-111, it reduces to

$$U_{\max} = \frac{EI}{2}\int_0^l \left(\frac{d^2y}{dx^2}\right)^2 dx + \frac{EA}{8}\int_0^l \left(\frac{dy}{dx}\right)^4 dx \tag{2-112}$$

The maximum kinetic energy is given by

$$T_{\max} = \frac{\gamma}{2}\int_0^l (y\omega_n)^2\, dx \tag{2-113}$$

in which γ is the mass per unit length of the beam. Substituting Eqs. 2-112 and 2-113 into $T_{\max} = U_{\max}$, we find that the natural circular frequency of an axially constrained beam is given by

$$\omega_n = \sqrt{\frac{EI\int_0^l (d^2y/dx^2)^2\, dx + (EA/4)\int_0^l (dy/dx)^4\, dx}{\gamma \int_0^l y^2\, dx}} \tag{2-114}$$

In Example 2-13 we shall see that the natural frequency of an axially constrained beam depends upon the amplitude at which it is vibrating.

EXAMPLE 2-13

To analyze the effects of axial beam constraints, let us consider the simply supported beam of rectangular cross section shown in Fig. 2-38, which is pinned at the ends to prevent their axial motion.

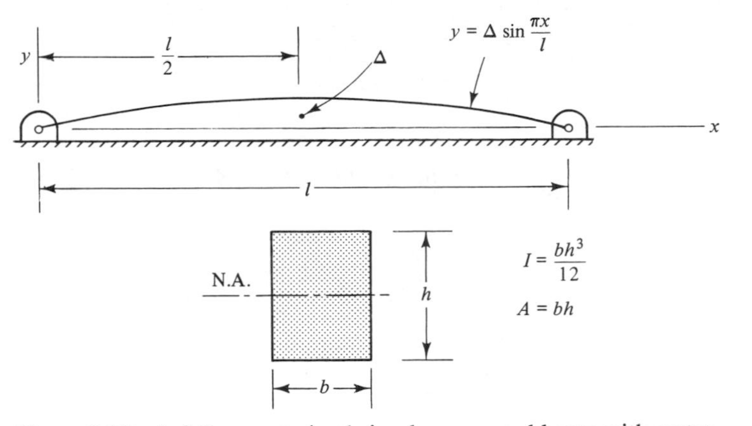

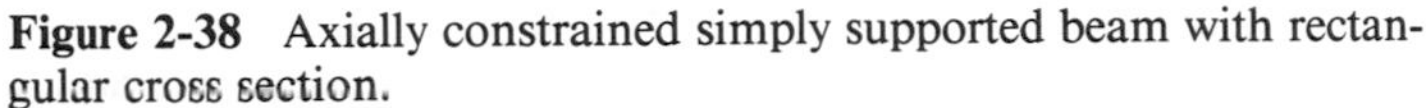

Figure 2-38 Axially constrained simply supported beam with rectangular cross section.

We wish to determine the lowest natural circular frequency of the beam in terms of the ratio Δ/h. In this ratio Δ is the amplitude of vibration of the beam at $x = l/2$, and h is the depth of the beam. The significance of this ratio will be seen later.

Solution. With axial constraints having little effect on the general shape of a beam's deflection curve, the trigonometric function

$$y = \Delta \sin \frac{\pi x}{l}$$

shown in the figure is selected as the shape function since it satisfies the deflection, slope, and moment conditions at $x = 0$ and $x = l$ for a beam pinned at both ends.

We shall not determine T_{max} and U_{max} and equate them as we have in preceding examples. Rather, we determine ω_n from Eq. 2-114, which was derived from using $T_{max} = U_{max}$ in the discussion just preceding this example. To do this, all we need is the shape function shown earlier and its derivatives

$$\frac{dy}{dx} = \frac{\Delta\pi}{l} \cos \frac{\pi x}{l}$$

and

$$\frac{d^2y}{dx^2} = \frac{-\Delta\pi^2}{l^2} \sin \frac{\pi x}{l}$$

to obtain the following:

$$EI \int_0^l \left(\frac{d^2y}{dx^2}\right)^2 dx = \frac{EI\Delta^2\pi^4}{l^4} \int_0^l \sin^2 \frac{\pi x}{l}\, dx = \frac{EI\Delta^2\pi^4}{2l^3}$$

$$\frac{EA}{4} \int_0^l \left(\frac{dy}{dx}\right)^4 dx = \frac{EA}{4}\left(\frac{\Delta\pi}{l}\right)^4 \int_0^l \cos^4 \frac{\pi x}{l}\, dx = \frac{3EA(\Delta\pi)^4}{32l^3}$$

$$\gamma \int_0^l y^2\, dx = \gamma\Delta^2 \int_0^l \sin^2 \frac{\pi x}{l} = \frac{\gamma\Delta^2 l}{2}$$

Substituting the above into Eq. 2-114, we find that

$$\omega_n = \sqrt{\frac{(EI\Delta^2\pi^4)/2l^3 + 3EA(\Delta\pi)^4/32l^3}{\gamma\Delta^2 l/2}}$$

or simplified,

$$\omega_n = \left(\frac{\pi}{l}\right)^2 \sqrt{\frac{EI}{\gamma}} \left(1 + \frac{3A\Delta^2}{16I}\right)^{1/2} \tag{2-115}$$

Referring to Fig. 2-38 to see that $I = bh^3/12$ and $A = bh$, Eq. 2-115 can be expressed in terms of the ratio Δ/h as

$$\omega_n = \left(\frac{\pi}{l}\right)^2 \sqrt{\frac{EI}{\gamma}} \left[1 + \frac{9}{4}\left(\frac{\Delta}{h}\right)^2\right]^{1/2} \tag{2-116}$$

The natural circular frequency of a beam pinned at both ends, but not constrained axially, is given by the first term in the bracket in Eq. 2-116. Thus, the second term in the bracket indicates the effects of the axial constraint. It shows that one effect of axially constraining a beam is to raise its fundamental natural frequency. Recalling that Δ is the amplitude of vibration at $x = l/2$, it also indicates that the natural frequency of the beam depends upon the amplitude of its vibration.

However, if the amplitude of the vibration is small compared with the depth of the beam, $\Delta = h/10$, for example, the increase in the magnitude of ω_n is only 1.1 percent, which is not a significant increase. Since the term

$$\frac{9}{4}\left(\frac{\Delta}{h}\right)^2$$

in Eq. 2-116 resulted in part from the ratio A/I in Eq. 2-115, it should be apparent that the *shape* of the beam cross section also has an effect on ω_n.

PROBLEMS

Problems 2-1 through 2-22 (Sections 2-1 through 2-3)

2-1. Determine the differential equation of motion for the undamped free vibration of the mass m. Express the undamped natural circular frequency ω_n in terms of the system parameters k and m.

Ans: $\omega_n = \sqrt{\dfrac{3k}{m}}$

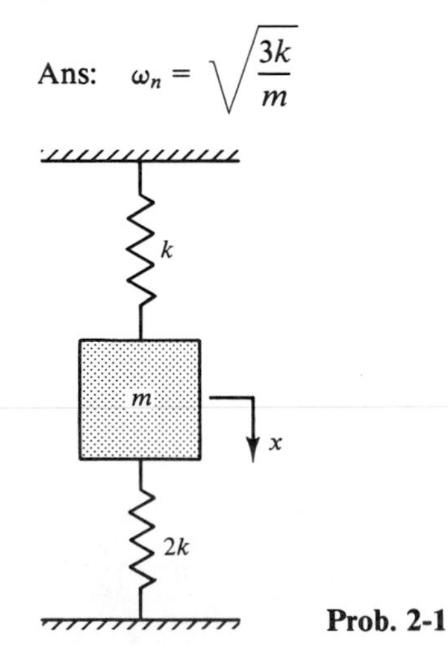

Prob. 2-1

2-2. A wooden cylinder of weight W and cross-sectional area A sinks to a depth h in a fluid that has a weight density ρ. Assume viscous damping with a damping coefficient c and determine the differential equation of motion. Express ω_n and the damping factor ζ in terms of the system parameters (c, ρ, A, and h).

Ans: $\omega_n = \sqrt{\dfrac{g}{h}}$

$$\zeta = \frac{c}{2\rho A}\sqrt{\frac{g}{h}}$$

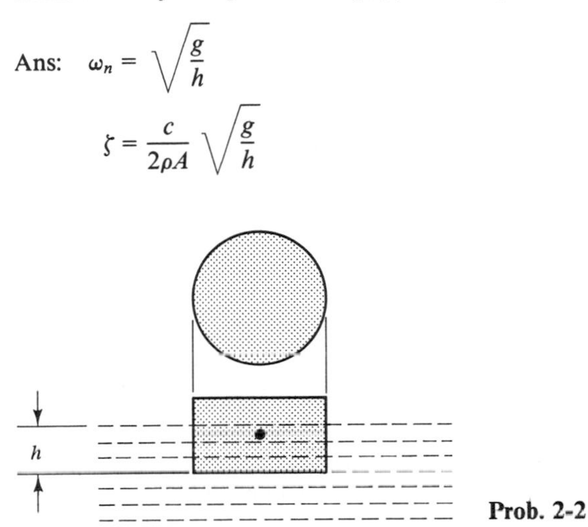

Prob. 2-2

2-3. The two pulleys are keyed together and have a combined mass moment of inertia $\bar{I}$ about the axis of rotation. Neglect damping and determine the differential equation of motion in terms of θ by (a) using Newton's second law and (b) using the concept of dynamic equilibrium. Express the undamped circular frequency ω_n in terms of the system parameters.

Ans: $\ddot{\theta} + \dfrac{kr^2}{\bar{I} + mR^2}\theta = 0$

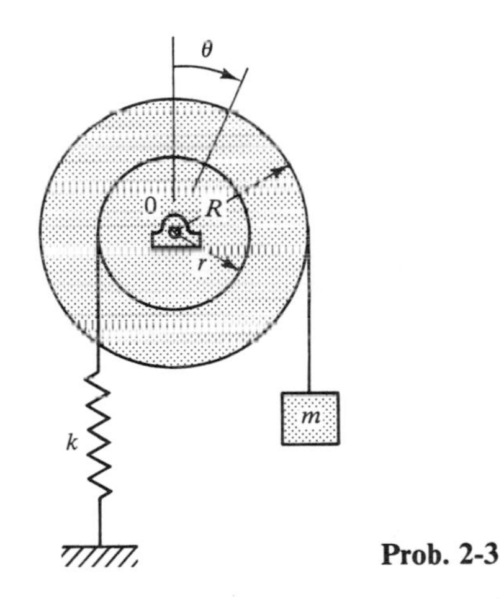

Prob. 2-3

2-4. The uniform slender rod AB of length l and weight W is attached to a torsional spring CD that has a torsional spring constant k_t. Coil springs of stiffness k are attached to the ends of the rod AB as shown. Determine the differential equation of motion for small angular oscillations of the rod by using the concept of dynamic equilibrium. Neglect damping. The centroidal mass moment of inertia $\bar{I}$ of a slender rod is $ml^2/12$. Express the undamped natural frequency f_n (Hz) in terms of the system parameters.

$$\text{Ans:}\quad f_n = \frac{1}{2\pi}\sqrt{\frac{48}{7ml^2}\left(\frac{5kl^2}{8} + k_t\right)}$$

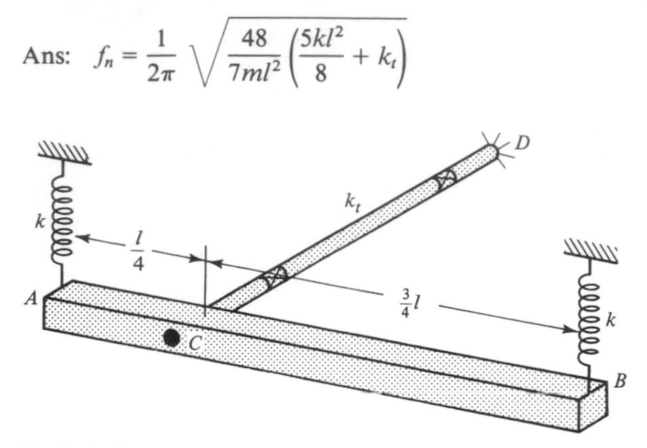

Prob. 2-4

2-5. A sphere of weight W and radius r rolls without slipping on a cylindrical surface as shown. For small angles of θ, determine (a) the differential equation of motion using the concept of dynamic equilibrium and (b) the undamped natural frequency f_n in terms of the parameters. Note that point A on the sphere moves to A' and that the angle ϕ is the absolute angular displacement of the sphere. The centroidal mass moment of inertia $\bar{I}$ of a sphere is $2mr^2/5$.

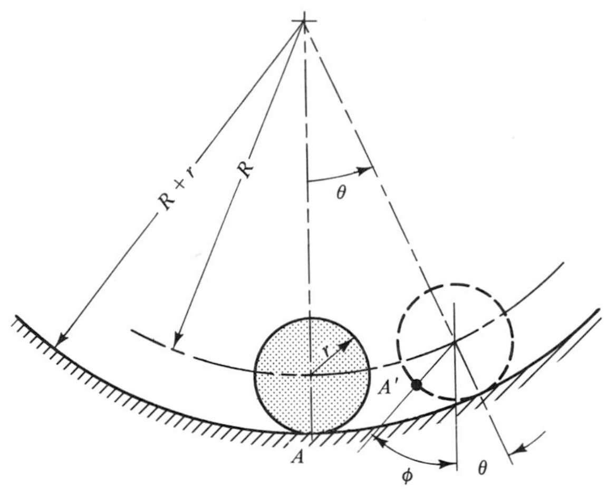

Prob. 2-5

2-6. The two disks A and B are driven at the same constant angular velocity with the senses shown in the accompanying figure. A uniform rod of weight W with center of gravity at G as shown rides in the grooves of the disks. If the center of gravity of the rod is displaced a distance x from the centerline between the two disks, the normal and friction forces between disk B and the rod are greater than the same forces between disk A and the rod. Since the friction forces that the disks exert on the rod are thus a function of x, the rod will oscillate back and forth with a frequency f_n. Assume that the coefficient of kinetic friction μ between the disks and rod is a constant and determine the differential equation of motion of the rod. Determine the frequency of oscillation if $l = 18$ in. and $\mu = 0.4$.

Partial ans: $\ddot{x} + \frac{\mu g}{l} x = 0$

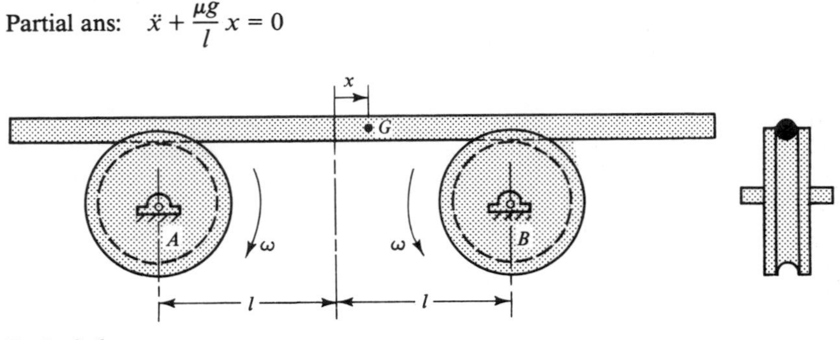

Prob. 2-6

2-7. The uniform slender rod is pinned to a support at A and to the center of a cylinder that rolls without slipping on a cylindrical surface of radius $l + r$. The rod has a length l, weight W_A, moment of inertia about G of $\bar{I}_G = W_A l^2/12g$, and moment of inertia about A of $I_A = W_A l^2/3g$. The cylinder has a weight of W_B and a centroidal moment of inertia of $\bar{I}_B = W_B r^2/2g$. For small angles of θ determine (a) the differential equation of motion of the system using dynamic equilibrium, and (b) the undamped natural frequency f_n in terms of the given parameters.

Partial ans: $f_n = \frac{1}{2\pi} \sqrt{\frac{(W_A l/2) + W_B l}{I_A + I_B(l/r)^2 + W_B l^2/g}}$

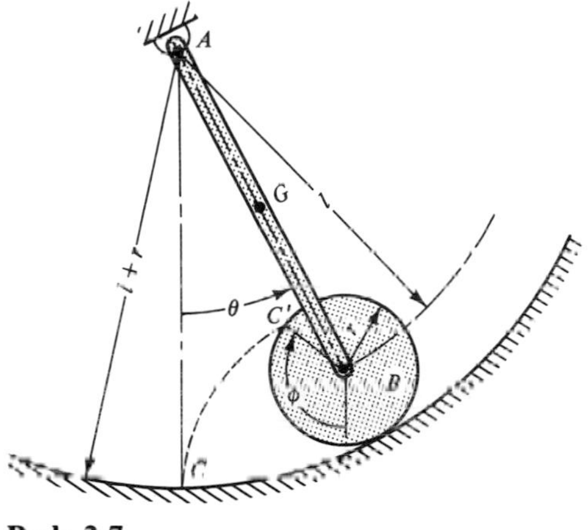

Prob. 2-7

2-8. The system shown in the accompanying figure consists of a cylinder of radius r that rolls without slipping and two slender rods each of length $l = 2r$. The mass of the cylinder is m, and the mass m_r of each rod is $m_r = m/4$. Derive the differential equation of motion for small oscillations and show that

$$\omega_n = \sqrt{\frac{12k}{11m}}$$

and

$$\zeta = \frac{3c}{\sqrt{132km}} \qquad \text{(damping factor)}$$

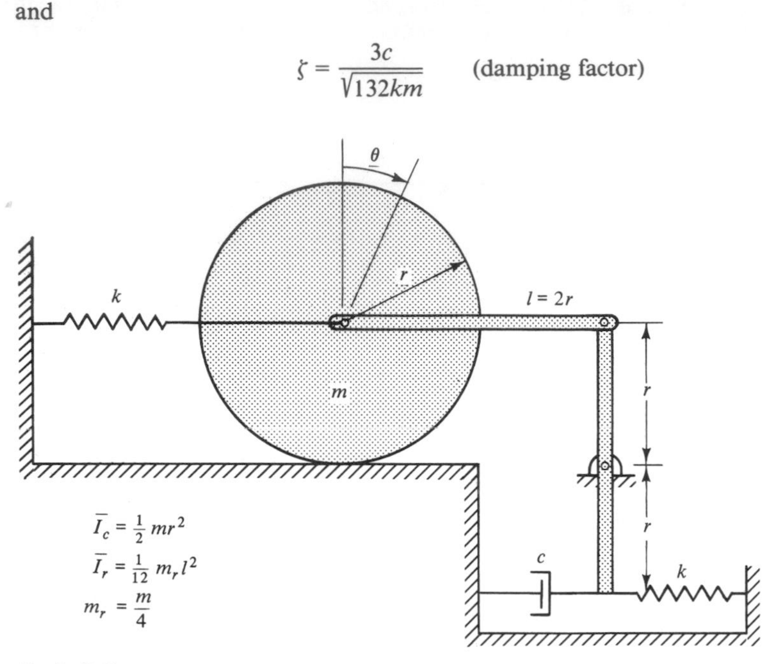

Prob. 2-8

2-9. A bicycle wheel and tire are supported so that they are free to rotate about their centroidal axis through the hub of the wheel. A small weight W is taped to the tire as shown in the accompanying figure at a distance R from the axis of rotation. When this weight is displaced slightly from the vertical axis shown, the wheel is observed to oscillate 3 cycles every 10 s. If $R = 0.28$ m, and $W = 3.34$ N, determine the centroidal mass moment of inertia $\bar{I}$ of the wheel and tire.

Ans: $\bar{I} = 0.237 \text{ N} \cdot \text{m} \cdot \text{s}^2$

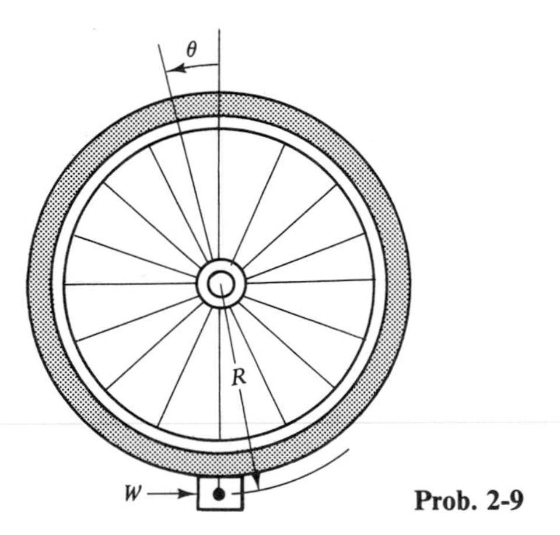

Prob. 2-9

2-10. A connecting rod weighing 10.21 lb is suspended on a knife-edge as shown in the accompanying figure. When the connecting rod is displaced slightly from a vertical axis, a stopwatch reveals that it takes 34.25 s to go through 30 cycles of oscillation. Determine the mass moment of inertia $\bar{I}$ of the rod about its mass center G, which is located 6.75 in. below the pivot point of the knife-edge.

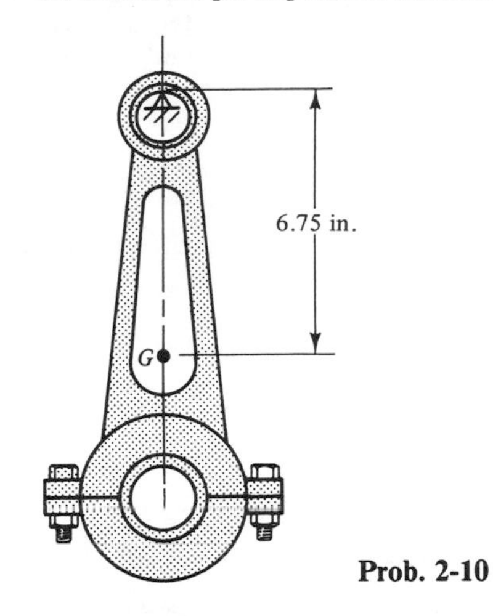

Prob. 2-10

2-11. A four-bladed aluminum propeller of a wind turbine is supported on a small shaft so that it is free to rotate about its centroidal axis. A weight $W = 20$ lb is taped to one of the blades at a distance $R = 72$ in. from the axis of rotation as shown. When the unbalanced weight W is displaced slightly from the vertical axis shown, the propeller is found to oscillate 10 cycles in 50.5 s. Determine the centroidal mass moment of inertia $\bar{I}$ of the propeller.

Ans: $\bar{I} = 661.6\ \text{lb} \cdot \text{in.} \cdot \text{s}^2$

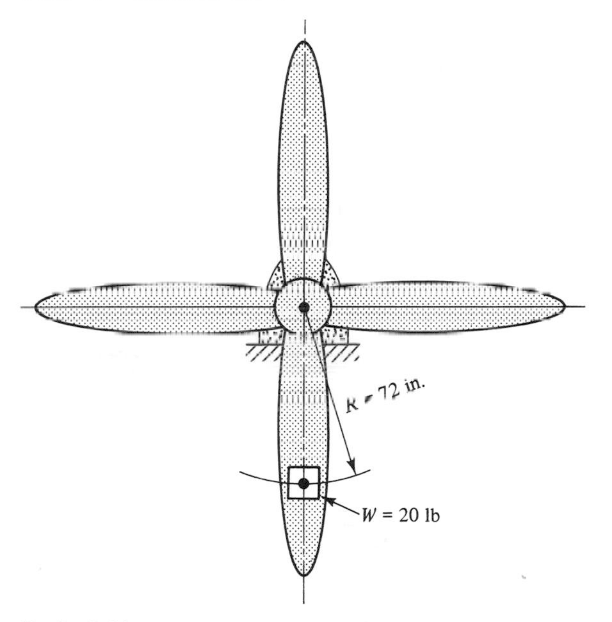

Prob. 2-11

2-12. A small space capsule containing instruments, and supporting the two solar panels shown, is suspended by a steel rod that acts as a torsional spring. The rod has a length of 1.0 m and a diameter of 12.5 mm. The mass moment of inertia of the capsule about the z axis shown depends upon the angle ϕ that the solar panels make with the z axis. When $\phi = 60°$, a stopwatch reveals that 10 cycles of oscillation takes 16.21 s. Determine the corresponding mass moment of inertia $\bar{I}_z$. [The shear modulus G for the steel of the rod is $82.7(10)^9$ N/m^2.]

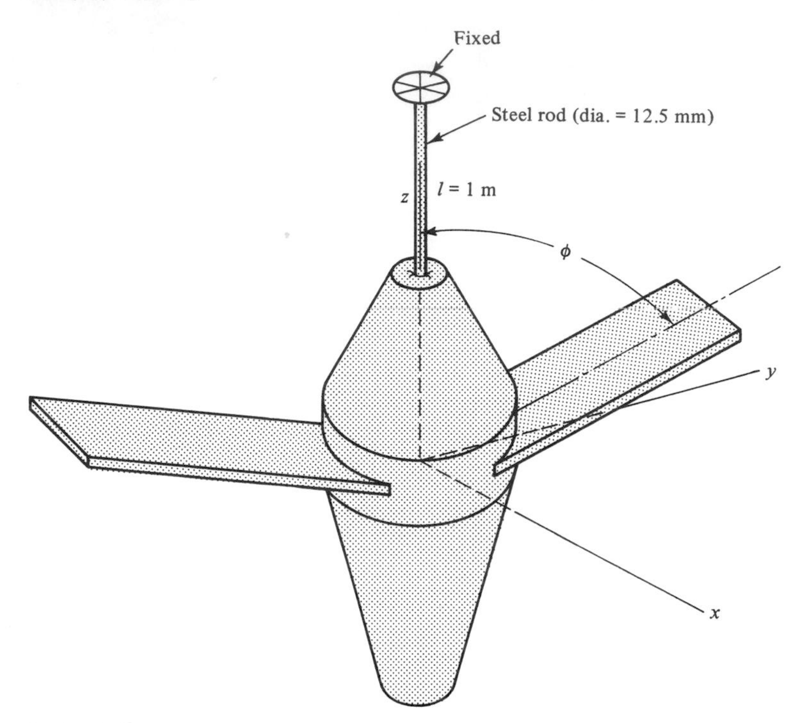

Prob. 2-12

2-13. The device shown in the accompanying figure is proposed as a means of determining the weight W and centroidal mass moment of inertia $\bar{I}$ of rubber-tired wheels of various sizes by recording the frequency of oscillation of the wheels both when free to rotate about an axle fixed to the horizontal bar AB and when locked to the bar through the axle. Determine the differential equations of motion for both a free and a locked wheel, and see if it is possible to determine W and $\bar{I}$ for a wheel by measuring the respective frequencies.

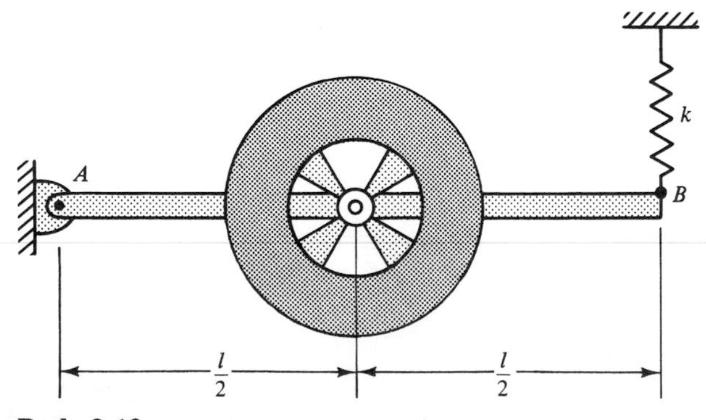

Prob. 2-13

2-14. Refer to Example 2-3 and determine the time t_m when y is a maximum for part a of the example in which damping was neglected. That is, determine t_m when $y = y_m$ for $\zeta = 0$.

2-15. The weight W and plate P are released from rest and fall freely through the distance h before the plate strikes the rigid support, which suddenly stops the upper end of the spring. Assuming that the plate does not rebound from the rigid support, and knowing that the spring is at its free length at the instant the plate makes contact with the rigid support, determine the maximum extension Δ_m of the spring if $W = 60$ lb, $k = 150$ lb/ft, and $h = 1$ ft. Also determine the time t_m at which this occurs. Neglect damping. (The reference for x shown in the figure is the static-equilibrium position).

Ans: $\Delta_m = 1.38$ ft

$t_m = 0.22$ s

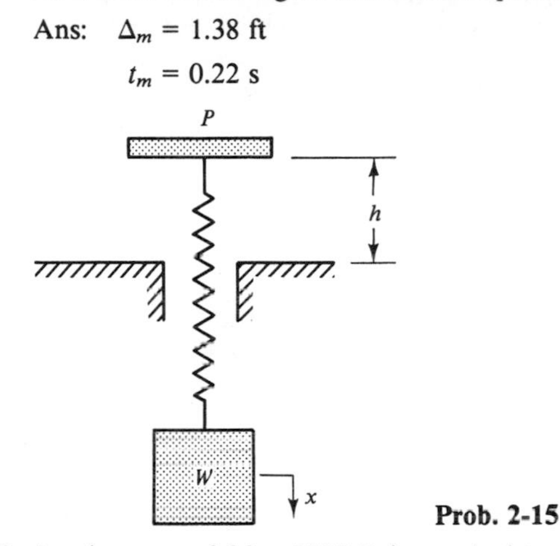

Prob. 2-15

2-16. An elevator weighing 8000 lb is attached to a steel cable that is wrapped around a drum rotating with a constant angular velocity of 3 rad/s. The radius of the drum is 1 ft. The cable has a net cross-sectional area of 1 in.^{2} and an effective modulus of elasticity of $E = 12(10)^6$ psi. A malfunction in the motor drive system of the drum causes the drum to stop suddenly when the elevator is moving down and the length l of the cable is 50 ft. Neglect damping and determine the maximum stress σ_m in the cable.

Ans: $\sigma_m = 31{,}180$ psi

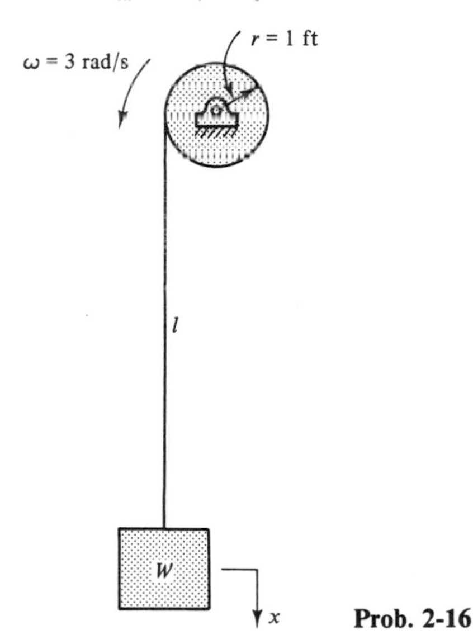

Prob. 2-16

2-17. Determine the maximum stress in the elevator cable of Prob. 2-16 if the damping is assumed to be such that $\zeta = 0.05$.

2-18. Consider that the elevator of Prob. 2-16 is moving upward instead of downward as in Prob. 2-16 when the drum suddenly stops at the instant $l = 50$ ft and determine the time t_0 at which the stress σ in the cable is zero. Neglect damping.

Ans: $t_0 = 0.0113$ s

2-19. The protection of packaged electronic equipment that is subject to damage when dropped accidentally is an important problem involving vibration theory.[15] The accompanying figure shows a TV set of mass m that is to be protected by some type of light cushioning material and dropped a distance h onto a concrete floor. The damage to the internal components of the TV set, if any, will depend upon the maximum acceleration experienced by the TV set and the corresponding maximum force determined from $F_{max} = m\ddot{x}_{max}$. Assume that the TV set and protective cushioning material can be modeled as a spring-and-mass system as shown in the accompanying figure, and that during the free fall the force in the spring is zero. Neglect damping and show that the differential equation of motion of m is

$$\ddot{x} + \frac{k}{m}x = g$$

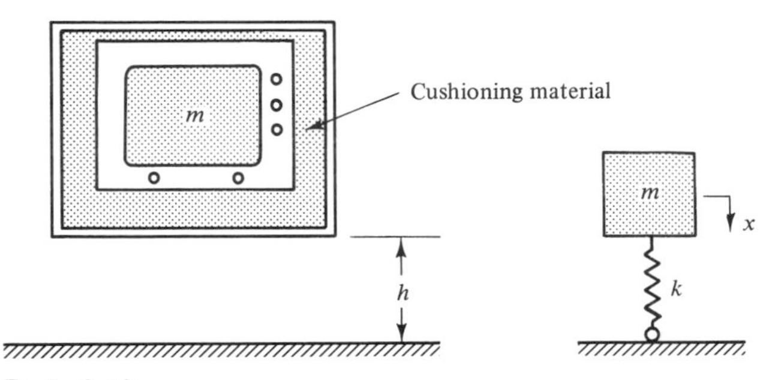

Prob. 2-19

in which x is the displacement of m following contact of the spring (cushioning material) with the floor. Determine a solution of this differential equation and show that the ratio of the maximum acceleration $\ddot{x}_{max}$ to the acceleration of gravity g is

$$\frac{|\ddot{x}_{max}|}{g} = \sqrt{1 + \frac{2h}{\Delta_s}}$$

in which $\Delta_s = W/k$ = static deflection.

2-20. An instrument package of mass m is mounted on a support structure inside a rocket as shown in the accompanying figure. The stiffness of the support structure is k. The firing of the rocket motor gives the rocket a sudden acceleration of $\ddot{y} = 4$ g's. Denoting the displacement of the rocket as y, and the displacement of m *relative* to the rocket as x, the absolute acceleration a_m of the instrument package is

$$a_m = \ddot{x} + \ddot{y}$$

[15] L. W. Gammell and J. L. Gretz, "Report on Effect of Drop Test Orientation on Impact Accelerations," Physical Test Laboratory, Texfoam Division, B. F. Goodrich Springs Products Division of B. F. Goodrich Co., Shelton, Conn., 1955.

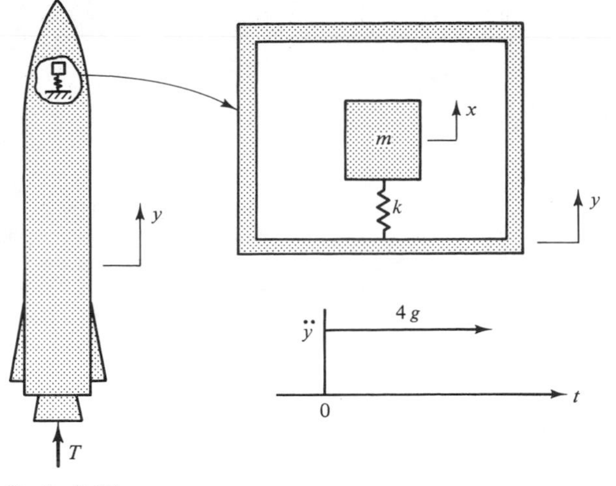

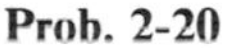
Prob. 2-20

Show that the differential equation of motion of m relative to the rocket is

$$\ddot{x} + \omega_n^2 x = -4g$$

Obtain a solution of this differential equation of relative motion and show that the absolute acceleration a_m of the instrument package is

$$a_m = 4g(1 - \cos \omega_n t)$$

2-21. A 0.4-ounce bullet having a horizontal velocity of 2000 ft/s becomes embedded in the mass A of the ballistic pendulum shown in the accompanying figure. If the block A weighs 32.2 lb and the two wires are each of length 2 ft, determine the maximum value θ that the wires make with the vertical using vibration theory.

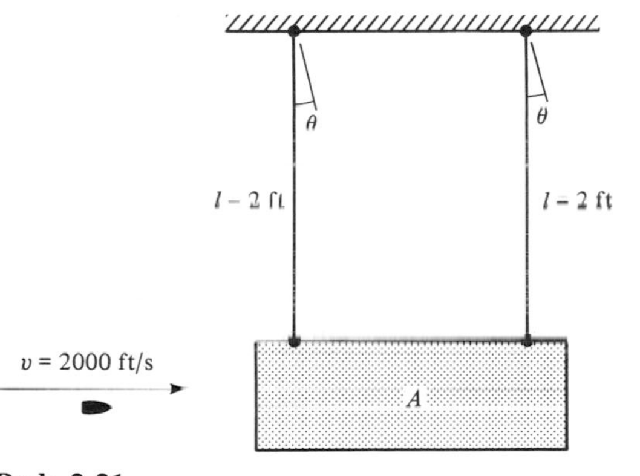

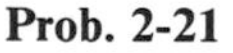
Prob. 2-21

2-22. The barrel of a large naval gun has an initial velocity of $\dot{x}_0$ as it begins its recoil after firing. The recoil mechanism consists of a spring and dashpot having parameters such that the gun is critically damped. Show that the maximum displacement during recoil is equal to $(\dot{x}_0 e^{-1})/\omega_n$ and occurs at time $t = 1/\omega_n$ regardless of the magnitude of the initial velocity $\dot{x}_0$.

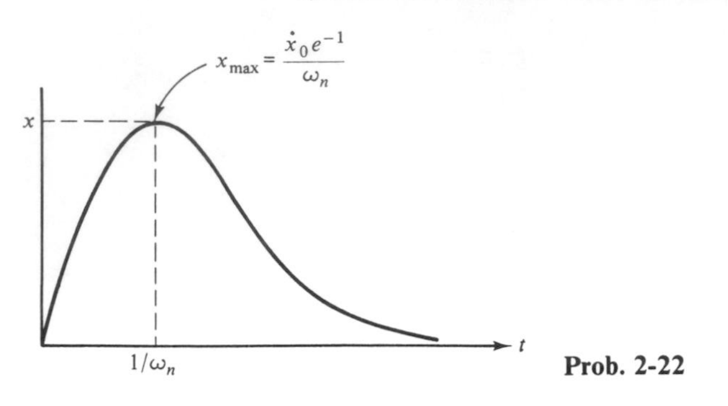

Prob. 2-22

Problems 2-23 through 2-34 (Sections 2-4 through 2-7)

2-23. A loaded railroad car weighing 35,000 lb is rolling at a constant velocity of v_0 when it couples with a spring and dashpot bumper system. If the recorded displacement-time curve of the loaded railroad car after coupling is as shown, determine (a) the damping factor ζ, (b) the spring constant k of the bumper system, and (c) the damping factor ζ of the system when the railroad car is *empty*. The unloaded railroad car weighs 8000 lb.

Ans: (a) $\zeta = 0.182$
(b) $k = 25{,}653$ lb/in.
(c) $\zeta = 0.38$

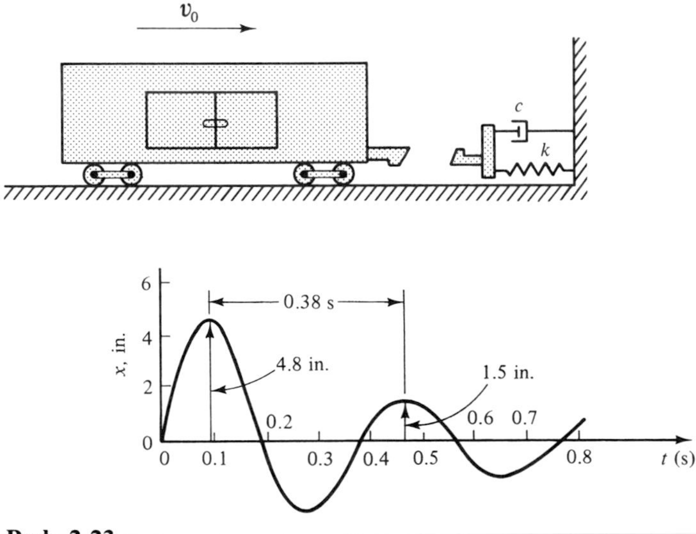

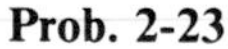
Prob. 2-23

2-24. The semilog plot of the amplitudes versus elapsed cycles s shown in the accompanying figure was made from an oscillograph record of a small steel structure. Determine the damping factor ζ and logarithmic decrement δ.

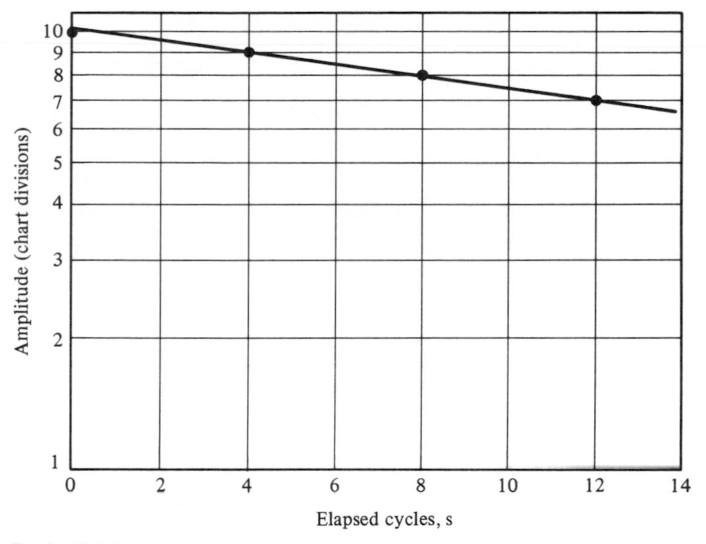

Prob. 2-24

2-25. Some years ago an end-zone stand was added to an existing football stadium located in the midwest. During the first use of the new stand many of the fans near the top had the feeling that the stand was swaying several inches, and some fans left the game fearing that it might collapse. While a subsequent visual inspection of the supporting columns did not show any damage, it was proposed to instrument the stand to determine its natural frequency f_n and damping factor ζ utilizing damped free-vibration measurements. One proposal was to excite the empty stand to obtain such measurements. Another was to obtain these measurements by monitoring the stand's motion during a game with the fans (approximately 10,000) providing the excitation. The latter procedure was used to obtain oscillograph records of the damped free vibration, and revealed that the natural frequency of $f_n = 1$ Hz and damping of 1.5 percent ($\zeta = 0.015$) were dangerously low. Considering that the empty stadium can be modeled as a simple spring-and-mass system as shown in part a of the accompanying figure, and that the spring-and-mass system for the stand full of people shown in part b is identical to that shown in part a except that the effective mass consists of that of both stand and fans, would the damping factor and frequency determined for an empty stand be the same as determined for the stand occupied by the 10,000 people?

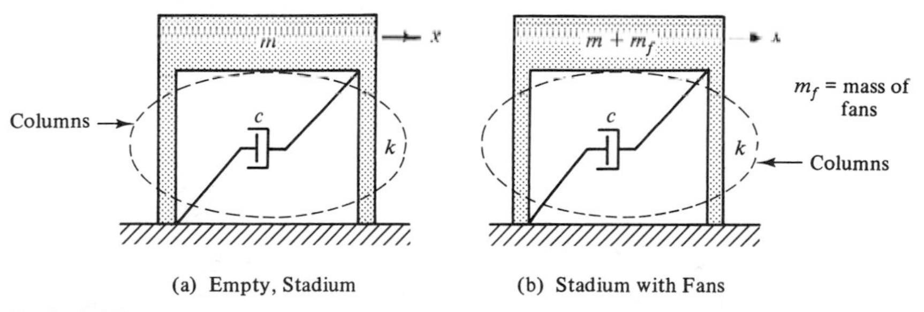

(a) Empty, Stadium (b) Stadium with Fans

Prob. 2-25

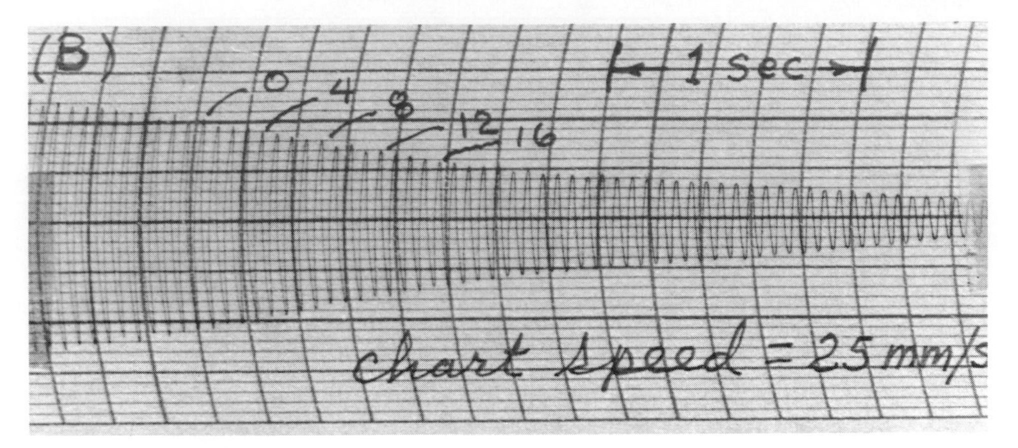

Prob. 2-26 (Aluminum)

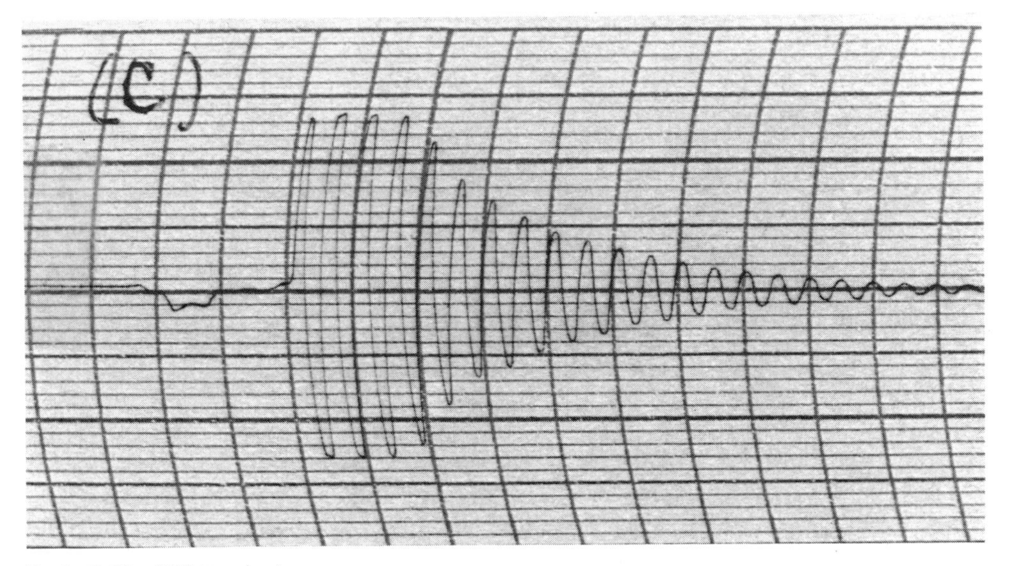

Prob. 2-27 (White pine)

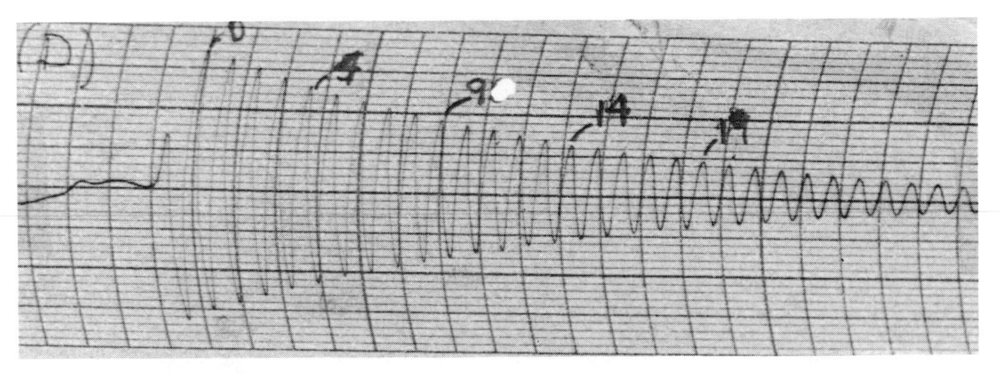

Prob. 2-28 (Composite)

2-26 to 2-29. The four oscillograph records in the accompanying photographs show the damped free vibrations of four small cantilever beams made of aluminum, white pine, a composite material consisting of Kevlar, and steel. The records were obtained from the output of a small accelerometer attached to the free end of each beam. Plot the amplitudes of vibration on semilog paper, and determine the damping factor and logarithmic decrement for each of the beams. Also determine whether the damping characteristics of each beam appear to conform to those of a viscous-damping model.

Prob. 2-26 Partial ans: $\zeta = 0.0056$
Prob. 2-28 Partial ans: $\zeta = 0.011$

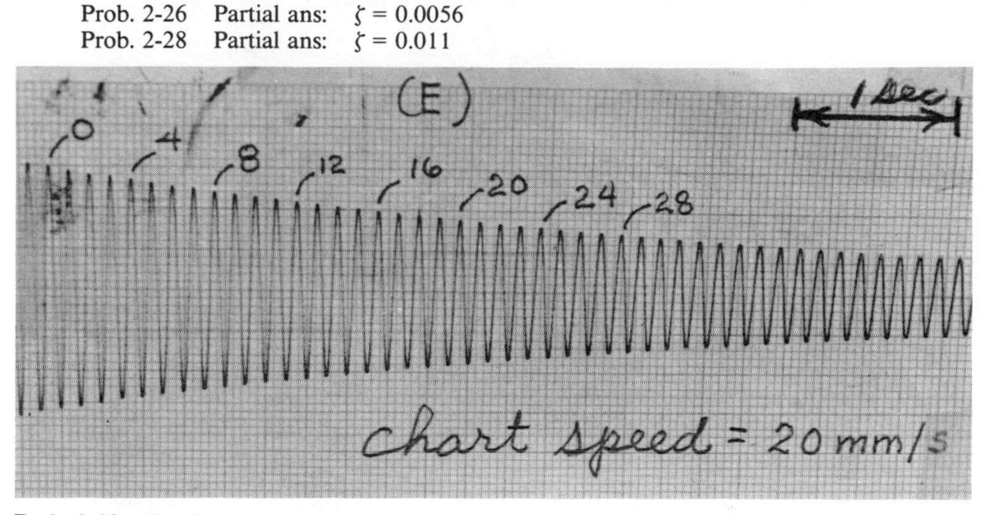

Prob. 2-29 (Steel)

2-30. A steel sphere of mass m_s and an aluminum sphere of mass m_a are immersed in a fluid of specific weight w as shown in part a of the accompanying figure. The radius of each sphere is r, and the specific weights of the steel and aluminum used are $w_s = 0.284$ lb/in.^{3} and $w_a = 0.10$ lb/in.^{3}, respectively. Consider that the vibration of a sphere in a fluid has the same characteristics as those of the spring-and-mass system shown in part b of the figure, and determine the ratio of the damping factor ζ_a of the aluminum sphere to that of the steel sphere ζ_s (ζ_a/ζ_s).

Ans: $\dfrac{\zeta_a}{\zeta_s} = 1.685$

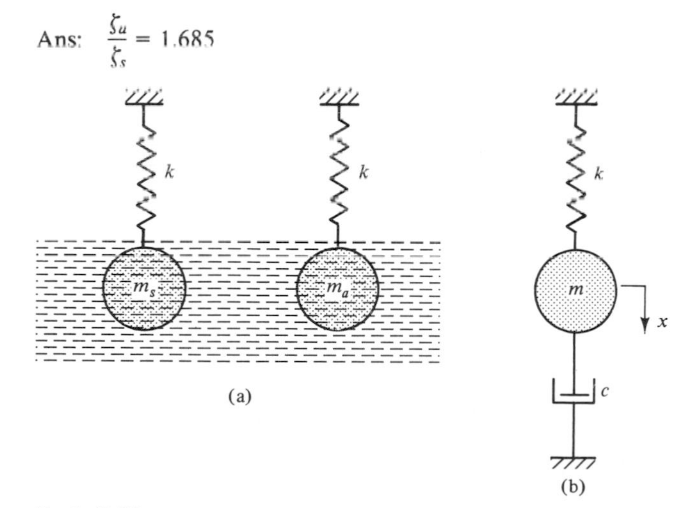

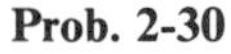

Prob. 2-30

2-31. The mass m is pinned to the end of a cantilever beam that has a bending stiffness factor of EI and a length of l. The spring constant k_b of the beam is $3EI/l^3$, and the spring constant of each of the two vertical springs is k. Determine the equivalent spring constant k_e of the system and its undamped natural circular frequency ω_n.

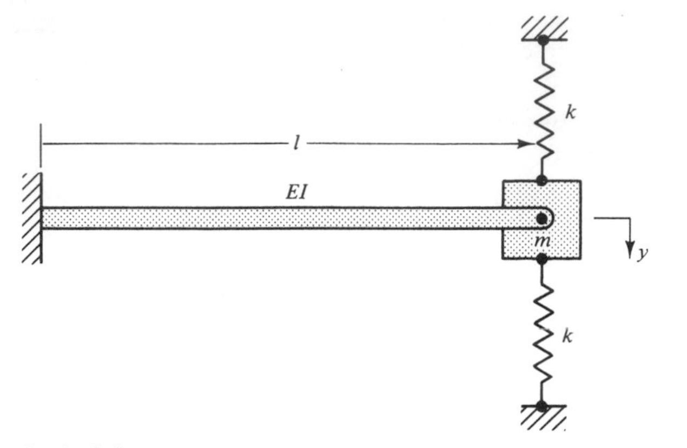

Prob. 2-31

2-32. The static deflection at $x = l/3$ of the fixed-fixed beam is 0.5 in. when a 1000-lb load is applied as shown in part a of the accompanying figure. A weight $W = 500$ lb is attached to a cable that is attached to the beam at $x = l/3$ as shown in b. Determine the natural frequency f_n of the system, if the spring constant k_c of the cable is $k_b/2$.

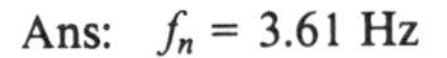
Ans: $f_n = 3.61$ Hz

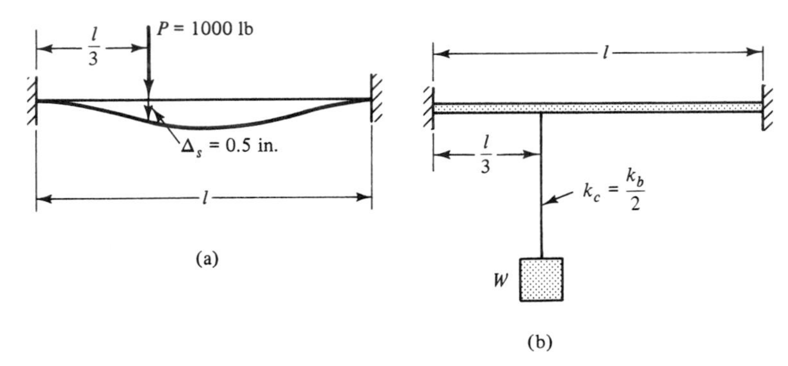

Prob. 2-32

2-33. The horizontal and vertical ground motions of earthquakes generally cause both bending and axial deformations of the columns of buildings. However, the major damage to the columns has been observed to come from bending deformations accompanying the lateral (horizontal) motion of the floors, with little damage resulting from the axial deformations. One of the reasons for this fact is that the columns are much stiffer axially than laterally. Thus, most buildings have much higher natural frequencies associated with axial deformations than with bending deformations. To illustrate this, consider the simple frame shown in the accompanying figure, and determine the ratio of the axial natural frequency f_a to the lateral natural frequency f_b. The effective length l of each of the two channels

(C10 × 130), which are welded to the large I beam, is 10 ft. Neglect the mass of the columns. The spring constants for each channel axially and laterally are

$$k_a = \frac{AE}{l}$$

and

$$k_b = \frac{12EI}{l^3}$$

where $E = 30(10)^6$, psi
$I = 103$ in.^{4} (moment of inertia of channel)
$A = 8.8$ in.^{2} (cross-sectional area of channel)

Ans: $f_a/f_b = 10.12$

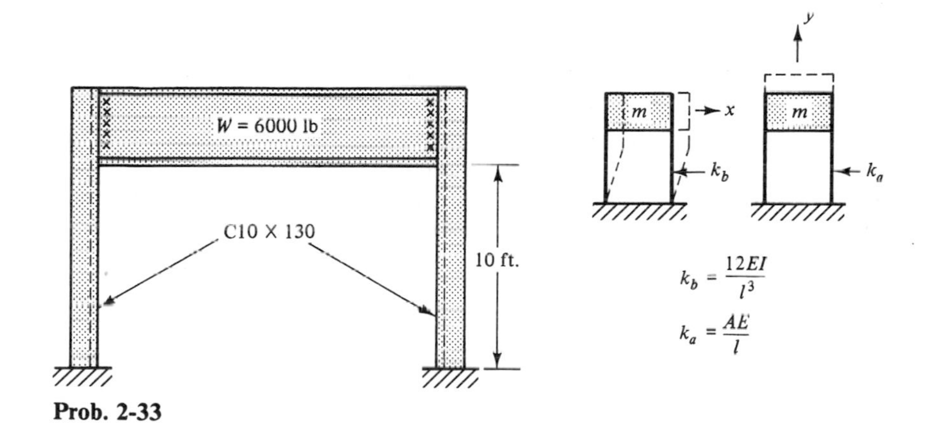

Prob. 2-33

2-34. A disk with a centroidal mass moment of inertia $\bar{I} = 0.03$ N · m · s^2 is attached to a stepped size shaft with torsional spring constants of $k_1 = 70$ N · m/rad and $k_2 = 280$ N · m/rad. Determine the equivalent torsional spring constant k_e for the shaft and the natural frequency of the system.

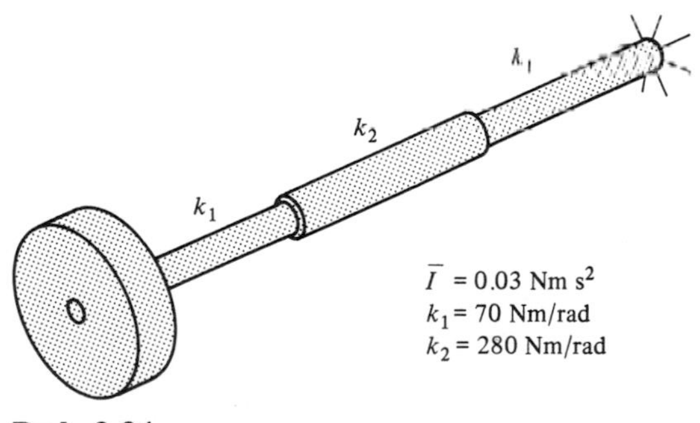

Prob. 2-34

Problems 2-35 through 2-50 (Sections 2-8 through 2-9)

2-35. A U-tube manometer contains a fluid of specific weight w which has a column length of l. Determine (a) the differential equation of motion of the fluid column using Newton's second law, (b) the undamped natural period τ_n, and (c) the undamped natural period τ_n using the energy method in which $T_{max} = U_{max}$.

Partial ans: $\tau_n = 2\pi \sqrt{\dfrac{l}{2g}}$

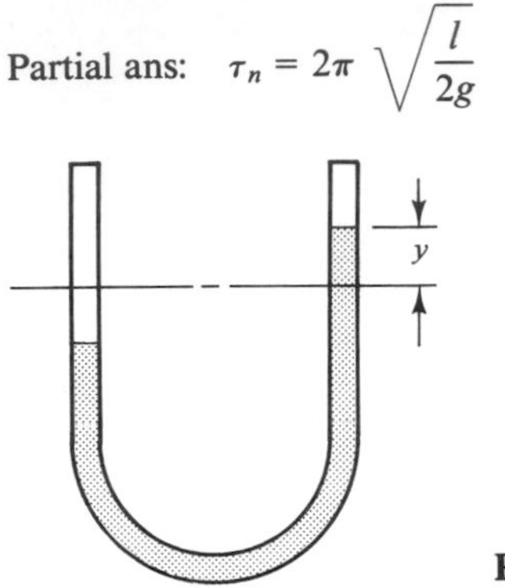

Prob. 2-35

2-36. Use the energy method, $T_{max} = U_{max}$, to determine the undamped natural frequency ω_n for the system shown in Prob. 2-8.

2-37. A uniform slender rod of length l and weight W is supported in two different ways as shown in the accompanying figure. Show that U_{max} for the rod when displaced θ_0 from its horizontal neutral equilibrium position in part a of the figure is

$$U_{max} = \frac{k}{2}(l\theta_0)^2$$

and that when displaced from a vertical neutral equilibrium position as shown in part b it is

$$U_{max} = \frac{k}{2}(l\theta_0)^2 + \frac{W}{4}l\theta_0^2$$

Why do these quantities differ?

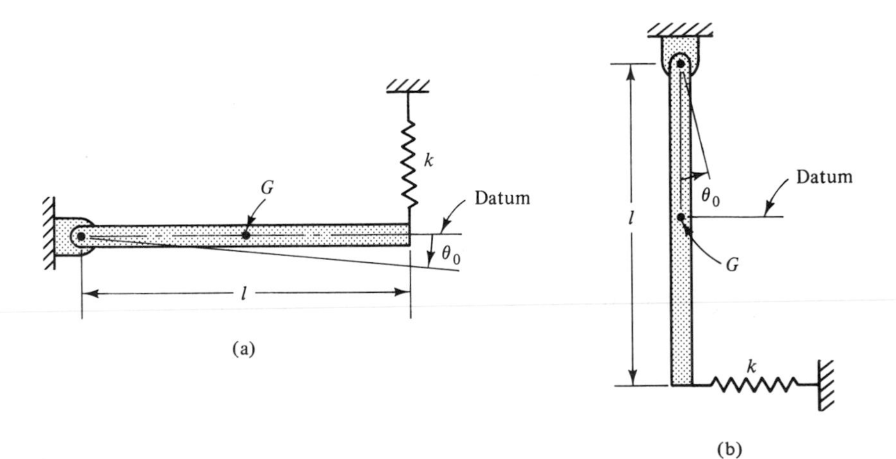

Prob. 2-37

2-38. A disk with a centroidal mass moment of inertia of $\bar{I}$ about the y axis shown is attached to the end of a circular shaft of length l, radius r, and mass per unit length γ. Assume a linear displacement function

$$\theta_s = \frac{\theta_0 y}{l}$$

for the maximum angular twist of any point along the shaft, and using $T_{max} = U_{max}$ show that

$$\omega_n = \sqrt{\frac{k}{\bar{I} + \bar{I}_s/3}}$$

where $\bar{I}_s$ = mass moment of inertia of the shaft about the longitudinal y axis
$k = GJ/l$ = torsional spring constant of shaft

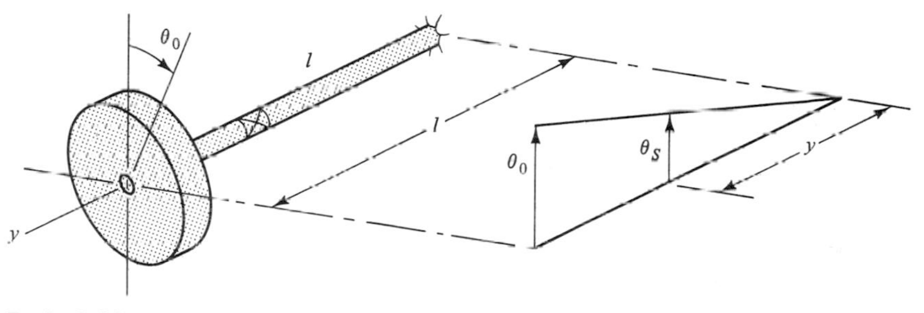

Prob. 2-38

2-39. Refer to Prob. 2-3 and use $T_{max} = U_{max}$ to determine the natural circular frequency of the system.

2-40. Refer to Prob. 2-4 and use $T_{max} = U_{max}$ to determine the natural circular frequency ω_n of the system.

2-41. A subsystem of a high-speed textile machine consists of a push rod, rocker arm, follower, and follower spring as shown in the accompanying figure. The subsystem is driven by a cam (assumed as rigid) as shown. The parameters of the system are

I_0 = mass moment of inertia of rocker arm about its axis of rotation
m_r = mass of push rod
m_f = mass of follower
m_s = mass of follower spring
k_s = spring constant of follower spring
k_r = spring constant of push rod

Use $T_{max} = U_{max}$ to determine the natural frequency of the subsystem in terms of its parameters. Include the effects of the masses m_s of the follower spring and m_r of the push rod (treat as a stiff spring). Assume that because of the type of linkage connection used the upper ends of the push rod and follower both move with the contacting portions of the rocker arm, and that the axial deformation of the push rod is negligible.

Ans: $$\omega_n = \sqrt{\frac{k_s b^2 + k_r a^2}{I_0 + m_r a^2/3 + (m_f + m_s/3)b^2}}$$

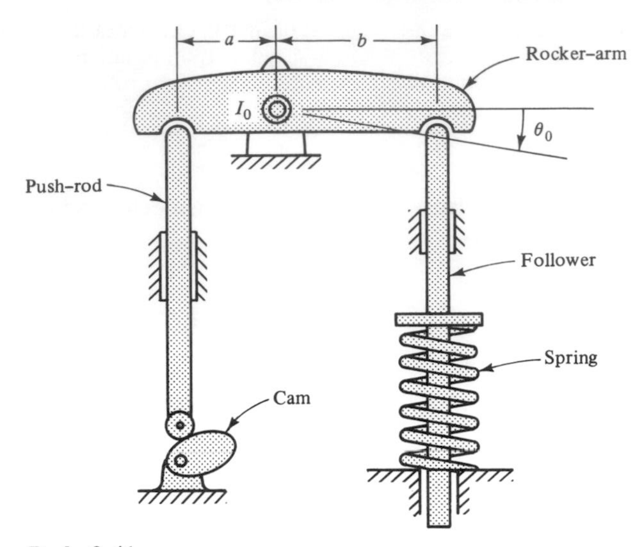

Prob. 2-41

2-42. A uniform fixed-fixed beam of mass per unit length γ, length l, and stiffness factor EI supports a uniform load of w lb per unit length as shown in the accompanying figure. Use the energy concept of $T_{max} = U_{max}$ to determine the natural circular frequency ω_n of the loaded beam. Use the shape function

$$y = A(-x^4 + 2lx^3 - l^2x^2)$$

which corresponds to the static-deflection curve of a uniformly loaded beam in which $A = w/(24EI)$. Compare the ω_n value found with the value determined in Example 2-9 for the beam alone, and also with the exact value given there.

Ans: $\omega_n = \dfrac{22.45}{l^2}\sqrt{\dfrac{EI}{\gamma + w/g}}$

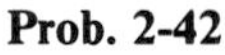

Prob. 2-42

2-43. Two identical disks of weight W are attached to a shaft that has a stiffness factor of EI as shown in part a of the accompanying figure. It is assumed that the bearings A and B offer no restraint to flexural vibrations (are self-aligning bearings), and that the disks are rigid bodies. The vibratory motion of the disks due to bending of the shaft consists of a translation y and a rotation corresponding to the slope dy/dx of the shaft at $x = l/3$ and $x = 2l/3$ (for small angles $\theta \cong \tan\theta = dy/dx$) as shown in part b of the accompanying figure. Neglect the mass of the shaft, and use the shape function $y = A\sin(\pi x/l)$ to determine (a) the natural circular frequency ω_n of the system if the rotary motion of the disks about their centroidal diametral axes (axes perpendicular to the plane of the paper through the centroids of the disks) is neglected, and (b) the natural circular frequency if both the translation and rotary motion of the disks are considered. (The mass moment of inertia of a disk of radius r about a centroidal diametral axis is $\bar{I} = Wr^2/4g$.)

Partial ans, part b: $\omega_n = \sqrt{\dfrac{g(24.35)EI}{Wl^3(0.75 + 0.617r^2/l^2)}}$

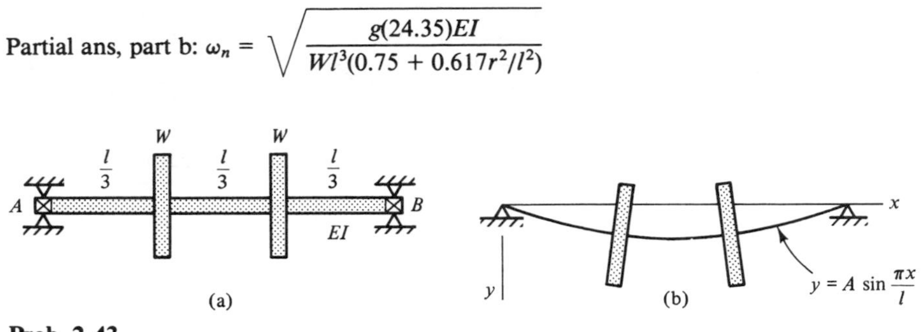

Prob. 2-43

2-44. A cable of length L and mass per unit length γ_c is attached to the end of a cantilever beam of length l and mass per unit length γ_b as shown in the accompanying figure. In selecting a shape function for the beam, the following two trigonometric functions are to be considered:

$$y = \Delta\left(1 - \cos\frac{\pi x}{l}\right)$$

$$y = \Delta\left(1 - \cos\frac{\pi x}{2l}\right)$$

Select the more appropriate of the two shape functions, and determine the natural circular frequency ω_n of the beam-and-cable system using the concept that $T_{max} = U_{max}$. Assume that the initial tension in the cable is sufficient to maintain the cable in tension at all times as the system vibrates. Express ω_n in terms of the following parameters:

m_b = mass of beam $(\gamma_b l)$
m_c = mass of cable $(\gamma_c L)$
k_b = spring constant of beam $(3EI/l^3)$
k_c = spring constant of cable (AE/L)

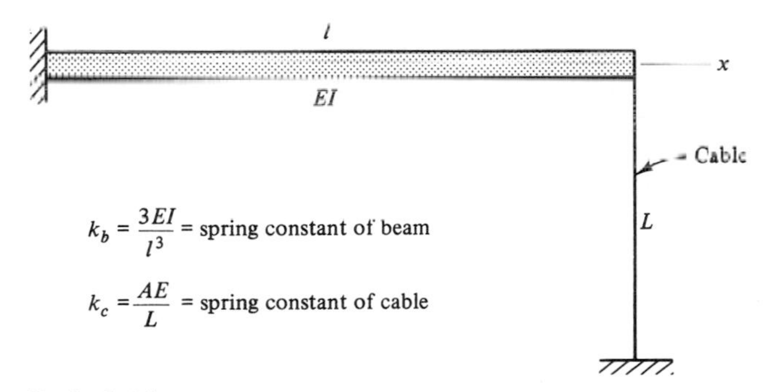

Prob. 2-44

2-45. Referring to Example 2-12, on page 106, consider that the steel plate is removed from the wooden beam, and determine f_n for the wood-only beam.

2-46. Referring to Example 2-12, on page 106, determine the natural frequency f_n of the wood-steel beam if a $\frac{1}{2}$- by 6-in. steel plate is added at the top of the beam.

Ans: $f_n = 23.29$ Hz

2-47. A simply supported reinforced-concrete beam 20 ft long is reinforced with four No. 7 steel bars ($\frac{7}{8}$ in. diameter) located as shown in the accompanying figure. The modulus of elasticity and specific weight of the concrete and steel materials are as follow:

	Specific weight, lb/ft^3	Modulus of elasticity, psi
Concrete	150	$5(10)^6$
Steel	490	$30(10)^6$

Consider that there are no cracks in the concrete of the beam, and determine its natural frequency f_n. It is common practice to assume that the area of the steel is concentrated along a line through the centroids of the bars, a distance of 15 in. from the top of the beam of this problem.

Ans: $f_n = 21.14$ Hz

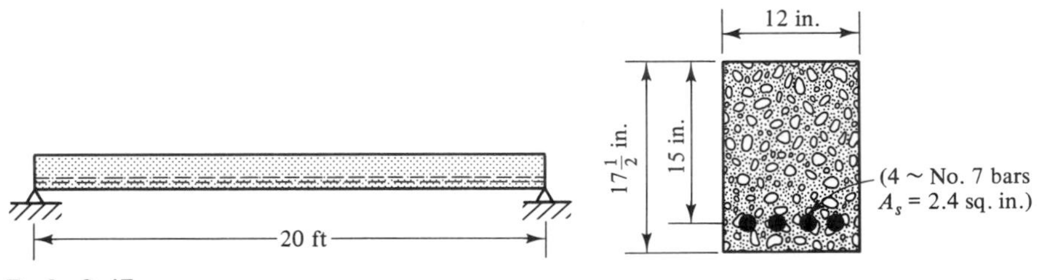

Prob. 2-47

2-48. Consider the beam of Prob. 2-47 as a concrete beam without reinforcing steel and determine its natural frequency f_n.

Ans: $f_n = 20.55$ Hz

2-49. Referring to Example 2-13, on page 110, determine the natural circular frequency of the pinned-pinned beam if the cross section is as shown in the accompanying figure.

$$\text{Ans:} \quad \omega_n = \frac{\pi^2}{l^2}\sqrt{\frac{EI}{\gamma}}\left[1 + 1.58\left(\frac{\Delta}{h}\right)^2\right]^{1/2}$$

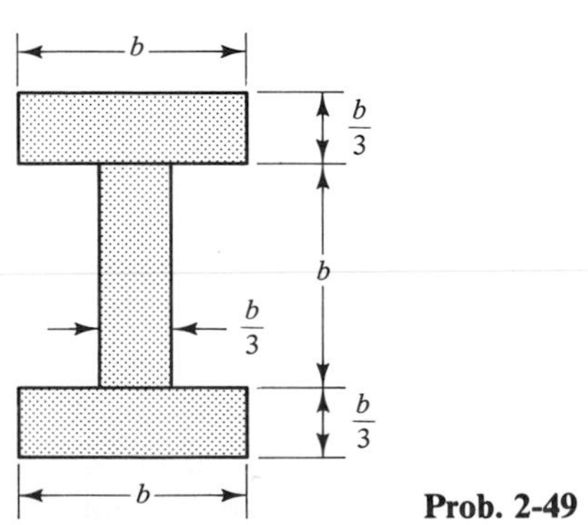

Prob. 2-49

2-50. Use the shape function $y = \Delta[1 - \cos(2\pi x/l)]$, which satisfies the deflection, slope, and moment conditions at the ends of a fixed-fixed beam, to determine the natural frequency of an axially constrained beam with the cross section shown in the accompanying figure. Express ω_n in terms of $(2\Delta)/h$, in which 2Δ is the amplitude of vibration at $x = l/2$ and h is the depth of the beam. The mass of the beam per unit length is γ.

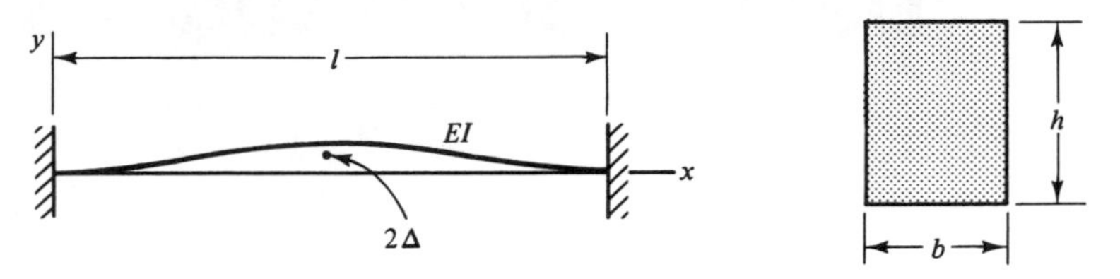

Prob. 2-50

Chapter 3

Harmonic Excitation of Single-Degree-of-Freedom Systems

3-1 INTRODUCTION

There are many sources of excitations that cause machines and structures to vibrate. They include unbalanced rotating devices, gusting winds, vortex shedding, moving vehicles, earthquakes, rough road surfaces, and so on.

The engineer is faced with a variety of problems when machines and structures are subjected to such forced vibrations. Unbalanced forces in the rotating parts of a machine can cause undesirable vibration of both the machine and the structure housing it, and in turn can be annoying and even harmful to the occupants of the structure. There are instruments that will not function properly or may be damaged, because of vibrating supports. When the frequency of an exciting force or support motion is the same as the natural frequency of the system upon which it is acting, a condition referred to as *resonance* occurs which can cause large displacements and accompanying stresses in machines and structures. These large alternating tensile and compressive stresses can lead to fatigue failures.

Although real systems are seldom excited by *purely* sinusoidally varying forces or support motions (only a single frequency component), analyses of the effects of such excitations on vibrating systems provide the foundation for the analysis of more complicated forms of excitation encountered in real systems, such as periodic excitations of various types containing multiple frequency components, random excitation, and so forth. The characteristics of transducers[1] that enable them to measure dynamic phenomena are developed by study of the sinusoidal excitations of simple spring-and-mass systems. The forced vibration of systems are usually caused by dynamic forces $F(t)$ or support motions $y(t)$ such as those shown acting on some typical single-degree-of-freedom systems in Fig. 3-1.

[1] Transducers are devices that transform energy from one form to another. A piezoelectric accelerometer, for example, changes mechanical energy into electrical energy.

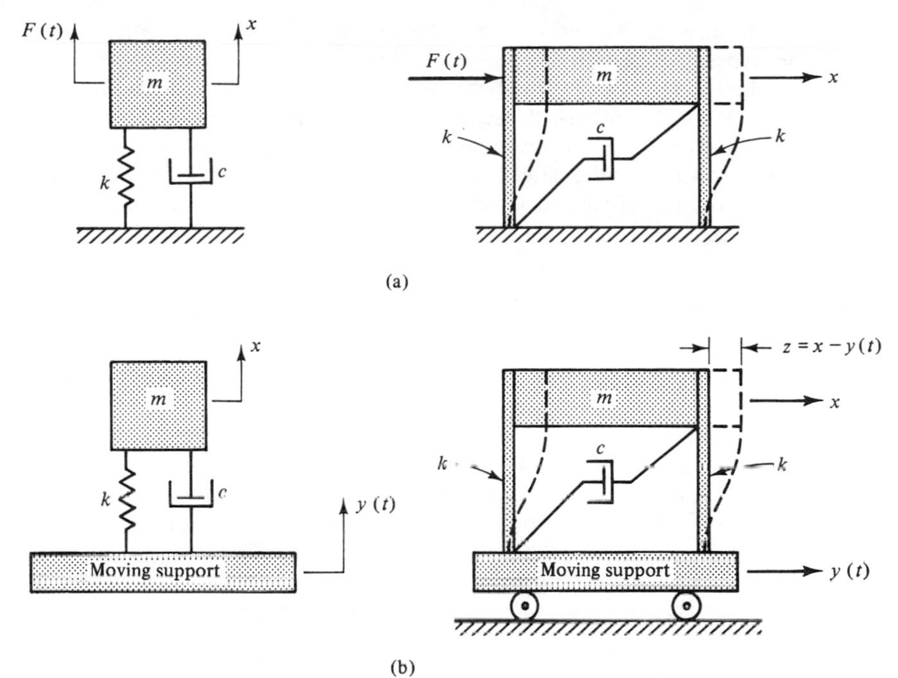

Figure 3-1 Two general types of excitation. (a) Force excitation. (b) Support excitation.

While all real systems have some damping, it is often neglected to simplify the analysis, and the theoretically undamped system is analyzed to gain insights into more complex systems. The terms associated with the vibration response of damped and theoretically undamped systems are discussed in the next few paragraphs, and the reader should study them carefully for future reference.

When a linear, constant-coefficient dynamic system is disturbed by certain types of exciting forces or support motions,[2] the resulting response of such a system can be considered to consist of the sum of two distinct components, which are:

1. The *forced response,* which resembles the exciting force or support motion in its mathematical form.
2. The *free response* (also referred to as *natural response*), which does not depend upon the characteristics of the exciting function, but only upon the physical parameters of the system itself. (The free response of a system can obviously also be induced by just initial conditions without the presence of an exciting function.)

In the mathematical solution used to determine the response of such a system, these two responses are usually obtained separately as the *particular* solution and the *complementary* or *homogeneous* solution. The sum of these two responses is then referred to as the *total response* of the system.

[2] These include sinusoidal excitations, decaying exponential excitations, and products of these two types, among others.

In any real system disturbed by a sinusoidal exciting force or support motion, or other types of excitation that repeat with time, the free response initiated when the excitation is applied dies out with time because of damping in the system, and eventually only the forced response remains. Because the free response of real systems dies out with time, it is often referred to as a *transient* response, and the forced response is known as the *steady-state* response. It is the latter two terms that we use in most instances in this text to refer to the two types of vibratory motion of real systems for which damping is considered.

3-2 EXCITATION DUE TO AN UNBALANCED ROTATING MASS

Unbalanced rotating masses in machines that subject them to periodic forces are one of the more common sources of excitation. A machine represented by a typical single-degree-of-freedom system is shown in Fig. 3-2a. It has a total mass m consisting of a translating mass M to which is attached a mass m_0 rotating about an axis through O. The mass of the slender rod attached to m_0 is considered negligible so that the total mass of the system is $m = M + m_0$. Since the center of gravity of m_0 does not coincide with its axis of rotation, it acts as an unbalanced rotating mass. The spring of constant k and dashpot of constant c represent the stiffness and damping, respectively, of the machine's mounting system.

Let us select up as the positive sense for x, $\dot{x}$, and $\ddot{x}$ as shown in Fig. 3-2, and note that x is the displacement of the nonrotating mass M from its static-equilibrium position. The total acceleration of the unbalanced mass m_0 is the vector sum of the normal acceleration $e\omega^2$ directed toward the axis of rotation O and the acceleration $\ddot{x}$ of the mass M.

Using D'Alembert's principle, the inertia effects $m_0 e\omega^2$, $m_0\ddot{x}$, and $M\ddot{x}$ are added to the free body by means of the dashed vectors shown in Fig. 3-2b. Each has a sense opposite to the assumed positive acceleration.

The senses of the forces acting on the free body are determined by giving each a

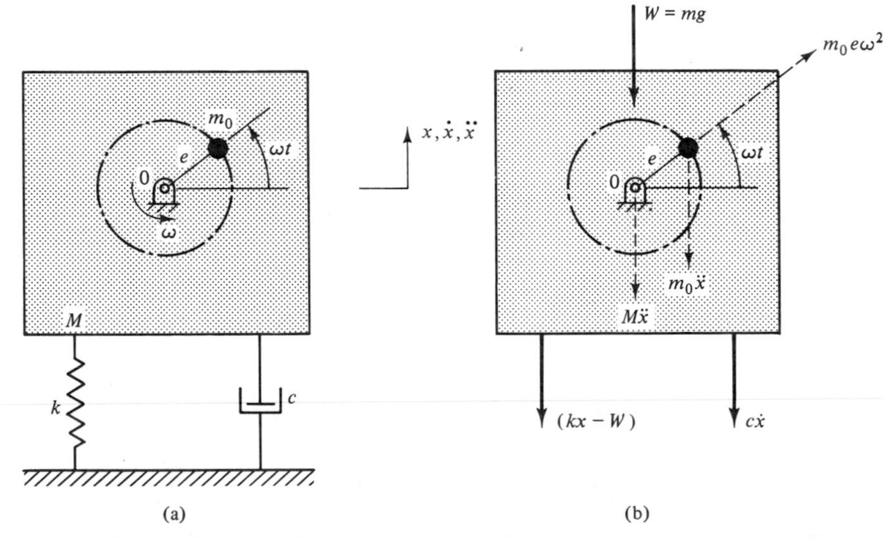

Figure 3-2 Single-degree-of-freedom system with unbalanced mass m_0. (a) System. (b) Free body.

sense consistent with the force it exerts on the mass M with the displacement, velocity, and acceleration of M all assumed positive. The spring force is $kx - W$ instead of kx since the system has an initial downward displacement in the static-equilibrium position with an accompanying force W in the spring.

We next sum the forces and inertia effects equal to zero, obtaining

$$\sum F_x = 0 = -W - (kx - W) - c\dot{x} - (M + m_0)\ddot{x} + m_0 e\omega^2 \sin \omega t \tag{3-1}$$

Letting $m = M + m_0$, and $m_0 e\omega^2 = F_0$, Eq. 3-1 can be written as

$$\ddot{x} + \frac{c}{m}\dot{x} + \frac{k}{m}x = \frac{F_0}{m}\sin \omega t \tag{3-2}$$

If we write the differential equation of motion in terms of the damping factor ζ and the undamped natural circular frequency ω_n of the system as explained in Sec. 2-2,

$$\ddot{x} + 2\zeta\omega_n\dot{x} + \omega_n^2 x = \frac{F_0}{m}\sin \omega t \tag{3-3}$$

in which $2\zeta\omega_n = c/m$ and $\omega_n^2 = k/m$.

Equation 3-3 is the general form of the *mathematical model* of any single-degree-of-freedom system subjected to a sinusoidal exciting force $F_0 \sin \omega t$. The complete solution of this equation is the sum of a *homogeneous* solution x_h and a *particular* solution x_p. That is, the total *response* is

$$x = x_h + x_p \tag{3-4}$$

The homogeneous solution of Eq. 3-3 for $\zeta < 1$ (known as the *free* or *transient response*) is

$$x_h = e^{-\zeta\omega_n t}(A \cos \omega_n\sqrt{1 - \zeta^2}t + B \sin \omega_n\sqrt{1 - \zeta^2}t) \tag{3-5}$$

(as derived in Sec. 2-3 and given by Eq. 2-20).

The particular solution (the *forced* or *steady-state response*) is best determined with the use of complex algebra, since the right-hand side of Eq. 3-3 is the sinusoidally varying term $(F_0/m)\sin \omega t$ (see Sec. 1-3). Thus, from Euler's equation,

$$e^{j\omega t} = \cos \omega t + j \sin \omega t \tag{3-6}$$

we can write that

$$\cos \omega t \overset{R}{=} e^{j\omega t}$$

and

$$\sin \omega t \overset{I}{=} e^{j\omega t}$$

in which R means the "real part of" and I means the "imaginary part of." With this in mind, we can express the right-hand side of Eq. 3-3 as $(F_0/m)e^{j\omega t}$, with the provision that only the imaginary part of the term will be used in the solution process. We then assume a particular solution having the form of

$$x_p = Xe^{j\omega t} \tag{3-7}$$

in which X is a *complex constant* to be determined so that it satisfies Eq. 3-3. Taking successive derivatives of Eq. 3-7, we obtain

and

$$\left.\begin{aligned}\dot{x}_p &= j\omega X e^{j\omega t}\\ \ddot{x}_p &= -\omega^2 X e^{j\omega t}\end{aligned}\right\} \tag{3-8}$$

Substituting Eqs. 3-7 and 3-8 into Eq. 3-3, and replacing $(F_0/m)\sin \omega t$ by $(F_0/m)e^{j\omega t}$, yields

$$(-\omega^2 + j\omega 2\zeta\omega_n + \omega_n^2)Xe^{j\omega t} = \frac{F_0}{m} e^{j\omega t} \tag{3-9}$$

Dividing this equation by ω_n^2, and noting that $m\omega_n^2 = k$, we see that

$$\left[1 - \left(\frac{\omega}{\omega_n}\right)^2 + j2\zeta\frac{\omega}{\omega_n}\right]X = \frac{F_0}{k} \tag{3-10}$$

The bracketed term in Eq. 3-10 can be written as

$$\left[1 - \left(\frac{\omega}{\omega_n}\right)^2 + j2\zeta\frac{\omega}{\omega_n}\right] = \sqrt{\left[1 - \left(\frac{\omega}{\omega_n}\right)^2\right]^2 + \left[2\zeta\frac{\omega}{\omega_n}\right]^2}\, e^{j\phi} \tag{3-11}$$

in which

$$\tan\phi = \frac{2\zeta(\omega/\omega_n)}{1 - (\omega/\omega_n)^2} \tag{3-12}$$

as shown in Fig. 3-3a.

Substituting the right-hand expression in Eq. 3-11 into Eq. 3-10,

$$X = \frac{(F_0/k)e^{-j\phi}}{\sqrt{[1 - (\omega/\omega_n)^2]^2 + [2\zeta(\omega/\omega_n)]^2}} \tag{3-13}$$

or

$$X = |X|e^{-j\phi} \tag{3-14}$$

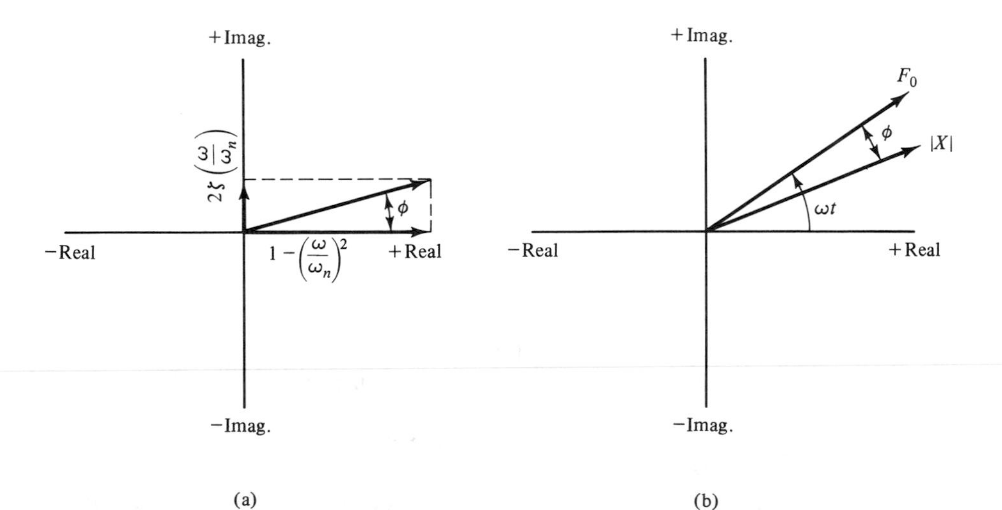

Figure 3-3 Vectors associated with steady-state response in the complex plane.

in which the *absolute value* of X is the amplitude

$$|X| = \frac{F_0/k}{\sqrt{[1 - (\omega/\omega_n)^2]^2 + [2\zeta(\omega/\omega_n)]^2}} \tag{3-15}$$

Thus, the particular solution of Eq. 3-3 is

$$x_p = |X|e^{-j\phi}e^{j\omega t} = |X|e^{j(\omega t - \phi)} \tag{3-16}$$

Using only the imaginary part of $e^{j(\omega t - \phi)}$,

$$x_p = |X|\sin(\omega t - \phi) \tag{3-17}$$

To obtain a more general picture of the particular solution just obtained, let us remember that we saw in Sec. 1-2 that it is possible to represent a sinusoidally varying quantity such as the displacement in Eq. 3-17, for example, by means of a rotating vector that generates a sine or cosine function as its projection on a vertical or horizontal axis. Our use of a complex quantity to represent the excitation caused by an inertia effect, and the resulting complex response in the preceding discussion, illustrates the same concept, but with rotating vectors expressed in complex notation.

Both the complex exciting force $F_0e^{j\omega t}$ and the resulting complex response $|X|e^{j(\omega t - \phi)}$ shown in Fig. 3-3b are rotating complex vectors (phasors) in the complex plane. They have magnitudes of F_0 and $|X|$, respectively, and their directions are given by the rotating complex *unit* vectors $e^{j\omega t}$ and $e^{j(\omega t - \phi)}$, respectively. They rotate with a common angular velocity ω, while always maintaining the same relative position. The angle of separation between them, referred to as the *phase angle,* is the angle by which the steady-state displacement lags the exciting force as shown in Fig. 3-3b.

With $\sin \omega t$ on the right side of Eq. 3-3, the exciting force and steady-state displacement at any time t are given by the projections of the respective complex vectors on the *imaginary* axis. If the right side of Eq. 3-3 had contained $\cos \omega t$ instead of $\sin \omega t$, the exciting force and steady-state displacement at any time t would be given by the projection of the respective complex vectors on the *real* axis.

Adding the homogeneous solution x_h given by Eq. 3-5 to the steady-state solution given by Eq. 3-17, the *complete* solution of Eq. 3-3 is

$$x = e^{-\zeta\omega_n t}(A \cos \omega_n\sqrt{1 - \zeta^2}t + B \sin \omega_n\sqrt{1 - \zeta^2}t) + |X|\sin(\omega t - \phi) \tag{3-18}$$

As can be seen, the vibratory motion described by Eq. 3-18 is a combination of two motions. One has a frequency of $\omega_n\sqrt{1 - \zeta^2}$ and an exponentially decreasing amplitude, while the other has a frequency of ω and a constant amplitude of $|X|$. The combined motion is shown in Fig. 3-4 for $\omega < \omega_n\sqrt{1 - \zeta^2}$.

As mentioned earlier, the decaying motion of Eq. 3-18 is a *transient* vibration that disappears with time, leaving just the *steady state* motion described by the particular solution. The constants A and B of the transient vibration depend upon the initial conditions, while the steady-state vibration depends only upon the forcing function and the physical parameters of the system.

To illustrate the use of the real part of $e^{j\omega t}$, consider the system shown in Fig. 3-5. The differential equation of motion of this system is

$$\ddot{x} + 2\zeta\omega_n\dot{x} + \omega_n^2 x = \frac{F_0}{m}\cos \omega t$$

in which

$$m = M + m_0$$

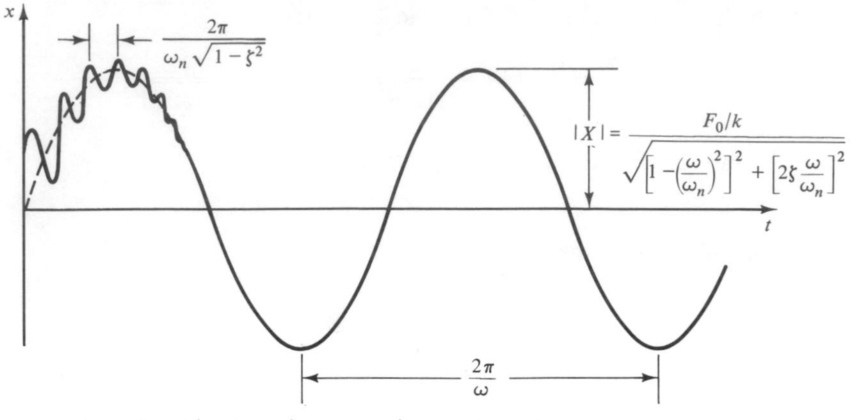

Figure 3-4 Combined motion—transient and steady-state.

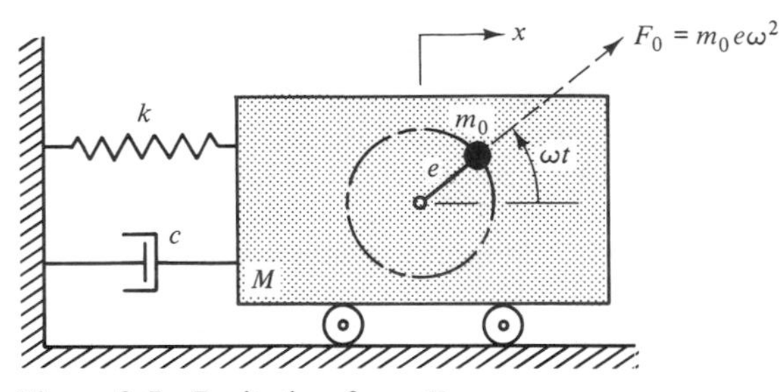

Figure 3-5 Excitation force $F_0 \cos \omega t$.

The excitation term includes $\cos \omega t$, which is the real part of $e^{j\omega t}$. The steady-state solution is then

$$x_p = |X|\cos(\omega t - \phi) \tag{3-19}$$

and the steady-state amplitude $|X|$ is the same as shown in Eq. 3-15, with the phase angle ϕ defined by Eq. 3-12.

Steady-State Vibration Characteristics

Equations 3-12 and 3-15 describe the steady-state behavior of a system subjected to a harmonic exciting force $F_0 \sin \omega t$ or $F_0 \cos \omega t$. Such an exciting force might come from an unbalanced rotating mass or vortex shedding, for example. Equation 3-15 defines the amplitude of the steady-state vibration, while Eq. 3-12 shows the phase relationship between the forcing function and the steady-state displacement. It is obvious from Eq. 3-15 that the amplitude of vibration increases with an increase in F_0, which would be expected.

To aid in determining the effects of the damping factor ζ and the frequency ratio ω/ω_n on the steady-state behavior of a vibrating system, Eq. 3-15 can be arranged in dimensionless form as

$$\frac{|X|}{F_0/k} = \frac{1}{\sqrt{[1-(\omega/\omega_n)^2]^2 + [2\zeta(\omega/\omega_n)]^2}} \tag{3-20}$$

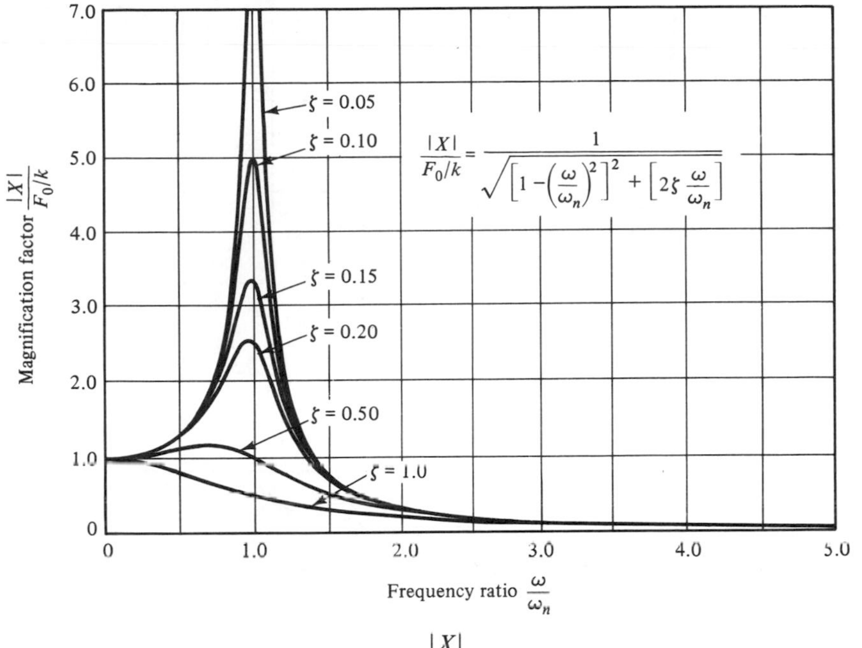

Figure 3-6 Plot of magnification factor $\dfrac{|X|}{F_0/k}$.

In this equation, F_0/k is the static displacement Δ_s that the system would have if the force F_0 were slowly applied to it, and $|X|$ is the amplitude of the steady-state vibration. The ratio $|X|/(F_0/k)$ is referred to as the *magnification factor* and, as evidenced by Eq. 3-20, is obviously a function of ζ and ω/ω_n. A plot of Eq. 3-20 for various magnitudes of damping is shown in Fig. 3-6. These curves reveal some important characteristics of the steady-state vibration of a system subjected to a sinusoidal excitation:

1. When $\omega/\omega_n \ll 1$, the magnification factor $|X|/(F_0/k)$ is very nearly 1, approaching the static-loading condition when $\omega/\omega_n = 0$.
2. When ω/ω_n approaches 1, and the damping factor ζ is small, the magnification factor can become very large (see Fig. 3-7 also).
3. When $\omega/\omega_n > 3$, the magnification factor is considerably less than 1, and the system approaches a motionless state as shown in the figure.
4. The damping factor ζ has a negligible effect on the magnification factor when $\omega/\omega_n \ll 1$, or when $\omega/\omega_n > 3$, but has a very significant effect in the region of $\omega/\omega_n \cong 1$.

With the magnitude of the damping inherent in most physical systems generally less than 20 percent ($\zeta < 0.2$), the magnification factor is a maximum when the exciting frequency ω is approximately equal to the natural frequency ω_n of the system.[3] This condition

[3] Setting the derivative of the right side of Eq. 3-20 with respect to ω/ω_n equal to zero yields $\omega/\omega_n = \sqrt{1 - 2\zeta^2}$, which shows that the magnification factor is actually maximum just short of $\omega/\omega_n = 1$, depending upon the magnitude of ζ. However, resonance is generally assumed as occurring when $\omega/\omega_n = 1$ for the magnitudes of damping encountered in most physical systems. For example, when $\zeta = 0.15$, $\omega/\omega_n = 0.977$.

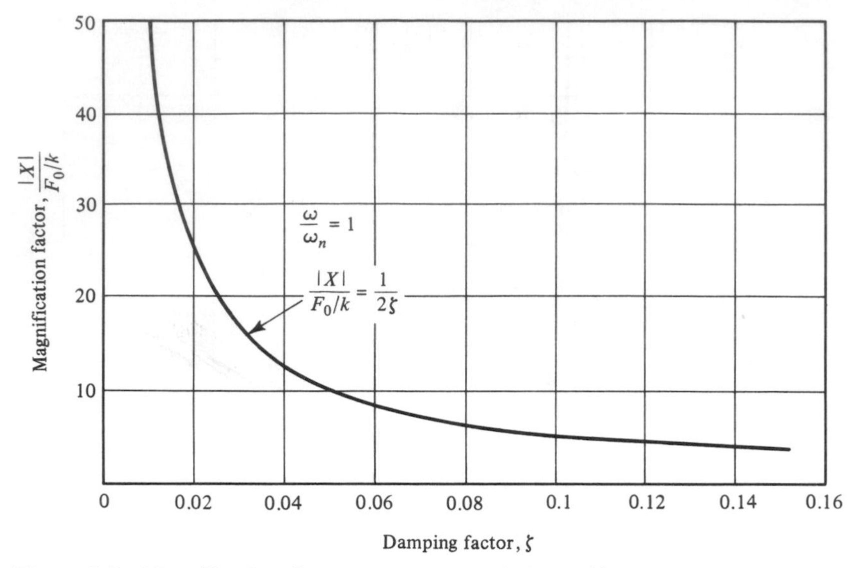

Figure 3-7 Magnification factors at resonance ($\omega/\omega_n = 1$).

is referred to as *resonance* and is obviously a condition that could be potentially dangerous to both machines and structures. When $\omega/\omega_n = 1$, Eq. 3-20 reduces to

$$\frac{|X|}{F_0/k} = \frac{1}{2\zeta} \tag{3-21}$$

which is the magnification factor at resonance.

The plot of Eq. 3-21 shown in Fig. 3-7 reveals that the magnification factor increases quite rapidly as the damping drops below 4 percent ($\zeta < 0.04$). A damping factor as low as 0.01 is not uncommon for steel and aluminum structural systems, and the substitution of $\zeta = 0.01$ into Eq. 3-21 produces a magnification factor of 50 at resonance for such systems. This means that the steady-state amplitude $|X|$ is 50 times greater than the static displacement F_0/k caused by applying F_0 statically.

If a system experiences very large amplitudes of vibration in the region of resonance, there are various means of reducing these amplitudes. High-damping materials or fluid dampers might be added to the system to increase the damping, or some type of vibration-absorbing device might be used. In many cases they can be reduced by simply changing the natural frequency of the system by changing its mass or stiffness. The response curve in Fig. 3-6 shows that when a system is vibrating near resonance, a relatively small change in ω/ω_n will move the system away from resonance and significantly reduce the amplitude of vibration. However, before any changes are made in the mass or stiffness, it must be known which side of the resonant peak the system is vibrating at to avoid changing ω_n in the wrong direction. For example, if the system were vibrating with a large amplitude at an ω/ω_n value greater than 1, a slight increase in the system's stiffness could make the situation worse, since this small increase would move the system even closer to resonance. The correct course in this instance would obviously be to either reduce the stiffness or increase the mass of the system, since either change would move the system away from resonance. The preceding discussion provides one reason that it is very important to be able to determine approximate natural frequencies of systems.

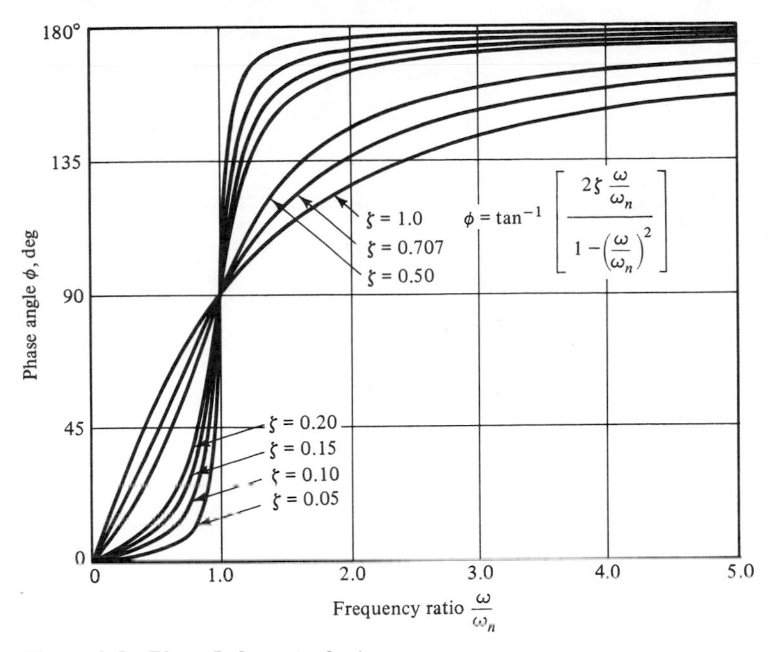

Figure 3-8 Plot of phase angle ϕ.

Equation 3-12 reveals that the phase angle ϕ is also a function of ζ and ω/ω_n, and Fig. 3-8 shows a family of curves obtained from plotting Eq. 3-12. Referring to this figure and Eq. 3-17, we can observe some interesting steady-state vibration characteristics associated with the phase angle ϕ:

1. When $\omega/\omega_n \ll 1$, the phase angle is quite small, which means that the exciting force $F_0 \sin \omega t$ is nearly in phase with the displacement since $x = |X|\sin(\omega t - \phi)$.
2. When $\omega/\omega_n = 1$, the phase angle $\phi = 90°$, and the exciting force $F_0 \sin \omega t$ is in phase with the velocity since

$$\dot{x} = |X|\omega \cos\left(\omega t - \frac{\pi}{2}\right) = |X|\omega \sin \omega t$$

3. When $\omega/\omega_n \gg 1$, the phase angle approaches 180°, and the sense of the exciting force $F_0 \sin \omega t$ is essentially opposite to that of the displacement since

$$x \cong |X|\sin(\omega t - \pi) = -|X|\sin \omega t$$

4. In the hypothetical case of zero damping ($\zeta = 0$), the phase angle is zero when $\omega/\omega_n < 1$, and 180° when $\omega/\omega_n > 1$.

EXAMPLE 3-1

The frame shown in Fig. 3-9 consists of a very stiff steel beam with a wide-flange section (W33 × 200), welded rigidly to two vertical channels (C8 × 11.5). The area moment of inertia I_c of each channel about its centroidal bending axis is 1.3 in.^{4} An eccentric exciter

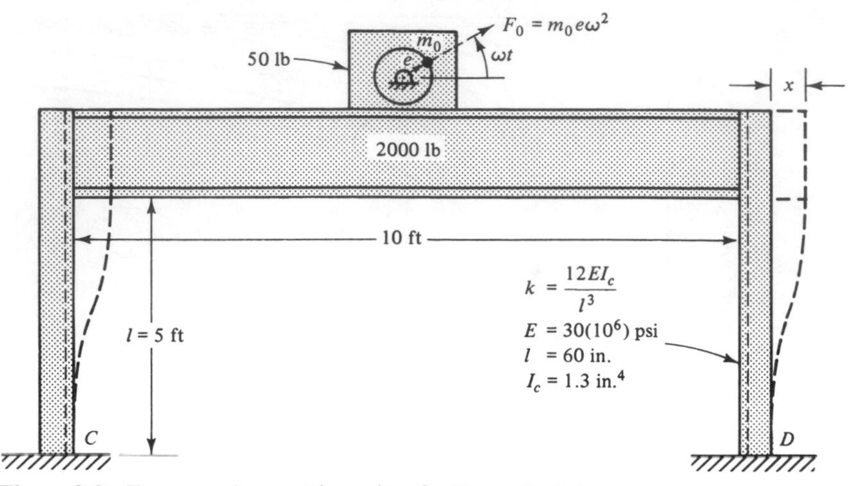

Figure 3-9 Frame and eccentric exciter for Example 3-1.

weighing 50 lb is attached to the beam, which weighs 2000 lb, and is used to excite the frame. The unbalanced weight of the exciter is 5 lb, and it has an eccentricity e of 2 in.

An LVDT (linear variable differential transformer)[4] is used to measure the lateral displacement of the frame by mounting the core of the LVDT horizontally at one of the upper corners of the frame and mounting the field windings, in which the core moves, to a rigid external body. By varying the rotational speed of the exciter until resonance occurs, the *maximum* horizontal amplitude of vibration of the frame can be found.

The maximum amplitude of the system was found to be $|X| = 0.15$ in. Assuming no bending in the beam, and considering the channels to be completely fixed at C and D, determine (a) the natural frequency f_n of the lateral vibration of the frame and exciter as a system; (b) the damping factor ζ of the system; (c) the magnification factor (M.F.) at resonance.

Solution. **a.** Since the beam has much more resistance to bending than the channels because of the depth of its section, it is reasonable to assume that the upper ends of the columns move horizontally with the beam without rotating as indicated in Fig. 3-9. With this assumption, the spring constant of each column is determined from part h of Table 2-1 on page 79. Since the columns act as parallel springs, the *equivalent* spring constant of the columns is

$$k_e = \frac{2(12EI_c)}{l^3} = \frac{2(12)30(10)^6(1.3)}{(60)^3} = 4333.3 \text{ lb/in.}$$

Neglecting the mass of the columns, the mass of the system is

$$m = \frac{2050}{386} = 5.31 \text{ lb} \cdot \text{s}^2/\text{in.}$$

[4] An LVDT is a variable inductance transducer that provides an output voltage that is linearly proportional to the displacement of its core relative to the transformer windings. A voltage to the primary winding provides the input to the transducer.

The natural circular frequency of the frame-and-exciter system is

$$\omega_n = \sqrt{\frac{k_e}{m}} = \sqrt{\frac{4333.33}{5.31}} = 28.57 \text{ rad/s}$$

so that

$$f_n = \frac{\omega_n}{2\pi} = 4.55 \text{ Hz}$$

b. At resonance, the exciting circular frequency ω equals the natural circular frequency ω_n determined in part a; so the exciting force can be computed as

$$F_0 = m_0 e \omega_n^2 = \frac{5}{386}(2)(28.57)^2 = 21.15 \text{ lb}$$

With $\omega/\omega_n = 1$,

$$\frac{|X|}{F_0/k} = \frac{1}{2\zeta}$$

and the damping factor is determined as

$$\zeta = \frac{21.15}{2(4333.3)(0.15)} = 0.016 = 1.6 \text{ percent}$$

c. The magnification factor at resonance is calculated as

$$\text{M.F.} = \frac{1}{2\zeta} = 31.25$$

EXAMPLE 3-2

A small wind turbine such as the one shown in Fig. 3-10 is supported by a cantilevered steel pole having a *torsional* stiffness of k (N · m/rad) and a torsional damping coefficient of c (N · m · s/rad). The center of gravity of the four-rotor-blade assembly of total mass m_0 has been found to be displaced a distance e from the axis of rotation of the assembly. The mass moment of inertia about the z axis of the complete turbine, including rotor assembly, housing pod, and contents, is I_z (kg · m^2). The total mass of the system is m (kg). The plane in which the blades rotate is located a distance d (meters) from the z axis as shown. As a part of a complete analysis of the vibration characteristics of the turbine system, we wish to determine: (a) the differential equation of motion of the torsional vibration of the system about the z axis; (b) the steady-state torsional response of the system using complex algebra. We shall neglect any effect of the mass and bending of the pole on the torsional response. We shall also neglect any gyroscopic effects of the rotating blades.

Solution. **a.** Figure 3-10 shows that the moment of the unbalanced rotating mass m_0 of the rotor blades about the z axis is

$$M(t) = (m_0 e \omega^2 \cos \omega t)d$$

or

$$M(t) = M_0 \cos \omega t$$

in which $M_0 = m_0 e \omega^2 d$. The pole also exerts a resisting torque of $k\theta$, and a torsional

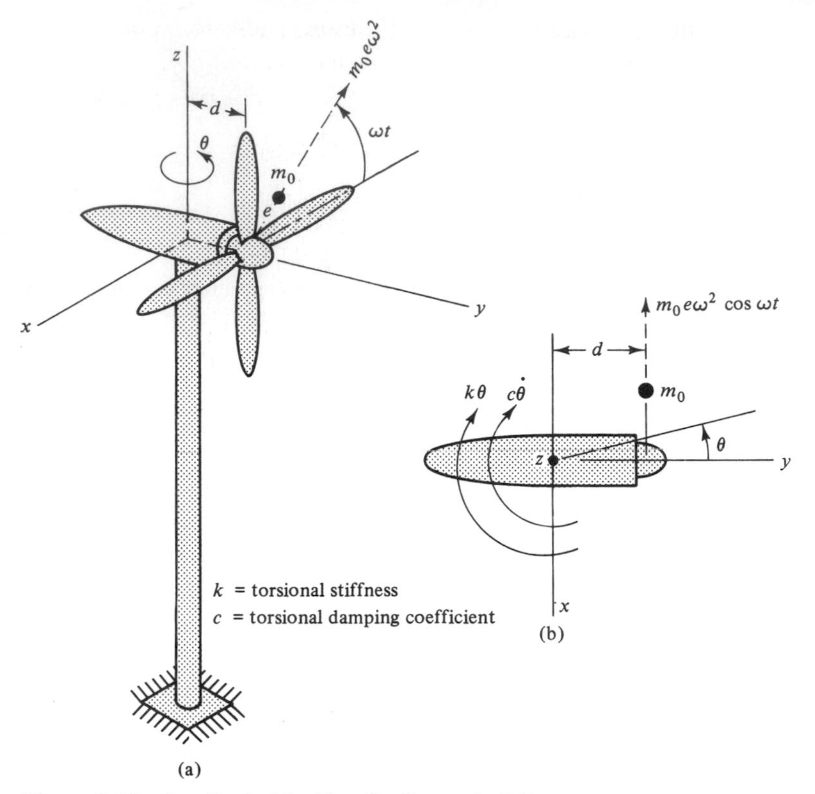

Figure 3-10 Small wind turbine for Example 3-2.

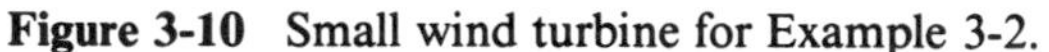

damping moment of $c\dot{\theta}$, as shown in Fig. 3-10b. Considering the moments about the z axis, we can write that

$$\sum M_z = I_z\ddot{\theta}$$

or that

$$-k\theta - c\dot{\theta} + M_0 \cos \omega t = I_z\ddot{\theta}$$

Dividing this expression by I_z, and collecting terms containing the dependent variable θ, yields

$$\ddot{\theta} + \frac{c\dot{\theta}}{I_z} + \frac{k\theta}{I_z} = \frac{M_0}{I_z} \cos \omega t \tag{3-22}$$

Letting $c/I_z = 2\zeta\omega_n$, and $k/I_z = \omega_n^2$, Eq. 3-22 can be written in the form of

$$\ddot{\theta} + 2\zeta\omega_n\dot{\theta} + \omega_n^2\theta = \frac{M_0}{I_z} \cos \omega t \tag{3-23}$$

where ζ = torsional-damping factor (dimensionless)
$\omega_n = \sqrt{k/I_z}$ = undamped natural circular torsional frequency (rad/s)
I_z = mass moment of inertia of wind-turbine system about the z axis ($\text{kg} \cdot \text{m}^2$)

b. Using complex algebra to obtain the steady-state solution of Eq. 3-23, cos ωt is replaced by the real part of $e^{j\omega t}$, and we assume a solution of the form

$$\theta = \Theta e^{j\omega t} \tag{3-24}$$

understanding that only the real part of $e^{j\omega t}$ is to be used in determining the complex constant Θ.

Substituting Eq. 3-24 and its derivatives into Eq. 3-23, and replacing cos ωt by $e^{j\omega t}$, yields

$$\left[1 - \left(\frac{\omega}{\omega_n}\right)^2 + j2\zeta\frac{\omega}{\omega_n}\right]\Theta = \frac{M_0}{I_z\omega_n^2} = \frac{M_0}{k}$$

or (see Fig. 3-3a)

$$\Theta = \frac{(M_0/k)e^{-j\phi}}{\sqrt{[1-(\omega/\omega_n)^2]^2 + [2\zeta(\omega/\omega_n)]^2}} = |\Theta|e^{-j\phi}$$

in which ϕ is the phase angle between the exciting moment of amplitude M_0 and the angular amplitude $|\Theta|$. This angle is given by

$$\tan\phi = \frac{2\zeta(\omega/\omega_n)}{1-(\omega/\omega_n)^2}$$

Thus, the complex form of the steady-state solution is

$$\theta = |\Theta|e^{j(\omega t - \phi)}$$

Taking the real part of $e^{j(\omega t-\phi)}$, the steady-state solution is

$$\theta = |\Theta|\cos(\omega t - \phi)$$

in which

$$|\Theta| = \frac{M_0/k}{\sqrt{[1-(\omega/\omega_n)^2]^2 + [2\zeta(\omega/\omega_n)]^2}}$$

It should be noted that if the above equation were rearranged algebraically in terms of the ratio $|\Theta|/(M_0/k)$, it would be identical in form to Eq. 3-20 and would yield the same curves as those shown in Fig. 3-6.

3-3 CRITICAL SPEED OF ROTATING SHAFTS

When a shaft that is rotating about its longitudinal axis bends about that axis (line AB in Fig. 3-11a), the bent shaft will *whirl* about its original axis of rotation as well as continuing to rotate about its longitudinal axis. Unbalanced disks, loose or worn bearings, and gyroscopic effects are examples of things that can cause shafts to whirl.

Let us consider the steady-state motion of the shaft shown in Fig. 3-11, which is whirling because the disk attached to it is not balanced. Because of imperfections in the disk, its mass center G has an eccentricity of e from the disk's geometric center C. Assuming that the center C of the disk is located at the origin O of the xyz axes when the shaft is aligned with line AB between the bearings, the lateral deflection of the shaft at the location of the disk is $r = OC$. The bent shaft and line AB form a rotating plane that makes an angle θ with the x axis as shown in Fig. 3-11b.

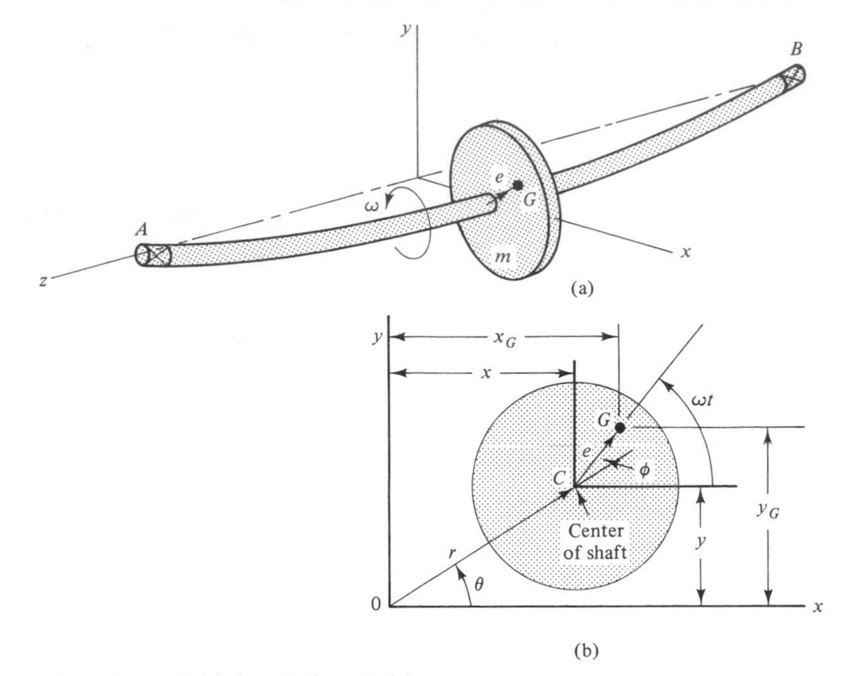

Figure 3-11 Whirling shaft and disk.

The angular velocity of the shaft-and-disk system with respect to the longitudinal axis of the shaft is ω, while the angular velocity of the rotating plane formed by the bent shaft and line AB is $\dot{\theta}$. The angle ϕ is the angle between the position vector $\mathbf{r}$ and the position vector $\mathbf{e}$ (e is the magnitude of the eccentricity). As we shall find subsequently, the shaft deflection r can become appreciable when the angular velocity ω becomes equal to the natural circular frequency ω_n of the lateral vibration of the system, that is, when the *critical* speed ω is such that $\omega/\omega_n = 1$. We shall also find that the magnitude of $\mathbf{r}$ is constant and that $\dot{\theta} = \omega$. This type of whirling motion is referred to as *synchronous whirl.*

We shall assume that the weight of the shaft is small compared with that of the attached disk. From Fig. 3-11b it can be seen that the coordinates of the mass center G are

$$\left.\begin{aligned} x_G &= x + e \cos \omega t \\ y_G &= y + e \sin \omega t \end{aligned}\right\} \tag{3-25}$$

in which x and y are the coordinates of the center C of the disk with respect to the axes xyz as shown in the figure.

Taking successive derivatives of Eq. 3-25, the acceleration components of the mass center of the disk are

$$\left.\begin{aligned} \ddot{x}_G &= \ddot{x} - e\omega^2 \cos \omega t \\ \ddot{y}_G &= \ddot{y} - e\omega^2 \sin \omega t \end{aligned}\right\} \tag{3-26}$$

For a circular shaft it is logical to assume that the stiffness and the damping of the shaft are the same regardless of the orientation of the shaft; that is, $k = k_x = k_y$ and $c =$

$c_x = c_y$. With this assumption, and using the acceleration components of Eq. 3-26, we can apply Newton's second law to obtain the equations

$$-kx - c\dot{x} = m(\ddot{x} - e\omega^2 \cos \omega t)$$

and

$$-ky - c\dot{y} = m(\ddot{y} - e\omega^2 \sin \omega t)$$

which can be reduced to the familiar forms of

$$\left.\begin{aligned} \ddot{x} + 2\zeta\omega_n\dot{x} + \omega_n^2 x &= e\omega^2 \cos \omega t \\ \ddot{y} + 2\zeta\omega_n\dot{y} + \omega_n^2 y &= e\omega^2 \sin \omega t \end{aligned}\right\} \tag{3-27}$$

in which

$$2\zeta\omega_n = \frac{c}{m} \qquad \text{and} \qquad \omega_n = \sqrt{\frac{k}{m}}$$

Assuming steady-state solutions of Eq. 3-27 in the complex form of

$$x = Xe^{j\omega t} \qquad \text{and} \qquad y = Ye^{j\omega t}$$

we can apply the procedure discussed in Sec. 3-2 to obtain

$$X = Y = \frac{e(\omega/\omega_n)^2 e^{-j\phi}}{\sqrt{[1 - (\omega/\omega_n)^2]^2 + [2\zeta(\omega/\omega_n)]^2}} \tag{3-28}$$

in which the amplitudes are

$$|X| = |Y| = \frac{e(\omega/\omega_n)^2}{\sqrt{[1 - (\omega/\omega_n)^2]^2 + [2\zeta(\omega/\omega_n)]^2}} \tag{3-29}$$

and the angle ϕ between the position vectors **r** and **e** can be determined from

$$\tan \phi = \frac{2\zeta(\omega/\omega_n)}{1 - (\omega/\omega_n)^2} \tag{3-30}$$

The rectangular displacement components of the disk center C are then

$$\left.\begin{aligned} x &= |X|\cos(\omega t - \phi) \\ y &= |Y|\sin(\omega t - \phi) \end{aligned}\right\} \tag{3-31}$$

Squaring the above components, we obtain

$$r = \sqrt{x^2 + y^2} = \sqrt{|X|^2\cos^2(\omega t - \phi) + |Y|^2\sin^2(\omega t - \phi)}$$

which, since $|X| = |Y|$, shows that

$$r = |X| = |Y| \qquad \text{(constant)} \tag{3-32}$$

Using the expressions for x and y in Eq. 3-31, and referring to Fig. 3-11b, we see that

$$\tan \theta = \frac{y}{x} = \frac{|Y|\sin(\omega t - \phi)}{|X|\cos(\omega t - \phi)} = \tan(\omega t - \phi)$$

which shows that

$$\theta = \omega t - \phi$$

and

$$\dot{\theta} = \omega \tag{3-33}$$

The above relationships show that the plane formed by the bent shaft and line AB whirls about line AB with an angular velocity $\dot{\theta}$ that is equal to the angular velocity ω of the shaft-and-disk system.

Figure 3-12 shows a plot of the dimensionless ratio $|X|/e$ versus the frequency ratio ω/ω_n for various magnitudes of damping, as obtained from Eq. 3-29. There are several ω/ω_n values that are of interest, as can be seen by observing the curves in Fig. 3-12, namely, $\omega/\omega_n \ll 1$, $\omega/\omega_n = 1$, and $\omega/\omega_n \gg 1$.

When $\omega/\omega_n \ll 1$, Fig. 3-12 shows that $|X|/e \cong 0$. Equation 3-30 shows that the phase angle $\phi \cong 0$, which means that the center of gravity G of the disk rotates on the *outside* of the path of the center C of the disk, with a radius of rotation of $OG \cong r + e$ as shown in Fig. 3-13a.

When $\omega/\omega_n = 1$ (the critical speed), Eqs. 3-29 and 3-32 show that

$$r = |X| = \frac{e}{2\zeta}$$

and Eq. 3-30 gives the phase angle as

$$\phi = 90°$$

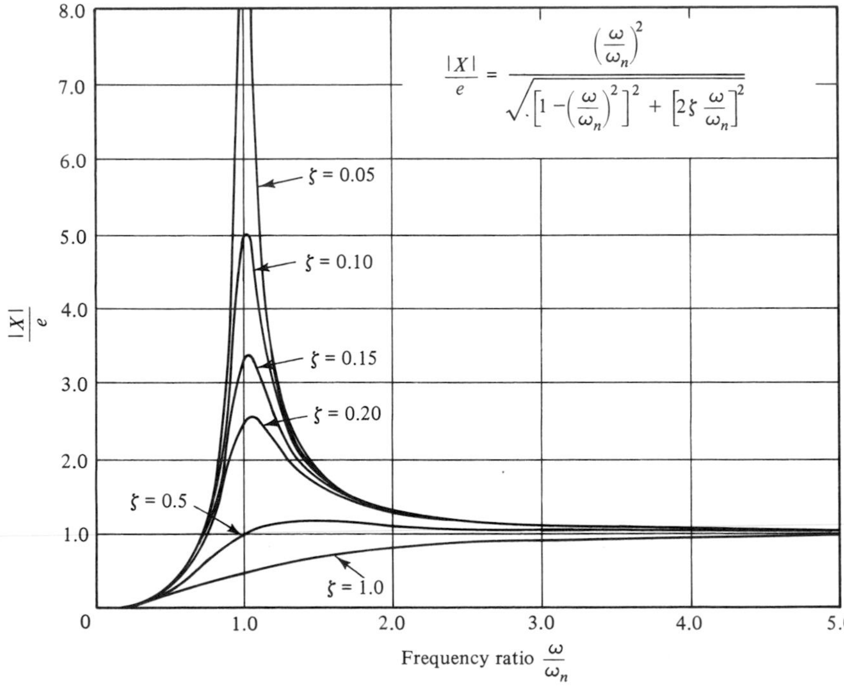

Figure 3-12 Plot of $|X|/e$ from Eq. 3-29.

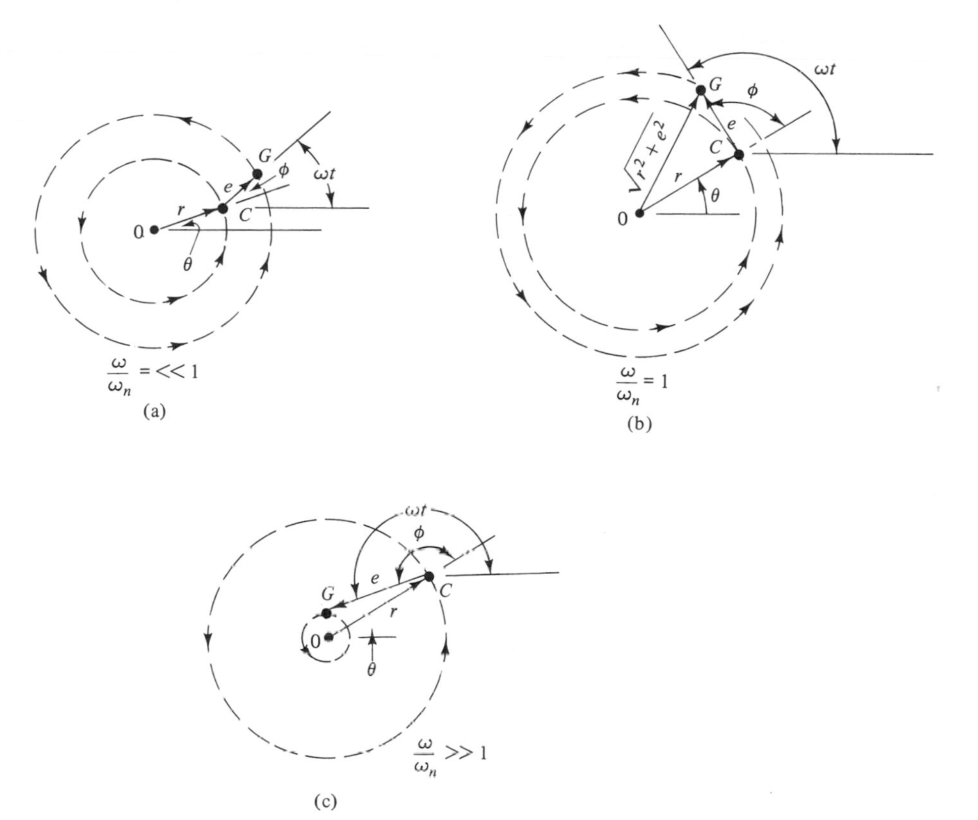

Figure 3-13 General whirling configurations for $\omega/\omega_n \ll 1$, $\omega/\omega_n = 1$, and $\omega/\omega_n \gg 1$.

The first of the above expressions indicates that the amplitude of the whirling motion can be quite large if the damping factor is small. Since $\phi = \pi/2$, the center of gravity G of the disk rotates in a circular path of radius $OG = \sqrt{r^2 + e^2}$ as shown in Fig. 3-13b.

When $\omega/\omega_n \gg 1$, Fig. 3-12 shows that

$$\frac{|X|}{e} \cong 1$$

Since the damping factor for a shaft-and-disk system is generally less than 10 percent ($\zeta < 0.1$), Eq. 3-30 shows that

$$\phi \cong 180^\circ$$

Thus, in this case, the center of gravity G of the disk rotates on the *inside* of the path of C with a radius of $OG \cong r - e$ as shown in Fig. 3-13c.

The effective spring constant k of the shaft appearing in the differential equations of motion depends not only upon the size of the shaft but also upon the degree of bending restraint provided by the bearings. For example, if the bearings are attached to a rigid support that prevents rotation of the bearings about any axis perpendicular to line AB, the k used would be the k for a fixed-fixed beam. However, if the bearings were free to rotate (were self-aligning), the k used would be that for a pinned-pinned beam. If several disks are attached to a rotating shaft, the natural frequencies (critical speeds) can be determined by procedures discussed in Chap. 5.

In starting and stopping rotating machines such as turbines that may operate at speeds above their natural frequencies, large amplitudes of vibration can build up as the machine passes through the critical speed ($\omega = \omega_n$). These can be minimized by passing through the critical speed as quickly as possible since the amplitude buildup does occur over some finite length of time. However, the problem can be avoided if it is possible to design the system with a natural frequency ω_n that is considerably above the operating frequency ω (see Fig. 3-12) when $\omega/\omega_n \ll 1$.

EXAMPLE 3-3

The shaft-and-disk system shown in Fig. 3-14 is supported by self-aligning bearings so the steel shaft can be considered as a simply supported beam for purposes of choosing a spring constant k. The rotating disk is fixed to the shaft midway between bearings A and B. The data for the system are as follows:

m = 12 kg (0.822 lb·s²/ft) (mass of disk)
l = 0.5 m (19.69 in.) (length of shaft)
d = 25.4 mm (1 in.) (diameter of shaft)
ρ = 7843 kg/m³ (15.22 lb·s²/ft⁴) (mass density of shaft)
E = 206.8 GPa [30(10)⁶ psi] (modulus of elasticity of shaft)

Since the system is to operate over a range of speeds varying from 2400 to 3600 rpm, there is some concern that the bearing forces R_A and R_B could become quite large at the critical speed if some imbalance exists in the rotating disk. Experience indicates that with precision machining, and using a homogeneous material, the eccentricity e of the disk can be kept to within 0.05 mm (0.002 in.) or less.

As part of the overall analysis, it is desired to determine (a) the critical speed ω_n of the disk-and-shaft system including the distributed mass of the shaft; (b) the maximum bearing forces R_A and R_B that could be anticipated in operating over a range of 2400 to 3600 rpm, with e = 0.05 mm and 2 percent damping assumed ($\zeta = 0.02$).

Solution. **a.** Since the distributed mass of the shaft is part of the system, we determine the natural circular frequency of the system (its critical speed) using Rayleigh's energy method, which is discussed in Sec. 2-9.

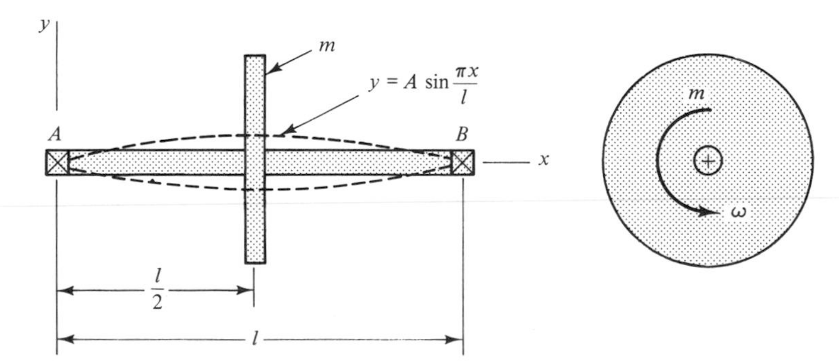

Figure 3-14 Disk-shaft system for Example 3-3.

Considering the shaft as a simply supported beam, the shape function

$$y = A \sin \frac{\pi x}{l} \tag{3-34}$$

satisfies the conditions of zero deflections and zero moments at the ends of the shaft and provides the slopes at each end that are consistent with the self-aligning features of the bearings. With the disk located at the center of the shaft, parameter A refers to both the displacement of the center of the shaft and the displacement of the disk.

The *maximum* kinetic energy of the shaft and disk is given by

$$T_{\max} = \frac{\gamma}{2} \int_0^l (y\omega_n)^2 \, dx + \frac{m}{2} (A\omega_n)^2$$

Substituting the shape function,

$$T_{\max} = \frac{\gamma A^2 \omega_n^2}{2} \int_0^l \sin^2 \frac{\pi x}{l} \, dx + \frac{m A^2 \omega_n^2}{2}$$

Upon integrating and substituting limits, we obtain

$$T_{\max} = \frac{A^2 \omega_n^2}{2} \left(\frac{\gamma l}{2} + m \right) \tag{3-35}$$

where γ = mass per unit length of shaft (kg/m)
$\gamma l = \rho(\pi d^2/4)l$ = mass of shaft (kg)
m = mass of disk (kg)

Using the SI system of units, the mass of the shaft is

$$\gamma l = 7843 \frac{\pi}{4} (0.0254)^2 (0.5) = 1.987 \text{ kg}$$

Substituting the value determined for γl and the value given for m into Eq. 3-35 gives the maximum kinetic energy as

$$T_{\max} = \frac{A^2 \omega_n^2}{2} (12.99) \tag{3-36}$$

Since y is measured from the static-equilibrium position of the system, the maximum *change* in the potential energy of the system from its potential energy in the static-equilibrium position is given by Eq. 2-70 as

$$U_{\max} = \frac{EI}{2} \int_0^l \left(\frac{d^2 y}{dx^2} \right)^2 dx$$

However, since the inertia effect due to the eccentricity of the disk's mass center has the same effect as a concentrated load at the center of the shaft, the maximum strain energy can be determined more simply and quite accurately using

$$U_{\max} = \tfrac{1}{2} k A^2 \tag{3-37}$$

in which k is the spring constant of a simply supported beam loaded at its center.

Referring to Table 2-1, the spring constant is

$$k = \frac{48EI}{l^3}$$

where $I = \dfrac{\pi d^4}{64} = \dfrac{\pi(0.0254)^4}{64} = 2.043(10)^{-8}\ \text{m}^4$

$E = 206.8(10)^9$ Pa

$l = 0.5$ m

Substituting these values into Eq. 3-37, we find that

$$U_{\text{max}} = \frac{A^2}{2}(1.622)(10)^6 \tag{3-38}$$

Substituting Eqs. 3-36 and 3-38 into $T_{\text{max}} = U_{\text{max}}$, and solving for ω_n, we find that

$$\omega_n = 353.4 \text{ rad/s}$$

so that

$$N = 3374 \text{ rpm}$$

which is the critical speed of the system.

b. Since the critical speed just determined is within the operating range of 2400 to 3600 rpm, the maximum displacement of the whirling shaft at the location of the disk will occur when $\omega/\omega_n = 1$. Thus, using Eq. 3-29,

$$r = |X| = \frac{e}{2\zeta} = \frac{0.05}{2(0.02)} = 1.25 \text{ mm}$$

Considering the shaft as a spring of stiffness k and deflection r as shown in Fig. 3-15, the *dynamic* bearing reactions at resonance are

$$R_A = R_B = \frac{kr}{2} = \frac{1.622(10)^6(0.00125)}{2} = 1013.8 \text{ N} = 227.9 \text{ lb}$$

Thus, it is apparent that a very small eccentricity, such as $e = 0.05$ mm in this example, can cause appreciable bearing forces.

It should be noted that while the dynamic bearing reactions just determined are constant in all directions at a particular whirling speed, the *total reactions,* which include the unidirectional static reactions due to the dead weight of the shaft and disk, vary with time in both magnitude and direction.

The ratio of the dynamic reactions to the static reactions can be determined as

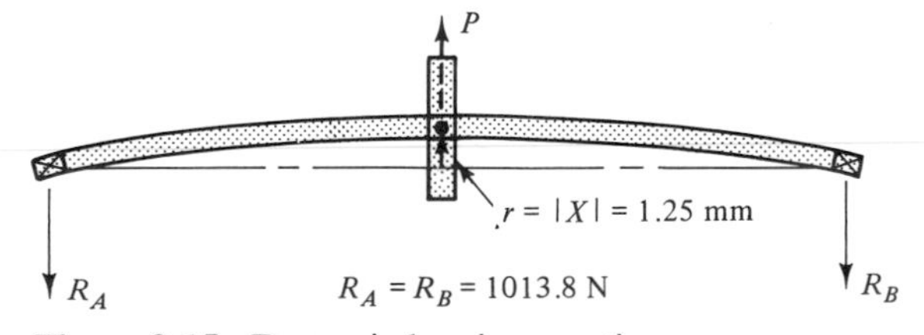

Figure 3-15 Dynamic bearing reactions.

$$\frac{R_A}{W_A} = \frac{1013.8}{[(12 + 1.99)/2]9.81} = 14.77$$

in which W_A is one-half the weight of the disk and shaft (the static reaction at A). This shows that the bearing force due to the unbalance of the disk as the system whirls is almost 15 times the force the bearing must carry in just supporting the weight of the shaft and disk in the static-equilibrium position.

3-4 VORTEX SHEDDING (FLOW-INDUCED VIBRATION)

Vibration that is caused by fluid flowing around or through objects is referred to as *flow-induced* vibration. Nuclear fuel rods, submarine periscopes, legs of offshore drilling platforms, transmission lines, and tall smokestacks are but a few examples of systems subjected to flow-induced vibration due to fluids flowing *around* structures. The catastrophic failure of the Tacoma Narrows bridge in 1940 is a well-known example of the results of wind-induced oscillations. Vibration caused by fluid flow *through* structures has been observed in oil pipelines, tubing in small pumping systems, air compressors, and so on.

There are various types of flow-induced vibration phenomena. The large, low-frequency (1- to 2-Hz) oscillation of ice-covered transmission lines, called *galloping,* occurs because of the lift and drag forces caused by air flowing over the ice formations on the lines. The "singing" of transmission lines and TV antennas at fairly high frequencies is caused by vortex shedding and is quite different from the phenomenon of galloping. The higher frequency in these instances is caused by the excitation of the higher harmonics of the system by the vortex shedding (see the discussion of the one-dimensional wave equation in Chap. 8).

As one might gather from the preceding discussion, the subject of flow-induced vibration is a rather broad one, and a comprehensive coverage of this important subject exceeds the intended scope of this text. Our discussion will be limited to some of the more important aspects of vibration caused by vortex shedding, since it is one of the most common sources of flow-induced vibrations. For a comprehensive coverage of flow-induced vibration, the reader is referred to an excellent book by R. D. Blevins.[5]

When fluid flows past a right-circular cylinder with sufficient velocity, vortices are formed in the wake of the fluid as shown in Fig. 3-16. Such vortices are frequently referred to as Kármán vortices, and shed in a regular pattern over a wide range of Reynolds numbers. The Reynolds number is given by

$$R = \frac{vd\rho}{\mu}$$

in which v is the fluid velocity, d is the diameter of the cylinder, ρ is the mass density of the fluid, and μ is the absolute viscosity of the fluid.

The vortices shed alternately from opposite sides of the cylinder with a frequency f. This causes an alternating pressure on each side of the cylinder, which acts as a sinusoidally varying force F *perpendicular* to the velocity of the fluid before its flow is disturbed as shown in Fig. 3-16. This force is given by

[5] R. D. Blevins, *Flow-Induced Vibration,* Van Nostrand Reinhold Co., New York, 1977.

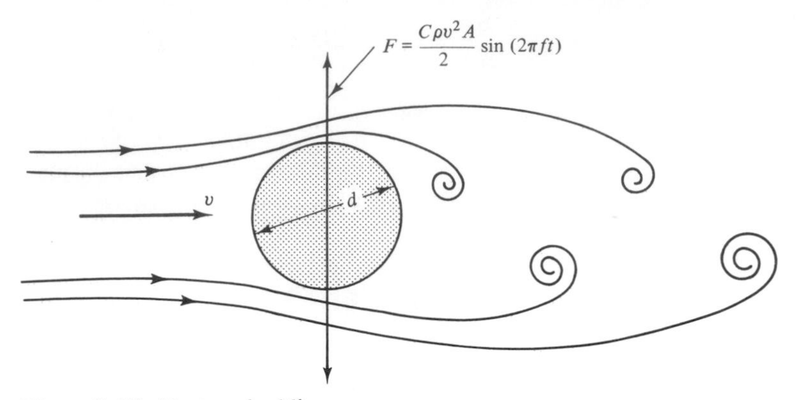

Figure 3-16 Vortex shedding.

$$F = \frac{C\rho v^2 A}{2} \sin(2\pi f t) \tag{3-39}$$

where C = drag coefficient (dimensionless) ($C \cong 1$ for a cylinder)
v = fluid velocity (ft/s, m/s)
A = projected area of cylinder perpendicular to v (ft^2, m^2)
ρ = mass density of fluid ($\text{lb} \cdot \text{s}^2/\text{ft}^4$, kg/m^3)

Experimental studies have confirmed that the frequency of the vortex shedding described above can be expressed as

$$f = \frac{Sv}{d} \tag{3-40}$$

where S = *Strouhal* number (dimensionless)
d = diameter of cylinder (ft, m)
v = velocity of fluid (ft/s, m/s)

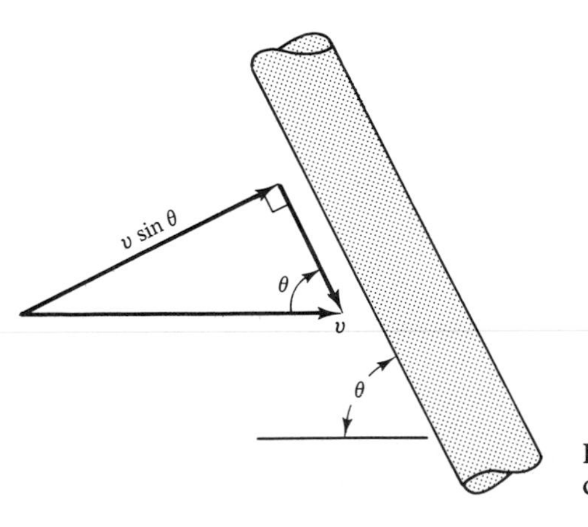

Figure 3-17 Velocity component perpendicular to structure.

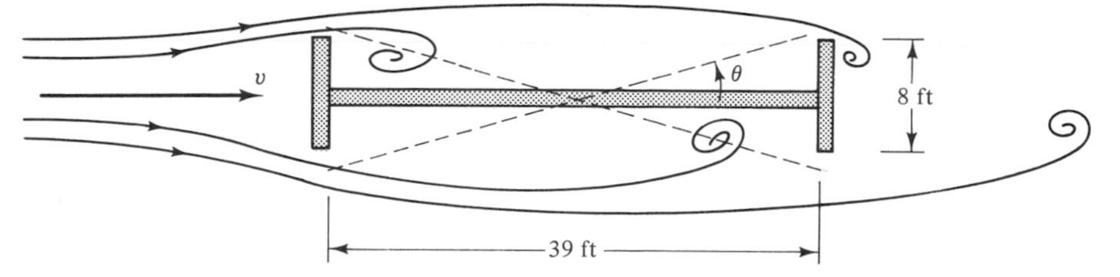

Figure 3-18 Bridge deck of Tacoma Narrows bridge.

Experimental studies have shown that the Strouhal number is very nearly constant over a wide range of Reynolds numbers, from $4(10)^2$ to $3(10)^5$. For this range a value of $S = 0.2$ is generally used for cylindrically shaped objects. When the vortex-shedding frequency approaches the natural frequency of the system being excited, large amplitudes of vibration can accompany this resonant condition.

If the flow velocity is not perpendicular to the structure interrupting the flow but has an angle of inclination θ as shown in Fig. 3-17, the vortex-shedding frequency becomes

$$f = \frac{Sv \sin \theta}{d} \tag{3-41}$$

in which $v \sin \theta$ is the component of velocity perpendicular to the structure.

The phenomenon of vortex shedding is not limited to cylindrically shaped bodies but occurs for all kinds of shapes, and Eqs. 3-39 and 3-40 are applicable for many different shapes when appropriate Strouhal numbers and drag coefficients are known for them. The Strouhal number for a variety of shapes can be found in the literature,[6] but drag-coefficient values are rather limited for shapes other than cylindrical or spherical. For cylindrically shaped bodies $C \cong 1.0$ for a large range of Reynolds numbers varying from $4(10)^2$ to $3(10)^5$. For spherically shaped bodies $C \cong 0.4$ over a range of Reynolds numbers from $(10)^3$ to $2(10)^5$.

Experimental studies at the University of Washington determined that the destruction of the Tacoma Narrows bridge in 1940 was caused by vortex shedding.[7] A steady wind of approximately 40 mph resulted in vortex shedding of a frequency coinciding with the natural *torsional* frequency of the H-shaped bridge deck (see Fig. 3-18). This resonant condition caused torsional displacements (angles of twist θ) along the bridge that reached amplitudes of approximately 45°. The bridge replacing the destroyed one was built with open side trusses between the bridge deck and the open bottom truss as shown in Fig. 3-19. This rectangular section is many times stiffer in torsion than the H-shaped section of the bridge that collapsed. Because it is stiffer, the natural frequency of the replacement bridge is considerably above any vortex-shedding frequency that might be encountered. In addition, the open side trusses and open slots in the bridge deck permit the passage of air through the structure, which minimizes the formation of unequal pressures on the top and bottom of the structure that develop from vortex shedding.

[6] Task Committee on Wind Forces, "Wind Forces on Structures," *Transactions of the American Society of Civil Engineers,* vol. 126, pp. 1124–1198, 1961. R. D. Blevins, *Flow-Induced Vibration,* Van Nostrand Reinhold Co., New York, 1977.

[7] University of Washington, *Engineering Experiment Station Bulletin* 116. J. P. Den Hartog, *Mechanical Vibrations,* 4th ed., pp. 305–308, McGraw-Hill Book Co., New York, 1956.

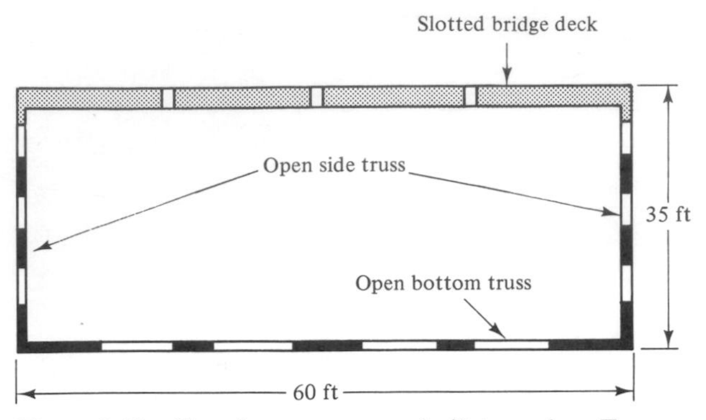

Figure 3-19 Closed-type structure built to replace Tacoma Narrows bridge destroyed in 1940.

Reducing Flow-Induced Vibration Caused by Vortex Shedding

The preceding paragraphs have been concerned with the vibrations induced by vortex shedding. Let us now discuss a few methods that have proved effective in reducing flow-induced vibration caused by vortex shedding.

One approach is to increase the damping in the system by introducing additional friction in the form of fluid dampers, or by cladding the structure with high-damping materials such as elastomers (rubberlike materials) and cork. The vibration of transmission lines has been effectively reduced by using a Stockbridge damper like that shown in Fig. 3-20. This damper consists of two masses attached to a short piece of stranded steel cable. The damper is in effect a spring-and-mass system that is "tuned" to the flow-induced vibration frequency by adjusting the length of the stranded cable and/or the weight of the attached masses. When the damper is attached to a transmission line at a point where the amplitude of vibration of the line is appreciable, the vibration of the damper dissipates energy because of the friction between the cable strands. In addition to dissipating energy, the spring-and-mass system acts as a vibration absorber (see Sec. 6-4). The latter phenomenon is very effective in reducing flow-induced vibration when properly tuned to the vortex-shedding frequency.

Another method is to change the surface of the system to break up the formation of the vortices. One such scheme that has proved effective for cylindrically shaped struc-

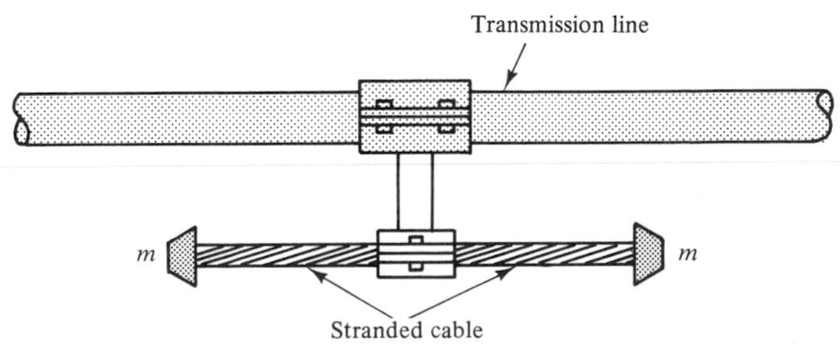

Figure 3-20 Stockbridge damper.

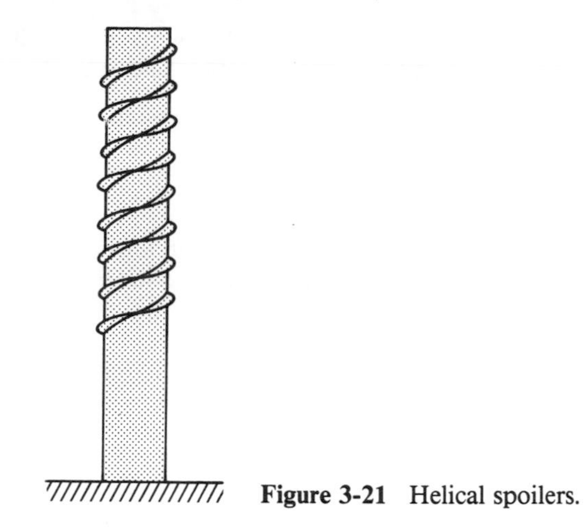

Figure 3-21 Helical spoilers.

tures such as industrial smokestacks is the use of *helical spoilers* such as those shown in Fig. 3-21.

Still another approach involves changing the natural frequency of the system away from the vortex-shedding frequency, thus avoiding resonance. As discussed previously, this was one of the modifications made in the design of the bridge that replaced the ill-fated Tacoma Narrows bridge.

EXAMPLE 3-4

The telescoping car antenna shown in Fig. 3-22 has a fundamental (lowest) natural frequency of 5.4 Hz when extended to a length of 2 ft. An analysis is required to investigate the frequencies of the vortex shedding around the antenna over a range of speeds from 5 to 65 mph to determine if a resonant condition might be encountered within that range. It is also desired to determine the Reynolds number for both 5 and 65 mph to see if they are within the Reynolds-number range of $400 \le R \le 3(10)^5$ for which the Strouhal number is 0.2.

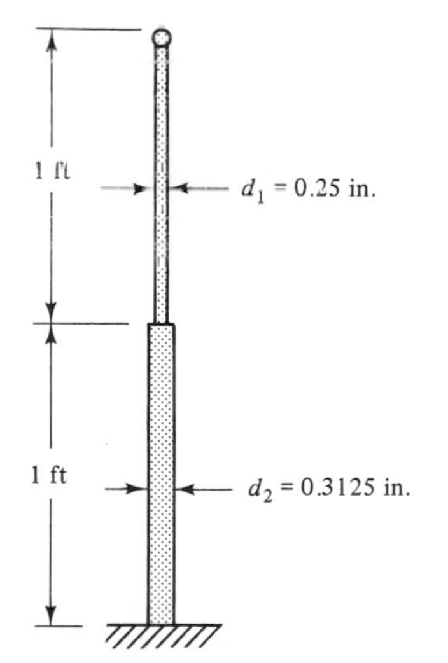

Figure 3-22 Car radio antenna.

The mass density of air is $\rho = 2.37(10)^{-3}$ slug/ft^3, while its absolute viscosity is $\mu = 3.8(10)^{-7}$ lb · s/ft^2. The two sections of the antenna have the diameters shown in Fig. 3-22.

Solution. The wind velocities relative to the antenna at 5 and 65 mph are

$$v_1 = 7.33 \text{ ft/s}$$
$$v_2 = 95.3 \text{ ft/s}$$

Assuming that the Reynolds number is in the region for which the Strouhal number is 0.2, we first use Eq. 3-40 to determine the vortex-shedding frequency when $v_1 = 7.33$ ft/s for the 0.25-in.-diameter section. It is determined as

$$f_1 = \frac{0.2v_1}{d_1} = \frac{0.2(7.33)}{0.25/12} = 70.4 \text{ Hz}$$

In similar fashion the shedding frequency of the 0.3125-in.-diameter section is found to be

$$f_1 = \frac{0.2(7.33)}{0.3125/12} = 56.3 \text{ Hz}$$

We next observe that at 5 mph, the *smallest* ratio of the excitation frequency f_1 to the fundamental frequency f_n is

$$\frac{f_1}{f_n} = \frac{\omega_1}{\omega_n} = \frac{56.3}{5.4} = 10.43$$

Figure 3-6 shows that when $\omega_1/\omega_n = 10.43$, the amplitude of vibration of the antenna would be negligible. As the speed increases above 5 mph, the ratio of the shedding frequency to the fundamental frequency moves farther to the right along the magnification-factor curve, so that resonance with the fundamental mode will not occur for speeds varying from 5 to 65 mph.

It should be mentioned here that the antenna has many natural frequencies above the fundamental frequency of 5.4 Hz, and it is thus possible for resonance to occur at one of these higher modes of vibration. As we shall see in Chap. 5, a number of the higher natural frequencies and corresponding mode shapes can be readily determined by modeling the antenna as a system of lumped masses.

The smallest value of Reynolds number will occur at 5 mph for the smaller diameter section, while the largest value occurs at 65 mph for the larger section. These values are computed as

$$\text{Smallest:} \qquad R = v_1 d_1 \frac{\rho}{\mu} = \frac{7.33(0.25/12)2.37(10)^{-3}}{3.8(10)^{-7}} = 952.4$$

$$\text{Largest:} \qquad R = v_2 d_2 \frac{\rho}{\mu} = \frac{95.3(0.3125/12)2.37(10)^{-3}}{3.8(10)^{-7}} = 15.5(10)^3$$

This shows that the Reynolds numbers for the 5 to 65 mph range are within the range of $400 \le R \le 3(10)^5$ for which $S = 0.2$.

3-5 BEATING

Beating is an interesting phenomenon that occurs when a system with very little damping is subjected to an excitation source that has a frequency very close to its natural frequency.

To begin our analysis of beating, let us consider Eq. 3-18 with $\omega/\omega_n \cong 1$ and the assumption of zero damping. With $\zeta = 0$, the phase angle ϕ is also zero, and Eq. 3-18 reduces to

$$x = A \cos \omega_n t + B \sin \omega_n t + |X| \sin \omega t \tag{3-42}$$

in which the amplitude $|X|$ of the forced response is

$$|X| = \frac{F_0/k}{1 - (\omega/\omega_n)^2}$$

as obtained from Eq. 3-15.

Using the initial conditions $x_0 = \dot{x}_0 = 0$, Eq. 3-42 yields

$$A = 0$$

and

$$B = -|X|\left(\frac{\omega}{\omega_n}\right)$$

Substituting these values into Eq. 3-42,

$$x = |X|\left(\sin \omega t - \frac{\omega}{\omega_n} \sin \omega_n t\right) \tag{3-43}$$

Considering $\omega/\omega_n \simeq 1$, Eq. 3-43 becomes

$$x = |X|(\sin \omega t - \sin \omega_n t) \tag{3-44}$$

Using the trigonometric identity

$$\sin \omega t - \sin \omega_n t = 2 \sin\left(\frac{\omega - \omega_n}{2}\right) t \cos\left(\frac{\omega + \omega_n}{2}\right) t$$

Eq. 3-44 may be written as

$$x = 2|X| \sin\left(\frac{\omega - \omega_n}{2}\right) t \cos\left(\frac{\omega + \omega_n}{2}\right) t \tag{3-45}$$

Equation 3-45 may be interpreted as a cosine wave

$$\cos\left(\frac{\omega + \omega_n}{2}\right) t$$

with a slowly varying amplitude of $2|X| \sin[(\omega - \omega_n)t/2]$ as shown in Fig. 3-23. This type of motion is referred to as beating.

Since the slowly varying amplitude $2|X| \sin[(\omega - \omega_n)t/2]$ goes through zero at multiples of π, the period of the beating motion is

$$\tau_b = \frac{2\pi}{\omega - \omega_n} = \frac{2\pi}{\Delta\omega} \qquad \text{(period of beating)} \tag{3-46}$$

As the figure shows, this period of beating is considerably longer than the vibration period

$$\tau = \frac{4\pi}{\omega + \omega_n} \qquad \text{(vibration period)}$$

since $\Delta\omega$ is a small quantity and $\omega + \omega_n \cong 2\omega_n$. It should be apparent that as the exciting frequency approaches the natural frequency of the system, the period of the beating increases, becoming theoretically infinite when $\omega/\omega_n = 1$.

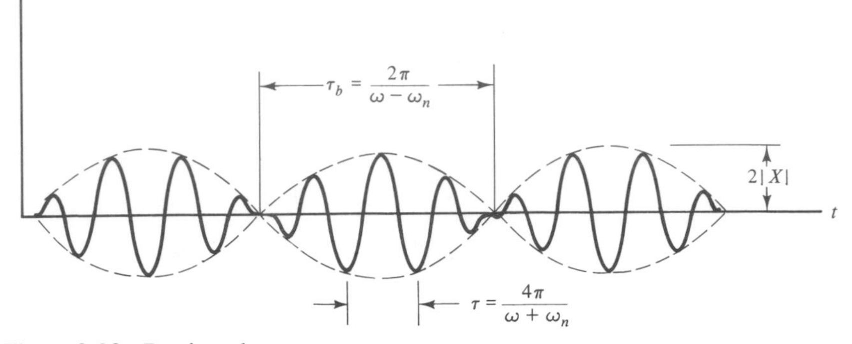

Figure 3-23 Beating phenomenon.

While Eq. 3-45 was derived with the assumption of zero damping, there is always some damping inherent in real systems, so the maximum beating amplitude *near* resonance will always be less than the steady-state amplitude $|X|$ *at* resonance.

3-6 SUPPORT EXCITATION

This section will be devoted to the study of sinusoidal support excitation of single-degree-of-freedom systems. Such a study is fundamental to understanding the basic principles of several important topics that will be discussed later in this chapter.

In general, the vibratory motion of a system subjected to support excitation may be analyzed in terms of the *absolute* motion[8] of its mass, or in terms of the motion of its mass *relative* to a moving support, depending upon which motion is of primary interest. For example, in isolating an instrument from a vibrating support, the absolute motion of the instrument would be of primary interest; while in using a vibration-measuring transducer, the output of which depends upon the motion of the transducer's mass element relative to the vibrating body to which it is attached, the relative motion would be of primary interest. In the discussion that follows, we analyze a system in terms of each of these motions, beginning with absolute motion.

Absolute Motion

Figure 3-24a shows a viscously damped spring-and-mass system. Our first step is to derive the differential equation of motion. As shown in the figure, x is the *absolute* displacement of the mass m from its static-equilibrium position, and the absolute displacement of the moving support is given by $y = Y \sin \omega t$. The displacement of m relative to the moving support is z. Assuming that x, y, and z and all their derivatives are positive upward, the *change* in the spring force from the static-equilibrium position is given by $k(x - y)$ and has the sense shown in the free-body diagram, which corresponds to a positive displacement of the mass with respect to the support. The viscous-damping force $c(\dot{x} - \dot{y})$ shown depends upon the relative velocity $(\dot{x} - \dot{y})$ between the mass and the support and has a sense corresponding to a positive *relative* velocity between the mass and the support.

Since the force in the spring in the static-equilibrium position is equal and opposite to the weight W, these forces cancel each other and are therefore omitted from the free-

[8] Absolute motion here refers to motion with respect to a coordinate system attached to the earth.

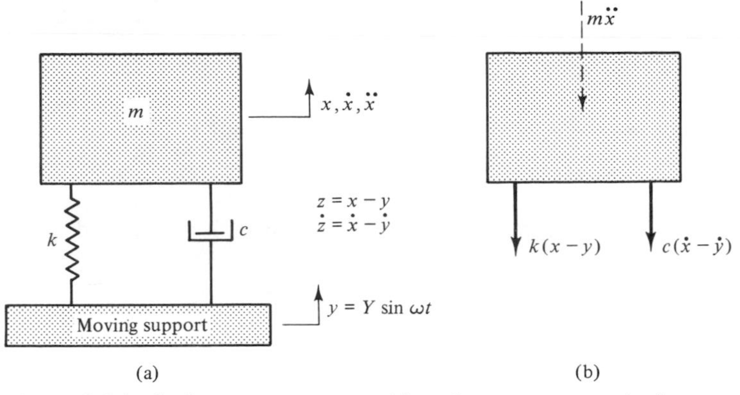

Figure 3-24 Spring-mass system subjected to support excitation $y = Y \sin \omega t$.

body diagram. Adding the inertia effect $m\ddot{x}$ with a sense opposite to that of the positive $\ddot{x}$ assumed, we can sum the forces and the inertia effect equal to zero as

$$\sum F_x = 0 = m\ddot{x} + c(\dot{x} - \dot{y}) + k(x - y) \tag{3-47}$$

Dividing through by m and separating the terms, we obtain

$$\ddot{x} + \frac{c}{m}\dot{x} + \frac{k}{m}x = \frac{c}{m}\dot{y} + \frac{k}{m}y$$

Letting $c/m = 2\zeta\omega_n$ and $k/m = \omega_n^2$, the differential equation becomes

$$\ddot{x} + 2\zeta\omega_n\dot{x} + \omega_n^2 x = 2\zeta\omega_n\dot{y} + \omega_n^2 y \tag{3-48}$$

and is in terms of the absolute motion of the mass.

Since the motion of the support is $y = Y \sin \omega t$, we assume a steady-state solution of Eq. 3-48 having the complex form $x = Xe^{j\omega t}$. Then

$$\left.\begin{aligned} x &= Xe^{j\omega t} \\ \dot{x} &= j\omega Xe^{j\omega t} \\ \ddot{x} &= -\omega^2 Xe^{j\omega t} \end{aligned}\right\} \tag{3-49}$$

Using the imaginary part of $Ye^{j\omega t}$ for $Y \sin \omega t$, the complex quantities for y and $\dot{y}$ are

$$\left.\begin{aligned} y &= Ye^{j\omega t} \\ \dot{y} &= j\omega Ye^{j\omega t} \end{aligned}\right\} \tag{3-50}$$

Substituting the terms in Eqs. 3-49 and 3-50 into Eq. 3-48, the result may be arranged as

$$\left[1 - \left(\frac{\omega}{\omega_n}\right)^2 + j2\zeta\frac{\omega}{\omega_n}\right]X = \left(1 + j2\zeta\frac{\omega}{\omega_n}\right)Y \tag{3-51}$$

Letting $a = 1 - (\omega/\omega_n)^2$ and $b = 2\zeta(\omega/\omega_n)$, Eq. 3-51 can be written more simply as

$$(a + jb)X = (1 + jb)Y \tag{3-52}$$

The ratio X/Y is then

$$\frac{X}{Y} = \frac{1 + jb}{a + jb} = \frac{\sqrt{1 + b^2}\, e^{j\phi_1}}{\sqrt{a^2 + b^2}\, e^{j\phi_2}} \tag{3-53}$$

or

$$\frac{X}{Y} = \frac{\sqrt{1 + b^2}}{\sqrt{a^2 + b^2}} e^{-j\phi} \tag{3-54}$$

where $\phi = \phi_2 - \phi_1$ = phase angle between X and Y
$\tan \phi_1 = b$
$\tan \phi_2 = b/a$

From Eq. 3-54 it can be seen that the dimensionless amplitude ratio is

$$\left|\frac{X}{Y}\right| = \frac{\sqrt{1 + [2\zeta(\omega/\omega_n)]^2}}{\sqrt{[1 - (\omega/\omega_n)^2]^2 + [2\zeta(\omega/\omega_n)]^2}} \tag{3-55}$$

A plot of Eq. 3-55 is shown in Fig. 3-25. The ratio $|F_T|/|F_0|$ shown on the figure is pertinent to a later discussion of vibration isolation.

To determine the phase angle ϕ between X and Y which is $\phi_2 - \phi_1$ as indicated above, the numerator and denominator of Eq. 3-53 are multiplied by $a - jb$, which is the *conjugate* of $a + jb$, to obtain

$$\frac{(1 + jb)(a - jb)}{(a + jb)(a - jb)}$$

which expanded is

$$\frac{a + b^2}{a^2 + b^2} + \frac{jb(a - 1)}{a^2 + b^2}$$

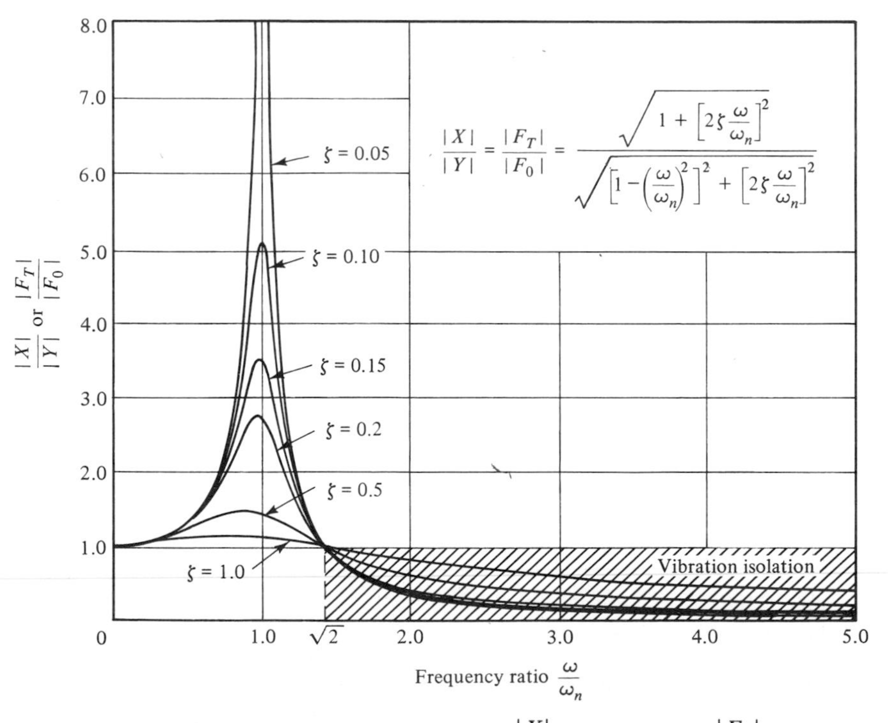

Figure 3-25 Plot of Eq. 3-55 (absolute motion) $\frac{|X|}{|Y|}$, and Eq. 3-79 $\frac{|F_T|}{|F_0|}$.

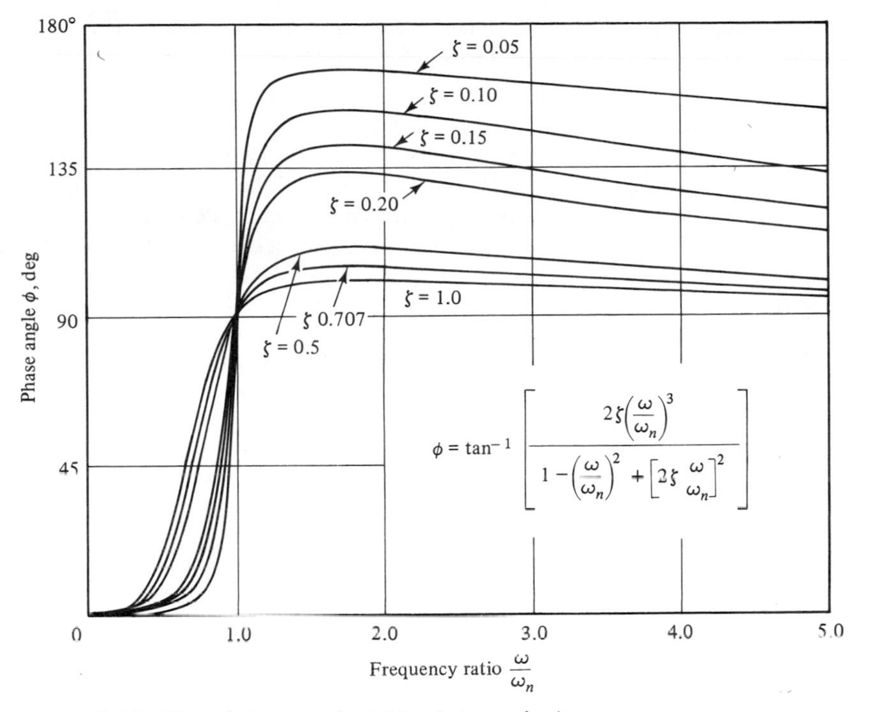

Figure 3-26 Plot of phase angle ϕ (absolute motion).

Dividing the imaginary part of the above expression by its real part gives

$$\tan(-\phi) = \frac{b(a-1)}{a+b^2}$$

or

$$\tan\phi = \frac{2\zeta(\omega/\omega_n)^3}{1-(\omega/\omega_n)^2+[2\zeta(\omega/\omega_n)]^2} \tag{3-56}$$

A plot of the phase angle ϕ as a function of ω/ω_n is shown in Fig. 3-26.

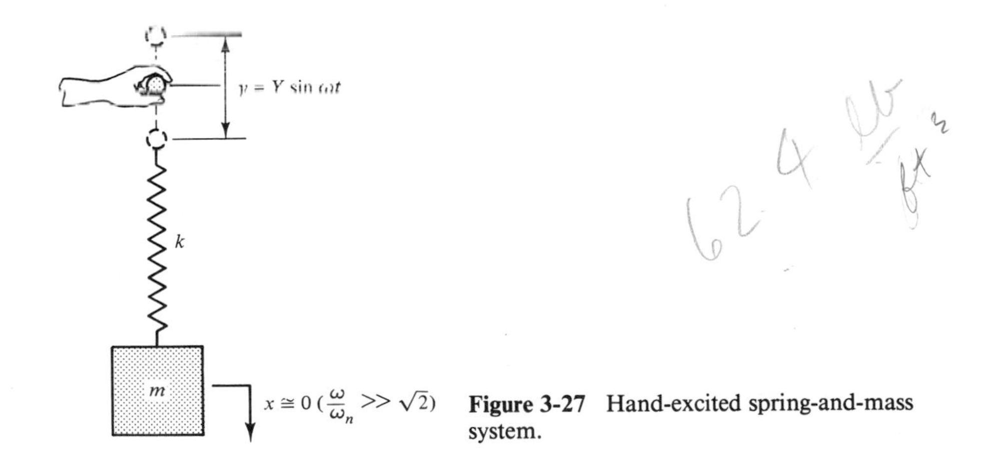

Figure 3-27 Hand-excited spring-and-mass system.

Referring to Fig. 3-25, let us note some of the characteristics of the steady-state response of a system subjected to support excitation:

1. The amplitude ratio $|X|/|Y| = 1$ for all values of damping when $\omega/\omega_n = \sqrt{2}$.
2. The ratio $|X|/|Y|$ is less than 1 when ω/ω_n is greater than $\sqrt{2}$, so $\omega/\omega_n = \sqrt{2}$ is the beginning of the region of *vibration isolation* (Sec. 3-8).
3. When $\omega/\omega_n \gg \sqrt{2}$, the ratio $|X|/|Y|$ is quite small, which means that the mass m is essentially stationary. This can be easily demonstrated by hand-exciting a low-frequency spring-and-mass system such as the one shown in Fig. 3-27.

Relative Motion

Referring again to Fig. 3-24, and remembering that x and y are absolute displacements, it can be seen that the displacement z of the mass m *relative* to the support motion $y = y(t)$ is

$$z = x - y$$

from which

$$\left.\begin{aligned} \dot{z} &= \dot{x} - \dot{y} \\ \ddot{z} &= \ddot{x} - \ddot{y} \end{aligned}\right\} \tag{3-57}$$

Substituting the above relationships into Eq. 3-47, we obtain

$$m(\ddot{z} + \ddot{y}) + c\dot{z} + kz = 0$$

Dividing the above by m and letting $c/m = 2\zeta\omega_n$ and $k/m = \omega_n^2$ gives us

$$\ddot{z} + 2\zeta\omega_n\dot{z} + \omega_n^2 z = -\ddot{y} \tag{3-58}$$

which is the differential equation of the system in terms of its *relative* motion.

To express the support excitation $y = Y \sin \omega t$, we take the imaginary part of $Ye^{j\omega t}$, from which $\ddot{y} = -\omega^2 Ye^{j\omega t}$. Assuming that $z = Ze^{j\omega t}$ is the steady-state solution, it follows that

$$\begin{aligned} \dot{z} &= j\omega Ze^{j\omega t} \\ \ddot{z} &= -\omega^2 Ze^{j\omega t} \end{aligned}$$

Substituting the complex expressions above into Eq. 3-58, and solving for the ratio Z/Y, we obtain

$$\frac{Z}{Y} = \frac{(\omega/\omega_n)^2 e^{-j\phi}}{\sqrt{[1 - (\omega/\omega_n)^2]^2 + [2\zeta(\omega/\omega_n)]^2}} \tag{3-59}$$

The phase angle between the support displacement and the relative displacement is given by

$$\tan \phi = \frac{2\zeta(\omega/\omega_n)}{1 - (\omega/\omega_n)^2} \tag{3-60}$$

A plot of the phase angle as a function of the frequency ratio is shown in Fig. 3-29.

The dimensionless amplitude ratio is

$$\left|\frac{Z}{Y}\right| = \frac{(\omega/\omega_n)^2}{\sqrt{[1 - (\omega/\omega_n)^2]^2 + [2\zeta(\omega/\omega_n)]^2}} \tag{3-61}$$

and a plot of this equation is shown in Fig. 3-28.

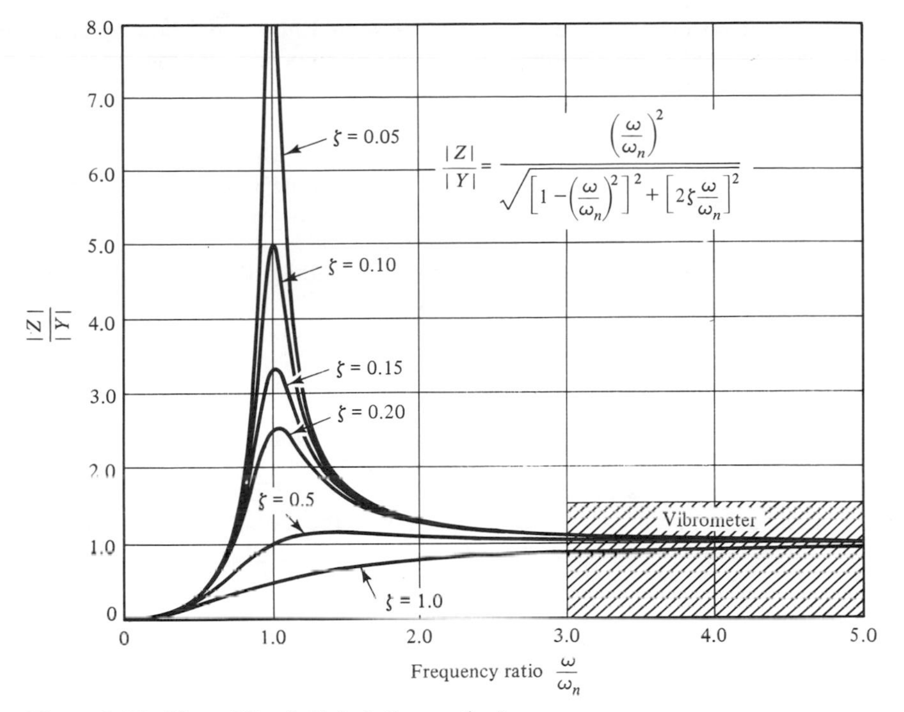

Figure 3-28 Plot of Eq. 3-61 (relative motion).

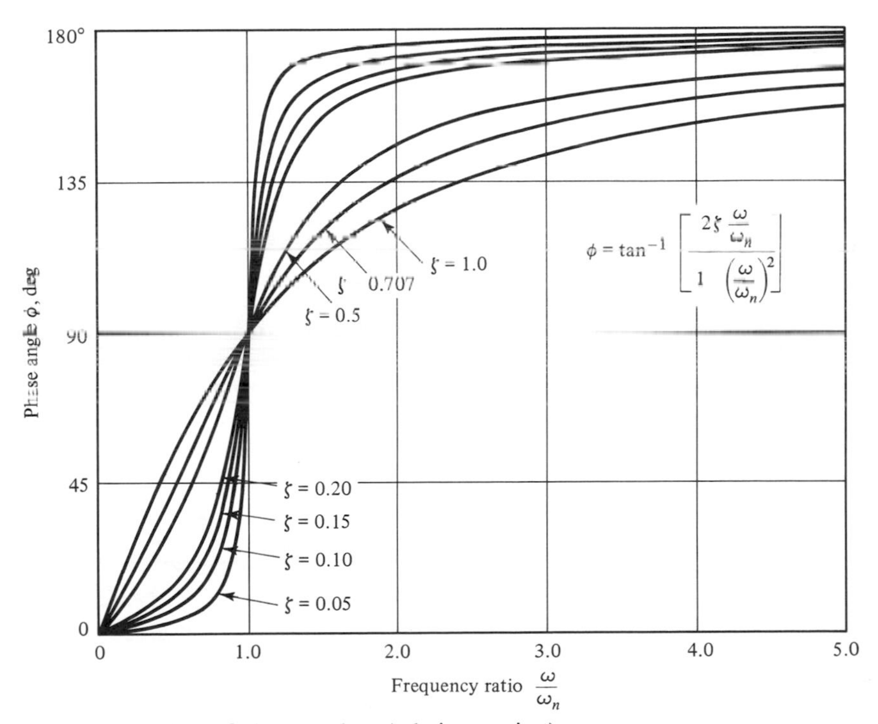

Figure 3-29 Plot of phase angle ϕ (relative motion).

Referring to Fig. 3-28, the reader should note the following characteristics of the *relative* steady-state response of a system to support excitation, as they will be pertinent to subsequent discussions:

1. When $\omega/\omega_n > 3$, the amplitude ratio $|Z|/|Y| \cong 1$, which indicates that the relative amplitude $|Z|$ is essentially the same as the amplitude $|Y|$ of the moving support. This is the principle upon which the vibrometer-type transducer operates in measuring vibratory motion (see Sec. 3-9).
2. When $|Z|/|Y| \cong 1$, which corresponds to the *absolute* ratio $|X|/|Y| \cong 0$, the mass m is essentially stationary.

EXAMPLE 3-5

The boat and trailer shown in Fig. 3-30a are being pulled over an undulating road at a velocity v. The contour of the road is such that it can be approximated fairly accurately by a sine wave having a wavelength of $l = 10$ ft and an amplitude of $Y = 0.5$ in. The total static deflection δ_s of the springs and tires of the trailer due to the weight of the boat and trailer has been measured as 1.5 in.

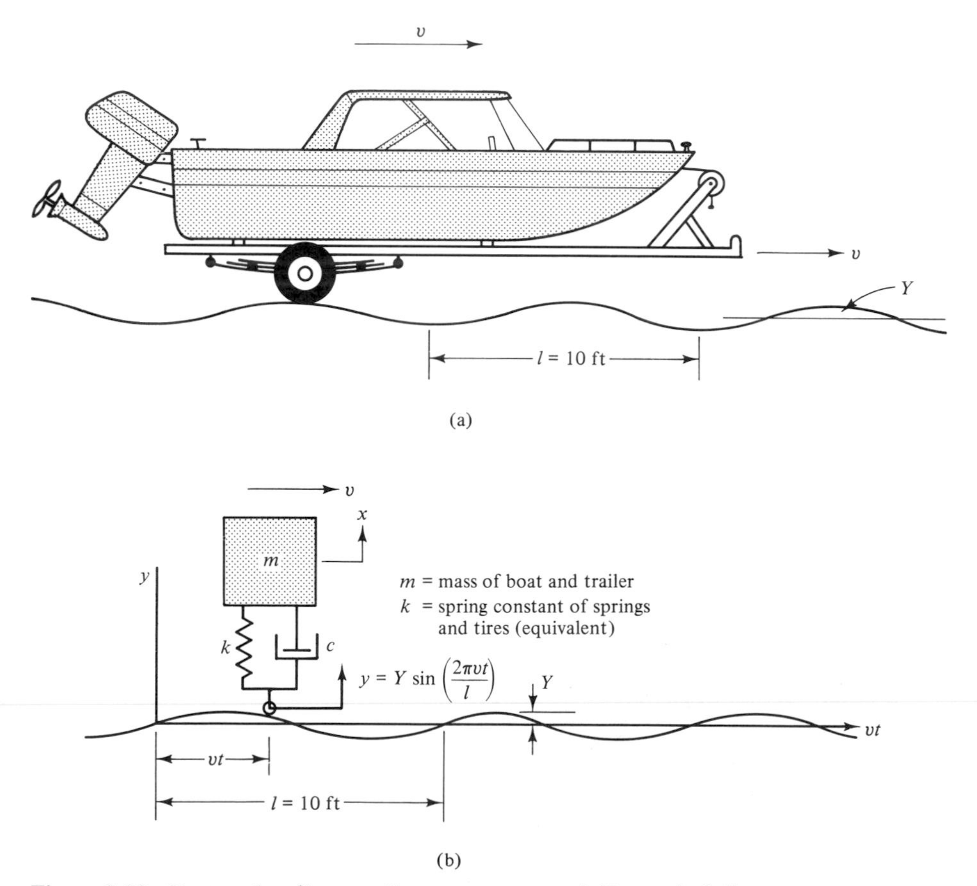

Figure 3-30 Boat and trailer traveling over wavy road (Example 3-5).

Assuming that the damping inherent in the system is viscous in nature and of such magnitude that $\zeta = 0.05$, determine: (a) the speed v at which the amplitude $|X|$ of the boat and trailer will be a maximum; (b) the value of the maximum amplitude referred to in part a; (c) the amplitude when the boat and trailer are traveling at the speed of 55 mph.

Solution. **a.** The boat and trailer are modeled by the spring-and-mass system shown in Fig. 3-30b. The contour of the road acts as a support excitation on the system, and is approximated by $y = Y \sin \omega t$.

The distance traveled is given by $S = vt$. Remembering that the period is related to ω by $\tau = 2\pi/\omega$, and that l is the distance traveled as the sine wave goes through one period, we can write that

$$l = v\tau = v\frac{2\pi}{\omega}$$

from which the exciting frequency is

$$\omega = \frac{2\pi v}{l} \tag{3-62}$$

Substituting Eq. 3-62 into the excitation function $y = Y \sin \omega t$, it becomes

$$y = Y \sin\left(\frac{2\pi vt}{l}\right)$$

The undamped natural circular frequency is given by

$$\omega_n = \sqrt{\frac{k}{m}} = \sqrt{\frac{g}{W/k}}$$

Remembering that $W/k = 1.5$ in. is the static deflection of the boat and trailer,

$$\omega_n = \sqrt{\frac{386}{1.5}} = 16.04 \text{ rad/s}$$

The amplitude $|X|$ will be maximum when $\omega/\omega_n = 1$. Thus,

$$\frac{2\pi v}{l} = 16.04$$

from which

$$v = \frac{16.04(10)}{2\pi} = 25.53 \text{ ft/s} = 17.41 \text{ mph}$$

b. Substituting $\zeta = 0.05$, $\omega/\omega_n = 1.0$, and $|Y| = 0.5$ in. into Eq. 3-55, the maximum amplitude at 17.41 mph is found to be

$$|X| = \frac{0.5\sqrt{1 + [2(0.05)]^2}}{2(0.05)} = 5.02 \text{ in.}$$

c. At 55 mph, the ratio of the circular excitation frequency ω to the natural circular frequency ω_n is also the ratio of the speeds, so that

$$\frac{\omega}{\omega_n} = \frac{55}{17.41} = 3.16$$

Therefore, referring to Eq. 3-55,

$$|X| = \frac{0.5\sqrt{1 + [2(0.05)(3.16)]^2}}{\sqrt{[1 - (3.16)^2]^2 + [2(0.05)(3.16)]^2}}$$
$$= 0.06 \text{ in. (at 55 mph)}$$

or

$$\left|\frac{X}{Y}\right| = \frac{0.06}{0.5} = 0.12$$

Referring to Fig. 3-25, it can be seen that $|X|/|Y|$ at 55 mph should be considerably less than 1 as found above, since $\omega/\omega_n = 3.16$ for 55 mph.

3-7 GENERAL PERIODIC EXCITATION/STEADY-STATE RESPONSE

In discussing forced vibrations up to this point, the disturbing excitations from either forces or support motions have been assumed to be simple harmonic functions of $\sin \omega t$ or $\cos \omega t$. However, force or support-motion excitations in general consist of several frequencies, or even possibly a theoretically infinite number of frequencies. Recalling the discussion of Fourier series in Sec. 1-5, *periodic functions* can be represented in the *time domain* by combinations of sines and/or cosines of the frequency components present in the waveform.

The steady-state response of a single-degree-of-freedom system subjected to an excitation containing multiple frequency components is simply the *sum of the particular solutions obtained* in considering each frequency component present in the excitation individually. That is, if the functions $x_{p_1}, x_{p_2}, \ldots, x_{p_n}$ are particular solutions of the linear differential equations

$$\begin{aligned} \ddot{x} + 2\zeta\omega_n\dot{x} + \omega_n^2 x &= f_1(t) \\ \ddot{x} + 2\zeta\omega_n\dot{x} + \omega_n^2 x &= f_2(t) \\ \vdots \qquad\qquad & \quad \vdots \\ \ddot{x} + 2\zeta\omega_n\dot{x} + \omega_n^2 x &= f_n(t) \end{aligned}$$

respectively, then

$$x(t) = x_{p_1} + x_{p_2} + \cdots + x_{p_n} \tag{3-63}$$

is a particular solution of the differential equation

$$\ddot{x} + 2\zeta\omega_n\dot{x} + \omega_n^2 x = f_1(t) + f_2(t) + \cdots + f_n(t)$$

As a simple example of using Eq. 3-63, assume that the support motion of the spring-and-mass system shown in Fig. 3-24a is

$$y = Y_1 \sin \omega_1 t + Y_2 \sin \omega_2 t \tag{3-64}$$

so that it consists of two frequencies ω_1 and ω_2. Reviewing the discussion of Eqs. 3-48 through 3-55, it should be apparent that the steady-state response to an excitation frequency ω is given by

$$x = |X| \sin(\omega t - \phi)$$

in which $|X|$ and ϕ are defined by Eqs. 3-55 and 3-56, respectively. It follows then that the steady-state response due to the support motion expressed by Eq. 3-64 is

$$x(t) = |X_1|\sin(\omega_1 t - \phi_1) + |X_2|\sin(\omega_2 t - \phi_2) \tag{3-65}$$

where

$$\left.\begin{aligned} |X_1| &= \frac{Y_1\sqrt{1 + [2\zeta(\omega_1/\omega_n)]^2}}{\sqrt{[1 - (\omega_1/\omega_n)^2]^2 + [2\zeta(\omega_1/\omega_n)]^2}} \\ \tan\phi_1 &= \frac{2\zeta(\omega_1/\omega_n)^3}{1 - (\omega_1/\omega_n)^2 + [2\zeta(\omega_1/\omega_n)]^2} \end{aligned}\right\}$$

$$\left.\begin{aligned} |X_2| &= \frac{Y_2\sqrt{1 + [2\zeta(\omega_2/\omega_n)]^2}}{\sqrt{[1 - (\omega_2/\omega_n)^2]^2 + [2\zeta(\omega_2/\omega_n)]^2}} \\ \tan\phi_2 &= \frac{2\zeta(\omega_2/\omega_n)^3}{1 - (\omega_2/\omega_n)^2 + [2\zeta(\omega_2/\omega_n)]^2} \end{aligned}\right\}$$

The reader should note that the *maximum* value of $x(t)$ is not apparent from a simple inspection of Eq. 3-65, since $\sin(\omega_1 t - \phi_1)$ and $\sin(\omega_2 t - \phi_2)$ become equal to 1 at different times. The maximum value of $x(t)$ could of course be obtained by computing Eq. 3-65 on a calculator or computer over a period of time sufficiently long to include the computation of the maximum value. However, if the steady-state response contains a large number of terms due to a large number of frequency components, it may be more practical to approximate the maximum value of $x(t)$ by calculating the *absolute maximum* response, which is merely the sum of the amplitudes obtained so that

$$x(t)_{\max} \cong |X_1| + |X_2| + \cdots + |X_n| \tag{3-66}$$

which is the upper bound of $x(t)$.

If either ω_1 or ω_2 in Eq. 3-65 is equal to ω_n, a condition of resonance will exist for that frequency component. For example, if $\omega_1/\omega_n = 1$, the first term of Eq. 3-66 would be dominant so that

$$x(t)_{\max} \cong |X_1|$$

It should be emphasized that the vibratory motion of a linear system consists of the frequency components present in the excitation source, and that the magnitude of a system's response depends both upon the damping present in the system and upon the ratios of the excitation frequencies to the natural frequency ω_n of the system.

Frequency, or spectrum, analyzers, which are discussed in Sec. 1-5, are often used to experimentally determine the frequency components present in the vibratory motion of a system. The data obtained can then be used to make modifications in the system, to either minimize or eliminate troublesome excitation frequencies, and/or to change the natural frequency of the system to avoid resonance with any of the exciting frequencies.[9]

EXAMPLE 3-6

Let us once again consider the boat and trailer of Example 3-5, but with a road contour that is represented by the sum of the three sine waves shown in Fig. 3-31a, which have wavelengths of 10, 6, and 4 ft. The amplitude of each of these waves is 0.5 in. as shown. The equations for y_1, y_2, and y_3 shown in the figure were determined as discussed in the beginning of Example 3-5. The first 10 ft of the *combined wave* is shown in Fig. 3-31b.

[9] Systems with n degrees of freedom have n natural frequencies so that any one of the natural frequencies may be susceptible to resonant excitation (see Chap. 5).

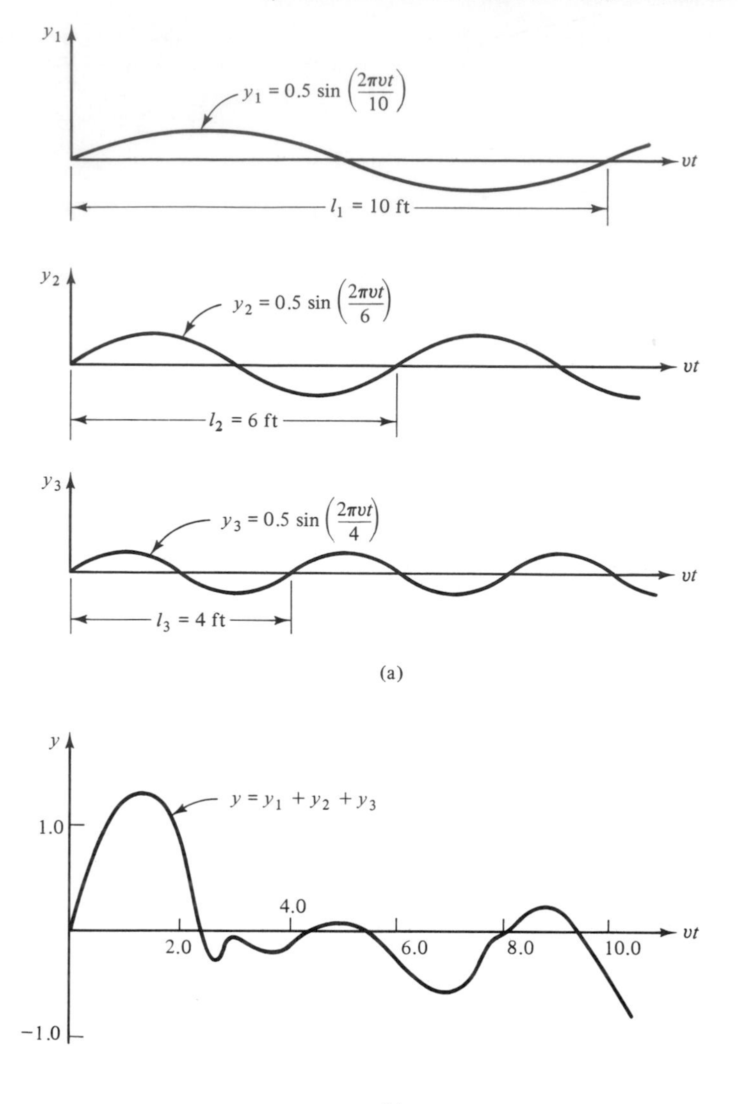

Figure 3-31 Road contour for Example 3-6. (a) Individual sine waves. (b) Combined-wave contour for first 10 ft.

Assuming that the boat and trailer have a static displacement of $\delta_s = 1.5$ in. and a viscous-damping factor of $\zeta = 0.05$, as in Example 3-5, determine (a) the critical speeds of the boat and trailer; (b) the maximum value of the response $x(t)$ of the boat and trailer for a speed of 20 mph using first an equation similar to Eq. 3-65 and then Eq. 3-66.

Solution. **a.** In Example 3-5, the natural circular frequency ω_n of the boat and trailer was found to be 16.04 rad/s. Using Eq. 3-62 with $\omega_n = 16.04$ rad/s in each case since $\omega/\omega_n = 1$ for critical speed, the three critical speeds are found to be

$$v_1 = \frac{16.04(10)}{2\pi} = 25.53 \text{ ft/s} \quad (17.41 \text{ mph})$$

$$v_2 = \frac{16.04(6)}{2\pi} = 15.32 \text{ ft/s} \quad (10.44 \text{ mph})$$

$$v_3 = \frac{16.04(4)}{2\pi} = 10.21 \text{ ft/s} \quad (6.96 \text{ mph})$$

b. At 20 mph ($v = 29.33$ ft/s), the three frequency ratios are

$$\frac{\omega_1}{\omega_n} = \frac{2\pi(29.33)}{10(16.04)} = 1.15$$

$$\frac{\omega_2}{\omega_n} = \frac{2\pi(29.33)}{6(16.04)} = 1.92$$

$$\frac{\omega_3}{\omega_n} = \frac{2\pi(29.33)}{4(16.04)} = 2.87$$

Substituting these frequency ratios, $\zeta = 0.05$, and $Y = 0.5$ into Eq. 3-55 yields the following steady-state amplitudes for 20 mph:

$$|X_1| = 1.47 \text{ in.}$$
$$|X_2| = 0.19 \text{ in.}$$
$$|X_3| = 0.07 \text{ in.}$$

Referring to Eq. 3-56, the phase angles are found to be

$$\phi_1 = 153.8° = 2.68 \text{ rad}$$
$$\phi_2 = 165.0° = 2.88 \text{ rad}$$
$$\phi_3 = 161.7° = 2.82 \text{ rad}$$

Substituting the above values into an equation similar to Eq. 3-65 gives

$$x(t) = 1.47 \sin(18.45t - 2.68) + 0.19 \sin(30.8t - 2.88) + 0.07 \sin(46.03t - 2.82)$$

Then by using very small increments of time over a sufficient interval of time, the maximum response of the boat and trailer is found as

$$x(t)_{\max} = 1.70 \text{ in.}$$

which occurs when $t = 1.59$ s.

Using the absolute-maximum approach,

$$x(t)_{\max} \cong 1.47 + 0.19 + 0.07 = 1.73 \text{ in.}$$

which is very close to the value obtained above.

EXAMPLE 3-7

The cam-follower system of Example 1-2 is shown again in Fig. 3-32a. The rotating cam subjects the viscously damped follower system of mass m to an excitation having the sawtooth shape shown in Fig. 3-32b. This displacement, or excitation, function has the form of

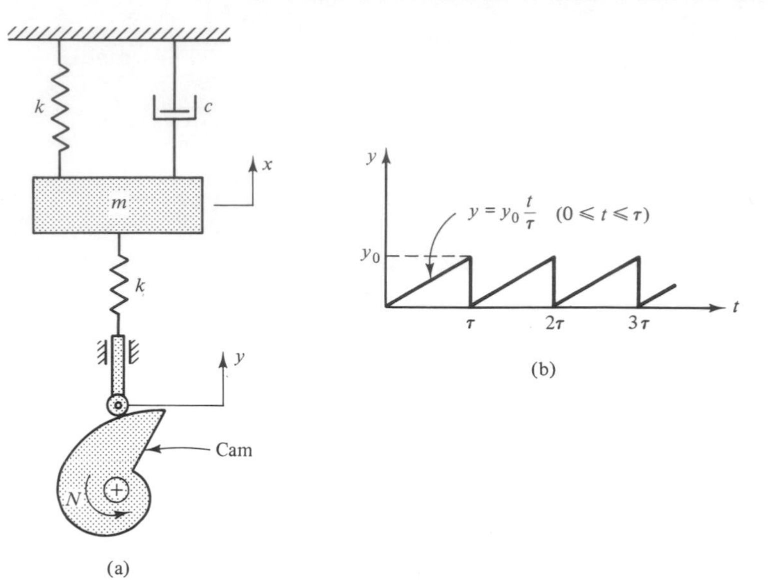

Figure 3-32 Cam-follower system with sawtooth excitation.

$$f(t) = y_0 \frac{t}{\tau} \qquad (0 \leq t \leq \tau)$$
$$f(t + \tau) = f(t)$$

in which y_0 is the amplitude of the sawtooth and τ is the period of the excitation function.

In Example 1-2, the Fourier series for the sawtooth displacement function was shown to be

$$y = y_0\left[\frac{1}{2} - \frac{1}{\pi} \sum_{i=1,2,\ldots}^{\infty} \frac{1}{i} \sin(i\omega_1 t)\right]$$

or

$$y = y_0\left[\frac{1}{2} - \frac{1}{\pi} \sum_{i=1,2,\ldots}^{\infty} \frac{1}{i} \sin(i2\pi f_1 t)\right] \qquad (3\text{-}67)$$

where $f_1 = 1/\tau$ (fundamental excitation frequency, Hz)
$\omega_1 = 2\pi f_1 = 2\pi N/60$ (fundamental circular excitation frequency, rad/s)
$N =$ (speed of rotating cam, rpm)

In analyzing the cam-follower system, we wish to: (a) determine the differential equation of motion of the system; (b) determine the steady-state response of the system; (c) obtain a sketch of the approximate steady-state response x versus the fundamental exciting frequency f_1 over a range of $0 \leq f_1 \leq 1.5 f_n$, in which f_n is the natural frequency of the follower system.

Solution. **a.** Summing the forces and inertia effect shown on the free body in Fig. 3-33, we obtain the differential equation of motion as

$$\ddot{x} + \frac{c}{m}\dot{x} + \frac{2k}{m}x = \frac{k}{m}y$$

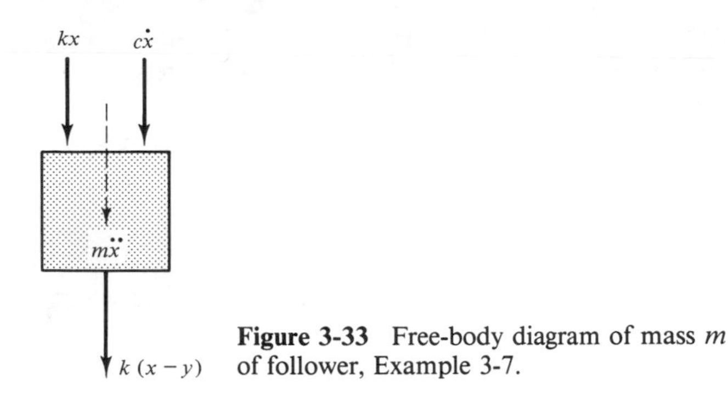

Figure 3-33 Free-body diagram of mass m of follower, Example 3-7.

or as

$$\ddot{x} + 2\zeta\omega_n\dot{x} + \omega_n^2 x = \frac{\omega_n^2}{2} y_0 \left[\frac{1}{2} - \frac{1}{\pi} \sum_{i=1,2,\ldots}^{\infty} \frac{1}{i} \sin(i\omega_1 t)\right] \tag{3-68}$$

in which $2\zeta\omega_n = c/m$, $\omega_n^2 = 2k/m$, and y has the form of the first expression in Eq. 3-67.

b. The steady-state response determined from Eq. 3-68 consists of two *particular solutions,* one that satisfies the constant term $\omega_n^2 y_0/4$, and a series solution in which *each term* of the series must satisfy the differential equation. Referring to the particular solution that satisfies the constant term as x_1, it can be seen from inspection of Eq. 3-68 that

$$x_1 = \frac{y_0}{4} \tag{3-69}$$

That is, when $x_1 = y_0/4$ and its derivatives are substituted into the left-hand side of Eq. 3-68, the result is the constant term $\omega_n^2 y_0/4$ since $\dot{x}_1 = \ddot{x}_1 = 0$.

To obtain a particular solution x_i for the ith term of the series in Eq. 3-68, we may write for any ith term that

$$\ddot{x} + 2\zeta\omega_n\dot{x} + \omega_n^2 x = -Y_i \sin(i\omega_1 t) \tag{3-70}$$

in which $Y_i = \omega_n^2 y_0/2i\pi$.

Assuming a steady-state solution having the complex form

$$x_i = X_i e^{j(i\omega_1 t)} \tag{3-71}$$

we find that

$$X_i = \frac{(Y_i/\omega_n^2)e^{-i\phi_i}}{\sqrt{[1 - (i\omega_1/\omega_n)^2]^2 + [2\zeta(i\omega_1/\omega_n)]^2}}$$

The phase angle ϕ_i between X_i and Y_i can be determined from

$$\tan \phi_i = \frac{2\zeta(i\omega_1/\omega_n)}{1 - (i\omega_1/\omega_n)^2}$$

The steady-state amplitudes of the series terms are then

$$|X_i| = \frac{-(y_0/i2\pi)}{\sqrt{[1 - (i\omega_1/\omega_n)^2]^2 + [2\zeta(i\omega_1/\omega_n)]^2}} \tag{3-72}$$

Adding the particular solution given by Eq. 3-69 and the particular solutions having the amplitudes given by Eq. 3-72, the steady-state response is given by

$$x = \frac{y_0}{4} + \sum_{i=1,2,\ldots}^{\infty} |X_i| \sin(i\omega_1 t - \phi_i) \tag{3-73}$$

Since $i\omega_1/\omega_n = if_1/f_n$, Eq. 3-73 can be written as

$$x = \frac{y_0}{2}\left\{\frac{1}{2} - \frac{1}{\pi}\sum_{i=1,2,\ldots}^{\infty} \frac{1}{i}\,\frac{\sin(i\omega_1 t - \phi_i)}{\sqrt{[1-(if_1/f_n)^2]^2 + [2\zeta(if_1/f_n)]^2}}\right\} \tag{3-74}$$

c. In sketching the steady-state response x as a function of the fundamental exciting frequency f_1, the following should first be noted:

1. A condition of resonance occurs for an ith term when $if_1 = f_n$, and it then becomes the dominant term in the series.
2. When the rotational speed of the cam is such that $f_1 = f_n$ ($i = 1$), the first term is dominant, and the remaining terms in the series contribute little to the response, since $2f_1$, $3f_1$, and so forth, are all greater than f_n. They correspond to ω/ω_n ratios in a region well past resonance (see Fig. 3-25). Thus, as the rotational speed of the cam increases so that $f_1 > f_n$, none of the terms in Eq. 3-74 can approach a condition of resonance, since they correspond to even larger ω/ω_n ratios.
3. It should be apparent from item 2 above that as the rotational speed of the cam decreases to where f_1 becomes less than f_n, resonance can occur for $i = 2$, $i = 3$, and so on. For example, when $f_1 = f_n/2$, the second term ($i = 2$) will be dominant since $2f_1/f_n = \omega_i/\omega_n = 1$.
4. The magnitudes of the resonant amplitudes are inversely proportional to i.

With the aid of these observations, a sketch of the steady-state response as a function of the cam speed f_1 can be made as shown in Fig. 3-34. The response exhibits peak amplitudes (in units of $y_0/2\pi$) at $f_1 = f_n, f_n/2, f_n/3$, and so on, as shown in the figure.

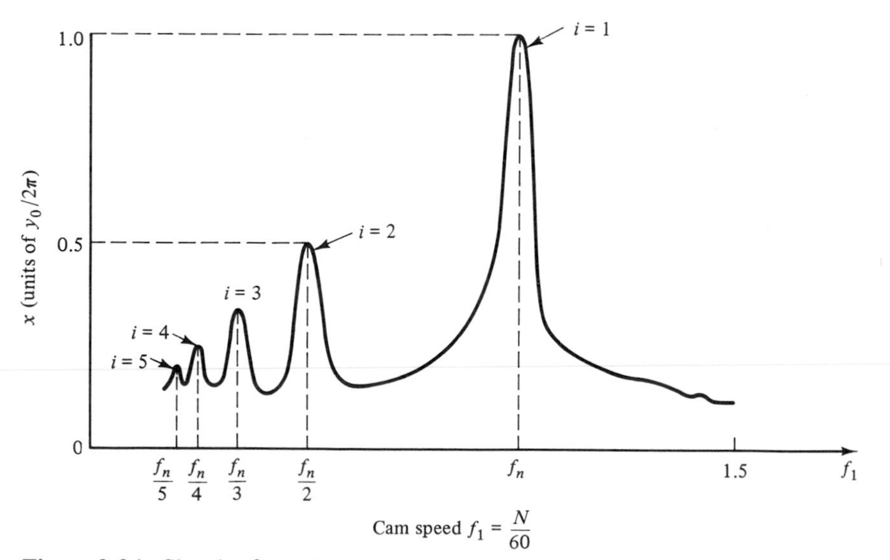

Figure 3-34 Sketch of steady-state response versus cam speed.

It is important for the reader to realize at this point that the general response characteristics of the cam-follower system that we have found thus far have all been determined by means of a theoretical analysis that has not involved any numerical calculations. This illustrates the amount of information that can be determined by such an analysis. Unfortunately, in this age of computers there seems to be an ill-advised tendency in many instances to overlook the use of theoretical analysis in favor of obtaining numerical solutions using the computer. Both are very valuable analytical tools and should be used complementally.

3-8 VIBRATION ISOLATION

In general, vibration-isolation analyses are concerned with reducing the magnitudes of forces transmitted from moving components of machines to supporting foundations as shown in Fig. 3-35a, or with reducing the support motion transmitted to instruments or equipment as shown in Fig. 3-35b.

When forces generated by moving components of a machine are transmitted to the foundation supporting the machine, undesirable vibrations are often induced in the entire structure housing the machine. The transmission of such forces can be reduced considerably by mounting the machine on so-called *isolation mounts,* which are basically pads of rubber or some type of elastomer such as neoprene. These pads act as springs with inherent damping, and are usually modeled by a spring and dashpot as shown in Fig. 3-35.

The motor-compressor units in refrigerators are supported on isolation mounts to minimize the transmission of forces to the refrigerator frame, and in turn to the floor upon which the refrigerator sits. The engines in gasoline-powered chain saws are isolated from the main frame to reduce the vibration transmitted to the hands of the operator.

Instruments and equipment can malfunction or even suffer serious damage if not properly isolated from vibrating supports upon which they are mounted. For example, an electron microscope housed in a building that is close to a street carrying heavy traffic would undoubtedly need to be isolated from the floor of the building. Circuit boards in electronic equipment are highly susceptible to fatigue failure if not isolated from vibrating support structures.

Many different types and configurations of vibration-isolation mounts are available commercially. As mentioned earlier, most of these are composed of rubber or some type of elastomer such as neoprene, and are modeled as spring-and-dashpot systems. Since many isolation systems consist of multiple isolation mounts, the stiffness k and the damping coefficient c shown in Fig. 3-35 represent *equivalent* stiffnesses and damping coefficients for the system.

Transmissibility of Forces

If we assume that the motion of the supporting foundation shown in Fig. 3-35a is zero, we see that the force F_T transmitted to the supporting foundation through the isolation system is

$$F_T = kx + c\dot{x} \tag{3-75}$$

in which x and $\dot{x}$ are the displacement and velocity, respectively, of the machine caused by the unbalanced force F_0. To determine the steady-state amplitude of F_T, we recall from Sec. 3-2 that the steady-state amplitude is given by Eq. 3-15, which is repeated here as

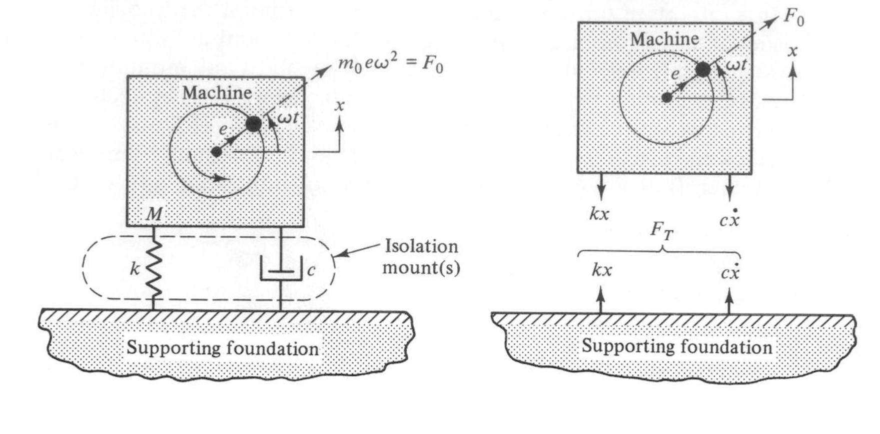

$$\left|\frac{F_T}{F_0}\right| = \frac{\sqrt{1 + \left[2\zeta \frac{\omega}{\omega_n}\right]^2}}{\sqrt{\left[1 - \left(\frac{\omega}{\omega_n}\right)^2\right]^2 + \left[2\zeta \frac{\omega}{\omega_n}\right]^2}}$$

(a)

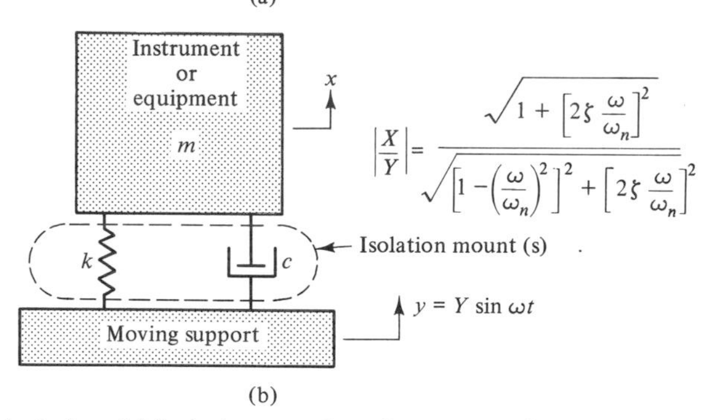

(b)

Figure 3-35 Vibration isolation. (a) Isolation to reduce force transmitted to support foundation. (b) Isolation to reduce vibration of equipment or instruments from support motion.

$$|X| = \frac{F_0/k}{\sqrt{[1 - (\omega/\omega_n)^2]^2 + [2\zeta(\omega/\omega_n)]^2}} \tag{3-15}$$

Equation 3-16 gives the steady-state displacement as

$$x = |X| e^{j(\omega t - \phi)} \tag{3-16}$$

The steady-state velocity is then

$$\dot{x} = j\omega |X| e^{j(\omega t - \phi)} \tag{3-76}$$

Substituting the last two equations above into Eq. 3-75 shows that

$$F_T = (k + jc\omega)|X| e^{j(\omega t - \phi)}$$

or

$$F_T = k\left(1 + \frac{jc\omega}{k}\right)|X|e^{j(\omega t - \phi)} \tag{3-77}$$

The steady-state amplitude of the transmitted force F_T is thus

$$|F_T| = k\sqrt{1 + \left(\frac{c\omega}{k}\right)^2}\,|X| \tag{3-78}$$

Referring to Eqs. 3-2 and 3-3, we see that

$$c = 2\zeta\omega_n m = \frac{2\zeta k}{\omega_n}$$

and

$$\frac{c}{k} = \frac{2\zeta}{\omega_n}$$

Rearranging Eq. 3-15 into the following form:

$$k|X| = \frac{F_0}{\sqrt{[1 - (\omega/\omega_n)^2]^2 + [2\zeta(\omega/\omega_n)]^2}}$$

we substitute it and the expression above for c/k into Eq. 3-78 to obtain

$$TR = \left|\frac{F_T}{F_0}\right| = \frac{\sqrt{1 + [2\zeta(\omega/\omega_n)]^2}}{\sqrt{[1 - (\omega/\omega_n)^2]^2 + [2\zeta(\omega/\omega_n)]^2}} = \left|\frac{X}{Y}\right| \tag{3-79}$$

which is in the form we desire.

Equation 3-79 is used to determine what portion of the exciting-force amplitude F_0 is transmitted to the foundation, and the ratio $|F_T|/|F_0|$ which delineates this is referred to as *transmissibility* (*TR*). For example, if an unbalanced rotating machine part were causing a force of amplitude F_0, and Eq. 3-79 showed that $|F_T|/|F_0| = 0.2$, the isolation system would be said to have 20 percent transmissibility.

The ratio $|X|/|Y|$ is also shown in Eq. 3-79, since this equation can also be used to determine what portion of the support-motion amplitude $|Y|$ is being transmitted to the system being excited by the support motion. The reader might refer to Eq. 3-55 at this point, to see that it is identical to Eq. 3-79.

Basic Concepts of Vibration Isolation

Equation 3-79 provides the basic parametric relationships that affect the transmissibility of forces and support motions. Referring to Fig. 3-25, which is a plot of Eq. 3-79, we can see that:

1. The region of vibration isolation begins at $\omega/\omega_n > \sqrt{2}$, since either ratio of Eq. 3-79 must be less than 1 for vibration isolation. Thus, for a given excitation frequency ω, the isolation mounts must be selected so that the natural frequency ω_n of the resulting system is less than $\omega/\sqrt{2}$. Since $\omega_n = \sqrt{k/m}$, and the mass of the mounts is generally much less than the mass of the system, appropriate isolation mounts are usually selected on the basis of their stiffness. However, there are

certain systems for which isolation is accomplished by adding mass to the system when the exciting frequency ω is very low.

2. Since the transmissibility of an exciting force or support motion decreases as ω/ω_n increases in the isolation region, the less stiff the isolation mounts the greater the efficiency of the isolation system. Although damping tends to reduce the efficiency of an isolation system, some damping must be present to minimize the peak response when the system passes through resonance during start-up or shutdown.
3. When $\omega/\omega_n > 3$, the response curves are about the same for different values of damping below 20 percent ($\zeta < 0.2$). This shows that in this region the transmissibility of a force or support motion is relatively unaffected by changing the damping. This is a fortunate feature of vibration isolation, since accurate values of ζ are generally not known.

Since the transmissibility is relatively unaffected by damping in the isolation region as just mentioned above, it is common practice to neglect damping in Eq. 3-79 when isolating a system.

With damping neglected, the transmissibility as expressed by Eq. 3-79 becomes

$$\text{TR} = \frac{1}{(\omega/\omega_n)^2 - 1} \tag{3-80}$$

in which the negative root of Eq. 3-79 has been used so that Eq. 3-80 will yield positive transmissibility values (ω/ω_n is always greater than $\sqrt{2}$ in the isolation region for which Eq. 3-80 is applicable).

The reduction R in transmissibility is given by

$$R = 1 - \text{TR} \tag{3-81}$$

and is used to indicate the efficiency of an isolation system. For example, if the transmissibility of a force through an isolation system is 20 percent, the reduction is 80 percent ($R = 0.80$). From Eqs. 3-80 and 3-81,

$$1 - R = \frac{1}{(\omega/\omega_n)^2 - 1}$$

from which

$$\frac{\omega}{\omega_n} = \sqrt{\frac{2 - R}{1 - R}} \tag{3-82}$$

Equation 3-82 can be used to determine the required stiffness k of an isolation system to accomplish a desired reduction R in the transmissibility of an excitation source of circular frequency ω acting on a system of mass m since $\omega_n = \sqrt{k/m}$. An even more usable form of Eq. 3-82 can be obtained by expressing the circular frequency ω of the excitation in terms of N (rpm or cpm) and the natural circular frequency ω_n in terms of the static deflection $\delta_s\,(W/k)$. Expressing ω as

$$\omega = \frac{2\pi N}{60}$$

and ω_n as

$$\omega_n = \sqrt{\frac{k}{m}} = \sqrt{\frac{kg}{W}}$$

we obtain

$$N = \frac{30}{\pi}\sqrt{\frac{kg\,(2-R)}{W\,(1-R)}} \qquad \text{(rpm or cpm)} \tag{3-83}$$

Substituting

$$W = k\delta_s$$

into Eq. 3-83 yields an alternate form of this equation as

$$N = \frac{30}{\pi}\sqrt{\frac{g\,(2-R)}{\delta_s\,(1-R)}} \qquad \text{(rpm or cpm)} \tag{3-83a}$$

In these equations:

k = stiffness of isolation system (lb/in. or kN/mm)
g = acceleration of gravity (386 in./s^2 or 9810 mm/s^2)
$\delta_s = W/k$ = static deflection (in. or mm)
W = weight of machine or structure (lb or kN)

Equation 3-83 can be used to determine the stiffness k of an isolation system that will accomplish a desired reduction R in the transmissibility of an excitation source of frequency N acting on a machine or structure of weight W.

Further examination of Eq. 3-83 reveals that for a given excitation speed N, the smaller the magnitude of k the larger the reduction R in the transmissibility (the smaller TR is).

Equation 3-83 can be plotted on log-log paper to facilitate the design of isolation systems by providing a graph such as the one shown in Fig. 3-36. Taking the logarithm of both sides of Eq. 3-83a gives

$$\underbrace{\log N}_{y} = \underbrace{-\frac{1}{2}\log \delta_s}_{mx} + \underbrace{\log\left[\frac{30}{\pi}\sqrt{\frac{g(2-R)}{(1-R)}}\right]}_{b} \tag{3-84}$$

which has the form of the equation of a straight line, $y = mx + b$, and plots as such on log-log paper. The upper part of Fig. 3-36 shows a family of curves obtained from this equation with R as a parameter. These curves can be used to determine the stiffness k that a system must have for a specified reduction in transmissibility. The lower family of curves shows the magnification of the exciting force or support motion that occurs at values of ω/ω_n less than 1 (to the left of resonance on a response curve such as shown in Fig. 3-25 if the damping were zero).

It is usually difficult to provide isolation at *very low* excitation frequencies. At these frequencies, the static deflection δ_s can become so large that isolation becomes impractical (see Example 3-9).

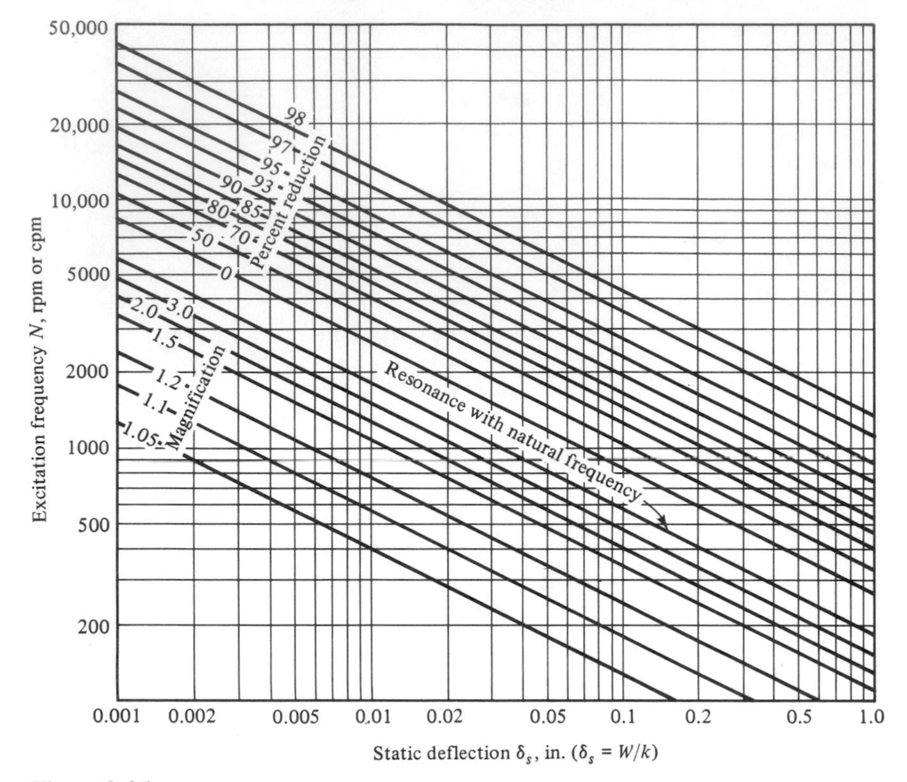

Figure 3-36

When it becomes necessary to provide a highly efficient isolation system ($R \geq 90$ percent) at *fairly low* excitation frequencies, the machine or instrument to be isolated is sometimes attached to, or rested upon, a rather large mass M such as that shown in Fig. 3-37. By using a fairly large mass such as a block of concrete, for example, a system's natural frequency becomes

$$f_n = \frac{\sqrt{k/(m + M)}}{2\pi}$$

and frequencies as low as 1.5 Hz can be obtained in this manner without undue difficulty. Equation 3-82 reveals that with $f_n = 1.5$ Hz, a 90 percent reduction ($R = 0.9$) can be obtained with a low excitation frequency of around 5 Hz. Equation 3-83a shows that with the above values, the system would have a static deflection of $\delta_s = 4.3$ in., which is a practical one.

It is often necessary to consider isolating a system for more than one excitation frequency. For example, in reciprocating engines there is a *primary* excitation frequency ω_1 and a *secondary* excitation $2\omega_1$. In some instances, the speed of a machine may vary during its operation and in so doing produces a *range* of excitation frequencies. In such instances, it should be apparent from Fig. 3-25 that the *lowest* excitation frequency is the one of primary importance, since $|X|/|Y|$ or $|F_T|/|F_0|$ decreases as ω/ω_n increases. That is, the reduction R for an excitation frequency ω_2 would be even greater than that for ω_1 when $\omega_1 < \omega_2$.

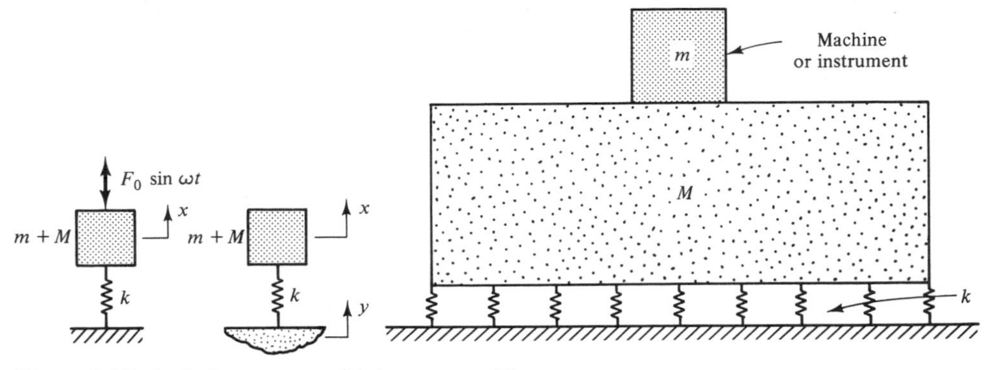

Figure 3-37 Isolation system with large mass M.

EXAMPLE 3-8

The machine shown in Fig. 3-38 has an armature with a small imbalance that is causing a force F_T to be transmitted to the foundation upon which the machine rests.

The machine weighs 4000 lb and has an operating speed of 2000 rpm. It is desired to reduce the amplitude of the transmitted force by 80 percent, using isolation pads represented by the springs shown in the figure.

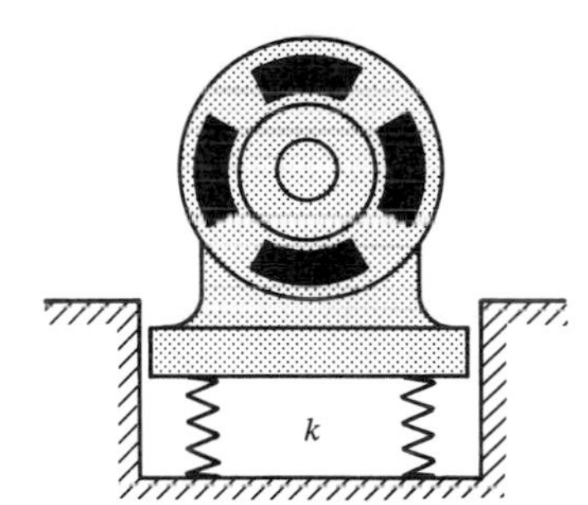

Figure 3-38 Machine with unbalanced rotating element

Solution. Figure 3-36 shows that the 80 percent reduction line intersects the horizontal 2000-rpm line at a point having an abscissa of $\delta_s = W/k = 0.05$ in. Thus, the desired stiffness of the isolation system is

$$k = \frac{4000}{0.05} = 80{,}000 \text{ lb/in.}$$

EXAMPLE 3-9

To illustrate the problems encountered in isolating systems from low-frequency excitations, let us consider the motion of a very large ship. The general motion of such a ship is a combination of motions about three mutually perpendicular axes. These motions, referred to as yaw, pitch, and roll, occur about the x, y, and z axes, respectively, as shown in Fig. 3-39a.

In calm seas the amplitudes of both pitch and yaw for a large ship are relatively small. However, rolling motion is generally always present, and is bothersome to passengers prone to motion sickness.

Therefore, it is desired to make a preliminary analysis of the feasibility of isolating

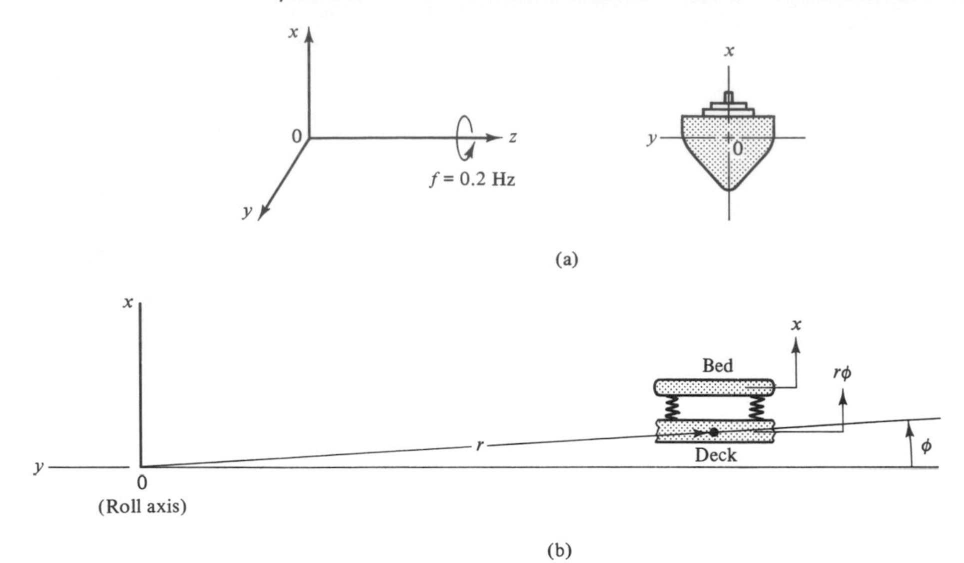

Figure 3-39 Ship of Example 3-9.

the beds on the ship from the rolling motion of the ship. The complete motion of the beds due to the roll of the ship consists of both translation and rotation. However, in this preliminary analysis, we neglect the angular motion of the beds and treat them as particles to simplify our analysis, and determine only the feasibility of isolating the *vertical* motion of the beds.

The roll frequency has been recorded as averaging approximately 0.2 Hz, and it is felt that eliminating 80 percent of the vertical motion would be sufficient.

Solution. Since 0.2 cps corresponds to 12 rpm, and $R = 0.8$, Eq. 3-83a becomes

$$12 = \frac{30}{\pi}\sqrt{\frac{386}{\delta_s}\frac{(2-0.8)}{(1-0.8)}}$$

from which

$$\delta_s = 1466.6 \text{ in.} = 122.2 \text{ ft}$$

Quite obviously such an isolation system would be impractical because of the space required to allow the springs or mounting pads such extreme static deflections.

Since the complete motion of the beds due to roll of the ship consists of both translation and rotation as mentioned earlier, a more complete analysis would include both motions (see Prob. 6-6).

EXAMPLE 3-10

As an example of changing the mass of a system to aid in isolating it, consider the compressor shown in Fig. 3-40a. The compressor weighs 2060 lb (9163 N), and operates at 1800 rpm. At this operating speed, undesirable vibration occurs when the compressor is attached directly to the floor of a small building in a manufacturing plant.

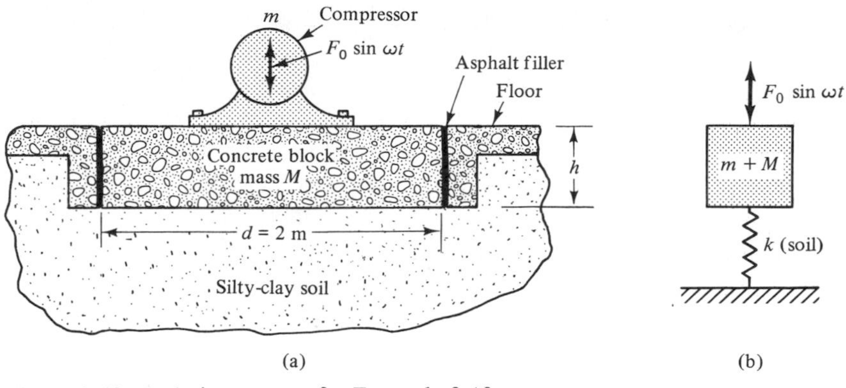

Figure 3-40 Isolation system for Example 3-10.

To reduce the annoying, and potentially damaging, vibration of the concrete floor that is resting on a silty-clay soil, it is proposed to isolate the compressor by mounting it on a square concrete block separated from the rest of the concrete floor as shown in Fig. 3-40a. The density of the concrete in the block has been measured as 23,563 N/m^3, and the *vertical compression* coefficient k_c of the silty-clay soil has been experimentally determined to be

$$k_c = 20.36(10)^6 \text{ N/m}^3$$

The geometry of the compressor leads to choosing a block 2 by 2 m, and our analysis is concerned with determining the depth h of the block that will yield a 75 percent reduction in the force transmitted by the compressor-block system to the supporting soil.

Solution. A study of the system leads to the assumption that it can be modeled as a simple spring-and-mass system as shown in Fig. 3-40b, with an effective mass equal to that of the compressor and block together, and with the soil acting as a spring of stiffness k.

The effective mass is thus

$$m_e = m + M = \frac{9163 + 2(2)(23{,}563)h}{9.81} = (934.1 + 9607.7h) \text{ kg}$$

The spring constant k of the soil is obtained by multiplying the bearing area by the compression coefficient k_c to obtain

$$k = k_c A = 20.36(10)^6(4) = 81.44(10)^6 \text{ N/m}$$

For $R = 0.75$, Eq. 3-82 gives

$$\frac{\omega}{\omega_n} = \sqrt{\frac{2 - 0.75}{1 - 0.75}} = 2.24$$

Since the operational speed of the compressor is 1800 rpm,

$$\omega_n = \frac{\omega}{2.24} = \frac{1800(2\pi)}{60(2.24)} = 84.15 \text{ rad/s}$$

Knowing the natural circular frequency, we can write that

$$84.15 = \sqrt{\frac{k}{m_e}} = \sqrt{\frac{81.44(10)^6}{934.1 + 9607.7h}} \quad \text{rad/s}$$

from which

$$h = 1.1 \text{ m} = 3.61 \text{ ft}$$

Thus, in supporting the compressor on a block of concrete weighing 103.7 kN, separating this block from the rest of the floor, and letting the block rest on the silty-clay soil, the undesirable vibrations in the floor of the building are reduced considerably.

3-9 TRANSDUCERS (FORCE AND VIBRATORY MOTION)

Transducers are generally described as devices that transfer energy from one form to another, and in so doing can have many different configurations. Our discussion is concerned with transducers that are used to measure dynamic forces and vibratory motion. Such transducers are usually modeled as simple spring-and-mass systems with viscous damping.

In using transducers to measure dynamic phenomena, the "flat response" frequency region of the transducer must be known if reliable data are to be obtained. It is only in this region that the response (the output) of the transducer is essentially independent of the frequency components present in the dynamic phenomenon being measured. As we shall see later, this "flat response" region of a transducer depends upon the ratios formed by the circular frequencies ω_i present in the dynamic phenomenon being measured and the natural circular frequency ω_n of the transducer. This fact dictates particular transducers for particular phenomena; so it is very helpful if the user has some idea of the frequency range of the phenomenon being measured to minimize the trial and error required in selecting an appropriate transducer.

Force Transducers

Load cells or pressure transducers that are used to measure dynamic forces or pressures frequently utilize resistance strain gages bonded to the elastic elements of the transducer to sense the strains resulting from the forces the transducer experiences.

The dynamic characteristics of force transducers can be determined by observing the characteristics of a spring-and-mass system subjected to a force $F_0 \sin \omega t$ as shown in Fig. 3-41. The output of the transducer is proportional to the strain of the elastic element of the transducer, and corresponds to the displacement x of the mass shown.

The magnification factor of a viscously damped spring-and-mass system is given by Eq. 3-20, and is repeated here as

$$\frac{|X|}{F_0/k} = \frac{1}{\sqrt{[1 - (\omega/\omega_n)^2]^2 + [2\zeta(\omega/\omega_n)]^2}} \tag{3-20}$$

Thinking in terms of a transducer, ω is a circular frequency of the system being measured, and ω_n is the natural circular frequency of the transducer.

The plot of Eq. 3-20 shown in Fig. 3-6 reveals that $|X|/(F_0/k)$ is very nearly *unity*

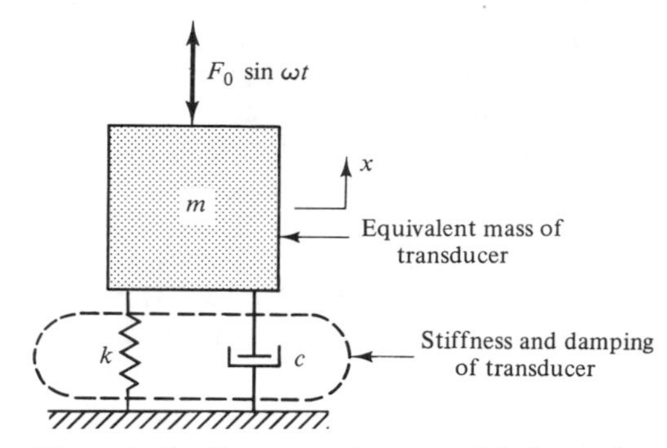

Figure 3-41 Force transducer modeled as spring-and-mass system.

for small values of ω/ω_n. This means that the amplitude $|X|$ of the response is essentially the same as the *static* displacement F_0/k resulting from a statically applied force of magnitude F_0 in this "flat-response" frequency region.

To examine this region in more detail, values of $|X|/(F_0/k)$ for $0 \leq \omega/\omega_n \leq 0.4$ are shown in Fig. 3-42. The curves plotted in this figure indicate that:

1. If $\omega/\omega_n \leq 0.2$, the response of the transducer is essentially independent of the excitation frequency ω, and $|X|/(F_0/k) \cong 1$ in this flat-response region.
2. The maximum error in the deviation of $|X|/(F_0/k)$ from unity in the "flat-response" region is less than 5 percent, regardless of the magnitude of damping $(0 \leq \zeta \leq 1.0)$.

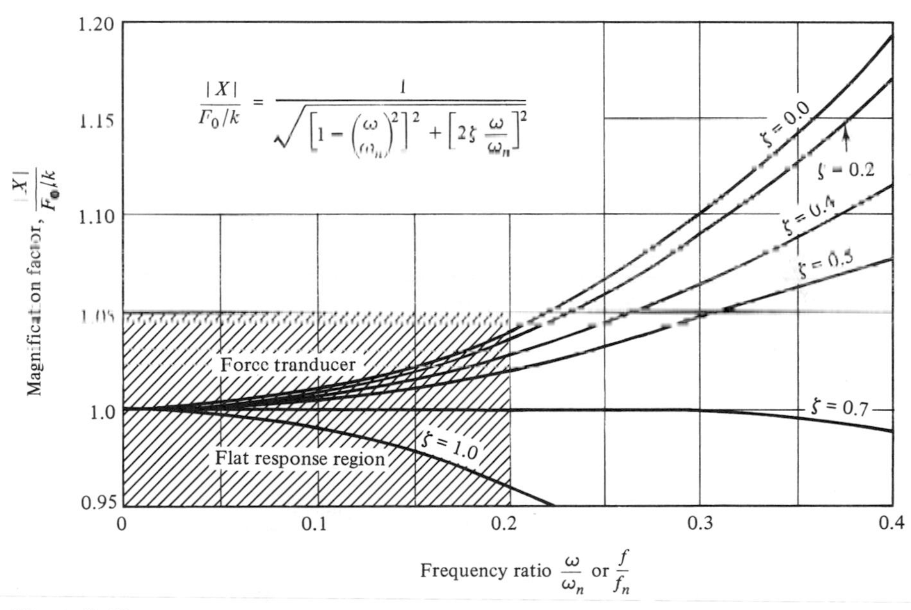

Figure 3-42

These observations lead us to the conclusion that accurate force measurements can be obtained from a force transducer when it is used to measure forces having frequencies up to approximately 20 percent of the fundamental natural frequency of the transducer. For example, a transducer having a fundamental natural frequency of 1000 Hz would yield accurate measurements of a dynamic force if the frequency of that force were no greater than 200 Hz.

Dynamic forces generally contain more than one frequency component, and when such multiple frequencies $f_1 < f_2 < \cdots < f_i$ are present, the highest frequency component f_i of the force should be less than 20 percent of the natural frequency f_n of the transducer since $f_1/f_n < f_2/f_n < \cdots < f_i/f_n \leq 0.2$.

From the above discussion it should be apparent that a high fundamental natural frequency is an important characteristic of force transducers. The static calibrations and fundamental natural frequencies of commercial force and pressure transducers are furnished by the manufacturer.

Force transducers that utilize a *piezoelectric* material (usually a polarized ferroelectric ceramic) as the sensing element can have fundamental natural frequencies well above 100,000 Hz. This is the type of transducer used in the impact hammer discussed in Sec. 1-7.

The deformation of the piezoelectric element shown in Fig. 3-43a produces a charge q on the pole faces that is proportional to the force $F(t)$. Since the voltage V (volts), charge q (coulombs), and capacitance C (farads) are related by $V = q/C$, the *sensitivity* of piezoelectric force transducers can be measured either in terms of picocoulombs (pC $= 10^{-12}$ coulomb) per unit of force (lb or N) or in terms of millivolts per unit of force.

Since the capacitance C includes the capacitance of both the piezoelectric element and the cable connecting the transducer to some instrument such as an oscilloscope or

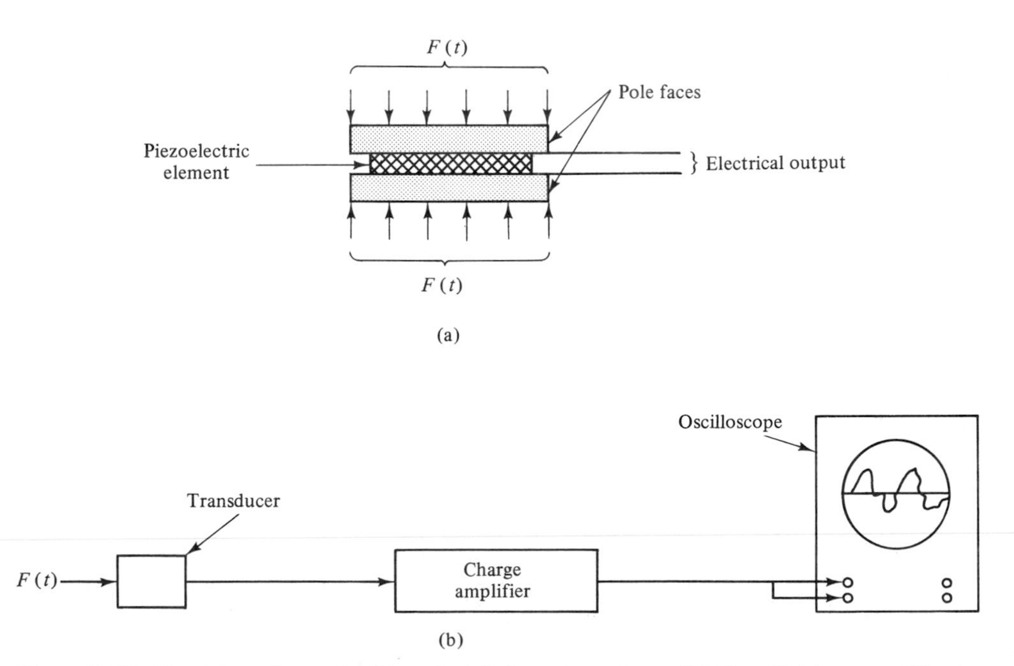

Figure 3-43 Force transducer. (a) Piezoelectric force transducer. (b) Use of charge amplifier to eliminate cable effect.

oscillograph, the apparent voltage V will be reduced considerably by the capacitance of the cable. However, this loss can be eliminated by adding a charge amplifier (signal conditioner) as shown in Fig. 3-43b, which is a common practice.

Phase Distortion

Phase distortion causes changes in the shape of a wave in the time domain. If a transducer is to accurately measure or follow a force $F(t)$ that has multiple frequency components, the phase shift of all the frequency-response components must be either zero or equal.

For example, consider that the force $F(t)$ in Fig. 3-44 is described by

$$F(t) = F_1 \sin \omega_1 t + F_3 \sin \omega_3 t \tag{3-85}$$

in which $\omega_3 = 3\omega_1$. Assume that the transducer has an ω_n such that $\omega_3/\omega_n \leq 0.2$, which ensures that the response is in the "flat-response" range since ω_3 is the higher-frequency component. From Eq. 3-17, the transducer would then have the steady-state response

$$x(t) = \frac{F_1}{k} \sin(\omega_1 t - \phi_1) + \frac{F_3}{k} \sin(\omega_3 t - \phi_3) \tag{3-86}$$

where

$$F_1/k = |X_1|$$

$$F_3/k = |X_3|$$

$$\tan \phi_1 = \frac{2\zeta(\omega_1/\omega_n)}{1 - (\omega_1/\omega_n)^2}$$

$$\tan \phi_3 = \frac{2\zeta(\omega_3/\omega_n)}{1 - (\omega_3/\omega_n)^2}$$

It should be apparent from Eqs. 3-85 and 3-86 that the response $x(t)$, which is the output of the transducer, would have the same waveform as the applied force $F(t)$ if $\phi_1 = \phi_3 = 0$, which would be theoretically true for zero damping. Fortunately, the damping inherent in force transducers is very small, so that they do approach the condition of zero phase shift. The damping factor ζ is generally less than 0.002 for metal-element transducers, and less than 0.001 for those with piezoelectric elements.

Referring again to the force $F(t)$ given by Eq. 3-85, if we assume that $\omega_3/\omega_n = 0.2$, which is the upper limit of the "flat-response" range, and also assume an upper value of $\zeta = 0.002$, we can use the appropriate expression given above for $\tan \phi_3$ to find that ϕ_3 is only 0.00083 rad (0.0475°). Since ω_1/ω_n would be less than 0.2, the phase angle ϕ_1 would be even smaller than ϕ_3, which illustrates that phase distortion is generally not a problem in using force transducers.

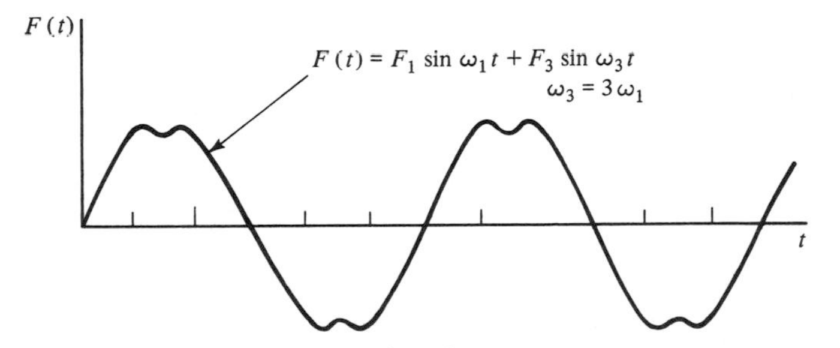

Figure 3-44 Force with multiple frequency components.

As mentioned previously, phase distortion can also be eliminated when all the frequency components shift equally. If Eq. 3-86 is written as

$$x(t) = \frac{F_1}{k} \sin \omega_1\left(t - \frac{\phi_1}{\omega_1}\right) + \frac{F_3}{k} \sin \omega_3\left(t - \frac{\phi_3}{\omega_3}\right)$$

it can be seen that each frequency component is shifted in the time domain by an equal amount if $\phi_1/\omega_1 = \phi_3/\omega_3$. This shows that equal shifting will occur if each phase angle is proportional to its corresponding excitation frequency. Referring to Fig. 3-8, it can be seen that for the particular value of $\zeta = 0.707$, with $\omega/\omega_n \leq 1.0$, the phase angle ϕ is essentially a linear function of ω/ω_n. That is,

$$\phi = \frac{\pi}{2}\frac{\omega}{\omega_n} \qquad \left(\frac{\omega}{\omega_n} \leq 1.0\right)$$

Since phase distortion is negligible in most force transducers because of their low inherent damping as discussed previously, there is no practical reason to make $\zeta = 0.707$ in designing force transducers.

Vibration-Measuring Transducers

Two general types of seismic transducers are used for vibration measurements. One is the *vibrometer* (sometimes referred to as a seismometer), which is designed with a low natural frequency so that its natural circular frequency ω_n is low compared with the circular frequency ω of the vibratory motion to be measured. The other is the *accelerometer,* which is designed to have a high natural frequency, so that its natural circular frequency ω_n is high compared with the circular frequency ω of the vibratory motion to be measured. As a result, vibrometers are generally thought of as low-frequency transducers, while accelerometers are referred to as high-frequency transducers.

Most vibrometers are electromagnetic transducers, which consist of a moving mass m within a coil and a permanent magnet fixed to the case as shown in Fig. 3-45a. In some

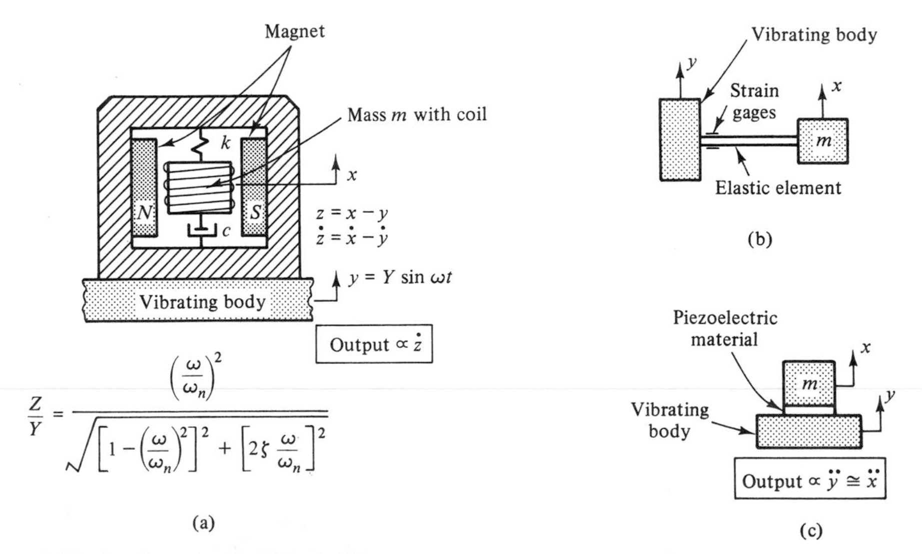

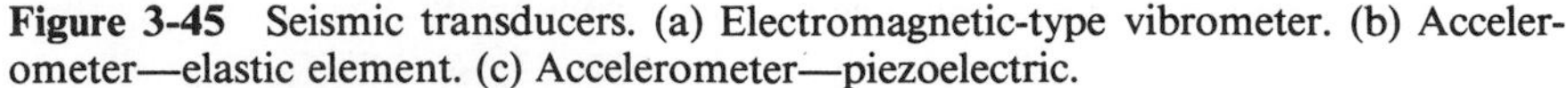
Figure 3-45 Seismic transducers. (a) Electromagnetic-type vibrometer. (b) Accelerometer—elastic element. (c) Accelerometer—piezoelectric.

electromagnetic transducers, the moving mass is the permanent magnet and the coil is fixed to the transducer case. In either case, the voltage output from the coil is proportional to the rate at which the magnetic flux lines are cut. This means that the voltage output is proportional to the *relative velocity* between the mass m and the vibrating body.

One type of transducer used for measuring accelerations consists basically of a mass m attached to some type of elastic element such as the small cantilever beam shown in Fig. 3-45b. When it is mounted on a vibrating body, its output is proportional to the *absolute acceleration* of the mass m, which is in turn essentially equal to the acceleration of the vibrating body (the support acceleration). Electrical-resistance strain gages are sometimes used to sense this acceleration since the strain in the elastic element caused by the inertia of the mass m is proportional to the absolute acceleration of the mass m.

At present, the most widely used accelerometer for measuring shock and vibration is the *piezoelectric* accelerometer, which is similar in many ways to the piezoelectric force transducer shown in Fig. 3-43a. The basic part of such a transducer is of course a piezoelectric material that generates an electrical charge q when stressed in either tension or compression, or in shear.

The basic theory needed to understand and use any of the vibration-measuring transducers shown in Fig. 3-45 can be acquired by analyzing simple spring-and-mass systems, which we shall proceed to do for each type of transducer shown, beginning with the vibrometer.

Vibrometer (Low-Frequency Transducer)

To determine the characteristics of this type of transducer and its useful frequency range, we refer to Eq. 3-61, which is applicable to the spring-and-mass system shown in Fig. 3-45a. It is repeated here for convenience as

$$\frac{|Z|}{|Y|} = \frac{(\omega/\omega_n)^2}{\sqrt{[1 - (\omega/\omega_n)^2]^2 + [2\zeta(\omega/\omega_n)]^2}} \tag{3-61}$$

Referring to Fig. 3-28, which is a plot of Eq. 3-61, we can see that when $\omega/\omega_n > 3$, the steady-state amplitude ratio $|Z|/|Y| \cong 1$ for a wide range of damping factors. Therefore, in this frequency range,

$$|Z| \cong |Y| \tag{3-87}$$

and since

$$|Z| = |X - Y|$$

$$X \cong 0$$

This means that the mass m remains essentially stationary as the case moves with the vibrating body.

Multiplying the first of the expressions in the paragraph above by the circular frequency ω of the vibrating body, we obtain the velocity relationship

$$|Z|\omega \cong |Y|\omega \qquad \text{(velocity amplitudes)} \tag{3-88}$$

which means that

$$\dot{z} \cong \dot{y} \qquad \text{(velocities)}$$

Therefore, the output voltage of an electromagnetic-type transducer, which is proportional to the relative velocity $\dot{z}$, is also essentially proportional to the velocity $\dot{y}$ of the vibration being measured. A vibration meter used with this type of transducer containing integrating and differentiating networks will yield direct readings of the displacement y and the acceleration $\ddot{y}$ of the vibrating body to which the transducer is attached.

The usable frequency range of a vibrometer depends upon its natural frequency ω_n, the damping present, and the accuracy desired in the approximation $|Z|\omega \cong |Y|\omega$. Without some type of damping mechanism in addition to the damping inherent in the vibrometer, its natural frequency ω_n should be no greater than one-third the frequency ω of the system being measured ($\omega/\omega_n > 3$). Electrical damping is frequently introduced into this type of transducer to make $\zeta \cong 0.7$, since this magnitude of damping extends the lower end of the "flat-response" range down to $\omega/\omega_n \cong 2$ (see Fig. 3-28). In addition to extending the frequency range, this magnitude of damping also eliminates phase distortion, as explained previously in the discussion of a force transducer.

When such a highly damped transducer is used for measuring ground motion, it would typically have a natural frequency of approximately 1 Hz and would be capable of accurately measuring ground motion with frequency components as low as 2 Hz.

Accelerometer (High-Frequency Transducer)

The beam and mass shown in Fig. 3-45b comprise a simple spring-and-mass system that can be used as an accelerometer as mentioned earlier. If the natural circular frequency ω_n of this system is considerably greater than the circular frequency ω of the vibrating body it is attached to, the ratio ω/ω_n is small and Eq. 3-61 reduces to

$$\frac{|Z|}{|Y|} \cong \left(\frac{\omega}{\omega_n}\right)^2$$

or

$$|Z| \cong \frac{|Y|\omega^2}{\omega_n^2} = \frac{\text{acceleration amplitude of vibrating body}}{\omega_n^2} \tag{3-89}$$

which is good for $\omega/\omega_n \leq 0.2$.

Equation 3-89 reveals that the relative *displacement* amplitude $|Z|$ is proportional to the *acceleration* amplitude $|Y|\omega^2$ of the vibrating body to which the accelerometer is attached. However, examination of Eq. 3-55, which is repeated below for convenience, shows that the displacement amplitude $|X|$ of the accelerometer mass is essentially equal to the displacement amplitude $|Y|$ of the vibrating body when $\omega/\omega_n \ll 1$. It follows then that $|X|\omega^2 = |Y|\omega^2$, and Eq. 3-89 can be written as

$$|Z| \cong \frac{|X|\omega^2}{\omega_n^2} = \frac{\text{acceleration amplitude of accelerometer mass } m}{\omega_n^2}$$

Since the acceleration amplitude of the mass element of the accelerometer is essentially equal to the acceleration amplitude of the vibrating body, the acceleration of the vibrating body is proportional to the strain accompanying the deflection of the beam due to the inertia effect $m\ddot{x}$.

The largest error in the approximation $|X| \cong |Y|$ occurs when the damping is theoretically zero, and Eq. 3-55, which is repeated here

$$\frac{|X|}{|Y|} = \frac{\sqrt{1 + [2\zeta(\omega/\omega_n)]^2}}{\sqrt{[1 - (\omega/\omega_n)^2]^2 + [2\zeta(\omega/\omega_n)]^2}} \tag{3-55}$$

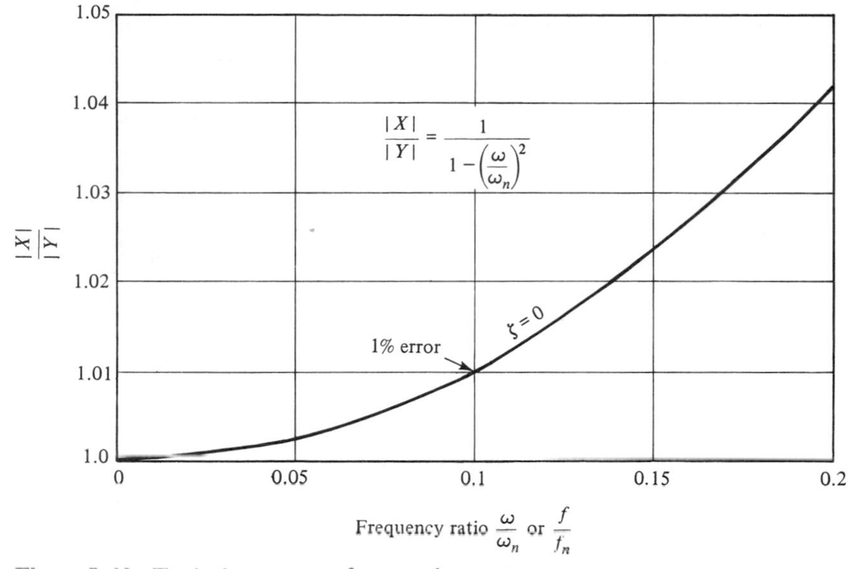

Figure 3-46 Typical response of an accelerometer.

reduces to

$$\frac{|X|}{|Y|} = \frac{1}{1 - (\omega/\omega_n)^2} \tag{3-90}$$

A plot of Eq. 3-90 shown in Fig. 3-46 shows the accuracy of the approximation as a function of ω/ω_n for $\zeta = 0$. It is apparent from this plot that the steady-state amplitude $|X|$ of the mass element m of the accelerometer is about 4 percent greater than the amplitude $|Y|$ of the vibrating body to which it is attached when $\omega/\omega_n = 0.2$ but is greater by only 1 percent or less when $\omega/\omega_n \leq 0.1$.

Let us now assume that the base of an accelerometer is subjected to the periodic motion

$$y = Y_1 \sin \omega_1 t + Y_2 \sin \omega_2 t + Y_3 \sin \omega_3 t$$

in which $\omega_1 < \omega_2 < \omega_3$. Differentiating this equation gives the base acceleration as

$$\ddot{y} = -Y_1\omega_1^2 \sin \omega_1 t - Y_2\omega_2^2 \sin \omega_2 t - Y_3\omega_3^2 \sin \omega_3 t \tag{3-91}$$

Considering the error to be negligible in assuming that $X_1 = Y_1$, $X_2 = Y_2$, and $X_3 = Y_3$ when $\omega_3/\omega_n \leq 0.1$, we can write Eq. 3-91 as

$$\ddot{y} = -X_1\omega_1^2 \sin \omega_1 t - X_2\omega_2^2 \sin \omega_2 t - X_3\omega_3^2 \sin \omega_3 t \tag{3-92}$$

Since the right side of Eq. 3-92 is $\ddot{x}$, the acceleration $\ddot{x}$ of the mass element m of the accelerometer is essentially the same as the base acceleration $\ddot{y}$, so that the acceleration of the mass element of the accelerometer can be used as an output representing the acceleration of the base to which it is attached (the vibrating body).

Piezoelectric Accelerometer

As mentioned previously, the most widely used transducer for shock and vibration measurements at present is the piezoelectric accelerometer. There are basically two types of

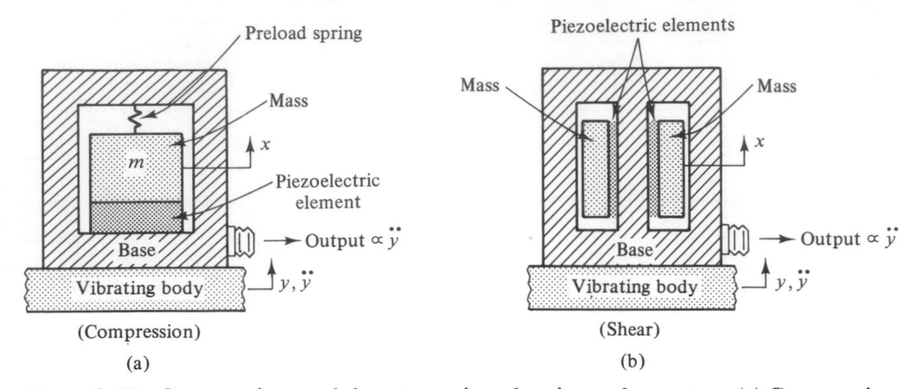

Figure 3-47 Compression- and shear-type piezoelectric accelerometers. (a) Compression. (b) Shear.

this accelerometer, as shown in Fig. 3-47. The *shear* type is less affected by airborne vibration (sound) than is the *compression* type. There is also an *isoshear* type not shown, which consists of multiple piezoelectric elements in shear. Characteristics of this type include high sensitivity and high signal-to-noise ratio.

Both compression and shear piezoelectric elements produce an electrical charge q that is proportional to the base acceleration $\ddot{y}$ when subjected to a force F ($F = m\ddot{x} \cong m\ddot{y}$).

These accelerometers are small, rugged, and reliable transducers that have quite stable characteristics over long periods of time. Their small size (typically 0.25 to 0.75 in. diameter) facilitates their use in small confined areas, and their light weight (typically 0.2 to 20 g) permits their use on lightweight test objects without appreciably affecting the vibration characteristics being measured.

For most acceleration measurements (10 g's or less), piezoelectric accelerometers are easily mounted using a wax-type material such as beeswax. When the acceleration is above 10 g's, other means should be employed, such as an epoxy cement or threaded mounting studs.

Another advantage of piezoelectric accelerometers over other types of transducers is that they are self-generating and as such require no external power supply. Small ones with correspondingly small mass elements can have natural frequencies of well over 100,000 Hz and will provide accurate measurements for frequency components of up to around 10,000 Hz, with negligible phase distortion. Natural frequencies of 30,000 to 50,000 Hz are typical for general-purpose piezoelectric accelerometers, depending upon the design.

Charge amplifiers were mentioned earlier in the discussion of piezoelectric force transducers, and they are also used with piezoelectric accelerometers to reduce sensitivity loss due to the capacitance in the cable connecting the transducer to such instruments as oscilloscopes and frequency analyzers. The sensitivity is usually given as picocoulombs per g of acceleration (pC/g), and the output of the charge amplifier is volts/g or millivolts/g.

3-10 STEADY-STATE VIBRATION MEASUREMENTS

The use of damped-free-vibration records as a means of experimentally measuring system damping was discussed in Sec. 2-4. Let us now consider some basic techniques used to obtain measurements pertinent to the steady-state dynamic behavior of structural systems.

The measurement of natural frequencies and damping is of primary importance in analyzing the steady-state characteristics of systems. The response measurements of small structural components subjected to an excitation are commonly used to study the characteristics of materials. For example, a beam will exhibit changes in its fundamental natural frequency and/or damping when it is subjected over a period of time to a hostile environment that damages the material of which the beam is composed.

The decrease observed in the natural frequency of concrete beam specimens has been used for years as a criterion for determining the resistance of concrete materials to alternating freezing and thawing cycles (ASTM Method C215). The formation of microscopic cracks in the concrete specimens resulting from the freezing and thawing causes a decrease in the modulus of elasticity of the concrete. Since the frequency of a beam depends upon its modulus of elasticity (see Sec. 2-9), the decrease in natural frequency observed can be used to measure the damage (the deterioration) caused by the freezing and thawing.

Vibration tests are also used to determine the environmental effects of temperature, moisture, and so forth on materials such as elastomers and materials used in composites such as Kevlar, graphite, and glass. In recent years material damping has received considerable attention in studies concerned with fatigue failure.[10]

To illustrate the measurements of natural frequencies and damping by means of forced-vibration tests, let us consider a typical test setup such as the one shown in Fig. 3-48a. The beam specimen is supported on soft rubber pads at the *nodal points* (points of zero displacement) of the fundamental mode of a free-free beam. It can be shown that these points are located at a distance of $0.23l$ from each end of the beam (see Table 8-1 on page 608 for the shape function of a free-free beam).

The specimen is subjected to a sinusoidal driving force, $F = F_0 \sin 2\pi ft$, at the center of the beam by a small electromechanical driver. The transverse vibration of the beam is monitored by an accelerometer attached to the beam as shown in Fig. 3-48a.

The input voltages to the horizontal and vertical axes of the oscilloscope are

$$\left.\begin{aligned} e_h &= C_1 \sin(2\pi ft) && \text{(horizontal axis—from driver)} \\ e_v &= C_2 \sin(2\pi ft - \phi) && \text{(vertical axis—from accelerometer)} \end{aligned}\right\} \tag{3-93}$$

in which C_1 and C_2 are the amplitudes of the inputs, and ϕ is the phase angle between e_h and e_v.

At resonance, the frequency f of the driving force is equal to the fundamental frequency f_n of the beam. When $f = f_n$, the response of the beam is a maximum and the phase angle $\phi = \pi/2$. Thus, at resonance Eq. 3-93 shows that the input voltages to the oscilloscope are

$$\left.\begin{aligned} e_h &= C_1 \sin(2\pi ft) \\ e_v &= -C_2 \cos(2\pi ft) \end{aligned}\right\} \tag{3-94}$$

Squaring each of these expressions and then adding them, we find that

$$\left(\frac{e_h}{C_1}\right)^2 + \left(\frac{e_v}{C_2}\right)^2 = 1$$

which is the equation of an ellipse. This ellipse, which is often referred to as a Lissajous figure, is shown in Fig. 3-48b.

The resonant frequency f_n of the beam is determined by adjusting the frequency f of

[10] P. W. Whaley, P. S. Chen, and G. M. Smith, "Continuous Measurement of Material Damping During Fatigue Tests," *Experimental Mechanics,* vol. 24, no. 4, pp. 342–348, December 1984.

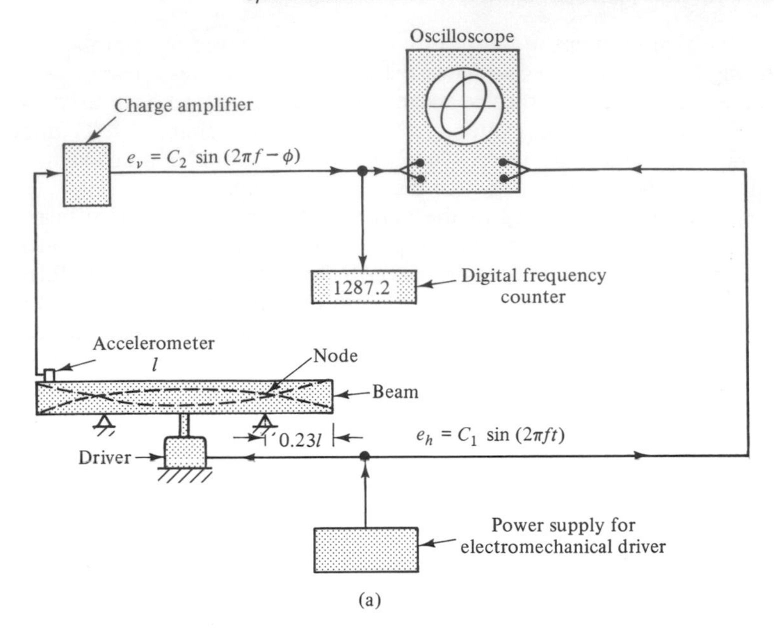

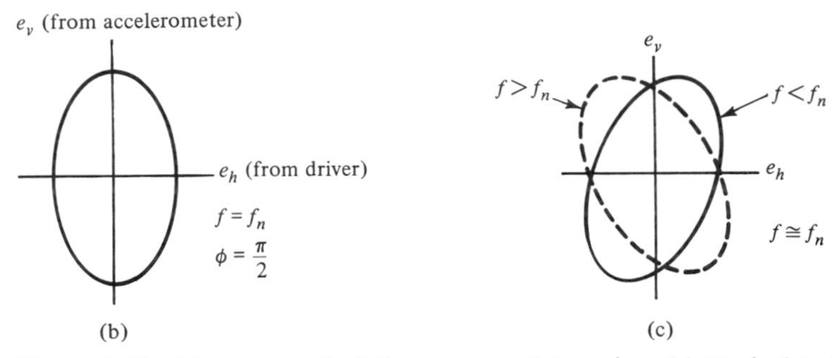

Figure 3-48 Measurement of frequency and damping. (a) Typical test setup. (b) Elliptical pattern at resonance. (c) Elliptical pattern near resonance.

the driver until the elliptical pattern on the oscilloscope is vertical. When the driving frequency is slightly above or below f_n, the elliptical pattern is slightly tilted from the vertical as shown in Fig. 3-48c. The best results are obtained by adjusting the power to the driver and/or the oscilloscope gains so that the horizontal (semiminor) axis of the ellipse is small in comparison with the vertical (semimajor) axis. That is, the natural frequency f_n is most readily detected by observing a very slender vertical ellipse. A *digital* frequency counter should be used if available to determine the excitation frequency f accurately, since dial readings are generally not very accurate.

Assuming that we have determined the natural frequency f_n of the free-free beam using the procedure discussed above, and that we know the weight of the beam and its dimensions, the modulus of elasticity E of the beam material can be calculated from (see Table 8-1 on page 608)

$$f_n = \frac{22.37}{2\pi}\sqrt{\frac{EIg}{Wl^3}}$$

or

$$E = \left(\frac{2\pi f_n}{22.37}\right)^2 \frac{Wl^3}{Ig} \tag{3-95}$$

where f_n = fundamental (first mode) natural frequency (Hz)
E = modulus of elasticity (lb/in.^{2} or N/m^2)
I = area moment of inertia of beam cross section (in.^{4} or m^4)
W = weight of beam (lb or N)
l = length of beam (in. or m)
g = acceleration of gravity (in./s^2 or m/s^2)

Frequency-Phase Method of Measuring Damping

The test setup shown in Fig. 3-48a may also be used to measure damping experimentally by providing data for a method referred to as the *frequency-phase* method.[11] The development of this method begins by assuming that the damping in the beam can be represented by a viscous-damping model and that the beam corresponds to a single-degree-of-freedom system when vibrating in a natural mode. With these assumptions, we may write from Eq. 3-12 that

$$\tan\phi = \frac{2\zeta(f/f_n)}{1-(f/f_n)^2}$$

from which

$$\zeta = \frac{1}{2}\left(\frac{f_n}{f} - \frac{f}{f_n}\right)\tan\phi \tag{3-96}$$

In these equations f is the driving frequency, f_n is the fundamental natural frequency of the beam, and ϕ is the phase angle between the acceleration of the beam and the sinusoidal driving force.

Equation 3-96 reveals that if f, f_n, and ϕ are known, the damping factor ζ can be calculated. A digital frequency counter should be used to measure both f and f_n, since the more accurately they are determined, the more accurate the value of ζ obtained.

The phase angle ϕ can be measured by means of a phase meter or by using measurements obtained from a Lissajous figure on the oscilloscope. To explain the latter method, let us refer to the elliptical patterns shown in Fig. 3-49. At resonance with $f = f_n$ and $\phi = \pi/2$, the elliptical pattern is vertical as shown in Fig. 3-49a. If the driving frequency is then decreased just slightly below the natural frequency, the elliptical pattern tilts from the vertical as shown in Fig. 3-49b.

Observing Fig. 3-49b and the terms in Eq. 3-93, it can be seen that when $2\pi ft = 0$, or multiples of π, $e_h = 0$ and $e_v = \mp C_2 \sin\phi$. The vertical distance $y_1/2$ is thus

$$\frac{y_1}{2} = e_v = C_2 \sin\phi \tag{3-97}$$

[11] G. M. Smith and H. D. Berns, "Frequency Phase Method for Measuring Material Damping," *Materials Research and Standards*, vol. 4, no. 5, pp. 225–227, May 1964.

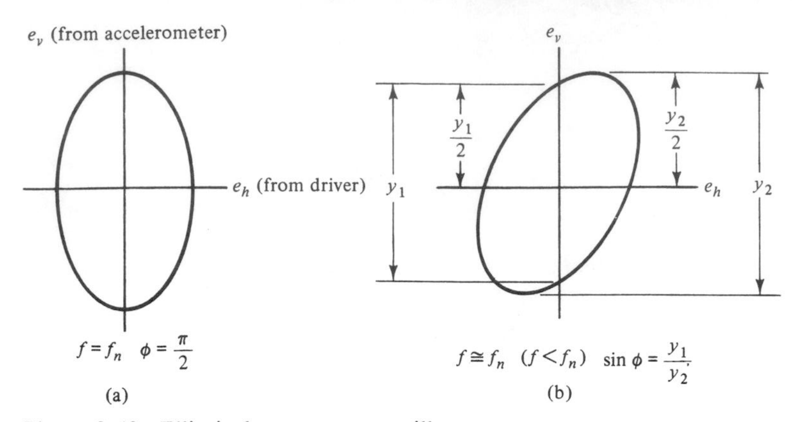

Figure 3-49 Elliptical patterns on oscilloscope screen.

Similarly, when $(2\pi ft - \phi) = \pi/2$, or odd multiples of $\pi/2$, e_v is at its maximum value ($\pm$), and the vertical distance $y_2/2$ is

$$\frac{y_2}{2} = (e_v)_{\max} = C_2 \tag{3-98}$$

Combining Eqs. 3-97 and 3-98,

$$\sin \phi = \frac{y_1}{y_2}$$

or

$$\phi = \sin^{-1} \frac{y_1}{y_2} \tag{3-99}$$

The values for y_1 and y_2 are obtained by measuring the *relative* magnitudes of the vertical distances y_1 and y_2 by any convenient means. One is to merely count the number of grid lines on the oscilloscope screen and use their ratio.

As mentioned above, the y_1/y_2 ratios are generally obtained from an elliptical pattern for which the driving frequency is just slightly below the natural frequency of the beam. While Eq. 3-96 does not indicate any such requirement, there are several reasons for using a frequency ratio close to resonance. One is the increased accuracy inherent in using larger numbers in the y_1/y_2 ratios. More importantly, it minimizes the energy losses at the supports.

The energy losses that occur at supports during vibration are a problem in all methods used to measure damping, and it is difficult to distinguish between the energy lost at the supports and that dissipated by the body internally. Therefore, it is essential that the support energy be reduced as much as possible so that the damping measured is predominantly that of the material.

Frequency-phase test results obtained from a 3- by 4- by 16-in. aluminum beam supported on rubber pads at the nodal points showed that the support damping was negligible when measurements were made with $y_1/y_2 \geq 0.8$. As the excitation frequency deviated further from the fundamental frequency, the higher modes, which do not have nodal points at the same location as the first mode, tended to increase the relative movement between the specimen and the supports, which increased the amount of support energy lost.

Quality Factor as a Measure of Damping

The quality factor Q is frequently used as a measure of damping with Q defined as

$$Q = \frac{1}{2\zeta} \tag{3-100}$$

Since a small damping factor ζ corresponds to a large Q value, a lightly damped system would be referred to as having a high Q factor. Equation 3-100 is associated with the excitation frequencies f_1 and f_2 that correspond to a *magnification factor* of $0.707/2\zeta$ as shown in Fig. 3-50. The frequency f_1 is below the resonant frequency f_n, and f_2 is above it. Recalling Eq. 3-21, we see that at resonance ($f = f_n$), the magnification factor is

$$\frac{|X|}{F_0/k} = \frac{1}{2\zeta} \tag{3-21}$$

The frequencies f_1 and f_2 are referred to as the *half-power* points, and $\Delta f = f_2 - f_1$ is referred to as the *bandwidth* of the system.

By measuring f_n and the bandwidth Δf at the half-power points, an approximate measure of Q can be determined from

$$Q \cong \frac{f_n}{\Delta f} = \frac{f_n}{f_2 - f_1} \tag{3-101}$$

To prove the approximation in Eq. 3-101, we substitute $0.707/2\zeta$ for $|X|/(F_0/k)$ in Eq. 3-20. We then square both sides of Eq. 3-20 and replace ω/ω_n with f/f_n to obtain

$$\frac{1}{2}\left(\frac{1}{4\zeta^2}\right) = \frac{1}{[1 - (f/f_n)^2]^2 + [2\zeta(f/f_n)]^2}$$

which, expanded, yields

$$\left(\frac{f}{f_n}\right)^4 + (4\zeta^2 - 2)\left(\frac{f}{f_n}\right)^2 + (1 - 8\zeta^2) = 0$$

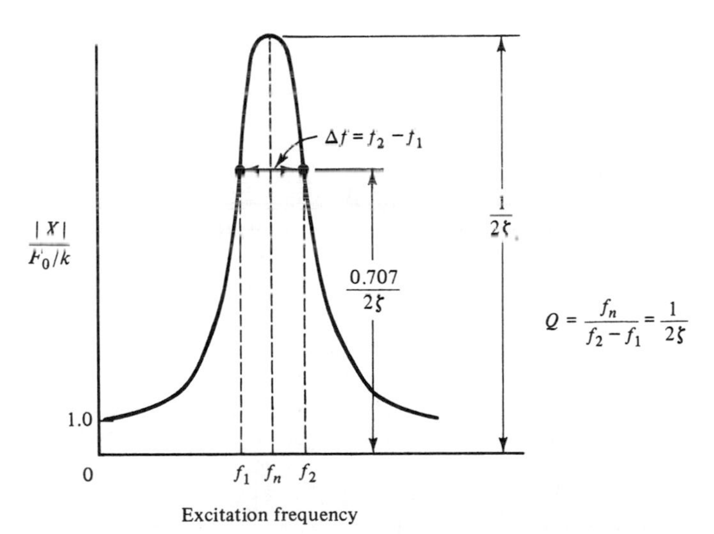

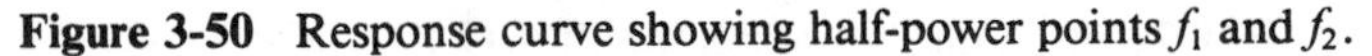

Figure 3-50 Response curve showing half-power points f_1 and f_2.

which is a quadratic in terms of $(f/f_n)^2$. Solving the quadratic shows that

$$\left(\frac{f}{f_n}\right)^2 = (1 - 2\zeta^2) \pm 2\zeta\sqrt{1 + \zeta^2} \tag{3-102}$$

If we now assume that $\zeta \ll 1$, and neglect the higher-order terms (ζ^2), Eq. 3-102 reduces to

$$\left(\frac{f}{f_n}\right)^2 \cong 1 \pm 2\zeta$$

Since f in this equation corresponds to the half-power points f_1 and f_2, we may write that

$$\left(\frac{f_2}{f_n}\right)^2 \cong 1 + 2\zeta \tag{3-103}$$

and that

$$\left(\frac{f_1}{f_n}\right)^2 \cong 1 - 2\zeta \tag{3-104}$$

Subtracting Eq. 3-104 from Eq. 3-103, we obtain

$$\frac{(f_2)^2 - (f_1)^2}{f_n^2} \cong 4\zeta$$

or

$$\frac{(f_2 - f_1)(f_2 + f_1)}{f_n^2} \cong 4\zeta \tag{3-105}$$

Since $(f_2 + f_1) \cong 2f_n$, Eq. 3-105 can be written as

$$\frac{1}{2\zeta} \cong \frac{f_n}{f_2 - f_1} = \frac{f_n}{\Delta f} \tag{3-106}$$

Comparing Eq. 3-106 with the definition of Q expressed by Eq. 3-100 reveals that

$$Q = \frac{1}{2\zeta} \cong \frac{f_n}{\Delta f}$$

which verifies the validity of Eq. 3-101.

There are disadvantages in measuring damping using Eq. 3-101 when compared with using the frequency-phase method discussed earlier. In the first place, the amplitude F_0 of the driving force must be held constant while obtaining the response amplitudes corresponding to the frequencies f_n, f_1, and f_2. To accomplish this, a force transducer would have to be added to the instrumentation shown in Fig. 3-48 to monitor the driving force. In addition, it is difficult to obtain accurate values for f_1 and f_2 corresponding to the response amplitude of $0.707/2\zeta$, since the response curve is very steep in the region of the half-power points. When using an accelerometer to obtain the responses, the acceleration amplitudes may be converted to amplitudes of displacement $|X|$ using the relationship

$$a_0 = |X|\omega^2$$

in which a_0 is the acceleration amplitude.

Since $Q = 1/2\zeta$, it should be apparent that Q may also be determined using the frequency-phase method for determining ζ. Substituting this relationship into Eq. 3-96 shows that

$$\frac{1}{Q} = 2\zeta = \left(\frac{f_n}{f} - \frac{f}{f_n}\right) \tan \phi$$

Thus, Q can be determined quite easily by the frequency-phase method, which does not require monitoring of the driving force and which avoids the difficulties mentioned above in obtaining accurate measurements at the half-power points.

Summary

Because of the complexity of damping mechanisms and the high sensitivity of the response near resonance to minute changes in damping, measured quantities for damping are generally more qualitative than quantitative. Furthermore, amplitude magnitudes, frequencies, specimen sizes and shapes, and support conditions are but a few of the many factors that can affect experimental results when measuring damping.

However, a general knowledge of the approximate magnitudes of damping that can be anticipated for various types of structures, materials, and constraints is essential in making the dynamic analyses required in designing systems. As an example, reinforced-concrete structures may have damping in the neighborhood of 5 to 10 percent, while the damping in welded steel structures is generally much lower, around 1 percent or less.

It is also possible to study the effects that environments have on materials by observing the relative changes in damping that occur with time and with environmental changes. For example, when microscopic cracks that lead to fatigue failure occur in a test specimen, the specimen may indicate these cracks by exhibiting a significant increase in damping.

As mentioned in Sec. 2-2, there are various types of damping mechanisms that occur in different types of systems, including fluid damping, Coulomb friction, and solid or material damping. Since the actual energy-dissipation mechanisms in most systems are only partially understood or even completely unknown, they are obviously difficult to model mathematically.

Fortunately, the vibration characteristics of many real systems with complicated or unknown damping mechanisms are quite similar to those observed for real and theoretical systems having viscous-damping mechanisms. As a result, the damping in such systems can be modeled as viscous damping by using an *equivalent viscous-damping factor* ζ_e. The modeling of damping using equivalent viscous damping is discussed in Sec. 3-11.

The test setup shown in Fig. 3-48 contains some of the more basic equipment that can be used for vibration measurements. However, with the scope of vibration analysis as broad as it is, vibration measurements generally require much more extensive and elaborate equipment than that shown in the figure. For example, digital equipment such as FFT frequency analyzers (see Sec. 1-5) is now commonly used for a variety of vibration measurements.[12]

3-11 EQUIVALENT VISCOUS DAMPING

In prior discussions concerning damped vibrations, a viscous-damping model was used in which the damping force $c\dot{x}$ was assumed to be proportional to the velocity and was generally

[12] P. W. Whaley and P. S. Chen, "Experimental Measurement of Material Damping Using Digital Equipment," *Shock and Vibration Bulletin,* vol. 53, part 4, p. 41, May 1983.

represented schematically by a dashpot. Such a simple viscous-damping model provides ordinary linear differential equations that are solved quite easily mathematically and that enable us to analyze and interpret various vibration phenomena.

Energy-dissipating mechanisms in materials such as internal molecular friction, fluid resistance, and so on are very complicated phenomena and difficult to model mathematically. Such modeling is further complicated by the fact that energy dissipation, or damping, can result from combinations of different types of damping mechanisms.

An equivalent-viscous-damping factor ζ_e can be determined for a nonviscous-damping mechanism by equating the energy dissipated per cycle by a viscous-damping mechanism to that dissipated by the nonviscous-damping mechanism. As we shall see later, the general characteristics of many systems damped by nonviscous mechanisms can be determined by using equivalent viscous damping.

We shall use equivalent viscous damping to investigate the characteristics of some systems having the following nonviscous forms of damping:

1. Velocity-squared damping (fluid damping)
2. Coulomb damping (dry friction damping)
3. Structural damping (internal damping in materials, sometimes referred to as solid or hysteretic damping)

Since the major effect of damping on a forced vibration occurs at or near resonance, equivalent-viscous-damping factors are usually determined at resonance. Coulomb damping is an exception, since as we shall see later, the amplitude of vibration theoretically goes to infinity when $\omega/\omega_n = 1$ for a system with this type of damping.

Energy Dissipated by Viscous Damping

In Sec. 3-2 it was shown that the steady-state response of a viscously damped system is

$$x = |X|\sin(\omega t - \phi)$$

Differentiating, the velocity is

$$\dot{x} = |X|\omega\cos(\omega t - \phi) \tag{3-107}$$

in which ϕ is the phase angle between F_0 and the displacement x, and the amplitude is

$$|X| = \frac{F_0/k}{\sqrt{[1 - (\omega/\omega_n)^2]^2 + [2\zeta(\omega/\omega_n)]^2}} \tag{3-15}$$

The energy dissipated by a viscous-damping force $F_d = c\dot{x}$ for 1 cycle is determined from

$$E_d = \oint c\dot{x}\,dx = c\int_0^{\tau} \dot{x}\left(\frac{dx}{dt}\right)dt$$

or

$$E_d = c\int_0^{\tau} \dot{x}^2\,dt \tag{3-108}$$

in which $\tau = 2\pi/\omega$ (the period of 1 cycle). Substituting Eq. 3-107 into Eq. 3-108,

$$E_d = c|X|^2\omega^2 \int_0^{2\pi/\omega} \cos^2(\omega t - \phi)\,dt$$

or

$$E_d = \frac{c|X|^2\omega^2}{2} \int_0^{2\pi/\omega} [1 + \cos 2(\omega t - \phi)]\, dt \tag{3-109}$$

The integration of Eq. 3-109 yields

$$E_d = \pi c|X|^2\omega \tag{3-110}$$

Remembering that $c/m = 2\zeta\omega_n$ (Sec. 2-2), and noting that $\omega = \omega_n = \sqrt{k/m}$ at resonance, Eq. 3-110 becomes

$$E_d = \pi 2\zeta k|X|^2 \tag{3-111}$$

which gives the energy *dissipated per cycle* by viscous damping at resonance.

Velocity-Squared Damping (Fluid Damping)

Velocity-squared damping is commonly used to describe the damping mechanism of a system vibrating in a fluid medium. The damping force is assumed proportional to the square of the velocity and can be approximated by

$$F_d = \pm\left(\frac{C\rho A}{2}\right)\dot{x}^2 = -\alpha|\dot{x}|\dot{x} \tag{3-112}$$

where $\dot{x}$ = velocity of vibrating body relative to fluid medium (ft/s or m/s)
$|\dot{x}|$ = absolute value of $\dot{x}$
C = drag coefficient (dimensionless)
A = projected area of body perpendicular to $\dot{x}$ (ft^2 or m^2)
ρ = mass density of fluid ($lb \cdot s^2/ft^4$ or kg/m^3)
$\alpha = C\rho A/2$

The damping force is put in the form of the product of $|\dot{x}|$ and $\dot{x}$ as shown in Eq. 3-112 so that it will always be 180° out of phase with the velocity when used in the differential equation of motion that follows.

The differential equation of motion of a single-degree-of-freedom system with velocity-squared damping is *nonlinear* and can be written as

$$\ddot{x} + \frac{\alpha|\dot{x}|\dot{x}}{m} + \frac{k}{m}x = \frac{F_0}{m}\sin \omega t \tag{3-113}$$

in which the damping force $\alpha|\dot{x}|\dot{x}$ always opposes motion of the mass in removing energy from the system. Fortunately, the important characteristics of a system with velocity-squared damping can be determined without having to solve the nonlinear differential equation, by using equivalent viscous damping.

The energy dissipated in 1 cycle by the damping force of Eq. 3-112 is

$$E_d = 2\alpha \int_0^{\tau/2} \dot{x}^2 \frac{dx}{dt}\, dt = 2\alpha \int_0^{\pi/\omega} \dot{x}^3\, dt \tag{3-114}$$

in which the integration is made over just one-half of the period τ and then multiplied by 2 to obtain the energy dissipated over a full cycle. This must be done since $\dot{x}^2$ is 180° out of phase with the velocity over only one-half of each cycle as shown in Fig. 3-51a. The damping force of $F_d = -\alpha|\dot{x}|\dot{x}$ has the form shown in Fig. 3-51b, in which it is always 180° out of phase with the velocity.

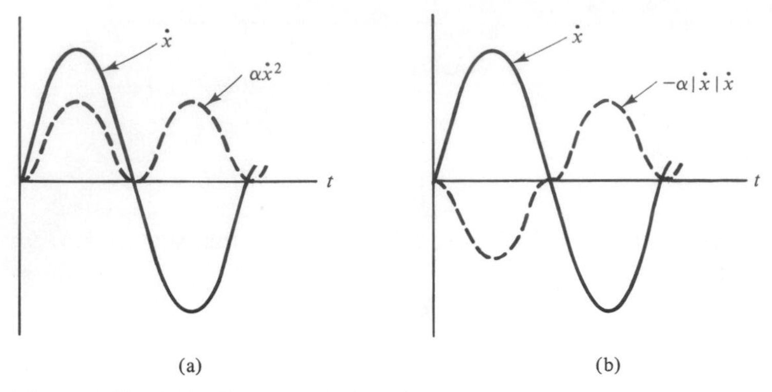

Figure 3-51 Velocity-squared damping.

Substituting the expression for $\dot{x}$ from Eq. 3-107 into Eq. 3-114 gives

$$E_d = 2\alpha|X|^3\omega^3 \int_0^{\pi/\omega} \sin^3(\omega t - \phi)\, dt$$

which yields

$$E_d = \tfrac{8}{3}\,\alpha|X|^3\omega^2$$

The energy dissipated per cycle for velocity-squared damping at resonance ($\omega = \omega_n$) is then

$$E_d = \frac{8}{3}\,\alpha|X|^3\omega_n^2 = \frac{8}{3}\,\alpha|X|^3\,\frac{k}{m} \tag{3-115}$$

Equating Eqs. 3-111 and 3-115 gives

$$\pi 2\zeta_e k|X|^2 = \frac{8}{3}\,\alpha|X|^3\,\frac{k}{m}$$

from which the equivalent-viscous-damping factor is determined as

$$\zeta_e = \frac{4\alpha|X|}{3\pi m}$$

Noting that $\alpha = C\rho A/2$, the equivalent-viscous-damping factor may also be written as

$$\zeta_e = \frac{2}{3}\,\frac{C\rho A|X|}{\pi m} \tag{3-116}$$

Assuming that the drag coefficient C and the mass density ρ of the fluid are constant, Eq. 3-116 reveals two important characteristics of velocity-squared damping:

1. It is proportional to the vibration amplitude $|X|$.
2. It is proportional to the ratio of the projected area A to the mass m (A/m).

Since ζ_e increases as the ratio of A/m increases, the inherent damping, even in a low-viscosity fluid such as air, can have a significant effect on the response of lightweight bodies having large cross-sectional areas.

EXAMPLE 3-11

Figure 3-52 shows a solid sphere and a hollow sphere, both of which are forced to vibrate at resonance with equal amplitudes $|X|$. Both spheres are made from the same material. The hollow sphere has an *outside* radius equal to the radius r of the solid sphere, and a wall thickness t of $0.01r$.

Assuming that the damping the spheres experience in moving through air is proportional to the square of their velocity, we are interested in determining (a) the ratio of an equivalent-viscous-damping factor ζ_h for the hollow sphere to an equivalent-viscous-damping factor ζ_s for the solid sphere; (b) the magnitude of the equivalent-viscous-damping factor for the sphere most affected by the fluid damping of the air as determined in part a, in terms of the amplitude $|X|$, the radius r, and the mass density γ of the material. Assume that $C \cong 1$ for a sphere and that the mass density of air is $2.4(10)^{-3}$ lb $\cdot$ s^2/ft^4.

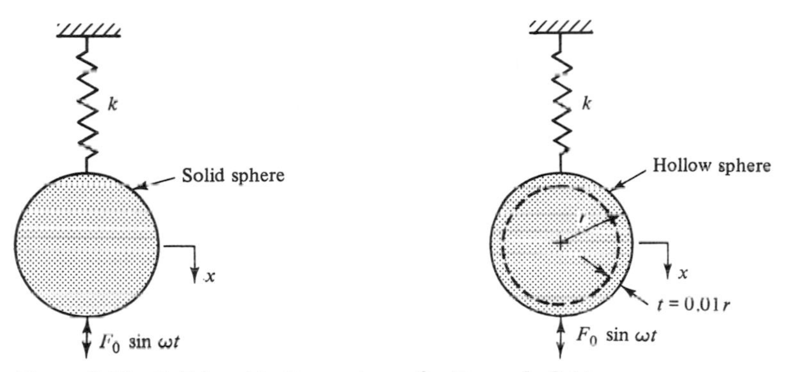

Figure 3-52 Solid and hollow spheres for Example 3-11.

Solution. **a.** Since the two spheres have the same outside geometry and vibrate in the same medium, C and ρ are the same for both spheres. With the amplitude $|X|$ and projected area A the same for both spheres, we can write from Eq. 3-116 that

$$\frac{\zeta_h}{\zeta_s} = \frac{\text{mass of solid sphere}}{\text{mass of hollow sphere}} = \frac{m_s}{m_h}$$

Since the spheres are made of the same material, the ratio ζ_h/ζ_s can also be obtained from the ratio V_s/V_h of the volumes of the two spheres. Thus,

$$\frac{\zeta_h}{\zeta_s} = \frac{\frac{4}{3}\pi r^3}{\frac{4}{3}\pi r^3[1-(0.99)^3]} = 33.7$$

in which the *inside* radius of the hollow sphere is $0.99r$. The above ratio shows that the equivalent-viscous-damping factor for the hollow sphere is approximately 34 times that of the solid sphere.

b. The ratio obtained in part a indicates that the damping due to air resistance is the largest for the hollow sphere. Using Eq. 3-116, the equivalent-viscous-damping factor for the hollow sphere is

$$\zeta_h = \frac{2(1)2.4(10)^{-3}\pi r^2|X|}{3\pi\gamma(\frac{4}{3})\pi r^3[1-(0.99)^3]}$$

which reduces to

$$\zeta_h = \frac{0.0129|X|}{\gamma r}$$

in which γ is the mass density of the material. The latter equation shows that for velocity-squared damping, a sphere formed from a low-mass-density material such as a plastic will have a larger damping factor than a similar one made from a heavier material such as steel. It is also of interest to note that the damping factor varies inversely as the radius of the sphere.

Coulomb Damping (Dry Friction)

Referring to Fig. 3-53 to see that the frictional damping force $F_d = \pm\mu W$ acts through a distance of $4|X|$ for each cycle, the energy dissipated in 1 cycle is

$$E_d = 4|X|F_d \tag{3-117}$$

To determine the *equivalent-viscous-damping coefficient* c_e, we equate Eqs. 3-110 and 3-117 to find that

$$c_e = \frac{4F_d}{\pi|X|\omega} \tag{3-118}$$

in which $\omega \neq \omega_n$. Recalling from an earlier discussion of viscous damping that $c_e = 2\zeta_e\omega_n m$, its substitution into Eq. 3-118 yields

$$\zeta_e = \frac{2F_d}{\omega_n m\pi|X|\omega} \tag{3-119}$$

which is the equivalent-viscous-damping factor for Coulomb damping.

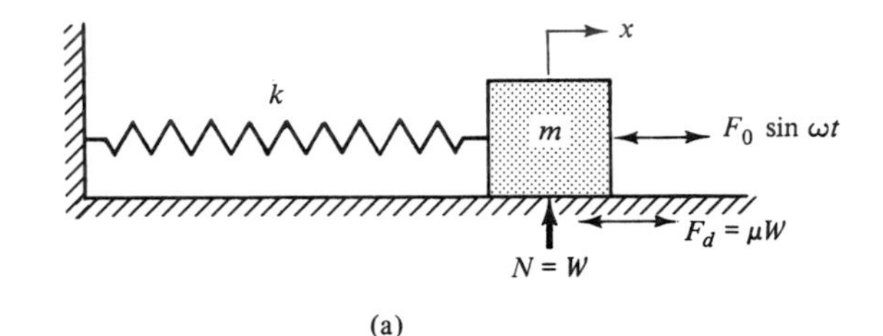

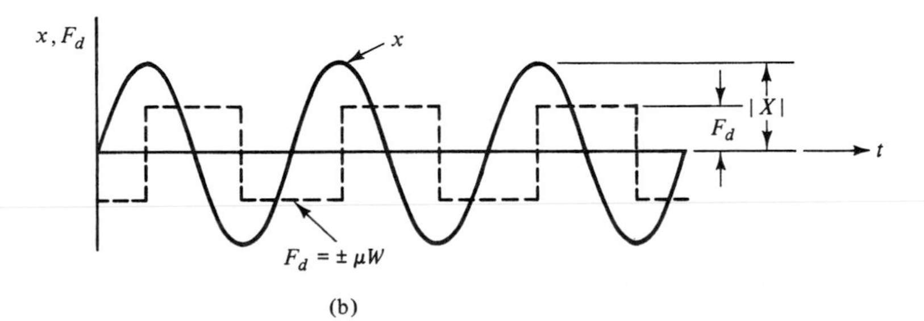

Figure 3-53 System with Coulomb damping. (a) Spring-and-mass system with Coulomb damping. (b) Displacement and Coulomb friction force as functions of time.

To investigate the response of a Coulomb-damped system at resonance, we multiply Eq. 3-119 by $2(\omega/\omega_n)$ to obtain

$$2\zeta_e \frac{\omega}{\omega_n} = \frac{4F_d}{\pi k|X|} \tag{3-120}$$

We then substitute the right side of Eq. 3-120 into the expression for the magnification factor of Eq. 3-15 to obtain

$$\frac{|X|}{F_0/k} = \frac{1}{\sqrt{[1-(\omega/\omega_n)^2]^2 + (4F_d/\pi k|X|)^2}} \tag{3-121}$$

Solving Eq. 3-121 for the amplitude $|X|$ yields

$$|X| = \frac{F_0}{k}\frac{\sqrt{1-(4F_d/\pi F_0)^2}}{1-(\omega/\omega_n)^2} \tag{3-122}$$

Equation 3-122 shows that the amplitude $|X|$ is theoretically infinite at resonance ($\omega = \omega_n$) for a Coulomb-damped system. The reader should note that for $|X|$ to be real, the quantity $4F_d/\pi F_0$ must be less than 1.

Structural Damping (Complex Stiffness)

The energy dissipation caused by cyclic stress and strain within a structural material is often referred to as structural damping. Other terms commonly used to refer to this type of energy dissipation are *hysteretic* damping, *solid* damping, and *displacement* damping. Experimental studies have shown that the energy dissipation per cycle for metals such as aluminum and steel which are common structural materials is approximately proportional to the square of the strain amplitude but is essentially independent of frequency.[13]

Structural damping can be formulated in terms of *complex stiffness* quantities, known as *complex spring constants* and *complex elastic moduli.* The use of these quantities yields mathematical models that are readily analyzed *if the excitation is harmonic.*

The complex spring constant k^* is expressed as

$$k^* = k(1 + j\eta) \tag{3-123}$$

where k = a spring constant giving the stiffness of the elastic element
η = a loss factor resulting from the deformation of the elastic element lagging the applied force during sinusoidal motion
$j = \sqrt{-1}$

The complex modulus of elasticity E^* has the same form as the complex spring constant and is defined as

$$E^* = E(1 + j\eta) \tag{3-124}$$

where E = Young's modulus for a material
η = a loss factor resulting from the normal strain ϵ lagging the normal stress σ during sinusoidal motion

[13] A. L. Kimball, "Vibration Damping Including the Case of Solid Friction," *Transactions of the ASME, APM,* vol. 51, no. 21, 1929.

The complex modulus is a damping model that can be used to determine the dynamic properties of materials.[14]

In a linear system, the strain amplitude is proportional to the vibration amplitude $|X|$. The energy dissipated per cycle for structural damping can thus be written simply as

$$E_d = \beta |X|^2 \tag{3-125}$$

in which β is a constant having the units of force/displacement like a spring constant. The magnitude of β can vary with the size, shape, material, and temperature of the structural system.

Equating Eqs. 3-125 and 3-110, the equivalent-viscous-damping coefficient c_e for structural damping is found to be

$$c_e = \frac{\beta}{\pi\omega} \tag{3-126}$$

If this equivalent coefficient replaces c in Eq. 3-2, which is the differential equation of motion of a viscously damped system, we obtain

$$\ddot{x} + \frac{\beta}{m\pi\omega}\dot{x} + \frac{k}{m}x = \frac{F_0}{m}\sin \omega t \tag{3-127}$$

If the motion is harmonic, we know that the steady-state solution is

$$x = |X| e^{j(\omega t - \phi)}$$

which may be differentiated to obtain the velocity as

$$\dot{x} = j\omega |X| e^{j(\omega t - \phi)} = j\omega x \tag{3-128}$$

Equation 3-127 can then be written as

$$\ddot{x} + \frac{k}{m}\left(1 + j\frac{\beta}{\pi k}\right)x = \frac{F_0}{m}e^{j\omega t}$$

or as

$$\ddot{x} + \frac{k}{m}(1 + j\eta)x = \frac{F_0}{m}e^{j\omega t} \tag{3-129}$$

in which $\eta = \beta/\pi k$ is the structural-damping factor and $k(1 + j\eta)$ is the complex spring constant k^*.

Assuming that the steady-state solution of Eq. 3-129 is $x = Xe^{j\omega t}$, and remembering that $\omega_n^2 = k/m$, the steady-state solution of this equation yields

$$\frac{|X|}{F_0/k} = \frac{1}{\sqrt{[1 - (\omega/\omega_n)^2]^2 + \eta^2}} \tag{3-130}$$

and

$$\tan \phi = \frac{\eta}{1 - (\omega/\omega_n)^2} \tag{3-131}$$

[14] "The Measurement of the Dynamic Properties of Elastomers and Elastomeric Mounts," ed. by B. M. Hillberry, Symposium sponsored by SAE and ASTM, Detroit, MI (Jan. 8–12, 1973). G. M. Smith, Y. C. Pao, and J. D. Fickes, "Determination of a Dynamic Model for Urethane Prosthetic Compounds," *Experimental Mechanics,* vol. 18, no. 9, October 1978. G. M. Smith, R. L. Bierman, and S. J. Zitek, "Determination of Dynamic Properties of Elastomers over Broad Frequency Range," *Experimental Mechanics,* vol. 23, no. 2, June 1983.

Comparing Eq. 3-130 with Eq. 3-15 with $\omega/\omega_n = 1$, it can be seen that

$$2\zeta_e = \eta \tag{3-132}$$

Let us now discuss further the complex modulus of elasticity E^*, which is defined in Eq. 3-124. For sinusoidal motion, the normal stress σ and strain ϵ can be represented in the complex plane as shown in Fig. 3-54, and since the strain must lag the stress for energy dissipation to occur, the strain is shown lagging the stress by the phase angle ψ. The stress and strain can be expressed as

$$\sigma = \sigma_0 e^{j\omega t} \tag{3-133}$$

and

$$\epsilon = \epsilon_0 e^{j(\omega t - \psi)} \tag{3-134}$$

in which σ_0 and ϵ_0 are the stress and strain *amplitudes,* respectively.

Remembering that the elastic modulus of a material is

$$E = \frac{\sigma}{\epsilon} \qquad \text{(static loading)}$$

it follows that the complex modulus of elasticity, which is a dynamic elastic modulus, may be defined as the ratio of the *dynamic* stress and strain. Thus, dividing Eq. 3-133 by Eq. 3-134, the complex modulus is

$$E^* = \frac{\sigma_0}{\epsilon_0} e^{j\psi} = \frac{\sigma_0}{\epsilon_0}(\cos\psi + j\sin\psi)$$

or

$$E^* = \frac{\sigma_0}{\epsilon_0}\cos\psi\left(1 + j\frac{\sin\psi}{\cos\psi}\right) \tag{3-135}$$

Referring to Fig. 3-55, it can be seen that $\sigma_0 \cos\psi$ is the component of stress that is in phase with the strain, and that the stress component $\sigma_0 \sin\psi$ leads the strain by 90°. These observations lead to the concept of a *storage modulus* E_1 and a *loss modulus* E_2, which are defined as

$$\left.\begin{aligned} E_1 &= \frac{\sigma_0}{\epsilon_0}\cos\psi \qquad \text{(storage modulus)} \\ E_2 &= \frac{\sigma_0}{\epsilon_0}\sin\psi \qquad \text{(loss modulus)} \end{aligned}\right\} \tag{3-136}$$

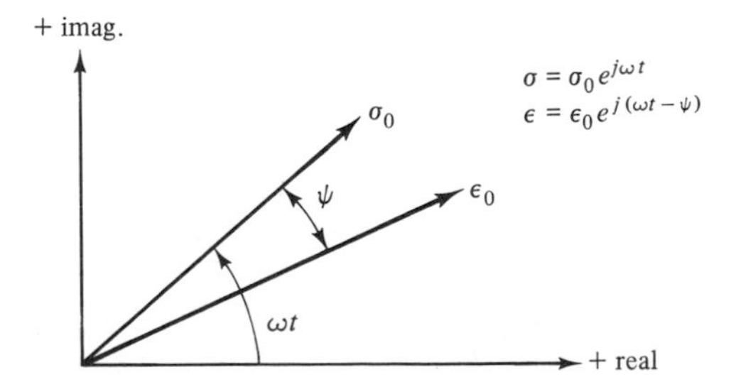

Figure 3-54 Harmonic stress and strain phasors in the complex plane.

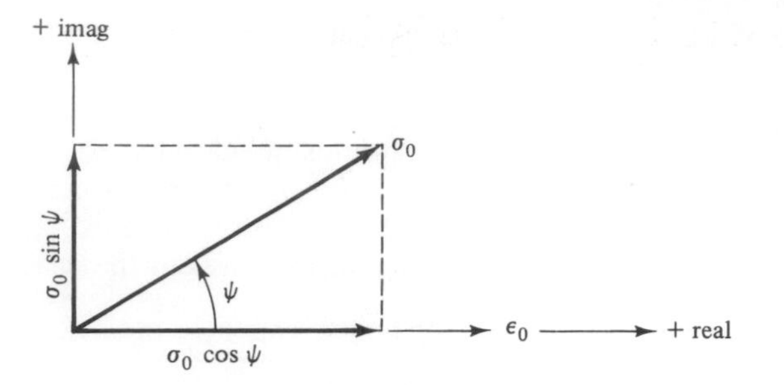

Figure 3-55 Stress components.

Thus, Eq. 3-135 can be written in the form of

$$E^* = E_1\left(1 + j\frac{E_2}{E_1}\right)$$

or more simply as

$$E^* = E_1(1 + j\eta) \tag{3-137}$$

in which the loss factor $\eta = E_2/E_1 = \tan \psi$. The storage modulus E_1 is actually a dynamic elastic-stiffness property of the material. If the loss factor $\eta = \tan \psi \leq 0.2$, which is true for most metals, it is common practice to assume that the storage modulus E_1 is equal to Young's modulus E since $\cos \psi \cong 1$.

With the assumption that $E_1 = E$, Eq. 3-137 becomes

$$E^* = E(1 + j\eta) \qquad (\text{for } \eta \leq 0.2)$$

which is the same as Eq. 3-124. However, there are materials, such as elastomers (rubberlike materials), that have loss factors greater than $\eta = 0.2$. For such materials the storage modulus E_1, which is a dynamic modulus, may differ considerably from the modulus of elasticity E, which is a static stress-strain ratio.

In the preceding paragraphs, only normal stresses and strains were discussed. However, the discussion is also germane to the case of shear stresses τ and shear strains γ, since it can be shown in a similar manner that the *complex shear modulus* G^* is

$$G^* = G_1(1 + j\eta) \tag{3-138}$$

where $\eta = G_2/G_1 = \tan \psi = \text{shear loss factor}$

$$G_1 = \frac{\tau_0}{\gamma_0} \cos \psi \quad (\text{shear storage modulus})$$

$$G_2 = \frac{\tau_0}{\gamma_0} \sin \psi \quad (\text{shear loss modulus})$$

As noted before, if $\eta = \tan \psi \leq 0.2$, the complex shear modulus reduces to

$$G^* = G(1 + j\eta) \tag{3-139}$$

in which G is the shear modulus of elasticity (modulus of rigidity). Recalling from elementary mechanics of materials that E, G, and Poisson's ratio ν are related by

$$G = \frac{E}{2(1 + \nu)}$$

the complex shear modulus can also be expressed in terms of Young's modulus E as

$$G^* = \frac{E}{2(1+\nu)}(1 + j\eta) \tag{3-140}$$

PROBLEMS

Problems 3-1 through 3-16 (Sections 3-1 through 3-5)

3-1. Determine (a) the differential equation of motion of the uniform slender rod if the damping is sufficient to keep the oscillations small for all values of the excitation frequency ω, (b) the damped natural frequency in terms of the system parameters, (c) the value of the damping coefficient c for critical damping, and (d) the amplitude of the steady-state response using complex algebra. The mass moment of inertia of the rod about an axis through O is $I_0 = ml^2/3$.

Partial ans: $\ddot{\theta} + \frac{c}{I_0}\left(\frac{l}{2}\right)^2\dot{\theta} + \frac{kl^2}{I_0}\theta = \frac{F_0 l}{I_0}\sin\omega t$

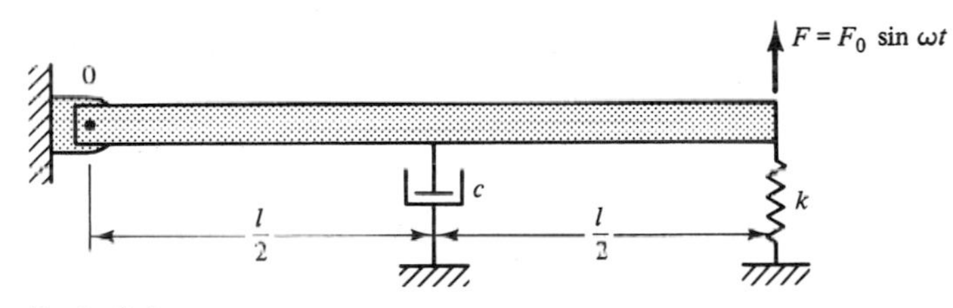

Prob. 3-1

3-2. Assume that the rod referred to in Prob. 3-1 was steel and that the system had a magnification factor (M.F.) of 2.5 when $\omega = \omega_n$. Then replace the steel rod with an aluminum one of identical length and cross section. Assuming that c and k are the same for both systems, show that

$$(\text{M.F.})_a = 2.5\sqrt{\frac{w_a}{w_s}} = 1.47$$

when $\omega = \omega_n$ for the system with the aluminum rod. In the above equation:

$(\text{M.F.})_a$ = magnification factor with the aluminum rod

w_a = specific weight of aluminum (169 lb/ft^3)

w_s = specific weight of steel (490 lb/ft^3)

3-3. A disk having a centroidal mass moment of inertia of $\bar{I}$ is attached to a shaft that has a torsional spring constant k_t and a torsional damping coefficient c. Determine the differential equation of motion of the disk and an expression for its steady-state amplitude $|\Theta|$.

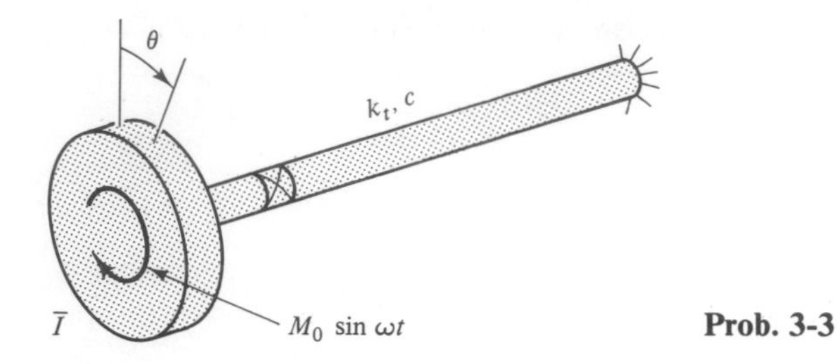

Prob. 3-3

3-4. Referring to Eq. 3-20, show that the peak magnification factor $|X|/(F_0/k)$ occurs when $\omega/\omega_n = \sqrt{1 - 2\zeta^2}$.

3-5. A critically damped spring-and-mass system is subjected to a sinusoidal excitation $F_0 \sin \omega t$ for which the excitation frequency ω is equal to the undamped natural circular frequency ω_n of the system. If the system starts from rest, show that the total response is given by

$$x = \frac{F_0}{2k}\left[(1 + \omega_n t)e^{-\omega_n t} - \cos \omega_n t\right]$$

3-6. A damping factor of $\zeta = 0.12$ is determined experimentally from its damped free vibrations for the system shown in the accompanying figure, as explained in Sec. 2-4. Assuming that there is no slippage between the light cable and the cylinder and that the cylinder is subjected to the moment $M_0 \sin \omega t$ shown, determine (a) the differential equation of motion of the system and (b) the amplitude $|X|$ of the forced response of the mass m if $M_0 = 10$ N·m and $\omega = 10$ rad/s.

Partial ans: $|X| = 29.7$ mm

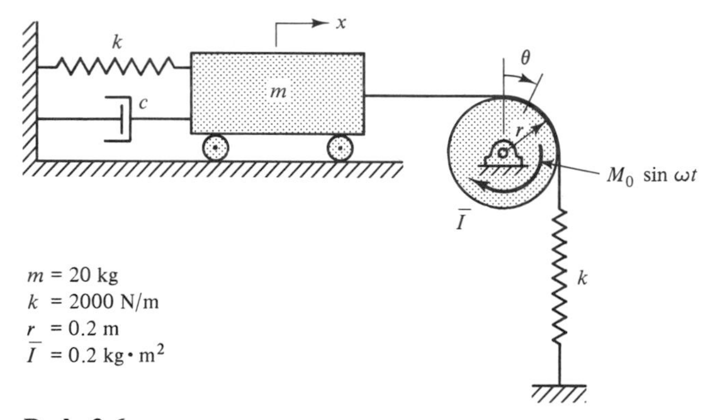

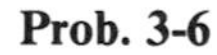
Prob. 3-6

3-7. The end of the tailpipe of an automobile is observed to vibrate with large amplitudes when the engine is idling. Considering the tailpipe as a beam, discuss the feasibility of reducing this undesirable vibration by (a) stiffening the tailpipe by adding an extra bracket, (b) repositioning the existing bracket, or (c) adding a mass at the end of the tailpipe.

3-8. In Example 3-1 the mass of the vertical channels (acting as columns) was neglected. Determine an appropriate shape function (see the discussion in Sec. 2-9) that will take into account the effect of the mass of the channels. If each channel weighs 11.5 lb/ft, determine (a) the natural frequency f_n of the system, (b) the damping factor ζ, and (c) the magnification factor M.F.

3-9. Show that Eq. 3-15 can be reduced to the nondimensional form

$$\frac{m|X|}{em_0} = \frac{(\omega/\omega_n)^2}{\sqrt{[1-(\omega/\omega_n)^2]^2 + [2\zeta(\omega/\omega_n)]^2}}$$

3-10. A Lazan fatigue-testing machine consists of a heavy frame mounted on soft springs (not shown) which enables the frame to remain essentially stationary during operation. Within the machine is a moving head of mass M mounted on two springs having a total stiffness of k lb/in. Attached to the moving head is an eccentric mass m_0 that is driven at ω rad/s. During operation, lateral motion of the moving head is restrained by guide arms that offer essentially no resistance to the vertical motion of the head. The test specimen of stiffness k_s is subjected to a sinusoidal force $F_s = k_s|X|\sin \omega t$ due to the sinusoidal motion of the head. Considering the equivalent spring-and-mass model of the test machine given in the accompanying figure, show that the amplitude of the sinusoidal force transmitted to the specimen will be $F_0 = m_0e\omega^2$ if $k = \omega^2(M + m_0)$. That is, M is adjusted by adding or removing small weights to "tune" the fatigue-testing machine to resonance at the operating speed ω in the absence of the test specimen, so that $F_s = k_s|X| = m_0e\omega^2$.

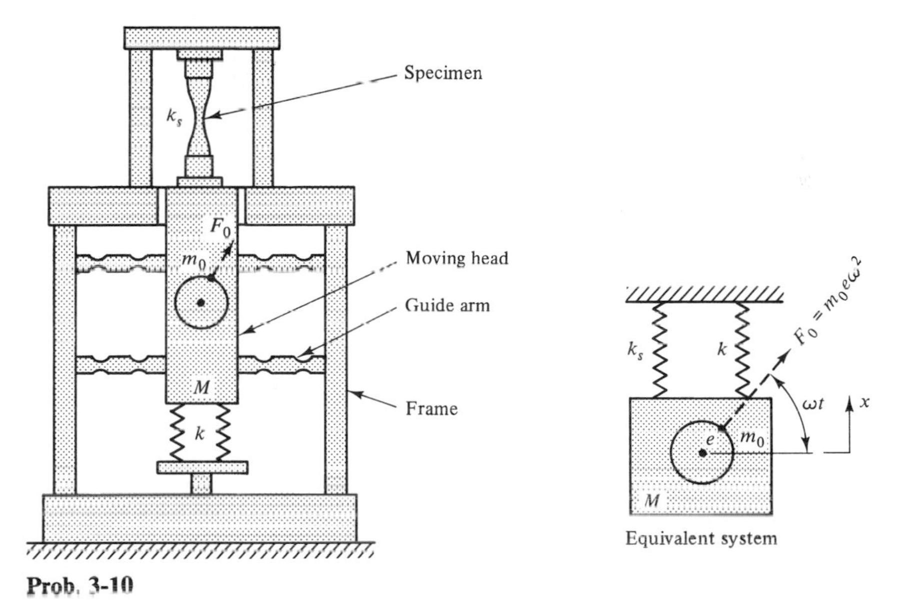

Prob. 3-10

3-11. In Example 3-3 the use of a 25.4-mm-diameter steel shaft resulted in rather high dynamic bearing reactions ($R_A = R_B = 1013.8$ N = 227.9 lb). If that shaft is replaced by another steel shaft having a diameter of 19 mm (0.75 in.), determine (a) the critical speed of the shaft, (b) the dynamic bearing reactions R_A and R_B at the critical speed, and (c) the ratio of the dynamic bearing reaction R_A to the static reaction W_A. The values for the eccentricity e, damping factor ζ, length l, and mass m of the disk are the same as those given in Example 3-3.

3-12. A 20-lb turbine rotor is mounted on a 1-in.-diameter steel shaft that is supported by the nonaligning bearings A and B shown in the accompanying figure. Assuming that the shaft is a fixed-fixed beam of 15 in. length, determine the dynamic bearing reactions R_A and R_B for an operating speed of 7200 rpm, an eccentricity of $e = 0.002$ in., and a damping factor of $\zeta = 0.02$. The specific weight of steel is 490 lb/ft^3. Select an appropriate shape function to determine ω_n using the concept of $T_{max} = U_{max}$. Use Eq. 3-37 to calculate U_{max}, and obtain k from Table 2-1 on page 80.

Partial ans: $R_A = 50.3$ lb

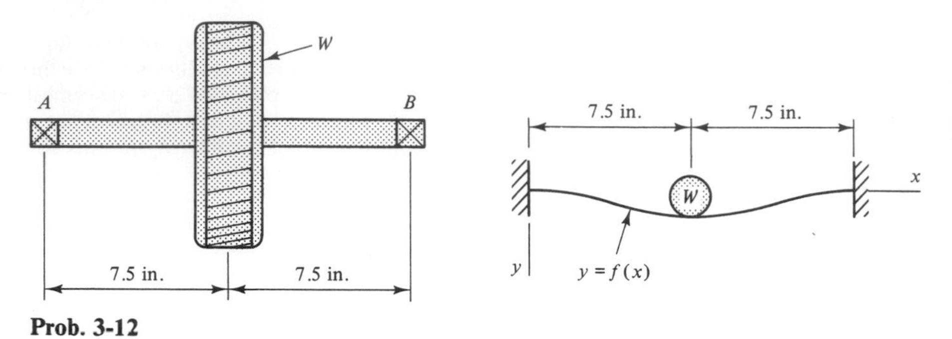

Prob. 3-12

3-13. Work Prob. 3-12 with the bearings A and B replaced by self-aligning bearings and all the other parameters and data unchanged.

3-14. The sensor of a color-selector unit that provides information to fishermen to help them determine the best color of lure to use in an area is lowered until about 8 ft of cable is out with the boat moving at around 2 mph. As this is done, the upper end of the cable can be felt to vibrate. If the diameter of the sensor is approximately 1.25 in., and the sensor and cable drag in the water at an angle of $\theta = 60°$ as shown in the accompanying figure, determine the approximate frequency that the fisherman feels. Although this is not a problem for fishermen using this device, the vibration of similar devices such as hydrophones being towed behind ships does occasionally cause problems.

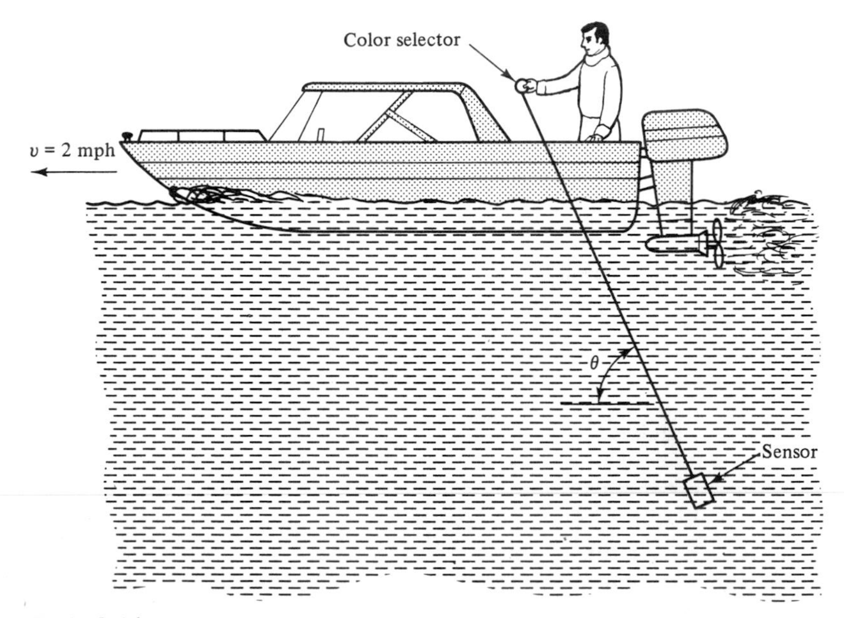

Prob. 3-14

3-15. The accompanying figure shows a tower structure supporting a crossarm that carries power-transmission lines. An inspection of such crossarms along a power line in a midwestern area revealed hairline cracks in the crossarms that appeared to be the result of fatigue, and it was speculated that the cracks could be due to vibration caused by vortex shedding. To check the vibration characteristics of the crossarms, a crossarm 36 ft long made of circular tubing with a mean radius of $r = 6$ in. and a wall thickness of $t = 0.25$ in. was considered. By modeling it as eight lumped masses (see Chap. 5), the first four natural frequencies of the crossarm (with the mass of the power lines neglected) were found to be $f_1 = 15.79$ Hz, $f_2 = 28.84$ Hz, $f_3 = 50.18$ Hz, and $f_4 = 115.52$ Hz. Would wind velocities ranging from 10 to 80 mph result in vortex-shedding frequencies corresponding to any of these first four natural frequencies?

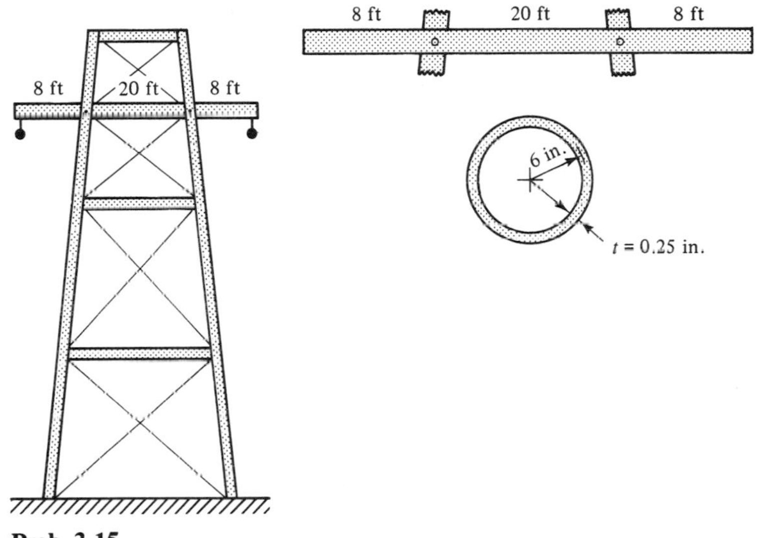

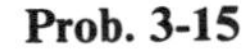

Prob. 3-15

3-16. A vertical suction pipe 3.25 m long with an outside diameter of 0.15 m is submerged to a depth of 1.25 m in a river that is flowing at 5 mph as shown in the accompanying figure. It is attached to a horizontal pipe that has a torsional spring constant of k_t = 15,000 N · m/rad. A simplified model of the two pipes is shown in part b of the figure. Vortex shedding subjects the system to a moment M_z about the z axis shown due to the force

$$F = \frac{C\rho v^2 A}{2} \sin(2\pi f t)$$

Determine the steady-state amplitude of vibration θ_0 of the suction pipe if the fundamental natural frequency of the system is 3.1 Hz and the damping factor is $\zeta = 0.25$. Recalling that $k_t = GJ/l$ for the horizontal pipe, it is suggested that the length of the horizontal pipe be increased from l to $2l$ as a means of reducing the amplitude of vibration. Do you agree with this suggestion?

Partial ans: $\theta_0 = 0.17$ rad

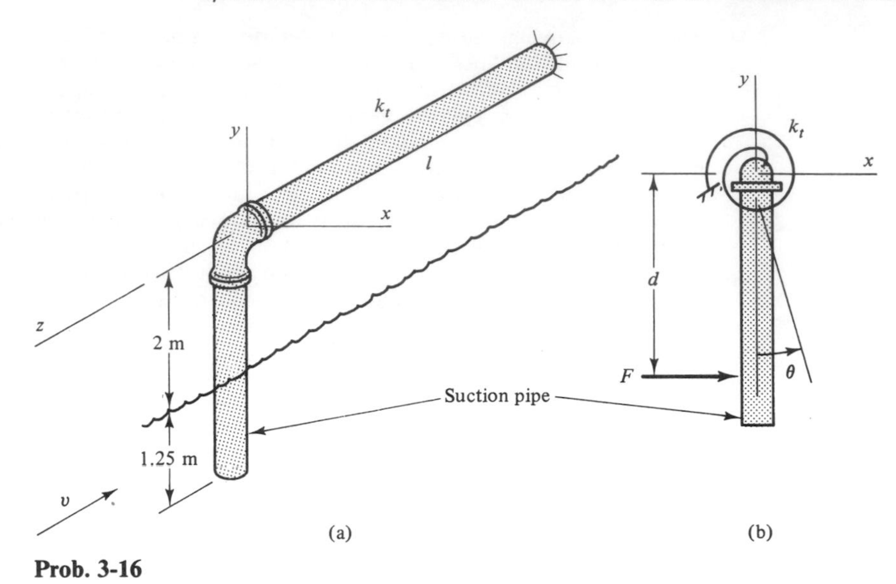

Prob. 3-16

Problems 3-17 through 3-26 (Sections 3-6 and 3-7)

3-17. Derive the differential equation of motion for the system shown, and determine the steady-state amplitude $|X|$ in terms of the system parameters.

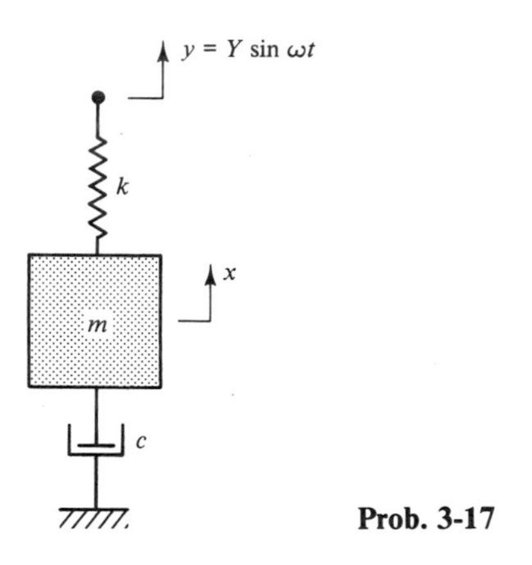

Prob. 3-17

3-18. The $\frac{1}{4}$- by 4- by 8-in. steel plate is attached to a spring of stiffness k and is forced to oscillate in an oil-filled chamber by the motion $y = Y \sin \omega t$ of the upper end of the spring as shown in the accompanying figure. The clearance h between the plate sides and the walls of the oil chamber is 0.2 in. Considering that the shear stress τ in the oil in contact with the plate is given by $\tau = \mu \dot{x}/h$, in which μ is the viscosity of the oil and $\dot{x}$ is the velocity of the plate, determine (a) the differential equation of motion of the system in terms of its parameters and (b) the viscosity μ of the oil if $|X|/|Y| = 40$ when $f/f_n = 1$ and $f_n = 4.2$ Hz with the plate vibrating freely in air.

Partial ans: $\mu = 1.21(10)^{-5}$ lb·s/in.^{2}

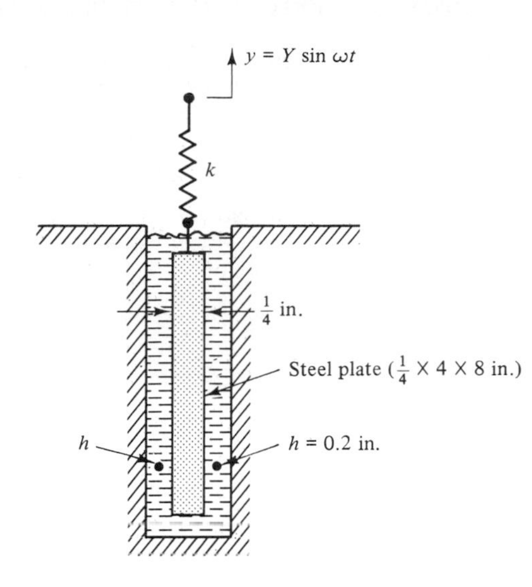

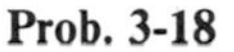

Prob. 3-18

3-19. A weight of $W = 500$ lb is attached to the end of an overhanging steel beam that is pinned at A and B as shown in the accompanying figure. The stiffness factor of the beam is $EI = 90(10)^6$ lb · in.^{2} Neglect damping and the mass of the beam and determine (a) the differential equation of motion of the system and (b) the steady-state amplitude of the weight W if the vertical motion of the pin at A is $y_A = Y \sin 2\pi f t$ in which $Y = 0.1$ in. and the frequency is $f = 5$ Hz. The support at B is stationary.

Ans: $\ddot{y}_c + \omega_n^2 y_c = \omega_n^2 \dfrac{Y}{2} \sin \omega t$

$|A| = 0.15$ in.

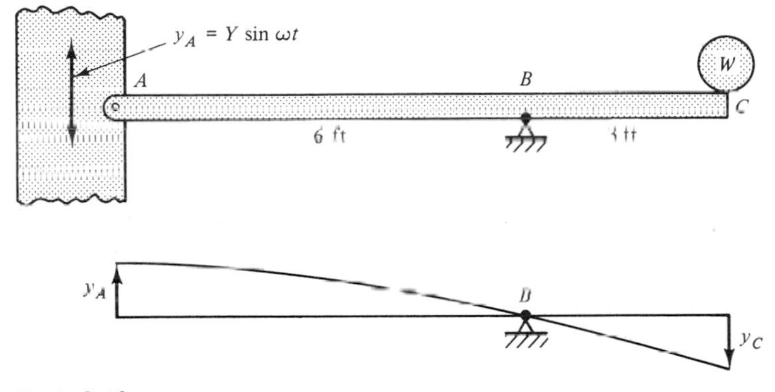

Prob. 3-19

3-20. The crude-oil-pumping rig shown in the accompanying figure is driven at 20 rpm, and the stroke of the top of the pump rod at A is 45 in. The inside diameter of the well pipe is 2 in., and the diameter of the pump rod is 0.75 in. The length of the pump rod and the length of the column of oil lifted during each stroke are essentially the same, and equal to 6000 ft. During the downward stroke a valve at the lower end of the pump rod opens to let a quantity of oil into the well pipe, and the column of oil is then lifted to obtain a discharge into the connecting pipeline at B. Thus, the amount of oil pumped in a given time depends upon the stroke of the lower end of the pump rod. Assuming that the motion of the upper end of the rod is essentially sinusoidal with a stroke of 45 in., and that the damping factor for the system is $\zeta = 0.5$, determine the output of the well in barrels per hour (bbl/h). The following data are pertinent to the solution:

$$\begin{aligned}\text{Specific gravity of the crude oil} &= 0.9\\ \text{Specific weight of steel} &= 490\ \text{lb/ft}^3\\ 1\ \text{bbl of oil} &= 42\ \text{gal}\\ 1\ \text{ft}^3 &= 7.481\ \text{gal}\end{aligned}$$

Ans: 21.7 bbl/h

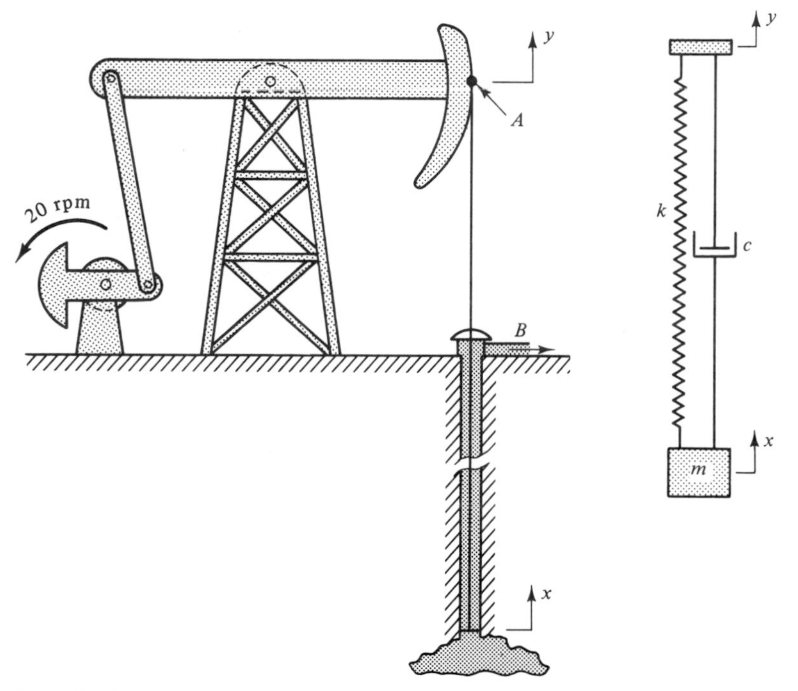

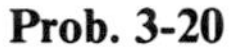
Prob. 3-20

3-21. The propeller and long steel shaft systems of large ships are susceptible to vibration problems. One source of axial excitation contributing to the vibration of such systems is the generation of pulse-type forces $F(t)$ that result from a propeller blade passing the restricted area between the propeller and the hull of the ship. Consider that the axial forcing function $F(t)$ can be represented by the rectangular pulse train shown in part c of the accompanying figure in which

$$\tau = \frac{1}{N\ \text{times number of propeller blades}}$$

and N is the rpm of the propeller at cruising speed. Use the Fourier series derived for the pulse train in Sec. 1-7, and determine whether a three- or four-bladed propeller should be used to avoid undesirable vibration if the following data pertain:

$N = 325$ rpm
$d_o = 20$ in. (outside diameter of shaft)
$d_i = 10$ in. (inside diameter of shaft)
$l = 175$ ft (length of shaft)
$E = 30(10)^6$ psi (modulus of elasticity of steel)
$w = 490$ lb/ft^3 (specific weight of steel)
$W = 22{,}500$ lb (weight of either a three- or four-bladed propeller)

Ans: Three-bladed propeller

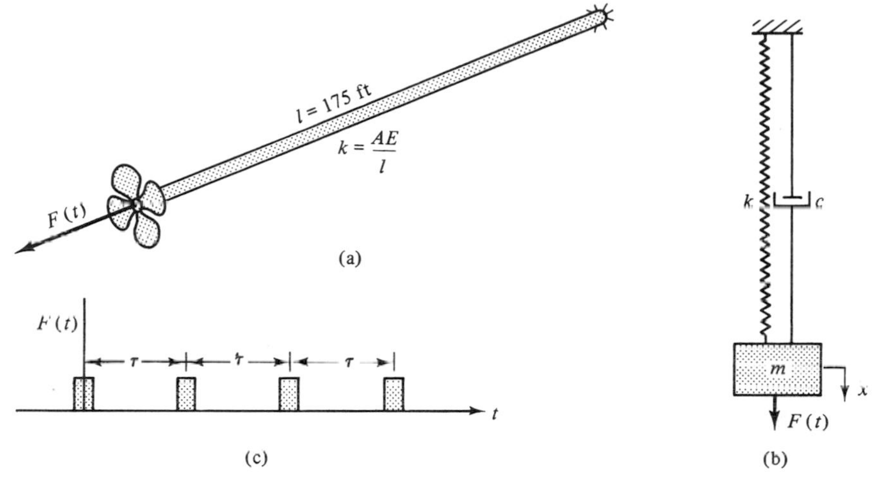

Prob. 3-21

3-22. Determine the differential equation of motion and the equation for the steady-state amplitude $|X|$ of the system shown in the accompanying figure.

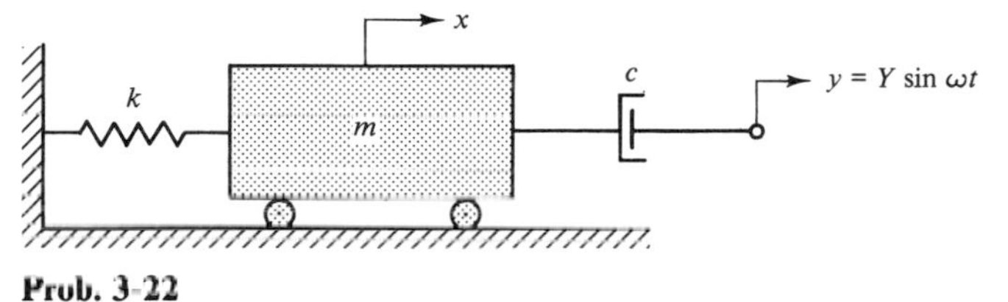

Prob. 3-22

3-23. A control box of mass m is attached to the end of a cantilever beam of stiffness $k = 3EI/l^3$. The sinusoidal motion of the support structure causes large flexural stresses in the beam, which results in fatigue failure of the beam when the support motion has a frequency of $f = 10$ Hz and the natural frequency of the beam system is $f_n = 12$ Hz. Which of the following would you recommend as a means of preventing such a fatigue failure? (a) Increase the length l of the beam. (b) Decrease the length l of the beam. (c) Add mass to the control box. Briefly explain the reason for your answer.

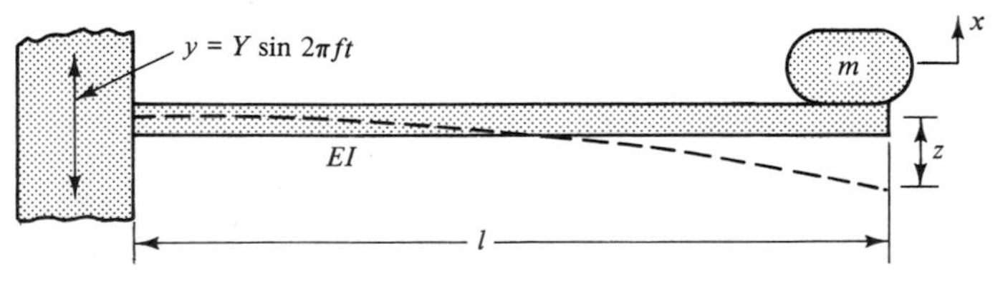

Prob. 3-23

3-24. To analyze the vibration of a single-cylinder engine such as the one shown in part a of the accompanying figure, it is necessary to determine the acceleration of the piston. For such engines, the ratio of the crank radius r to the length l of the connecting rod (r/l) is considerably less than 1. Considering that $r/l \ll 1$, and that the angular velocity ω of the crank is constant, show that the absolute acceleration a_p of the piston can be approximated by the equation

$$a_p = \ddot{x} - r\omega^2\left(\cos \omega t + \frac{r}{l}\cos 2\omega t\right)$$

in which $\ddot{x}$ is the acceleration of point O on the engine block, and the sum of the last two terms is the acceleration $\ddot{y}$ of the piston relative to O. That is,

$$a_p = \ddot{x} + a_{p/O} = \ddot{x} + \ddot{y}$$

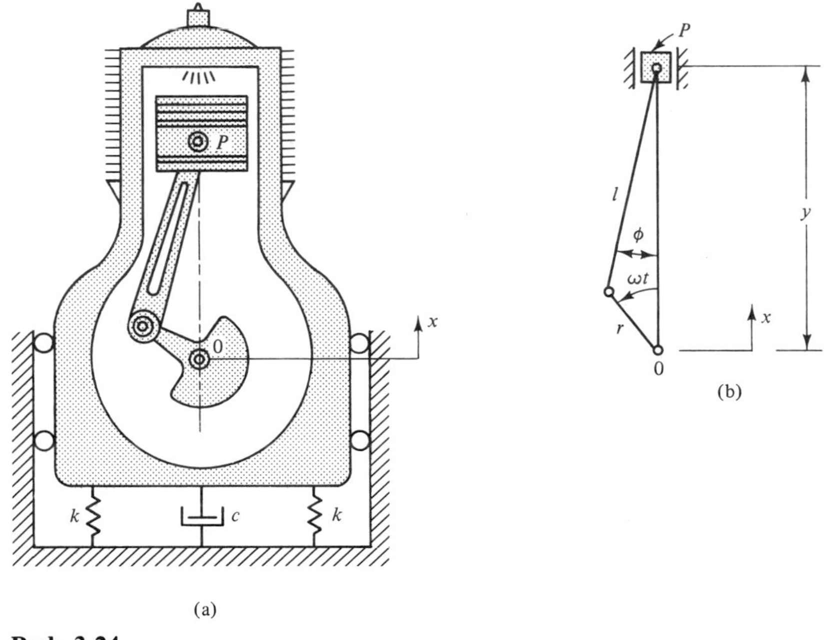

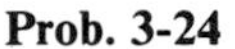
Prob. 3-24

3-25. An approximation for the *absolute* acceleration a_p of the piston of the single cylinder of Prob. 3-24 was shown to be

$$a_p = \ddot{x} - r\omega^2 \cos \omega t - l\left(\frac{r}{l}\right)^2 \omega^2 \cos 2\omega t$$

where $\ddot{x}$ = absolute acceleration of engine block (point O)
r = radius of crank
l = length of connecting rod
ω = engine speed, rad/s

Expressing the effective mass of the piston as M_p (the mass of the piston plus a portion of the connecting rod) and the total mass of the engine as m, show that the differential equation of motion of the engine is

$$\ddot{x} + \frac{c}{m}\dot{x} + \frac{2k}{m}x = \frac{M_p}{m}\omega^2\left[r\cos\omega t + l\left(\frac{r}{l}\right)^2\cos 2\omega t\right]$$

Assume that the crank and the remaining portion of the connecting-rod mass, which is considered concentrated at the crankpin, is balanced so that its center of gravity is at the axis of rotation O. It is suggested that the concept of dynamic equilibrium be used to derive the differential equation of motion.

3-26. In Prob. 3-25, the differential equation of motion of the single-cylinder engine (see Prob. 3-24) was shown to be

$$\ddot{x} + \frac{c}{m}\dot{x} + \frac{2k}{m}x = \frac{M_p}{m}\omega^2\left[r\cos\omega t + l\left(\frac{r}{l}\right)^2\cos 2\omega t\right]$$

Determine the absolute maximum steady-state response of the engine if the numerical data for the single-cylinder engine of Probs. 3-24 and 3-25 are as follows:

$W_p = 5$ lb (effective weight of piston and portion of connecting rod)
$W_e = 200$ lb (total weight of engine)
$N = 600$ rpm (engine speed)
$r = 3$ in. (crank radius)
$l = 12$ in. (length of connecting rod)
$k = 1200$ lb/in. (spring constant of each spring)
$\zeta = 0.02$ (damping factor)

Ans: $x(t)_{max} = 0.82$ in.

Problems 3-27 through 3-39 (Sections 3-8 through 3-11)

3-27. An instrument that weighs 100 lb is to be mounted on a support that is vibrating with an amplitude of 0.2 in. at a frequency of 1800 cpm by means of four identical springs. Determine (a) the spring constant of each spring for 80 percent isolation and (b) the resulting amplitude of vibration of the instrument.

Ans: $k = 384.6$ lb/in.
$|X| = 0.04$ in.

3-28. A fragile instrument and its housing, which together weigh 10 lb, are mounted on four symmetrically located identical springs as a means of isolating the instrument from the vibration of the table supporting it. Measurements indicate that the table reaches a maximum amplitude of 0.004 in. somewhere in the frequency range of 20 to 30 Hz. Determine the spring constant k of each spring so that the amplitude of vibration of the instrument will be limited to 0.0002 in. as a maximum.

Ans: $k = 4.87$ lb/in.

3-29. A large machine that weighs 30,000 lb is found to be transmitting a force of 500 lb to its supporting foundation when running at a speed of 1200 rpm. The total (equivalent) spring constant of the supporting springs is $k = 150{,}000$ lb/in. Determine the magnitude of the unbalanced force F_0 developed by the machine. What is the approximate amplitude of vibration of the machine?

Ans: $F_0 = 3590$ lb, $|X| = 0.003$ in.

3-30. A machine weighing 644 lb is mounted on a vibrating support that has a sinusoidal motion with a frequency of 30 Hz and an amplitude of 0.05 in. by means of four identical springs. Determine the equivalent spring constant k_e of the four springs that will limit the maximum acceleration of the machine to 0.25 g's.

Ans: $k_e = 36{,}612$ lb/ft

3-31. A fragile instrument of mass $m = 100$ kg is used on a table that is vibrating because of its proximity to rotating machines that are running in the area. Since the vibration of the table affects the operation of the instrument, it is desired to use an accelerometer and a frequency analyzer to obtain the rms values of the frequency components of the acceleration (in g's) of the table. The acceleration of the table is shown in part a of the accompanying figure. To reduce the acceleration of the instrument to an acceptable limit of 0.02 g's, it is proposed to isolate the instrument from the table by means of several 25-mm-thick rubber pads as shown in part b of the figure. Determine the minimum number of pads required to reduce the acceleration of the instrument to 0.02 g's or less if the stiffness k of each pad is 250 N/mm. It is assumed that the damping is 10 percent ($\zeta = 0.1$) regardless of the number of pads used.

Ans: Four pads

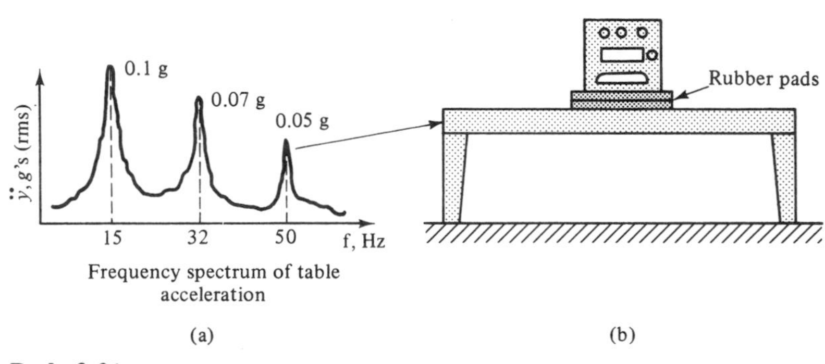

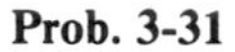

Prob. 3-31

3-32. Before working this vibration-isolation problem, it is suggested that the reader refer to Probs. 3-24, 3-25, and 3-26 for background information for a typical single-cylinder engine. In Prob. 3-25 the differential equation of motion of a single-cylinder engine was given as

$$\ddot{x} + \frac{c}{m}\dot{x} + \frac{2k}{m}x = \frac{M_P}{m}\omega^2\left[r\cos\omega t + l\left(\frac{r}{l}\right)^2\cos 2\omega t\right]$$

where $2k = k_e$ = equivalent spring constant
m = total mass of engine
M_P = effective mass of piston plus portion of connecting rod
r = crank radius
l = length of connecting rod
ω = engine speed, rad/s

Neglecting damping, and using the numerical data given in Prob. 3-26, determine (a) the maximum force that the engine transmits to the foundation supporting it when bolted directly to it as shown in part a of the accompanying figure and (b) the equivalent spring constant k_e of a mounting system that will reduce the transmitted force determined in part a by 80 percent. Compare the k_e determined with the value of $2k = 2400$ lb/in. of Prob. 3-26.

Ans: $F_{max} = 191.8$ lb, $k_e = 287.7$ lb/in.

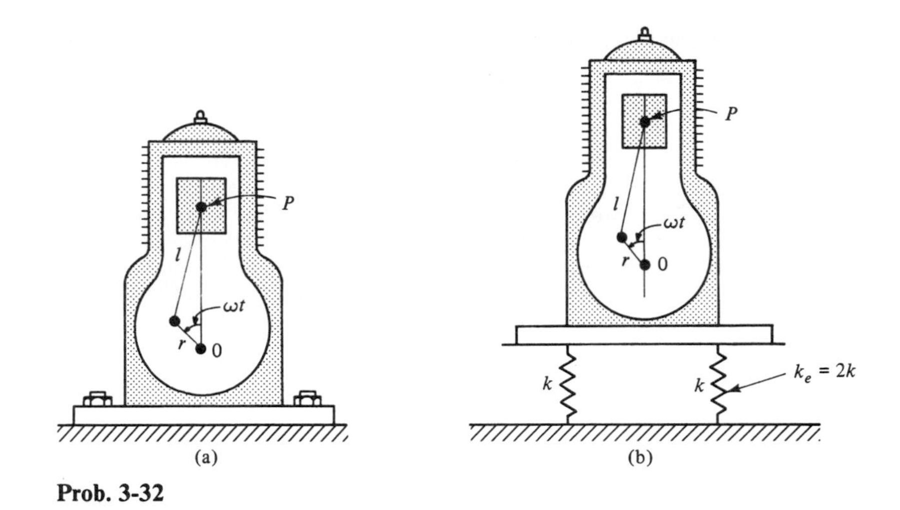

Prob. 3-32

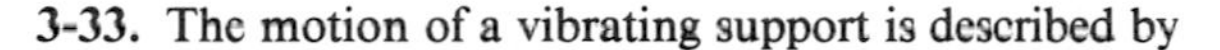

3-33. The motion of a vibrating support is described by

$$y = Y_1 \sin \omega_1 t + Y_2 \sin \omega_2 t + Y_3 \sin \omega_3 t$$

in which $\omega_1 < \omega_2 < \omega_3$. (a) Which ω is of primary importance in the design of isolation mounts? (b) Which ω is of primary importance in the use of a vibrometer-type transducer to measure the support motion? (c) Which ω is of primary importance in the use of an accelerometer-type transducer to measure the support motion?

3-34. The transducer shown in the accompanying figure is designed to measure gas pressures that fluctuate with frequencies lying between 100 and 600 cpm. It consists of a rod-and-piston assembly, a coil spring, and a linear-variable-differential transformer (LVDT). The displacement of the piston caused by the gas pressure produces a corresponding displacement of the core of the LVDT, and the voltage generated by the LVDT is fed to an appropriate recording device. When calibrated with static pressures, the static displacements of the core and corresponding voltages produced by the LVDT are proportional to the pressures. If the mass of the rod-and-piston assembly is 0.12 kg, determine an appropriate value for the spring constant k such that $\omega/\omega_n \leq 0.1$ for the range of pressure frequencies

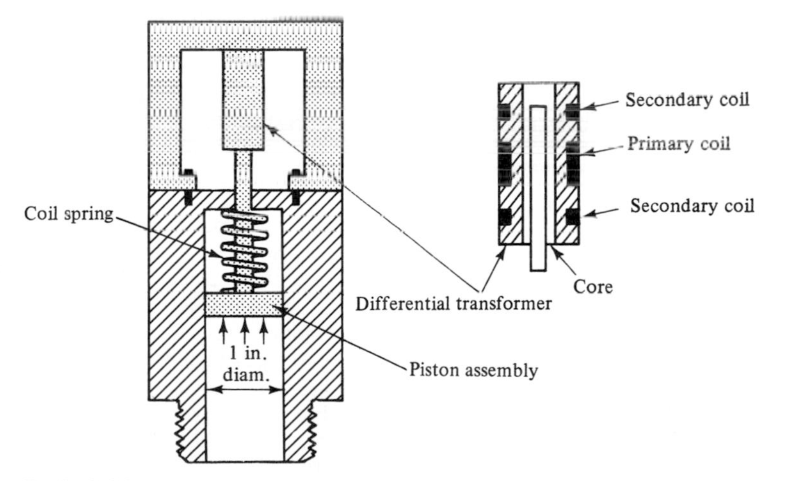

Prob. 3-34

encountered, so that the static calibration corresponds to dynamic pressure measurements.

Ans: $k = 47.37$ N/mm

3-35. The instrument shown in the accompanying figure is attached to the 85th floor of the Empire State Building to measure the lateral oscillation characteristics of the building caused by strong gusting winds. The weight W is rigidly attached to the two vertical elastic elements, each of which has a stiffness of k. The vertical penholder is pinned to the frame of the instrument at A and to the weight W at B. Previous observations have shown that the lateral oscillations of the 85th floor can have amplitudes as large as 2 ft when it is vibrating at approximately 0.2 Hz. Should the instrument shown in the figure be designed as an accelerometer or as a vibrometer? Give a brief explanation of the reason for your answer.

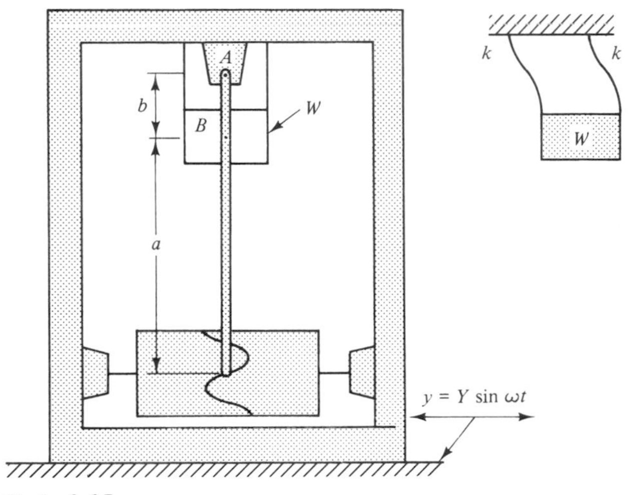

Prob. 3-35

3-36. Refer to Prob. 3-35 and the accompanying figure, and assume that the instrument was designed as an accelerometer for which $W = 0.4$ N and $a = 9b$. Determine (a) the value of the spring constant k for each elastic element for $\omega/\omega_n = 0.1$ and (b) the amplitude Y of the oscillation of the 85th floor when the chart amplitude reading on the drum is 50 mm with a period of $\tau = 5$ s ($f = 0.2$ Hz).

Ans: $k = 3.22$ N/m, $Y = 0.5$ m

3-37. A small trailer with hard rubber tires is to be pulled behind an automobile at a speed of 55 mph to evaluate the surface characteristics of a highway. The output from the electrical strain gages attached to the leaf springs of the trailer is to be recorded on magnetic tape for use in subsequent analyses. The deflection and accompanying strain ϵ of the springs is proportional to the relative displacement $z = x - y$, in which x is the vertical displacement of the trailer of mass m and y is the vertical displacement of the trailer's wheels. Since the tires are of hard rubber, the displacement y of the wheels should be essentially the same as the contour of the road surface. A trial run of a trailer built from a preliminary design furnished the frequency spectrum shown in the accompanying figure, which was obtained by feeding the output from the strain gages into a spectrum analyzer. Determine the minimum static deflection that the springs of the trailer must have if $z \cong y$ as the trailer moves along the road.

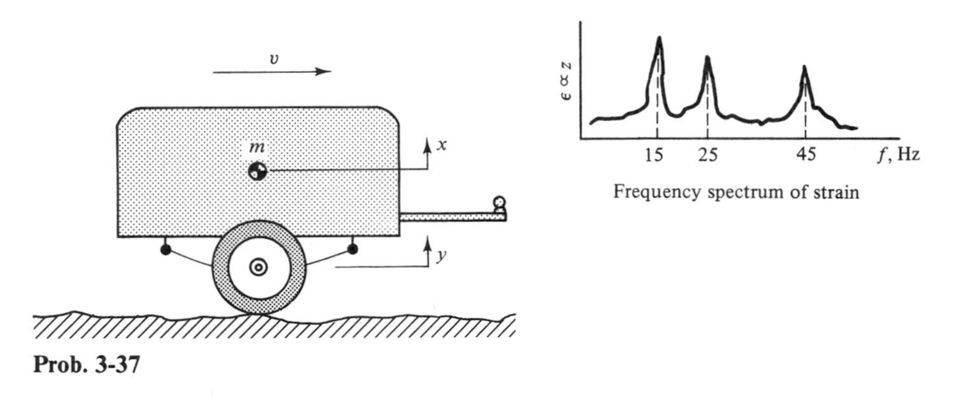

Prob. 3-37

3-38. The lateral oscillations of very tall buildings caused by gusting winds are usually very low in frequency and fairly high in amplitude. The small cantilever beam of length l that has a weight W and two strain gages attached to it, as shown in the accompanying figure, is to be designed as a transducer to monitor the lateral motion (more specifically the acceleration) of the upper floors of very tall buildings. One specification of the transducer is that the strain ϵ in the beam element at its outer fibers is to be at least 50 μin. when the lateral acceleration of the floor to which it is attached has an amplitude of $0.1g$. Another specification requires that the transducer have a frequency response that is essentially flat up to 0.5 Hz. The beam is to be made of aluminum having a modulus of elasticity of $E = 10(10)^6$ psi. Determine the length l of the beam if $W = 0.2$ lb, $b = \frac{1}{2}$ in., and $h = \frac{1}{16}$ in. as shown in the accompanying figure.

Ans: $l = 8.14$ in.

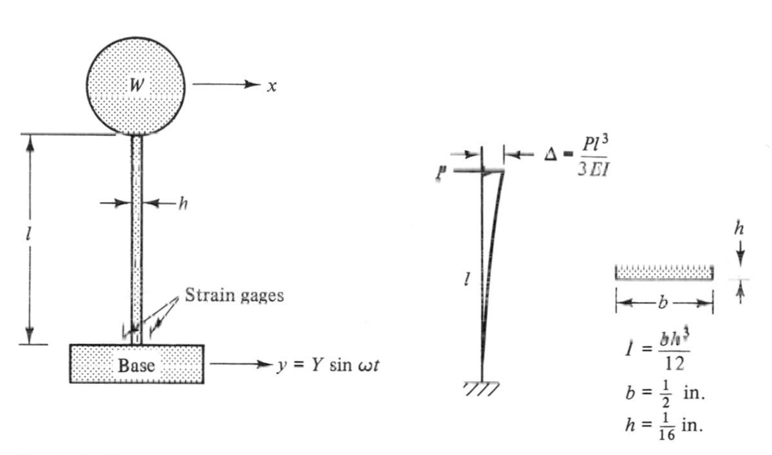

Prob. 3-38

3-39. The steel sphere shown in the accompanying figure is forced to oscillate in a water medium by the sinusoidal motion $y = Y \sin \omega t$ of the upper end of the spring of stiffness k. When $\omega/\omega_n = 1$ and $Y = 0.1$ in., the steady-state amplitude $|X|$ of the sphere is recorded as 0.5 in. Assuming velocity-squared damping, what would be the steady-state amplitude $|X_a|$

of an aluminum sphere of identical size with the excitation frequency ω adjusted so that $\omega/\omega_n = 1$ with $Y = 0.1$?

Ans: $|X_a| = 0.30$ in.

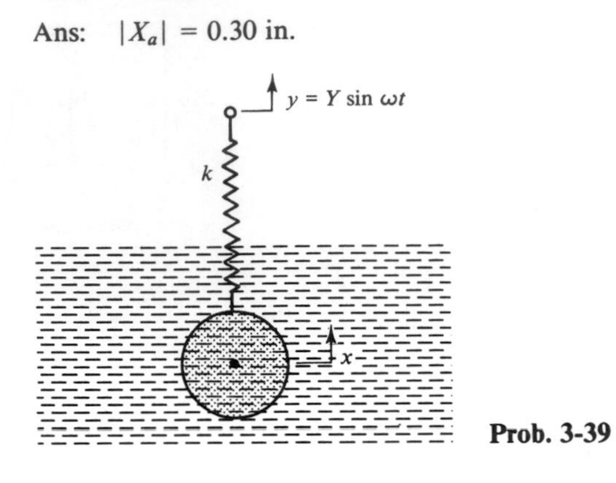

Prob. 3-39

Chapter 4

Transient Vibration from Nonharmonic Excitation

4-1 INTRODUCTION

In Chap. 3 we discussed the steady-state response of single-degree-of-freedom systems subjected to sinusoidal (harmonic) excitations. In this chapter we discuss the responses of single-degree-of-freedom systems subjected to *nonharmonic* excitations. Such responses consist of *transient* vibrations rather than steady-state oscillations.

If a nonharmonic excitation consists of a force or other disturbance that is applied suddenly (during a time that is short compared with the natural period of oscillation) or that undergoes sudden changes in magnitude, it is referred to as a *shock* excitation.

Several typical examples of nonharmonic forces are shown in Fig. 4-1. The exciting forces illustrated in Fig. 4-1a and d do not persist indefinitely, and there is obviously no steady-state response for a system upon which either of them acts; the total response of the system is thus its transient response. These forces are referred to as transient nonharmonic exciting forces.

On the other hand, the exciting forces shown in Fig. 4-1b and c are not transient but remain indefinitely, and although the *vibration* of a real system subjected to the excitation of one of these forces would be transient, the *total response* of the system would not go to zero but would become constant and persist indefinitely. Precisely speaking, then, the total response of the system is not transient. However, it is common practice to refer to any total response as transient if it does not include a steady-state *oscillation.* Using such a definition, the total response of a system subjected to either the transient or nontransient nonharmonic forces shown in Fig. 4-1 would be transient, and it is in this sense that the term transient is used in this text.

Figure 4-2b shows a structural system subjected to a typical nonharmonic *support excitation* $\ddot{y}$ (acceleration), which can result from ground motion such as that caused by earthquakes, heavy moving vehicles, blasting, and pile driving.

Observing the displacements shown in Fig. 4-2b, it should be apparent that the stresses caused in a structural system by support excitations are proportional to the *relative dis-*

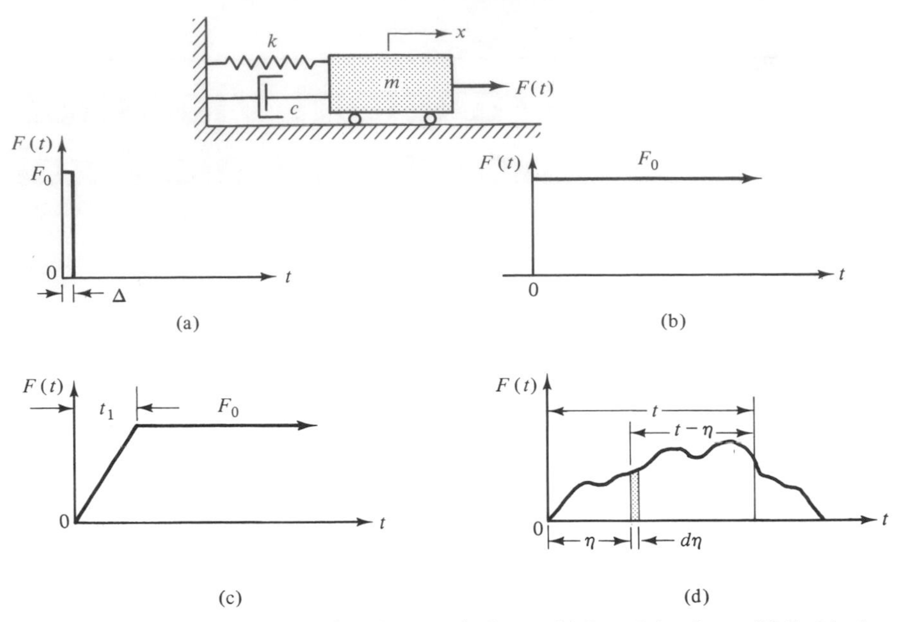

Figure 4-1 Typical examples of nonharmonic forces. (a) Impulsive force. (b) Suddenly applied force (step function). (c) Step function with rise time t_1. (d) General arbitrary force.

placement z, which is the displacement the mass m has with respect to the moving support. This usually makes the relative displacement of more practical importance than the absolute displacement x.

The acceleration $\ddot{y}$ of the moving support appears in the differential equation of motion of such a system, which is derived in terms of the relative displacement z as shown by Eq. 3-58. Thus, the acceleration of the moving support is an important variable in the analysis of vibration caused by support excitation.

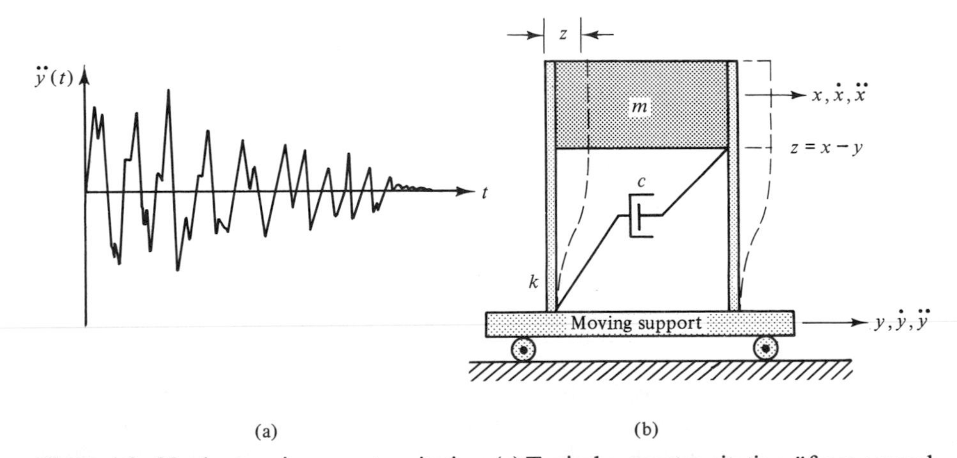

Figure 4-2 Nonharmonic support excitation. (a) Typical support excitation $\ddot{y}$ from ground motion. (b) Simple structural system subjected to support excitation.

A linear system subjected to a transient nonharmonic excitation will generally oscillate during the excitation with a varying amplitude, and at a frequency that corresponds to the natural frequency of the system. However, because of damping the oscillations will disappear after some interval of time. The peak response of a transient vibration is usually difficult to obtain analytically, except for a few simple excitations, of which the suddenly applied force shown in Fig. 4-1b is an example. Thus, numerical methods are widely used at present to obtain response spectrums with the use of microcomputers. Such spectrums facilitate the design of systems that are subjected to nonharmonic, or shock-type, excitations. As we shall see later, the response spectrum of a system subjected to a specified excitation is a graphical plot of the peak response of the system as a function of the system's natural period or frequency. The *modal-analysis* method discussed later in Chap. 6 will show how the response of a multiple-degree-of-freedom system can be determined by the use of the response spectrum of a single-degree-of-freedom system subjected to the same excitation.

4-2 NONHARMONIC FORCE EXCITATION

If we sketch a free-body diagram of the single-degree-of-freedom system shown in Fig. 4-1, and use either Newton's second law or the concept of dynamic equilibrium, we can obtain the differential equation of motion of the system as

$$\ddot{x} + 2\zeta\omega_n\dot{x} + \omega_n^2 x = \frac{F(t)}{m} \tag{4-1}$$

in which $F(t)$ represents any force that varies with time.

In this section the time-varying forces we are interested in are nonharmonic force excitations; we discuss several such excitations and the solutions of differential equations of motion containing them.

Impulsive Force

Let us first consider $F(t)$ as being a large force that occurs over a short interval of time such that

$$F(t) = \begin{Bmatrix} F_0 & (0 < t < \Delta) \\ 0 & (0 > t > \Delta) \end{Bmatrix} \tag{4-2}$$

as shown in Fig. 4-1a. If we then consider the interval Δ over which this pulse $F_0\Delta$ acts as being very small in comparison with the period τ_n of the system, we can replace this pulse with the "ideal" impulse

$$F(t) = C\delta(t) \tag{4-3}$$

discussed in Chap. 1. Equation 4-3 indicates an impulse of magnitude C occurring at time $t = 0$.

Using this expression in Eq. 4-1, we obtain the differential equation of motion of a damped system subjected to an impulsive force as

$$\ddot{x} + 2\zeta\omega_n\dot{x} + \omega_n^2 x = \frac{C}{m}\,\delta(t) \tag{4-4}$$

The solution of Eq. 4-4 can be obtained with the use of Laplace transforms (see Appendix A). With the initial conditions of $x_0 = \dot{x}_0 = 0$, the Laplace transform of Eq. 4-4 yields

$$\bar{x} = \mathcal{L}(x) = \frac{C/m}{s^2 + 2\zeta\omega_n s + \omega_n^2}$$

in which $\bar{x}$ indicates the Laplace transform of x. Then, using the table of Laplace transforms, the solution is found as

$$x = \frac{Ce^{-\zeta\omega_n t}}{m\omega_n\sqrt{1-\zeta^2}} \sin \omega_n\sqrt{1-\zeta^2}\, t \tag{4-5}$$

The reader should note from Eq. 4-5 that the response of a system subjected to an impulsive force depends primarily upon the *magnitude* of the impulse rather than upon its *shape*.

Alternate Solution

The response of a system to an impulsive force can also be obtained by considering that the impulse produces an *instantaneous change* in the *momentum* of the system before any appreciable displacement occurs. Using this concept, we can consider the damped free vibration response of the system using the initial conditions

$$\left.\begin{aligned} x_0 &= 0 \\ \dot{x}_0 &= \frac{C}{m} \end{aligned}\right\} \tag{4-6}$$

with $m\dot{x}_0$ representing the instantaneous change in momentum that the impulse of magnitude C causes.

Let us now again write Eq. 2-20, which is the solution of the differential equation of motion of damped free vibration,

$$x = e^{-\zeta\omega_n t}(A \cos \omega_n\sqrt{1-\zeta^2}\, t + B \sin \omega_n\sqrt{1-\zeta^2}\, t) \tag{2-20}$$

Using the initial conditions specified in Eq. 4-6, we find that

$$A = 0$$

and

$$B = \frac{C}{m\omega_n\sqrt{1-\zeta^2}}$$

Substituting these quantities into Eq. 2-20 repeated above, we obtain

$$x = \frac{Ce^{-\zeta\omega_n t}}{m\omega_n\sqrt{1-\zeta^2}} \sin \omega_n\sqrt{1-\zeta^2}\, t$$

which is the same as Eq. 4-5, which was obtained by the Laplace-transform method.

Arbitrary Force

The system response obtained for an impulsive force can be used to obtain its response to any arbitrary force $F(t)$ by considering the arbitrary force to be the sum of a sequence of impulses, one of which is shown in Fig. 4-1d. The response of a linear system to $F(t)$ is then found as the superposition of its responses to the individual impulses.

Written in terms of the independent variable η shown in Fig. 4-1d, $F(t)$ becomes $F(\eta)$. If we now consider an impulse of magnitude $F(\eta)\,d\eta$ acting at time η (the shaded area in Fig. 4-1d), the *differential displacement* dx of the system at any later time t due to the impulse at time η can be written directly from Eq. 4-5 as

$$dx = \frac{F(\eta)\,d\eta\, e^{-\zeta\omega_n(t-\eta)}}{m\omega_d} \sin \omega_d(t-\eta)$$

in which $\omega_d = \omega_n\sqrt{1-\zeta^2}$ and $(t-\eta)$ is the time elapsed after the impulse at time η. The total displacement x at any later time t due to the *sum* of all the impulses $F(\eta)\,d\eta$ of the excitation function $F(\eta)$ in the interval from 0 to t is

$$x = \frac{1}{m\omega_d}\int_0^t F(\eta)e^{-\zeta\omega_n(t-\eta)} \sin \omega_d(t-\eta)\, d\eta \tag{4-7}$$

Known as *Duhamel's integral* (the convolution integral), this represents a *particular* solution of the differential equation of motion of a damped system subjected to any excitation function $F(\eta)$. Unless $F(\eta)$ is a simple analytical function, such as a step function, the integral will usually have to be evaluated by some numerical method.

If the initial conditions are not zero, the homogeneous solution x_h of the differential equation of motion (see Eq. 2-20) must be added to the particular solution given by Eq. 4-7 to obtain the complete solution. The constants A and B in the homogeneous solution are then determined by applying the initial conditions to the complete solution. Since it can be shown that the particular solution and its derivative with respect to time are both equal to zero at $t = 0$, we can determine the constants A and B in such a case by applying the initial conditions ($x = x_0$, $\dot{x}_0 = \dot{x}$, at $t = 0$) to the homogeneous solution alone to find that

$$x_h = e^{-\zeta\omega_n t}\left[x_0 \cos \omega_d t + \left(\frac{\dot{x}_0 + \zeta\omega_n x_0}{\omega_d}\right)\sin \omega_d t\right] \tag{4-8}$$

in which once again $\omega_d = \omega_n\sqrt{1-\zeta^2}$. Thus, the complete solution that yields the total response of a single-degree-of-freedom system at any time t is given by

$$x = e^{-\zeta\omega_n t}\left[x_0 \cos \omega_d t + \left(\frac{\dot{x}_0 + \zeta\omega_n x_0}{\omega_d}\right)\sin \omega_d t\right] + \frac{1}{m\omega_d}\int_0^t F(\eta)e^{-\zeta\omega_n(t-\eta)} \sin \omega_d(t-\eta)\, d\eta \tag{4-9}$$

When it is difficult to obtain an analytical solution of the transient response using Duhamel's integral, an easier approach is to solve the differential equation of motion by a numerical integration method.[1]

[1] The numerical integration of Duhamel's integral to obtain large amounts of data, such as for plotting a response curve, is not recommended on a microcomputer since it is much less efficient than a program such as the one given in Example 4-4 (the former took 100 times the computing time of the latter to obtain the same amount of data).

Suddenly Applied Force (Step Function)

When a system is subjected to a suddenly applied force $F(t) = F_0$ as shown in Fig. 4-1b, Eq. 4-1 gives the differential equation of motion as

$$\ddot{x} + 2\zeta\omega_n\dot{x} + \omega_n^2 x = \frac{F_0}{m} \tag{4-10}$$

The solution of Eq. 4-10 can be obtained in a straightforward manner by adding the particular solution x_p to the homogeneous solution given by Eq. 2-20. Since the right side of Eq. 4-10 is a constant, an inspection of the equation shows that

$$x_p = \frac{F_0}{m\omega_n^2} = \frac{F_0}{k} \tag{4-11}$$

Thus, the complete solution of Eq. 4-10 is

$$x = e^{-\zeta\omega_n t}(A\cos\omega_d t + B\sin\omega_d t) + \frac{F_0}{k} \tag{4-12}$$

in which $\omega_d = \omega_n\sqrt{1-\zeta^2}$.

If the initial conditions are $x_0 = \dot{x}_0 = 0$, we can find from Eq. 4-12 that

$$A = \frac{-F_0}{k}$$

and

$$B = \frac{-F_0\zeta}{k\sqrt{1-\zeta^2}}$$

Substituting these expressions into Eq. 4-12 yields

$$\frac{x}{F_0/k} = 1 - e^{-\zeta\omega_n t}\left(\cos\omega_d t + \frac{\zeta}{\sqrt{1-\zeta^2}}\sin\omega_d t\right) \tag{4-13}$$

in which F_0/k is the static displacement δ_s that would occur if the force F_0 were applied statically.

Recalling that

$$\omega_d = \omega_n\sqrt{1-\zeta^2} = \frac{2\pi}{\tau_n}\sqrt{1-\zeta^2}$$

and that the two trigonometric terms in Eq. 4-13 can be combined into a single trigonometric function and phase angle ϕ, as shown in Fig. 4-3, Eq. 4-13 can be written in the form of

$$\frac{x}{F_0/k} = 1 - \frac{e^{-\zeta 2\pi t/\tau_n}}{\sqrt{1-\zeta^2}}\left[\cos\left(\frac{2\pi}{\tau_n}\sqrt{1-\zeta^2}\,t - \phi\right)\right] \tag{4-14}$$

in which

$$\tan\phi = \frac{\zeta}{\sqrt{1-\zeta^2}}$$

If the damping is theoretically zero, Eq. 4-14 reduces to

$$\frac{x}{F_0/k} = 1 - \cos\frac{2\pi t}{\tau_n} \tag{4-15}$$

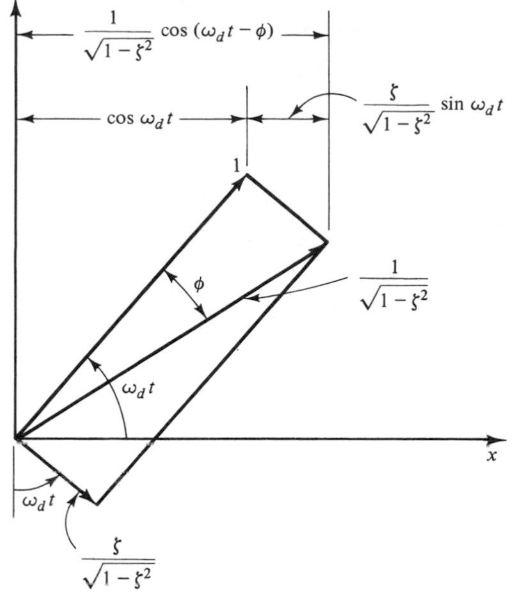

Figure 4-3

which shows that $x/(F_0/k)$ reaches a maximum value of 2 when $t/\tau_n = \frac{1}{2}$. That is, a suddenly applied force of a given magnitude produces a peak response (displacement) that is *twice* that of a slowly applied force of the same magnitude. The family of curves from Eq. 4-14 shown in Fig. 4-4 reveals that as the damping increases, the peak response decreases, moving further and further below the theoretically undamped peak of $2(F_0/k)$.

The reader should note that the family of curves shown in Fig. 4-4 is applicable to an infinite number of systems of varying τ_n. For example, the response $x/(F_0/k)$ at time $t = 0.2$ s for a system with $\tau_n = 0.1$ s would be the same as that of another system with $\tau_n = 0.2$ s when $t = 0.4$ s, and so forth.

Equation 4-15 can also be obtained quite easily from Duhamel's integral given by Eq. 4-7. With $\zeta = 0$ and $F(\eta) = F_0$, it becomes

$$x = \frac{F_0}{m\omega_n} \int_0^t \sin \omega_n(t - \eta)\, d\eta$$

which when integrated yields

$$x = \frac{F_0}{m\omega_n^2} \cos \omega_n(t - \eta)\Big|_0^t$$

Substituting limits, and noting that $m\omega_n^2 = k$, we obtain

$$\frac{x}{F_0/k} = 1 - \cos \omega_n t$$

which is the same as Eq. 4-15.

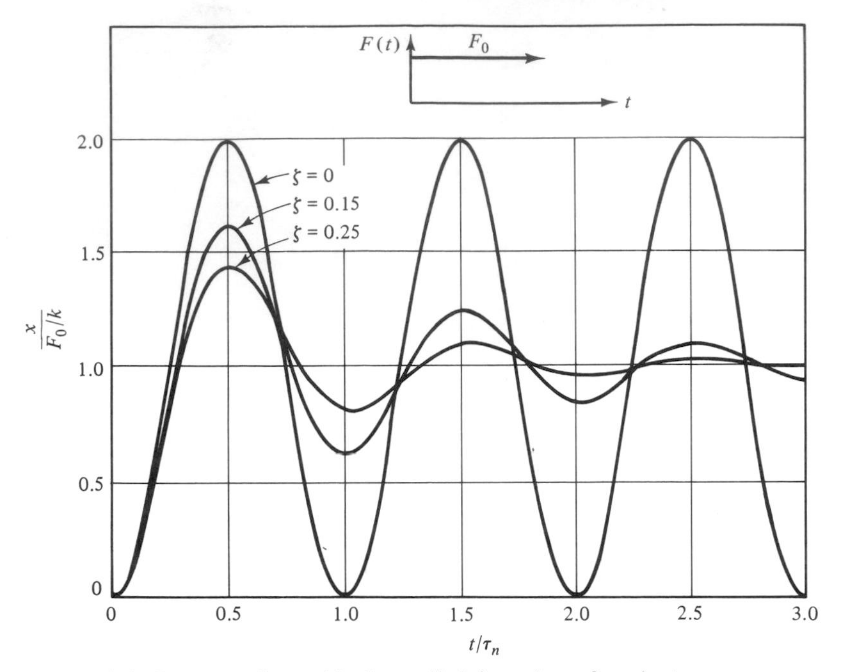

Figure 4-4 Response for suddenly applied force (step function).

Step Function (with Rise Time)

The step function shown in Fig. 4-1c with a rise time t_1 is an excitation that is more realistic than the suddenly applied force just discussed in which the rise time was theoretically zero. The former occurs quite commonly in real systems, and the application of rapidly applied loads with various rise times is frequently used in tests to evaluate the effects of dynamic loading on structural systems.

In our previous discussion of a suddenly applied force with zero rise time acting on a system with no damping, the maximum response of the system was found to be twice that caused by a force of the same magnitude applied statically. When a step function with rise time t_1 acts on a system, the maximum response of the system varies according to the ratio of the rise time to the undamped natural period of the system, t_1/τ_n. However, as might be expected, the maximum dynamic response approaches that caused by the suddenly applied force with zero rise time when t_1/τ_n is very, very small, and approaches that caused by a static load when t_1/τ_n becomes sufficiently large. For example, we shall see later that when $t_1/\tau_n > 4$, the dynamic response caused by a step function is essentially the same as that caused by a statically applied load having the same magnitude as the maximum value of the step function.

With this in mind, the initial analysis in the design of a structural system might well be made by approaching the problem as one in statics, and then checking to see if a dynamic analysis is necessary. That is, if a system has a rather high natural frequency f_n so that its natural period τ_n is very short, the system response might be essentially the same as the static response, even though the rise time t_1 was very short.

It is somewhat difficult to obtain an analytical solution yielding the response of a damped system subjected to a step function with a finite rise time. Therefore, let us assume that the damping is theoretically zero in the discussions that immediately follow, so that we can obtain analytical solutions. System damping will be included later when we discuss a numerical integration method for obtaining the transient response of such systems.

Solution by Duhamel's Integral

If $\zeta = 0$, and the initial conditions are assumed to be zero, Duhamel's integral given by Eq. 4-7 can be used without difficulty to obtain a solution of Eq. 4-1, in which $F(t)$, or $F(\eta)$, is defined by

$$F(\eta) = \frac{F_0 \eta}{t_1} \qquad 0 \le \eta \le t_1$$

$$F(\eta) = F_0 \qquad \eta \ge t_1$$

With these conditions, Duhamel's integral gives the response at any time $t \ge t_1$ as

$$x = \frac{F_0}{m\omega_n t_1} \int_0^{t_1} \eta \sin \omega_n(t - \eta)\, d\eta + \frac{F_0}{m\omega_n} \int_{t_1}^{t} \sin \omega_n(t - \eta)\, d\eta \tag{4-16}$$

which, upon integrating and substituting limits, yields

$$x = \frac{F_0}{m\omega_n^2} \left[1 - \frac{\sin \omega_n t}{\omega_n t_1} + \frac{\sin \omega_n (t - t_1)}{\omega_n t_1} \right]$$

or since $\omega_n = 2\pi/\tau_n$ and $k = m\omega_n^2$,

$$x = \frac{F_0}{k} \left[1 - \frac{\tau_n}{2\pi t_1} \sin \frac{2\pi t}{\tau_n} + \frac{\tau_n}{2\pi t_1} \sin \frac{2\pi}{\tau_n}(t - t_1) \right] \qquad (t \ge t_1) \tag{4-17}$$

Direct Solution of Differential Equation of Motion

The step function with rise time t_1 can be considered as the sum of two *ramp functions* such as those shown in Fig. 4-5. Such a consideration facilitates a solution of the differential equation of motion of a system subjected to this type of excitation.

The first ramp function starts at $t = 0$, and the second starts after a time delay equal to the rise time t_1 of the step function. With $\zeta = 0$ and $F(t) = (F_0/t_1)t$, Eq. 4-1 becomes

$$\ddot{x} + \omega_n^2 x = \frac{F(t)}{m} = \frac{F_0 t}{m t_1} \qquad (t \le t_1) \tag{4-18}$$

The complete solution of Eq. 4-18 is the sum of the homogeneous solution $x_h = A \cos \omega_n t + B \sin \omega_n t$ and the particular solution x_p. The latter can be determined from an inspection of Eq. 4-18 as $x_p = F_0 t/m\omega_n^2 t_1$. That is,

$$x = A \cos \omega_n t + B \sin \omega_n t + \frac{F_0 t}{m\omega_n^2 t_1} \tag{4-19}$$

Using the initial conditions $x_0 = \dot{x}_0 = 0$, we find from Eq. 4-19 that

$$A = 0$$

and

$$B = \frac{-F_0}{m\omega_n^3 t_1} = \frac{-F_0}{k\omega_n t_1}$$

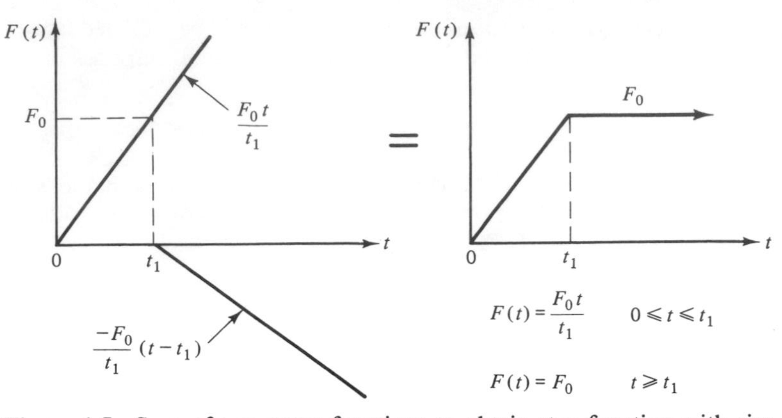

Figure 4-5 Sum of two ramp functions to obtain step function with rise time t_1.

The response of the system for $t \leq t_1$ is then found to be

$$x = \frac{F_0}{k}\left(\frac{t}{t_1} - \frac{1}{\omega_n t_1}\sin \omega_n t\right) \qquad (t \leq t_1) \tag{4-20}$$

The system response caused by a *negative* ramp function starting at time t_1 can be written from an inspection of Eq. 4-20 and noting that the *elapsed* time for this function is $t - t_1$ instead of t. Thus,

$$x = -\frac{F_0}{k}\left[\frac{t - t_1}{t_1} - \frac{1}{\omega_n t_1}\sin \omega_n(t - t_1)\right] \qquad (t > t_1) \tag{4-21}$$

Since Eq. 4-20 is also valid for $t > t_1$, the sum of Eqs. 4-20 and 4-21 gives the response for $t > t_1$ as

$$x = \frac{F_0}{k}\left[1 - \frac{1}{\omega_n t_1}\sin \omega_n t + \frac{1}{\omega_n t_1}\sin \omega_n(t - t_1)\right]$$

or

$$\frac{x}{F_0/k} = 1 - \frac{\tau_n}{2\pi t_1}\sin \frac{2\pi t}{\tau_n} + \frac{\tau_n}{2\pi t_1}\sin \frac{2\pi}{\tau_n}(t - t_1) \qquad (t > t_1) \tag{4-22}$$

which is the same as Eq. 4-17 obtained using Duhamel's integral.

The effect of the ratio t_1/τ_n on the peak response is not apparent from a cursory inspection of Eq. 4-22. Since the response will naturally lag the excitation $F(t)$, it is logical to assume that the peak response occurs at a time $t = t_m$, which is greater than the rise time t_1. To determine this peak-response time, Eq. 4-22 can be written in a more convenient form using the identity

$$\sin u - \sin v = 2 \sin \tfrac{1}{2}(u - v) \cdot \cos \tfrac{1}{2}(u + v)$$

Then, choosing

$$u = \frac{2\pi}{\tau_n}(t - t_1)$$

and

$$v = \frac{2\pi t}{\tau_n}$$

Eq. 4-22 can be written as

$$\frac{x}{F_0/k} = 1 - \frac{\tau_n}{\pi t_1}\left[\sin\frac{\pi t_1}{\tau_n}\cdot\cos\frac{2\pi}{\tau_n}\left(t - \frac{t_1}{2}\right)\right] \qquad (t > t_1) \tag{4-23}$$

Since the maximum displacement will occur when the velocity $\dot{x}$ is zero, we can differentiate Eq. 4-23 to obtain

$$\dot{x} = 0 = \sin\frac{2\pi}{\tau_n}\left(t_m - \frac{t_1}{2}\right) \tag{4-24}$$

in which t_m is the time at which the response is a maximum.

It now follows from Eq. 4-24 that

$$\frac{2\pi}{\tau_n}\left(t_m - \frac{t_1}{2}\right) = i\pi \qquad i = 1, 2, 3, \ldots$$

from which

$$t_m = \frac{i\tau_n}{2} + \frac{t_1}{2} \qquad i = 1, 2, 3, \ldots$$

Thus, the *cosine* term in Eq. 4-23 becomes

$$\cos\frac{2\pi}{\tau_n}\left(\frac{i\tau_n}{2} + \frac{t_1}{2} - \frac{t_1}{2}\right) = \cos i\pi \qquad i = 1, 2, 3, \ldots$$

so that the maximum response is found to be

$$\frac{x_m}{F_0/k} = 1 - \frac{\tau_n}{\pi t_1}\sin\frac{\pi t_1}{\tau_n}(\cos i\pi) \qquad i = 1, 2, 3, \ldots \tag{4-25}$$

The *sine* term in Eq. 4-25 can be positive or negative depending upon the magnitude of t_1/τ_n, while the *cosine* term can be positive or negative depending upon the integer i. The maximum response will obviously occur when the signs are such that the product of the two trigonometric terms is negative. Therefore, in determining the maximum response, the maximum value of $x_m/(F_0/k)$, the following should be noted:

$$\frac{x_m}{F_0/k} = 1 + \frac{\tau_n}{\pi t_1}\sin\frac{\pi t_1}{\tau_n} \qquad 0 < \frac{t_1}{\tau_n} \le 1 \qquad i = 1$$

$$\frac{x_m}{F_0/k} = 1 + \frac{\tau_n}{\pi t_1}\left|\sin\frac{\pi t_1}{\tau_n}\right| \qquad 1 \le \frac{t_1}{\tau_n} \le 2 \qquad i = 2$$

$$\frac{x_m}{F_0/k} = 1 + \frac{\tau_n}{\pi t_1}\sin\frac{\pi t_1}{\tau_n} \qquad 2 \le \frac{t_1}{\tau_n} \le 3 \qquad i = 3$$

. . . .

From these relationships, we can conclude that Eq. 4-25 can also be written as

$$\frac{x_m}{F_0/k} = 1 + \frac{\tau_n}{\pi t_1}\left|\sin\frac{\pi t_1}{\tau_n}\right| \qquad \frac{t_1}{\tau_n} > 0 \tag{4-26}$$

for computing the maximum response.

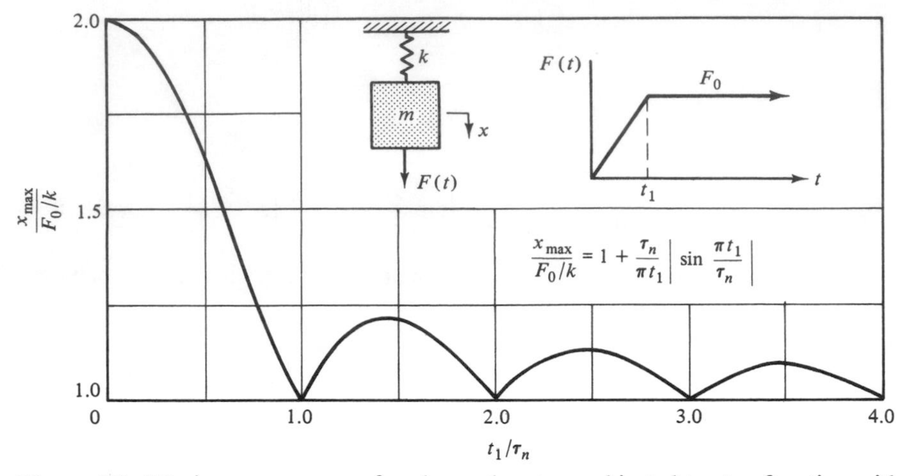

Figure 4-6 Maximum response of undamped system subjected to step function with rise time t_1.

Figure 4-6 shows a plot of the response as a function of t_1/τ_n. From this plot it is apparent that as t_1/τ_n increases, the maximum dynamic response decreases, approaching the response caused by a statically applied force of magnitude F_0 when $t_1/\tau_n > 4$.

EXAMPLE 4-1

A wooden beam as shown in Fig. 4-7a is fabricated by gluing four rough-cut planks together as shown in Fig. 4-7c. Each plank is 2 in. thick. The modulus of elasticity of the wood is $E = 1.5(10)^6$ psi, and its specific weight is $w = 0.022$ lb/in.^{3}

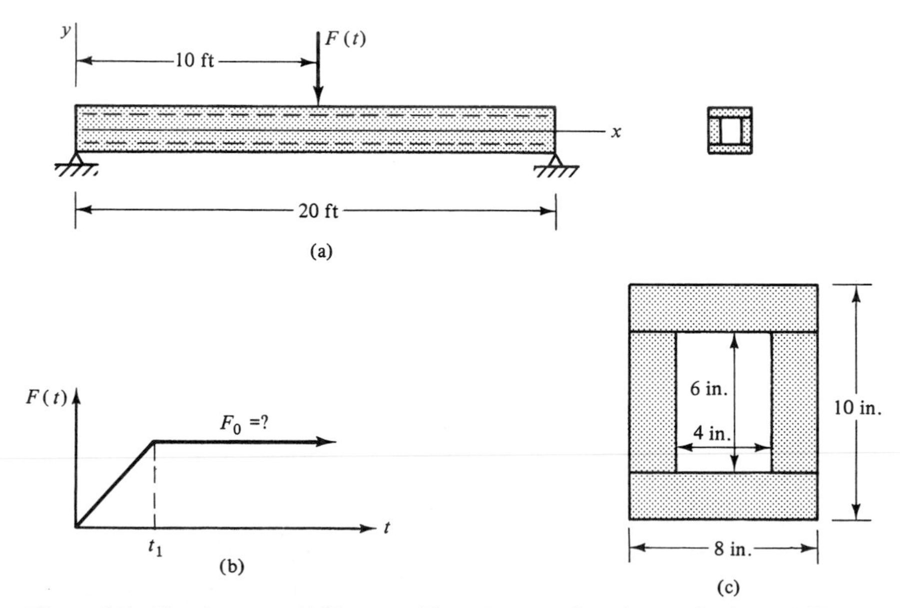

Figure 4-7 Simply supported beam subjected to step-function excitation. (a) Simply-supported wooden beam. (b) Step function. (c) Cross section.

Knowing the allowable stress for both glue and wood, and with both the shear stress in the glued joints and the maximum normal stress in the wood determined in terms of a static load P_0 applied at the center of the beam, it is found from principles of elementary mechanics of materials that the magnitude of the design load P_0 depends upon the shear strength of the glue rather than upon the strength of the wood. For the glue used, the allowable shear stress is reached when the design load $P_0 = 6000$ lb.

To analyze the behavior of the glued joints under dynamic loading, it is proposed to subject the beam to a step-function force with rise time t_1 as shown in Fig. 4-7b. The force $F(t)$ is to be applied by means of a servohydraulic testing machine, programmed to control both the rise time t_1 and the force magnitude F_0.

If the rise time is selected as 30 ms, what should the force magnitude F_0 be limited to, so that the *dynamic* deflection of the center of the beam is essentially the same as the static deflection caused by a slowly applied design load of 6000 lb? Assume that the beam behaves as a simple single-degree-of-freedom system having only the fundamental natural frequency f_n and that it is theoretically undamped ($\zeta = 0$).[2]

Solution. The fundamental undamped natural frequency of the beam can be determined by the energy method explained in Sec. 2-9 in which $T_{max} = U_{max}$. Using the shape function

$$y = \Delta \sin \frac{\pi x}{l} \tag{4-27}$$

which is the exact shape function for the fundamental mode of vibration of a simply supported beam, the maximum kinetic energy of the beam is given by

$$T_{max} = \frac{\gamma}{2} \int_0^l (\omega_n y)^2 \, dx = \frac{\gamma \omega_n^2 \Delta^2}{2} \int_0^l \sin^2 \frac{\pi x}{l} \, dx$$

After integrating and substituting limits,

$$T_{max} = \frac{\gamma \omega_n^2 \Delta^2}{2} \left(\frac{l}{2}\right) \tag{4-28}$$

where γ = mass of beam per unit length (lb · s²/in.² or N · s²/m²)
l = length of beam (in. or m)

The maximum change in the potential energy from the potential energy in the static-equilibrium position is given by

$$U_{max} = \frac{EI}{2} \int_0^l \left(\frac{d^2 y}{dx^2}\right)^2 dx = \frac{EI \Delta^2 \pi^4}{2 l^4} \int_0^l \sin^2 \frac{\pi x}{l} \, dx$$

which yields

$$U_{max} = \frac{EI \Delta^2 \pi^4}{2 l^4} \left(\frac{l}{2}\right) \tag{4-29}$$

[2] Although the beam has many degrees of freedom, a transient solution for the deflection with the beam assumed as a single-degree-of-freedom system will be reasonably accurate since, as shown in the more rigorous analyses of this type of problem discussed in Chaps. 6 and 8, the major portion of the total response obtained from the superposition of modes comes from the fundamental mode in most situations.

Substituting Eqs. 4-28 and 4-29 into $T_{max} = U_{max}$, we obtain

$$\omega_n = \frac{\pi^2}{l^2}\sqrt{\frac{EI}{\gamma}}$$

from which the fundamental natural frequency is found to be

$$f_n = \frac{\omega_n}{2\pi} = \frac{\pi}{2l^2}\sqrt{\frac{EI}{\gamma}} \quad \text{Hz} \tag{4-30}$$

Referring to the beam cross section shown in Fig. 4-7c, and knowing the specific weight of the wood, we can calculate the following:

$$\gamma = \frac{(80 - 24)(0.022)}{386} = 3.19(10)^{-3} \text{ lb}\cdot\text{s}^2/\text{in.}^2$$

$$I = \frac{8(10)^3}{12} - \frac{4(6)^3}{12} = 594.7 \text{ in.}^4$$

Substituting these values, along with $l = 240$ in. and $E = 1.5(10)^6$ psi, into Eq. 4-30, we find that the fundamental natural frequency is

$$f_n = \frac{\pi}{2(240)^2}\sqrt{\frac{1.5(10)^6(594.7)}{3.19(10)^{-3}}} = 14.42 \text{ Hz}$$

and the fundamental natural period is

$$\tau_n = \frac{1}{f_n} = 0.0693 \text{ s}$$

Thus,

$$\frac{t_1}{\tau_n} = \frac{0.030}{0.0693} = 0.43$$

Substituting this value into Eq. 4-26 gives

$$\frac{y_{max}}{F_0/k} = 1.72 \tag{4-31}$$

This result can be checked approximately by referring to Fig. 4-6.

Since F_0/k in Eq. 4-31 corresponds to the static deflection of the beam for the 6000-lb design load, the magnitude F_0 of the step function with rise time $t_1 = 0.03$ s is

$$F_0 = \frac{6000}{1.72} = 3488 \text{ lb} \tag{4-32}$$

Using $F_0 = 3488$ lb and $k = 48EI/l^3$, the maximum deflection at the center of the beam using Eq. 4-31 is found to be

$$y_{max} = 1.72\frac{F_0}{k} = \frac{1.72(3488)(240)^3}{48(1.5)(10)^6(594.7)} = 1.94 \text{ in.}$$

This result, which is based upon the beam being modeled as a single-degree-of-freedom system, is within 1 percent of a modal-analysis solution in which the beam is modeled as five lumped masses (see Prob. 6-24).

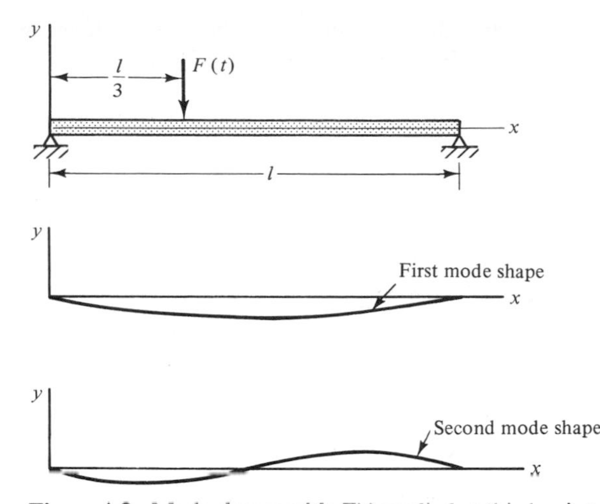

Figure 4-8 Mode shapes with $F(t)$ applied at third point.

In the preceding example, if $F(t)$ were applied at say $x = l/3$, as shown in Fig. 4-8, instead of at the center of the beam, such a solution based upon just the fundamental mode shape and frequency would not be very accurate. With $F(t)$ applied at a third point, the shape of the dynamic response would no longer be symmetrical about the center of the beam, primarily because of the excitation of the second mode shown in Fig. 4-8. This mode is not present in the response when $F(t)$ is applied at the center of the beam.

Section 5-12, which is concerned with the response of beams modeled as lumped-mass systems, will provide the reader with a better understanding of how a beam's natural frequencies and corresponding mode shapes affect its dynamic response.

4-3 NONHARMONIC SUPPORT EXCITATION

As discussed in Sec. 3-6, the differential equation of motion of the system shown in Fig. 4-2b can be expressed either in terms of the *absolute* displacement x of the mass m or in terms of the *relative* displacement z of the mass with respect to the moving support.

As can be seen from Fig. 4-2b, the distortion, or deformation, of each of the elastic elements is $x - y$. The damping force acting on the mass is $c(\dot{x} - \dot{y})$ since the velocity of the dashpot components with respect to each other is $\dot{x} - \dot{y}$. With k as the combined stiffness of the two elastic elements, the free-body diagram of the mass is as shown schematically in Fig. 4-9.

Applying Newton's second law,

$$-k(x - y) - c(\dot{x} - \dot{y}) = m\ddot{x} \tag{4-33}$$

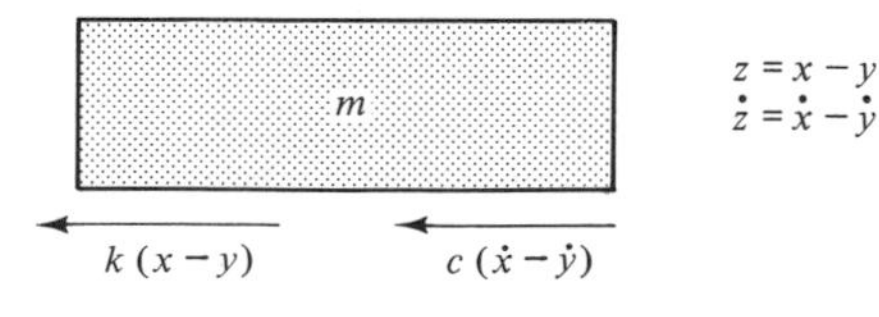

$z = x - y$
$\dot{z} = \dot{x} - \dot{y}$

Figure 4-9 Free-body diagram of system shown in Fig. 4-2b.

in which x, $\dot{x}$, and $\ddot{x}$ are the *absolute* displacement, velocity, and acceleration, respectively, of the mass m, and y and $\dot{y}$ are the absolute displacement and velocity, respectively, of the moving support.

Since y and $\dot{y}$ are both excitation functions, Eq. 4-33 is written as

$$\ddot{x} + \frac{c}{m}\dot{x} + \frac{k}{m}x = \frac{c}{m}\dot{y} + \frac{k}{m}y$$

or as

$$\ddot{x} + 2\zeta\omega_n\dot{x} + \omega_n^2 x = 2\zeta\omega_n\dot{y} + \omega_n^2 y \tag{4-34}$$

Equation 4-34 is the differential equation of motion of the system in terms of the *absolute* motion of the mass, with the excitation appearing in terms of both y and $\dot{y}$.

Noting that $z = x - y$, $\dot{z} = \dot{x} - \dot{y}$, and $\ddot{x} = \ddot{z} + \ddot{y}$, Eq. 4-33 can be written as

$$-kz - c\dot{z} = m(\ddot{z} + \ddot{y})$$

from which we obtain

$$\ddot{z} + 2\zeta\omega_n\dot{z} + \omega_n^2 z = -\ddot{y} \tag{4-35}$$

Equation 4-35 is the differential equation of motion of the system in terms of the motion of the mass relative to the moving support. In this equation, the excitation is the negative of the absolute acceleration of the moving support.

As mentioned previously, the relative response z is of more practical interest than the absolute response x since the stresses induced in structural systems subjected to support excitation depend upon relative displacements. Although the relative response can be determined by solving Eq. 4-34, such a solution may be complicated by the fact that excitation involves both y and $\dot{y}$ of the moving support. However, solving Eq. 4-35 to obtain the relative response z involves only the acceleration $\ddot{y}$ of the moving support. When the excitation is due to ground motion such as that which occurs during earthquakes, the acceleration $\ddot{y}$ is the quantity most generally monitored since accelerometers provide an accurate and convenient means of doing so. Therefore, to avoid the complications mentioned, our discussion of the transient responses due to support excitation will deal primarily with Eq. 4-35.

If $\ddot{y}$ is a simple analytical function, a direct solution of Eq. 4-35 can usually be obtained in a straightforward manner. However, it may be difficult to determine the peak response of z from such an analytical solution, particularly for systems in which damping is considered. In instances in which such a solution is difficult to obtain, or when it is difficult to determine the maximum response from such a solution, a numerical integration of Eq. 4-35 can be used quite readily to obtain the desired results. A procedure for formulating the differential equation of motion in terms of parameters pertinent to its solution by numerical integration using the fourth-order Runge-Kutta method will be discussed in Sec. 4-6.

Another approach to obtaining the relative response z is by the use of Duhamel's integral (Eq. 4-7), which is easily modified to include the arbitrary support acceleration $\ddot{y}(\eta)$ shown in Fig. 4-10. Comparing Eqs. 4-1 and 4-35, it can be seen that $F(\eta)/m$ in Eq. 4-7 may be replaced by $-\ddot{y}(\eta)$ to obtain

$$z = \frac{-1}{\omega_d}\int_0^t \ddot{y}(\eta)e^{-\zeta\omega_n(t-\eta)} \sin \omega_d(t - \eta)\, d\eta \tag{4-36}$$

which represents a *particular* solution of Eq. 4-35. Unless $\ddot{y}(\eta)$ is a simple analytical function,

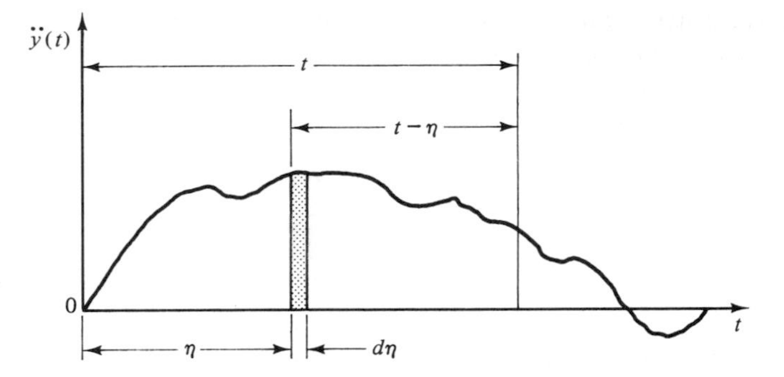

Figure 4-10 Arbitrary support acceleration.

the integral will usually have to be evaluated by some numerical method. If the initial conditions are not zero, the homogeneous solution of Eq. 4-35 must be added to the particular solution given by Eq. 4-36 to obtain a complete solution, as was done in obtaining Eq. 4-9.

EXAMPLE 4-2

The displacement of the support of the simple structural system shown in Fig. 4-2 is the *sine pulse* shown in Fig. 4-11a. Considering the sine pulse as the superposition of two sine waves as shown in Fig. 4-11b, and considering the damping as theoretically zero, we wish to determine the relative response z of the system as a function of time.

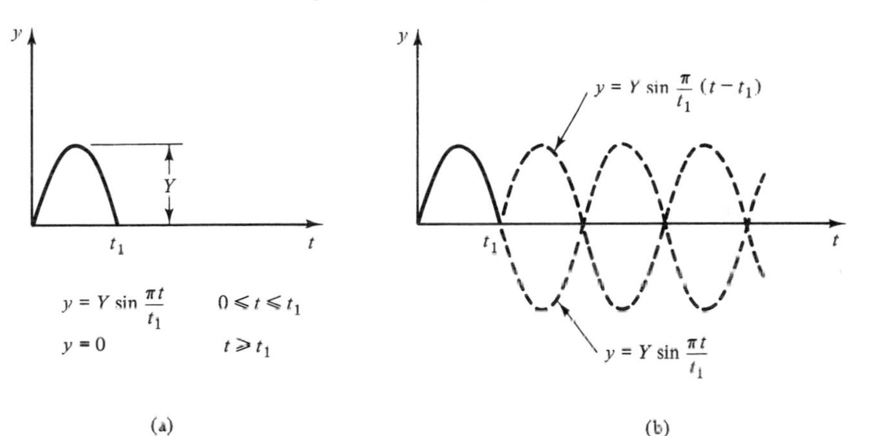

Figure 4-11 Sine pulse of Example 4-2. (a) Sine pulse of duration t_1. (b) Superposition of two sine waves to obtain the sine pulse.

Solution. With the sine wave starting at $t = 0$,

$$y = Y \sin \frac{\pi t}{t_1} = Y \sin \omega t$$

Two successive differentiations yield the acceleration of the moving support as

$$\ddot{y} = \frac{-Y\pi^2}{t_1^2} \sin \frac{\pi t}{t_1} = -Y\omega^2 \sin \omega t \tag{4-37}$$

in which $\omega = \pi/t_1$. Since $\zeta = 0$, Eq. 4-35 reduces to

$$\ddot{z} + \omega_n^2 z = Y\omega^2 \sin \omega t \tag{4-38}$$

The homogeneous solution of Eq. 4-38 is

$$z_h = A \cos \omega_n t + B \sin \omega_n t \tag{4-39}$$

in which $\omega_n = \sqrt{k/m}$. Assuming a particular solution of Eq. 4-38 as

$$z_p \overset{I}{=} Ze^{j\omega t}$$

we obtain

$$z_p = \frac{Y(\omega/\omega_n)^2}{1 - (\omega/\omega_n)^2} \sin \omega t \tag{4-40}$$

Adding Eqs. 4-39 and 4-40, we find that

$$z = A \cos \omega_n t + B \sin \omega_n t + \frac{Y(\omega/\omega_n)^2}{1 - (\omega/\omega_n)^2} \sin \omega t \tag{4-41}$$

Assuming that the initial conditions are $x_0 = \dot{x}_0 = 0$, the initial conditions for the *relative* motion are

$$\left.\begin{aligned} z_0 &= x_0 - y_0 = 0 \\ \dot{z}_0 &= \dot{x}_0 - \dot{y}_0 = -Y\omega \end{aligned}\right\} \tag{4-42}$$

The reader should realize that it would have been incorrect to assume that $\dot{z}_0 = 0$.

Using these initial conditions in Eq. 4-41, we find that

$$\left.\begin{aligned} A &= 0 \\ B &= \frac{-Y(\omega/\omega_n)}{1 - (\omega/\omega_n)^2} \end{aligned}\right\} \tag{4-43}$$

Substituting the above into Eq. 4-41, we obtain

$$z_1 = \frac{Y}{1 - (\omega/\omega_n)^2}\left[\left(\frac{\omega}{\omega_n}\right)^2 \sin \omega t - \frac{\omega}{\omega_n} \sin \omega_n t\right] \tag{4-44}$$

in which z_1 indicates a solution for the sine-wave excitation starting when $t = 0$. Noting that $\omega = \pi/t_1$ and that $\omega_n = 2\pi/\tau_n$, Eq. 4-44 can be written as

$$z_1 = \frac{Y}{1 - (\tau_n/2t_1)^2}\left[\left(\frac{\tau_n}{2t_1}\right)^2 \sin \frac{\pi t}{t_1} - \left(\frac{\tau_n}{2t_1}\right) \sin \frac{2\pi t}{\tau_n}\right] \qquad (t \leq t_1) \tag{4-45}$$

The solution for the second sine wave, which starts at time t_1, can be written from Eq. 4-45 by merely replacing t by $t - t_1$. Thus,

$$z_2 = \frac{Y}{1 - (\tau_n/2t_1)^2}\left[\left(\frac{\tau_n}{2t_1}\right)^2 \sin \frac{\pi}{t_1}(t - t_1) - \left(\frac{\tau_n}{2t_1}\right) \sin \frac{2\pi}{\tau_n}(t - t_1)\right] \qquad (t \geq t_1) \tag{4-46}$$

With Eq. 4-45 giving the response for $t \leq t_1$, we can see from Fig. 4-11b that adding Eqs. 4-45 and 4-46 will give the response for $t \geq t_1$ as

$$z = \frac{Y}{1 - (\tau_n/2t_1)^2} \left\{ \left(\frac{\tau_n}{2t_1} \right)^2 \left[\sin \frac{\pi t}{t_1} + \sin \frac{\pi}{t_1} (t - t_1) \right] - \frac{\tau_n}{2t_1} \left[\sin \frac{2\pi t}{\tau_n} + \sin \frac{2\pi}{\tau_n} (t - t_1) \right] \right\} \qquad (t \geq t_1) \qquad \text{(4-47)}$$

4-4 ROTATIONAL TRANSIENT RESPONSE

Previous discussions concerning the transient response of simple spring-and-mass systems involving linear motion are also applicable to systems that involve rotation. For example, the differential equation of motion of the system shown in Fig. 4-12a is

$$\ddot{\theta} + 2\zeta\omega_n\dot{\theta} + \omega_n^2\theta = \frac{M(t)}{\bar{I}} \qquad \text{(4-48)}$$

where $2\zeta\omega_n = \dfrac{cr^2}{\bar{I}}$

ζ = rotational damping factor

$\omega_n = \sqrt{kr^2/\bar{I}}$ = undamped natural circular frequency

I = mass moment of inertia of rotating element

$M(t)$ = moment excitation

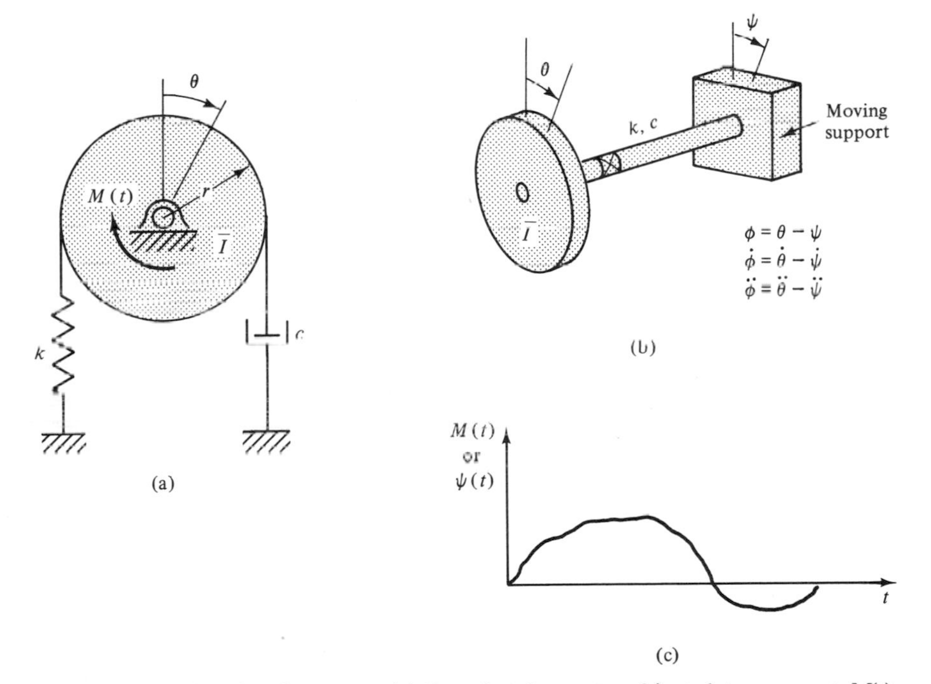

Figure 4-12 Rotational systems. (a) Rotational system subjected to moment $M(t)$. (b) Rotational system subjected to support excitation. (c) Arbitrary moment or support-excitation functions.

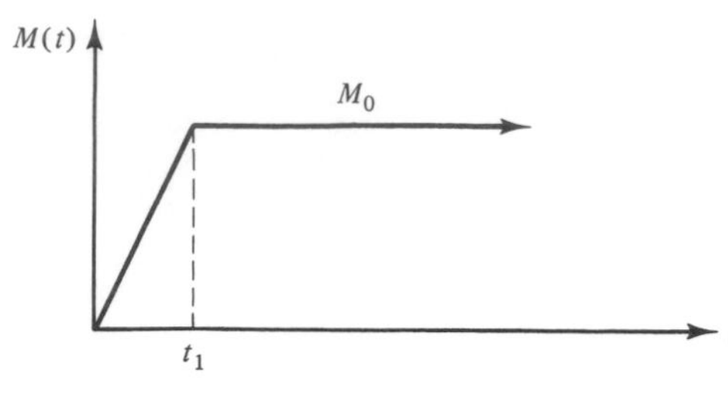

Figure 4-13 Step function moment with rise time t_1.

Equation 4-48 is analogous to Eq. 4-1, in which $M(t)/\bar{I}$ replaces $F(t)/m$ and θ replaces x. If $M(t)$ is a step function of magnitude M_0 with a rise time t_1 as shown in Fig. 4-13, the dimensionless term $x_{max}/(F_0/k)$ in Fig. 4-6 is replaced by $\theta_{max}/(M_0/kr^2)$. The denominator of the latter corresponds to the static angular displacement θ_s that would occur if the moment increased very slowly up to a magnitude of M_0. In this instance $\dot{\theta}$ and $\ddot{\theta}$ in Eq. 4-48 are essentially zero, and Eq. 4-48 reduces to

$$\omega_n^2 \theta_s = \frac{M_0}{\bar{I}}$$

from which

$$\theta_s = \frac{M_0}{\bar{I}\omega_n^2} = \frac{M_0}{kr^2}$$

Thus, Fig. 4-6, with its ordinate in terms of $\theta_{max}/(M_0/kr^2)$, is applicable to the system shown in Fig. 4-12a when $\zeta = 0$.

When a system is subjected to an angular support excitation $\psi(t)$ such as that shown in Fig. 4-12b, the differential equation of motion of the system in terms of the *relative* angular displacement ϕ (the displacement of the disk relative to the support) is found to be

$$\ddot{\phi} + 2\zeta\omega_n\dot{\phi} + \omega_n^2\phi = -\ddot{\psi} \tag{4-49}$$

where $2\zeta\omega_n = c/\bar{I}$
ζ = rotational damping factor
$\omega_n = \sqrt{k/\bar{I}}$
k = torsional spring constant
$\bar{I}$ = mass moment of inertia of rotating element about axis of rotation
$\ddot{\psi}$ = angular acceleration of support

Comparing Eqs. 4-49 and 4-35, it can be seen that they are analogous, with ϕ corresponding to the relative linear displacement z and $\ddot{\psi}$ corresponding to the linear acceleration $\ddot{y}$.

4-5 RESPONSE SPECTRUMS

A response spectrum (sometimes referred to as a shock spectrum) is a plot of the *maximum* response of a large number of single-degree-of-freedom systems of varying undamped natural frequencies f_n (or periods τ_n) to a shock excitation such as the transient-type one shown in Fig. 4-11, for example. The plot of $x_{max}/(F_0/k)$ versus t_1/τ_n shown in Fig. 4-6 is a typical example of a response spectrum plotted for a step-function excitation with a rise time of

t_1. Each point on this response spectrum represents the maximum response of a single-degree-of-freedom system having a particular natural frequency f_n or period τ_n. Although a particular excitation has a unique response spectrum, it is possible for two different types of excitations to have very similar response spectrums.

With varying values of the damping factor ζ, the response spectrum for a specified excitation is a *family of curves,* with each curve describing the response for a particular magnitude of damping. As we shall see later, the maximum *transient* response is generally not as sensitive to damping as is the maximum steady-state response discussed in Chap. 3. Thus, an undamped single-degree-of-freedom system is usually used as the standard model for response spectrums.

Since most equipment, appliances, instruments, and machine parts are subjected to shocks of some type at one time or another, the effects of shock excitation should be considered in the design of most devices. Response spectra provide pertinent information for designing experimental shock tests for devices using shock-testing machines.[3] In such tests the duration of a pulse or the rise time of a step excitation, for example, can be made a value noted from a particular response spectrum to provide the most severe shock to the device for a given excitation magnitude.

As we shall see in Chap. 6, which is concerned with modal analysis, the response spectrums of single-degree-of-freedom systems provide a powerful means of determining the transient response of multiple-degree-of-freedom systems.

The analytical solutions necessary for computing response spectrums are usually difficult to obtain for most shock excitations. Various techniques of data reduction, and the numerical computation of response spectrums, may be found in the literature.[4] Since microcomputers are widely used at present to numerically integrate differential equations, we shall be concerned at this point with discussing some of the basic concepts pertinent to obtaining differential equations of motion in terms of key parameter values that will facilitate the numerical computations necessary for obtaining response spectrums.

4-6 NUMERICAL INTEGRATION FOR OBTAINING RESPONSE SPECTRUMS

The maximum response for a given excitation can be plotted in terms of displacement, velocity, or acceleration. Which is used depends upon how the spectrum is used. It is generally desirable to express the response in terms of a dimensionless quantity such as $x_{max}/(F_0/k)$ as was done in Fig. 4-6. When the excitation is characterized by a time parameter such as the rise time t_1 of a step function (Fig. 4-6) or the duration t_1 of a sine pulse (Fig. 4-11a), it is convenient to compute and plot the maximum response as a function of t_1/τ_n as shown in Fig. 4-6.

To illustrate the selection of parameters that facilitate the numerical integration of differential equations of motion as a means of obtaining response spectrums, let us again write Eq. 4-1, which is

$$\ddot{x} + 2\zeta\omega_n\dot{x} + \omega_n^2 x = \frac{F(t)}{m} \tag{4-1}$$

[3] I. Vigness, "Shock Testing Machines," chap. 26, *Shock and Vibration Handbook,* vol. 2, ed. by C. M. Harris and C. E. Crede, McGraw-Hill Book Co., New York, 1961.

[4] S. Rubin, "Concepts in Shock Data Analysis," chap. 23, *Shock and Vibration Handbook,* vol. 2, ed. by C. M. Harris and C. E. Crede, McGraw-Hill Book Co., New York, 1961.

For purposes of discussion, we consider that $F(t)$ is the step function shown in Fig. 4-6, and that we wish to obtain a response spectrum for various damping-factor values with $x_{max}/(F_0/k)$ plotted as a function of t_1/τ_n.

If we let $\tau_n = 1$, then

$$\left.\begin{aligned} \frac{t_1}{\tau_n} &= t_1 \\ \omega_n &= \frac{2\pi}{\tau_n} = 6.28319 \\ \omega_n^2 &= 39.47848 \end{aligned}\right\} \tag{4-50}$$

With $\tau_n = 1$, the value of t_1/τ_n is simply equal to the value of t_1.

Assuming that $F(t)$ is applied very slowly, $\dot{x}$ and $\ddot{x}$ in Eq. 4-1 are equal to zero for all practical purposes, so that the static displacement due to a static load F_0 is

$$x_s = \frac{F_0}{m\omega_n^2} = \frac{F_0}{k} \tag{4-51}$$

To facilitate our computations, let us use a static deflection $x_s = 1$ and a unit mass of $m = 1$. Equation 4-51 then becomes

$$F_0 = \omega_n^2 = 39.47848 \tag{4-52}$$

and the value of $x_{max}/(F_0/k)$ is simply the value of x_{max}.

Upon substituting the above values into Eq. 4-1, we obtain

$$\ddot{x} + 2\zeta(6.28319)\dot{x} + 39.47848x = F(t) \tag{4-53}$$

where $F(t) = \dfrac{F_0 t}{t_1} = 39.47848\,\dfrac{t}{t_1} \qquad 0 \le t \le t_1$

$F(t) = F_0 = 39.47848 \qquad t \ge t_1$

Thus, by selecting particular values for the key parameters τ_n, m, and x_s, the differential equation of motion given by Eq. 4-53 is in a form that can be numerically integrated to yield results in term of *dimensionless quantities*.

The following general comments should be noted for programming Eq. 4-53 for solution on a digital computer to obtain values of $(x_{max})/(F_0/k)$ as a function of t_1/τ_n:

1. A value of $x_{max}/(F_0/k)$ is equal to the value of x_{max} since $F_0/k = 1$.
2. The value of t_1/τ_n is equal to the value of t_1, since $\tau_n = 1$.
3. For each damping factor ζ used, the differential equation is integrated for a number of values of t_1/τ_n sufficient to obtain a reasonable plot of $x_{max}/(F_0/k)$ as a function of t_1/τ_n. This means that for n values of t_1/τ_n, Eq. 4-53 must be integrated n times over the necessary time interval t_{max} for each damping factor (ζ) value desired.
4. To obtain the $x_{max}/(F_0/k)$ value that occurs during each integration interval ($0 < t \le t_{max}$), the absolute value of x_{i+1} calculated at $(t_i + \Delta t)$ is compared with the value of XMAX (a variable name having a value assigned to it that is equal to the maximum absolute value of x previously determined), that is, the maximum value of x_i calculated at t_i and all previously calculated values of x. If $|x_{i+1}|$ is greater than XMAX, $|x_{i+1}|$ is assigned to XMAX. If $|x_{i+1}|$ is less than XMAX, the value of XMAX remains unchanged at that step. Using this procedure, XMAX will have a value corresponding to the largest value of $x_{max}/(F_0/k)$ at the end of

the integration time t_{max} and is available for printout or a computer graphics plot.

5. Since the maximum response of x may occur when $t > t_1/\tau_n$, the integration time t_{max} should be greater than t_1/τ_n, to ensure that the maximum response is obtained during the time interval of the integration. In the FORTRAN program shown later, the integration time was chosen as 2.5 s for all integrations, and t_1/τ_n was varied from 0 to 2, using increments of 0.1.
6. The step size specified by the time increment Δt must be small enough to obtain a reasonably accurate integration. In general, the accuracy of a numerical integration increases as Δt decreases. However, if Δt is too small, the roundoff error can be excessive. Therefore, some criteria must be established for selecting a suitable value of Δt. Considering that a transient response will oscillate with a frequency corresponding to the natural frequency f_n of the system, we can use the natural period τ_n as the criterion for selecting a value. The authors have found that a value of Δt selected within a range of $\tau_n/100$ to $\tau_n/50$ will generally yield excellent results. That is,

$$\frac{\tau_n}{100} \leq \Delta t \leq \frac{\tau_n}{50}$$

Thus, a good value of Δt for the numerical integration of Eq. 4-53 is in the range of $0.01 \leq \Delta t \leq 0.02$, since $\tau_n = 1$.

Fourth-Order Runge-Kutta Method/FORTRAN Programs

One of the most widely used numerical methods for integrating differential equations is the fourth-order Runge-Kutta method, since it is self-starting, has good accuracy, and is easy to program in the FORTRAN language.

To write a FORTRAN program for numerically integrating Eq. 4-53, the differential equation is written in terms of its highest-order derivative as $\ddot{x} = f(t, x, \dot{x})$, yielding

$$\ddot{x} = -2\zeta(6.28319)\dot{x} - 39.47848x + F(t) \tag{4-54}$$

where
$$F(t) = \frac{F_0 t}{t_1} = 39.47848\,\frac{t}{t_1} \qquad 0 \leq t \leq t_1$$
$$F(t) = F_0 = 39.47848 \qquad t > t_1$$

The Runge-Kutta recurrence formulas for solving Eq. 4-54 in terms of the step size $h = \Delta t$ are[5]

$$x_{i+1} = x_i + h\dot{x}_i + \frac{h}{6}(k_1 + k_2 + k_3)$$

and

$$\dot{x}_{i+1} = \dot{x}_i + \frac{1}{6}(k_1 + 2k_2 + 2k_3 + k_4) \tag{4-55}$$

[5] M. L. James, G. M. Smith, and J. C. Wolford, *Applied Numerical Methods for Digital Computation,* 3d ed., chap. 6, Harper & Row, New York, 1985.

$$
\left.
\begin{aligned}
\text{where}\quad k_1 &= (h)f(t_i, x_i, \dot{x}_i) \\
k_2 &= (h)f\left(t_i + \frac{h}{2}, x_i + \frac{h}{2}\dot{x}_i, \dot{x}_i + \frac{k_1}{2}\right) \\
k_3 &= (h)f\left(t_i + \frac{h}{2}, x_i + \frac{h}{2}\dot{x}_i + \frac{h}{4}k_1, \dot{x}_i + \frac{k_2}{2}\right) \\
k_4 &= (h)f\left(t_i + h, x_i + h\dot{x}_i + \frac{h}{2}k_2, \dot{x}_i + k_3\right)
\end{aligned}
\right\} \tag{4-56}
$$

The numerical solution begins with the substitution of the initial values of x and $\dot{x}$ into Eq. 4-54 to obtain a value of the function $f(t_i, x_i, \dot{x}_i)$ for use in determining k_1. The successive k values are then determined for use in the recurrence formulas of Eq. 4-55 to

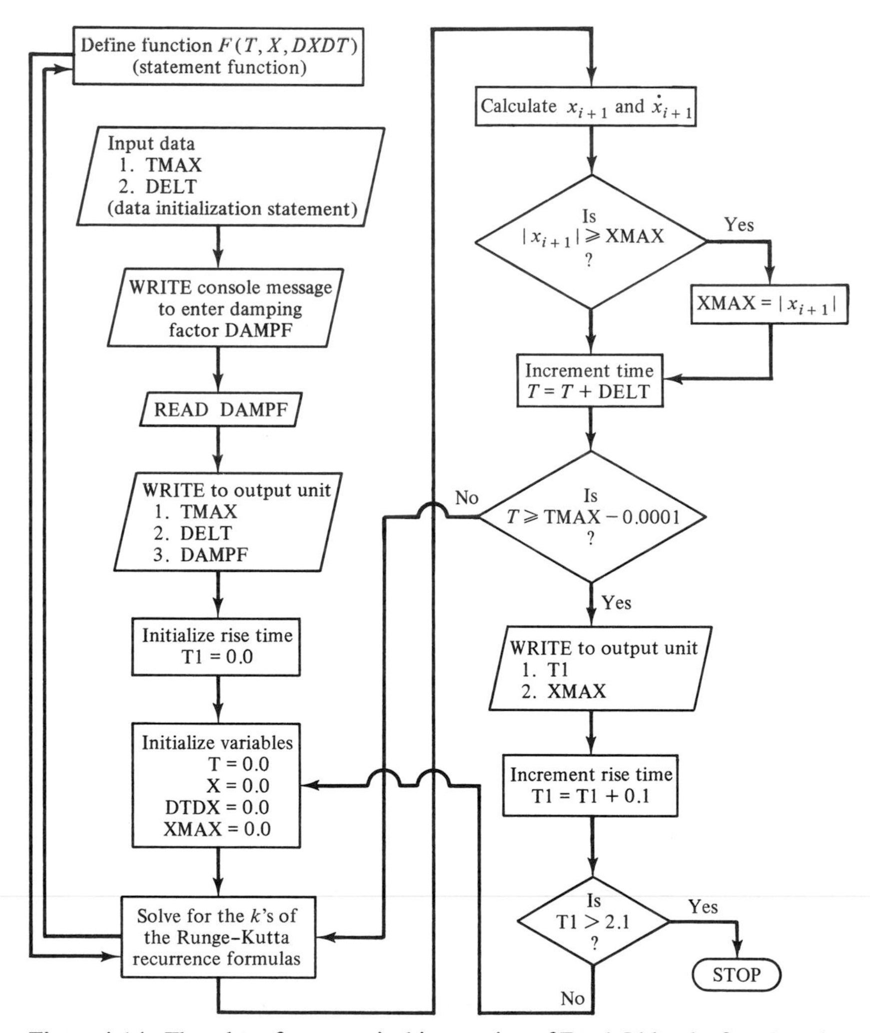

Figure 4-14 Flowchart for numerical integration of Eq. 4-54 by the fourth-order Runge-Kutta method.

obtain values of x_{i+1} and $\dot{x}_{i+1}$. The latter are then used in Eqs. 4-54 and 4-56 to obtain new k values for substitution into Eq. 4-55 to obtain x_{i+2} and $\dot{x}_{i+2}$, and so on. An outline of the procedure is shown in the flowchart of Fig. 4-14. The initial conditions used are

$$x_0 = 0$$
$$\dot{x}_0 = 0$$

The FORTRAN variable names and the quantities that they represent are as follows:

Variable name	Quantity
TMAX	Maximum time t_{max} for integration, 2.5 s
DELT	Step size time increment Δt, 0.01 s
T	Independent variable time, t
T1	Rise time t_1 ranging from 0 to 2 in increments of 0.1
X	Displacement, x
XMAX	Maximum response of x
DXDT	Velocity $dx/dt = \dot{x}$
DAMPF	Damping factor ζ (entered as data when called for on computer console)
STEP(T,T1)	Name of *Function Subprogram* to generate $F(t)$
F(T,X,DXDT)	Statement Function for Eq. 4-54
AK1, AK2, AK3, AK4	Quantities k_1, k_2, k_3, and k_4 of Runge-Kutta formulas

The FORTRAN program (STEP.FOR) used to implement the procedures outlined in the flowchart of Fig. 4-14 is shown below.[6] The program was run on a microcomputer having a RAM (random-access memory) of 64K. As the program is written, the input required is a value for the damping factor ζ, which is entered from the keyboard of the console when the message of the FORMAT statement 2 appears on the console screen. The Function Subprogram STEP(T,T1) generates the step function with a rise time t_1, which is referenced in the Statement Function in the second line of the program.

```
C   STEP.FOR
C   THIS PROGRAM OBTAINS THE MAXIMUM RESPONSE XMAX OF A SINGLE-
C   DEGREE-OF-FREEDOM SYSTEM SUBJECTED TO A STEP FUNCTION WITH
C   A RISE TIME OF T1.  XMAX/XSTATIC = XMAX SINCE XSTATIC = 1.
C   T1/TAU = T1 SINCE TAU = 1.  THE ONLY INPUT REQUIRED FOR A
C   COMPUTER RUN IS A VALUE FOR THE DAMPING FACTOR (DAMPF).
C   EACH COMPUTER RUN CONSISTS OF DETERMINING XMAX FOR EACH
C   VALUE OF T1 WHICH RANGES FROM 0.0 TO 2. IN INCREMENTS OF 0.1.
C   DEVICE * IN READ AND WRITE STATEMENTS IS THE CONSOLE.
C   DEVICE 2 IN WRITE STATEMENTS IS THE PRINTER.
C   * IN THE PLACE OF A FORMAT STATEMENT NUMBER MEANS FREE FORMAT.
      DATA TMAX,DELT/2.5,0.01/
      F(T,X,DXDT)=-2.*DAMPF*6.28319*DXDT-39.47848*X+STEP(T,T1)
      OPEN (2, FILE='PRN')
      WRITE(*,2)
2     FORMAT(' ','ENTER DAMPING FACTOR VALUE DAMPF (F10.0)',/)
      READ(*,4)DAMPF
```

[6] The FORTRAN version used here is compatible with IBM personal computers. In using other versions of FORTRAN it may be necessary to modify some of the statements slightly.

```
4       FORMAT(F10.0)
        WRITE(2,6)TMAX,DELT,DAMPF
6       FORMAT(' ',5X,'TMAX = ',F4.2,3X,'DELT = ',F5.3,3X,
       *'DAMPF = ',F4.2,/)
        T1 = 0.0
8       T = 0.0
        X = 0.0
        DXDT = 0.0
        XMAX = 0.0
9       AK1=DELT*F(T,X,DXDT)
        AK2=DELT*F(T+DELT/2.,X+DELT*DXDT/2.,DXDT+AK1/2.)
        AK3=DELT*F(T+DELT/2.,X+DELT*(DXDT/2.+AK1/4.),DXDT+AK2/2.)
        AK4=DELT*F(T+DELT,X+DELT*(DXDT+AK2/2.),DXDT+AK3)
        X = X + DELT*(DXDT+(AK1+AK2+AK3)/6.)
        DXDT = DXDT+(AK1+2.*AK2+2.*AK3+AK4)/6.
        IF(ABS(X) .GE. XMAX) XMAX=ABS(X)
        T = T + DELT
        IF(T .GE. TMAX-.0001)GO TO 10
        GO TO 9
10      WRITE(2,12)T1,XMAX
12      FORMAT(' ',11X,'T1 = ',F5.3,5X,'XMAX = ',F6.4)
        T1 = T1 + 0.1
        IF(T1 .GT. 2.1)GO TO 14
        GO TO 8
14      STOP
        END

        FUNCTION STEP(T,T1)
        IF(T1 .EQ. 0.0)GO TO 2
        IF(T .GT. T1)GO TO 2
        STEP = 39.47848*T/T1
        RETURN
2       STEP = 39.47848
        RETURN
        END
```

The printout obtained from the program in which $\zeta = 0.05$ is as follows:

```
TMAX = 2.50    DELT =  .010    DAMPF =  .05

       T1 = 0.000     XMAX = 1.8545
       T1 =  .100     XMAX = 1.8405
       T1 =  .200     XMAX = 1.7993
       T1 =  .300     XMAX = 1.7334
       T1 =  .400     XMAX = 1.6465
       T1 =  .500     XMAX = 1.5443
       T1 =  .600     XMAX = 1.4317
       T1 =  .700     XMAX = 1.3158
       T1 =  .800     XMAX = 1.2029
```

```
T1 =  .900    XMAX = 1.1013
T1 = 1.000    XMAX = 1.0396
T1 = 1.100    XMAX = 1.0758
T1 = 1.200    XMAX = 1.1201
T1 = 1.300    XMAX = 1.1494
T1 = 1.400    XMAX = 1.1620
T1 = 1.500    XMAX = 1.1589
T1 = 1.600    XMAX = 1.1423
T1 = 1.700    XMAX = 1.1157
T1 = 1.800    XMAX = 1.0833
T1 = 1.900    XMAX = 1.0515
T1 = 2.000    XMAX = 1.0343
```

The data from the numerical integration of Eq. 4-54 for $\zeta = 0$, 0.05, and 0.1 were used to obtain the curves shown in the response spectrum of Fig. 4-15. The plot of the data for $\zeta = 0$ corresponds to the analytical results shown in Fig. 4-6.

If it is desired to obtain the response x as a function of time t, computed values of x at time t could be printed out by a WRITE statement. However, such a *time-domain* plot of x versus t would not yield any additional information for engineering purposes beyond that shown in the response spectrum of Fig. 4-15.

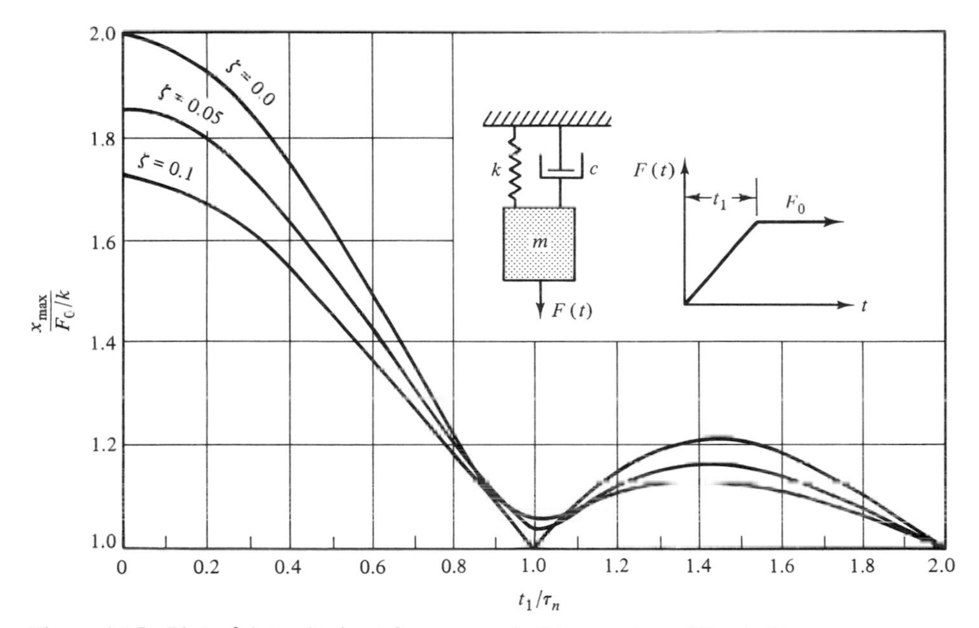

Figure 4-15 Plot of data obtained from numerical integration of Eq. 4-54.

EXAMPLE 4-3

The curves shown in Fig. 4-16 are the response spectrum of the spring-and-mass system shown in Fig. 4-16a for various degrees of damping when the system was subjected to the sine-pulse force excitation shown in Fig. 4-16b. On page 251 is a FORTRAN computer program STEP.FOR that was used to obtain similar curves for the same system when it

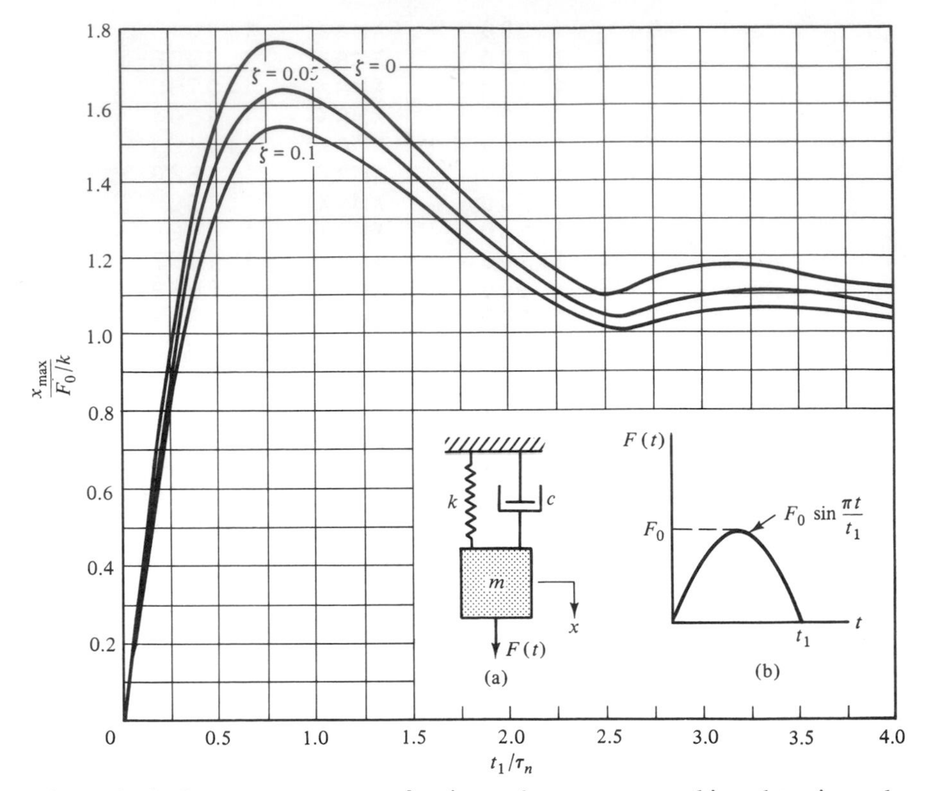

Figure 4-16 Response spectrum of spring-and-mass system subjected to sine-pulse force excitation.

was subjected to a step-function excitation with a rise time of t_1. We wish to modify that program so that the sine-pulse excitation replaces the step-function excitation, and so that it will yield data enabling us to duplicate the curves shown in Fig. 4-16.

The pulse-duration time is to vary from 0.2 to 4 s, in increments of 0.2 s.

Solution. As explained previously, values are selected for the key parameters τ_n, x_s, and m, so that the value of $x_{max}/(F_0/k)$ is equal to the value of x_{max} and the value of t_1/τ_n is equal to t_1. That is,

$$\tau_n = 1$$

$$\omega_n = \frac{2\pi}{\tau_n} = 6.28319$$

$$\omega_n^2 = 39.47848$$

$$m = 1$$

$$x_s = \frac{F_0}{m\omega_n^2} = \frac{F_0}{k} = 1$$

$$F_0 = \omega_n^2 = 39.47848$$

With these values, the equation in the Statement Function is identical to Eq. 4-54, with the exception of the function $F(t)$. That is,

$$\ddot{x} = -2\zeta(6.28319)\dot{x} - 39.47848x + F(t)$$

where $F(t) = 39.47848 \sin \frac{\pi t}{t_1} \qquad (0 \le t \le t_1)$

$F(t) = 0 \qquad (t \ge t_1)$

To compute values of x_{max} corresponding to values of t_1/τ_n up to 4 as shown in Fig. 4-16, the maximum integration time TMAX should be greater than 4. Thus, the DATA statement in the first line of the program is changed to

```
DATA TMAX,DELT/5.,0.01/
```

in which a value of 5 for TMAX should ensure that the maximum value of x is obtained during the integration.

If we arbitrarily select FUNCT as the name of the Function Subprogram replacing STEP, the Statement Function in the second line of the program should be changed to

```
F(T,X,DXDT)=-2.*DAMPF*6.28319*DXDT
                    -39.47848*X+FUNCT(T,T1)
```

and the Function Subprogram used to generate the sine pulse should be

```
          FUNCTION FUNCT(T,T1)
          IF(T .GT. T1)GO TO 2
          IF(T .LE. T1)GO TO 1
1         FUNCT = 39.47848*SIN(3.14159*T/T1)
          RETURN
2         FUNCT = 0.0
          RETURN
          END
```

The initial value of t_1 (T1 = 0.0 in the original program) must be changed to a value other than zero, to avoid division by zero in the Function Subprogram. Thus, the statement following the FORMAT statement 6 is changed to

```
T1 = 0.2
```

Since t_1 is to vary from 0.2 to 4 s, in increments of 0.2 s, the two statements following the FORMAT statement 12 must also be changed to

```
T1 = T1 + 0.2
IF(T1 .GT. 4.1)GO TO 14
```

With the preceding modifications, we now have the desired program, which will give us data for duplicating the curves shown in Fig. 4-16.

EXAMPLE 4-4

The curves shown in Fig. 4-17 are the response spectrum of the spring-and-mass system shown in Fig. 4-17a for various degrees of damping, when the system was subjected to the sine-pulse *support excitation* (acceleration) shown in Fig. 4-17b. In these plots, the maximum acceleration of the mass m in g's is plotted as a function of t_1/τ_n. The amplitude of the support acceleration is 386 in./s^2, which is 1 g.

It is desired to modify the computer program STEP.FOR on page 251 to enable us

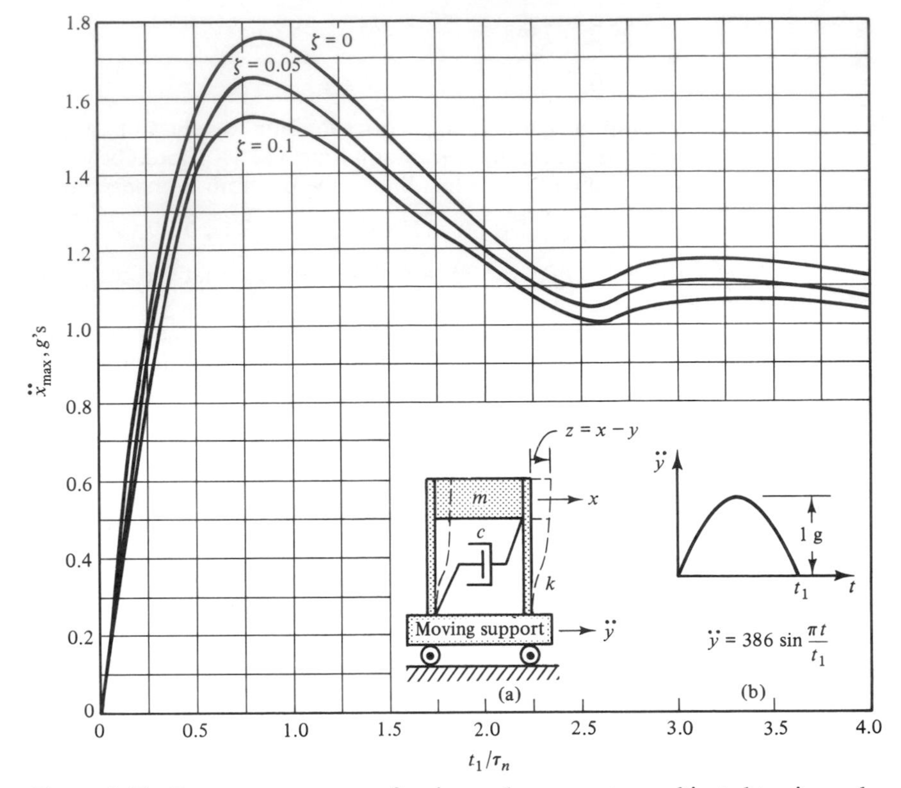

Figure 4-17 Response spectrum of spring-and-mass system subjected to sine-pulse support excitation (acceleration).

to obtain the data necessary to duplicate the curves shown in Fig. 4-17. The pulse-duration time is to vary from 0.2 to 4 s in increments of 0.2 s.

Solution. Using the differential equation for *relative* motion, given by Eq. 4-35, the equation for the Statement Function of the FORTRAN program is

$$\ddot{z} = -2\zeta\omega_n\dot{z} - \omega_n^2 z - \ddot{y} \tag{4-57}$$

in which

$$\ddot{y} = 386 \sin \frac{\pi t}{t_1} \qquad (0 \le t \le t_1)$$

and

$$\ddot{y} = 0 \qquad (t \ge t_1)$$

To calculate the maximum value of $\ddot{x}$ for each value of t_1, we note from Eq. 4-57 that since $\ddot{z} = \ddot{x} - \ddot{y}$,

$$\ddot{x} = \ddot{z} + \ddot{y} = -2\zeta\omega_n\dot{z} - \omega_n^2 z \tag{4-58}$$

Again letting $\tau_n = 1$, so that the value of t_1/τ_n is equal to the value of t_1,

$$\tau_n = 1 \text{ s}$$
$$\omega_n = \frac{2\pi}{\tau_n} = 6.28319 \text{ rad/s}$$
$$\omega_n^2 = 39.47848$$

Substituting these values into Eq. 4-57, and using the variable names Z and DZDT for z and $\dot{z}$, respectively, and FUNCT for the Function Subprogram name, the FORTRAN Statement Function becomes

```
F(T,Z,DZDT)=-2.*DAMPF*6.28319*DZDT
                          -39.47848*Z-FUNCT(T,T1)
```

From Eq. 4-58, the FORTRAN statement for computing $\ddot{x}$ (XDD) in g's is

```
XDD = (-2.*DAMPF*6.28319*DZDT-39.47848*Z)/386.
```

in which $g = 386$ in./s^2.

The initial conditions must next be determined. The support excitation is

$$\ddot{y} = 386 \sin \frac{\pi t}{t_1}$$

Assuming that $\dot{y}_0 = y_0 = 0$, we find that

and

$$\left.\begin{aligned} z_0 &= x_0 - y_0 = 0 \\ \dot{z}_0 &= \dot{x}_0 - \dot{y}_0 = 0 \end{aligned}\right\} \tag{4-59}$$

In the program that follows, these *relative initial conditions* are specified by

```
   Z = 0.0
DZDT = 0.0
```

The term XDDM in the program is the name used for $\ddot{x}_{max}$ measured in g's. The modified program is as follows:

```
C   SINEBASE.FOR
C   THIS PROGRAM OBTAINS THE MAX. ACCELERATION XDDM IN g'S FOR
C   A SDOF SYSTEM SUBJECTED TO A SINE PULSE OF 1 g. XDDM/INPUT =
C   XDDM SINCE INPUT = 1.0.  SINCE THE PERIOD TAU OF THE SYSTEM
C   IS EQUAL TO 1.0, THE RATIO T1/TAU IS SIMPLY T1.  THE ONLY
C   INPUT REQUIRED FOR A COMPUTER RUN IS A VALUE FOR THE DAMP-
C   ING FACTOR (DAMPF).  XDDM IS DETERMINED FOR EACH VALUE OF T1
C   WHICH RANGES FROM 0 TO 4.0 IN INCREMENTS OF 0.2.
C   DEVICE * IN READ AND WRITE STATEMENTS IS THE CONSOLE.
C   DEVICE 2 IN WRITE STATEMENTS IS THE PRINTER.
C   * IN THE PLACE OF A FORMAT STATEMENT NUMBER MEANS FREE FORMAT.
      DATA TMAX,DELT/5.,0.01/
      F(T,Z,DZDT)=-2.*DAMPF*6.28319*DZDT-39.47848*Z-FUNCT(T,T1)
      OPEN (2,FILE='PRN')
      WRITE(*,2)
```

```
2       FORMAT(' ','ENTER DAMPING FACTOR VALUE DAMPF (F10.0)',/)
        READ(*,4)DAMPF
4       FORMAT(F10.0)
        WRITE(2,6)TMAX,DELT,DAMPF
6       FORMAT(' ',5X,'TMAX = ',F4.2,3X,'DELT = ',F5.3,3X,
       *'DAMPF = ',F4.2,/)
        T1 = 0.2
8       T = 0.0
        Z = 0.0
        DZDT = 0.0
        XDDM = 0.0
9       AK1=DELT*F(T,Z,DZDT)
        AK2=DELT*F(T+DELT/2.,Z+DELT*DZDT/2.,DZDT+AK1/2.)
        AK3=DELT*F(T+DELT/2.,Z+DELT*(DZDT/2.+AK1/4.),DZDT+AK2/2.)
        AK4=DELT*F(T+DELT,Z+DELT*(DZDT+AK2/2.),DZDT+AK3)
        Z = Z + DELT*(DZDT+(AK1+AK2+AK3)/6.)
        DZDT = DZDT+(AK1+2.*AK2+2.*AK3+AK4)/6.
        XDD = (-2.*DAMPF*6.28319*DZDT-39.47848*Z)/386.
        IF(ABS(XDD) .GE. XDDM) XDDM=ABS(XDD)
        T = T + DELT
        IF(T .GE. TMAX-.0001)GO TO 10
        GO TO 9
10      WRITE(2,12)T1,XDDM
12      FORMAT(' ',11X,'T1 = ',F5.3,5X,'XDDM = ',F6.4)
        T1 = T1 + 0.2
        IF(T1 .GT. 4.01)GO TO 14
        GO TO 8
14      STOP
        END

        FUNCTION FUNCT(T,T1)
        IF(T .GT. T1)GO TO 2
        IF(T .LE. T1)GO TO 1
1       FUNCT = 386.0*SIN(3.141593*T/T1)
        RETURN
2       FUNCT = 0.0
        RETURN
        END
```

The printout obtained from the program in which $\zeta = 0.05$ is as follows:

```
TMAX = 5.00    DELT =  .010    DAMPF =  .05

           T1 =  .200     XDDM =  .7175
           T1 =  .400     XDDM = 1.2791
           T1 =  .600     XDDM = 1.5738
           T1 =  .800     XDDM = 1.6510
           T1 = 1.000     XDDM = 1.6235
           T1 = 1.200     XDDM = 1.5525
```

```
T1 = 1.400        XDDM = 1.4659
T1 = 1.600        XDDM = 1.3770
T1 = 1.800        XDDM = 1.2916
T1 = 2.000        XDDM = 1.2123
T1 = 2.200        XDDM = 1.1396
T1 = 2.400        XDDM = 1.0733
T1 = 2.600        XDDM = 1.0432
T1 = 2.800        XDDM = 1.0810
T1 = 3.000        XDDM = 1.1016
T1 = 3.200        XDDM = 1.1095
T1 = 3.400        XDDM = 1.1083
T1 = 3.600        XDDM = 1.1005
T1 = 3.800        XDDM = 1.0881
T1 = 4.000        XDDM = 1.0725
```

Response Spectrums Associated with Earthquakes

The ground accelerations caused by earthquakes are very erratic, as shown in Fig. 4-18, which shows an accelerogram recorded during the 1971 San Fernando, California, earthquake at the Pacoima Dam.[7] The response spectrums obtained from earthquakes are frequently plotted in the log form shown in Fig. 4-19, which was determined using the accelerogram shown in Fig. 4-18.[8]

Such response spectrums can be determined using values of the maximum *relative* displacement z computed for a number of undamped natural frequencies. Considering the damping as small so that $\sqrt{1-\zeta^2} \cong 1$ and $\omega_n \cong \omega_d$, and assuming that the relative motion is a sine wave with a varying amplitude as shown in Fig. 4-20, values for plotting the relative maximum velocities and accelerations as functions of the natural frequency f_n can be calculated from the computed values of the maximum relative displacement z_{max} using[9]

$$\left.\begin{aligned} \dot{z}_{max} &= z_{max}(\omega_n) = z_{max}(2\pi f_n) \\ \ddot{z}_{max} &= z_{max}(\omega_n)^2 = z_{max}(2\pi f_n)^2 \end{aligned}\right\} \qquad (4\text{-}60)$$

Since the peak response usually occurs during the earthquake excitation, the motion is not purely harmonic and the values of $\dot{z}_{max}$ and $\ddot{z}_{max}$ determined from Eq. 4-60 are not exact. However, experience has shown that such a procedure yields reasonably good approximations of the exact values, which makes it possible to plot a response spectrum in the form shown in Fig. 4-19. The reader might find it of interest to locate the intersection of a particular value of z_{max} with a particular frequency in Fig. 4-19, and check the corresponding values of $\dot{z}_{max}$ and $\ddot{z}_{max}$ obtained against the respective values obtained using Eq. 4-60.

As another example of this type of response spectrum, a normalized-average-response spectrum for the 1940 El Centro, California, earthquake is shown in Fig. 4-21, in which z_{max}, $\dot{z}_{max}$, and $\ddot{z}_{max}$ are plotted as functions of the undamped natural period τ_n. This

[7] Computer plot of data obtained from California Institute of Technology, "Strong Motion Earthquake Accelerograms," Report EERL 71-50, September 1971.

[8] M. Maheshwari, "Correlation between Linear and Nonlinear Seismic Response of Structures," Ph.D. Dissertation, Department of Engineering Mechanics, University of Nebraska, 1974.

[9] G. W. Housner, "Vibration of Structures Induced by Seismic Waves," chap. 50, *Shock and Vibration Handbook,* vol. 3, ed. by C. M. Harris and C. E. Crede, McGraw-Hill Book Co., New York, 1961.

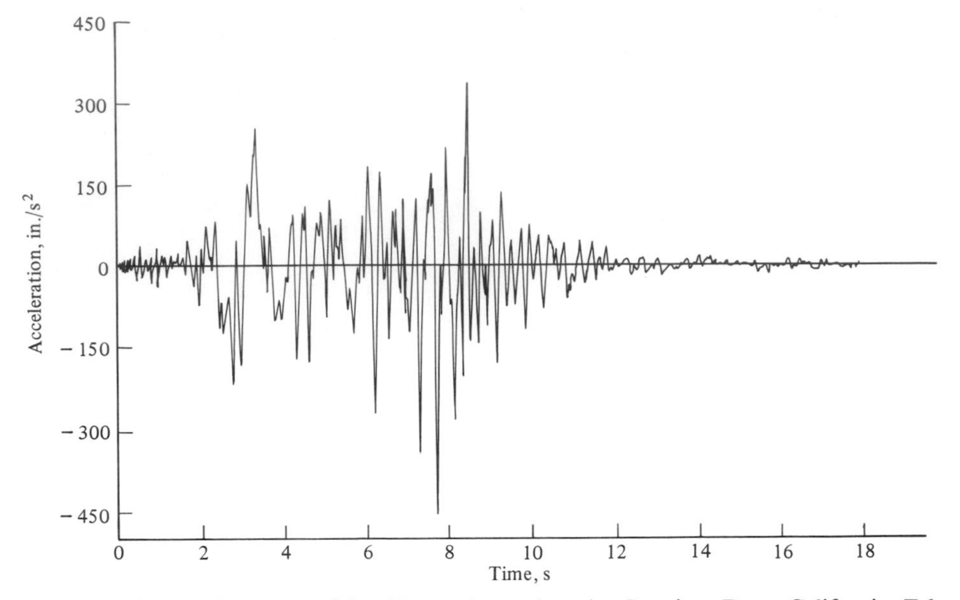

Figure 4-18 Accelerogram of San Fernando earthquake, Pacoima Dam, California, Feb. 9, 1971, component S16E.

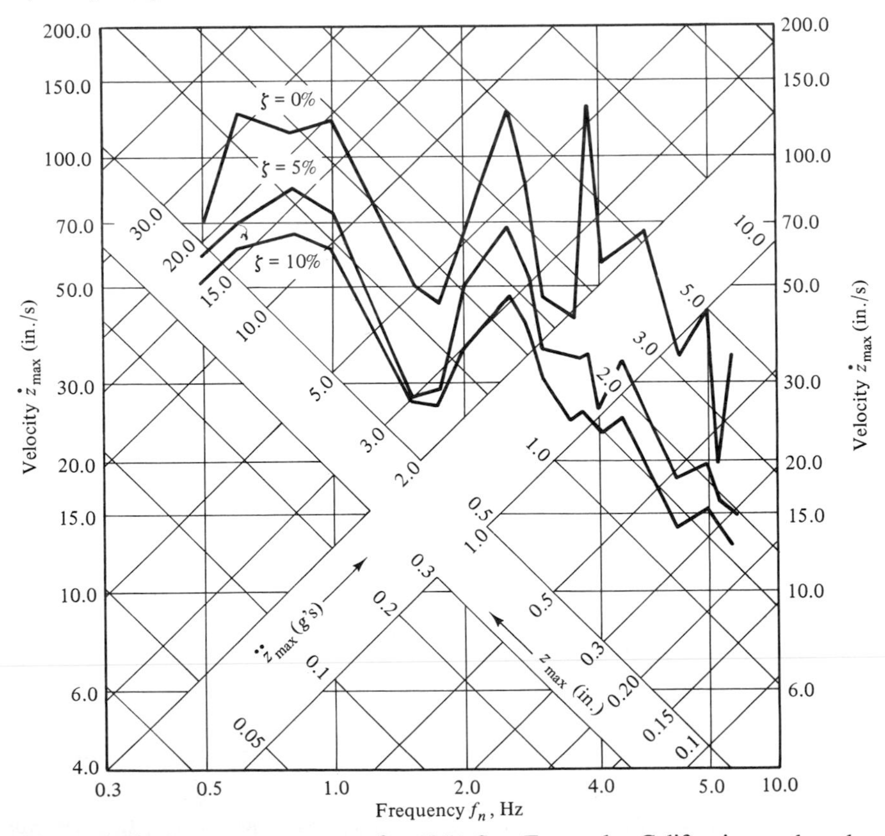

Figure 4-19 Response spectrum for 1971 San Fernando, California, earthquake (Pacoima Dam, component S16E).

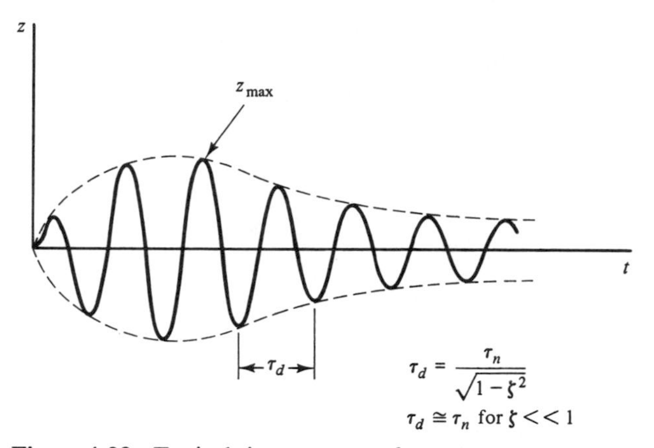

Figure 4-20 Typical time response from shock excitation.

response spectrum is used frequently in the analysis of vibration problems involving nuclear reactors.[10]

The response spectrums shown in Figs. 4-19 and 4-21 will be used in Chap. 6 to illustrate a general procedure for determining the response of a multiple-degree-of-freedom system subjected to an earthquake.

To obtain the maximum relative response as a function of the undamped natural frequency f_n by means of numerical integration, Eq. 4-57 is written as

$$\ddot{z} = -2\zeta(2\pi f_n)\dot{z} - (2\pi f_n)^2 z - \ddot{y} \tag{4-61}$$

while if it is to be plotted as a function of the undamped natural period τ_n, it is written in the form of

$$\ddot{z} = -2\zeta\frac{2\pi}{\tau_n}\dot{z} - \left(\frac{2\pi}{\tau_n}\right)^2 z - \ddot{y} \tag{4-62}$$

A computer program for obtaining a numerical solution of either Eq. 4-61 or 4-62 using the Runge-Kutta method discussed earlier is similar to the programs shown earlier. However, when $\ddot{y}$ corresponds to an arbitrary function such as that occasioned by an earthquake, the Function Subprogram must include some type of arbitrary function generator to generate $\ddot{y}$ for input to the Function Statement of the main program.

Arbitrary Function Generator

The following Function Subprogram can be used to generate arbitrary types of excitation forces or support motions:

```
FUNCTION FUNCT(T)
COMMON TDAT(129), YDAT(129)
IMAX=129
IMIN=1
DO 3 L = 1,7
IMED=(IMAX+IMIN)/2
```

[10] U.S. Atomic Energy Commission, "Nuclear Reactors and Earthquakes," TID-7024, Office of Technical Services, Washington, D.C., 1963.

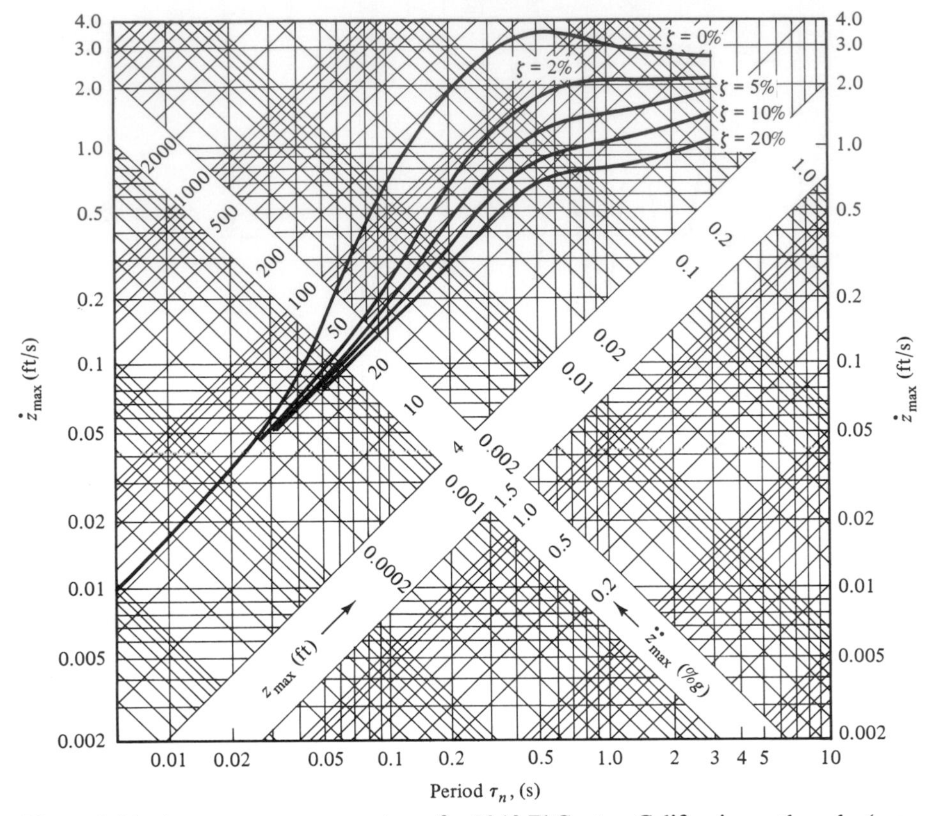

Figure 4-21 Average response spectrum for 1940 El Centro, California, earthquake (maximum ground acceleration 33 percent of *g*).

```
      IF(T .LE. TDAT(IMED))GO TO 2
      IMIN=IMED
      GO TO 3
2     IMAX=IMED
3     CONTINUE
      SLOPE=(YDAT(IMAX)-YDAT(IMIN))/(TDAT(IMAX)-TDAT(IMIN))
      FUNCT=YDAT(IMIN)+SLOPE*(T-TDAT(IMIN))
      RETURN
      END
```

Initially, IMAX has a value equal to the maximum number of data points permitted in representing the arbitrary function, and IMIN has a value of 1, which is the subscript for the first data point. In the program above, the maximum number of data values permitted is 129.

The data values for the independent variable time *T* are stored in the array TDAT, and the corresponding values for the dependent variable are stored in the array YDAT.

The DO loop beginning with DO 3 L = 1,7 executes a *bisection* search technique to determine which TDAT values the current value of the independent variable T lies between. The test value 7 of the DO loop is the required number of bisections for 129 data points. The maximum number of data points permitted can be extended by changing this test value. The test values and the corresponding maximum number of data points permitted are as follows:

Test value	Maximum number of data points
7	129
8	257
9	519
10	1025
11	2049

If the actual number of data points used to represent an arbitrary function is less than the maximum number permitted by the test value, TDAT must be filled with increasing values (arbitrary) and YDAT filled with zeros. The preceding discussion will be illustrated in the FORTRAN program of Example 4-5.

The SLOPE and FUNCT statements obtain the value of FUNCT, which is then returned to the Statement Function of the main program as the value of function FUNCT for the current value of its argument.

EXAMPLE 4-5

The single-degree-of-freedom system shown in Fig. 4-22 is subjected to the support acceleration $\ddot{y}$, which can be represented by the 13 data points shown in Fig. 4-23. We wish to write a FORTRAN program using the Function Subprogram as an arbitrary function generator that will compute the maximum *relative* response z_{max} for a range of undamped natural frequencies of $0.4 \leq f_n \leq 10$. The program is to be run with a damping factor of $\zeta = 0.05$, and results obtained for plotting a response spectrum for z_{max} versus f_n.

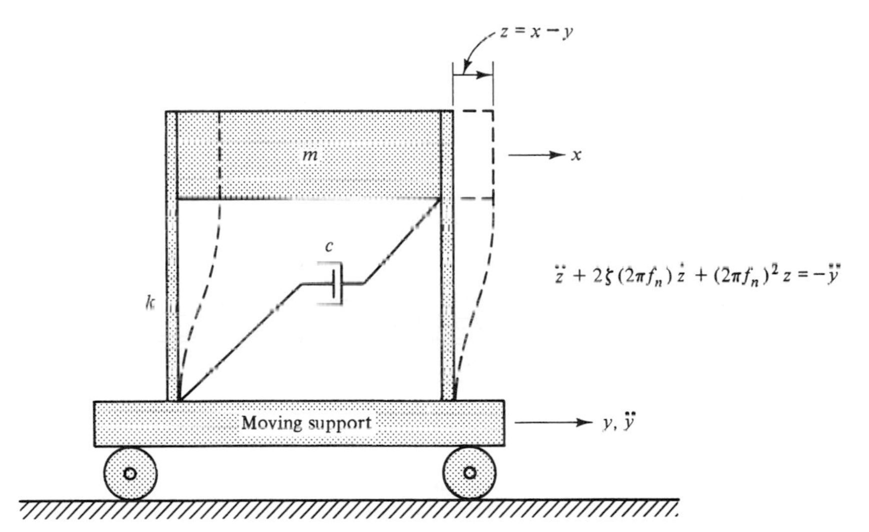

Figure 4-22 System subjected to the support excitation shown in Fig. 4-23.

Solution. The maximum relative response is determined by integrating Eq. 4-61 for each value of f_n over the range of 0.4 to 10 Hz. Incrementing f_n in steps of 0.2 should provide sufficient data to plot a fairly smooth response spectrum.

The FORTRAN program for this example, called ARBGEN.FOR, is quite

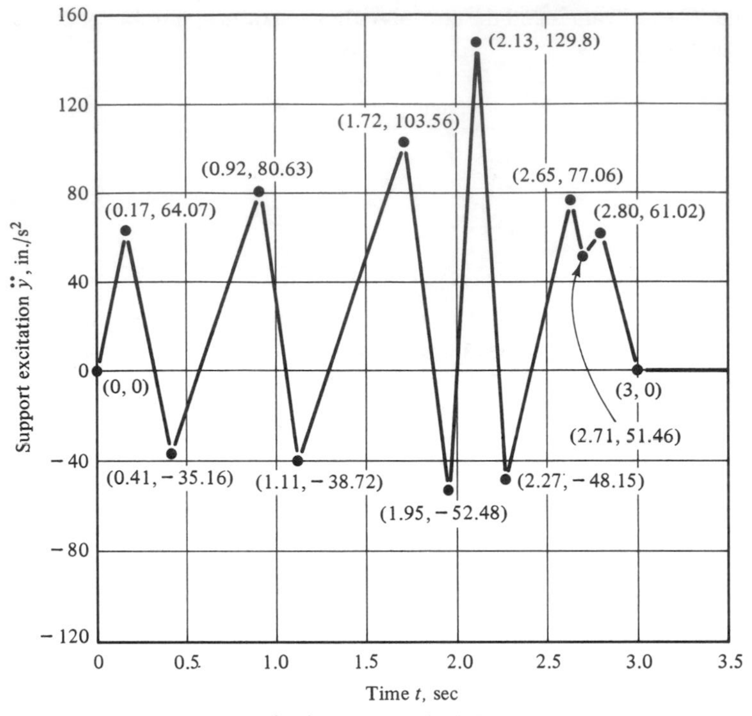

Figure 4-23 Support excitation, Example 4-5.

similar to the program given in Example 4-4. The FORTRAN variable names and the quantities they represent are as follows:

Variable name	Quantity
TMAX	Maximum time for integration, 5.0 s
DELT	Step size time increment Δt, 0.01 s
T	Independent variable time t
Z	Relative displacement, z
ZMAX	Maximum response, z_{max}
DZDT	Relative velocity, $\dot{z}$
DAMPF	Damping factor ζ
FUNCT	Name of Function Subprogram to generate $\ddot{y}$
F(T,Z,DZDT)	Statement Function for Eq. 4-61
FN	Undamped natural frequency ranging from 0.4 to 10 Hz in increments of 0.2
N	Number of data points (13 in this case)
I	DO variable
TDAT(I)	Array-storing independent variable time data values
YDAT(I)	Array-storing dependent variable data values for support excitation $\ddot{y}$
AK1, AK2, AK3, AK4	Quantities k_1, k_2, k_3, and k_4 of Runge-Kutta formulas

The program, as it is written, will permit up to 129 data points. The 13 pairs of data values shown in Fig. 4-23 are stored as 13 records in the Data File FORT10.DAT, which is transmitted from the disk on which it is stored to the main program by the statement

```
READ(10,*)TDAT(I), YDAT(I)
```

which is in the DO loop beginning with DO 6 I = 1,N.

```
C   ARBGEN.FOR
C   THIS PROGRAM USES AN ARBITRARY FUNCTION GENERATOR TO
C   GENERATE SUPPORT EXCITATION YDD TO OBTAIN THE MAXIMUM
C   RELATIVE RESPONSE ZMAX.  THE DATA FOR YDD IS STORED
C   IN A DATA FILE (FORT10.DAT).  THE PROGRAM AS WRITTEN CAN
C   HANDLE A MAXIMUM OF 129 DATA POINTS FOR YDD VERSUS TIME T
C   BUT CAN EASILY BE EXTENDED TO 257 OR 513 OR LARGER AS
C   EXPLAINED IN TEXT.  ZMAX IS DETERMINED FOR EACH NATURAL
C   FREQUENCY FN WHICH RANGES FROM 0.4 TO 10 HZ IN INCREMENTS
C   OF 0.2.  DATA INPUT REQUIRED FOR A COMPUTER RUN ARE: NUMBER
C   OF DATA POINTS N, DAMPING FACTOR (DAMPF), AND DATA STORED IN
C   THE DATA FILE FOR THE ARBITRARY SUPPORT EXCITATION.
C   DEVICE * IN READ AND WRITE STATEMENTS IS THE CONSOLE.
C   DEVICE 2 IN WRITE STATEMENTS IS THE PRINTER.
C   DEVICE 10 IN READ STATEMENTS IS DATA FILE 'FORT10.DAT'
C   * IN THE PLACE OF A FORMAT STATEMENT NUMBER MEANS FREE FORMAT.
      COMMON TDAT(129),YDAT(129)
      DATA TMAX,DELT/5.,0.01/
      F(T,Z,DZDT)=-2.*DAMPF*6.28319*FN*DZDT-39.47848*FN*FN*Z
     *-FUNCT(T)
      OPEN (2,FILE='PRN')
      OPEN (10,FILE='FORT10.DAT',STATUS='OLD')
C
C   ENTER FROM CONSOLE THE NUMBER OF DATA POINTS TO BE USED IN
C   THE LINEAR FUNCTION GENERATOR
C
      WRITE(*,1)
1     FORMAT(' ','ENTER NUMBER OF DATA POINTS N (I3)',/)
      READ(*,2)N
2     FORMAT(I3)
      WRITE(*,3)
3     FORMAT(' ','ENTER DAMPING FACTOR VALUE DAMPF (F10.0)',/)
      READ(*,4)DAMPF
4     FORMAT(F10.0)
      WRITE(2,5)TMAX,DELT,DAMPF
5     FORMAT(' ',5X,'TMAX = ',F4.2,3X,'DELT = ',F5.3,3X,
     *'DAMPF = ',F4.2,/)
C
C   READ DATA FROM DATA FILE (FORT10.DAT)
C
      DO 6 I = 1,N
      READ(10,*)TDAT(I),YDAT(I)
6     CONTINUE
C
C   FILL TDAT ARRAY WITH ARBITRARY INCREASING VALUES AND FILL
C   YDAT ARRAY WITH ZERO VALUES
C
```

```
      NP1 = N+1
      DO 7 I = NP1,129
      TDAT(I) = TDAT(I-1) + 1.
      YDAT(I) = 0.0
7     CONTINUE
      FN = 0.4
8     T = 0.0
      Z = 0.0
      DZDT = 0.0
      ZMAX = 0.0
9     AK1=DELT*F(T,Z,DZDT)
      AK2=DELT*F(T+DELT/2.,Z+DELT*DZDT/2.,DZDT+AK1/2.)
      AK3=DELT*F(T+DELT/2.,Z+DELT*(DZDT/2.+AK1/4.),DZDT+AK2/2.)
      AK4=DELT*F(T+DELT,Z+DELT*(DZDT+AK2/2.),DZDT+AK3)
      Z = Z + DELT*(DZDT+(AK1+AK2+AK3)/6.)
      DZDT = DZDT+(AK1+2.*AK2+2.*AK3+AK4)/6.
      IF(ABS(Z) .GE. ZMAX) ZMAX = ABS(Z)
      T = T + DELT
      IF(T .GE. TMAX-.0001)GO TO 10
      GO TO 9
10    WRITE(2,12)FN,ZMAX
12    FORMAT(' ',11X,'FN = ',F6.3,5X,'ZMAX = ',F6.4)
      FN = FN + 0.2
      IF(FN .GT. 10.01)GO TO 14
      GO TO 8
14    STOP
      END

      FUNCTION FUNCT(T)
      COMMON TDAT(129), YDAT(129)
      IMAX=129
      IMIN=1
      DO 3 L = 1,7
      IMED=(IMAX+IMIN)/2
      IF(T .LE. TDAT(IMED))GO TO 2
      IMIN=IMED
      GO TO 3
2     IMAX=IMED
3     CONTINUE
      SLOPE=(YDAT(IMAX)-YDAT(IMIN))/(TDAT(IMAX)-TDAT(IMIN))
      FUNCT=YDAT(IMIN)+SLOPE*(T-TDAT(IMIN))
      RETURN
      END
```

A partial printout of the results obtained from the computer program is as follows:

```
TMAX = 5.00    DELT =  .010    DAMPF =  .05

         FN =  .400       ZMAX = 7.5231
         FN =  .600       ZMAX = 4.7469
         FN =  .800       ZMAX = 3.7580
```

```
FN = 1.000        ZMAX = 4.2434
FN = 1.200        ZMAX = 4.3737
FN = 1.400        ZMAX = 3.5842
FN = 1.600        ZMAX = 2.6270
FN = 1.800        ZMAX = 1.7764
FN = 2.000        ZMAX = 1.7387
FN = 2.200        ZMAX = 1.5724
FN = 2.400        ZMAX = 1.4413
FN = 2.600        ZMAX = 1.2446
FN = 2.800        ZMAX =  .8453
FN = 3.000        ZMAX =  .6526
FN = 3.200        ZMAX =  .5996
FN = 3.400        ZMAX =  .4989
FN = 3.600        ZMAX =  .4201
FN = 3.800        ZMAX =  .3213
```

A plot of z_{max} versus f_n on log-log scales results in the response spectrum shown in Fig. 4-24.

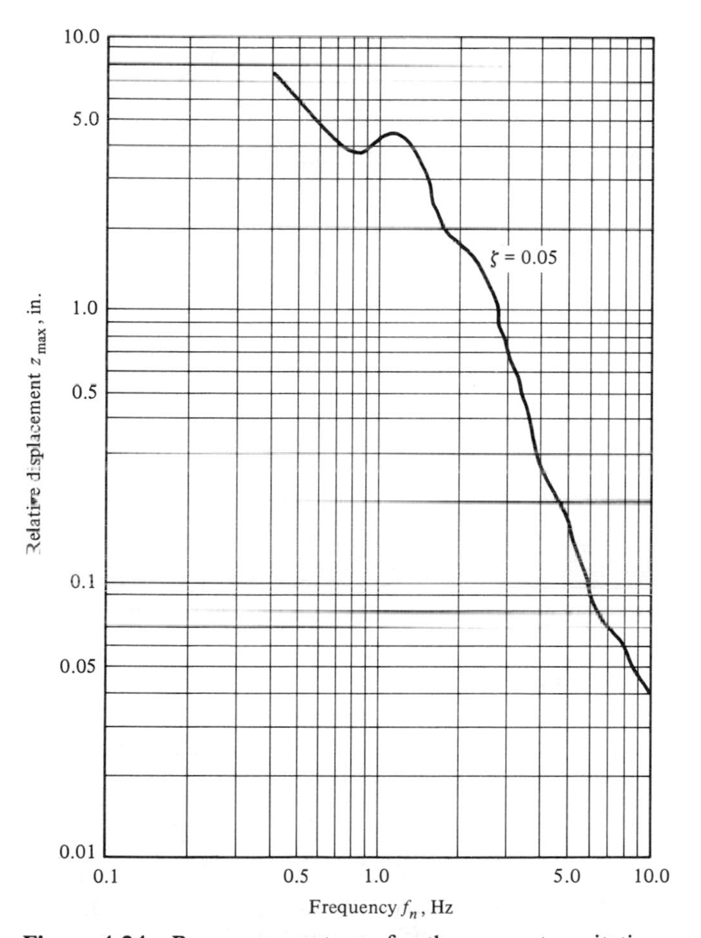

Figure 4-24 Response spectrum for the support excitation shown in Fig. 4-23 ($\zeta = 0.05$).

The program in the preceding example, which includes the arbitrary function generator discussed, can be used with microcomputers to obtain the response of systems subjected to earthquakes. A typical earthquake requires approximately 2000 data values to represent the ground acceleration; so a test value of 11, allowing up to 2049 data values, would be used in the DO loop in the Function Subprogram that supplies the arbitrary function generator. Such data should be stored in a disk file for transmission to the main program by means of a READ statement, as was done in Example 4-5. The time required for a microcomputer to compute data for plotting a response spectrum for an earthquake, however, might be considerable.

PROBLEMS

Problems 4-1 through 4-13 (Sections 4-1 through 4-4)

4-1. Show that the equation giving the time t_m at which the peak response occurs for a spring-and-mass system subjected to an impulsive force is

$$\tan(\omega_n \sqrt{1 - \zeta^2}\, t_m) = \frac{\sqrt{1 - \zeta^2}}{\zeta}$$

4-2. The spring-and-mass system shown in part a of the accompanying figure is subjected to the impulsive force $F(t)$ shown in part b of the figure. If $k = 2500$ N/m, $m = 2.5$ kg, and $c = 10$ N·s/m, determine the maximum response of the system.

Ans: $x_m = 57.6$ mm

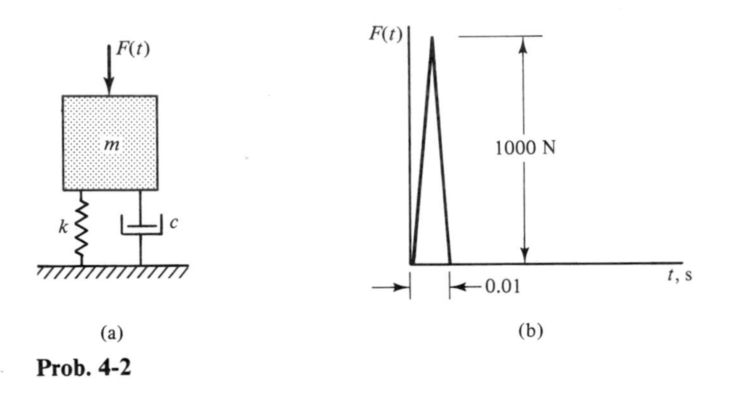

Prob. 4-2

4-3. Referring to Prob. 4-2, determine the velocity at the end of the impulsive force ($t = 0.01$ s) by considering Eq. 4-5. Compare the result obtained with $\dot{x}_0 = C/m$ shown in Eq. 4-6.

4-4. Neglect damping and show that the response of a spring-and-mass system subjected to the excitation force $F(t)$ shown in the accompanying figure for $t > t_1$ is

$$x = \frac{F_0}{k}\left[-\frac{\sin \omega_n t}{\omega_n t_1} + \frac{\sin \omega_n (t - t_1)}{\omega_n t_1} + \cos \omega_n (t - t_1)\right]$$

(Hint: Consider the solution given for a step function with rise time t_1 and the solution given for a step function with zero rise time.)

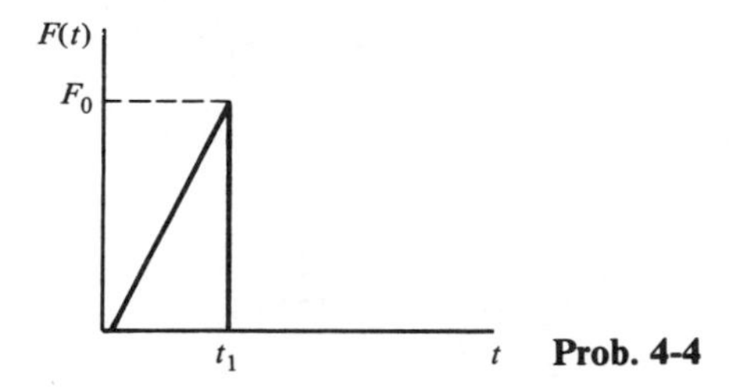

Prob. 4-4

4-5. The simply supported beam shown in part a of the accompanying figure is modeled as a single-degree-of-freedom system in which the distributed mass of the beam is lumped as a single mass m that is equal to one-half the total mass of the beam and that is located at the center of the beam as shown in part b of the figure. A pneumatic loading device is used to subject the beam to a simulated blast-type load

$$F(t) = F_0 e^{-bt}$$

which decreases exponentially with time as shown in part c of the figure. Show that the response of the simplified model of the beam at $x = l/2$ is

$$y = \frac{F_0 l^3}{48EI[1 + (b/\omega_n)^2]}\left(e^{-bt} - \cos \omega_n t + \frac{b}{\omega_n} \sin \omega_n t\right)$$

where EI = stiffness factor of beam
l = length of beam
$\omega_n = \sqrt{96EI/m_b l^3}$

The total mass of the beam is m_b = 2 m.

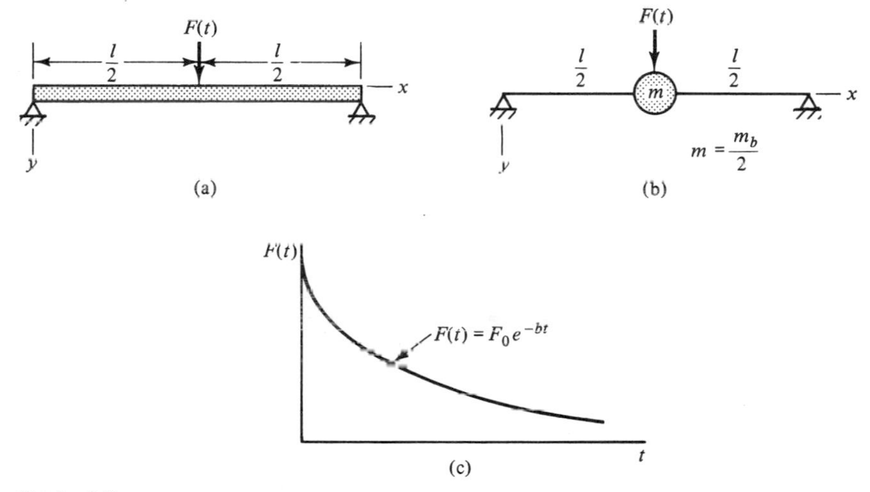

Prob. 4-5

4-6. Show that the response spectrum for a spring-and-mass system subjected to the rectangular pulse shown in part a of the accompanying figure with $\zeta = 0$ is described by

$$\frac{x_{max}}{F_0/k} = 2 \sin \frac{\pi t_1}{\tau_n} \qquad (t_1/\tau_n \leq 0.5)$$

$$\frac{x_{max}}{F_0/k} = 2 \qquad (t_1/\tau_n \geq 0.5)$$

as shown graphically in part b of the figure.

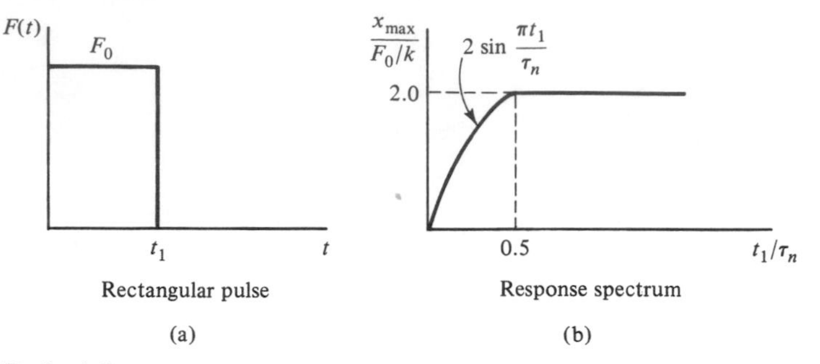

Prob. 4-6

4-7. If the differential equation of motion of the system shown in part a of the accompanying figure is given by Eq. 4-34, determine the excitation functions y and $\dot{y}$ in that equation for $0 \leq t \leq t_1$ if the acceleration of the moving support is described by

$$\ddot{y} = 386 \sin \frac{\pi t}{t_1} \qquad (0 \leq t \leq t_1)$$

$$\ddot{y} = 0 \qquad (t \geq t_1)$$

which is shown graphically in part b of the figure. Assume initial values of $\dot{y}_0 = y_0 = 0$.

Partial ans: $y = \frac{386 t_1}{\pi}\left(t - \frac{t_1}{\pi} \sin \frac{\pi t}{t_1}\right) \qquad (0 \leq t \leq t_1)$

m
c
k
x
Moving support
$y, \ddot{y}$
(a)

$\ddot{y}$
1 g
t_1
t
$\ddot{y} = 386 \sin \frac{\pi t}{t_1} \quad 0 \leqslant t \leqslant t_1$
$\ddot{y} = 0 \quad t \geqslant t_1$
(b)

Prob. 4-7

4-8. The recording device shown in part a of the accompanying figure consists of a mass m fastened to two identical springs and a drum that rotates about a vertical axis as shown. A pen fastened to the mass traces a curve on the drum that represents the relative motion z between the rotating drum and the suspended mass. The frame housing the drum and mass is attached to a support that is given a sudden constant acceleration $\ddot{y} = a_0$. Neglecting damping, determine (a) the differential equation of motion in terms of the relative motion z and (b) the solution of the differential equation of motion if the initial conditions are $z_0 = \dot{z}_0 = 0$. Sketch a portion of the trace of z that appears on the drum.

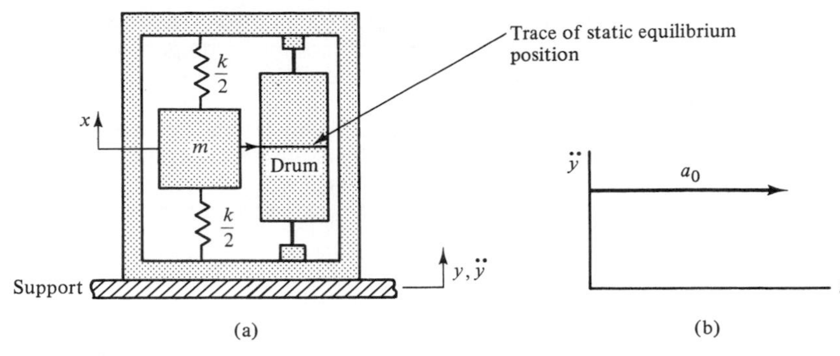

Prob. 4-8

4-9. Work Prob. 4-8 with the spring-and-mass assembly of the recorder considered as being viscously damped.

4-10. Equation 4-49 is the differential equation of motion of the system shown in part b of Fig. 4-12 in terms of the relative angular displacement ϕ ($\phi = \theta - \psi$). Verify this equation.

4-11. The simply supported composite beam formed of wood and steel portions as shown in part a of the accompanying figure is the same beam that is shown in Fig. 2-35 of Example 2-12, in which its undamped frequency f_n was found to be 16.22 Hz. Considering the transformed cross section representing an equivalent steel beam as shown in part c of the figure and neglecting damping, determine (a) the deflection of the beam at midspan if $F(t)$ is a step function with a rise time of $t_1 = 0.5$ s and a peak magnitude of $F_0 = 5000$ lb as shown in part b of the figure, (b) the maximum stresses in the wood and in the steel if the ratio of the moduli of elasticity of the two materials is $E_s/E_w = 20$ as used in Example 2-12, and (c) whether the values determined for the deflection and stresses are very accurate.

Ans: $y|_{x=l/2} = 0.72$ in., $\sigma_s = 13{,}010$ psi,

$\sigma_w = 1721$ psi

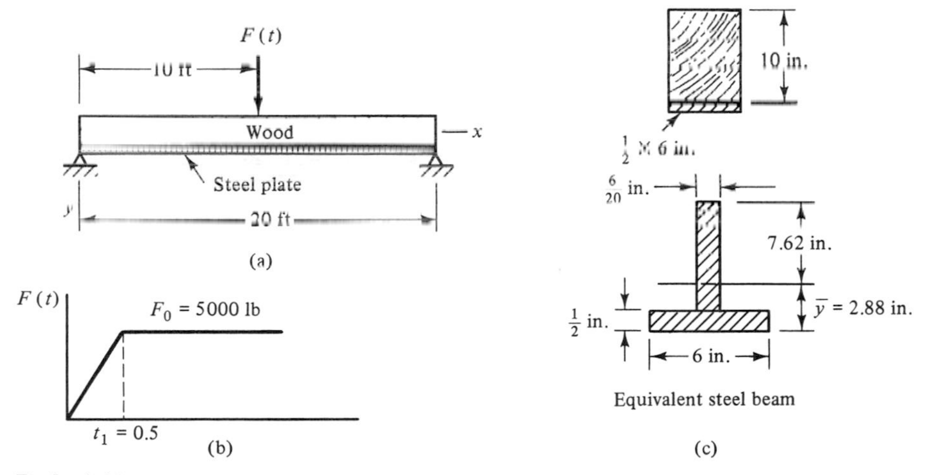

Prob. 4-11

4-12. Work Prob. 4-11 if the rise time t_1 is 30 ms instead of 0.5 s.

4-13. An instrument package of mass m is mounted on a support structure of stiffness k in the nose of a rocket as shown in part a of the accompanying figure. The support structure, which could involve springs, beams, plates, and so forth, is represented schematically by the spring-and-mass system shown in part b of the figure. Assuming that the nose cone to which the support structure is rigidly fastened experiences the acceleration shown in part c of the figure, and neglecting damping, determine (a) an expression for the relative displacement z of the instrument package with respect to the rocket and (b) the absolute acceleration $\ddot{x}$ of the instrument package as a function of time. If z is to be kept small, should k/m be large or small?

Partial ans: $\ddot{x} = a_0 t\left(1 - \dfrac{1}{\omega_n t}\sin \omega_n t\right)$

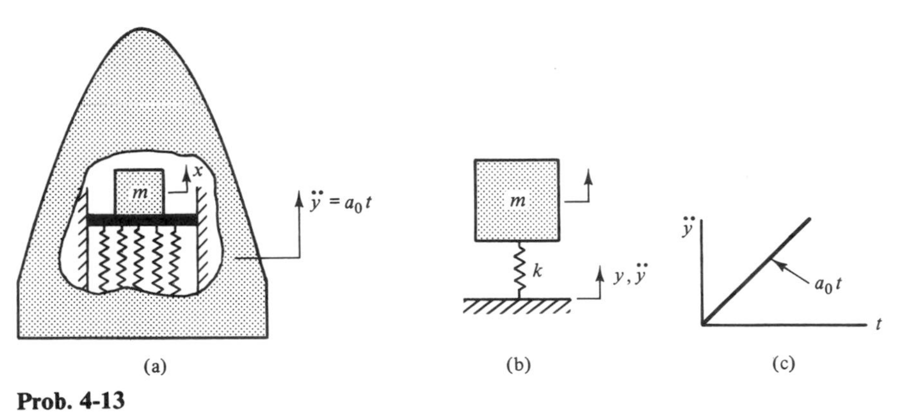

Prob. 4-13

Problems 4-14 through 4-21 (Sections 4-5 through 4-6)

4-14. The program STEP.FOR on page 251 was written to obtain the curves making up the response spectrum of a viscously damped spring-and-mass system subjected to a step-function excitation with a rise time of t_1. In Example 4-3 this program was modified to obtain the curves making up the response spectrum of the same spring-and-mass system subjected to the sine-pulse excitation shown in Fig. 4-16. In the present problem we wish to obtain the curves making up the response spectrum of the same viscously damped spring-and-mass system subjected to the excitation $F(t)/m$ shown in the insert of the accompanying figure (see Eq. 4-1 for the differential equation of motion). Since this excitation has the units of acceleration, it can be considered as an *input* excitation instead of a force excitation as was considered in Example 4-3. The curves shown in the main portion of the accompanying figure give the maximum acceleration $\ddot{x}_{max}$ of the mass m in g's when $F(t)/m$ is a sine pulse having a magnitude of 1 g as shown in the figure insert. Modify the STEP.FOR program so that it will obtain the data necessary for plotting the curves of the spectrum shown in the accompanying figure.

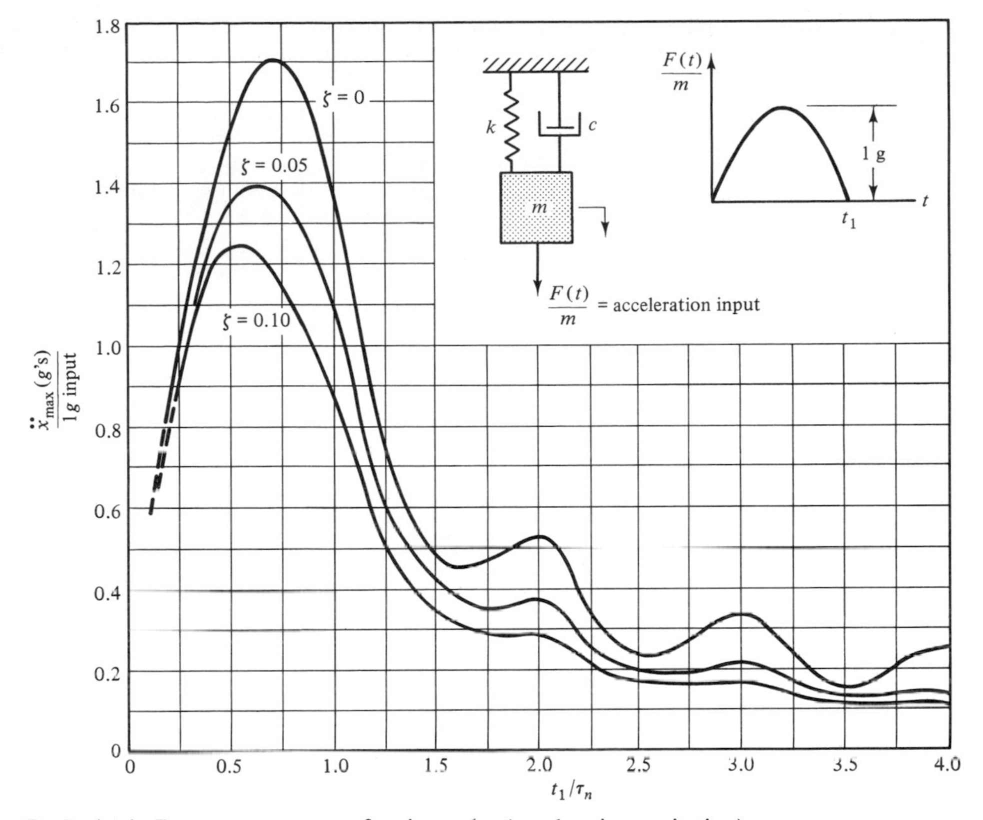

Prob. 4-14 Response spectrum for sine pulse (acceleration excitation).

4-15. The function $F(t)$ shown in the accompanying figure consists of a ramp function with a rise time of t_1 and an exponentially decaying function when $t \geq t_1$. That is,

$$F(t) = F_0 \frac{t}{t_1} \qquad (0 \leq t \leq t_1)$$

$$F(t) = F_0 e^{-b(t-t_1)} \qquad (t_1 \leq t)$$

in which the value of b is such that $F(2t_1) = F_0/2$. Write a Function Subprogram named FUNCT to generate the above function, and modify the program STEP.FOR given on page 251 to obtain $x_{max}/(F_0/k)$ for an interval of $0.1 \leq t_1 \leq 2.0$ in increments of 0.1 s with TMAX = 2.5.

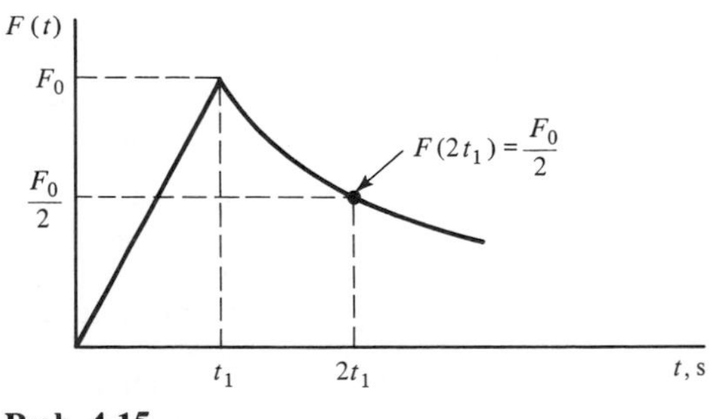

Prob. 4-15

4-16. Use the modified STEP.FOR program determined in Prob. 4-15 to obtain a response spectrum consisting of the curves of x_{max} versus t_1/τ_n for $\zeta = 0.0, 0.05$, and 0.1.

4-17. The acceleration of the upper end of the elevator cable shown in graphical form in part a of the accompanying figure is to be used to determine the relative motion z_{max} of the elevator of mass m as it travels between floors of a very tall building as shown schematically in part b of the figure. The natural frequency f_n of the elevator system is 5 Hz when the length l of the cable is 50 ft. Modify the ARBGEN.FOR program given in Example 4-5 so that it can be used to generate the $\ddot{y}$ function shown in part a of the figure, and obtain a response spectrum consisting of the curves of z_{max} versus f_n for $\zeta = 0$ with the cable length varying from 50 to 1000 ft. Neglect the mass of the cable, and assume that its length is essentially constant over a travel interval of one floor. Is this assumption a reasonable one? If z_{max} is to be limited to a magnitude of 1.0 in. for the system, what is the maximum permissible length l_m of the cable?

Ans: $l_m = 432.5$ ft

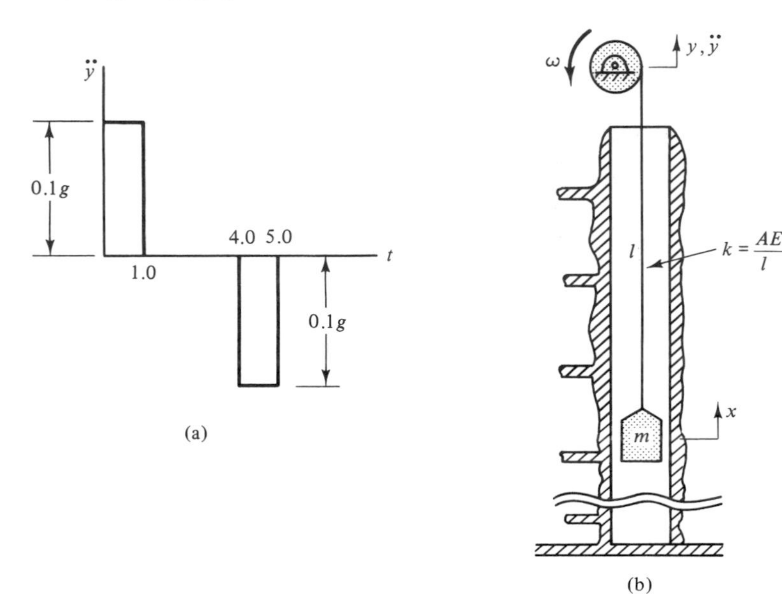

Prob. 4-17

4-18. Two designs have been proposed for the supporting columns of a storage tank at an oil refinery that is located in a region in which earthquakes occur. The design shown in part a of the accompanying figure consists of four steel pipes, each of which has an outside diameter of 6.62 in. (nominal 6 in.) and a length of 10 ft. The design shown in part b of the figure employs 14 steel pipes, each of which has an outside diameter of 4.5 in. (nominal 4 in.) and a length of 10 ft. It is desired to evaluate the two designs to determine which would be the better in resisting damage when subjected to an earthquake such as the 1940 El Centro, California, earthquake. Assuming that the pipes are fixed at both ends, the spring constant of each is $12\ EI/l^3$, in which I is the area moment of inertia of the pipe about a diametral axis. The equivalent spring constant k_e of each system is shown in the figure, and because of the relationship between the number of pipes and the magnitude of I in each, they are approximately equal in magnitude. Since the stiffness is essentially the same for each design, the natural frequency f_n of each will also be essentially the same. Depending upon how full the tank is, the natural frequency can be expected to vary from 2 to 5 Hz based upon data from existing tanks. With damping assumed to be $\zeta = 0.02$, evaluate the two designs by using the response spectrum of the 1940 El Centro, California, earthquake shown in Fig. 4-21, and determine whether or not the column distortions calculated exceed the 60,000 psi elastic limit of the column material [which is accompanied by a strain of $2000(10)^{-6}$ in./in.]. (Hint: Express the strain ϵ in terms of the relative displacement z, outside pipe radius r, and pipe length l.)

Ans: Elastic limit is exceeded in 6-in. pipes but not in 4-in. pipes.

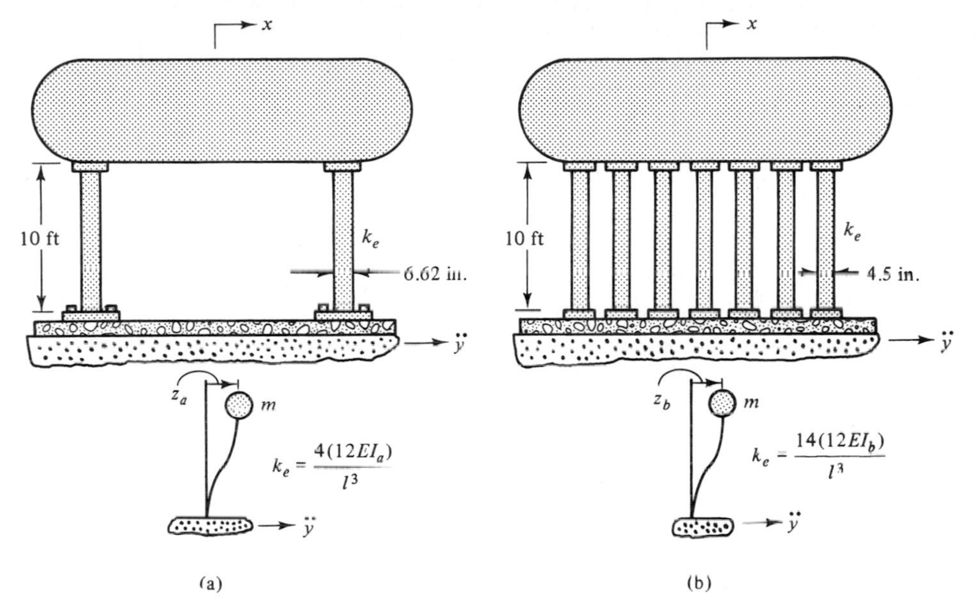

Prob. 4-18

4-19. Refer to Prob. 4-18 and evaluate the two designs using the same data, parameter values, and design criteria, but using the response spectrum of the 1971 San Fernando, California earthquake shown in Fig. 4-19 instead of the one used in Prob. 4-18.

4-20. Refer to Example 4-5 and use the Function Subprogram for the arbitrary function generator shown there to obtain the maximum relative response z_{max} for the support acceleration excitation shown in the accompanying figure. Obtain data for plotting a response spectrum for $\zeta = 0.05$ for a natural frequency range of $0.4 \leq f_n \leq 10$ in increments of 0.2 Hz as was done in Example 4-5.

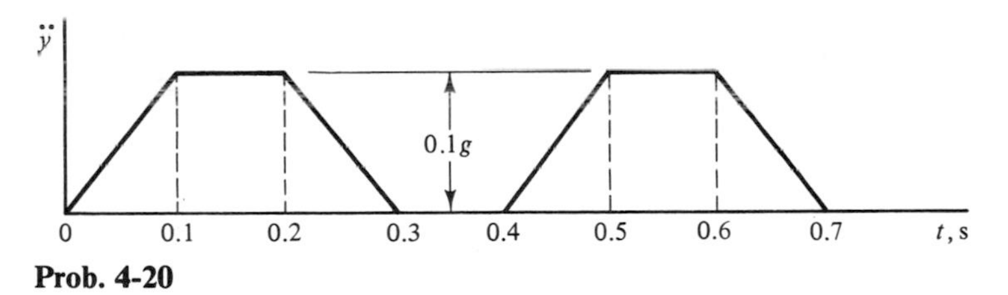

Prob. 4-20

4-21. The characteristics of explosive-generated ground motions such as blasting in rock quarries are quite similar to those of earthquakes. Consequently, such ground motions have been used in studying the response of structures to earthquakes. A typical ground-acceleration record from a large blast explosion is shown in part a of the accompanying figure. Part b of the figure gives the peaks of the record and the times at which they occur in tabular form. Use the ARBGEN.FOR program given on page 265, and obtain the maximum relative response z_{max} due to the ground acceleration produced by the blast explosion. Run the program to obtain data for plotting curves for a response spectrum for $\zeta = 0.0$, 0.05, and 0.10 for a natural frequency range of $0.4 \leq f_n \leq 10$ in increments of 0.2 Hz as was done in Example 4-5.

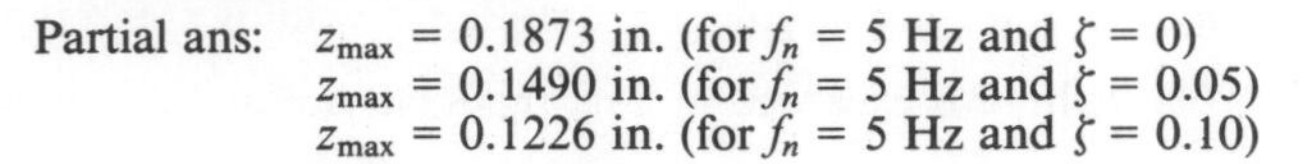

Partial ans: $z_{max} = 0.1873$ in. (for $f_n = 5$ Hz and $\zeta = 0$)
$z_{max} = 0.1490$ in. (for $f_n = 5$ Hz and $\zeta = 0.05$)
$z_{max} = 0.1226$ in. (for $f_n = 5$ Hz and $\zeta = 0.10$)

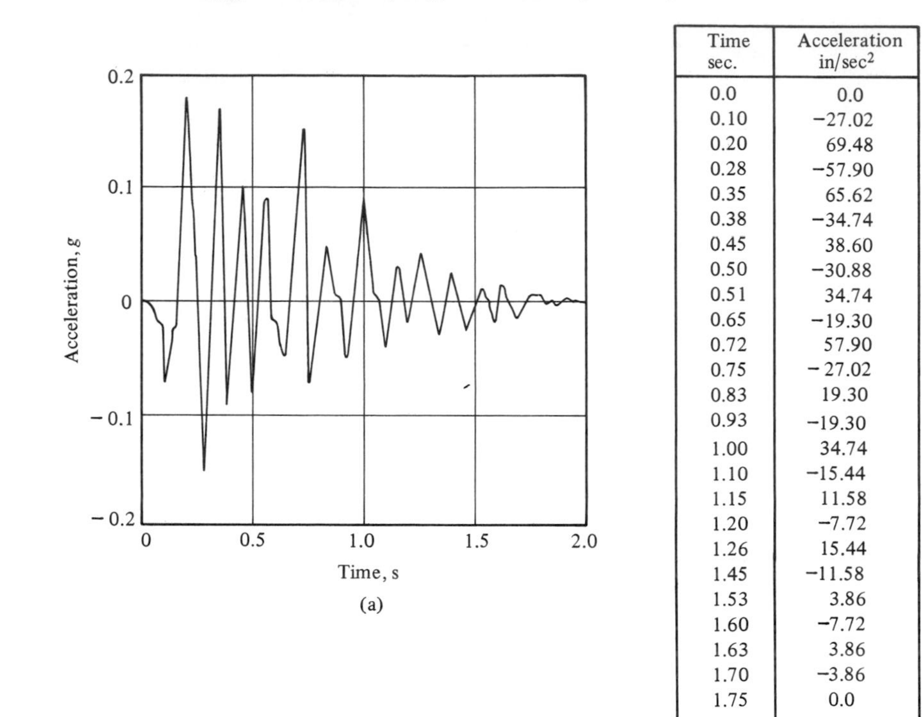

(a)

Time sec.	Acceleration in/sec^2
0.0	0.0
0.10	−27.02
0.20	69.48
0.28	−57.90
0.35	65.62
0.38	−34.74
0.45	38.60
0.50	−30.88
0.51	34.74
0.65	−19.30
0.72	57.90
0.75	−27.02
0.83	19.30
0.93	−19.30
1.00	34.74
1.10	−15.44
1.15	11.58
1.20	−7.72
1.26	15.44
1.45	−11.58
1.53	3.86
1.60	−7.72
1.63	3.86
1.70	−3.86
1.75	0.0

(b)

Prob. 4-21

Chapter 5

Free Vibration of Multiple-Degree-of-Freedom Systems (Discrete and Lumped-Mass Systems)

5-1 INTRODUCTION

Discussions in preceding chapters have been confined to single-degree-of-freedom systems for which a single coordinate was sufficient to describe the motion of the system. In this chapter we discuss the free vibration of discrete and lumped-mass systems that have more than one degree of freedom and consequently require more than one coordinate to describe their motion. In general, discrete and lumped-mass systems have finite numbers of degrees of freedom, while continuous systems (systems of continuous mass) theoretically have an infinite number of degrees of freedom.[1]

An n-degree-of-freedom system inherently has n natural frequencies and n normal modes of vibration describing its configurations as it vibrates at the different frequencies. Thus, when a system vibrates freely at a particular natural frequency, the normal mode corresponding to that frequency describes the configuration of the system for that particular mode, and there is a particular relationship between the amplitudes of the coordinates defining the motion of the system.

5-2 DIFFERENTIAL EQUATIONS OF MOTION

The differential equations of motion of multiple-degree-of-freedom systems can be derived (1) by the direct application of Newton's second law, (2) by using the concept of dynamic

[1] A discrete system is one in which the inertial, elastic, and damping properties are all clearly described by distinct masses, springs, and damping mechanisms, respectively. When the distributed properties of beams, rods, plates, and so on are discretized by modeling them as systems composed of lumped masses, lumped elastic elements, and modal damping (Eq. 5-116), such systems are generally referred to as lumped-mass systems or as lumped-parameter systems. These discrete systems should not be confused with discrete-time systems, for which the variables are determined at distinct instants of time, and which are described by difference equations rather than by differential equations.

equilibrium, (3) by utilizing the definitions of influence coefficients, and (4) by the use of Lagrange's equations. The first three of these approaches will be discussed in this chapter, while the use of Lagrange's equations for this purpose will be deferred until Chap. 7 in which they are also used in developing the finite-element method for use in vibration analysis. For the sake of simplicity we use systems having only two or three degrees of freedom to illustrate procedures using Newton's second law directly for deriving the differential equations of motion of systems having any number of degrees of freedom. We then discuss the conversion of these equations to matrix form in terms of mass, stiffness, and damping matrices.

Generalized Coordinates

Before we can derive the differential equations of motion of a multiple-degree-of-freedom system, we must specify a sufficient number of *independent* coordinates to completely describe the configurations of the system. There is one coordinate associated with each degree of freedom, and any one of these coordinates can change without necessitating a change in the others, and is thus *independent* of the others. Such coordinates are referred to as *generalized coordinates,* and there is thus obviously one such coordinate for each degree of freedom of the system.

Care must be exercised in selecting the generalized coordinates. For example, we could write for the double pendulum shown in Fig. 5-1 that

$$x_1 = l_1 \sin \theta_1 \qquad x_2 = l_1 \sin \theta_1 + l_2 \sin \theta_2$$
$$y_1 = l_1 \cos \theta_1 \qquad y_2 = l_1 \cos \theta_1 + l_2 \cos \theta_2$$

However, x_1, x_2, y_1, and y_2 are not independent coordinates since the four of them can be expressed in terms of the two coordinates θ_1 and θ_2. Thus, there are only two independent coordinates for the system, and the double pendulum has but two degrees of freedom.

Deriving Differential Equations of Motion Using Newton's Second Law

In deriving the differential equations of motion using this method, it is very important that the following steps be used:

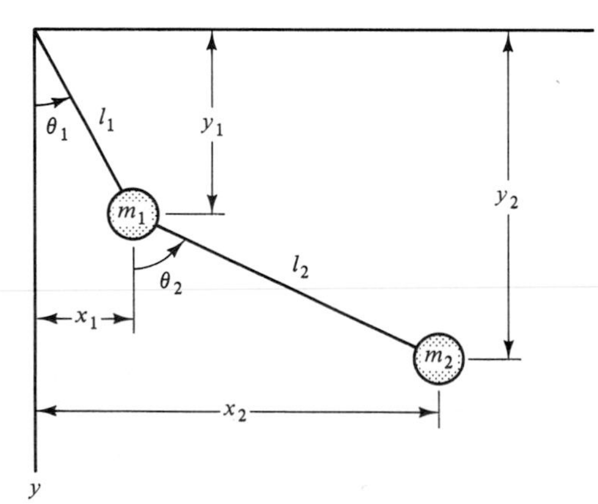

Figure 5-1 Double pendulum.

1. Select a *common* positive sense for the displacements, velocities, and accelerations associated with the generalized coordinates used.
2. Whether the generalized coordinate represents a linear or angular displacement, measure it from the static-equilibrium position of the system.
3. Sketch a free-body diagram of *each* mass element of the system. Elastic and damping forces acting on a mass element are shown on the free-body diagram with senses consistent with the forces they exert *on* the mass element when it has an assumed positive displacement and velocity.
4. Apply Newton's second law to the free body of each mass element. This will result in n differential equations of motion for an n-degree-of-freedom system. As we shall see later, these equations are generally not independent of each other, since a particular coordinate can appear in more than one of them. The equations are then said to be *coupled* equations.

With the preceding steps in mind, Newton's second law applied to the ith mass element yields

$$\sum F_x = m_i \ddot{x}_i \tag{5-1}$$

in which $\ddot{x}_i$ is the acceleration of the mass center of the ith mass element. In the summation any force acting in the same sense as the positive sense assumed for x_i is given a plus sign, and any force acting in the opposite sense is given a negative sign.

For a particular coordinate θ_i, describing the angular displacement of the ith element, we write that

$$\sum M_0 = I_0 \ddot{\theta}_i \tag{5-2}$$

in which $\sum M_0$ is the summation of moments of the forces about an axis through 0, and I_0 is the mass moment of inertia of the element about the same axis. As discussed in elementary dynamics, the point 0 must be fixed or moving with constant velocity. However, if the axis used coincides with an axis through the mass center G of the element, then

$$\sum M_G = \bar{I} \ddot{\theta}_i \tag{5-3}$$

Equation 5-3 is always valid, regardless of the motion of the mass center. In writing either of the above equations, a moment acting with the same sense as the assumed positive sense for θ_i is given a positive sign, while moments of opposite sense are assigned a negative sign.

EXAMPLE 5-1

The two-degree-of-freedom system shown in Fig. 5-2a consists of a pulley, a mass m, two springs, a dashpot, and a cable connecting them as shown. The pulley's centroidal mass moment of inertia is $\bar{I}$, and the dashpot has a coefficient of c. There is sufficient friction between the cable and pulley to prevent the cable from slipping.

We wish to determine the differential equations of motion of the system when the pulley is subjected to a time-varying moment $M(t)$ as shown in the figure.

Solution. The generalized coordinates θ and x shown in Fig. 5-2a are selected to describe the displacements of the system elements from their static-equilibrium positions. Arbitrarily assuming that $x > r\theta$, the forces exerted by the springs on the pulley and mass m accompanying *positive* displacements θ and x from the static-equilibrium positions are as shown on the free-body diagrams of Fig. 5-2b. The

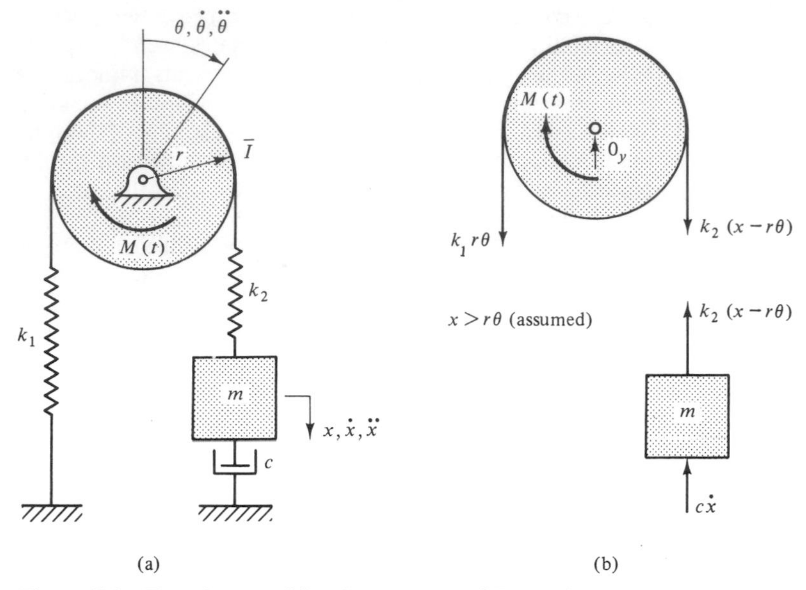

Figure 5-2 Two-degree-of-freedom system of Example 5-1.

damping force $c\dot{x}$, acting on the mass m, has a sense opposing the positive downward velocity $\dot{x}$ of the mass m, since damping forces always oppose motion.

Using the free-body diagram of the pulley, we can write that

$$\sum M_0 = \bar{I}\ddot{\theta}$$

This summation yields

$$-k_1 r^2\theta + k_2(x - r\theta)r + M(t) = \bar{I}\ddot{\theta}$$

or

$$\bar{I}\ddot{\theta} + (k_1 + k_2)r^2\theta - k_2 r x = M(t) \tag{5-4}$$

Applying Newton's second law to the free-body diagram of the mass m

$$\sum F_x = m\ddot{x}$$

we obtain

$$-k_2(x - r\theta) - c\dot{x} = m\ddot{x}$$

or

$$m\ddot{x} + c\dot{x} + k_2 x - k_2 r\theta = 0 \tag{5-5}$$

Equations 5-4 and 5-5 are the differential equations of motion of the system. If the excitation moment $M(t)$ were zero, the system would have damped free vibration when either or both of the mass elements were disturbed from the static-equilibrium position.

The free-body diagrams shown in Fig. 5-2b are based upon the assumption that $x > r\theta$. It is suggested that the reader sketch the free-body diagrams with the assumption that $x < r\theta$ and verify that they will also yield Eqs. 5-4 and 5-5.

It should also be pointed out that the gravity force mg was not shown on the free-body diagram for the mass m, since the forces in the springs necessary to keep the system in its static-equilibrium position cancel the effect of mg as explained in Chap. 2.

EXAMPLE 5-2

We wish to write the differential equations of motion of Example 5-1 in matrix form (the basic concepts and operations of matrix algebra are given in Appendix B for the reader who feels a need to review the subject).

Solution. The matrix form of Eqs. 5-4 and 5-5 is given by

$$\begin{bmatrix} \bar{I} & 0 \\ 0 & m \end{bmatrix} \begin{Bmatrix} \ddot{\theta} \\ \ddot{x} \end{Bmatrix} + \begin{bmatrix} 0 & 0 \\ 0 & c \end{bmatrix} \begin{Bmatrix} \dot{\theta} \\ \dot{x} \end{Bmatrix} + \begin{bmatrix} (k_1 + k_2)r^2 & -k_2 r \\ -k_2 r & k_2 \end{bmatrix} \begin{Bmatrix} \theta \\ x \end{Bmatrix} = \begin{Bmatrix} M(t) \\ 0 \end{Bmatrix} \tag{5-6}$$

which can be verified by multiplying the first row of the square matrices by the corresponding column matrices to obtain Eq. 5-4, and then multiplying the second row of the square matrices by the corresponding column matrices to obtain Eq. 5-5.

Mass, Stiffness, and Damping Matrices

The matrix form of equations for multiple-degree-of-freedom systems will generally involve a mass matrix **M**, a stiffness matrix **K**, and a damping matrix **C**. For example, referring to Eq. 5-6, these matrices are

$$\mathbf{M} = \begin{bmatrix} \bar{I} & 0 \\ 0 & m \end{bmatrix} \quad \text{(mass matrix)}$$

$$\mathbf{K} = \begin{bmatrix} (k_1 + k_2)r^2 & -k_2 r \\ -k_2 r & k_2 \end{bmatrix} \quad \text{(stiffness matrix)}$$

$$\mathbf{C} = \begin{bmatrix} 0 & 0 \\ 0 & c \end{bmatrix} \quad \text{(damping matrix)}$$

The general form of the matrix equation for an n-degree-of-freedom system subjected to excitation forces and/or moments is

$$\begin{bmatrix} m_{11} & m_{12} & \cdots & m_{1n} \\ m_{21} & m_{22} & \cdots & m_{2n} \\ \cdot & \cdot & \cdots & \cdot \\ m_{n1} & m_{n2} & \cdots & m_{nn} \end{bmatrix} \begin{Bmatrix} \ddot{x}_1 \\ \ddot{x}_2 \\ \vdots \\ \ddot{x}_n \end{Bmatrix} + \begin{bmatrix} c_{11} & c_{12} & \cdots & c_{1n} \\ c_{21} & c_{22} & \cdots & c_{2n} \\ \cdot & \cdot & \cdots & \cdot \\ c_{n1} & c_{n2} & \cdots & c_{nn} \end{bmatrix} \begin{Bmatrix} \dot{x}_1 \\ \dot{x}_2 \\ \vdots \\ \dot{x}_n \end{Bmatrix}$$
$$+ \begin{bmatrix} k_{11} & k_{12} & \cdots & k_{1n} \\ k_{21} & k_{22} & \cdots & k_{2n} \\ \cdot & \cdot & \cdots & \cdot \\ k_{n1} & k_{n2} & \cdots & k_{nn} \end{bmatrix} \begin{Bmatrix} x_1 \\ x_2 \\ \vdots \\ x_n \end{Bmatrix} = \begin{Bmatrix} F_1 \\ F_2 \\ \vdots \\ F_n \end{Bmatrix} \tag{5-7}$$

in which $x_1, x_2, \ldots, x_n$ are generalized coordinates that can be either linear or angular displacements, and $F_1, F_2, \ldots, F_n$ are excitation forces or moments.

Equation 5-7 can be written in simple, compact form as

$$\mathbf{M\ddot{X}} + \mathbf{C\dot{X}} + \mathbf{KX} = \mathbf{F} \tag{5-8}$$

in which **F** is a column matrix involving excitation forces or moments. It should be noted that Eq. 5-8, which may represent a system with many degrees of freedom, has the same general appearance as the differential equation of motion of a single-degree-of-freedom system.

For undamped free vibration, Eq. 5-8 reduces to

$$\mathbf{M\ddot{X}} + \mathbf{KX} = \mathbf{0} \tag{5-9}$$

in which **0** is a *null* column matrix containing only zero elements. As we shall see later, this equation is fundamental to the formation of equations pertinent to determining the natural frequencies and normal-mode configurations of n-degree-of-freedom systems.

A unique characteristic of the mass, stiffness, and damping matrices of linear systems with small displacements is that they are all symmetric. That is,

$$m_{ij} = m_{ji} \quad \text{(elements of mass matrix)}$$
$$c_{ij} = c_{ji} \quad \text{(elements of damping matrix)}$$
$$k_{ij} = k_{ji} \quad \text{(elements of stiffness matrix)}$$

The mass matrix is usually a diagonal matrix, in which all the elements are zero except for those on the diagonal, as in the mass matrix of Eq. 5-6, for example. However, in the case of *dynamic coupling,* which is discussed in Sec. 5-4, the mass matrix is non-diagonal and will be of the general form shown in Eq. 5-7. The method of *finite elements,* discussed in Chap. 7, also involves nondiagonal mass matrices.

In general, many of the elements of the stiffness matrix are zero. In fact, as we shall soon see, the elements of such a matrix for an n-degree-of-freedom system are generally zero everywhere, except for those which are on the diagonal and immediately above and below it. The stiffness matrix is thus referred to as a *banded* matrix.

EXAMPLE 5-3

We wish to determine the differential equations of motion in matrix form for the undamped free vibration of the three-story building shown schematically in Fig. 5-3a.

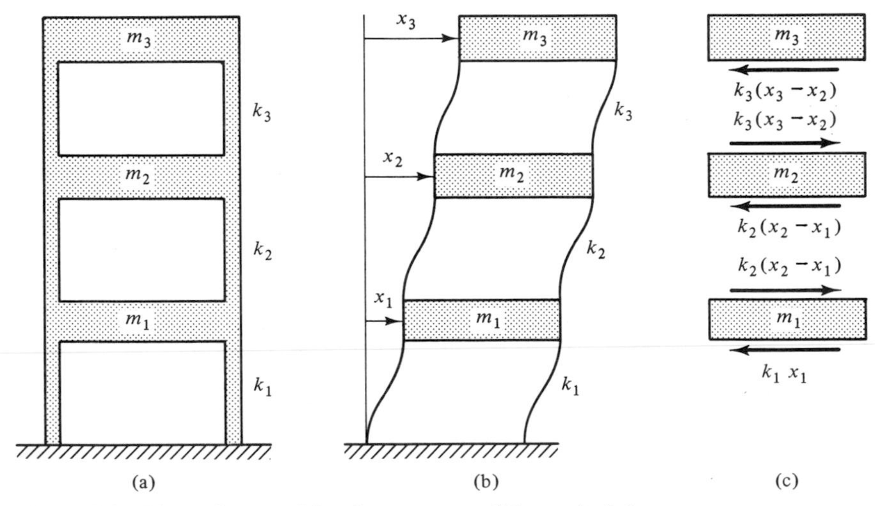

Figure 5-3 Three-degree-of-freedom system of Example 5-3.

We assume that the mass distribution of the building can be represented by the lumped masses at the different levels, shown in the figure. We assume further that the girders of the structure are infinitely rigid in comparison with the supporting columns, so that the general configuration of the columns will be as shown in Fig. 5-3b.

The spring constants shown in the figure are equivalent constants, representing the aggregate stiffness of the number of columns supporting a given floor. This is done by considering the columns as springs in parallel.

Solution. The coordinates x_1, x_2, and x_3 and the velocities and accelerations associated with them are assumed positive to the right. Arbitrarily assuming that $x_3 > x_2 > x_1$, the free-body diagrams in Fig. 5-3c show the forces that the columns exert on the masses. In applying Newton's second law, the forces having senses the same as the positive sense assumed for the acceleration are summed positive; the others are summed negative.

Applying Newton's second law to each free body of the three masses, we obtain

$$-k_1x_1 + k_2(x_2 - x_1) - m_1\ddot{x}_1$$
$$-k_2(x_2 - x_1) + k_3(x_3 - x_2) = m_2\ddot{x}_2$$
$$-k_3(x_3 - x_2) = m_3\ddot{x}_3$$

Collecting and rearranging terms in each of the above equations, we obtain

$$\left.\begin{aligned} m_1\ddot{x}_1 + (k_1 + k_2)x_1 - k_2x_2 &= 0 \\ m_2\ddot{x}_2 - k_2x_1 + (k_2 + k_3)x_2 - k_3x_3 &= 0 \\ m_3\ddot{x}_3 - k_3x_2 + k_3x_3 &= 0 \end{aligned}\right\} \tag{5-10}$$

Inspecting Eq. 5-10, it should be apparent that the differential equations of motion for the undamped free vibration of the three-story building can be written in matrix form as

$$\begin{bmatrix} m_1 & 0 & 0 \\ 0 & m_2 & 0 \\ 0 & 0 & m_3 \end{bmatrix} \begin{Bmatrix} \ddot{x}_1 \\ \ddot{x}_2 \\ \ddot{x}_3 \end{Bmatrix} + \begin{bmatrix} (k_1 + k_2) & -k_2 & 0 \\ -k_2 & (k_2 + k_3) & -k_3 \\ 0 & -k_3 & k_3 \end{bmatrix} \begin{Bmatrix} x_1 \\ x_2 \\ x_3 \end{Bmatrix} = \begin{Bmatrix} 0 \\ 0 \\ 0 \end{Bmatrix} \tag{5-11}$$

As observed before, Eq. 5-11 can also be written in the more compact form as

$$\mathbf{M\ddot{X}} + \mathbf{KX} = \mathbf{0}$$

in which the mass matrix **M** and the stiffness matrix **K** are as shown in Eq. 5-11.

It should be noted that the stiffness matrix **K** is symmetrical and that the nonzero elements appear on the diagonal and immediately above and below it. That is, $k_{13} = k_{31} = 0$.

5-3 INFLUENCE COEFFICIENTS (STIFFNESS, FLEXIBILITY, DAMPING)

In the preceding section it was shown that the differential equations of motion of a system can be written in the form of a mass matrix **M**, a stiffness matrix **K**, and a damping matrix **C**. As we shall see later, the differential equations of motion can also be written in terms of a flexibility matrix **A**, which is the inverse of the stiffness matrix **K**. That is, $\mathbf{K}^{-1} = \mathbf{A}$, or $\mathbf{A}^{-1} = \mathbf{K}$.

The elements k_{ij}, a_{ij}, and c_{ij} of the stiffness, flexibility, and damping matrices, respectively, are referred to as *influence coefficients.*

The use of influence coefficients facilitates the vibration analysis of systems having many degrees of freedom. In fact, as we shall soon see, the differential equations of some such systems and the equations for determining their natural circular frequencies can be written by simply inspecting the system and applying the definition of the influence coefficient used.

Stiffness Coefficients

Designating the coordinates $q_1, q_2, \ldots, q_n$ of an n-degree-of-freedom system as the generalized coordinates that define the linear and/or angular displacements of the masses comprising the system, we can define the stiffness coefficient k_{ij} as the force or moment required to hold a particular coordinate q_i fixed when a coordinate q_j is given a *unit* linear (or angular) displacement, with all other coordinates held fixed. The stiffness coefficient k_{jj} $(i = j)$ then becomes the static force or moment required to give the coordinate q_j a unit linear or angular displacement.

To determine the k_{ij}'s of a system using this definition, we first note that each mass in the system will have a generalized coordinate for each type of motion that it has, and that the subscripts of the k_{ij}'s refer to these generalized coordinates. With this in mind we can state that for a system configuration consisting of a positive *unit* linear (or angular) displacement q_j of one of the masses, the k_{ij}'s for that particular q_j are the forces and/or moments that must be applied to the masses associated with the coordinates q_i $(i = 1, 2, \ldots, n)$ to maintain the system in that configuration, and are determined in total by applying the definition of k_{ij} in turn for all $q_j (j = 1, 2, \ldots, n)$.

Thus, the stiffness coefficients $k_{1j}, k_{2j}, \ldots, k_{nj}$ in the stiffness matrix **K** are the elements of the jth column and are the forces or moments that must be applied to the masses associated with the coordinates $q_1, q_2, \ldots, q_n$ to hold the system in a configuration consisting of a positive unit linear or angular displacement q_j of one of the masses, with all the other mass displacements held at zero. For example, if the mass associated with a translation coordinate q_1 and a rotation coordinate q_2 is given a positive unit linear displacement $(j = 1)$, and its angular displacement and all displacements of the other masses are held at zero, the resulting configuration yields the first column of the **K** matrix. Similarly, if the same mass is given a positive unit angular displacement $(j = 2)$, and its linear displacement and all displacements of the other masses are held at zero, the resulting configuration yields the elements of the second column of **K**. For n degrees of freedom (n generalized coordinates), the above procedure is continued through $j = n$ to obtain all the elements of the **K** matrix.

As will be shown in the examples that follow, the stiffness coefficients are expressed in terms of the spring constants and can be either positive or negative depending upon how they act. If a stiffness coefficient (force or moment) acts in the same sense as the sense assumed positive for the associated generalized coordinate, it will be positive. If it acts in the opposite sense, it will be negative.

EXAMPLE 5-4

It is desired to determine the stiffness matrix **K** for the three-story building of Example 5-3, which is shown again in this example in Fig. 5-4a.

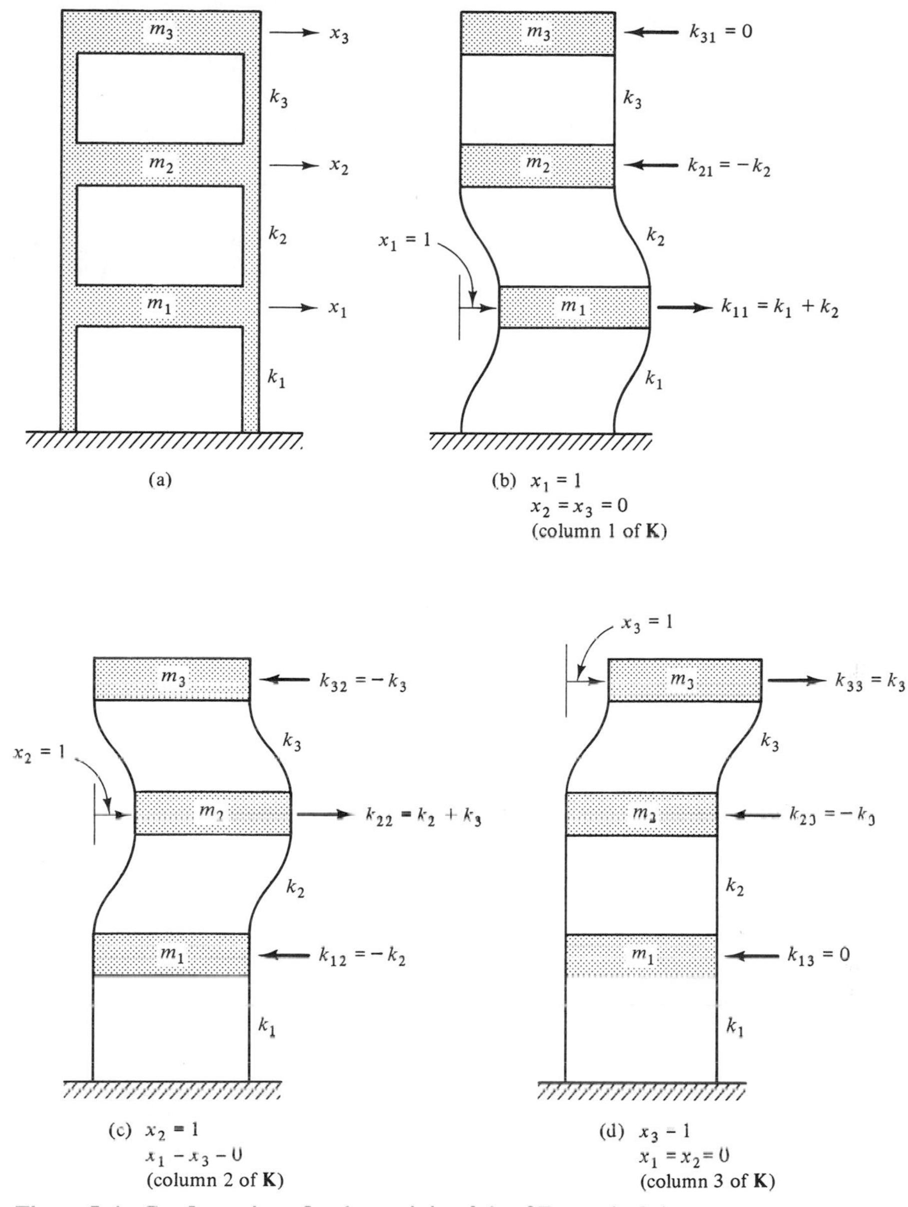

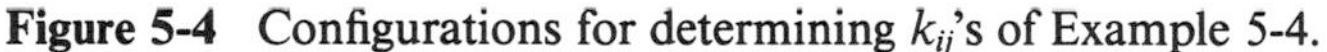
Figure 5-4 Configurations for determining k_{ij}'s of Example 5-4.

Solution. We obtain the elements of **K** by referring to Fig. 5-4 and using the definition of stiffness coefficients. To determine the first column of **K**, the mass m_1 is given a positive unit displacement ($x_1 = 1$), with m_2 and m_3 held fixed, so that $x_2 = x_3 = 0$ as shown in Fig. 5-4b. It can be seen from the figure that to maintain this configuration, the required forces must be

$$\left.\begin{aligned} k_{11} &= k_1 + k_2 \\ k_{21} &= -k_2 \\ k_{31} &= 0 \end{aligned}\right\} \qquad \text{(first column of } \mathbf{K}\text{)}$$

It can also be seen, as mentioned earlier, that these forces are in terms of the spring constants and are forces per unit displacement. Thus, k_{11} is the force necessary to deform the elastic elements k_1 and k_2 by an amount $x_1 = 1$ in giving the mass m_1 that unit displacement. To keep the displacement of m_2 zero requires a force k_{21} having a sense opposite to that of a positive x_2, as shown in Fig. 5-4b. With m_2 held fixed by k_{21}, there is no tendency for m_3 to displace, so k_{31} is zero.

To determine the second column of **K**, m_2 is given a positive unit displacement ($x_2 = 1$), with m_1 and m_3 held fixed as shown in Fig. 5-4c. Again, from inspection of Fig. 5-4c, it can be seen that the forces required to maintain this configuration are

$$\left.\begin{aligned} k_{12} &= -k_2 \\ k_{22} &= k_2 + k_3 \\ k_{32} &= -k_3 \end{aligned}\right\} \qquad \text{(second column of } \mathbf{K}\text{)}$$

In similar fashion, the third column of **K** is obtained from the forces required to maintain the configuration shown in Fig. 5-4d, in which $x_3 = 1$ and $x_1 = x_2 = 0$, so that

$$\left.\begin{aligned} k_{13} &= 0 \\ k_{23} &= -k_3 \\ k_{33} &= k_3 \end{aligned}\right\} \qquad \text{(third column of } \mathbf{K}\text{)}$$

Thus, the stiffness matrix is

$$[k_{ij}] = \mathbf{K} = \begin{bmatrix} (k_1 + k_2) & -k_2 & 0 \\ -k_2 & (k_2 + k_3) & -k_3 \\ 0 & -k_3 & k_3 \end{bmatrix}$$

which is the same matrix **K** as that expressed by Eq. 5-11 of Example 5-3.

It should be pointed out that the k_{ij}'s we have found are equal and opposite in sense to the forces that the elastic elements exert *on the masses.* For example, k_{11} is equal and opposite in sense to the forces that the elastic elements k_1 and k_2 exert on m_1, when it has a positive unit displacement. Therefore, it should be apparent that the application of Newton's second law would yield

$$-[k_{ij}]\begin{Bmatrix} x_1 \\ x_2 \\ x_3 \end{Bmatrix} = \begin{bmatrix} m_1 & 0 & 0 \\ 0 & m_2 & 0 \\ 0 & 0 & m_3 \end{bmatrix}\begin{Bmatrix} \ddot{x}_1 \\ \ddot{x}_2 \\ \ddot{x}_3 \end{Bmatrix}$$

or

$$-\mathbf{KX} = \mathbf{M\ddot{X}}$$

in which $-\mathbf{KX}$ are the forces that the elastic elements exert on the masses. Conversely, $\mathbf{KX}$ are the forces that the masses exert on the elastic elements.

EXAMPLE 5-5

A shaft with three disks attached to it is fixed at one end as shown in Fig. 5-5a. The mass moments of inertia of the disks are I_1, I_2, and I_3. The torsional spring constant of each of the three shaft intervals of length l is $k = GJ/l$, in which G is the shear modulus and J is the polar moment of inertia of the shaft.

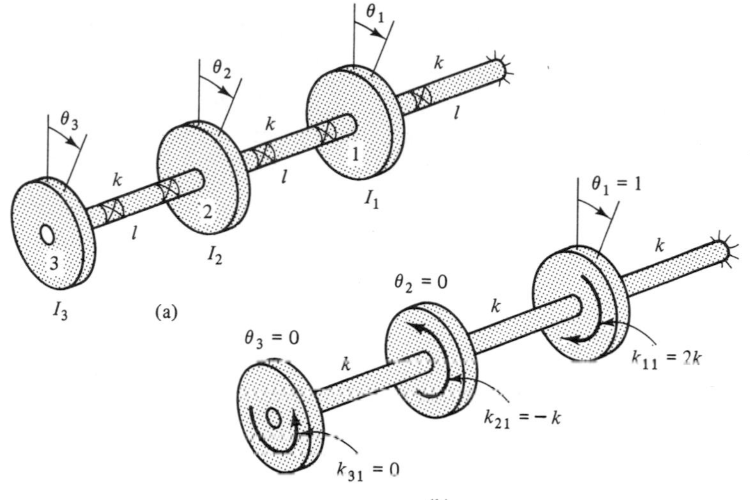

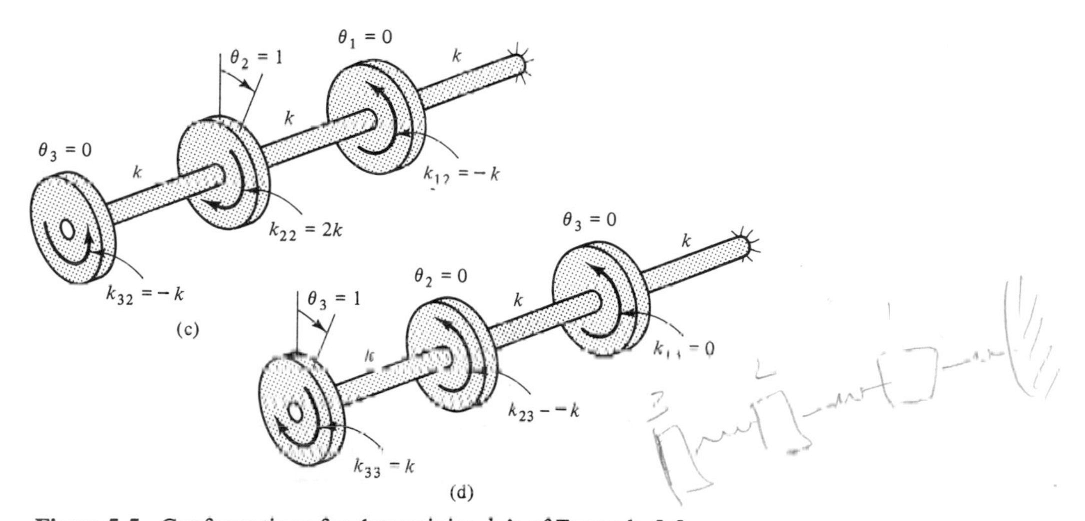

Figure 5-5 Configurations for determining k_{ij}'s of Example 5-5.

Using the angular displacements θ_1, θ_2, and θ_3 (the generalized coordinates) of the disks shown in Fig. 5-5a, and the definition of k_{ij}, determine (a) the stiffness matrix **K**, (b) the differential equations of motion of the undamped free vibration of the system in matrix form.

Solution. **a.** To determine the first column of **K**, disk 1 (station 1) is given a unit angular displacement ($\theta = 1$), with disks 2 and 3 (stations 2 and 3) held fixed, so that $\theta_2 = \theta_3 = 0$ as shown in Fig. 5-5b. The moment k_{11} required to rotate disk 1 through the angle $\theta = 1$ is $2k$, and the moment k_{21} required to keep disk 2 fixed is

$-k$. Since disk 2 is held fixed by the moment k_{21}, there is no tendency for disk 3 to rotate, so $k_{31} = 0$. Thus,

$$\left.\begin{aligned} k_{11} &= 2k \\ k_{21} &= -k \\ k_{31} &= 0 \end{aligned}\right\} \qquad \text{(first column of } \mathbf{K}\text{)}$$

Similarly, giving disk 2 a unit angular displacement ($\theta_2 = 1$), with disks 1 and 3 held fixed ($\theta_1 = \theta_3 = 0$), it can be seen from Fig. 5-5c that the moments required to maintain this configuration are

$$\left.\begin{aligned} k_{12} &= -k \\ k_{22} &= 2k \\ k_{32} &= -k \end{aligned}\right\} \qquad \text{(second column of } \mathbf{K}\text{)}$$

Finally, with $\theta_3 = 1$ and $\theta_1 = \theta_2 = 0$, the required moments for the configuration shown in Fig. 5-5d are

$$\left.\begin{aligned} k_{13} &= 0 \\ k_{23} &= -k \\ k_{33} &= k \end{aligned}\right\} \qquad \text{(third column of } \mathbf{K}\text{)}$$

b. The general form of the matrix equation expressing the equations of motion of the undamped free vibration is

$$\mathbf{M}\ddot{\mathbf{\Theta}} + \mathbf{K}\mathbf{\Theta} = \mathbf{0}$$

Since the mass matrix $\mathbf{M}$ consists simply of the mass moments of inertia on the diagonal, the matrix equation expressing the differential equations of motion of the undamped free vibration of the disk-and-shaft system is given by

$$\begin{bmatrix} I_1 & 0 & 0 \\ 0 & I_2 & 0 \\ 0 & 0 & I_3 \end{bmatrix} \begin{Bmatrix} \ddot{\theta}_1 \\ \ddot{\theta}_2 \\ \ddot{\theta}_3 \end{Bmatrix} + \begin{bmatrix} 2k & -k & 0 \\ -k & 2k & -k \\ 0 & -k & k \end{bmatrix} \begin{Bmatrix} \theta_1 \\ \theta_2 \\ \theta_3 \end{Bmatrix} = \mathbf{0} \tag{5-12}$$

The reader might note that this system with its three disks is analogous to the three-story building of Example 5-4.

The differential equations of motion can, of course, also be obtained by applying Newton's second law to the free-body diagrams of the three disks. However, it should be apparent at this point that it is easier to obtain the differential equations of motion by determining the stiffness coefficients than by sketching free bodies and using Newton's second law. This is particularly true for systems having many degrees of freedom. After some practice, and with the definition of the stiffness coefficients k_{ij} well in mind, the reader should be able to look at a system such as the one shown in Fig. 5-5a and write down the stiffness coefficients of the matrix $\mathbf{K}$ by inspection, without even finding it necessary to sketch the configurations shown in Fig. 5-5.

EXAMPLE 5-6

Figure 5-6 shows a six-degree-of-freedom system, which is a simplified model of a five-story building, with a spring-and-mass system (k_4 and m_4) attached to the fourth floor (the lumped mass m_3).

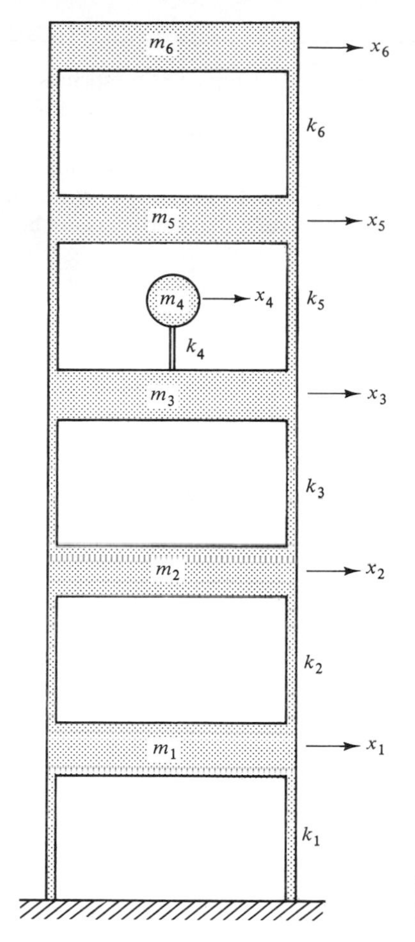

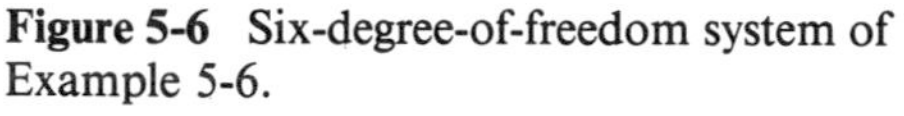

Figure 5-6 Six-degree-of-freedom system of Example 5-6.

We wish to determine the fourth column of the stiffness matrix **K** of the system by inspecting the system and using the definition of the stiffness coefficients k_{ij}.

Solution. In determining the fourth column of **K**, we visualize the horizontal forces required to maintain the configuration of $x_4 = 1$ with all the other masses (stations) held fixed, by inspecting Fig. 5-6 and mentally giving a unit linear displacement to m_4. Such an inspection shows that

$$\left.\begin{aligned} k_{14} &= 0 \\ k_{24} &= 0 \\ k_{34} &= -k_4 \\ k_{44} &= k_4 \\ k_{54} &= 0 \\ k_{64} &= 0 \end{aligned}\right\} \quad \text{(fourth column of } \mathbf{K}\text{)}$$

It is left as an exercise for the reader to determine the remaining columns of the stiffness matrix (see Prob. 5-3).

EXAMPLE 5-7

The system in Fig. 5-7a consists of two identical bars supported by springs as shown. Each bar has a length l, a mass m, and a centroidal mass moment of inertia of $\bar{I}$. The upper bar moves with general plane motion (a combination of translation and rotation), and the generalized coordinates q_1 and q_2 define the vertical translation and rotation of the bar, respectively. The lower bar rotates about the pin O located at its mass center, with q_3 defining this motion. The springs have free lengths such that both bars are horizontal when the system is in static equilibrium.

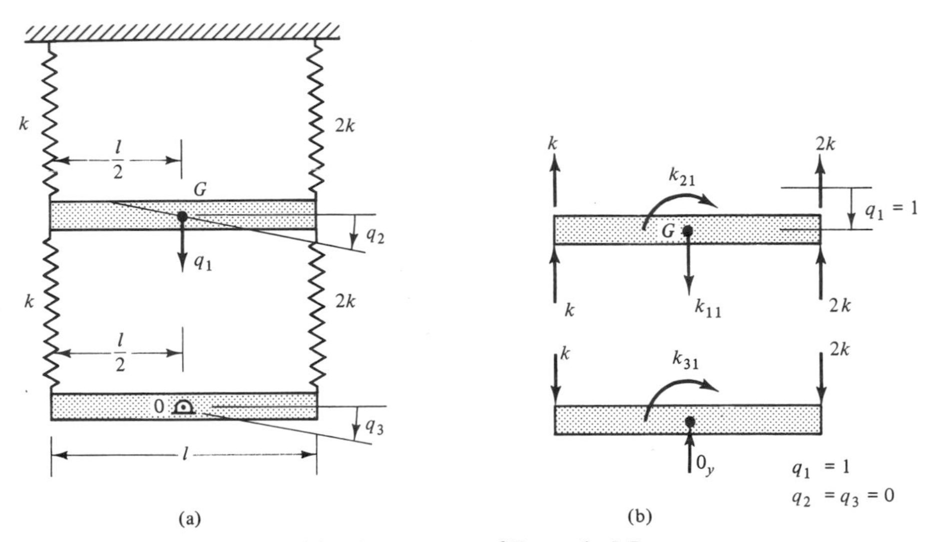

Figure 5-7 Three-degree-of-freedom system of Example 5-7.

The differential equations of motion of the undamped free vibration of this three-degree-of-freedom system can be expressed as

$$\begin{bmatrix} m & 0 & 0 \\ 0 & \bar{I} & 0 \\ 0 & 0 & \bar{I} \end{bmatrix} \begin{Bmatrix} \ddot{q}_1 \\ \ddot{q}_2 \\ \ddot{q}_3 \end{Bmatrix} + \mathbf{K} \begin{Bmatrix} q_1 \\ q_2 \\ q_3 \end{Bmatrix} = \mathbf{0}$$

It is desired to determine the first column of the **K** matrix.

Solution. In analyzing this system, it is not as easy to determine the senses of the stiffness coefficients by a simple inspection as it was in prior examples. This is because the upper mass has two motions and the spring forces are involved. A good procedure for a system such as this one is to sketch the free-body diagrams of the mass elements for each system configuration (there will be a set of free-body diagrams for each column of **K**). The forces and moments shown acting on the free bodies are forces and moments *per unit displacement*.

Before sketching these free-body diagrams, let us discuss the subscripting of the stiffness coefficients for this type of system. In systems considered earlier, there was always just one generalized coordinate associated with each mass. In the system of this example, however, while the lower bar requires just the one generalized coordinate q_3 associated with its rotation, the upper mass requires the two generalized coordinates q_1 and q_2 to define its translation and rotation, respectively (see Fig.

5-7a). Thus, the subscripts of the stiffness coefficients refer to *particular motions (particular coordinates) of particular masses.*

With the preceding in mind, and referring to the definition of stiffness coefficients on page 284, let us proceed to determining the first column of **K**. The necessary free bodies are shown in Fig. 5-7b. The force k_{11} is applied at the mass center of the upper bar and is of such magnitude and sense that it gives the bar a positive unit translation ($q_1 = 1$). The moment k_{21} shown is the moment required to keep the upper bar from rotating ($q_2 = 0$) when k_{11} is applied, and the moment k_{31} keeps the lower bar from rotating ($q_3 = 0$) when k_{11} is applied.

The reader should note that each of the k_{ij}'s has been given a sense the same as the sense assumed positive for the generalized coordinates as shown in Fig. 5-7. The spring forces have senses consistent with the forces that they exert on the two bars for the system configuration shown. The correct signs for the k_{ij}'s are obtained using this sign convention when the forces and moments are summed equal to zero. Performing these summations on the upper free body, we find that

$$\sum F_v = 0 = k_{11} - 6k$$

and

$$\sum M_G = 0 = k_{21} - 4k\frac{l}{2} + 2k\frac{l}{2}$$

from which

$$k_{11} = 6k$$

and

$$k_{21} = kl$$

Referring to the free body of the lower bar and summing moments about the pin O,

$$\sum M_0 = 0 = k_{31} + 2k\frac{l}{2} - k\frac{l}{2}$$

from which

$$k_{31} = \frac{-kl}{2}$$

The first column of **K** is then

$$\left\{\begin{matrix} 6k \\ kl \\ \dfrac{-kl}{2} \end{matrix}\right\} \qquad \text{(first column of } \mathbf{K}\text{)}$$

To obtain the second column of **K**, two more free bodies are sketched. A moment k_{22} is applied to the upper bar to give it a positive unit displacement ($q_2 = 1$), along with a force k_{12} to keep it from translating ($q_1 = 0$). A moment k_{32} is applied to the lower bar to keep it from rotating ($q_3 = 0$). The pin and spring forces consistent with this configuration are then added.

To obtain the third column of **K**, another set of free bodies is sketched. A moment k_{33} is applied to the lower bar to give it a positive unit angular displacement ($q_3 = 1$). A force k_{13} is applied to the upper bar to keep it from translating ($q_1 = 0$), along with a moment k_{23} to keep it from rotating ($q_2 = 0$). The pin and spring forces consistent with this configuration are then added.

It is left as an exercise for the reader to obtain the second and third columns of **K** by sketching the appropriate free bodies discussed in the preceding two paragraphs and by performing the appropriate summations of forces and moments (see Prob. 5-5).

EXAMPLE 5-8

Let us now consider a system in which the stiffness coefficients involve forces not only from the elastic elements of the system but from gravity as well. Two identical bars are suspended vertically from pins A and B as shown in Fig. 5-8a. Each rod has a length l, a mass m, and a mass moment of inertia I_0 about the pinned end.

Using the generalized coordinates θ_1 and θ_2 shown in the figure, we wish to determine the stiffness matrix **K** and the differential equations of motion for the undamped free vibration of the bars.

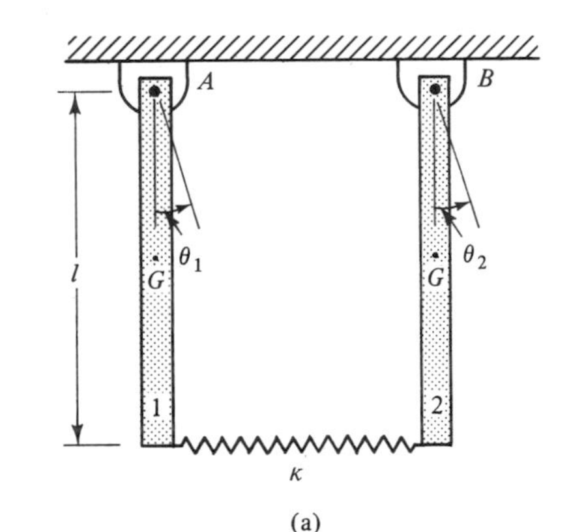

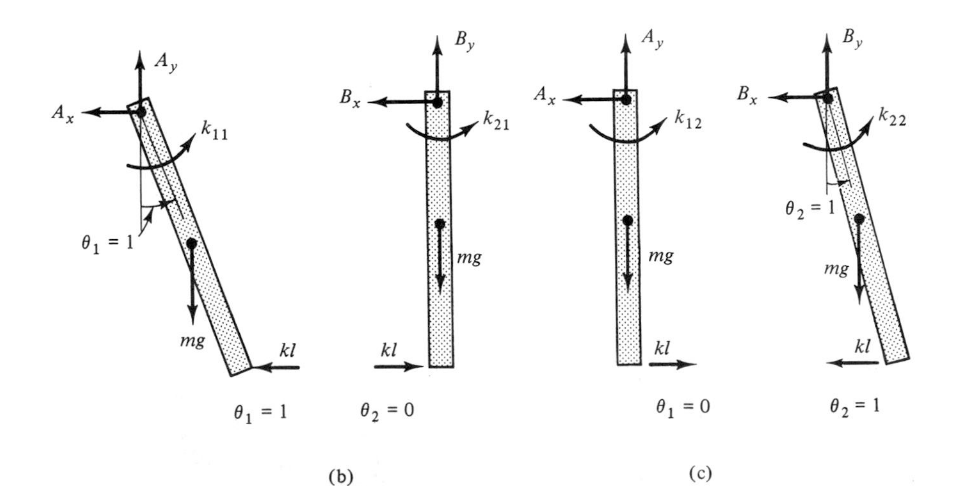

Figure 5-8 Two-degree-of-freedom system of Example 5-8.

Solution. As in Example 5-7, we use free-body diagrams of the bars. They are shown in Fig. 5-8b and c. We first note that in the vertical static-equilibrium position the force in the spring is zero. Therefore, as the bars are rotated from their vertical positions, the gravity forces of magnitude mg must be included on the free-body diagrams, since there is no initial elastic force in the spring (no initial strain energy in the system) to cancel the change in potential energy caused by the change in position.

The free bodies for determining the first column of **K** are shown in Fig. 5-8b. The moment k_{11} is the moment required to give the left bar a positive unit angular displacement ($\theta_1 = 1$), while k_{21} is the moment on the right bar required to keep it fixed ($\theta_2 = 0$). Note that both of these moments are given the same sense as the sense selected as positive for the generalized coordinates θ_1 and θ_2. With this convention, we minimize the chance of making sign errors for the stiffness coefficients. The spring force kl shown is based upon the assumption that θ_1 is small, so that

$$kl \sin \theta_1 \cong kl\theta_1 = kl$$

The sense of the spring force is consistent with the force that the spring exerts on the left bar when the bar is given a unit angular displacement.

Again assuming only small amplitudes of oscillation, we can use the free-body diagrams shown in Fig. 5-8b to obtain

$$\sum M_A = 0 = k_{11} - mg\frac{l}{2} - kl^2$$
$$\sum M_B = 0 = k_{21} + kl^2$$

from which

$$\left.\begin{aligned} k_{11} &= kl^2 + mg\frac{l}{2} \\ k_{21} &= -kl^2 \end{aligned}\right\} \qquad \text{(first column of } \mathbf{K}\text{)}$$

The two stiffness coefficients obtained above are moments per unit of angular displacement and usually have units of lb-in./rad or N · m/rad.

To obtain the second column of **K**, we use the free bodies shown in Fig. 5-8c. The moment k_{22} is the moment necessary to give the right bar a positive unit angular displacement ($\theta_2 = 1$), while the moment k_{12} is the moment on the left bar required to maintain it fixed ($\theta_1 = 0$). Note that both of these moments have been given the same sense as the sense selected as positive for the generalized coordinates, and that the spring force kl has a sense consistent with the force it exerts on the right bar due to the bar's positive unit angular displacement. From these free bodies we write that

$$\sum M_A = 0 = k_{12} + kl^2$$

and

$$\sum M_B = 0 = k_{22} - mg\frac{l}{2} - kl^2$$

from which

$$\left.\begin{aligned} k_{12} &= -kl^2 \\ k_{22} &= kl^2 + mg\frac{l}{2} \end{aligned}\right\} \qquad \text{(second column of } \mathbf{K}\text{)}$$

Thus, the differential equations of motion in matrix form for the undamped free vibration of the two vertical bars are given by

$$\begin{bmatrix} I_0 & 0 \\ 0 & I_0 \end{bmatrix} \begin{Bmatrix} \ddot{\theta}_1 \\ \ddot{\theta}_2 \end{Bmatrix} + \begin{bmatrix} \left(kl^2 + \dfrac{mgl}{2}\right) & -kl^2 \\ -kl^2 & \left(kl^2 + \dfrac{mgl}{2}\right) \end{bmatrix} \begin{Bmatrix} \theta_1 \\ \theta_2 \end{Bmatrix} = \mathbf{0}$$

Flexibility Coefficients

Designating the coordinates $q_1, q_2, \ldots, q_n$ of an n-degree-of-freedom system as the generalized coordinates that define the linear and/or angular displacements of the masses comprising the system, we can define the flexibility coefficient a_{ij} as the linear or angular displacement that occurs for a particular coordinate q_i when a *unit* force (or moment) is applied at the location of a coordinate q_j, with all other coordinates *free* to displace.

To determine the a_{ij}'s of a system using this definition, we first note that each mass in the system will have a generalized coordinate for each type of motion that it has, and that the subscripts of the a_{ij}'s refer to these generalized coordinates. With this in mind, we can state that when a positive *unit* force (or moment) is applied to a system mass with coordinate q_j, with all other coordinates free to displace, the a_{ij}'s for that particular q_j are the linear and/or angular displacements of the masses associated with the coordinates q_i ($i = 1, 2, \ldots, n$) and are determined in total by applying the definition of a_{ij} in turn for all q_j ($j = 1, 2, \ldots, n$).

To observe the general form of the differential equations of motion in terms of the flexibility coefficients, let us consider the three-story building shown in Fig. 5-9a. It is assumed that the distributed mass of the building can be represented effectively by the lumped masses at the different levels shown in the figure. It is further assumed that the girders of the structure are infinitely rigid in comparison with the stiffness of the columns, so that the general configuration of the columns is as shown in Fig. 5-9b, with x_1, x_2, and x_3 being the positive horizontal displacements of the masses.

In using flexibility coefficients to obtain the differential equations of motion of the system, let us use the concept of dynamic equilibrium. We first select to the right as positive for the displacement, velocity, and acceleration of each generalized coordinate. In sketching

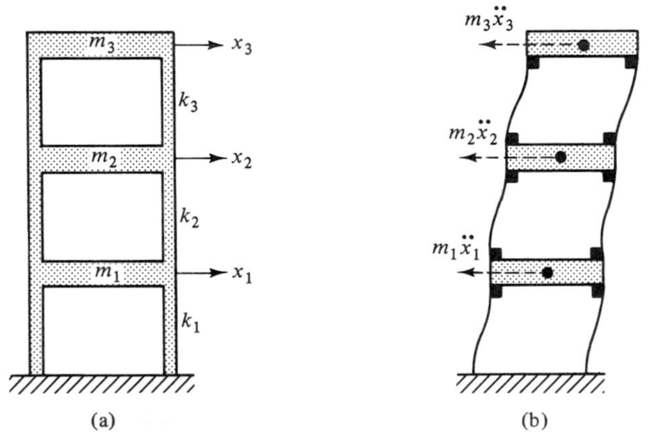

Figure 5-9 Model of three-story building.

the free bodies we assume that the displacement, velocity, and acceleration of each mass are positive. We then add the inertia effects shown by the dashed vectors in Fig. 5-9b with their senses opposite to the assumed positive sense of the acceleration of each mass.

These inertia effects are equal in magnitude to the forces that the various *masses exert on the columns* and can thus be treated as if they were *forces* acting on the structure. Noting that the total displacement of a particular mass is equal to the sum of the displacements caused by each inertia effect (force) acting on the columns (the elastic elements), and remembering the definition of a flexibility coefficient, we can write that

$$\left.\begin{aligned} x_1 &= -a_{11}m_1\ddot{x}_1 - a_{12}m_2\ddot{x}_2 - a_{13}m_3\ddot{x}_3 \\ x_2 &= -a_{21}m_1\ddot{x}_1 - a_{22}m_2\ddot{x}_2 - a_{23}m_3\ddot{x}_3 \\ x_3 &= -a_{31}m_1\ddot{x}_1 - a_{32}m_3\ddot{x}_3 - a_{33}m_3\ddot{x}_3 \end{aligned}\right\} \tag{5-13}$$

These are the differential equations of motion for the undamped free vibration of the three-story building in terms of the flexibility coefficients. The reader is reminded that the actual acceleration of the structure is always toward the static-equilibrium position for any configuration. Thus, although the accelerations were assumed positive to the right for the configuration shown in Fig. 5-9b, the actual acceleration of each mass for this configuration is to the left, which is negative. It follows then that the insertion of negative values for $\ddot{x}_1$, $\ddot{x}_2$, and $\ddot{x}_3$ in Eq. 5-13 will yield the positive displacements shown in the figure.

Equation 5-13 can be put in matrix form as

$$\begin{Bmatrix} x_1 \\ x_2 \\ x_3 \end{Bmatrix} + \begin{bmatrix} a_{11} & a_{12} & a_{13} \\ a_{21} & a_{22} & a_{23} \\ a_{31} & a_{32} & a_{33} \end{bmatrix}\begin{bmatrix} m_1 & 0 & 0 \\ 0 & m_2 & 0 \\ 0 & 0 & m_3 \end{bmatrix}\begin{Bmatrix} \ddot{x}_1 \\ \ddot{x}_2 \\ \ddot{x}_3 \end{Bmatrix} = \mathbf{0} \tag{5-14}$$

and the general form of the matrix equation for the undamped free vibration of an n-degree-of-freedom system in terms of its flexibility coefficients is

$$\begin{Bmatrix} x_1 \\ x_2 \\ \vdots \\ x_n \end{Bmatrix} + \begin{bmatrix} a_{11} & a_{12} & \cdots & a_{1n} \\ a_{21} & a_{22} & \cdots & a_{2n} \\ \cdot & \cdot & \cdots & \cdot \\ a_{n1} & a_{n2} & \cdots & a_{nn} \end{bmatrix}\begin{bmatrix} m_{11} & m_{12} & \cdots & m_{1n} \\ m_{21} & m_{22} & \cdots & m_{2n} \\ \cdot & \cdot & \cdots & \cdot \\ m_{n1} & m_{n2} & \cdots & m_{nn} \end{bmatrix}\begin{Bmatrix} \ddot{x}_1 \\ \ddot{x}_2 \\ \vdots \\ \ddot{x}_n \end{Bmatrix} = \mathbf{0}$$

or more simply

$$\mathbf{X} + \mathbf{A}\mathbf{M}\ddot{\mathbf{X}} = \mathbf{0} \tag{5-15}$$

In the latter form, **A** and **M** are the flexibility and mass matrices, respectively. As mentioned previously, the mass matrix **M** is a diagonal matrix, unless the coordinates selected lead to dynamic coupling (see Sec. 5-4). When **M** is a diagonal matrix, Eq. 5-14 shows that $\mathbf{M}\ddot{\mathbf{X}}$ is simply a column matrix, so that

$$\begin{Bmatrix} x_1 \\ x_2 \\ \vdots \\ x_n \end{Bmatrix} + \mathbf{A}\begin{Bmatrix} m_1\ddot{x}_1 \\ m_2\ddot{x}_2 \\ \vdots \\ m_n x_n \end{Bmatrix} = \mathbf{0} \tag{5-16}$$

If Eq. 5-15 is premultiplied by $\mathbf{A}^{-1}$ (the inverse of **A**), we obtain

$$\mathbf{M}\ddot{\mathbf{X}} + \mathbf{A}^{-1}\mathbf{X} = \mathbf{0} \tag{5-17}$$

Comparing Eq. 5-17 with Eq. 5-9, it can be seen that

$$\mathbf{A}^{-1} = \mathbf{K} \tag{5-18}$$

Thus, the inverse of the flexibility matrix **A** is the stiffness matrix **K**. Conversely,

$$\mathbf{K}^{-1} = \mathbf{A} \tag{5-19}$$

The reciprocal theorem can be used to show that **A** is a symmetric matrix by proving that $a_{ij} = a_{ji}$.[2] It then follows from Eq. 5-18 that if **A** is a symmetric matrix, the stiffness matrix **K** is also symmetric since the inverse of a symmetric matrix is also symmetric.

EXAMPLE 5-9

We wish to determine the flexibility matrix **A** and the differential equations of motion of the three-story building shown in Fig. 5-9 by using the definition of flexibility coefficients. The equivalent spring constants for the columns supporting the first, second, and third masses are $3k$, $2k$, and k, respectively, as shown in Fig. 5-10, which shows the three respective configurations associated with the unit forces F_1, F_2, and F_3.

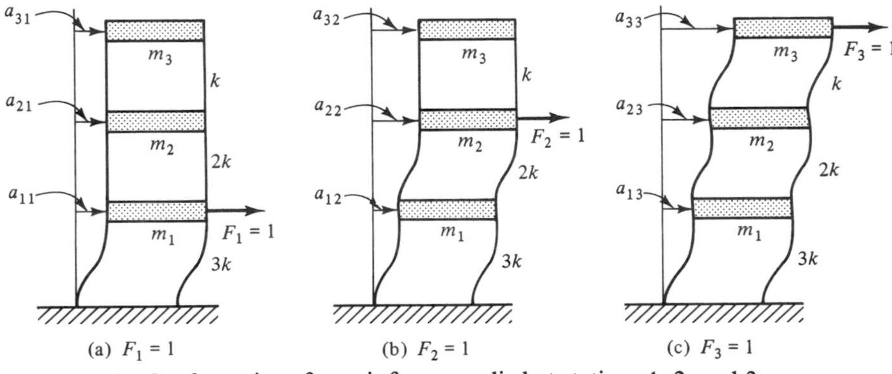

Figure 5-10 Configurations for unit forces applied at stations 1, 2, and 3.

Solution. To determine the first column of **A**, a unit force $F_1 = 1$ is applied at m_1 (station 1), which deforms the structure as shown in Fig. 5-10a. Assuming that all displacements are positive to the right, we obtain from elementary statics and inspection of the configuration that

$$F_1 = (3k)a_{11}$$

so that

$$a_{11} = \frac{1}{3k}$$

The coefficient a_{11} is positive since m_1 has a positive displacement due to the application of F_1. Since the horizontal forces on the columns supporting the upper two masses are zero, it should be apparent from Fig. 5-10a that $a_{21} = a_{31} = a_{11}$. Thus, the first column of **A** is

[2] See almost any text on framed structures.

$$\left.\begin{aligned} a_{11} &= \frac{1}{3k} \\ a_{21} &= \frac{1}{3k} \\ a_{31} &= \frac{1}{3k} \end{aligned}\right\} \quad \text{(first column of } \mathbf{A}\text{)}$$

A positive unit force $F_2 = 1$ is next applied to m_2 (station 2), as shown in Fig. 5-10b, to determine the second column of **A**. Again from statics and an inspection of the building configuration, it can be seen that the columns in each of the first two stories are subjected to the unit force F_2. Thus,

$$\left.\begin{aligned} a_{12} &= \frac{1}{3k} \\ a_{22} &= \frac{1}{3k} + \frac{1}{2k} = \frac{5}{6k} \\ a_{32} &= a_{22} = \frac{5}{6k} \end{aligned}\right\} \quad \text{(second column of } \mathbf{A}\text{)}$$

Similarly, a unit force F_3 is applied to m_3, resulting in the configuration shown in Fig. 5-10c. Once again considering static equilibrium, it can be seen that the columns supporting all three masses are subjected to the force F_3, so that

$$\left.\begin{aligned} a_{13} &= \frac{1}{3k} \\ a_{23} &= \frac{1}{3k} + \frac{1}{2k} = \frac{5}{6k} \\ a_{33} &= \frac{1}{3k} + \frac{1}{2k} + \frac{1}{k} = \frac{11}{6k} \end{aligned}\right\} \quad \text{(third column of } \mathbf{A}\text{)}$$

Having determined the three columns of **A**, the differential equations of motion for the undamped free vibration of the building in terms of the flexibility matrix **A** are

$$\begin{Bmatrix} x_1 \\ x_2 \\ x_3 \end{Bmatrix} + \frac{1}{6k}\begin{bmatrix} 2 & 2 & 2 \\ 2 & 5 & 5 \\ 2 & 5 & 11 \end{bmatrix}\begin{Bmatrix} m_1\ddot{x}_1 \\ m_2\ddot{x}_2 \\ m_3\ddot{x}_3 \end{Bmatrix} = \mathbf{0}$$

EXAMPLE 5-10

A disk of mass m and centroidal moment of inertia $\bar{I}$ about a z axis is attached to the end of a cantilever beam of length l, stiffness factor EI, and a mass that is small compared with that of the disk. The disk has general plane motion (it both translates and rotates) as shown in Fig. 5-11b, so that it has the two degrees of freedom that are defined by the generalized coordinates $q_1 = y$ and $q_2 = \theta$ shown in the figure.

With only the mass of the disk to consider (there is only one station), the differential equations of motion for the undamped free vibration of the system in terms of the flexibility coefficients are

$$\begin{Bmatrix} y \\ \theta \end{Bmatrix} + \begin{bmatrix} a_{11} & a_{12} \\ a_{21} & a_{22} \end{bmatrix}\begin{Bmatrix} m\ddot{y} \\ \bar{I}\ddot{\theta} \end{Bmatrix} = \mathbf{0}$$

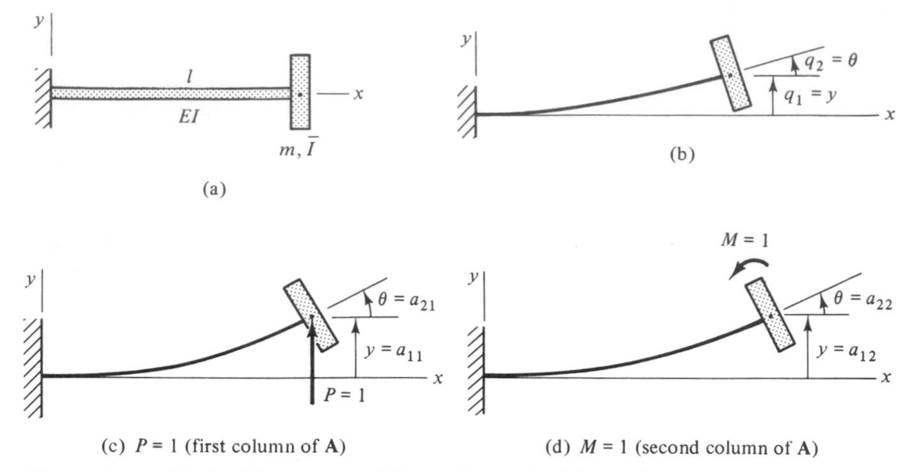

Figure 5-11 Shaft-disk system of Example 5-10 with configurations to determine flexibility coefficients.

In terms of the stiffness coefficients they are

$$\begin{Bmatrix} m\ddot{y} \\ \bar{I}\ddot{\theta} \end{Bmatrix} + \begin{bmatrix} k_{11} & k_{12} \\ k_{21} & k_{22} \end{bmatrix} \begin{Bmatrix} y \\ \theta \end{Bmatrix} = \mathbf{0}$$

Remembering that the mass of the beam is negligible, (a) determine the flexibility matrix **A**, (b) determine the stiffness matrix **K**, and (c) using the results of parts a and b, and considering that $\mathbf{A}^{-1} = \mathbf{K}$, show that

$$\mathbf{AK} = \mathbf{I} = \begin{bmatrix} 1 & 0 \\ 0 & 1 \end{bmatrix}$$

in which **I** is the *identity* matrix.

Solution. **a.** The system of this example consists of a single mass having two motions, a linear displacement and an angular displacement as shown in Fig. 5-11b. The generalized coordinates associated with these displacements are $q_1 = y$ and $q_2 = \theta$, respectively, and their positive senses are assumed as up and counterclockwise, respectively.

Remembering that flexibility coefficients are linear or angular displacement per unit force or moment by definition, we apply a unit force ($P = 1$) as shown in Fig. 5-11c, and determine from elementary beam theory that the deflection y and rotation θ of the disk due to the application of this unit force are

$$\left.\begin{aligned} y = a_{11} &= \frac{l^3}{3EI} \\ \theta = a_{21} &= \frac{l^2}{2EI} \end{aligned}\right\} \quad \text{(first column of } \mathbf{A}\text{)}$$

in which $\theta = dy/dx$ is the slope of the disk end of the beam.[3]

[3] The deflection and slope can be determined by various methods discussed in elementary texts on mechanics of materials.

We next apply a positive unit moment ($M = 1$) as shown in Fig. 5-11d, and determine that the translation and rotation of the disk caused by the application of this moment are

$$\left.\begin{aligned} y = a_{12} &= \frac{l^2}{2EI} \\ \text{and} \qquad \theta = a_{22} &= \frac{l}{EI} \end{aligned}\right\} \qquad \text{(second comumn of } \mathbf{A}\text{)}$$

Thus, the flexibility matrix is determined as

$$\mathbf{A} = \frac{l}{6EI}\begin{bmatrix} 2l^2 & 3l \\ 3l & 6 \end{bmatrix} \tag{5-20}$$

b. Referring to Fig. 5-12a, we see that the force k_{11} is the force necessary to give the disk a positive unit translation, while the moment k_{21} is the moment required to keep the disk from rotating when the force k_{11} is applied. Note that k_{11} is applied with the same sense as that assumed positive for y, while k_{21} is applied with the sense necessary to maintain $\theta = 0$. From elementary beam theory

$$\left.\begin{aligned} P = k_{11} &= \frac{12EI}{l^3} \\ M = k_{21} &= \frac{-6EI}{l^2} \end{aligned}\right\} \qquad \text{(first column of } \mathbf{K}\text{)}$$

In determining the second column of **K**, we see from Fig. 5-12b that the moment k_{22} applied to the disk is the moment required to give it a positive unit rotation ($\theta = 1$), while the force k_{12} is the force necessary to keep the disk from translating when k_{22} is applied. Note that the moment k_{22} has the same sense as that assumed positive for θ, while the force k_{12} has the sense necessary to maintain $y = 0$.

Again using elementary beam theory, we find that

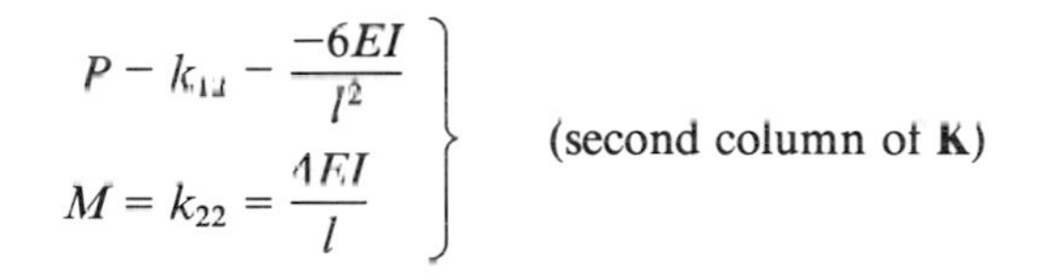

$$\left.\begin{aligned} P = k_{12} &= \frac{-6EI}{l^2} \\ M = k_{22} &= \frac{4EI}{l} \end{aligned}\right\} \qquad \text{(second column of } \mathbf{K}\text{)}$$

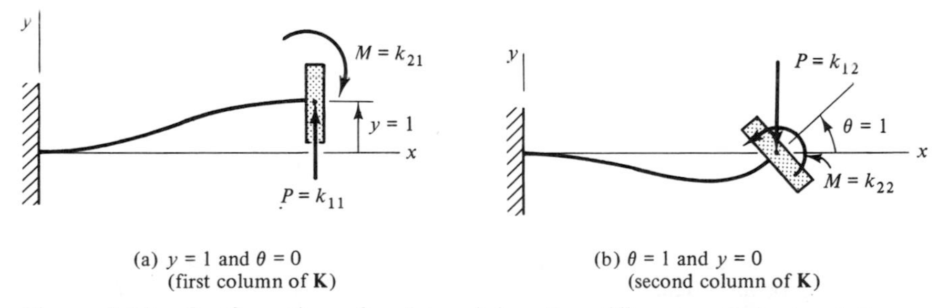

(a) $y = 1$ and $\theta = 0$ (first column of **K**)

(b) $\theta = 1$ and $y = 0$ (second column of **K**)

Figure 5-12 Configurations for determining the stiffness coefficients of Example 5-10.

The stiffness matrix is thus

$$\mathbf{K} = \frac{6EI}{l}\begin{bmatrix} 2/l^2 & -1/l \\ -1/l & 2/3 \end{bmatrix} \tag{5-21}$$

c. To show that **AK** = **I**, we premultiply Eq. 5-18 by **A** to obtain

$$\mathbf{AA}^{-1} = \mathbf{I} = \mathbf{AK}$$

Then, postmultiplying the matrix of Eq. 5-20 by the matrix of Eq. 5-21, we find that

$$\mathbf{AK} = \left(\frac{l}{6EI}\right)\left(\frac{6EI}{l}\right)\begin{bmatrix} 2l^2 & 3l \\ 3l & 6 \end{bmatrix}\begin{bmatrix} 2/l^2 & -1/l \\ -1/l & 2/3 \end{bmatrix} = \begin{bmatrix} 1 & 0 \\ 0 & 1 \end{bmatrix} \tag{5-22}$$

Choice of Stiffness or Flexibility Matrix

In some problems it is easier to determine the flexibility coefficients a_{ij} than the stiffness coefficients k_{ij}, while in others the reverse is true. In modeling beams as lumped-mass systems it is generally easier to determine the flexibility coefficients than the stiffness coefficients. For example, the three-lumped-mass model of an overhanging beam shown in Fig. 5-13a becomes statically indeterminate to the second degree when stiffness coefficients

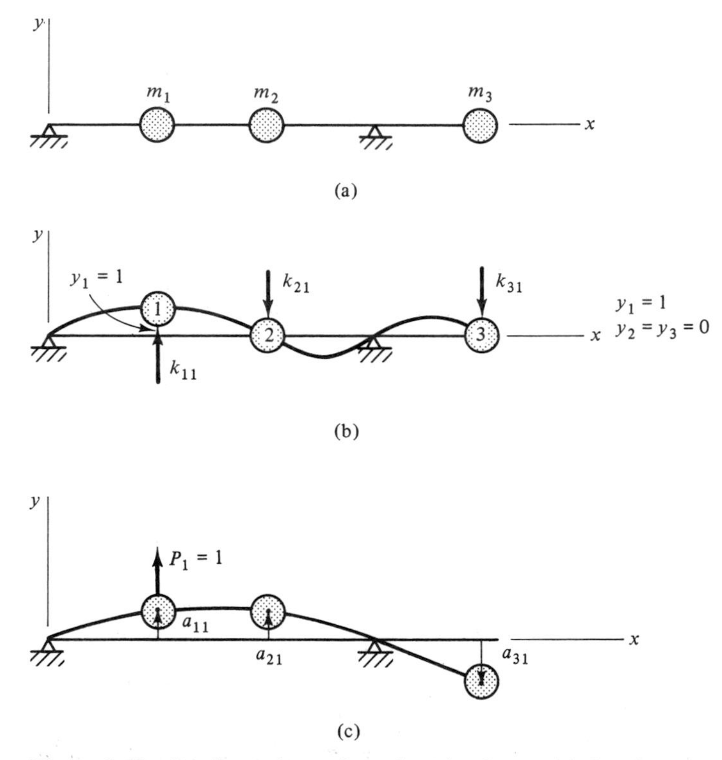

Figure 5-13 Configurations of overhanging beam. (a) Overhanging beam. (b) Statically indeterminate configuration for determining first column of **K** matrix. (c) Statically determinate configuration for determining first column of **A** matrix.

are considered. This can be seen by observing the configuration in Fig. 5-13b for determining the first column of **K** in which k_{21} and k_{31} act the same as unknown pin reactions in keeping m_2 and m_3 fixed when k_{11} is applied, as dictated by the definition of a stiffness coefficient ($y_1 = 1$, $y_2 = y_3 = 0$). On the other hand, it is relatively easy to determine the flexibility coefficients since the beam configuration shown in Fig. 5-13c for determining the first column of **A** is for a statically determinate beam.

In the case of systems such as those shown in Fig. 5-14, the stiffness coefficients can be determined by simple inspection. For example, if $F_1 = k_{11}$ is applied to station 1 (mass 1), so that $x_1 = 1$, and m_2 and m_3 are held fixed by k_{21} and k_{31}, respectively, a simple inspection, and a consideration of statics, shows that

$$
\begin{aligned}
k_{11} &= k_1 + k_2 \\
k_{21} &= -k_2 \\
k_{31} &= 0
\end{aligned}
$$

On the other hand, if a unit force ($F_1 = 1$) is applied at station 1, with m_1, m_2, and m_3 free to displace, it is found necessary to write three equations of equilibrium for m_1, m_2, and m_3, which must be solved simultaneously to obtain the flexibility coefficients a_{11}, a_{21}, and a_{31}. It is left as an exercise for the reader to determine the flexibility matrix **A** by this procedure, with $k_1 = k_2 = k_3 = k_4 = k$ (see Prob. 5-12).

When systems are unconstrained (free-free) as illustrated by the two systems shown in Fig. 5-15, the flexibility coefficients are infinite. For example, if a unit force ($F_1 = 1$) is applied to m_1 of the system in Fig. 5-15a, the displacements (the a_{ij}'s) become $a_{11} = a_{21} = a_{31} = \infty$, since there are no other external forces to resist this statically applied force. These free-free systems are referred to as *semidefinite* systems, and their equations of motion must be determined using stiffness coefficients rather than flexibility coefficients. As we shall see later, one of the natural frequencies of such systems will be zero, corresponding to the zero root of the frequency equation when the system moves as a rigid body. The

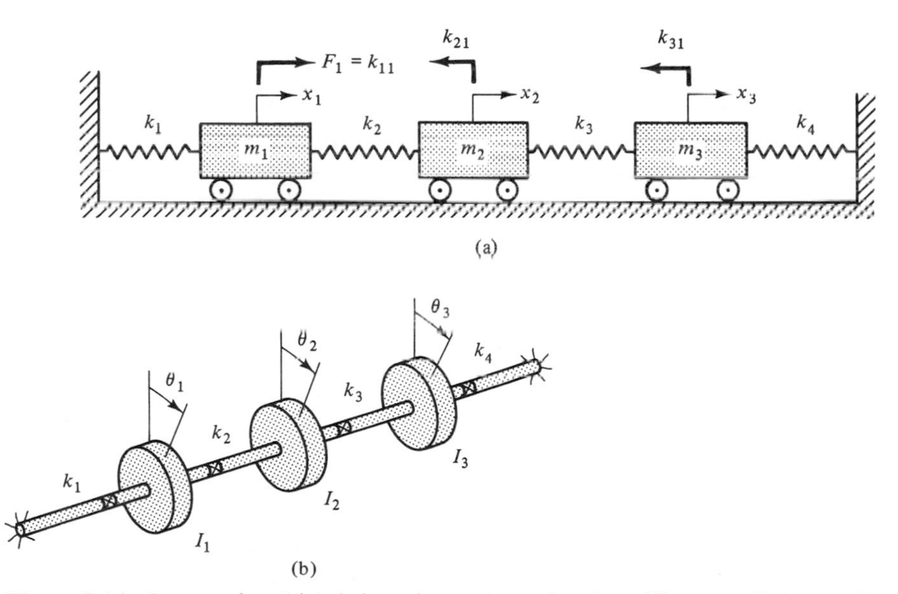

Figure 5-14 Systems in which it is easier to determine the stiffness coefficients. (a) Rectilinear motion. (b) Rotational motion.

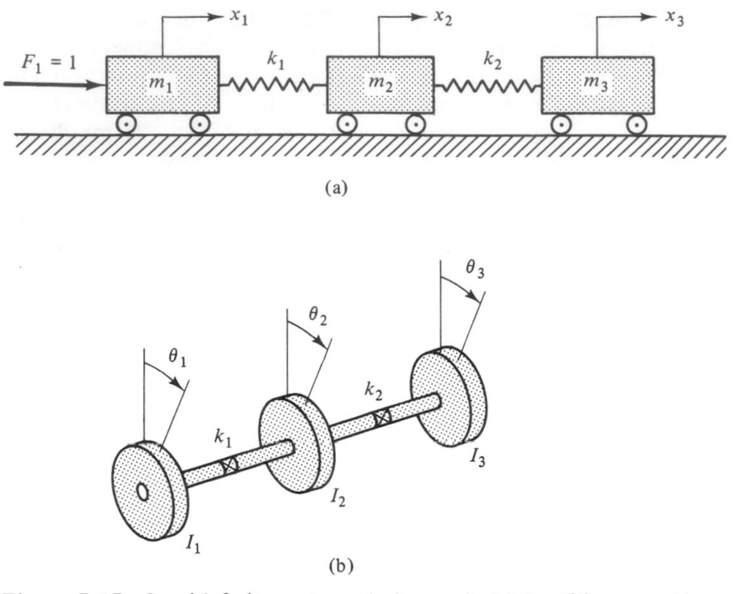

Figure 5-15 Semidefinite systems (a_{ij}'s = ∞). (a) Rectilinear motion. (b) Rotational motion.

elements of the *inverse* of **K** are all infinite for semidefinite systems, since they correspond to the elements of the flexibility matrix **A**.

As we shall see later, the eigenvalues (natural frequencies) and corresponding eigenvectors (mode shapes) can be determined from equations written in terms of either stiffness or flexibility matrices, with the one used depending upon the information desired and which matrix is the more easily determined. For example, the sweeping method used in Sec. 5-10 to determine the lower eigenvalues and eigenvectors works best when using equations written in terms of the flexibility matrix **A.** However, if it were easier to obtain the stiffness matrix **K** in such an instance, the flexibility matrix could still be used in the sweeping method by using the FORTRAN computer program that follows to invert the **K** matrix to obtain **A**.

Program for Inverting Matrices (INVERT.FOR)[4]

The program presented here uses a Gauss-Jordan algorithm, which overlays the original matrix in memory with the final inverted matrix. This reduces the storage required in memory, which makes it feasible to invert fairly large matrices by means of microcomputers, even when used in conjunction with other programs. Partial pivoting is used in the program to circumvent division by zero and to improve the accuracy obtained.

The variable names used in the program represent the following quantities:

[4] M. L. James, G. M. Smith, and J. C. Wolford, *Applied Numerical Methods,* 3d ed., pp. 190–198, Harper & Row, New York, 1985.

FORTRAN name	Quantity
A(I,J)	Elements of the original and inverted matrices
N	Order of matrix to be inverted
K	The pivot row number
JJ	A variable that takes on the value of the row having the largest pivot element
BIG	Takes on values of the elements in the *k*th column (column with same number as pivot row) which are possible pivot elements, eventually taking on the value of the pivot element used (largest element in absolute value in the pivot column that can serve as a pivot element)
TEMP	Temporary name used for the elements of the row selected to become the pivot row before interchange is made
INTER	An *N* by 2 array that indicates which columns of the final matrix must be exchanged, and in what order this must be done, in order to compensate for the fact that the original matrix was not augmented by an identity matrix and, as a result, no row interchange of these missing elements could take place. Starting from the bottom row of this matrix, the two elements of the row indicate which columns of the final matrix must be exchanged. If the two elements have the same value, no column interchange should be made corresponding to that row. The order in which the interchange of the final matrix must be done is controlled by progressing up the array INTER from the bottom row to the top.

The FORTRAN program is as follows:

```
C   INVERT.FOR
C   MATRIX INVERSION USING GAUSS-JORDAN REDUCTION. INVERTED
C   MATRIX OVERLAYS ORIGINAL MATRIX IN MEMORY. PARTIAL PIVOTING
C   IS USED IN THIS PROGRAM TO AVOID DIVISION BY ZERO AND TO
C   IMPROVE ACCURACY.  THE ONLY INPUT REQUIRED TO RUN THE
C   PROGRAM IS THE ORDER N AND VALUES OF THE ELEMENTS OF THE
C   MATRIX TO BE INVERTED.
C   DEVICE * IN READ AND WRITE STATEMENTS IS THE CONSOLE.
C   DEVICE 2 IN WRITE STATEMENTS IS THE PRINTER.
C   * IN THE PLACE OF A FORMAT STATEMENT NUMBER MEANS FREE FORMAT.
      DIMENSION A(15,15),INTER(15,2)
      OPEN (2,FILE='PRN')
      WRITE(*,1)
1     FORMAT(' ','READ IN ORDER OF MATRIX (I2)',/)
      READ(*,2)N
2     FORMAT(I2)
      WRITE(*,20)
20    FORMAT(' ','READ IN MATRIX TO BE INVERTED',/)
      DO 3 I = 1,N
      DO 3 J = 1,N
      WRITE(*,40)I,J
      READ(*,50)A(I,J)
3     CONTINUE
40    FORMAT(' ','ROW = ',I2,3X,'COLUMN = ',I2,'  (E14.7)',/)
50    FORMAT(E14.7)
C
```

```
C  WRITE OUT MATRIX TO BE INVERTED
C
      WRITE(2,60)
60    FORMAT(' ',5X,'MATRIX TO BE INVERTED IS AS FOLLOWS',/)
65    FORMAT(' ',5X,'A(',I2,',',I2,') = ',E14.7)
      DO 70 I = 1,N
      DO 70 J = 1,N
      WRITE(2,65)I,J,A(I,J)
70    CONTINUE
C
C  CYCLE K, THE PIVOT ROW NUMBER, FROM 1 TO N
      DO 12 K = 1,N
      JJ = K
      IF(K .EQ. N)GO TO 6
      KP1 = K + 1
      BIG = ABS(A(K,K))
C
C  SEARCH FOR LARGEST PIVOT ELEMENT
C
      DO 5 I = KP1,N
      AB = ABS(A(I,K))
      IF(BIG-AB)4,5,5
4     BIG = AB
      JJ = I
5     CONTINUE
C
C  MAKE DECISION ON NECESSITY OF ROW INTERCHANGE AND
C  STORE THE NUMBERS OF THE TWO ROWS INTERCHANGED DURING KTH
C  REDUCTION. IF NO INTERCHANGE, THE NUMBERS STORED BOTH EQUAL K
C
6     INTER(K,1) = K
      INTER(K,2) = JJ
      IF(JJ-K)7,9,7
C
C  DO ROW INTERCHANGE
C
7     DO 8 J = 1,N
      TEMP = A(JJ,J)
      A(JJ,J) = A(K,J)
8     A(K,J) = TEMP
C
C  CALCULATE ELEMENTS OF REDUCED MATRIX
C  FIRST CALCULATE ELEMENTS OF PIVOT ROW
C
9     DO 10 J = 1,N
      IF(J .EQ. K)GO TO 10
      A(K,J) = A(K,J)/A(K,K)
10    CONTINUE
C
C  CALCULATE ELEMENT REPLACING PIVOT ELEMENT
```

```
C
      A(K,K) = 1./A(K,K)
C
C  CALCULATE NEW ELEMENTS NOT IN PIVOT ROW OR COLUMN
      DO 11 I = 1,N
      IF(I .EQ. K)GO TO 11
      DO 110 J = 1,N
      IF(J .EQ. K)GO TO 110
      A(I,J) = A(I,J) - A(K,J)*A(I,K)
110   CONTINUE
11    CONTINUE
C
C  CALCULATE REPLACEMENT ELEMENTS FOR PIVOT COLUMN-EXCEPT PIVOT
C  ELEMENT
C
      DO 120 I = 1,N
      IF(I .EQ. K)GO TO 120
      A(I,K) = -A(I,K)*A(K,K)
120   CONTINUE
12    CONTINUE
C
C  REARRANGE COLUMNS OF FINAL MATRIX OBTAINED
C
      DO 13 L = 1,N
      K = N-L+1
      KROW = INTER(K,1)
      IROW = INTER(K,2)
      IF(KROW .EQ. IROW)GO TO 13
      DO 130 I = 1,N
      TEMP = A(I,IROW)
      A(I,IROW) = A(I,KROW)
      A(I,KROW) = TEMP
130   CONTINUE
13    CONTINUE
C
C  WRITE OUT INVERTED MATRIX
C
      WRITE(2,14)
14    FORMAT('0',5X,'INVERTED MATRIX'/)
15    FORMAT(' ',5X,'A(',I2,',',I2,') = ',E14.7)
      DO 16 I = 1,N
      DO 16 J = 1,N
      WRITE(2,15)I,J,A(I,J)
16    CONTINUE
      STOP
      END
```

Damping Coefficients

Damping coefficients c_{ij} must be determined when system damping is included in terms of a damping matrix **C** as shown in Eq. 5-8. If a particular coordinate q_j is given a positive linear or angular velocity ($\dot{q}_j = 1$) with all other coordinates held fixed, the c_{ij}'s ($i = 1, 2, \ldots, n$) are the forces or moments required to maintain the specified velocity configuration.

Recalling the definition of a stiffness coefficient, it can be seen that the damping coefficient is similar, except that the c_{ij}'s are associated with forces and moments and corresponding unit velocities, while the k_{ij}'s are associated with forces and moments and corresponding unit displacements. As a result, the procedure for obtaining the damping matrix **C** is similar to the procedure for obtaining the stiffness matrix **K**, which we discussed earlier. Like the stiffness matrix, the damping matrix is also symmetric. That is, $c_{ij} = c_{ji}$. The determination of a damping matrix is illustrated in Example 5.11.

EXAMPLE 5-11

It is desired to determine the damping matrix **C** for the three-story building shown in Fig. 5-16 by considering the definition of c_{ij} while observing the velocity configurations shown in the figure. The damping coefficients are c_1, c_2, and c_3, and are damping forces per unit of velocity (lb · s/in. or N · s/m).

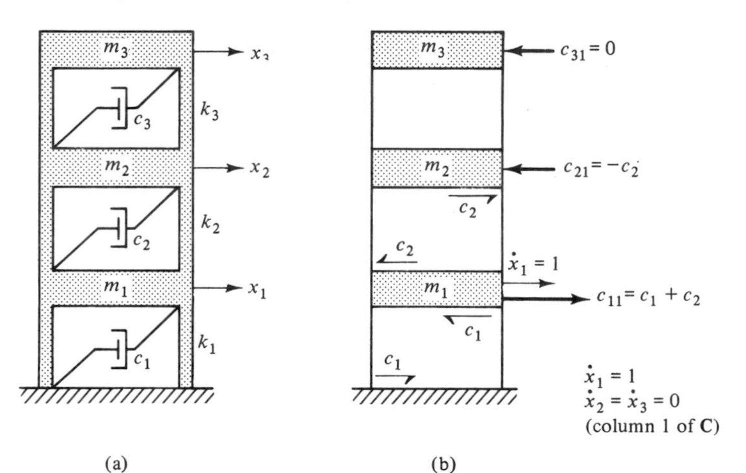

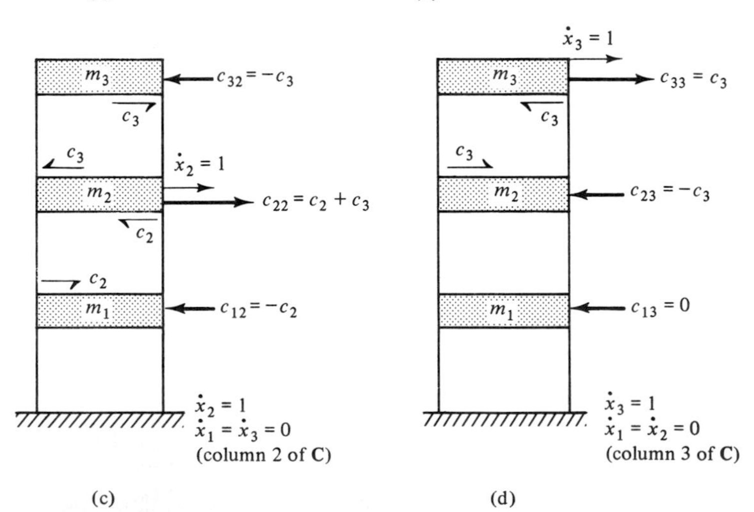

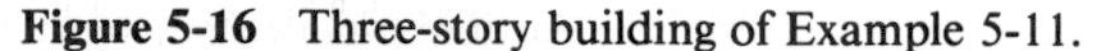

Figure 5-16 Three-story building of Example 5-11.

Solution. To determine the first column of **C**, the mass m_1 (station 1) is given a positive unit velocity ($\dot{x}_1 = 1$), with m_2 and m_3 held fixed so that $\dot{x}_2 = \dot{x}_3 = 0$ as shown in Fig. 5-16b. Recalling that a damping force always opposes velocity, and with the displacement and velocity of the building assumed as positive to the right, the viscous-damping forces per unit of velocity acting on m_1 and m_2 when $\dot{x}_1 = 1$ are c_1 and c_2, as shown in Fig. 5-16b. Inspection of this figure shows the required forces to be

$$\left.\begin{aligned} c_{11} &= c_1 + c_2 \\ c_{21} &= -c_2 \\ c_{31} &= 0 \end{aligned}\right\} \qquad \text{(first column of } \mathbf{C}\text{)}$$

The coefficient c_{21} is negative since it acts with a sense opposite to that assumed positive for x_2.

To determine the second column of **C**, the mass m_2 is given a unit velocity ($\dot{x}_2 = 1$), with m_1 and m_3 held motionless so that $\dot{x}_1 = \dot{x}_3 = 0$. The damping forces acting on the three masses for this case are shown in Fig. 5-16c. Inspection of this figure reveals that

$$\left.\begin{aligned} c_{12} &= -c_2 \\ c_{22} &= c_2 + c_3 \\ c_{32} &= -c_3 \end{aligned}\right\} \qquad \text{(second column of } \mathbf{C}\text{)}$$

In a similar manner, the third column of **C** is obtained by observing the forces required to give the mass m_3 a unit velocity ($\dot{x}_3 = 1$) with $\dot{x}_1 = \dot{x}_2 = 0$, as shown in Fig. 5-16d. From this figure

$$\left.\begin{aligned} c_{13} &= 0 \\ c_{23} &= -c_3 \\ c_{33} &= c_3 \end{aligned}\right\} \qquad \text{(third column of } \mathbf{C}\text{)}$$

Using the damping coefficients found above, the damping matrix is

$$\mathbf{C} = \begin{bmatrix} (c_1 + c_2) & -c_2 & 0 \\ -c_2 & (c_2 + c_3) & -c_3 \\ 0 & -c_3 & c_3 \end{bmatrix} \tag{5-23}$$

The matrix equation for damped free vibration of the three-degree-of-freedom system shown in Fig. 5-16 is

$$\mathbf{M}\begin{Bmatrix} \ddot{x}_1 \\ \ddot{x}_2 \\ \ddot{x}_3 \end{Bmatrix} + \mathbf{C}\begin{Bmatrix} \dot{x}_1 \\ \dot{x}_2 \\ \dot{x}_3 \end{Bmatrix} + \mathbf{K}\begin{Bmatrix} x_1 \\ x_2 \\ x_3 \end{Bmatrix} = \mathbf{0}$$

or more simply

$$\mathbf{M}\ddot{\mathbf{X}} + \mathbf{C}\dot{\mathbf{X}} + \mathbf{K}\mathbf{X} = \mathbf{0} \tag{5-24}$$

In these equations **C** is the damping matrix of Eq. 5-23, and **M** and **K** are the mass and stiffness matrices, respectively.

In Example 5-11, it was fairly easy to determine the correct sense of the damping influence coefficients by inspection. In instances in which the senses are not readily apparent, a good procedure is to sketch free-body diagrams of the mass elements and show the c_{ij}'s acting with the same sense as the positive sense assumed for the generalized coordinates.

The correct signs for the c_{ij} values will then result when the forces or moments are summed to zero. This procedure was used in Example 5-7 to obtain the k_{ij} values of the stiffness matrix, since their senses were not apparent from a simple inspection.

There is no practical means of determining accurate values for the c_{ij} elements of the damping matrix for structural systems.

5-4 COORDINATE COUPLING (DYNAMIC AND STATIC)

Differential equations of motion are said to be *dynamically coupled* if the coordinates used to describe the motion of the system result in a nondiagonal mass matrix $\mathbf{M}$, and *statically coupled* if they result in a nondiagonal stiffness matrix $\mathbf{K}$. Since it is possible to write the differential equations of motion of some systems in terms of more than one set of generalized coordinates, dynamic and/or static coupling can sometimes be eliminated when it occurs with one set of coordinates by selecting a different set of coordinates. We examine these two types of coupling in turn.

Dynamic Coupling

The two-degree-of-freedom system shown in Fig. 5-17 vibrates with general plane motion, which is a combination of translation and rotation. Two of many possible sets of coordinates for describing the motion of this system are shown in Fig. 5-17b and c. Let us first derive the equations of motion of the system using the set of coordinates shown in Fig. 5-17b. In this set x is the vertical displacement of the left end of the bar and θ is the rotation of the bar as viewed from a set of axes translating with the left end of the bar. The centroidal mass moment of inertia of the bar is $\bar{I}$, and for the time being we designate the acceleration of the mass center of the bar as $\ddot{x}_G$. The positive sense assumed for θ, $\dot{\theta}$, and $\ddot{\theta}$ is counterclockwise, with up being positive for x, $\dot{x}$, and $\ddot{x}$.

In using the concept of dynamic equilibrium to derive the differential equations of motion of the system, we add the inertia couple $\bar{I}\ddot{\theta}$ with a sense opposite that assumed positive for the angular acceleration $\ddot{\theta}$ and the linear inertia effect $m\ddot{x}_G$ with a sense opposite that assumed positive for the acceleration $\ddot{x}_G$ of the mass center of the bar. We then add the spring forces with senses consistent with the forces the springs exert on the bar when it has a positive displacement from its static-equilibrium position. This yields the free-body diagram shown in Fig. 5-17b. Using this free-body diagram, we write the summation of the forces and inertia effect as

$$\sum F_x = 0 = k_1 x + k_2(x + l\theta) + m\ddot{x}_G \tag{5-25}$$

Assuming that the oscillations are small, we can let

$$x_G = x + a\theta$$

from which

$$\ddot{x}_G = \ddot{x} + a\ddot{\theta}$$

When the right side of the latter expression above is substituted into Eq. 5-25, one of the equations of motion is obtained as

$$m\ddot{x} + ma\ddot{\theta} + (k_1 + k_2)x + k_2 l\theta = 0 \tag{5-26}$$

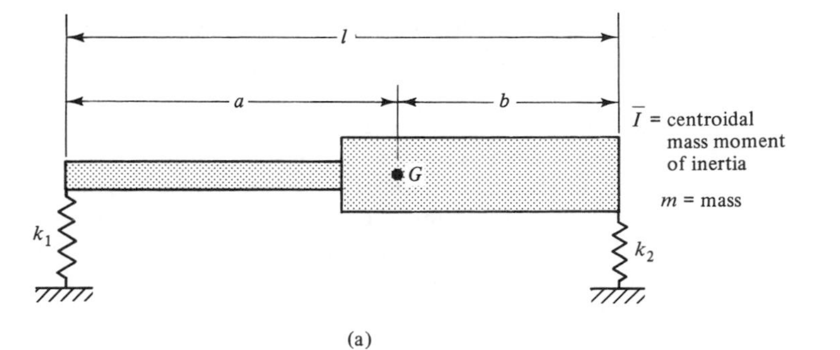

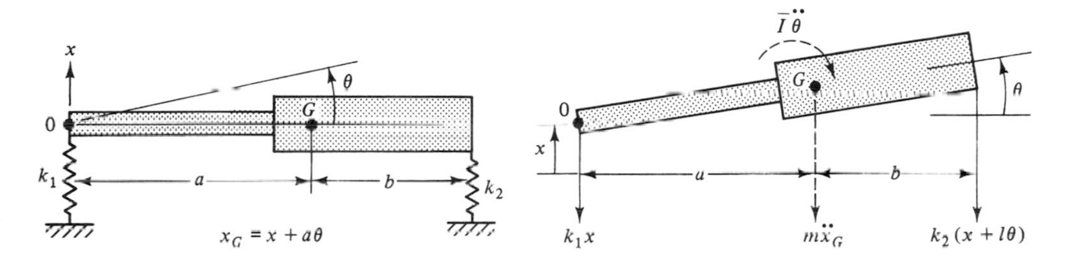

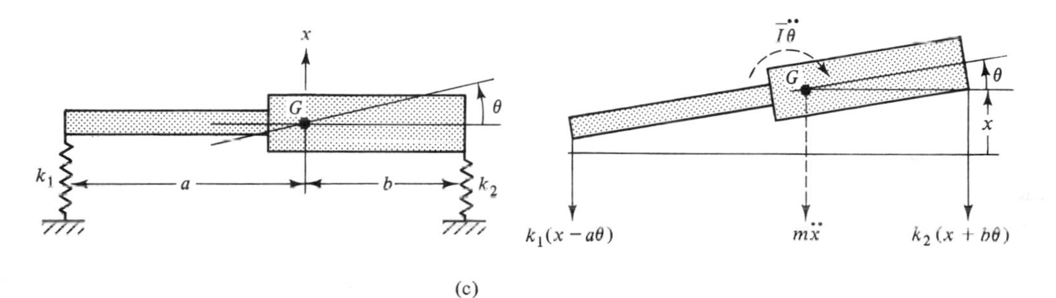

Figure 5-17 Two-degree-of-freedom system described by two different coordinate systems. (b) Coordinates that result in dynamic coupling. (c) Coordinates that eliminate dynamic coupling.

Summing moments of the forces and inertia effects equal to zero about an axis through O at the left end of the bar, the other equation is

$$\sum M_0 = 0 = ma\ddot{x}_G + \bar{I}\ddot{\theta} + k_2(x + l\theta)l$$

or

$$ma\ddot{x} + (ma^2 + \bar{I})\ddot{\theta} + k_2xl + k_2l^2\theta = 0 \tag{5-27}$$

Equations 5-26 and 5-27 are the equations of motion, and appear in matrix form as

$$\begin{bmatrix} m & ma \\ ma & (ma^2 + \bar{I}) \end{bmatrix} \begin{Bmatrix} \ddot{x} \\ \ddot{\theta} \end{Bmatrix} + \begin{bmatrix} (k_1 + k_2) & k_2 l \\ k_2 l & k_2 l^2 \end{bmatrix} \begin{Bmatrix} x \\ \theta \end{Bmatrix} = \mathbf{0} \tag{5-28}$$

Inspection of Eq. 5-28 shows that the mass matrix is nondiagonal, which follows from both $\ddot{x}$ and $\ddot{\theta}$ appearing in Eqs. 5-26 and 5-27; so we have dynamic coupling.

Let us next derive the differential equations of motion of the same system using the set of coordinates shown in Fig. 5-17c. In this set x is now the vertical displacement of the mass center of the bar while θ is the rotation of the bar as viewed from a set of axes translating with the mass center. The free-body diagram shown in Fig. 5-17c is obtained in the same manner discussed for the one shown in Fig. 5-17b. Referring to the former and summing the forces and inertia effect, we obtain

$$\sum F_x = 0 = k_1(x - a\theta) + k_2(x + b\theta) + m\ddot{x}$$

or

$$m\ddot{x} + (k_1 + k_2)x + (k_2 b - k_1 a)\theta = 0 \tag{5-29}$$

Summing moments of the forces and inertia effects about G equal to zero,

$$\sum M_G = 0 = \bar{I}\ddot{\theta} + k_2(x + b\theta)b - k_1(x - a\theta)a$$

or

$$\bar{I}\ddot{\theta} + (k_2 b^2 + k_1 a^2)\theta + (k_2 b - k_1 a)x = 0 \tag{5-30}$$

The matrix equation for the differential equations of motion, Eqs. 5-29 and 5-30, is

$$\begin{bmatrix} m & 0 \\ 0 & \bar{I} \end{bmatrix} \begin{Bmatrix} \ddot{x} \\ \ddot{\theta} \end{Bmatrix} + \begin{bmatrix} (k_1 + k_2) & (k_2 b - k_1 a) \\ (k_2 b - k_1 a) & (k_2 b^2 + k_1 a^2) \end{bmatrix} \begin{Bmatrix} x \\ \theta \end{Bmatrix} = 0 \tag{5-31}$$

Inspection of Eq. 5-31 reveals that since the mass matrix contains no elements off of the diagonal, dynamic coupling has been eliminated. This is confirmed by noting that only $\ddot{x}$ appears in Eq. 5-29 and only $\ddot{\theta}$ in Eq. 5-30.

We can generalize from the preceding discussion that when the linear coordinate x is the displacement of the mass center, and the displacements of other points on the body are expressed in terms of this coordinate and the angular rotation θ of the body, dynamic coupling will usually be avoided. Example 5-7, in which the linear coordinate q_1 and rotational coordinate q_2 were selected in this manner as shown in Fig. 5-7 to eliminate dynamic coupling, further illustrates this generalization.

A rather quick way of determining whether the coordinates selected lead to dynamic coupling is to examine the expression obtained for the kinetic energy of a system. If the expression contains products of the coordinates, there will be dynamic coupling. If it does not, there will be no dynamic coupling. To illustrate this, the use of the coordinates x and θ in Fig. 5-17b results in the kinetic-energy expression

$$T = \tfrac{1}{2}m(\dot{x} + a\dot{\theta})^2 + \tfrac{1}{2}\bar{I}\dot{\theta}^2$$

which indicates dynamic coupling because of the product $2\dot{x}a\dot{\theta}$ that appears upon expanding the term in parentheses. However, using the coordinates shown in Fig. 5-17c, the kinetic-energy expression is

$$T = \tfrac{1}{2}m\dot{x}^2 + \tfrac{1}{2}\bar{I}\dot{\theta}^2$$

which indicates no dynamic coupling since there are no coordinate products.[5]

In general, the choice of coordinates is arbitrary as far as determining the dynamic characteristics of a system is concerned. The dynamic coupled equations given by Eqs. 5-26 and 5-27 will yield the same natural frequencies, for example, as will the uncoupled equations given by Eqs. 5-29 and 5-30. (The reader should note that the x coordinate in the first set of these equations is not the same as the x in the second set.) However, there is an advantage in selecting coordinates that lead to uncoupled equations of motion, since a diagonal mass matrix is more easily inverted than a nondiagonal one. It is easy to demonstrate that $\mathbf{M}^{-1}$ (the inverse of $\mathbf{M}$) for a diagonal matrix is simply another diagonal matrix whose elements are the reciprocals of the corresponding elements of $\mathbf{M}$. For example, if the mass matrix is

$$\mathbf{M} = \begin{bmatrix} m_1 & 0 & 0 \\ 0 & m_2 & 0 \\ 0 & 0 & m_3 \end{bmatrix}$$

then the inverse of $\mathbf{M}$ is simply

$$\mathbf{M}^{-1} = \begin{bmatrix} 1/m_1 & 0 & 0 \\ 0 & 1/m_2 & 0 \\ 0 & 0 & 1/m_3 \end{bmatrix} \tag{5-32}$$

This is an important feature, for as we shall see in the next section, the inversion of the mass matrix simplifies the form of the homogeneous algebraic equations from which the natural frequencies and mode shapes of systems are determined.

Static Coupling

The coordinates shown in Fig. 5-17b and c both result in static coupling since the stiffness matrices in both Eqs. 5-28 and 5-31 contain off-diagonal elements. This can be eliminated by letting the coordinate x be the displacement of a point on the body where a force acting in conjunction with the spring forces on the body would produce only translation of the body (a point through which the resultant of the spring forces acts). However, as we have seen, this will result in dynamic coupling unless the resultant of the spring forces passes through the mass center of the body; so it is not particularly advantageous.

In Sec. 5-7 we discuss a general procedure for decoupling differential equations of motion that are coupled dynamically and/or statically. This procedure, referred to as *modal analysis,* leads to a set of independent equations in terms of the *principal coordinates* of a system, each of which can be solved independently of the others.

5-5 NATURAL FREQUENCIES AND MODE SHAPES (EIGENVALUES AND EIGENVECTORS)

An n-degree-of-freedom system has n natural frequencies, and for each natural frequency there is a corresponding normal mode shape that defines a distinct relationship between

[5] The basis for this check for dynamic coupling will become evident when the derivation of the differential equations of motion of systems using Lagrange's equations is discussed in Sec. 7-2.

the amplitudes of the generalized coordinates for that mode. The squares of the natural circular frequencies and corresponding sets of coordinate values describing the normal mode shapes are referred to as *eigenvalues* and *eigenvectors,* respectively, and they are of fundamental importance in the analysis of the free or forced vibration of multiple-degree-of-freedom systems.

It was shown in Chap. 2 that the damped natural circular frequency ω_d is related to the undamped natural circular frequency ω_n by

$$\omega_d = \omega_n\sqrt{1 - \zeta^2}$$

and that they are essentially the same in magnitude for many real systems that have damping of less than 20 percent ($\zeta < 0.2$). For this reason, the eigenvalues and eigenvectors of multiple-degree-of-freedom systems are usually determined considering undamped free vibration.

To formulate the general algebraic equations from which the eigenvalues and eigenvectors of a system are determined, we first *premultiply* Eq. 5-9 by $\mathbf{M}^{-1}$ (the inverse of $\mathbf{M}$) to obtain

$$\ddot{\mathbf{X}} + \mathbf{M}^{-1}\mathbf{K}\mathbf{X} = \mathbf{0} \qquad (5\text{-}33)$$

which can also be written as

$$\begin{Bmatrix} \ddot{x}_1 \\ \ddot{x}_2 \\ \vdots \\ \ddot{x}_n \end{Bmatrix} + \mathbf{M}^{-1}\mathbf{K} \begin{Bmatrix} x_1 \\ x_2 \\ \vdots \\ x_n \end{Bmatrix} = \mathbf{0}$$

For undamped free vibration, we can assume harmonic motion for each mass, so that

$$\left.\begin{aligned} x_i &= X_i e^{j\omega t} \\ \ddot{x}_i &= -\omega^2 X_i e^{j\omega t} \end{aligned}\right\} \qquad (5\text{-}34)$$

in which $i = 1, 2, \ldots, n$. Substituting the expressions of Eq. 5-34 into Eq. 5-33, we obtain

$$-\omega^2\mathbf{X} + \mathbf{M}^{-1}\mathbf{K}\mathbf{X} = \mathbf{0}$$

or

$$[\mathbf{M}^{-1}\mathbf{K} - \omega^2\mathbf{I}]\mathbf{X} = \mathbf{0} \qquad (5\text{-}35)$$

in which the identity matrix $\mathbf{I}$ indicates that ω^2 appears in the terms on the diagonal of the square matrix

$$[\mathbf{M}^{-1}\mathbf{K} - \omega^2\mathbf{I}]$$

Considering that $\mathbf{M}$ is a diagonal matrix, as discussed in Sec. 5-4, the inverse of $\mathbf{M}$ is simply (see Eq. 5-32)

$$M^{-1} = \begin{bmatrix} 1/m_1 & 0 & \cdot & \cdot & \cdot & 0 \\ 0 & 1/m_2 & \cdot & \cdot & \cdot & 0 \\ \cdot & \cdot & \cdot & \cdot & \cdot & \cdot \\ 0 & 0 & \cdot & \cdot & \cdot & 1/m_n \end{bmatrix}$$

With this in mind, it can be shown that $\mathbf{M}^{-1}\mathbf{K}$ in Eq. 5-35 has the form of

$$\mathbf{M}^{-1}\mathbf{K} = \begin{bmatrix} \frac{k_{11}}{m_1} & \frac{k_{12}}{m_1} & \cdots & \frac{k_{1n}}{m_1} \\ \frac{k_{21}}{m_2} & \frac{k_{22}}{m_2} & \cdots & \frac{k_{2n}}{m_2} \\ \cdot & \cdot & \cdots & \cdot \\ \frac{k_{n1}}{m_n} & \frac{k_{n2}}{m_n} & \cdots & \frac{k_{nn}}{m_n} \end{bmatrix}$$

which is often referred to as the *dynamic matrix,* since it describes the dynamic properties of a system. It should be noted that the dynamic matrix is, in general, nonsymmetric. Finally, letting $\omega^2 = \lambda$, Eq. 5-35 has the form of

$$\begin{bmatrix} \left(\frac{k_{11}}{m_1} - \lambda\right) & \frac{k_{12}}{m_1} & \cdots & \frac{k_{1n}}{m_1} \\ \frac{k_{21}}{m_2} & \left(\frac{k_{22}}{m_2} - \lambda\right) & \cdots & \frac{k_{2n}}{m_2} \\ \cdots & \cdots & \cdots & \cdots \\ \frac{k_{n1}}{m_n} & \frac{k_{n2}}{m_n} & \cdots & \left(\frac{k_{nn}}{m_n} - \lambda\right) \end{bmatrix} \begin{Bmatrix} X_1 \\ X_2 \\ \vdots \\ X_n \end{Bmatrix} = \mathbf{0} \qquad (5\text{-}36)$$

Equation 5-36 consists of a set of *homogeneous algebraic* equations, in which there are n values of λ that will satisfy the set of equations. These n values of λ are the *eigenvalues,* and they correspond to the squares of the natural circular frequencies of the system. The values of λ are obtained by setting the *determinant* $|D|$ of the coefficient matrix of Eq. 5-36 equal to zero. That is,

$$|D| = \begin{vmatrix} \left(\frac{k_{11}}{m_1} - \lambda\right) & \frac{k_{12}}{m_1} & \cdots & \frac{k_{1n}}{m_1} \\ \frac{k_{21}}{m_2} & \left(\frac{k_{22}}{m_2} - \lambda\right) & \cdots & \frac{k_{2n}}{m_2} \\ \cdots & \cdots & \cdots & \cdots \\ \frac{k_{n1}}{m_n} & \frac{k_{n2}}{m_n} & \cdots & \left(\frac{k_{nn}}{m_n} - \lambda\right) \end{vmatrix} = 0 \qquad (5\text{-}37)$$

The expansion of the above determinant results in an nth-degree polynomial

$$\lambda^n + b_1\lambda^{n-1} + b_2\lambda^{n-2} + \cdots + b_n = 0 \qquad (5\text{-}38)$$

which is referred to as the *frequency* or *characteristic* equation. The roots of this equation are the λ_i values (eigenvalues) that make the determinant equal to zero. The constants b_1, $b_2, \ldots, b_n$ depend upon the values of the stiffness coefficients and of the masses. Since the roots $\lambda_i = \omega_i^2$, an n-degree-of-freedom system has n natural frequencies, $\omega_1, \omega_2, \omega_3, \ldots,$ ω_n, thus the reference to the polynomial as the frequency equation.

After the eigenvalues are determined, they can be substituted one value at a time back into Eq. 5-36, to obtain the eigenvectors, which are corresponding sets of relationships between the unknown X_i's. The eigenvectors describe the normal mode shapes, or mode configurations, corresponding to the natural frequencies.

As demonstrated in several of the examples in Sec. 5-3, the stiffness coefficients can often be determined quite simply by an inspection of a physical model. The k_{ij}/m_i elements of the determinant of Eq. 5-37 can be obtained in this manner, as is shown in Example 5-12.

EXAMPLE 5-12

Two identical disks of centroidal mass moment of inertia I are attached to a steel shaft that is fixed at the right end as shown in Fig. 5-18. Each section of shaft has a diameter d, a length l, and a torsional spring constant k. Using the system data,

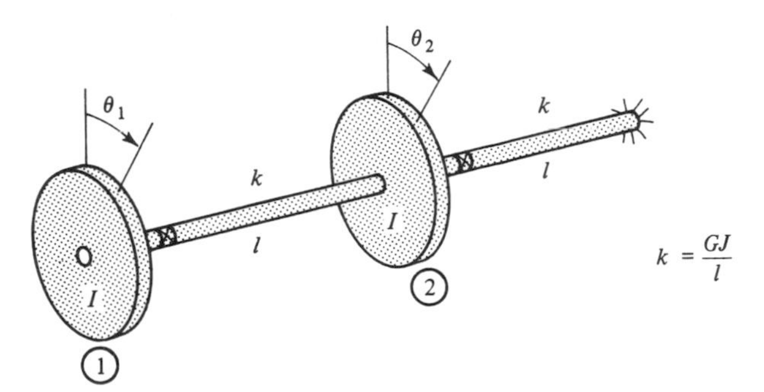

Figure 5-18 Two-degree-of-freedom system of Example 5-12.

l = 24 in. (length of each section of shaft)
d = 1.25 in. (diameter of shaft)
$G = 12(10)^6$ psi (shear modulus)
$I = 12$ lb · in. · s^2 (centroidal mass moment of inertia of disk)

determine (a) the eigenvalues of the system, (b) the natural frequencies of the system (Hz), (c) the eigenvectors (normal-mode shapes) of the system.

Solution. **a.** Using the coordinates θ_1 and θ_2 shown in Fig. 5-18, it can be seen by inspecting Fig. 5-18 and recalling the definition of stiffness coefficients that k_{11} is the moment necessary to give disk 1 a unit rotation, while k_{21} is the moment that must be applied to disk 2 to keep it from rotating when k_{11} is applied. That is, for $\theta_1 = 1$ and $\theta_2 = 0$,

$$k_{11} = k$$
$$k_{21} = -k$$

Similarly, for $\theta_2 = 1$ and $\theta_1 = 0$,

$$k_{12} = -k$$
$$k_{22} = 2k$$

From the above, the stiffness matrix is

$$\mathbf{K} = \begin{bmatrix} k & -k \\ -k & 2k \end{bmatrix} \tag{5-39}$$

The mass matrix is

$$\mathbf{M} = \begin{bmatrix} I & 0 \\ 0 & I \end{bmatrix} \tag{5-40}$$

Referring to Eqs. 5-37, 5-39, and 5-40, we can write that

$$|D| = \begin{vmatrix} \left(\dfrac{k}{I} - \lambda\right) & \dfrac{-k}{I} \\ \dfrac{-k}{I} & \left(\dfrac{2k}{I} - \lambda\right) \end{vmatrix} = 0 \tag{5-41}$$

from which we obtain the frequency (characteristic) equation as

$$\left(\frac{k}{I} - \lambda\right)\left(\frac{2k}{I} - \lambda\right) - \left(\frac{k}{I}\right)^2 = 0$$

or

$$\lambda^2 - \frac{3k}{I}\lambda + \left(\frac{k}{I}\right)^2 = 0 \tag{5-42}$$

in which $\lambda = \omega^2$. The roots of Eq. 5-42 are the eigenvalues, and they are found to be

$$\left.\begin{aligned} \lambda_1 &= \omega_1^2 = \frac{k}{I}\left[\frac{3 - \sqrt{5}}{2}\right] \\ \lambda_2 &= \omega_2^2 = \frac{k}{I}\left[\frac{3 + \sqrt{5}}{2}\right] \end{aligned}\right\} \tag{5-43}$$

b. Referring to part c of Table 2-1 on page 78, and to the data given at the beginning of the example, the torsional spring constant is computed as

$$k = \frac{GJ}{l} = \frac{12(10)^6}{24}\left[\frac{\pi(1.25)^4}{32}\right] = 120(10)^3 \text{ lb}\cdot\text{in./rad}$$

and

$$\frac{k}{I} = \frac{120(10)^3}{12} = 10(10)^3 \text{ s}^{-2}$$

Substituting the value of k/I into the expressions of Eq. 5-43, we find that

$$\lambda_1 = \omega_1^2 = 3.82(10)^3$$
$$\lambda_2 = \omega_2^2 = 26.18(10)^3$$

from which the two natural frequencies are found to be

$$f_1 = \frac{\omega_1}{2\pi} = \frac{\sqrt{3.82(10)^3}}{2\pi} = 9.84 \text{ Hz}$$

$$f_2 = \frac{\omega_2}{2\pi} = \frac{\sqrt{26.18(10)^3}}{2\pi} = 25.75 \text{ Hz}$$

c. To determine the eigenvectors (normal mode shapes) corresponding to the eigenvalues determined, we refer to Eq. 5-36, and write that

$$\begin{bmatrix} \left(\frac{k}{I} - \lambda\right) & \frac{-k}{I} \\ \frac{-k}{I} & \left(\frac{2k}{I} - \lambda\right) \end{bmatrix} \begin{Bmatrix} \Theta_1 \\ \Theta_2 \end{Bmatrix} = \mathbf{0} \tag{5-44}$$

in which Θ_1 and Θ_2 are the eigenvector components. From Eq. 5-44, we can write that

$$\left.\begin{aligned} \left(\frac{k}{I} - \lambda\right)\Theta_1 - \frac{k}{I}\Theta_2 = 0 \quad \text{(a)} \\ \frac{-k}{I}\Theta_1 + \left(\frac{2k}{I} - \lambda\right)\Theta_2 = 0 \quad \text{(b)} \end{aligned}\right\} \tag{5-45}$$

The relationships between the eigenvector components for λ_1 and λ_2 can be determined from either Eq. 5-45a or 5-45b. Using the first of these,

$$\frac{\Theta_2}{\Theta_1} = \frac{k/I - \lambda}{k/I} \tag{5-46}$$

in which $k/I = 10(10)^3$. Substituting the values for λ_1 and λ_2, in turn, into Eq. 5-46, yields

$$\left.\frac{\Theta_2}{\Theta_1}\right)_1 = \frac{10(10)^3 - 3.82(10)^3}{10(10)^3} = 0.62 \qquad \text{(first mode)}$$

and

$$\left.\frac{\Theta_2}{\Theta_1}\right)_2 = \frac{10(10)^3 - 26.18(10)^3}{10(10)^3} = -1.62 \qquad \text{(second mode)}$$

Although there are an infinite number of values of Θ_1 and Θ_2 that will satisfy the above ratios, a convenient procedure is to normalize the ratios by arbitrarily assigning a value of *unity* to one of the eigenvector components. For example, letting $\Theta_1 = 1$, we obtain the eigenvectors as

$$\begin{Bmatrix} \Theta_1 \\ \Theta_2 \end{Bmatrix}_1 = \begin{Bmatrix} 1 \\ 0.62 \end{Bmatrix} \qquad \text{(eigenvector for first mode)}$$

and

$$\begin{Bmatrix} \Theta_1 \\ \Theta_2 \end{Bmatrix}_2 = \begin{Bmatrix} 1 \\ -1.62 \end{Bmatrix} \qquad \text{(eigenvector for second mode)}$$

The eigenvector for the first mode describes the undamped free-vibration configuration of the system when it is vibrating at a frequency of 9.84 Hz, in which $\Theta_2 = 0.62\Theta_1$ at any instant in time. Since the ratio is positive, the disks are vibrating *in phase* with each other.

Similarly, the eigenvector for the second mode describes the system configuration for the undamped free-vibration mode when the system is vibrating at a frequency of 25.75 Hz, with the disks 180° *out of phase* with each other, as indicated by the negative ratio ($\Theta_2 = -1.62\Theta_1$).

The twist configurations of the shaft (the mode shapes) are shown in Fig.

5-19. Note that for the second mode, the angular displacement of the shaft is zero at the *node* (determined from the similar triangles shown in the figure).

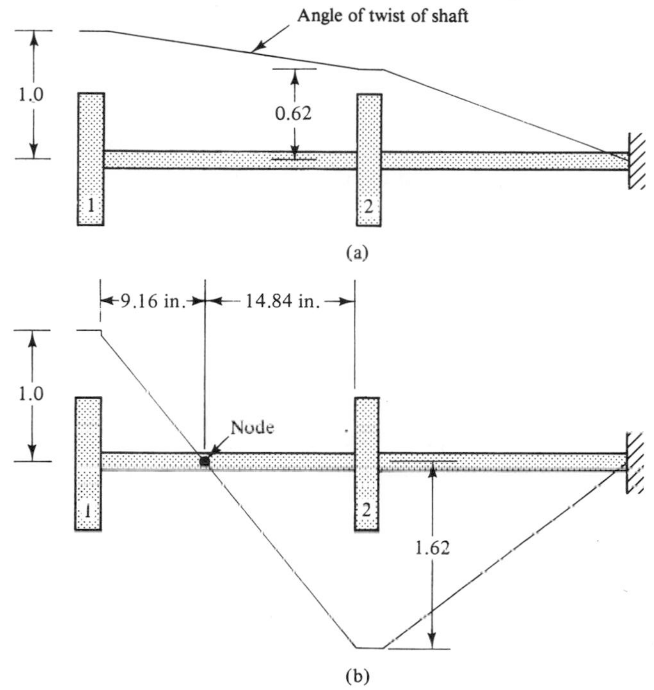

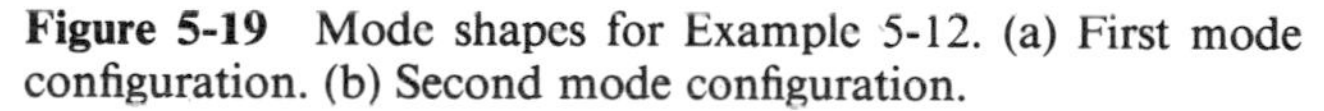

Figure 5-19 Mode shapes for Example 5-12. (a) First mode configuration. (b) Second mode configuration.

EXAMPLE 5-13

Figure 5-20 shows a free-free system, which is a semidefinite system having a zero natural frequency that corresponds to the system rotating as a rigid body. We wish to determine (a) the natural frequencies of the system in terms of the torsional-spring constant k and the mass moment of inertia I, (b) the normal mode shapes (the eigenvectors).

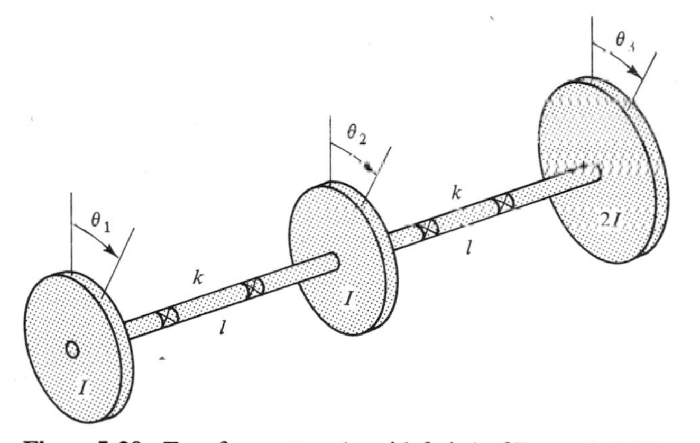

Figure 5-20 Free-free system (semidefinite) of Example 5-13.

Solution. **a.** Inspecting Fig. 5-20, and recalling the definition of stiffness coefficients, the homogeneous algebraic equations corresponding to Eq. 5-36 are found to be

$$\begin{bmatrix} \left(\frac{k}{I} - \lambda\right) & \frac{-k}{I} & 0 \\ \frac{-k}{I} & \left(\frac{2k}{I} - \lambda\right) & \frac{-k}{I} \\ 0 & \frac{-k}{2I} & \left(\frac{k}{2I} - \lambda\right) \end{bmatrix} \begin{Bmatrix} \Theta_1 \\ \Theta_2 \\ \Theta_3 \end{Bmatrix} = \mathbf{0} \tag{5-47}$$

To satisfy these equations, the determinant

$$|D| = \begin{vmatrix} \left(\frac{k}{I} - \lambda\right) & \frac{-k}{I} & 0 \\ \frac{-k}{I} & \left(\frac{2k}{I} - \lambda\right) & \frac{-k}{I} \\ 0 & \frac{-k}{2I} & \left(\frac{k}{2I} - \lambda\right) \end{vmatrix} = 0 \tag{5-48}$$

must be equal to zero, as indicated by Eq. 5-48.

Expanding this determinant, and simplifying, gives the frequency equation as

$$\lambda^3 - \frac{7}{2}\left(\frac{k}{I}\right)\lambda^2 + 2\left(\frac{k}{I}\right)^2 \lambda = 0$$

or

$$\lambda\left[\lambda^2 - \frac{7}{2}\frac{k}{I}\lambda + 2\left(\frac{k}{I}\right)^2\right] = 0 \tag{5-49}$$

in which $\lambda = \omega^2$. The roots (the eigenvalues) of Eq. 5-49 are found to be

$$\left.\begin{aligned} \lambda_1 &= 0 \\ \lambda_2 &= 0.72\,\frac{k}{I} \\ \lambda_3 &= 2.78\,\frac{k}{I} \end{aligned}\right\} \tag{5-50}$$

Using the above, the natural frequencies are then found to be

$$\left.\begin{aligned} f_1 &= \frac{\omega_1}{2\pi} = 0 \\ f_2 &= \frac{\omega_2}{2\pi} = \frac{1}{2\pi}\sqrt{0.72\,\frac{k}{I}}\ \text{Hz} \\ f_3 &= \frac{\omega_3}{2\pi} = \frac{1}{2\pi}\sqrt{2.78\,\frac{k}{I}}\ \text{Hz} \end{aligned}\right\} \tag{5-51}$$

b. The first and third of Eqs. 5-47 are

$$\left(\frac{k}{I} - \lambda\right)\Theta_1 - \frac{k}{I}\Theta_2 = 0$$

and

$$\frac{-k}{2I}\Theta_2 + \left(\frac{k}{2I} - \lambda\right)\Theta_3 = 0$$

from which it can be seen that both Θ_2 and Θ_3 can be expressed in terms of Θ_1 as

$$\left.\begin{aligned} \frac{\Theta_2}{\Theta_1} &= \frac{k/I - \lambda}{k/I} \\ \frac{\Theta_3}{\Theta_1} &= \frac{k/I - \lambda}{2(k/2I - \lambda)} \end{aligned}\right\} \tag{5-52}$$

Substituting $\lambda_1 = 0$ into Eq. 5-52 yields

$$\left.\begin{aligned} \left.\frac{\Theta_2}{\Theta_1}\right)_1 &= 1 \\ \left.\frac{\Theta_3}{\Theta_1}\right)_1 &= 1 \end{aligned}\right\} \quad \text{(rigid body motion)}$$

which corresponds to the system rotating as a rigid body, with $\Theta_1 = \Theta_2 = \Theta_3$.

Substituting $\lambda_2 = 0.72\ k/I$ into Eq. 5-52, we find that

$$\left.\begin{aligned} \left.\frac{\Theta_2}{\Theta_1}\right)_2 &= \frac{k/I - 0.72(k/I)}{k/I} = 0.28 \\ \left.\frac{\Theta_3}{\Theta_1}\right)_2 &= \frac{k/I - 0.72(k/I)}{2[k/2I - 0.72(k/I)]} = -0.64 \end{aligned}\right\} \quad \text{(first vibration mode)}$$

Similarly, substituting $\lambda_3 = 2.78\ k/I$ into Eq. 5-52 shows that

$$\left.\begin{aligned} \left.\frac{\Theta_2}{\Theta_1}\right)_3 &= -1.78 \\ \left.\frac{\Theta_3}{\Theta_1}\right)_3 &= 0.39 \end{aligned}\right\} \quad \text{(second vibration mode)}$$

An arbitrary value can be assigned to any one of the Θ's, with the other two then expressed in terms of it. For example, if we let $\Theta_1 = 1$, the eigenvectors become

$$\begin{Bmatrix} \Theta_1 \\ \Theta_2 \\ \Theta_3 \end{Bmatrix}_1 = \begin{Bmatrix} 1.0 \\ 1.0 \\ 1.0 \end{Bmatrix} \qquad \begin{Bmatrix} \Theta_1 \\ \Theta_2 \\ \Theta_3 \end{Bmatrix}_2 = \begin{Bmatrix} 1.0 \\ 0.28 \\ -0.64 \end{Bmatrix} \qquad \begin{Bmatrix} \Theta_1 \\ \Theta_2 \\ \Theta_3 \end{Bmatrix}_3 = \begin{Bmatrix} 1.0 \\ -1.78 \\ 0.39 \end{Bmatrix}$$

in which the components of the eigenvectors are said to be *normalized* with respect to Θ_1. It should be pointed out that the components of the eigenvectors can also be normalized with respect to either Θ_2 or Θ_3 in the same manner. The magnitude of the referencing component is referred to as the *normalizing factor.* The eigenvector can also be normalized by giving it a unit magnitude, in which case the normalizing factor is the square root of the sum of the squares of the eigenvector components.

The rigid-body motion corresponding to $\lambda_1 = 0$ is not actually vibratory motion. Therefore, the free-free system shown in Fig. 5-20 has only two natural frequencies and two corresponding mode shapes as shown in Fig. 5-21, corresponding to the elastic vibratory motion.

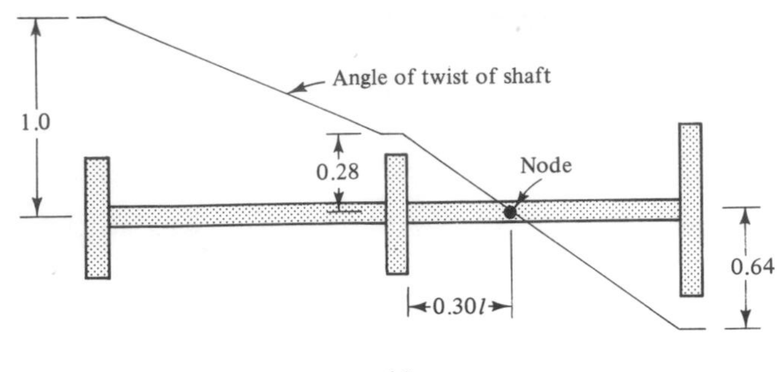

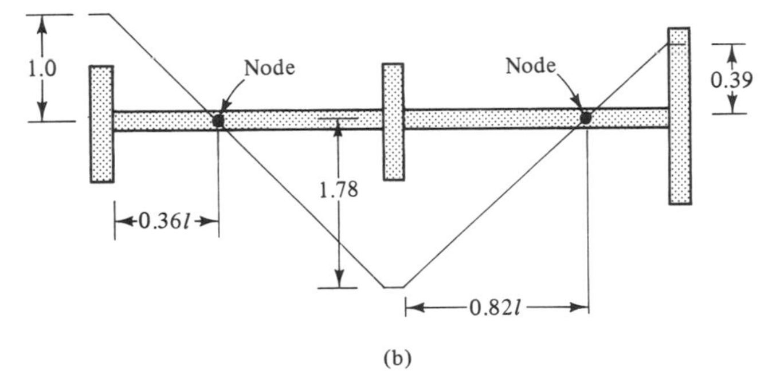

Figure 5-21 Vibration mode configurations of Example 5-13. (a) First vibration mode corresponding to $\lambda_2 = 0.72\ k/I$. (b) Second vibration mode corresponding to $\lambda_3 = 2.78\ k/I$.

EXAMPLE 5-14

We wish to determine the natural frequencies and corresponding normal-mode configurations of the three-story building shown schematically in Fig. 5-22, which has the mass and stiffness values given in the figure.

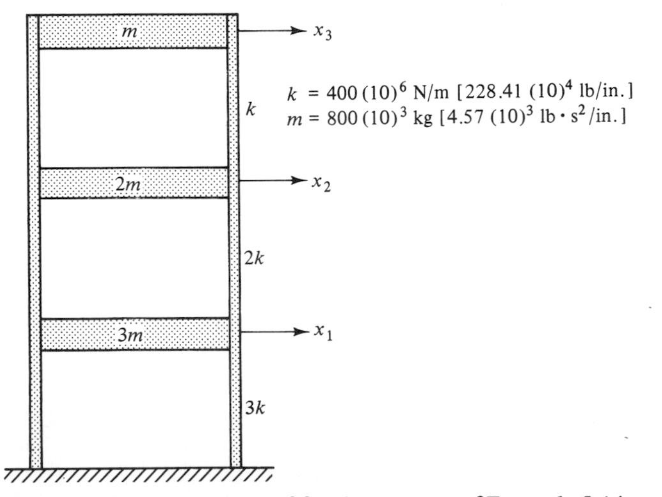

Figure 5-22 Three-degree-of-freedom system of Example 5-14.

Solution. Remembering the definition of stiffness coefficients k_{ij}, and inspecting Fig. 5-22, we find the following stiffness coefficients:

$$\begin{array}{lll} k_{11} = 5k & k_{12} = -2k & k_{13} = 0 \\ k_{21} = -2k & k_{22} = 3k & k_{23} = -k \\ k_{31} = 0 & k_{32} = -k & k_{33} = k \end{array}$$

Noting that $m_1 = 3m$, $m_2 = 2m$, and $m_3 = m$, we can see from Eq. 5-37 that

$$|D| = \begin{vmatrix} \left(\dfrac{5k}{3m} - \lambda\right) & \dfrac{-2k}{3m} & 0 \\ \dfrac{-k}{m} & \left(\dfrac{3k}{2m} - \lambda\right) & \dfrac{-k}{2m} \\ 0 & \dfrac{-k}{m} & \left(\dfrac{k}{m} - \lambda\right) \end{vmatrix} = 0 \tag{5-53}$$

Expanding this determinant, and simplifying, the frequency equation is

$$\lambda^3 - \frac{25}{6}\left(\frac{k}{m}\right)\lambda^2 + \frac{27}{6}\left(\frac{k}{m}\right)^2\lambda - \left(\frac{k}{m}\right)^3 = 0 \tag{5-54}$$

Substituting the values $k = 400(10)^6$ N/m and $m = 800(10)^3$ kg into Eq. 5-54 yields the frequency equation

$$\lambda^3 - 20.83(10)^2\lambda^2 + 11.25(10)^5\lambda - 12.50(10)^7 = 0 \tag{5-55}$$

Although various numerical methods are available for obtaining the roots of polynomials of order higher than 2, the roots of the third-degree polynomial above can be obtained quite efficiently by the *incremental-search* method.[6] In this method, values of the polynomial $f(\lambda)$ are determined for successive values of λ, in increments of $\Delta\lambda$, until a sign change occurs for $f(\lambda)$, which indicates that a root has been passed (see Fig. 5-23). A closer approximation of the root can then be obtained by returning to the last value of λ preceding the sign change, and beginning with this value, again determining values of $f(\lambda)$ for successive values of λ, using a smaller increment $\Delta\lambda$ than was used initially, until the sign of $f(\lambda)$ again changes. This procedure is repeated, with progressively smaller increments of λ, until a sufficiently accurate value of the root is obtained. Using this method, the roots of Eq. 5-55 are found to be

$$\left.\begin{array}{l} \lambda_1 = \omega_1^2 = 149.6 \\ \lambda_2 = \omega_2^2 = 652.6 \\ \lambda_3 = \omega_3^2 = 1280.9 \end{array}\right\} \tag{5-56}$$

from which the natural frequencies are determined as

$$f_1 = \frac{\sqrt{149.6}}{2\pi} = 1.95 \text{ Hz}$$

$$f_2 = \frac{\sqrt{652.6}}{2\pi} = 4.07 \text{ Hz}$$

$$f_3 = \frac{\sqrt{1280.9}}{2\pi} = 5.70 \text{ Hz}$$

[6] M. L. James, G. M. Smith, and J. C. Wolford, *Applied Numerical Methods,* 3d ed., Chap. 2, Harper & Row, New York, 1985.

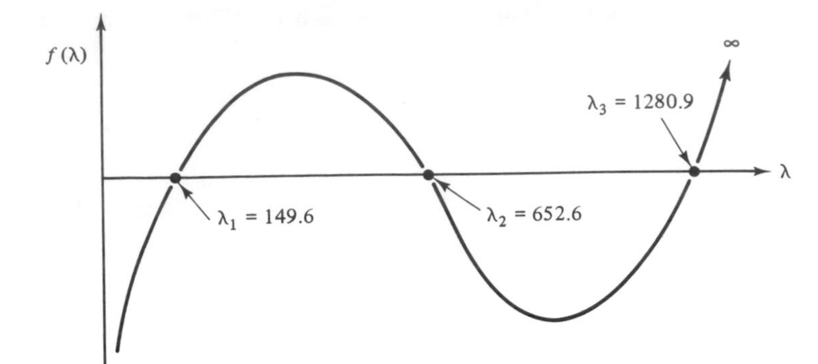

Figure 5-23 Approximate graph of the polynomial of Eq. 5-55 as a function of λ.

To determine the eigenvectors (the natural mode shapes), we note from Eq. 5-53 that

$$\begin{bmatrix} \left(\frac{5k}{3m} - \lambda\right) & \frac{-2k}{3m} & 0 \\ \frac{-k}{m} & \left(\frac{3k}{2m} - \lambda\right) & \frac{-k}{2m} \\ 0 & \frac{-k}{m} & \left(\frac{k}{m} - \lambda\right) \end{bmatrix} \begin{Bmatrix} X_1 \\ X_2 \\ X_3 \end{Bmatrix} = 0 \tag{5-57}$$

From the first and last equations of Eq. 5-57, the eigenvector components X_2 and X_3 can be written in terms of X_1 as

$$\left.\begin{aligned} \frac{X_2}{X_1} &= \frac{5k/3m - \lambda}{2k/3m} \\ \frac{X_3}{X_1} &= \left(\frac{3}{2}\right)\frac{5k/3m - \lambda}{k/m - \lambda} \end{aligned}\right\} \tag{5-58}$$

Calculating that $k/m = 500$ from the data given, and using the values of λ shown in Eq. 5-56, we determine from Eq. 5-58 that

First mode:

$$\left.\frac{X_2}{X_1}\right)_1 = 2.05 \qquad \left.\frac{X_3}{X_1}\right)_1 = 2.93$$

Second mode:

$$\left.\frac{X_2}{X_1}\right)_2 = 0.54 \qquad \left.\frac{X_3}{X_1}\right)_2 = -1.78$$

Third mode:

$$\left.\frac{X_2}{X_1}\right)_3 = -1.34 \qquad \left.\frac{X_3}{X_1}\right)_3 = 0.86$$

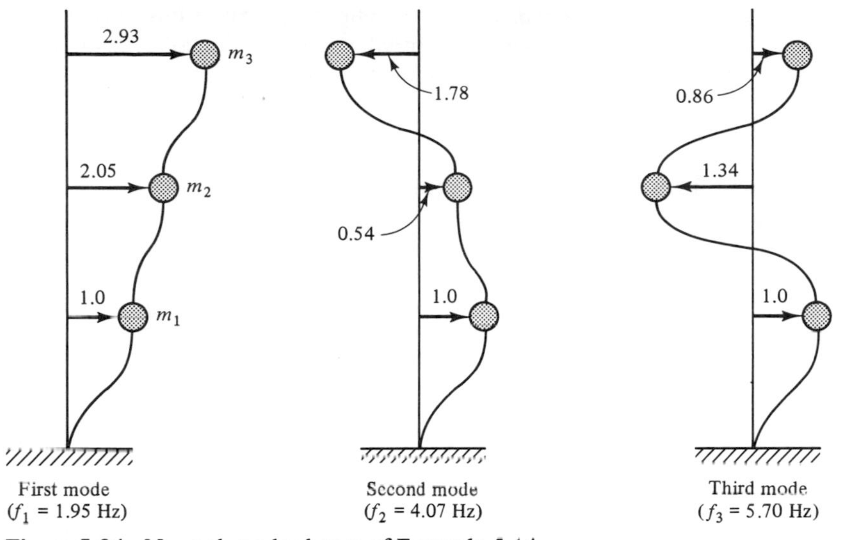

Figure 5-24 Normal-mode shapes of Example 5-14.

If we normalize the preceding relationships with respect to X_1 by assuming that $X_1 = 1$, the normal-mode shapes (the eigenvectors) are

$$\begin{Bmatrix} X_1 \\ X_2 \\ X_3 \end{Bmatrix}_1 = \begin{Bmatrix} 1.0 \\ 2.05 \\ 2.93 \end{Bmatrix} \qquad \begin{Bmatrix} X_1 \\ X_2 \\ X_3 \end{Bmatrix}_2 = \begin{Bmatrix} 1.0 \\ 0.54 \\ -1.78 \end{Bmatrix} \qquad \begin{Bmatrix} X_1 \\ X_2 \\ X_3 \end{Bmatrix}_3 = \begin{Bmatrix} 1.0 \\ -1.34 \\ 0.86 \end{Bmatrix}$$

The normal-mode shapes are shown in Fig. 5-24.

Modal Matrix [*u*]

The modal matrix, which we shall refer to as $[u]$, is simply a square matrix, in which the columns correspond to the eigenvectors (the normal-mode shapes) of a system. The modal matrix for an *n*-degree-of-freedom system has the general form of

$$[u] = \left[\begin{Bmatrix} \\ \\ \\ \end{Bmatrix}_1 \begin{Bmatrix} \\ \\ \\ \end{Bmatrix}_2 \begin{Bmatrix} \\ \\ \\ \end{Bmatrix}_3 \cdots \begin{Bmatrix} \\ \\ \\ \end{Bmatrix}_n \right] \tag{5-59}$$

Modes

As a specific example, the modal matrix of the system of Example 5-14 is

$$[u] = \begin{bmatrix} 1.0 & 1.0 & 1.0 \\ 2.05 & 0.54 & -1.34 \\ 2.93 & -1.78 & 0.86 \end{bmatrix}$$

In the process of *modal analysis,* which is discussed later in both Secs. 5-7 and 6-6, coupled equations of motion of a system can be uncoupled by a linear transformation that involves the modal matrix of the system. The uncoupling of the equations of motion provides a set of *independent* equations of motion in terms of *principal* or *normal* coordinates, which considerably simplifies the vibration analysis of multiple-degree-of-freedom systems. Therefore, the modal matrix is of fundamental importance in the study and analysis of such systems.

Determining Eigenvalues and Eigenvectors Using Flexibility Coefficients

The frequency (characteristic) equation expressed in Eq. 5-38 was obtained from equations of motion that were formulated in terms of the stiffness coefficients k_{ij}. As mentioned earlier, the frequency equation and in turn the eigenvalues and eigenvectors of a system can also be determined from equations formulated in terms of the flexibility coefficients a_{ij}.

Assuming that the mass matrix **M** is a diagonal matrix (not coupled dynamically), the equations of motion expressed in Eq. 5-15 can be written for an n-degree-of-freedom system in terms of the flexibility coefficients, as

$$\begin{Bmatrix} x_1 \\ x_2 \\ \vdots \\ x_n \end{Bmatrix} + \begin{bmatrix} a_{11}m_1 & a_{12}m_2 & \cdots & a_{1n}m_n \\ a_{21}m_1 & a_{22}m_2 & \cdots & a_{2n}m_n \\ \cdots & \cdots & \cdots & \cdots \\ a_{n1}m_1 & a_{n2}m_2 & \cdots & a_{nn}m_n \end{bmatrix} \begin{Bmatrix} \ddot{x}_1 \\ \ddot{x}_2 \\ \vdots \\ \ddot{x}_n \end{Bmatrix} = \mathbf{0} \tag{5-60}$$

Assuming harmonic motion for each mass, so that $x_i = X_i e^{j\omega t}$ and $\ddot{x}_i = -\omega^2 X_i e^{j\omega t}$, Eq. 5-60 becomes

$$\begin{Bmatrix} X_1 \\ X_2 \\ \vdots \\ X_n \end{Bmatrix} + \begin{bmatrix} a_{11}m_1 & a_{12}m_2 & \cdots & a_{1n}m_n \\ a_{21}m_1 & a_{22}m_2 & \cdots & a_{2n}m_n \\ \cdots & \cdots & \cdots & \cdots \\ a_{n1}m_1 & a_{n2}m_2 & \cdots & a_{nn}m_n \end{bmatrix} \begin{Bmatrix} -\omega^2 X_1 \\ -\omega^2 X_2 \\ \vdots \\ -\omega^2 X_n \end{Bmatrix} = \mathbf{0} \tag{5-61}$$

which can also be written as

$$\begin{bmatrix} \left(a_{11}m_1 - \dfrac{1}{\omega^2}\right) & a_{12}m_2 & \cdots & a_{1n}m_n \\ a_{21}m_1 & \left(a_{22}m_2 - \dfrac{1}{\omega^2}\right) & \cdots & a_{2n}m_n \\ \cdots & \cdots & \cdots & \cdots \\ a_{n1}m_1 & a_{n2}m_2 & \cdots & \left(a_{nn}m_n - \dfrac{1}{\omega^2}\right) \end{bmatrix} \begin{Bmatrix} X_1 \\ X_2 \\ \vdots \\ X_n \end{Bmatrix} = \mathbf{0} \tag{5-62}$$

or more compactly as

$$\left[\mathbf{AM} - \frac{1}{\omega^2}\mathbf{I}\right]\mathbf{X} = \mathbf{0} \tag{5-63}$$

in which the identity matrix **I** indicates that $1/\omega^2$ appears in terms of the diagonal of the square matrix

$$\left[\mathbf{AM} - \frac{1}{\omega^2}\mathbf{I}\right]$$

Thus, the eigenvalues of Eq. 5-62 are $1/\omega^2$. The eigenvalues are determined by setting the determinant of the coefficient matrix of Eq. 5-62 equal to zero with $\lambda = 1/\omega^2$,

$$|D| = \begin{vmatrix} (a_{11}m_1 - \lambda) & a_{12}m_2 & \cdot & \cdot & \cdot & a_{1n}m_n \\ a_{21}m_1 & (a_{22}m_2 - \lambda) & \cdot & \cdot & \cdot & a_{2n}m_n \\ \cdot & \cdot & \cdot & \cdot & \cdot & \cdot \\ a_{n1}m_1 & a_{n2}m_2 & \cdot & \cdot & \cdot & (a_{nn}m_n - \lambda) \end{vmatrix} = 0 \qquad (5\text{-}64)$$

The expansion of this determinant results in the nth-degree polynomial equation

$$\lambda^n + b_1\lambda^{n-1} + b_2\lambda^{n-2} + \cdots + b_n = 0 \qquad (5\text{-}65)$$

for which the roots ($\lambda = 1/\omega^2$) are the eigenvalues. The constants $b_1, b_2, \ldots, b_n$ depend upon the values of the flexibility coefficients and the values of the masses. The eigenvalues obtained from Eq. 5-65 are thus the reciprocals of the squares of the undamped natural circular frequencies.

After the eigenvalues are determined, they can be substituted, one value at a time, back into Eq. 5-62 to obtain the corresponding eigenvectors, which are the normal-mode shapes.

EXAMPLE 5-15

It is desired to determine the natural frequencies and the normal-mode shapes of the system shown in Fig. 5-18 of Example 5-12 using flexibility coefficients.

Solution. Remembering the definition of a_{ij} while inspecting Fig. 5-18, we find that

$$a_{11} = \frac{1}{k} + \frac{1}{k} = \frac{2}{k}$$

$$a_{12} = \frac{1}{k}$$

$$a_{21} = \frac{1}{k}$$

$$a_{22} = \frac{1}{k}$$

Substituting the above quantities into Eq. 5-64, we see that

$$|D| = \begin{vmatrix} \left(\frac{2I}{k} - \lambda\right) & \frac{I}{k} \\ \frac{I}{k} & \left(\frac{I}{k} - \lambda\right) \end{vmatrix} = 0$$

Expanding this determinant,

$$\left(\frac{2I}{k} - \lambda\right)\left(\frac{I}{k} - \lambda\right) - \left(\frac{I}{k}\right)^2 = 0$$

or

$$\lambda^2 - 3\frac{I}{k}\lambda + \left(\frac{I}{k}\right)^2 = 0$$

The roots of this quadratic are found to be

$$\left.\begin{aligned}\lambda_1 &= \frac{1}{\omega_1^2} = \frac{I}{k}\left(\frac{3+\sqrt{5}}{2}\right)\\ \lambda_2 &= \frac{1}{\omega_2^2} = \frac{I}{k}\left(\frac{3-\sqrt{5}}{2}\right)\end{aligned}\right\} \tag{5-66}$$

Using the parameter values

$$I = 12 \text{ lb}\cdot\text{in.}\cdot\text{s}^2$$
$$k = 120(10)^3 \text{ lb}\cdot\text{in./rad}$$

given in Example 5-12, we find from Eq. 5-66 that

$$\left.\begin{aligned}\lambda_1 &= 2.62(10)^{-4}\text{ s}^2\\ \lambda_2 &= 0.382(10)^{-4}\text{ s}^2\end{aligned}\right\} \tag{5-67}$$

Remembering that $\lambda = 1/\omega^2$,

$$f_1 = \frac{\omega_1}{2\pi} = \frac{1}{2\pi\sqrt{\lambda_1}} = 9.84\text{ Hz}$$

$$f_2 = \frac{\omega_2}{2\pi} = \frac{1}{2\pi\sqrt{\lambda_2}} = 25.75\text{ Hz}$$

These frequency values are, of course, the same as those determined in Example 5-12 in which stiffness coefficients were used.

To determine the normal-mode shapes (the eigenvectors), we substitute the a_{ij} values determined into Eq. 5-62 to obtain

$$\begin{bmatrix}\left(\frac{2I}{k}-\lambda\right) & \frac{I}{k}\\ \frac{I}{k} & \left(\frac{I}{k}-\lambda\right)\end{bmatrix}\begin{Bmatrix}\Theta_1\\ \Theta_2\end{Bmatrix} = \mathbf{0} \tag{5-68}$$

The relationship between Θ_1 and Θ_2 can be obtained from either the first or second equation of Eq. 5-68. Using the second one,

$$\frac{\Theta_2}{\Theta_1} = \frac{-I/k}{I/k-\lambda} \tag{5-69}$$

Substituting the given values of I and k and the values of λ in Eq. 5-67 into Eq. 5-69 shows that

$$\left.\frac{\Theta_2}{\Theta_1}\right)_1 = \frac{-10^{-4}}{10^{-4}-2.62(10)^{-4}} = 0.62 \qquad \text{(first mode)}$$

and

$$\left.\frac{\Theta_2}{\Theta_1}\right)_2 = \frac{-10^{-4}}{10^{-4}-0.382(10)^{-4}} = -1.62 \qquad \text{(second mode)}$$

which are the same values as those determined in Example 5-12.

Normalizing the above relationships with respect to $\Theta_1 = 1$ the modal matrix is

$$[u] = \begin{bmatrix} 1.0 & 1.0 \\ 0.62 & -1.62 \end{bmatrix}$$

It has been shown in the preceding paragraphs that the natural frequencies and normal-mode shapes of a system can be determined from equations formulated in terms of either the stiffness or the flexibility coefficients. The pertinent equations involving the use of each of these types of influence coefficients are summarized for convenience in the two paragraphs that follow.

Equations Involving Stiffness Coefficients

$$[\mathbf{M}^{-1}\mathbf{K} - \lambda\mathbf{I}]\mathbf{X} = \mathbf{0} \qquad \text{(see Eq. 5-36 for general form)}$$
$$|D| = |\mathbf{M}^{-1}\mathbf{K} - \lambda\mathbf{I}| = 0 \qquad \text{(see Eq. 5-37 for general form)}$$
$$\lambda^n + b_1\lambda^{n-1} + b_2\lambda^{n-2} + \cdots + b_n = 0 \qquad \text{(see Eq. 5-38)}$$

in which $\lambda = \omega^2$ and n indicates the number of degrees of freedom.

Equations Involving Flexibility Coefficients

$$[\mathbf{AM} - \lambda\mathbf{I}]\mathbf{X} = \mathbf{0} \qquad \text{(see Eq. 5-62 for general form)}$$
$$|D| = |\mathbf{AM} - \lambda\mathbf{I}| = 0 \qquad \text{(see Eq. 5-64 for general form)}$$
$$\lambda^n + b_1\lambda^{n-1} + b_2\lambda^{n-2} + \cdots + b_n = 0 \qquad \text{(see Eq. 5-65)}$$

in which $\lambda = 1/\omega^2$ and n again indicates the number of degrees of freedom.

The eigenvalues and eigenvectors were determined in a straightforward manner for the two- and three-degree-of-freedom systems analyzed in the preceding examples. However, the computations required to determine them become increasingly more difficult as the number of degrees of freedom becomes greater than three. As a result, various numerical methods are used with the digital computer to facilitate the determination of the eigenvalues and eigenvectors of systems having a large number of degrees of freedom. Some of these methods with accompanying computer programs are presented in Secs. 5-9, 5-10, and 5-11.

5-6 ORTHOGONALITY PROPERTIES OF THE NORMAL MODES

Basically, there are two orthogonality relationships involving the normal modes (the eigenvectors). One involves the mass matrix **M**, the other the stiffness matrix **K**.[7]

To develop these relationships, we begin with the matrix equation for undamped free vibration, Eq. 5-9, which is repeated here as

$$\mathbf{M\ddot{X}} + \mathbf{KX} = \mathbf{0} \tag{5-9}$$

[7] A familiarity with the orthogonality properties of the normal modes (the eigenvectors) is the key to understanding the concepts of modal analysis that are discussed in Secs. 5-7 and 6-6. As will be seen there, these concepts greatly simplify the response analysis of both the free and forced vibration of multiple-degree-of-freedom systems.

where $\mathbf{M} = [m_{ij}]$ (mass matrix)
$\mathbf{X} = \{x_i\}$ (column matrix of coordinates)
$\ddot{\mathbf{X}} = \{\ddot{x}_i\}$ (column matrix of accelerations)
$\mathbf{K} = [k_{ij}]$ (stiffness matrix)

Let us initially assume that the mass matrix is nondiagonal and symmetric. With this assumption, and further assuming harmonic motion, so that $x_i = X_i e^{j\omega t}$, we obtain from Eq. 5-9 that

$$\omega^2\mathbf{MX} = \mathbf{KX} \tag{5-70}$$

Equation 5-70 must hold for any normal mode of vibration; so designating $\mathbf{X}_r$ and $\mathbf{X}_s$ as the eigenvectors for the rth and sth modes, respectively, we can see from Eq. 5-70 that

$$\omega_r^2\mathbf{MX}_r = \mathbf{KX}_r \qquad (r\text{th mode}) \tag{5-71}$$

and

$$\omega_s^2\mathbf{MX}_s = \mathbf{KX}_s \qquad (s\text{th mode}) \tag{5-72}$$

in which ω_r^2 and ω_s^2 are the squares of the undamped natural circular frequencies of the rth and sth modes, respectively.

Postmultiplying the *transpose* of Eq. 5-71 by $\mathbf{X}_s$ gives

$$\omega_r^2[\mathbf{MX}_r]^T\mathbf{X}_s = [\mathbf{KX}_r]^T\mathbf{X}_s \tag{5-73}$$

From matrix algebra (see Appendix B), we know that

$$\left.\begin{aligned}[\mathbf{MX}_r]^T &= \mathbf{X}_r^T\mathbf{M}^T\\ [\mathbf{KX}_r]^T &= \mathbf{X}_r^T\mathbf{K}^T\end{aligned}\right\} \tag{5-74}$$

Substituting these equalities into Eq. 5-73, we obtain

$$\omega_r^2\mathbf{X}_r^T\mathbf{M}^T\mathbf{X}_s = \mathbf{X}_r^T\mathbf{K}^T\mathbf{X}_s \tag{5-75}$$

Since $\mathbf{M}$ and $\mathbf{K}$ are symmetric matrices, $\mathbf{M}^T = \mathbf{M}$ and $\mathbf{K}^T = \mathbf{K}$, so that Eq. 5-75 can be written as

$$\omega_r^2\mathbf{X}_r^T\mathbf{MX}_s = \mathbf{X}_r^T\mathbf{KX}_s \tag{5-76}$$

We next *premultiply* Eq. 5-72 by $\mathbf{X}_r^T$ to obtain

$$\omega_s^2\mathbf{X}_r^T\mathbf{MX}_s = \mathbf{X}_r^T\mathbf{KX}_s \tag{5-77}$$

Subtracting Eq. 5-77 from Eq. 5-76 yields

$$(\omega_r^2 - \omega_s^2)\mathbf{X}_r^T\mathbf{MX}_s = \mathbf{0} \tag{5-78}$$

Considering r and s as two distinct modes, $\omega_r^2 \neq \omega_s^2$, Eq. 5-78 shows that

$$\mathbf{X}_r^T\mathbf{MX}_s = \mathbf{0} \qquad (r \neq s) \tag{5-79}$$

Equation 5-79 expresses the orthogonality relationship between *any* two eigenvectors, $\mathbf{X}_r$ and $\mathbf{X}_s$, relative to the mass matrix $\mathbf{M}$.

Since $\mathbf{X}_r$ is a column matrix, $\mathbf{X}_r^T$ is a *row* matrix. Thus, for an n-degree-of-freedom system, Eq. 5-79 can also be written as

$$[X_1 \quad X_2 \quad X_3 \cdots X_n]_r \mathbf{M} \begin{Bmatrix} X_1 \\ X_2 \\ X_3 \\ \vdots \\ X_n \end{Bmatrix}_s = \mathbf{0} \tag{5-80}$$

If the mass matrix **M** is a diagonal matrix (no dynamic coupling), which is generally the case, then Eq. 5-80 becomes

$$[X_1 \quad X_2 \quad X_3 \cdots X_n]_r \begin{Bmatrix} m_1 X_1 \\ m_2 X_2 \\ m_3 X_3 \\ \vdots \\ m_n X_n \end{Bmatrix}_s = \mathbf{0} \tag{5-81}$$

or, more compactly,

$$\sum_{i=1}^{n} m_i (X_i)_r (X_i)_s = 0 \tag{5-81a}$$

In the case of dynamic coupling in which **M** is not a diagonal matrix, the general form of Eq. 5-80 is

$$[X_1 \quad X_2 \cdots X_n]_r \begin{Bmatrix} m_{11}X_1 + m_{12}X_2 + \cdots + m_{1n}X_n \\ m_{21}X_1 + m_{22}X_2 + \cdots + m_{2n}X_n \\ \cdot \quad \cdot \quad \cdot \quad \cdot \quad \cdot \quad \cdot \quad \cdot \\ m_{n1}X_1 + m_{n2}X_2 + \cdots + m_{nn}X_n \end{Bmatrix}_s = \mathbf{0} \tag{5-82}$$

or, more compactly,

$$\sum_{i=1}^{n} \sum_{j=1}^{n} m_{ij} (X_i)_r (X_j)_s = \mathbf{0} \tag{5-82a}$$

Substituting Eq. 5-79 into either Eq. 5-76 or Eq. 5-77, we find that

$$\mathbf{X}_r' \mathbf{K} \mathbf{X}_s = \mathbf{0} \qquad (r \neq s) \tag{5-83}$$

which expresses the orthogonality relationship between *any* two eigenvectors, $\mathbf{X}_r$ and $\mathbf{X}_s$, relative to the stiffness matrix **K**.

As we shall soon see, the orthogonality relationships expressed by Eqs. 5-79 and 5-83 are fundamental to decoupling equations of motion. They are also useful in checking the accuracy of eigenvectors (normal modes) determined by the numerical procedures required in the analysis of systems having a large number of degrees of freedom. Since **M** is usually a diagonal matrix, it is somewhat more efficient to check the accuracy of the eigenvectors by using Eq. 5-81 rather than Eq. 5-83.

Although the term "orthogonal" means mutually perpendicular, it is more of a mathematical than a geometric concept in most vibration problems. Mathematically, two vectors are said to be orthogonal if their scalar product is zero. One of the few cases in which the orthogonality of the normal modes of vibration can be given a simple geometric interpretation is shown in Fig. 5-25. Assuming that the particle of mass m has the two degrees of freedom described by the x and y coordinates, it is possible for the mass to vibrate along some line at a frequency corresponding to the first mode. The vibration

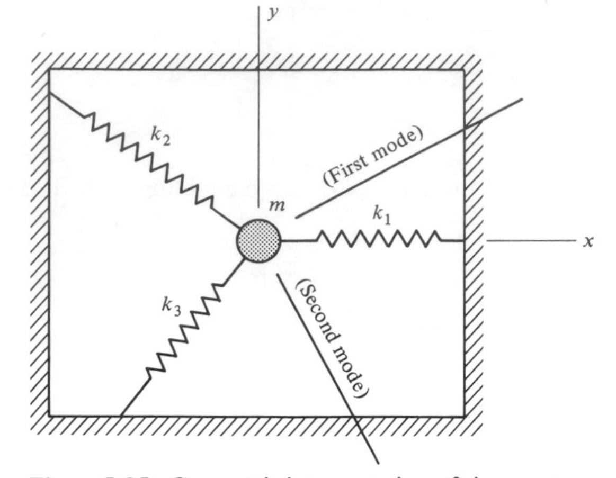

Figure 5-25 Geometric interpretation of eigenvectors.

corresponding to the second mode would then be along a line perpendicular to that of the first mode, at the second-mode frequency.

Generalized Mass and Stiffness Matrices

If we consider that $r = s$, Eqs. 5-79 and 5-83 are not equal to zero. Thus, when $r = s$ and $\mathbf{M}$ is a diagonal matrix, we can write from Eqs. 5-79 and 5-81a that

$$\mathbf{M}_r = \mathbf{X}_r^T\mathbf{M}\mathbf{X}_r = \sum_{i=1}^{n} m_i(X_i^2)_r \qquad (r = 1, 2, 3, \ldots, n) \tag{5-84}$$

in which $\mathbf{M}_r$ is referred to as the *generalized mass* for the rth mode.

In the case of dynamic coupling, in which $\mathbf{M}$ is a nondiagonal matrix, Eq. 5-82a with $r = s$ shows that the generalized mass is

$$\mathbf{M}_r = \sum_{i=1}^{n}\sum_{j=1}^{n} m_{ij}(X_i)_r(X_j)_r \qquad (r = 1, 2, 3, \ldots, n) \tag{5-84a}$$

In a similar manner, with $r = s$, we can write from Eq. 5-83 that

$$\mathbf{K}_r = \mathbf{X}_r^T\mathbf{K}\mathbf{X}_r \qquad (r = 1, 2, 3, \ldots, n) \tag{5-85}$$

in which $\mathbf{K}_r$ is referred to as the *generalized stiffness* for the rth mode.

It now follows from Eqs. 5-76, 5-84, and 5-85 that

$$\left.\begin{aligned} \omega_r^2\mathbf{X}_r^T\mathbf{M}\mathbf{X}_r &= \mathbf{X}_r^T\mathbf{K}\mathbf{X}_r \\ \text{or} \qquad \omega_r^2\mathbf{M}_r &= \mathbf{K}_r \end{aligned}\right\} \tag{5-86}$$

The reader should note that $\mathbf{M}_r$ and $\mathbf{K}_r$ are 1×1 matrices, and although matrices do not have numerical values, Eqs. 5-84 and 5-85 are used to calculate the values of the single elements of these matrices.

EXAMPLE 5-16

In Example 5-12, the eigenvalues and eigenvectors were found to be

$$\lambda_1 = \omega_1^2 = 3.82(10)^3$$
$$\lambda_2 = \omega_2^2 = 26.18(10)^3$$

and

$$\begin{Bmatrix} \Theta_1 \\ \Theta_2 \end{Bmatrix}_1 = \begin{Bmatrix} 1.0 \\ 0.62 \end{Bmatrix} \qquad \begin{Bmatrix} \Theta_1 \\ \Theta_2 \end{Bmatrix}_2 = \begin{Bmatrix} 1.0 \\ -1.62 \end{Bmatrix}$$

respectively.

Noting that the mass moment of inertia of each disk is $I = 12$ lb · in. · s², (a) check the orthogonality relationship of the two modes, (b) determine the elements of the generalized mass matrix, and (c) determine the elements of the generalized stiffness matrix.

Solution. **a.** The orthogonality relationship of the two modes can be checked by using either Eq. 5-81 or Eq. 5-83. Using the former, and noting that $m_1 = m_2 = I = 12$, we obtain

$$[1.0 \quad 0.62]_1 \begin{Bmatrix} 12 & (1.0) \\ 12 & (-1.62) \end{Bmatrix}_2 = -0.053$$

which is not quite zero because of the roundoff in the computations used to determine the eigenvalues and eigenvectors in Example 5-12.

b. The generalized mass terms M_r are determined using Eq. 5-84 and are found to be

$$M_1 = \sum_{i=1}^{2} m_i(\Theta_i^2)_1 = 12(1)^2 + 12(0.62)^2 = 16.61 \text{ lb} \cdot \text{in.} \cdot \text{s}^2$$

and

$$M_2 = \sum_{i=1}^{2} m_i(\Theta_i^2)_2 = 12(1)^2 + 12(-1.62)^2 = 43.49 \text{ lb} \cdot \text{in.} \cdot \text{s}^2$$

c. The generalized stiffness terms K_r can be determined from Eq. 5-85, or more simply from Eq. 5-86, since ω_1^2 and ω_2^2 are known. Using the latter, we find that

$$K_1 = \omega_1^2 M_1 = 3.82(10)^3(16.61) = 63.45(10)^3 \text{ lb} \cdot \text{in./rad}$$

and

$$K_2 = \omega_2^2 M_2 = 26.18(10)^3(43.49) = 11.39(10)^5 \text{ lb} \cdot \text{in./rad}$$

EXAMPLE 5-17

In Example 5-14, the eigenvalues and eigenvectors were determined using a pocket calculator, and the calculations were rounded to two places after the decimal point. In Sec. 5-10 a computer program utilizing an iteration method determined the same eigenvalues and eigenvectors more accurately as

$$\lambda_1 = \omega_1^2 = 149.560029$$
$$\lambda_2 = \omega_2^2 = 652.108901$$
$$\lambda_3 = \omega_3^2 = 1281.664096$$

$$\begin{Bmatrix} X_1 \\ X_2 \\ X_3 \end{Bmatrix}_1 = \begin{Bmatrix} 1.0 \\ 2.051321 \\ 2.926780 \end{Bmatrix} \qquad \begin{Bmatrix} X_1 \\ X_2 \\ X_3 \end{Bmatrix}_2 = \begin{Bmatrix} 1.0 \\ 0.543674 \\ -1.787118 \end{Bmatrix} \qquad \begin{Bmatrix} X_1 \\ X_2 \\ X_3 \end{Bmatrix}_3 = \begin{Bmatrix} 1.0 \\ -1.344993 \\ 0.860339 \end{Bmatrix}$$

We wish to (a) evaluate the accuracy of the eigenvalues and eigenvectors by checking the orthogonality relationships among the three modes, (b) determine the generalized mass elements M_1, M_2, and M_3 of the three modes.

Solution. **a.** Referring to Example 5-14, the lumped masses modeling the building are

$$m_1 = 24(10)^5 \text{ kg} \qquad m_2 = 16(10)^5 \text{ kg} \qquad m_3 = 8(10)^5 \text{ kg}$$

Using Eq. 5-81, the orthogonality check of the first and second modes ($r = 1, 2$) shows that

$$[1.0 \quad 2.051321 \quad 2.926780]_1 (10)^5 \begin{Bmatrix} 24(1.0) \\ 16(0.543674) \\ 8(-1.787118) \end{Bmatrix}_2 = -1.1$$

Similarly, for $r = 1, 3$ and $r = 2, 3$, we obtain

$$[1.0 \quad 2.051321 \quad 2.926780]_1 (10)^5 \begin{Bmatrix} 24(1.0) \\ 16(-1.344993) \\ 8(0.860339) \end{Bmatrix}_3 = -1.4$$

and

$$[1.0 \quad 0.543674 \quad -1.787118]_2 (10)^5 \begin{Bmatrix} 24(1.0) \\ 16(-1.344993) \\ 8(0.860339) \end{Bmatrix}_3 = -2.2$$

b. Utilizing Eq. 5-84, the generalized mass elements M_r are found to be

$$M_1 = 10^5[24(1.0)^2 + 16(2.051321)^2 + 8(2.926780)^2] = 159.86(10)^5$$
$$M_2 = 10^5[24(1.0)^2 + 16(0.543674)^2 + 8(-1.787118)^2] = 54.28(10)^5$$
$$M_3 = 10^5[24(1.0)^2 + 16(-1.344993)^2 + 8(0.860339)^2] = 58.87(10)^5$$

The reader might question the accuracy of the eigenvectors, since the orthogonality checks resulted in numbers that do not appear to be zero. However, the results of -1.1, -1.4, and -2.2 obtained in the computations above do indicate that the values of the eigenvectors are quite accurate, for two reasons. First, these numbers are quite small compared with the numbers in the computations that involve magnitudes on the order of 10^5 as shown above. Second, the results are very small compared with the values determined for the generalized mass elements, which involve magnitudes on the order of 10^7, as shown above.

As we shall see in the following section, the quantities -1.1, -1.4, and -2.2 correspond to the off-diagonal elements of a 3×3 matrix, while the generalized mass elements M_1, M_2, and M_3 correspond to elements on the diagonal of the matrix. Thus, for practical purposes this 3×3 matrix can be written as

$$\begin{bmatrix} M_1 & 0 & 0 \\ 0 & M_2 & 0 \\ 0 & 0 & M_3 \end{bmatrix} = \begin{bmatrix} 159.86(10)^5 & 0 & 0 \\ 0 & 54.28(10)^5 & 0 \\ 0 & 0 & 58.87(10)^5 \end{bmatrix}$$

5-7 MODAL ANALYSIS—PRINCIPAL COORDINATES

Modal analysis basically involves the uncoupling of differential equations of motion as a means of obtaining independent equations to facilitate the response analysis of multiple-degree-of-freedom systems. The independent equations that result from the decoupling process are expressed in terms of coordinates that are referred to as *principal coordinates,* and in terms of normal-mode parameters that include the natural frequencies and modal-damping properties of the system. In this process there is one independent equation for each normal (natural) mode of vibration, and it can be solved as if it were the equation of a single-degree-of-freedom system. The system response is then obtained by superposition of the responses of the individual normal modes.

Although the greatest advantage of modal analysis lies in its use in determining the response of forced systems, it can also be used to some advantage in obtaining the free-vibration response of systems that comes from *initial conditions.* The discussion that follows will also be pertinent to the use of modal analysis in obtaining the total response of forced systems, which is discussed in Chap. 6.

Uncoupled Equations for Undamped Free Vibration

As explained in Sec. 5-4, differential equations of motion are dynamically coupled if the coordinates selected lead to nondiagonal mass matrices **M**, while static coupling is present when the stiffness matrix **K** is nondiagonal. Dynamic coupling can usually be avoided by a proper selection of coordinates, as explained in Sec. 5-4. However, it is generally quite difficult to select coordinates that eliminate static coupling, and as a result it is usually present in equations of motion.

As discussed previously, the equations of motion for undamped free vibration, with dynamic and/or static coupling, can be written in either of the following forms:

$$\left.\begin{aligned} \mathbf{M}\ddot{\mathbf{X}} + \mathbf{K}\mathbf{X} &= \mathbf{0} \\ \mathbf{M}\{\ddot{x}_i\} + \mathbf{K}\{x_i\} &= \mathbf{0} \end{aligned}\right\} \tag{5-87}$$

in which

$$\mathbf{X} = \{x_i\} = \text{column matrix of generalized coordinates}$$

and

$$\ddot{\mathbf{X}} = \{\ddot{x}_i\} = \text{column matrix of accelerations}$$

To decouple these equations of motion, we start with the *linear transformation* relationship

$$\{x_i\} = [u]\{\delta_i\} \tag{5-88}$$

from which

$$\{\ddot{x}_i\} = [u]\{\ddot{\delta}_i\} \tag{5-89}$$

In these equations, $[u]$ is the *modal matrix* (see Eq. 5-59), and the δ_i's are referred to as *principal (normal) coordinates*. In general, the principal coordinates do not have a geometric interpretation but are used only in a mathematical sense to transform equations from one form to another.

Substituting Eqs. 5-88 and 5-89 into Eq. 5-87 gives

$$\mathbf{M}[u]\{\ddot{\delta}_i\} + \mathbf{K}[u]\{\delta_i\} = \mathbf{0} \tag{5-90}$$

We next premultiply Eq. 5-90 by $[u]^T$ (the transpose of $[u]$) to obtain

$$[u]^T\mathbf{M}[u]\{\ddot{\delta}_i\} + [u]^T\mathbf{K}[u]\{\delta_i\} = \mathbf{0} \tag{5-91}$$

which we shall show reduces to uncoupled equations having the form

$$\begin{bmatrix} \diagdown & & \\ & M_r & \\ & & \diagdown \end{bmatrix} \begin{Bmatrix} \ddot{\delta}_1 \\ \ddot{\delta}_2 \\ \vdots \\ \ddot{\delta}_n \end{Bmatrix} + \begin{bmatrix} \diagdown & & \\ & \omega_r^2 M_r & \\ & & \diagdown \end{bmatrix} \begin{Bmatrix} \delta_1 \\ \delta_2 \\ \vdots \\ \delta_n \end{Bmatrix} = 0 \qquad (r = 1, 2, \ldots, n) \tag{5-92}$$

where M_r = generalized mass of rth mode (see Eq. 5-84)
ω_r = undamped natural circular frequency of rth mode

To verify Eq. 5-92, we first note that

$$[u] = \left[\begin{Bmatrix} X_1 \\ X_2 \\ \vdots \\ X_n \end{Bmatrix}_1 \begin{Bmatrix} X_1 \\ X_2 \\ \vdots \\ X_n \end{Bmatrix}_2 \cdots \begin{Bmatrix} X_1 \\ X_2 \\ \vdots \\ X_n \end{Bmatrix}_r \cdots \begin{Bmatrix} X_1 \\ X_2 \\ \vdots \\ X_n \end{Bmatrix}_s \cdots \begin{Bmatrix} X_1 \\ X_2 \\ \vdots \\ X_n \end{Bmatrix}_n \right] \tag{5-93}$$

Mode

The transpose of $[u]$ is then

$$[u]^T = \begin{bmatrix} [X_1 & X_2 & \cdot & \cdot & \cdot & X_n]_1 \\ [X_1 & X_2 & \cdot & \cdot & \cdot & X_n]_2 \\ \cdot & \cdot & \cdot & \cdot & \cdot & \cdot \\ [X_1 & X_2 & \cdot & \cdot & \cdot & X_n]_r \\ \cdot & \cdot & \cdot & \cdot & \cdot & \cdot \\ [X_1 & X_2 & \cdot & \cdot & \cdot & X_n]_s \\ \cdot & \cdot & \cdot & \cdot & \cdot & \cdot \\ [X_1 & X_2 & \cdot & \cdot & \cdot & X_n]_n \end{bmatrix} \tag{5-94}$$

Mode

in which r and s refer to any two eigenvectors (mode shapes) of an n-degree-of-freedom system. Considering the matrix-multiplication process of $[u]^T\mathbf{M}[u]$ in Eq. 5-91, which involves the rth *row* of Eq. 5-94 and the sth *column* of Eq. 5-93, we see that

$$[X_1 \quad X_2 \quad \cdot \quad \cdot \quad \cdot \quad X_n]_r \mathbf{M} \begin{Bmatrix} X_1 \\ X_2 \\ \vdots \\ X_n \end{Bmatrix}_s = \mathbf{X}_r^T\mathbf{M}\mathbf{X}_s \tag{5-95}$$

which is the orthogonality relationship of the rth and sth modes shown in Eq. 5-79. Thus,

$$\mathbf{X}_r^T\mathbf{M}\mathbf{X}_s = \mathbf{0} \qquad (r \neq s)$$

when $r \neq s$. This shows that all the off-diagonal elements of $[u]^T\mathbf{M}[u]$ are *zero*. However, when $r = s$, we should recall from Eq. 5-84 that Eq. 5-95 gives

$$\mathbf{X}_r^T\mathbf{M}\mathbf{X}_r = \mathbf{M}_r \qquad (r = 1, 2, 3, \ldots, n)$$

in which the generalized mass $\mathbf{M}_r$ corresponds to an element *on the diagonal* of $[u]^T\mathbf{M}[u]$.

Therefore, we can conclude that

$$[u]^T\mathbf{M}[u] = \begin{bmatrix} M_1 & 0 & 0 & \cdot & \cdot & \cdot & 0 \\ 0 & M_2 & 0 & \cdot & \cdot & \cdot & 0 \\ 0 & 0 & M_3 & \cdot & \cdot & \cdot & 0 \\ \cdot & \cdot & \cdot & \cdot & \cdot & \cdot & \cdot \\ 0 & 0 & 0 & \cdot & \cdot & \cdot & M_n \end{bmatrix} \tag{5-96}$$

Similarly, for $[u]^T\mathbf{K}[u]$ of Eq. 5-91, it can be seen that

$$[X_1 \quad X_2 \quad \cdot \quad \cdot \quad \cdot \quad X_n]_r \mathbf{K} \begin{Bmatrix} X_1 \\ X_2 \\ \vdots \\ X_n \end{Bmatrix}_s = \mathbf{X}_r^T\mathbf{K}\mathbf{X}_s \tag{5-97}$$

which is the orthogonality relationship shown in Eq. 5-83. Thus,

$$\mathbf{X}_r^T\mathbf{K}\mathbf{X}_s = 0 \qquad (r \neq s)$$

when $r \neq s$. This shows that all the off-diagonal elements of $[u]^T\mathbf{K}[u]$ are *zero*. However, when $r = s$, Eq. 5-97 corresponds to Eq. 5-85, so that

$$\mathbf{X}_r^T\mathbf{K}\mathbf{X}_r = \mathbf{K}_r \qquad (r = 1, 2, 3, \ldots, n)$$

in which the generalized stiffness $\mathbf{K}_r$ corresponds to an element on *the diagonal* of $[u]^T\mathbf{K}[u]$. Thus,

$$[u]^T\mathbf{K}[u] = \begin{bmatrix} K_1 & 0 & 0 & \cdot & \cdot & \cdot & 0 \\ 0 & K_2 & 0 & \cdot & \cdot & \cdot & 0 \\ 0 & 0 & K_3 & \cdot & \cdot & \cdot & 0 \\ \cdot & \cdot & \cdot & \cdot & \cdot & \cdot & \cdot \\ 0 & 0 & 0 & \cdot & \cdot & \cdot & K_n \end{bmatrix} \tag{5-98}$$

Recalling from Eq. 5-86 that

$$\omega_r^2\mathbf{M}_r = \mathbf{K}_r \qquad (r = 1, 2, 3, \ldots, n)$$

Eq. 5-98 can be written as

$$[u]^T\mathbf{K}[u] = \begin{bmatrix} \omega_1^2 M_1 & 0 & 0 & \cdot & \cdot & \cdot & 0 \\ 0 & \omega_2^2 M_2 & 0 & \cdot & \cdot & \cdot & 0 \\ 0 & 0 & \cdot & \cdot & \cdot & \cdot & \cdot \\ \cdot & \cdot & \cdot & \cdot & \omega_r^2 M_r & \cdot & \cdot \\ \cdot & \cdot & \cdot & \cdot & \cdot & \cdot & \cdot \\ 0 & 0 & 0 & \cdot & \cdot & \cdot & \omega_n^2 M_n \end{bmatrix} \tag{5-99}$$

where ω_r = undamped natural circular frequency of rth mode

$M_r = \sum_{i=1}^{n} m_i(X_i^2)_r$ = generalized mass of rth mode when mass matrix is diagonal

$M_r = \sum_{i=1}^{n} \sum_{j=1}^{n} m_{ij}(X_i)_r(X_j)_r$ = generalized mass of rth mode when mass matrix is nondiagonal

Referring to Eqs. 5-96 and 5-99, we can see that Eq. 5-91 reduces to

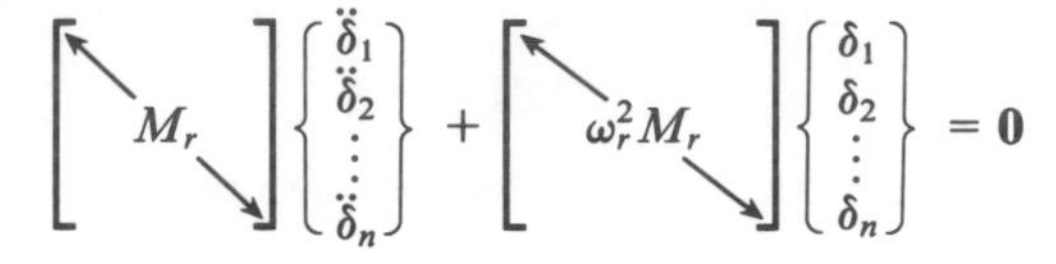

which is identical to Eq. 5-92. Thus, Eq. 5-92 shows that the n uncoupled (independent) equations of an n-degree-of-freedom system, in terms of the principal coordinates, are

$$\left.\begin{aligned} \ddot{\delta}_1 + \omega_1^2 \delta_1 &= 0 \\ \ddot{\delta}_2 + \omega_2^2 \delta_2 &= 0 \\ &\cdots \\ \ddot{\delta}_r + \omega_r^2 \delta_r &= 0 \\ &\cdots \\ &\cdots \\ \ddot{\delta}_n + \omega_n^2 \delta_n &= 0 \end{aligned}\right\} \tag{5-100}$$

The solution (see Eq. 2-21) for any rth mode of Eq. 5-100 is then simply

$$\delta_r = A_r \cos \omega_r t + B_r \sin \omega_r t \qquad (r = 1, 2, 3, \ldots, n) \tag{5-101}$$

in which the constants A_r and B_r depend upon the initial conditions of the problem.

Undamped Free Vibration Response

Looking at both Eqs. 5-88 and 5-101, it follows that the undamped free vibration of an n-degree-of-freedom system due to initial conditions can be determined using

$$\begin{Bmatrix} x_1 \\ x_2 \\ x_3 \\ \vdots \\ x_n \end{Bmatrix} = [u] \begin{Bmatrix} A_1 \cos \omega_1 t + B_1 \sin \omega_1 t \\ A_2 \cos \omega_2 t + B_2 \sin \omega_2 t \\ A_3 \cos \omega_3 t + B_3 \sin \omega_3 t \\ \cdots \\ A_n \cos \omega_n t + B_n \sin \omega_n t \end{Bmatrix} \tag{5-102}$$

in which $\{x_i\}$ are the generalized coordinates used to describe the vibratory motion of the system. The constants A_r and B_r are determined from the initial conditions:

$$\{x_i\}_{t=0} \qquad \text{(displacements at } t = 0\text{)}$$
$$\{\dot{x}_i\}_{t=0} \qquad \text{(velocities at } t = 0\text{)}$$

For a system with n degrees of freedom, the initial conditions will result in $2n$ algebraic equations containing n unknowns in A_r and n unknowns in B_r.

To facilitate the computations involved in initial-condition problems, it is sometimes convenient to write Eq. 5-102 in the following form:

$$\begin{Bmatrix} x_1 \\ x_2 \\ x_3 \\ \vdots \\ x_n \end{Bmatrix} = \delta_1 \begin{Bmatrix} X_1 \\ X_2 \\ X_3 \\ \vdots \\ X_n \end{Bmatrix}_1 + \delta_2 \begin{Bmatrix} X_1 \\ X_2 \\ X_3 \\ \vdots \\ X_n \end{Bmatrix}_2 + \cdots + \delta_r \begin{Bmatrix} X_1 \\ X_2 \\ X_3 \\ \vdots \\ X_n \end{Bmatrix}_r + \cdots + \delta_n \begin{Bmatrix} X_1 \\ X_2 \\ X_3 \\ \vdots \\ X_n \end{Bmatrix}_n \tag{5-103}$$

in which $\delta_1, \delta_2, \delta_3, \ldots, \delta_n$ correspond to Eq. 5-101.

EXAMPLE 5-18

The undamped natural frequencies and the modal matrix $[u]$ of the structural system in Example 5-14 were found to be

$$\omega_1 = \sqrt{149.6} = 12.23 \text{ rad/s}$$
$$\omega_2 = \sqrt{652.6} = 25.55 \text{ rad/s}$$
$$\omega_3 = \sqrt{1280.9} = 35.79 \text{ rad/s}$$

and

$$[u] = \begin{bmatrix} 1.0 & 1.0 & 1.0 \\ 2.05 & 0.54 & -1.34 \\ 2.93 & -1.78 & 0.86 \end{bmatrix}$$

respectively.

The system is released from rest at $t = 0$ with the displacements shown in Fig. 5-26, and we wish to determine the undamped free vibration of the system as a function of time.

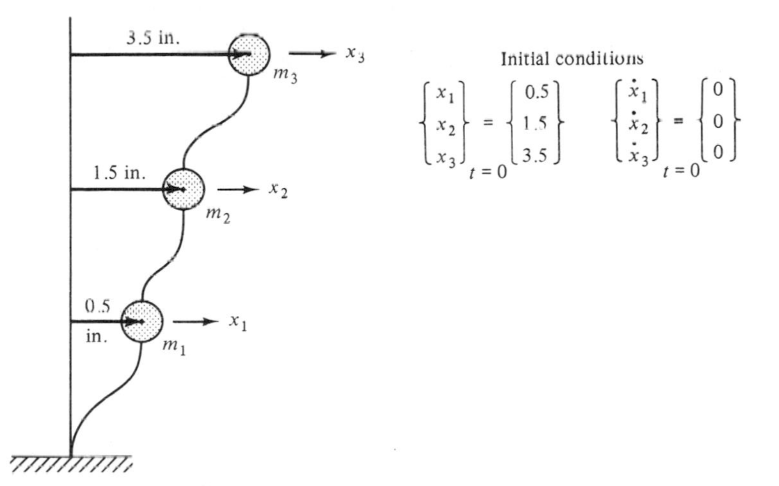

Figure 5-26 Configuration of system of Example 5-18 at $t = 0$.

Solution. Using the initial displacements shown in Fig. 5-26, Eq. 5-102 reduces to

$$\begin{Bmatrix} 0.5 \\ 1.5 \\ 3.5 \end{Bmatrix}_{t=0} = [u]\begin{Bmatrix} A_1 \\ A_2 \\ A_3 \end{Bmatrix} \tag{5-104}$$

Since the initial velocities are zero, it can be seen from Eq. 5-102 that

$$\begin{Bmatrix} 0.0 \\ 0.0 \\ 0.0 \end{Bmatrix}_{t=0} = [u]\begin{Bmatrix} B_1\omega_1 \\ B_2\omega_2 \\ B_3\omega_3 \end{Bmatrix} \tag{5-105}$$

which shows that $B_1 = B_2 = B_3 = 0$.

Substituting the modal matrix $[u]$ into Eq. 5-104 and carrying out the matrix multiplication will yield three algebraic equations that must be solved simultaneously to determine the unknowns A_1, A_2, and A_3. Although a number of methods are suitable for obtaining solutions of simultaneous algebraic equations, we shall use the *matrix-inversion* method, which is easily programmed for use on the digital computer.

If we premultiply Eq. 5-104 by $[u]^{-1}$ (the inverse of $[u]$), we see that

$$[u]^{-1}\begin{Bmatrix} 0.5 \\ 1.5 \\ 3.5 \end{Bmatrix} = \begin{Bmatrix} A_1 \\ A_2 \\ A_3 \end{Bmatrix} \tag{5-106}$$

This shows that if $[u]^{-1}$ is known, the unknown constants A_1, A_2, and A_3 can be determined by performing the simple matrix multiplication indicated in Eq. 5-106. Recalling the computer program INVERT.FOR, shown on page 303, we can use it to obtain $[u]^{-1}$. Using the values of the elements of $[u]$ as input data for the elements $A(I, J)$ of this program, we obtain the inverse of $[u]$ as

$$[u]^{-1} = \begin{bmatrix} 0.1495810 & 0.2055888 & 0.1464038 \\ 0.4430427 & 0.1611999 & -0.2639940 \\ 0.4073766 & -0.3667880 & 0.1175902 \end{bmatrix} \tag{5-107}$$

The constant values obtained from Eq. 5-106 using the above inverse matrix are found to be

$$A_1 = 0.896 \text{ in.}$$
$$A_2 = -0.461 \text{ in.}$$
$$A_3 = 0.065 \text{ in.}$$

Thus, the principal coordinates, as functions of time t, are

$$\left.\begin{aligned} \delta_1 &= 0.896 \cos(12.23\,t) \\ \delta_2 &= -0.461 \cos(25.55\,t) \\ \delta_3 &= 0.065 \cos(35.79\,t) \end{aligned}\right\} \tag{5-108}$$

If we then substitute the expressions of Eq. 5-108 and the eigenvectors of the modal matrix $[u]$ into Eq. 5-103, we obtain

$$\begin{aligned} x_1 &= 0.896 \cos(12.23\,t) - 0.461 \cos(25.55\,t) + 0.065 \cos(35.79\,t) \\ x_2 &= 1.837 \cos(12.23\,t) - 0.249 \cos(25.55\,t) - 0.087 \cos(35.79\,t) \\ x_3 &= 2.625 \cos(12.23\,t) + 0.821 \cos(25.55\,t) + 0.056 \cos(35.79\,t) \end{aligned}$$

which is the undamped time response of the system, resulting from the given initial conditions.

Program for Matrix Multiplication (MATMULT.FOR)

The FORTRAN program MATMULT.FOR can be used to perform the matrix multiplication involving $[u]^{-1}$, such as that shown in Eq. 5-106 of Example 5-18, to obtain the unknown constants A_r and B_r, for initial-condition problems of systems that have a large number of degrees of freedom.

The variable names and the quantities they represent are as follows:

FORTRAN name	Quantity
N	Order of square matrix
A(I,J)	Elements of square matrix **A**
X(I)	Elements of column matrix **X**
B(I)	Elements of column matrix resulting from the matrix multiplication of **AX**

The FORTRAN program as written will handle systems having up to 15 degrees of freedom. If there are more than 15 degrees of freedom, the DIMENSION statement must be changed. The program is as follows:

```
C   MATMULT.FOR
C   THIS PROGRAM MULTIPLIES A SQUARE MATRIX A BY A COLUMN
C   MATRIX X.  THE PRODUCT OF AX IS A COLUMN MATRIX NAMED B.
C   INPUT DATA REQUIRED IS ORDER N OF MATRIX A, ELEMENTS A(I,J)
C   OF MATRIX A, AND ELEMENTS X(I) OF COLUMN MATRIX X.
C   DEVICE * IN READ AND WRITE STATEMENTS IS THE CONSOLE.
C   DEVICE 2 IN WRITE STATEMENTS IS THE PRINTER.
C   * IN THE PLACE OF A FORMAT STATEMENT NUMBER MEANS FREE FORMAT.
      DIMENSION A(15,15),X(15),B(15)
      OPEN (2,FILE='PRN')
      WRITE(*,1)
1     FORMAT(' ','READ IN ORDER N OF MATRIX A (I2)',/)
      READ(*,2)N
2     FORMAT(I2)
      WRITE(*,20)
20    FORMAT(' ','READ IN MATRIX A',/)
      DO 3 I = 1,N
      DO 3 J = 1,N
      WRITE(*,40)I,J
      READ(*,50)A(I,J)
3     CONTINUE
40    FORMAT(' ','ROW = ',I2,3X,'COLUMN = ',I2,'     (E14.7)',/)
50    FORMAT(E14.7)
C
C   WRITE OUT MATRIX A
C
      WRITE(2,60)
60    FORMAT(' ',5X,'MATRIX A IS AS FOLLOWS',/)
      DO 4 I = 1,N
      DO 4 J = 1,N
      WRITE(2,65)I,J,A(I,J)
4     CONTINUE
65    FORMAT(' ',5X,'A(',I2,',',I2,') = ',E14.7)
C
C   READ IN COLUMN MATRIX X
C
      WRITE(*,70)
70    FORMAT(' ','READ IN COLUMN MATRIX X (E14.7)',/)
      DO 6 I = 1,N
      WRITE(*,75)I
      READ(*,80)X(I)
6     CONTINUE
75    FORMAT(' ',5X,'X(',I2,') =     (E14.7)',/)
80    FORMAT(E14.7)
C
```

```
C   WRITE OUT COLUMN MATRIX X
C
      WRITE(2,85)
85    FORMAT('0',5X,'COLUMN MATRIX X IS AS FOLLOWS',/)
      DO 8 I = 1,N
      WRITE(2,90)I,X(I)
8     CONTINUE
90    FORMAT(' ',5X,'X(',I2,') = ',E14.7)
C
C   MULTIPLY MATRIX A TIMES X
C
      DO 10 I = 1,N
      SUM = 0.0
      DO 10 J = 1,N
      SUM = SUM + A(I,J)*X(J)
      B(I) = SUM
10    CONTINUE
C
C   WRITE OUT MATRIX B OBTAINED FROM AX
C
      WRITE(2,95)
95    FORMAT('0',5X,'PRODUCT OF AX IS AS FOLLOWS',/)
100   FORMAT(' ','(',I2,') = ',E14.7)
      DO 12 I = 1,N
      WRITE(2,100)I,B(I)
12    CONTINUE
      STOP
      END
```

Uncoupled Equations for Damped Free Vibration

The matrix form of the differential equations of motion for damped free vibration is

$$\mathbf{M}\ddot{\mathbf{X}} + \mathbf{C}\dot{\mathbf{X}} + \mathbf{K}\mathbf{X} = \mathbf{0} \tag{5-109}$$

in which **C** is the damping matrix

$$\mathbf{C} = \begin{bmatrix} c_{11} & c_{12} & \cdot & \cdot & \cdot & c_{1n} \\ c_{21} & c_{22} & \cdot & \cdot & \cdot & c_{2n} \\ \cdot & \cdot & \cdot & \cdot & \cdot & \cdot \\ c_{n1} & c_{n2} & \cdot & \cdot & \cdot & c_{nn} \end{bmatrix} \tag{5-110}$$

With the addition of this matrix, the equations of motion can also be coupled by damping, as well as being dynamically and/or statically coupled. This occurs if there are off-diagonal damping coefficients c_{ij} in the damping matrix.

If we substitute the modal-matrix transformations

$$\{x_i\} = [u]\{\delta_i\}$$
$$\{\dot{x}_i\} = [u]\{\dot{\delta}_i\}$$
$$\{\ddot{x}_i\} = [u]\{\ddot{\delta}_i\}$$

into Eq. 5-109, and then premultiply the result by $[u]^T$, we obtain

$$[u]^T\mathbf{M}[u]\{\ddot{\delta}_i\} + [u]^T\mathbf{C}[u]\{\dot{\delta}_i\} + [u]^T\mathbf{K}[u]\{\delta_i\} = \mathbf{0} \tag{5-111}$$

The first and last terms of Eq. 5-111 reduce to diagonal matrices (see Eqs. 5-96 and 5-99) because of the orthogonality relationships between the eigenvectors relative to the mass matrix $\mathbf{M}$ and stiffness matrix $\mathbf{K}$. However, the second term

$$[u]^T\mathbf{C}[u]$$

does not usually reduce to a diagonal matrix. One practical approach to completely uncoupling the damped free-vibration equations is to introduce the concept of *proportional damping*.

In proportional damping, the damping matrix $\mathbf{C}$ is assumed to be proportional to the mass matrix $\mathbf{M}$ or to the stiffness matrix $\mathbf{K}$. For example, if we assume that

$$\mathbf{C} = \alpha\mathbf{M}$$

in which α is a constant, then

$$[u]^T\mathbf{C}[u] = \alpha[u]^T\mathbf{M}[u] = \begin{bmatrix} \nwarrow & & \\ & \alpha M_r & \\ & & \searrow \end{bmatrix}$$

in which M_r is the generalized mass of the rth mode. From this result, it is common practice to assume that the modal damping has the form of

$$2\zeta_r\omega_r M_r = \alpha M_r \tag{5-112}$$

Similarly, if we assume that

$$\mathbf{C} = \beta\mathbf{K}$$

in which β is a constant, we obtain

$$[u]^T\mathbf{C}[u] - \beta[u]^T\mathbf{K}[u] = \begin{bmatrix} \nwarrow & & \\ & \beta\omega_r^2 M_r & \\ & & \searrow \end{bmatrix} \tag{5-113}$$

in which case the modal damping can also be expressed as

$$2\zeta_r\omega_r M_r = \beta\omega_r^2 M_r \tag{5-114}$$

With the assumption of proportional damping that leads to either Eq. 5-112 or Eq. 5-114, the complete uncoupling of Eq. 5-111 reduces it to the form of

$$\begin{bmatrix} \nwarrow & & \\ & M_r & \\ & & \searrow \end{bmatrix}\begin{Bmatrix} \ddot{\delta}_1 \\ \ddot{\delta}_2 \\ \vdots \\ \ddot{\delta}_n \end{Bmatrix} + \begin{bmatrix} \nwarrow & & \\ & 2\zeta_r\omega_r M_r & \\ & & \searrow \end{bmatrix}\begin{Bmatrix} \dot{\delta}_1 \\ \dot{\delta}_2 \\ \vdots \\ \dot{\delta}_n \end{Bmatrix} + \begin{bmatrix} \nwarrow & & \\ & \omega_r^2 M_r & \\ & & \searrow \end{bmatrix}\begin{Bmatrix} \delta_1 \\ \delta_2 \\ \vdots \\ \delta_n \end{Bmatrix} = \mathbf{0} \tag{5-115}$$

Equation 5-115 reveals that instead of n coupled equations, we now have n uncoupled equations of the form

$$\ddot{\delta}_r + 2\zeta_r\omega_r\dot{\delta}_r + \omega_r^2\delta_r = 0 \qquad (r = 1, 2, 3, \ldots, n) \tag{5-116}$$

where ζ_r = modal damping factor of the rth mode
ω_r = undamped natural circular frequency of rth mode

Equation 5-116 for an rth mode has the same form as the differential equation of motion of a single-degree-of-freedom system, which has the solution (see Eq. 2-20)

$$\delta_r = e^{-\zeta_r \omega_r t}(A_r \cos \omega_r \sqrt{1 - \zeta_r^2}\, t + B_r \sin \omega_r \sqrt{1 - \zeta_r^2}\, t) \tag{5-117}$$

in which the constants A_r and B_r must be determined from the initial conditions.

Damped Free-Vibration Response

The damped free vibration of a system resulting from initial conditions can be obtained in a manner similar to the procedure presented for obtaining the undamped free-vibration response as illustrated by Example 5-18. In the damped case the equations defining δ_1, δ_2, $\delta_3, \ldots, \delta_n$ expressed by Eq. 5-117 are used in Eq. 5-103.

Rayleigh's Proportional Damping[8] The modal damping term $2\zeta_r \omega_r M_r$ in Eq. 5-116 was obtained by making the assumption that the damping matrix **C** was proportional to either the mass matrix **M** or the stiffness matrix **K**.

Lord Rayleigh considered proportional damping to have the form of

$$\mathbf{C} = \alpha \mathbf{M} + \beta \mathbf{K} \tag{5-118}$$

in which α and β are constants. If this form of proportional damping is substituted into Eq. 5-111, the modal damping assumes the form of

$$2\zeta_r \omega_r M_r = (\alpha + \beta \omega_r^2) M_r \tag{5-119}$$

In most vibration problems, it is immaterial whether the modal damping $2\zeta_r \omega_r M_r$ is based upon **C** being proportional to **M** or **K**, or a combination of **M** and **K** as in Eq. 5-118. Thus, from a practical point of view, the damping factor ζ_r and the undamped natural circular frequency ω_r can be interpreted as being properties that are inherent in the system. Furthermore, typical values and characteristics of the damping factor ζ_r for various types of structures are available from experimental studies. For example, a damping factor of $\zeta_r = 0.05$ (5 percent damping) is a reasonable value to use for the various modes in evaluating the *elastic* response of concrete structures subjected to earthquakes (see Example 6-10).

5-8 METHODS FOR SOLUTION OF EIGENVALUE PROBLEMS—GENERAL

Various methods and techniques have evolved through the years for solving eigenvalue problems because of their importance in solving problems in various disciplines such as engineering, science, and mathematics. The characteristic polynomial method was used in Sec. 5-5 to develop some insight into the nature of the eigenvalue problem encountered in vibration analysis. However, the use of this method to determine both the eigenvalues and the corresponding eigenvectors for large systems is quite cumbersome. Therefore, in the next three sections we discuss some of the more widely used methods for solving eigenvalue problems, including the power method, Hotelling's deflation method in conjunction with the power method, and Jacobi's method.

[8] Lord Rayleigh, *Theory of Sound,* vol. 1, Dover Publications, New York, 1945.

As a general rule, the most suitable method for solving a particular eigenvalue problem depends to some extent upon the size of the matrix (the degrees of freedom of a vibrating system), and upon how many eigenvalues and corresponding eigenvectors are required in the solution of the problem.

The power method presented in Sec. 5-9 is used frequently when only the smallest and/or the largest eigenvalue(s) are desired. An advantage of this method over most of the other methods is that the eigenvalues and corresponding eigenvectors are obtained simultaneously in the iteration process. Furthermore, this method can be performed quite easily on present-day pocket calculators for problems involving small matrices of order $n \leq 5$.

In Sec. 5-10 we present Hotelling's deflation method for "sweeping out" previously determined eigenvalues and corresponding eigenvectors, so that intermediate eigenvalues and eigenvectors can be determined by means of the power method. Used in conjunction with Hotelling's deflation method, the power method can be used to find all the eigenvalues and eigenvectors for matrices of order up to approximately 10. It is also particularly well suited for solving vibration problems involving matrices of order much larger than 10 when only a relatively small number of eigenvalues and eigenvectors are required (as in obtaining the forced response of systems subjected to transient excitations in Chap. 6). When this method is used for matrices of order greater than approximately 10, a loss of accuracy occurs in the higher eigenvalues because of roundoff error. In a 20-degree-of-freedom system, for example, perhaps the first 10 eigenvalues and eigenvectors would be accurate enough for use, while the rest would not.

Jacobi's method, which is presented in Sec. 5-11, is perhaps one of the most reliable methods but is relatively inefficient in comparison with some of the other methods for matrices of order larger than approximately 20. One advantage of this method is that it obtains all the eigenvalues and eigenvectors simultaneously, and with excellent and *uniform* accuracy. Jacobi's method also works well for vibration problems involving semidefinite (free-free) systems.

It is generally recognized at present that one of the most efficient and accurate methods for obtaining all the eigenvalues and eigenvectors of large systems (20 to several hundred degrees of freedom) is Householder's method used in conjunction with an algorithm called *QL*. This method is based upon an orthogonal transformation that produces a large number of zeros in a given row, yielding what is referred to as a *tridiagonalized* matrix. Various techniques, including the *QL* algorithm, are then used to determine the eigenvalues from the tridiagonalized matrix. A detailed discussion of Householder's method is beyond the intended scope of this text, and the interested reader is referred to a discussion of this method by Ortega.[9]

5-9 POWER METHOD FOR DETERMINING EIGENVALUES AND EIGENVECTORS

Various iterative methods are available for determining eigenvalues and eigenvectors. The iterative method most commonly used for this purpose in vibration problems is the *power* method, which is discussed in this section.

The power method is used most frequently when only the smallest and/or largest eigenvalue(s) are desired, but it can also be used when intermediate eigenvalues and ei-

[9] James Ortega, *Mathematical Methods for Digital Computers,* vol. II, pp. 94–115, ed. by A. Ralston and H. S. Wilf, John Wiley & Sons, New York, 1967.

genvectors are to be found. For determining the latter, a deflation technique that will be discussed later is used. An advantage of the power method is that the eigenvectors are obtained simultaneously with the eigenvalues, rather than requiring separate operations as in the polynomial method discussed earlier. In general, the power method is preferred over the polynomial method for obtaining the natural frequencies and normal-mode shapes of systems having more than three degrees of freedom.

It was shown in Sec. 5-5 that the matrix equation used to find the eigenvalues (from which the natural frequencies are determined) can be formulated in terms of either the stiffness matrix **K** or the flexibility matrix **A**. For use with the power method, these matrix equations are expressed as

$$\mathbf{M}^{-1}\mathbf{K}\mathbf{X} = \lambda\mathbf{X} \tag{5-120}$$

in which $\lambda = \omega^2$, and

$$\mathbf{A}\mathbf{M}\mathbf{X} = \lambda\mathbf{X} \tag{5-121}$$

in which $\lambda = 1/\omega^2$.

Using these equations, the iteration process converges to the *largest* eigenvalue λ and corresponding eigenvector **X**. The proof of such convergence can be found in the literature.[10]

Since $\lambda = \omega^2$, the iteration of Eq. 5-120, which is in terms of the stiffness matrix **K**, converges to a value of λ for obtaining the *largest* natural circular frequency ω. Conversely, the iteration of Eq. 5-121, which is in terms of the flexibility matrix **A**, converges to a value of λ that yields the *lowest* natural circular frequency, since $\lambda = 1/\omega^2$.

The steps involved in the iteration process, which are pertinent for use with either Eq. 5-120 or 5-121, are as follows:

1. Arbitrarily assume values for the components of the eigenvector $\mathbf{X} = (X_1, X_2, \ldots, X_n)$, and subsequently refer to it as $\mathbf{X}_0$. Choosing a value of unity for each component is generally satisfactory. However, an exception does occur in the case of semidefinite systems, for which the use of unit components causes immediate convergence to the rigid-body-mode frequency of zero. Substitute the assumed components of **X** into the left side of whichever of Eqs. 5-120 or 5-121 is being used, and carry out the matrix multiplication. This yields a first approximation of the right side of Eq. 5-120 or Eq. 5-121, with

$$\lambda\mathbf{X} = \begin{Bmatrix} \lambda X_1 \\ \lambda X_2 \\ \lambda X_3 \\ \vdots \\ \lambda X_n \end{Bmatrix} \tag{5-122}$$

2. Normalize the vector $\lambda\mathbf{X}$ obtained in step 1. This can be done by dividing the vector by the magnitude of any one of its components (referred to as the normalizing factor), or by normalizing the vector **X** to a unit magnitude, in which case the normalizing factor is the square root of the sum of the squares of the eigenvector components.
3. Use the components of the normalized vector as improved values of **X**, and substitute them into the left side of Eq. 5-120 (or 5-121). Then, carrying out the

[10] M. L. James, G. M. Smith, and J. C. Wolford, *Applied Numerical Methods,* 3d ed., p. 706, Harper & Row, New York, 1985.

matrix multiplication, another approximation of the right-hand side of Eq. 5-120 (or 5-121) is obtained.

4. Repeat steps 2 and 3 in turn, until Eq. 5-120 (or 5-121) is essentially satisfied, that is, until values of the eigenvector components found in step 2 in two consecutive iterations vary by less than some preassigned ϵ value. The final normalizing factor that the iteration process converges to will be the largest eigenvalue λ, and the elements of $\mathbf{X}$ will be the components of the eigenvector associated with it.

In carrying out the iteration procedure, we are in effect forming a sequence of vectors $\mathbf{BX}_0$, $\mathbf{B}^2\mathbf{X}_0$, $\mathbf{B}^3\mathbf{X}_0$, . . . , $\mathbf{B}^k\mathbf{X}_0$ in which $\mathbf{X}_0$ is the arbitrary vector initially assumed, and $\mathbf{B} = \mathbf{M}^{-1}\mathbf{K}$ (see Eq. 5-120). The reference to this iterative method as the *power method* comes from the fact that the sequence above consists of powers of the matrix $\mathbf{B}$.

EXAMPLE 5-19

As a means of illustrating the steps just outlined, let us determine the lowest natural circular frequency ω_1 of the system shown in Fig. 5-27. It consists of three masses and three springs, connected as shown.

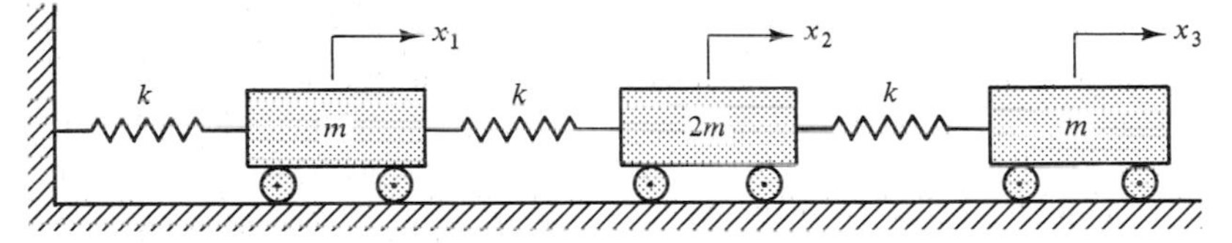

Figure 5-27 System of Example 5-19.

Solution. Since we wish to determine the lowest natural circular frequency of the system, the iteration procedure will be applied to Eq. 5-121, in which $\lambda_1 = 1/\omega_1^2$. Remembering the definition of flexibility coefficients while inspecting Fig. 5-27, we obtain the flexibility matrix as

$$\mathbf{A} = \begin{bmatrix} 1/k & 1/k & 1/k \\ 1/k & 2/k & 2/k \\ 1/k & 2/k & 3/k \end{bmatrix} \tag{5-123}$$

The mass matrix is simply

$$\mathbf{M} = \begin{bmatrix} m & 0 & 0 \\ 0 & 2m & 0 \\ 0 & 0 & m \end{bmatrix} \tag{5-124}$$

Referring to the general form of **AM**, as shown in Eq. 5-60, Eqs. 5-123 and 5-124 show that

$$\mathbf{AM} = \begin{bmatrix} m/k & 2m/k & m/k \\ m/k & 4m/k & 2m/k \\ m/k & 4m/k & 3m/k \end{bmatrix} = \frac{m}{k}\begin{bmatrix} 1 & 2 & 1 \\ 1 & 4 & 2 \\ 1 & 4 & 3 \end{bmatrix}$$

Thus, Eq. 5-121 becomes

$$\frac{m}{k}\begin{bmatrix} 1 & 2 & 1 \\ 1 & 4 & 2 \\ 1 & 4 & 3 \end{bmatrix}\begin{Bmatrix} X_1 \\ X_2 \\ X_3 \end{Bmatrix} = \lambda\begin{Bmatrix} X_1 \\ X_2 \\ X_3 \end{Bmatrix} \tag{5-125}$$

Carrying out steps 1 and 2, we find from the first iteration of Eq. 5-125 that

$$\frac{m}{k}\begin{bmatrix} 1 & 2 & 1 \\ 1 & 4 & 2 \\ 1 & 4 & 3 \end{bmatrix}\begin{Bmatrix} 1 \\ 1 \\ 1 \end{Bmatrix} = \frac{m}{k}\begin{Bmatrix} 4 \\ 7 \\ 8 \end{Bmatrix} = 4\,\frac{m}{k}\begin{Bmatrix} 1.0 \\ 1.75 \\ 2.0 \end{Bmatrix}$$

Then applying step 3, we find from the second iteration, that

$$\frac{m}{k}\begin{bmatrix} 1 & 2 & 1 \\ 1 & 4 & 2 \\ 1 & 4 & 3 \end{bmatrix}\begin{Bmatrix} 1.0 \\ 1.75 \\ 2.0 \end{Bmatrix} = \frac{m}{k}\begin{Bmatrix} 6.5 \\ 12.0 \\ 14.0 \end{Bmatrix} = 6.5\,\frac{m}{k}\begin{Bmatrix} 1.0 \\ 1.84 \\ 2.15 \end{Bmatrix}$$

The third iteration yields

$$\frac{m}{k}\begin{bmatrix} 1 & 2 & 1 \\ 1 & 4 & 2 \\ 1 & 4 & 3 \end{bmatrix}\begin{Bmatrix} 1.0 \\ 1.84 \\ 2.15 \end{Bmatrix} = 6.83\,\frac{m}{k}\begin{Bmatrix} 1.0 \\ 1.85 \\ 2.17 \end{Bmatrix}$$

The fourth iteration yields

$$\frac{m}{k}\begin{bmatrix} 1 & 2 & 1 \\ 1 & 4 & 2 \\ 1 & 4 & 3 \end{bmatrix}\begin{Bmatrix} 1.0 \\ 1.85 \\ 2.17 \end{Bmatrix} = 6.87\,\frac{m}{k}\begin{Bmatrix} 1.0 \\ 1.85 \\ 2.17 \end{Bmatrix}$$

Comparing the third and fourth iterations, it can be seen that the normalizing factor resulting from the fourth iteration varies little from that resulting from the third iteration and that there is no difference to two decimal places between the eigenvector components of the two iterations. Thus, a good approximation of the value of λ_1 is given by

$$\lambda_1 = 6.87\,\frac{m}{k}$$

Using this eigenvalue, the lowest natural circular frequency is found to be

$$\omega_1 = \frac{1}{\sqrt{\lambda_1}} = 0.38\sqrt{\frac{k}{m}}\ \text{rad/s}$$

The eigenvector $\mathbf{X}_1$ (the first natural mode shape), associated with ω_1 is

$$\mathbf{X}_1 = \begin{Bmatrix} 1.0 \\ 1.85 \\ 2.17 \end{Bmatrix}$$

A fifth iteration would of course yield a still better approximation of the value of λ_1 and the corresponding eigenvector.

EXAMPLE 5-20

A small airplane is modeled using the three lumped masses shown in Fig. 5-28.[11] It is assumed for this simplified model of the airplane that the wings are uniform cantilever beams of length l and stiffness factor EI, and that $m_1 = m_3$ and $m_2 = 4m_1$.

[11] The discretizing of distributed-mass systems by means of lumped-mass modeling is discussed in Sec. 5-12.

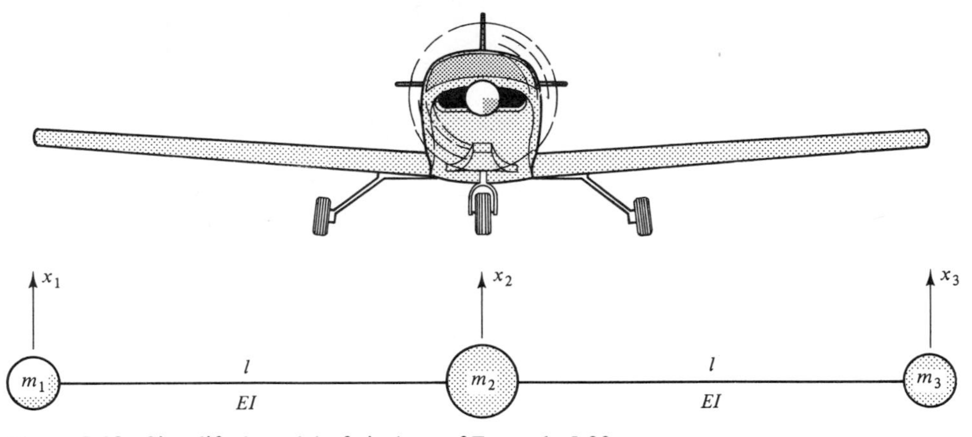

Figure 5-28 Simplified model of airplane of Example 5-20.

We wish to determine the largest natural circular frequency and corresponding normal mode of the free-free system represented by the airplane using the power method.

Solution. Recalling from the discussion in Sec. 5-3 that the flexibility coefficients of a semidefinite (free-free) system are infinite, the power method must be applied to the matrix equation expressed in terms of the stiffness matrix **K** (Eq. 5-120).

Considering the wings as simple cantilever beam elements, with each having a stiffness of $k = 3EI/l^3$ as shown in part d of Table 2-1 on page 78, the stiffness coefficients k_{ij} shown in Fig. 5-29 are determined. Thus, the stiffness matrix is

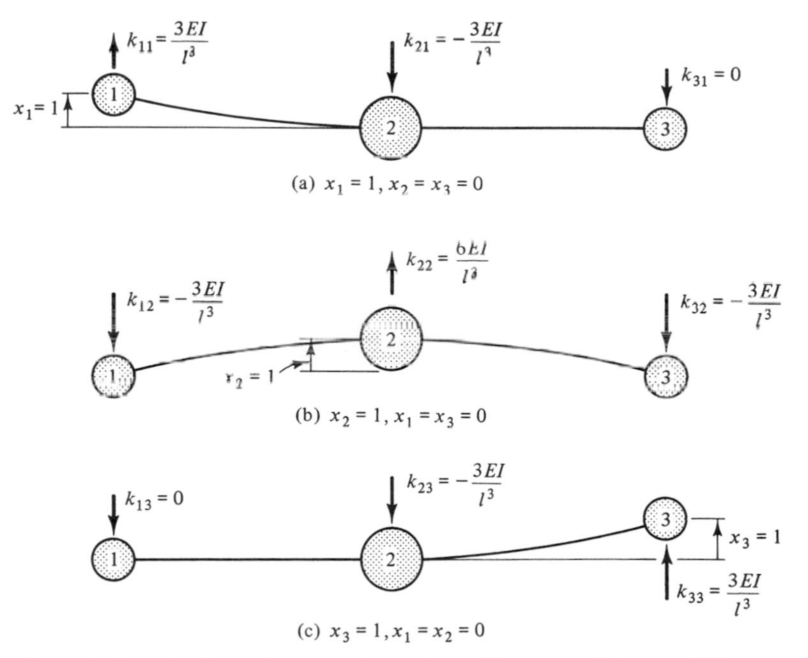

Figure 5-29 Configurations to determine stiffness coefficients of Example 5-20.

$$\mathbf{K} = \frac{EI}{l^3}\begin{bmatrix} 3 & -3 & 0 \\ -3 & 6 & -3 \\ 0 & -3 & 3 \end{bmatrix} \tag{5-126}$$

The mass matrix is

$$\mathbf{M} = m_1\begin{bmatrix} 1 & 0 & 0 \\ 0 & 4 & 0 \\ 0 & 0 & 1 \end{bmatrix} \tag{5-127}$$

Recalling the general form of $\mathbf{M}^{-1}\mathbf{K}$ when $\mathbf{M}$ is a diagonal matrix (see Sec. 5-5), we can write from Eqs. 5-126 and 5-127 that

$$\mathbf{M}^{-1}\mathbf{K} = \frac{EI}{m_1 l^3}\begin{bmatrix} 3 & -3 & 0 \\ -\frac{3}{4} & \frac{6}{4} & -\frac{3}{4} \\ 0 & -3 & 3 \end{bmatrix}$$

Using this equality, Eq. 5-120 is ready for iteration as

$$\frac{EI}{m_1 l^3}\begin{bmatrix} 3 & -3 & 0 \\ -\frac{3}{4} & \frac{6}{4} & -\frac{3}{4} \\ 0 & -3 & 3 \end{bmatrix}\begin{Bmatrix} X_1 \\ X_2 \\ X_3 \end{Bmatrix} = \lambda\begin{Bmatrix} X_1 \\ X_2 \\ X_3 \end{Bmatrix} \tag{5-128}$$

in which $\lambda = \omega^2$.

If we assume unit values for each of the eigenvector components as suggested in the iteration steps given earlier, inspection of Eq. 5-128 shows that it will yield only zero values for λX_1, λX_2, and λX_3, which is the rigid-body mode with $\lambda = 0$. To avoid this mode, let us assume that the components of $\mathbf{X}$ are each of unit magnitude but with alternating signs. Then carrying out steps 1 and 2 of the iteration procedure, the first iteration of Eq. 5-128 yields

$$\frac{EI}{m_1 l^3}\begin{bmatrix} 3 & -3 & 0 \\ -\frac{3}{4} & \frac{6}{4} & -\frac{3}{4} \\ 0 & -3 & 3 \end{bmatrix}\begin{Bmatrix} 1 \\ -1 \\ 1 \end{Bmatrix} = \frac{EI}{m_1 l^3}\begin{Bmatrix} 6 \\ -3 \\ 6 \end{Bmatrix} = \frac{6EI}{m_1 l^3}\begin{Bmatrix} 1.0 \\ -0.5 \\ 1.0 \end{Bmatrix}$$

Applying step 3 next, the second iteration provides

$$\frac{EI}{m_1 l^3}\begin{bmatrix} 3 & -3 & 0 \\ -\frac{3}{4} & \frac{6}{4} & -\frac{3}{4} \\ 0 & -3 & 3 \end{bmatrix}\begin{Bmatrix} 1.0 \\ -0.5 \\ 1.0 \end{Bmatrix} = \frac{EI}{m_1 l^3}\begin{Bmatrix} 4.5 \\ -2.25 \\ 4.5 \end{Bmatrix} = \frac{4.5EI}{m_1 l^3}\begin{Bmatrix} 1.0 \\ -0.5 \\ 1.0 \end{Bmatrix}$$

The third iteration gives us

$$\frac{EI}{m_1 l^3}\begin{bmatrix} 3 & -3 & 0 \\ -\frac{3}{4} & \frac{6}{4} & -\frac{3}{4} \\ 0 & -3 & 3 \end{bmatrix}\begin{Bmatrix} 1.0 \\ -0.5 \\ 1.0 \end{Bmatrix} = \frac{4.5EI}{m_1 l^3}\begin{Bmatrix} 1.0 \\ -0.5 \\ 1.0 \end{Bmatrix}$$

which is identical to the result obtained in the second iteration. Thus, the largest natural circular frequency is determined by

$$\omega_3 = \sqrt{\lambda_3} = \sqrt{\frac{4.5EI}{m_1 l^3}}$$

and the corresponding normal mode is

$$\mathbf{X}_3 = \begin{Bmatrix} 1.0 \\ -0.5 \\ 1.0 \end{Bmatrix}$$

as shown in Fig. 5-30.

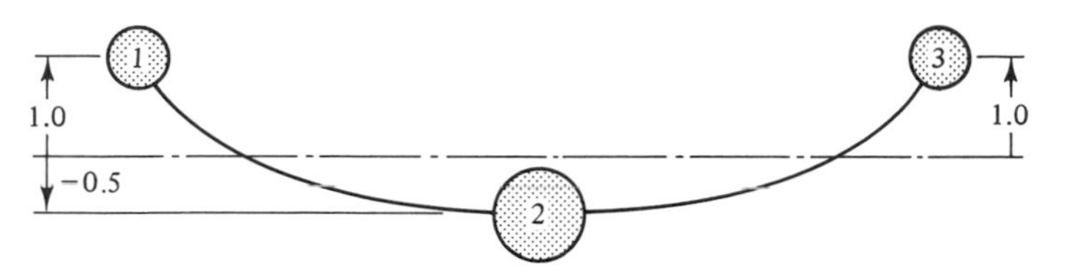

Figure 5-30 Normal-mode shape corresponding to largest natural frequency of Example 5-20.

It should be apparent from the discussion and examples presented thus far, that the power method is ideally suited for use with the digital computer in solving vibration problems.

If the largest and next-to-largest eigenvalues are nearly the same in magnitude, convergence may be slow. If the largest eigenvalue has a multiplicity of 2 ($\lambda_1 = \lambda_2$), convergence is to the largest eigenvalue. However, the components of the eigenvector will converge to any of the eigenvectors (which usually form a two-dimensional space) associated with λ_1, with the one obtained depending upon the vector assumed initially.[12] Other special cases, which are rarely encountered, are discussed by Faddeev and Faddeeva.[13]

5-10 ITERATION FOR INTERMEDIATE EIGENVALUES AND EIGENVECTORS—HOTELLING'S DEFLATION METHOD

The iteration procedure discussed in Sec. 5-9 can be extended to determine the intermediate eigenvalues and corresponding eigenvectors that lie between the smallest and largest eigenvalues. The process for obtaining these intermediate values, which will be discussed in this section, is known as Hotelling's deflation method. It can be used for systems having up to approximately 10 degrees of freedom and is particularly well suited for much larger systems when only a relatively small number of eigenvalues and eigenvectors is required (as in obtaining the total response of a forced-vibration system, for example). It will become apparent in Chap. 6 that a modal analysis involving only the first four or five normal modes is often sufficient to obtain the forced-vibration response of a 50-degree-of-freedom system.

Using the iteration procedure with Eq. 5-120, and in conjunction with Hotelling's deflation technique, will yield the eigenvalues λ in descending order, which determines the natural circular frequencies in descending order since $\lambda = \omega^2$. Conversely, while the same

[12] See Example 7-7, which illustrates the eigenvectors accompanying multiple eigenvalues using Jacobi's method.

[13] D. K. Faddeev and U. N. Faddeeva, *Computational Methods of Linear Algebra,* trans. Robert C. Williams, W. H. Freeman, San Francisco, 1963.

process with Eq. 5-121 will again yield the eigenvalues in descending order, the natural circular frequencies will be obtained in *ascending* order since $\lambda = 1/\omega^2$.

Deflation for Determining Natural Circular Frequencies in Ascending Order

Let us discuss first the use of the deflation technique in conjunction with Eq. 5-121. In this process, the original *flexibility* matrix will be referred to as $\mathbf{A}_1$ and will be used in the iteration procedure to obtain the largest eigenvalue λ_1 ($\lambda_1 > \lambda_2 > \cdots > \lambda_n$) and associated eigenvector $\mathbf{X}_1$. Since $\lambda = 1/\omega^2$, the iteration process converges to the lowest natural circular frequency ω_1. With λ_1 and $\mathbf{X}_1$ determined by the iteration process, a new matrix $\mathbf{A}_2\mathbf{M}$ is formed as

$$\mathbf{A}_2\mathbf{M} = \mathbf{A}_1\mathbf{M} - \lambda_1\mathbf{X}_1\mathbf{X}_1^T\mathbf{M} \tag{5-129}$$

To prove that the iteration process with this new matrix $\mathbf{A}_2\mathbf{M}$ converges to λ_2 and corresponding eigenvector $\mathbf{X}_2$, we postmultiply Eq. 5-129 by the eigenvector $\mathbf{X}_1$ to obtain

$$\mathbf{A}_2\mathbf{M}\mathbf{X}_1 = \mathbf{A}_1\mathbf{M}\mathbf{X}_1 - \lambda_1\mathbf{X}_1\mathbf{X}_1^T\mathbf{M}\mathbf{X}_1 \tag{5-130}$$

Recalling the orthogonality relationship expressed by Eq. 5-79, we can write that

$$\mathbf{X}_r^T\mathbf{M}\mathbf{X}_s = \begin{Bmatrix} 0 & r \neq s \\ 1 & r = s \end{Bmatrix} \tag{5-131}$$

in which the components of $\mathbf{X}_r$ are normalized to make $\mathbf{X}_r^T\mathbf{M}\mathbf{X}_r = 1$. From this, $\mathbf{X}_1^T\mathbf{M}\mathbf{X}_1 = 1$, and noting that after the iteration of Eq. 5-121 is completed,

$$\mathbf{A}_1\mathbf{M}\mathbf{X}_1 = \lambda_1\mathbf{X}_1$$

Eq. 5-130 reduces to

$$\mathbf{A}_2\mathbf{M}\mathbf{X}_1 = \mathbf{A}_1\mathbf{M}\mathbf{X}_1 - \lambda_1\mathbf{X}_1 = 0 \tag{5-132}$$

This shows that $\mathbf{A}_2\mathbf{M}\mathbf{X}_1 = 0$, from which $\lambda = 0$ and $\mathbf{X} = \mathbf{X}_1$ is a solution of the equation

$$\mathbf{A}_2\mathbf{M}\mathbf{X} = \lambda\mathbf{X} \tag{5-133}$$

Therefore, the eigenvalue λ_1 of $\mathbf{A}_1\mathbf{M}$ is not an eigenvalue associated with the eigenvector $\mathbf{X}_1$ of $\mathbf{A}_2\mathbf{M}$ but has been replaced by a *zero value.*

We next need to show that $\lambda_2, \lambda_3, \ldots, \lambda_n$ and the associated eigenvectors $\mathbf{X}_2, \mathbf{X}_3, \ldots, \mathbf{X}_n$ are eigenvalues and eigenvectors of $\mathbf{A}_2\mathbf{M}$. To demonstrate this, Eq. 5-129 is postmultiplied by $\mathbf{X}_2$ to obtain

$$\mathbf{A}_2\mathbf{M}\mathbf{X}_2 = \mathbf{A}_1\mathbf{M}\mathbf{X}_2 - \lambda_1\mathbf{X}_1\mathbf{X}_1^T\mathbf{M}\mathbf{X}_2 \tag{5-134}$$

Noting from Eq. 5-131 that $\mathbf{X}_1^T\mathbf{M}\mathbf{X}_2 = 0$, Eq. 5-134 reduces to

$$\mathbf{A}_2\mathbf{M}\mathbf{X}_2 = \mathbf{A}_1\mathbf{M}\mathbf{X}_2 \tag{5-135}$$

Recalling that $(\lambda_1, \mathbf{X}_1), (\lambda_2, \mathbf{X}_2), \ldots, (\lambda_n, \mathbf{X}_n)$ are solutions of

$$\mathbf{A}_1\mathbf{M}\mathbf{X} = \lambda\mathbf{X}$$

we can write that

$$\mathbf{A}_1\mathbf{M}\mathbf{X}_2 = \lambda_2\mathbf{X}_2$$

and thus conclude from Eq. 5-135 that

$$\mathbf{A}_2\mathbf{M}\mathbf{X}_2 = \lambda_2\mathbf{X}_2 \tag{5-136}$$

By postmultiplying Eq. 5-129 successively by $\mathbf{X}_3, \mathbf{X}_4, \ldots, \mathbf{X}_n$, we can show in a manner similar to that above that

$$\begin{aligned} \mathbf{A}_2\mathbf{M}\mathbf{X}_3 &= \lambda_3\mathbf{X}_3 \\ \mathbf{A}_2\mathbf{M}\mathbf{X}_4 &= \lambda_4\mathbf{X}_4 \\ &\vdots \\ \mathbf{A}_2\mathbf{M}\mathbf{X}_n &= \lambda_n\mathbf{X}_n \end{aligned}$$

Thus, the new matrix $\mathbf{A}_2\mathbf{M}$ has eigenvalues $\lambda_2, \lambda_3, \ldots, \lambda_n$ and associated eigenvectors $\mathbf{X}_2, \mathbf{X}_3, \ldots, \mathbf{X}_n$, and a zero eigenvalue associated with the eigenvector $\mathbf{X}_1$. The largest eigenvalue of $\mathbf{A}_2\mathbf{M}$ is λ_2, and the iteration of Eq. 5-133 will result in convergence to λ_2 and its associated eigenvector $\mathbf{X}_2$.

It can be shown in a similar manner that

$$\mathbf{A}_3\mathbf{M} = \mathbf{A}_2\mathbf{M} - \lambda_2\mathbf{X}_2\mathbf{X}_2^T\mathbf{M} \tag{5-137}$$

sweeps out λ_2 as an eigenvalue of $\mathbf{A}_3\mathbf{M}$, leaving λ_3 as its largest eigenvalue, so that the iteration of

$$\mathbf{A}_3\mathbf{M}\mathbf{X} = \lambda\mathbf{X}$$

converges to λ_3 and its associated eigenvector $\mathbf{X}_3$.

Reviewing the preceding discussion, the general procedure for determining the natural circular frequencies ω_i and the corresponding normal-mode shapes $\mathbf{X}_i$ in *ascending* order can be summarized as follows:

$$\mathbf{A}_1\mathbf{M}\mathbf{X} = \lambda\mathbf{X} \quad \text{(iteration for } \lambda_1 \text{ and } \mathbf{X}_1) \quad (\lambda_1 = 1/\omega_1^2) \tag{5-138a}$$

$$\mathbf{X}_i^T\mathbf{M}\mathbf{X}_i = 1 \quad \text{(normalization)} \tag{5-138b}$$

$$\mathbf{A}_{i+1}\mathbf{M} = \mathbf{A}_i\mathbf{M} - \lambda_i\mathbf{X}_i\mathbf{X}_i^T\mathbf{M} \quad \text{(formation of new matrix)} \tag{5-138c}$$

$$\mathbf{A}_{i+1}\mathbf{M}\mathbf{X} = \lambda\mathbf{X} \quad \text{(iteration for } \lambda_{i+1} \text{ and } \mathbf{X}_{i+1}) \tag{5-138d}$$

$$i = 1, 2, 3, \ldots, s$$

in which the original flexibility matrix used is $\mathbf{A}_1$ ($i = 1$) for the first iteration process involving Eq. 5-121. The formation of s new matrices for successive iterations will yield $s + 1$ eigenvalues and eigenvectors, including λ_1 and $\mathbf{X}_1$.

Assuming that the coordinates selected result in a diagonal mass matrix, the normalizing factor that satisfies Eq. 5-138b is obtained by noting that

$$\mathbf{X}_i^T\mathbf{M}\mathbf{X}_i = [X_1 \quad X_2 \quad \cdot \quad \cdot \quad \cdot \quad X_n]_i \begin{Bmatrix} m_1X_1 \\ m_2X_2 \\ \vdots \\ m_nX_n \end{Bmatrix}_i$$

or performing the matrix multiplication indicated,

$$\mathbf{X}_i^T\mathbf{M}\mathbf{X}_i = [m_1X_1^2 + m_2X_2^2 + \cdots + m_nX_n^2]_i$$

in which the component values of $\mathbf{X}_i$ obtained from *each* iteration process have been normalized with respect to X_1 (making $X_1 = 1$). The normalizing factor is then

$$\text{N.F.} = \sqrt{m_1(1)^2 + m_2X_2^2 + \cdots + m_nX_n^2} \tag{5-139}$$

The $\mathbf{X}_i$ values satisfying Eq. 5-138b that are used in Eq. 5-138c are then

$$X_1 = \frac{1}{\text{N.F.}}, X_2 = \frac{X_2}{\text{N.F.}}, \ldots, X_n = \frac{X_n}{\text{N.F.}}$$

A FORTRAN program referred to as DEFLATEA.FOR performs the computations indicated in Eqs. 5-138 and 5-139 to obtain the natural circular frequencies ω_i and corresponding normal modes $\mathbf{X}_i$ in *ascending order.* The features of the program and the program itself are presented in the paragraphs that follow.

Computer Program DEFLATEA.FOR

Some important features of the program are as follows:

1. The program is written with *double precision* to minimize the accumulation of roundoff errors that occur in the deflation and iteration processes.
2. The data input for $s + 1$ (integer) specifies the number of natural frequencies and normal modes to be determined. For example, if $N = 10$ for a 10-degree-of-freedom system, and only the first four modes are to be determined, then $s + 1 = 4$.
3. The orthogonality relationship $\mathbf{X}_1^T\mathbf{M}\mathbf{X}_{s+1} = 0$, between the first eigenvector $\mathbf{X}_1$ and the $s + 1$th eigenvector $\mathbf{X}_{s+1}$, is calculated and printed out, to indicate the accuracy attained.
4. The mass matrix $\mathbf{M}$ is considered to be a diagonal matrix.
5. The deflation process is done in the main program, and the iteration process explained in Sec. 5-9 is done by the subroutine subprogram POWERA.
6. Convergence of the iteration process in the subroutine subprogram POWERA is controlled by comparing the difference between two successive sets of eigenvector component values with the value prescribed for EPSI. A typical value would be EPSI $= 10^{-6}$.

The variable names and the quantities they represent in both the main program and the subroutine subprogram POWERA are as follows:

Variable name	Quantity
A(I,J)	Elements of the flexibility matrix. Also used in subroutine subprogram POWERA in which λ is the eigenvalue of matrix $\mathbf{A}$
LAMBDA(I)	λ of the equation $\mathbf{AMX} = \lambda\mathbf{X}$ where $\lambda = 1/\omega^2$. (Nonsubscripted LAMBDA is the dummy name for the eigenvalue in subroutine subprogram POWERA)
OMEGA(I)	Natural undamped circular frequency ω_i rad/s
M(I)	Elements of the diagonal mass matrix that for simplicity are stored as a one-dimensional array
NF	Normalizing factor (see Eq. 5-139)
SP1	$s + 1$ equals the total number of eigenvalues, or natural frequencies, and normal modes to be determined
AM(I,J)	Elements of the product $\mathbf{AM}$
X(I)	Eigenvector
XX(I,J)	A two-dimensional array in which the eigenvectors are stored as columns
XIXITM(I,J)	Elements of the matrix $X_iX_i^T\mathbf{M}$
ITER(I) and IT	A counter for the number of iterations to converge to a value of λ and corresponding eigenvector

Variable name	Quantity
N	Order of flexibility and mass matrices
EPSI	A small quantity such as 10^{-6} used to control convergence of the iteration procedure in subroutine subprogram POWERA
II	A FORTRAN DO variable that corresponds to subscript i of Eq. 5-138
I,J,K	DO variables
PROD	The product $X_i^T\mathbf{M}\mathbf{X}_{s+1}$ calculated as an indication of accuracy attained
D(I)	Name for the vector **AX** in subroutine POWERA (also equal to $\lambda\mathbf{X}$)
Z(I)	Name for the normalized D(I). D(I) is normalized so that D(1) = 1.0
DIFF	The difference between two successive eigenvector component values in the iterative procedure of subroutine POWERA

The FORTRAN program DEFLATEA.FOR is as follows:[14]

```
C   DEFLATEA.FOR
C   THIS PROGRAM FINDS S+1 VALUES OF OMEGA OF THE EQUATION
C   AMX = (1./OMEGA**2)X IN ASCENDING ORDER.  PROGRAM AS WRITTEN
C   CAN HANDLE A 10 DEGREE OF FREEDOM SYSTEM.  DIMENSION
C   STATEMENTS IN MAIN PROGRAM AND SUBROUTINE SUBPROGRAM
C   'POWERA' ARE DIMENSIONED (10,10).  THEREFORE, THE MATRIX
C   ORDER N CAN BE ENTERED WITH VALUES RANGING FROM 2 UP TO 10.
C   DEVICE * IN READ AND WRITE STATEMENTS IS THE CONSOLE.
C   DEVICE 2 IN WRITE STATEMENTS IS THE PRINTER.
C   * IN THE PLACE OF A FORMAT STATEMENT NUMBER MEANS FREE FORMAT.
      IMPLICIT REAL*8(A-H,O-Z)
      REAL*8 M,LAMBDA,LPRIME,NF
      INTEGER SP1
C
      DIMENSION A(10,10),AM(10,10),LAMBDA(10),X(10),XX(10,10)
      DIMENSION XIXITM(10,10),ITER(10),M(10),OMEGA(10)
      OPEN (2,FILE='PRN')
C
      WRITE(*,101)
101   FORMAT(' ','S+1 VALUE? MAT. ORDER N? EPSI? (2I2,F10.0)'/)
      READ(*,*)SP1,N,EPSI
      WRITE(*,103)
103   FORMAT(' ','READ IN UPPER TRIANGULAR ELEMENTS OF MATRIX A',/)
105   FORMAT(' ','ROW = ',I2,3X,'COLUMN = ',I2,'  (F20.0) ',/)
106   FORMAT(F20.0)
      DO 110 I = 1,N
      DO 110 J = I,N
      WRITE(*,105)I,J
      READ(*,106)A(I,J)
110   CONTINUE
C
```

[14] The FORTRAN version used here is compatible with IBM personal computers. In using other versions of FORTRAN it may be necessary to modify some of the statements slightly.

```
C  ASSIGN VALUES TO SYMMETRIC ELEMENTS BELOW MAIN DIAGONAL
C
      DO 3 I = 2,N
      IMIN1 = I-1
      DO 3 J = 1,IMIN1
3     A(I,J) = A(J,I)
C
C  WRITE OUT THE FLEXIBILITY MATRIX A(I,J)
C
      WRITE(2,4)
4     FORMAT(' ',5X,'THE FLEXIBILITY MATRIX IS:',//)
5     FORMAT(' ',5X,'A(',I2,',',I2,') = ',E14.7)
      DO 6 I = 1,N
      DO 6 J = 1,N
      WRITE(2,5)I,J,A(I,J)
6     CONTINUE
C
C  READ THE DIAGONAL ELEMENTS OF THE MASS MATRIX M (A ONE
C  DIMENSIONAL ARRAY).
C
      WRITE(*,127)
127   FORMAT(' ','READ IN ELEMENTS OF MASS MATRIX M',/)
125   FORMAT(' ',5X,'MASS NUMBER = ',I2,'  (F10.0) ', /)
126   FORMAT(F10.0)
      DO 120 I = 1,N
      WRITE(*,125)I
      READ(*,126)M(I)
120   CONTINUE
C
C  WRITE OUT THE DIAGONAL ELEMENTS OF THE MASS MATRIX
C
      WRITE(2,7)
7     FORMAT(///,' ',5X,'THE ELEMENTS OF THE DIAGONAL MASS',
     *' MATRIX ARE:',//)
8     FORMAT(' ',5X,'M(',I2,') = ',E14.7)
      DO 22 I = 1,N
      WRITE(2,8)I,M(I)
22    CONTINUE
C
C  DETERMINE THE PRODUCT OF THE A AND M MATRICES
C
      DO 9 I = 1,N
      DO 9 J = 1,N
9     AM(I,J) = A(I,J)*M(J)
C
C  BEGIN A MAJOR DO LOOP IN WHICH THE STEPS GIVEN BY EQUATIONS
C  5-138  ARE CARRIED OUT.  DO VARIABLE 'II' CORRESPONDS TO
C  LOWER CASE LETTER I OF EQUATIONS 5-138.
C
```

```
      DO 100 II = 1,SP1
      CALL POWERA(AM,LPRIME,X,N,EPSI,IT)
      ITER(II) = IT
      LAMBDA(II) = LPRIME
C
C  ASSIGN THE EIGENVECTOR COMPONENTS RETURNED FROM SUBROUTINE
C  'POWERA' AS A COLUMN OF THE TWO-DIMENSIONAL ARRAY XX.
C
      DO 10 I = 1,N
10    XX(I,II) = X(I)
      IF(II.EQ.SP1)GO TO 100
C
C  CALCULATE NORMALIZING FACTOR NF FOR NORMALIZING THE
C  EIGENVECTORS SO THAT (XTRANSPOSE)(M)X = 1.0.
C
      SUM = 0.0
      DO 11 I = 1,N
11    SUM = SUM + M(I)*X(I)**2
      NF = DSQRT(SUM)
C
C  DIVIDE EACH COMPONENT OF X BY NF.
C
      DO 12 I = 1,N
12    X(I) = X(I)/NF
C
C  CALCULATE ELEMENTS OF THE N BY N MATRIX X(XTRANSPOSE)(M).
C
      DO 13 I = 1,N
      DO 13 J = 1,N
13    XIXITM(I,J) = X(I)*X(J)*M(J)
C
C  GET THE NEW (A)(M) MATRIX.
C
      DO 14 I = 1,N
      DO 14 J = 1,N
14    AM(I,J) = AM(I,J)-XIXITM(I,J)*LAMBDA(II)
100   CONTINUE
C
C  WRITE OUT THE LAMBDAS,OMEGAS AND CORRESPONDING EIGENVECTORS.
C
      DO 18 I = 1,SP1
      OMEGA(I) = DSQRT(1./LAMBDA(I))
      WRITE(2,15)I,LAMBDA(I),ITER(I)
15    FORMAT(///,' ',5X,'LAMBDA(',I2,') =  ',D14.7,' USING', I4,
     *' ITERATIONS')
      WRITE(2,23)I,OMEGA(I)
      WRITE(2,16)
16    FORMAT(/,' ',5X,'THE EIGENVECTOR COMPONENTS ARE:'/)
      WRITE(2,17)(XX(K,I),K=1,N)
```

```
17    FORMAT(' ',18X,D17.10)
18    CONTINUE
23    FORMAT(' ',5X,'OMEGA(',I2,') =   ',D14.7,)
C
C  CHECK ON THE ORTHOGONALITY OF (XTRANSPOSE)(M)(X) USING
C  THE FIRST EIGENVECTOR FOR (XTRANSPOSE) AND THE (S+1)TH
C  EIGENVECTOR FOR X - AS AN INDICATION OF THE ACCURACY ATTAINED.
C
      PROD = 0.0
      DO 19 I = 1,N
      PROD = PROD + XX(I,1)*M(I)*XX(I,SP1)
19    CONTINUE
      WRITE(2,20)
20    FORMAT(///,' ',5X,'THE PRODUCT (XTRANSPOSE)(M)(X) USING')
      WRITE(2,21)PROD
21    FORMAT(' ',5X,'THE FIRST AND (S+1)TH VECTOR IS',F20.17)
      STOP
      END

      SUBROUTINE POWERA(A,LAMBDA,X,N,EPSI,IT)
C  SUBROUTINE POWERA FINDS THE LARGEST EIGENVALUE OF A MATRIX
C  AND THE ASSOCIATED EIGENVECTOR COMPONENTS BY THE ITERATIVE
C  PROCEDURE CALLED THE POWER METHOD.
C
      IMPLICIT REAL*8(A-H,O-Z)
      REAL*8 LAMBDA
      DIMENSION A(10,10),X(10),D(10),Z(10)
C  ASSIGN VALUES TO COMPONENTS OF VECTOR X.
      X(1) = 1.0
      DO 2 I = 2,N
2     X(I) = X(I-1) + 0.1
C  CALCULATE COMPONENTS OF THE VECTOR AX = LAMBDA*X.  THIS
C  VECTOR IS CALLED D(I).
      IT = 0.
3     DO 4 I = 1,N
      D(I) = 0.
      DO 4 J = 1,N
      D(I) = D(I) + A(I,J)*X(J)
4     CONTINUE
      IT = IT + 1
C  NORMALIZE THE VECTOR D(I) AND CALL IT Z(I)
      DO 5 I = 1,N
5     Z(I) = D(I)/D(1)
C  CHECK TO SEE IF REQUIRED ACCURACY HAS BEEN ATTAINED.
      DO 6 I = 1,N
      DIFF = X(I) - Z(I)
      IF(DABS(DIFF) - EPSI*DABS(Z(I)))6,6,7
6     CONTINUE
      GO TO 9
```

```
C   REQUIRED ACCURACY HAS NOT BEEN ATTAINED.  ASSIGN COMPONENTS
C   OF THE Z VECTOR TO X AND ITERATE AGAIN.
7       DO 8 I = 1,N
8       X(I) = Z(I)
        IF(IT .GE. 100)GO TO 10
        GO TO 3
C   REQUIRED ACCURACY HAS BEEN ATTAINED
9       DO 20 I = 1,N
20      X(I) = Z(I)
C   OR MAXIMUM NUMBER OF ITERATIONS HAS BEEN REACHED.
10      LAMBDA = D(1)
        RETURN
        END
```

EXAMPLE 5-21

The distributed mass of a tightly stretched steel cable is to be modeled by five lumped masses equally spaced along the cable as shown in Fig. 5-31a.[15] The tension T in the 120-ft-long cable is 10,000 lb, and the cable weighs 2.5 lb/ft.

We wish to determine which of the following lumped-mass models better represents the distributed mass of the cable:

Model 1:

$$m_1 = m_2 = m_3 = m_4 = m_5 = \frac{\text{total mass of cable}}{6}$$

Model 2:

$$m_1 = m_2 = m_3 = m_4 = m_5 = \frac{\text{total mass of cable}}{5}$$

Both models consist of five equally spaced lumped masses as shown in Fig. 5-31a. However, in the first model each lumped mass is assumed to consist of 10 ft of cable to each side of the lumped mass, so that each lumped mass of that model represents 20 ft of the 120-ft cable. With this assumption, one-half of the mass of the cable between m_1 and the left end of the cable and one-half of the mass of the cable between m_5 and the right end of the cable are not included as part of the lumped masses. Thus, the total mass of the five lumped masses is less than the total mass of the cable. In the second model, each lumped mass is arbitrarily assigned one-fifth of the total mass of the cable; so they obviously sum to the total mass of the cable.

As a basis for determining which model is better, let us determine the lowest three natural frequencies of each model and compare them with the corresponding exact frequencies as determined from the wave equation (see Sec. 8-2). The exact natural circular frequencies are given by

$$\omega_i = \frac{i\pi}{l}\sqrt{\frac{T}{\gamma}} = i\omega_1 \qquad i = 1, 2, 3, \ldots \tag{5-140}$$

[15] Lumped-mass modeling is discussed in Sec. 5-12.

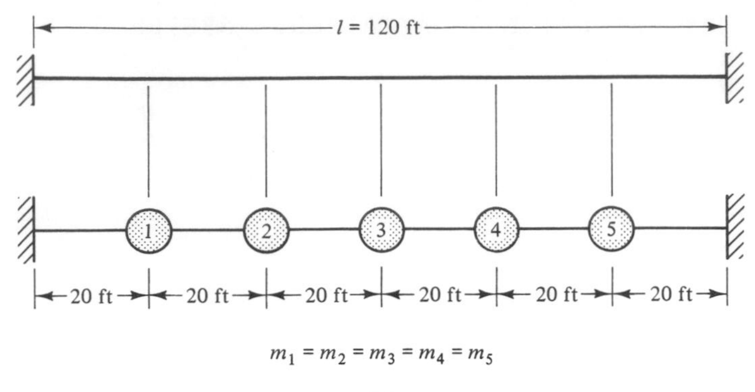

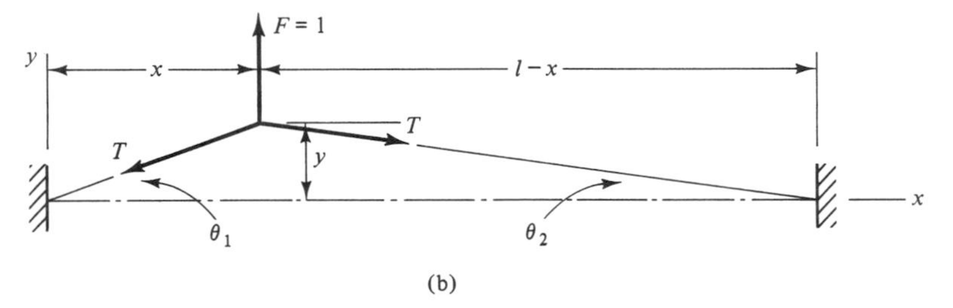

Figure 5-31 Tightly stretched cable modeled by five lumped masses. (a) Cable modeled by five lumped masses. (b) General configuration of cable due to unit force ($F = 1$).

where T = tension in cable = 10,000 lb
γ = mass/unit length = 2.5/[386(12)] = $5.4(10)^{-4}$ lb · s^2/in.^{2}
l = length of cable = 120(12) = 1440 in.

Use the FORTRAN program DEFLATEA.FOR to obtain the natural frequencies of the two lumped-mass models.

Solution. To determine the flexibility coefficients, we first consider a unit force ($F = 1$) as being applied at some point along the cable as shown in Fig. 5-31b. Considering this point to be in equilibrium, and assuming that the tension T remains the same everywhere in the cable since the displacement y shown is very small, we can sum forces to find that

$$\sum F_y = 0 = F - T \sin \theta_1 - T \sin \theta_2 \tag{5-141}$$

Also, since the displacement y is very small, we can assume that

$$\sin \theta_1 \cong \tan \theta_1 = \frac{y}{x}$$

$$\sin \theta_2 \cong \tan \theta_2 = \frac{y}{l - x}$$

Substituting these trigonometric relationships into Eq. 5-141, with $F = 1$, we find that

$$y = \frac{x(l - x)}{Tl} \tag{5-142}$$

Equation 5-142 gives the displacement of any point along the cable at which a unit force is applied, which is also the definition of the *diagonal* flexibility coefficients a_{ii}. For example, when a unit force is applied at $x = l/6$, the value of y obtained from Eq. 5-142 is the flexibility coefficient a_{11}. Thus,

$$a_{11} = \frac{(l/6)[l - (l/6)]}{Tl} = \frac{5l}{36T}$$

Similarly, using x values corresponding to the locations of the other lumped masses in Eq. 5-142, the remaining diagonal coefficients are found to be

$$a_{22} = \frac{2l}{9T}$$

$$a_{33} = \frac{l}{4T}$$

$$a_{44} = \frac{2l}{9T}$$

$$a_{55} = \frac{5l}{36T}$$

The off-diagonal flexibility coefficients can now be easily determined by using the values of the diagonal coefficients just determined and some simple geometry. For example, referring to Fig. 5-32a and using the value obtained above for a_{11}, we can use similar triangles to find the off-diagonal coefficients

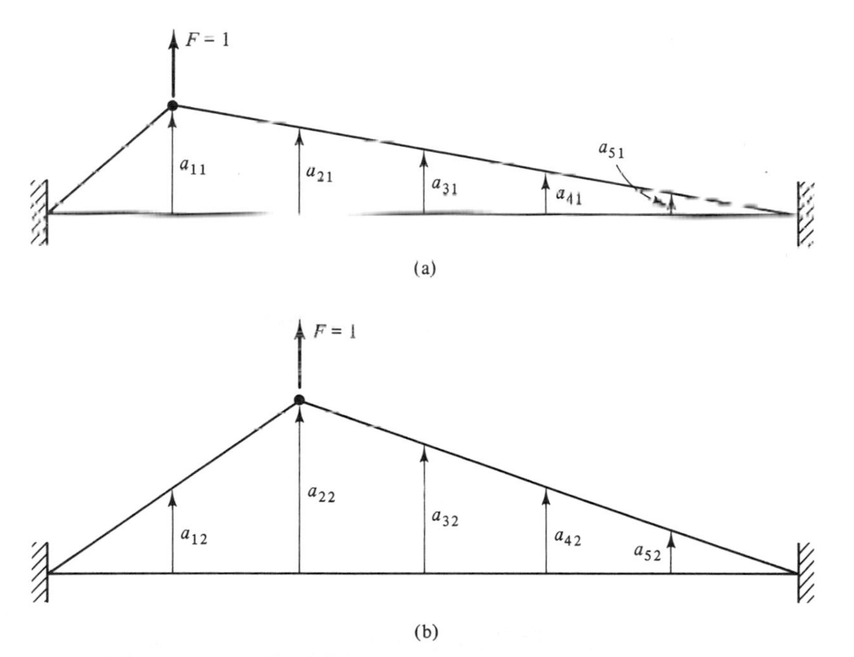

Figure 5-32 Determining off-diagonal flexibility coefficients.

$$a_{21} = \frac{4}{5} a_{11} = \frac{4}{5}\left(\frac{5l}{36T}\right) = \frac{l}{9T}$$

$$a_{31} = \frac{3}{5} a_{11} = \frac{l}{12T}$$

$$a_{41} = \frac{2}{5} a_{11} = \frac{l}{18T}$$

$$a_{51} = \frac{1}{5} a_{11} = \frac{l}{36T}$$

which are elements of the first column of the flexibility matrix. The coefficients a_{12}, $a_{32}, \ldots, a_{52}$, which are elements of the second column of the flexibility matrix, are found from the similar triangles shown in Fig. 5-32b with a_{22} known. Similar figures can be sketched using the other three diagonal coefficients to determine the off-diagonal coefficients of the remaining three columns of the flexibility matrix.

Input Data for DEFLATEA.FOR

Recalling that the flexibility matrix is a symmetric matrix with $a_{ij} = a_{ji}$, and noting that the upper triangular elements of the flexibility matrix are read in as data in the program DEFLATEA.FOR, the following coefficients are calculated for $T = 10{,}000$ lb and $l = 1440$ in.:

$$a_{11} = a_{55} = \frac{5l}{36T} = 0.0200$$

$$a_{12} = a_{24} = a_{45} = \frac{l}{9T} = 0.0160$$

$$a_{13} = a_{35} = \frac{l}{12T} = 0.0120$$

$$a_{14} = a_{25} = \frac{l}{18T} = 0.0080$$

$$a_{15} = \frac{l}{36T} = 0.0040$$

$$a_{22} = a_{44} = \frac{2l}{9T} = 0.0320$$

$$a_{23} = a_{34} = \frac{l}{6T} = 0.0240$$

$$a_{33} = \frac{l}{4T} = 0.0360$$

The data for the lumped masses of each of the models are:

Model 1:

$$m_i = \frac{2.5(20)}{386} = 0.1295 \text{ lb}\cdot\text{s}^2/\text{in.} \qquad i = 1, 2, \ldots, 5$$

Model 2:

$$m_i = \frac{2.5(120)}{386(5)} = 0.1554 \text{ lb} \cdot \text{s}^2/\text{in.}$$

Other input data:

$$\begin{aligned} S + 1 &= 3 \quad \text{(to obtain first three modes)} \\ N &= 5 \quad \text{(five degrees of freedom)} \\ \epsilon &= 0.000001 \end{aligned}$$

Using the data above pertinent to Model 1 as input to the program DEFLATEA.FOR, the following printout is obtained:

```
THE FLEXIBILITY MATRIX IS:


A( 1, 1) =  .2000000E-01
A( 1, 2) -  .1C00000E-01
A( 1, 3) =  .1200000E-01
A( 1, 4) =  .8000000E-02
A( 1, 5) =  .4000000E-02
A( 2, 1) =  .1600000E-01
A( 2, 2) =  .3200000E-01
A( 2, 3) =  .2400000E-01
A( 2, 4) =  .1600000E-01
A( 2, 5) =  .8000000E-02
A( 3, 1) =  .1200000E-01
A( 3, 2) =  .2400000E-01
A( 3, 3) =  .3600000E-01
A( 3, 4) =  .2400000E-01
A( 3, 5) =  .1200000E-01
A( 4, 1) =  .8000000E-02
A( 4, 2) =  .1600000E-01
A( 4, 3) =  .2400000E-01
A( 4, 4) =  .3200000E-01
A( 4, 5) =  .1600000E-01
A( 5, 1) =  .4000000E-02
A( 5, 2) =  .8000000E-02
A( 5, 3) =  .1200000E-01
A( 5, 4) =  .1600000E-01
A( 5, 5) =  .2000000E-01


THE ELEMENTS OF THE DIAGONAL MASS MATRIX ARE:


M( 1) =  .1295000E+00
M( 2) =  .1295000E+00
M( 3) =  .1295000E+00
M( 4) =  .1295000E+00
M( 5) =  .1295000E+00
```

```
LAMBDA( 1) =     .1159922D-01 USING  11 ITERATIONS
OMEGA( 1) =      .9285080D+01

THE EIGENVECTOR COMPONENTS ARE:

                .1000000000D+01
                .1732050883D+01
                .2000000205D+01
                .1732051088D+01
                .1000000205D+01


LAMBDA( 2) =     .3108000D-02 USING  44 ITERATIONS
OMEGA( 2) =      .1793740D+02

THE EIGENVECTOR COMPONENTS ARE:

                .1000000000D+01
                .1000000094D+01
                .2555917646D-06
               -.9999996509D+00
               -.9999997444D+00


LAMBDA( 3) =     .1554000D-02 USING  81 ITERATIONS
OMEGA( 3) =      .2536731D+02

THE EIGENVECTOR COMPONENTS ARE:

                .1000000000D+01
               -.4320931745D-09
               -.1000000001D+01
               -.4325499374D-09
                .1000000000D+01


THE PRODUCT (XTRANSPOSE)(M)(X) USING
THE FIRST AND (S+1)TH VECTOR IS -.00000000033591349
```

Figure 5-33 shows the first, second, and third mode configurations of Model 1.

The first three natural frequencies obtained for each of the models are tabulated below for comparison with the exact values shown, which were calculated from Eq. 5-140.

	Exact (Eq. 5-140)	Model 1	Model 2
First mode	9.39 rad/s	9.29 rad/s	8.48 rad/s
Second mode	18.78	17.93	16.37
Third mode	28.17	25.37	23.16

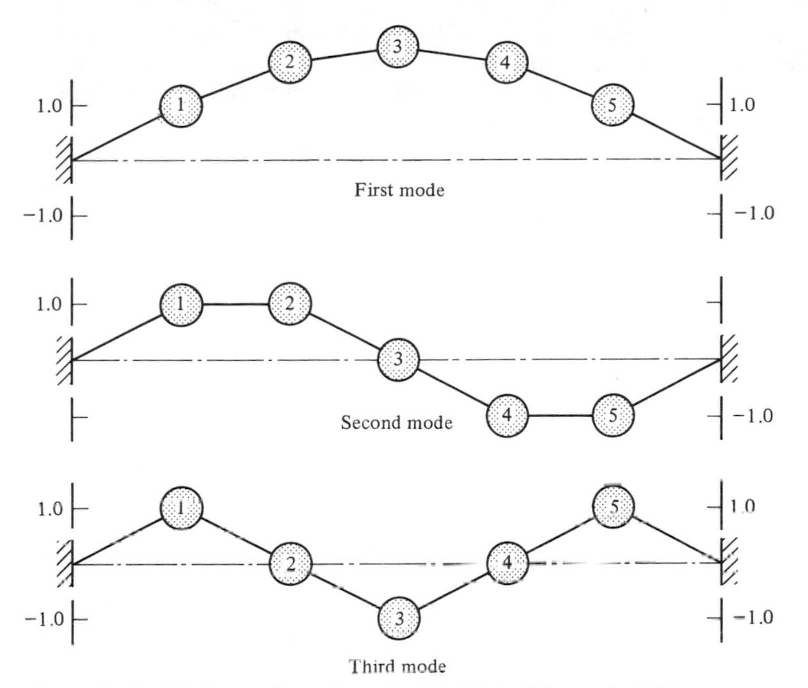

Figure 5-33 Mode configurations for Model 1 of Example 5-21.

The comparison shows that Model 1 is the better lumped-mass model, since its first natural circular frequency is within 1 percent of the exact value, while the natural circular frequency obtained using Model 2 is in error by more than 10 percent. It should be noted that the percentage error of the results obtained from the models increases for successively higher modes. For example, the errors in the second and third modes of Model 1 are approximately 5 and 11 percent, respectively. This occurs because the modal configurations become increasingly distorted for successive modes, as shown in Fig. 5-33. Therefore, more and more lumped masses are required to model the distributed mass for accurate results at successively higher modes. For instance, using a seven-lumped-mass model in the present example with each mass representing a 15-ft segment of the cable, we would expect excellent results for the first and second modes, and reasonably good results for the third mode.

Deflation Procedure for Determining Natural Circular Frequencies in Descending Order

The preceding paragraphs have discussed the process involved in determining the eigenvalues and corresponding eigenvectors of the system matrix equation

$$\mathbf{AMX} = \lambda\mathbf{X}$$

in descending order. In such systems, $\lambda = 1/\omega^2$, so the natural circular frequencies were obtained in *ascending* order. Equations 5-129 through 5-137 were used to prove the deflation procedure.

When the system matrix equation is formulated as

$$\mathbf{M}^{-1}\mathbf{KX} = \lambda\mathbf{X}$$

the use of the deflation procedure will again result in obtaining the eigenvalues in descending order, but since $\lambda = \omega^2$ in the above equation, the natural circular frequencies will also be obtained in *descending* order. Equations 5-129 through 5-137 can again be used to prove the deflation procedure by merely replacing **AM** by $\mathbf{M}^{-1}\mathbf{K}$.

The general procedure for determining the natural circular frequencies in descending order is then summarized as follows:

$$\left.\begin{aligned} &[\mathbf{M}^{-1}\mathbf{K}]_1\mathbf{X} = \lambda\mathbf{X} \qquad \text{(iteration for } \lambda_1 \text{ and } \mathbf{X}_1 \text{ with } \lambda_1 = \omega_1^2) && \text{(a)} \\ &\mathbf{X}_i^T\mathbf{M}\mathbf{X}_i = 1 \qquad \text{(normalization)} && \text{(b)} \\ &[\mathbf{M}^{-1}\mathbf{K}]_{i+1} = [\mathbf{M}^{-1}\mathbf{K}]_i - \lambda_i\mathbf{X}_i\mathbf{X}_i^T\mathbf{M} \qquad \text{(formation of new matrix)} && \text{(c)} \\ &[\mathbf{M}^{-1}\mathbf{K}]_{i+1}\mathbf{X} = \lambda\mathbf{X} \qquad \text{(iteration for } \lambda_{i+1} \text{ and } \mathbf{X}_{i+1}) && \text{(d)} \end{aligned}\right\} \qquad (5\text{-}143)$$

$$i = 1, 2, \ldots, s$$

in which $[\mathbf{M}^{-1}\mathbf{K}]_1$ involves the original stiffness matrix **K** for the first iteration process of Eq. 5-120. The formation of new matrices s times, which are used for successive iterations, will yield a total of $s + 1$ eigenvalues and eigenvectors, including λ_1 and $\mathbf{X}_1$.

The program DEFLATEA.FOR on page 353, which was used to obtain the natural circular frequencies in ascending order, is easily modified to perform the computations referred to in Eq. 5-143 for determining the natural circular frequencies in descending order. The upper triangular elements of the matrix **K** should be read in instead of the matrix **A**, and the heading for the matrix elements should indicate that they are stiffness-matrix elements rather than flexibility-matrix elements. In forming matrix $\mathbf{M}^{-1}\mathbf{K}$ rather than **AM**, statement 9 must be changed to

```
9 MIK(I,J) = K(I,J)/M(I)
```

in which the array name MIK is used for the matrix product $\mathbf{M}^{-1}\mathbf{K}$. The array name MIK would then replace the array name AM in the few places it appears throughout the remainder of the main program. The statement following

```
DO 18 I = 1,SP1
```

must also be changed to

```
OMEGA(I) = DSQRT(LAMBDA(I))
```

In the case of semidefinite systems, no difficulty is encountered in the iteration process for determining the largest eigenvalue and associated eigenvector. However, the use of the deflation procedure to obtain intermediate eigenvalues and eigenvectors is not recommended because of the difficulties encountered with the zero frequency of the rigid-body mode.

As we shall see in the next section, Jacobi's method works very well for semidefinite systems in obtaining all the natural frequencies and normal modes, including the zero frequency and corresponding values for all the eigenvector components for the rigid-body mode. Semidefinite systems with more than one rigid-body mode (more than one zero frequency) are also handled with facility by the Jacobi method as evidenced by Example 7-7. Some readers might prefer to use Jacobi's method for all problems involving the use of the equation

$$\mathbf{M}^{-1}\mathbf{K}\mathbf{X} = \lambda\mathbf{X} \qquad (5\text{-}120)$$

5-11 JACOBI'S METHOD FOR DETERMINING EIGENVALUES AND EIGENVECTORS

Jacobi's method is one of the most efficient and reliable methods of determining the eigenvalues and eigenvectors of systems having up to 20 degrees of freedom. One advantage of this method is that it obtains all the eigenvalues and associated eigenvectors simultaneously, and obtains them with excellent and uniform accuracy. It also works very well for semidefinite (free-free) systems, which is not always true for the iteration process utilizing Hotelling's deflation procedure discussed in the preceding section; it sometimes breaks down for free-free systems.

Since Jacobi's method provides a means of analyzing semidefinite systems, in which the influence coefficients are infinite, let us consider it as a means of obtaining the solution of the equation

$$\mathbf{M}^{-1}\mathbf{K}\mathbf{X} = \lambda\mathbf{X} \tag{5-120}$$

in which $\mathbf{K}$ is the stiffness matrix and the eigenvalues $\lambda_i = \omega_i^2$.

The basic Jacobi method solves the eigenvalue problem

$$\mathbf{A}\mathbf{X} = \lambda\mathbf{X} \tag{5-144}$$

in which $\mathbf{A}$ is a symmetric matrix. However, the matrix $\mathbf{M}^{-1}\mathbf{K}$ in Eq. 5-120 is not generally symmetric, but as we shall see, the eigenvalue problem represented by Eq. 5-120 can be reduced to a standard symmetric eigenvalue problem, which can then be solved.

Jacobi's method consists of performing a sequence of transformations on a sequence of matrices, beginning with a symmetric matrix $\mathbf{A}$ (referred to as $\mathbf{A}_1$ in the equations that follow) for which the eigenvalues and eigenvectors are to be determined. These transformations are of the form

$$\left.\begin{aligned} \mathbf{A}_2 &= \mathbf{R}_1(i,j)\mathbf{A}_1\mathbf{R}_1^T(i,j) \\ \mathbf{A}_3 &= \mathbf{R}_2(i,j)\mathbf{A}_2\mathbf{R}_2^T(i,j) \\ \mathbf{A}_4 &= \mathbf{R}_3(i,j)\mathbf{A}_3\mathbf{R}_3^T(i,j) \\ &\vdots \\ \mathbf{A}_{r+1} &= \mathbf{R}_r(i,j)\mathbf{A}_r\mathbf{R}_r^T(i,j) \\ &\vdots \end{aligned}\right\} \tag{5-145}$$

in which the $\mathbf{R}(i, j)$'s are *plane rotation matrices*, having rotation angles θ in the $x_i x_j$ plane of n-dimensional space (n is the order of the matrix $\mathbf{A}$). A plane rotation matrix converts the coordinates of a point in space (or the components of a vector from the origin to the point) to new coordinates resulting from a rotation of axes in one of the coordinate planes. For example, in Euclidean space

$$\begin{Bmatrix} x' \\ y' \\ z' \end{Bmatrix} = \begin{bmatrix} \cos\theta & \sin\theta & 0 \\ -\sin\theta & \cos\theta & 0 \\ 0 & 0 & 1 \end{bmatrix} \begin{Bmatrix} x \\ y \\ z \end{Bmatrix} \quad \text{or} \quad \mathbf{P}' = \mathbf{R}(x, y)\mathbf{P}$$

gives the new coordinates of a point due to a rotation of the x and y axes through the angle θ, and $\mathbf{R}(x, y)$ is referred to as the rotation matrix for this transformation of coordinates.[16]

It can be shown that the matrix $\mathbf{H}^{-1}\mathbf{A}\mathbf{H}$ has the same eigenvalues as the matrix $\mathbf{A}$, when $\mathbf{H}$ is any square matrix of the same order as $\mathbf{A}$. It can also be shown that the *transpose*

[16] For a detailed development of the method, see M. L. James, G. M. Smith, and J. C. Wolford, *Applied Numerical Methods,* 3d ed., pp. 254–274, Harper & Row, New York, 1985.

of a *rotation matrix* is equal to its inverse. Therefore, all the matrices $\mathbf{A}_2, \mathbf{A}_3, \ldots, \mathbf{A}_{r+1}, \ldots$, in Eq. 5-145 have the same eigenvalues as A. The rotation angle θ for the transformation $\mathbf{R}_r(i, j)\mathbf{A}_r\mathbf{R}_r^T(i, j)$ can be selected to make an off-diagonal element a_{ij} of $\mathbf{A}_{r+1}$ equal to zero. Because of symmetry, an a_{ji} will also be reduced to zero. A sequence of these transformations is performed with varying values of i and j, with each making a corresponding off-diagonal element equal to zero. When all the off-diagonal elements have been thus reduced to zero, one sweep is said to have been completed. A number of sweeps will be required, because when an off-diagonal element of $\mathbf{A}_{r+1}$ is made zero in a transformation, the off-diagonal element of $\mathbf{A}_r$ that was made zero in the previous transformation will generally become nonzero. However, all the off-diagonal elements tend toward zero as the sweeps progress, so that

$$\mathbf{A}_{\text{final}} = \mathbf{R}_{\text{final}} \cdots \underbrace{\mathbf{R}_3 \underbrace{\mathbf{R}_2 \underbrace{\mathbf{R}_1 \mathbf{A}_1 \mathbf{R}_1^T}_{\mathbf{A}_2} \mathbf{R}_2^T}_{\mathbf{A}_3} \mathbf{R}_3^T}_{\mathbf{A}_4} \cdots \mathbf{R}_{\text{final}}^T \qquad (5\text{-}146)$$

is a diagonal matrix, with all the off-diagonal elements essentially equal to zero.

It can be shown that the transformation $[X]^{-1}\mathbf{A}[X]$, in which $[X]$ is an $n \times n$ matrix, with the eigenvectors $\{X_i\}$ $(i = 1, 2, \ldots, n)$ as its columns, results in a diagonal matrix $[D]$ with the eigenvalues of $\mathbf{A}_1$ on its diagonal. It then follows that the diagonal matrix $\mathbf{A}_{\text{final}}$ of Eq. 5-146 is the matrix $[D]$, and its diagonal elements are the eigenvalues of $\mathbf{A}_1$. Furthermore, the square matrix

$$[R_1^T R_2^T R_3^T \cdots R_{\text{final}}^T]$$

is the square matrix $[X]$, which has the eigenvectors of $\mathbf{A}_1$ as its columns.

The eigenvalues on the diagonal of $\mathbf{A}_{\text{final}}$ are, in general, not ordered with respect to their algebraic magnitude as might be desired. A separate algorithm could be used to reorder the eigenvalues and associated eigenvectors, but in the Jacobi computer program given later a technique is used to order the eigenvalues in the process of making the off-diagonal elements equal to zero. In a sweep, j is always greater than i, so whenever a_{ii} is less than a_{jj}, these two diagonal elements are not properly ordered. They are interchanged in the process of making element a_{ij} zero by adding 90° to the rotation angle θ if it is negative, or by subtracting 90° from it if it is positive.

The Eigenvalue Problem $\mathbf{M}^{-1}\mathbf{KX} = \lambda\mathbf{X}$ ($\mathbf{KX} = \lambda\mathbf{MX}$)

Although the matrices $\mathbf{M}$ and $\mathbf{K}$ are both symmetric, the matrix $\mathbf{M}^{-1}\mathbf{K}$ is generally not, which means that the matrix equation

$$\mathbf{M}^{-1}\mathbf{KX} = \lambda\mathbf{X} \qquad (5\text{-}120)$$

cannot be solved directly by Jacobi's method. In many instances the mass matrix $\mathbf{M}$ is diagonal, and some computational effort can be saved by writing a computer program specifically for diagonal mass matrices. However, a full mass matrix is sometimes encountered (see Chap. 7), and the discussion that follows will be concerned with reducing Eq. 5-120 to a standard eigenvalue problem in which the mass matrix $\mathbf{M}$ may be either diagonal or full but must be symmetrical and *positive definite* in either case.[17]

[17] A matrix is positive definite if every principal minor of the matrix is positive. A principal minor of a matrix is the determinant of the matrix remaining after striking out a row and the same numbered column of the original matrix. (This is but one of several possible definitions.)

Since $\mathbf{M}$ is generally both symmetric and positive definite, it can be written as

$$\mathbf{M} = \mathbf{L}\mathbf{L}^T$$

in which $\mathbf{L}$ is a lower triangular matrix and $\mathbf{L}^T$ is the upper triangular transpose of $\mathbf{L}$. This decomposition of $\mathbf{M}$ is known as the Cholesky decomposition, and the equations for performing it will be given later in this discussion.

Writing Eq. 5-120 in the form

$$\mathbf{K}\mathbf{X} = \lambda\mathbf{M}\mathbf{X}$$

and replacing $\mathbf{M}$ with its decomposition, we obtain

$$\mathbf{K}\mathbf{X} = \lambda\mathbf{L}\mathbf{L}^T\mathbf{X}$$

Premultiplying both sides of this equation by $\mathbf{L}^{-1}$ yields

$$\mathbf{L}^{-1}\mathbf{K}\mathbf{X} = \lambda\mathbf{L}^{-1}\mathbf{L}\mathbf{L}^T\mathbf{X}$$

Since

$$(\mathbf{L}^T)^{-1}\mathbf{L}^T = \mathbf{I}$$

and

$$\mathbf{K}\mathbf{I} - \mathbf{K}$$

the above can be rewritten as

$$\mathbf{L}^{-1}\mathbf{K}\underbrace{(\mathbf{L}^T)^{-1}\mathbf{L}^T}_{\mathbf{I}}\mathbf{X} = \lambda\underbrace{\mathbf{L}^{-1}\mathbf{L}}_{\mathbf{I}}\mathbf{L}^T\mathbf{X} \tag{5-147}$$

Because $(\mathbf{L}^T)^{-1} = (\mathbf{L}^{-1})^T$, Eq. 5-147 can be expressed as

$$\left.\begin{aligned} [\mathbf{L}^{-1}\mathbf{K}(\mathbf{L}^{-1})^T]\{\mathbf{L}^T\mathbf{X}\} &= \lambda\{\mathbf{L}^T\mathbf{X}\} \\ \text{or} \qquad \mathbf{A}\mathbf{Y} &= \lambda\mathbf{Y} \end{aligned}\right\} \tag{5-148}$$

where $\mathbf{A} = \mathbf{L}^{-1}\mathbf{K}(\mathbf{L}^{-1})^T$

$\mathbf{Y} = \mathbf{L}^T\mathbf{X}$

in the latter form.

With $\mathbf{K}$ a symmetric matrix, it can be verified that $\mathbf{A}$ will also be symmetric, and Eq. 5-148 can thus be solved by the Jacobi method. The eigenvalues λ of Eq. 5-148 are identical to those of the equation in its original form (Eq. 5-120). However, it should be apparent from the second expression below Eq. 5-148 that the eigenvectors $\mathbf{X}$ of the original equation must be obtained from

$$\mathbf{X} = (\mathbf{L}^T)^{-1}\mathbf{Y} = (\mathbf{L}^{-1})^T\mathbf{Y} \tag{5-149}$$

The Cholesky Decomposition of a Symmetric Positive Definite Matrix

By equating a symmetric matrix $\mathbf{A}$ to the product of a lower triangular matrix and its transpose, such as

$$\begin{bmatrix} l_{11} & 0 & 0 & 0 \\ l_{21} & l_{22} & 0 & 0 \\ l_{31} & l_{32} & l_{33} & 0 \\ l_{41} & l_{42} & l_{43} & l_{44} \end{bmatrix} \begin{bmatrix} l_{11} & l_{21} & l_{31} & l_{41} \\ 0 & l_{22} & l_{32} & l_{42} \\ 0 & 0 & l_{33} & l_{43} \\ 0 & 0 & 0 & l_{44} \end{bmatrix} = \begin{bmatrix} a_{11} & a_{21} & a_{31} & a_{41} \\ a_{21} & a_{22} & a_{32} & a_{42} \\ a_{31} & a_{32} & a_{33} & a_{43} \\ a_{41} & a_{42} & a_{43} & a_{44} \end{bmatrix}$$

and then carrying out the multiplications in symbolic form to solve for the l's, the pattern of the equations for solving for these elements becomes

$$\left.\begin{array}{l}\left.\begin{array}{ll} l_{11} = \sqrt{a_{11}} & \\ l_i = a_i / l_{11} & (\text{for } i = 2, 3, \ldots, n) \end{array}\right\} \quad j = 1 \\ \left.\begin{array}{ll} l_{ij} = \sqrt{a_{ij} - \sum_{k=1}^{j-1} l_{ik} l_{jk}} & (\text{for } i = j) \\ l_{ij} = \dfrac{a_{ij} - \sum_{k=1}^{j-1} l_{ik} l_{jk}}{l_{jj}} & (\text{for } i = j+1, \quad j+2, \ldots, n) \end{array}\right\} \quad j = 2, 3, \ldots, n \end{array}\right\} \tag{5-150}$$

If matrix **A** is not positive definite, then at some stage of the computation one of the radicands in Eq. 5-150 will become negative, and computation will cease.

Computer Program—JACOBI.FOR

The Jacobi program that follows can be used to determine the undamped natural circular frequencies ω_i and corresponding eigenvectors $\mathbf{X}_i$ (normal-mode shapes) of a system having the matrix equation

$$\mathbf{M}^{-1}\mathbf{K}\mathbf{X} = \lambda \mathbf{X}$$

The solutions of this equation, λ_i, are the squares of the natural circular frequencies of the system. The eigenvalues and eigenvectors are all obtained in a final iteration and with equal accuracy.

Some important features of the program are as follows:

1. The determination of the sine and cosine of the rotation angle uses equations that ensure a high degree of accuracy. The program is also written in double precision to further ensure the accuracy of the results.
2. The only data-input values required are the order n of the stiffness matrix **K**, and the upper triangular elements of the matrices **K** and **M**.
3. The mass matrix **M** may be either a diagonal or full matrix but must be positive definite in either case (which it will be if no error has been made). If it is not positive definite, one of the radicands in the Cholesky algorithm will become negative at some stage of the computation, which will cause the computation to cease.
4. The program will handle zero eigenvalues (rigid-body modes) and eigenvalues having a multiplicity of two or more. All the eigenvalues should be positive or zero since they are the squares of the natural circular frequencies. However, because of small inaccuracies in the data, roundoff error, and the inherent imperfection of any iterative numerical method, a theoretically zero eigenvalue can turn out to have a small value instead. If this small value is negative, ω_i is simply set equal to zero in the program to avoid asking the computer to attempt the square root of a negative number.
5. The eigenvectors are normalized such that the first component of each is equal to 1.0, unless the first component is zero or very nearly so. In either of the latter instances, the eigenvector is normalized such that its *largest* component is equal to 1.0.

The important variable names and the quantities they represent in the JACOBI.FOR program are as follows:

Variable or array name	Quantity
K	Array name for the stiffness matrix **K**
N	The order of matrix **K**
M	Array name for the mass matrix **M**
L	Lower triangular matrix formed in the Cholesky decomposition of **M**. **L** is determined by the subprogram DECOMP
LT	The transpose of **L**, determined by the subprogram DECOMP
LINV	The inverse of **L**, determined by subprogram MATINV
LINVTR	The transpose of the inverse of **L**
RT	Originally an N by N identity matrix which eventually becomes the product matrix of all the transposed rotation matrices, $[\mathbf{R}_1^T\mathbf{R}_2^T\mathbf{R}_3^T \cdots \mathbf{R}_{\text{final}}^T]$
A	The array name for the matrix $\mathbf{L}^{-1}\mathbf{K}(\mathbf{L}^{-1})^T$
OMEGA	A one-dimensional array containing the natural circular frequencies ω_i
NSWEEP	A counter that counts the number of sweeps (iterations)
NRSKIP	A counter that counts the number of rotations skipped in a given sweep
AV	$(a_{ij} + a_{ji})/2$. Also used after convergence for the average of the absolute values of the eigenvector components when the first eigenvector component is zero
DIFF	$a_{ii} - a_{jj}$
RAD	$\sqrt{(a_{ii} - a_{jj})^2 + 4\left[\dfrac{a_{ij} + a_{ji}}{2}\right]^2}$
SINE	$\sin\theta$ (θ is the rotation angle)
COSINE	$\cos\theta$
PROD1	The scalar product of the first and last eigenvectors of the symmetric matrix $\mathbf{L}^{-1}\mathbf{K}(\mathbf{L}^{-1})^T$; used as a check on the accuracy of the results
PROD	An array name for the square matrix [LINVTR][RT]. The array RT after convergence contains the eigenvectors of $\mathbf{L}^{-1}\mathbf{K}(\mathbf{L}^{-1})^T$ as its columns. It must be premultiplied by $(\mathbf{L}^{-1})^T$ to yield a matrix containing the eigenvectors of $\mathbf{M}^{-1}\mathbf{K}$ as its columns
II	The row number of the largest element of an eigenvector whose first element is zero. When the first element is zero, normalization will make the largest element equal to 1.0. The value of II is determined by the subprogram SEARCH

The program JACOBI.FOR is as follows:

```
C   JACOBI.FOR
C   THIS DOUBLE-PRECISION PROGRAM DETERMINES THE EIGENVALUES
C   AND EIGENVECTORS SATISFYING THE EQUATION KX = (LAMBDA)MX
C   USING JACOBI'S METHOD, WHERE K IS THE STIFFNESS MATRIX
C   AND M IS A FULL (POSITIVE DEFINITE) MASS MATRIX.  THE PROGRAM
C   AS WRITTEN CAN HANDLE UP TO A 12 DEGREE OF FREEDOM SYSTEM.
C   THE DIMENSION STATEMENTS MUST BE CHANGED FOR LARGER SYSTEMS.
C   DEVICE * IN READ AND WRITE STATEMENTS IS THE CONSOLE.
C   DEVICE 2 IN WRITE STATEMENTS IS THE PRINTER.
C   * IN THE PLACE OF A FORMAT STATEMENT NUMBER MEANS FREE FORMAT.
      IMPLICIT REAL*8 (A-H,O-Z)
      REAL*8 K,M,L,LT,LINV,LINVTR
```

```
      DIMENSION K(12,12),RT(12,12),A(12,12)
      DIMENSION OMEGA(12),M(12,12),L(12,12),LT(12,12)
      DIMENSION LINV(12,12),LINVTR(12,12),PROD(12,12)
C  READ IN THE ORDER OF THE STIFFNESS MATRIX K.
      OPEN (2,FILE='PRN')
      WRITE(*,101)
101   FORMAT(' ','ENTER ORDER N OF THE K MATRIX.  (I2)',/)
      READ(*,*)N
C  READ IN THE UPPER TRIANGULAR ELEMENTS OF MATRIX K
103   FORMAT(' ENTER K(',I2,',',I2,') (F20.0)',/)
104   FORMAT(F20.0)
      DO 3 I = 1,N
      DO 3 J = I,N
      WRITE(*,103)I,J
      READ(*,104)K(I,J)
3     CONTINUE
C  ASSIGN VALUES TO ELEMENTS BELOW THE MAIN DIAGONAL OF THE
C  K MATRIX USING SYMMETRY.
      DO 4 I = 2,N
      IMIN1 = I-1
      DO 4 J = 1,IMIN1
4     K(I,J) = K(J,I)
C  WRITE OUT MATRIX K.
      WRITE(2,5)
5     FORMAT(' ',5X,'THE STIFFNESS MATRIX K IS:',/)
      DO 6 I = 1,N
      DO 6 J = 1,N
      WRITE(2,105)I,J,K(I,J)
105   FORMAT(' ',5X,'K(',I2,',',I2,') = ',E14.7)
6     CONTINUE
C  READ IN THE UPPER TRIANGLE ELEMENTS OF MATRIX M
      DO 7 I=1,N
      DO 7 J=I,N
      WRITE(*,106)I,J
      READ(*,107)M(I,J)
7     CONTINUE
106   FORMAT(' ENTER M(',I2,',',I2,') (F20.0)',/)
107   FORMAT(F20.0)
C  ASSIGN VALUES TO ELEMENTS BELOW MAIN DIAGONAL
      DO 108 I = 2,N
      IMIN1 = I-1
      DO 108 J = 1,IMIN1
108   M(I,J) = M(J,I)
C  WRITE OUT THE MASS MATRIX
      WRITE(2,109)
109   FORMAT(' ',5X,'THE MASS MATRIX M IS:',/)
      DO 202 I = 1,N
      DO 202 J = 1,N
202   WRITE(2,203)I,J,M(I,J)
```

```
203    FORMAT(' ',5X,'M(',I2,',',I2,') = ',E14.7)
C   DECOMPOSE THE MASS MATRIX USING CHOLESKY DECOMPOSITION
       CALL DECOMP(M,N,L,LT)
C   GET THE INVERSE OF LOWER TRIANGULAR MATRIX L
       CALL MATINV(L,LINV,N)
C   GET THE TRANSPOSE OF MATRIX LINV
       DO 204 I = 1,N
       DO 204 J = 1,N
204    LINVTR(I,J) = LINV(J,I)
C   GET THE MATRIX PRODUCT [LINV][K][LINVTR]
       CALL MATMPY(N,K,LINVTR,PROD)
       CALL MATMPY(N,LINV,PROD,A)
C   GENERATE AN N X N IDENTITY MATRIX RT WHICH WILL EVENTUALLY
C   BECOME THE MATRIX CONTAINING THE EIGENVECTORS.
       DO 14 I = 1,N
       DO 13 J = 1,N
       RT(I,J) = 0.0
13     CONTINUE
       RT(I,I) = 1.0
14     CONTINUE
C   INITIALIZE A COUNTER TO ZERO FOR COUNTING NUMBER OF SWEEPS.
       NSWEEP = 0
C   INITIALIZE A COUNTER TO ZERO FOR COUNTING NUMBER OF ROTATIONS
C   SKIPPED.
15     NRSKIP = 0
C   BEGIN A SWEEP WHICH WILL TRANSFORM EACH OFF-DIAGONAL
C   ELEMENT IN TURN TO ZERO.
       NMIN1 = N-1
       DO 25 I - 1,NMIN1
       IP1 = I+1
       DO 24 J = IP1,N
       AV - 0.5*(A(I,J) + A(J,I))
       DIFF = A(I,I) - A(J,J)
       RAD = DSQRT(DIFF*DIFF + 4.*AV*AV)
C   CHECK TO SEE IF RAD IS ZERO.  IF IT IS, NO ROTATION WILL
C   BE PERFORMED SINCE CALCULATION OF SINE AND COSINE WOULD
C   CAUSE OVERFLOW.
       IF(RAD .EQ. 0.0)GO TO 20
C   CHECK TO SEE IF DIFF IS NEGATIVE.  IF IT IS, THE ELEMENTS
C   A(I,I) AND A(J,J) ON THE DIAGONAL NEED TO BE INTERCHANGED,
C   AND THEREFORE A ROTATION WILL BE PERFORMED.
       IF(DIFF .LT. 0.0)GO TO 18
       IF(DABS(A(I,I)) .EQ. DABS(A(I,I))+100.*DABS(AV))GO TO 16
       GO TO 17
16     IF(DABS(A(J,J)) .EQ. DABS(A(J,J))+100.*DABS(AV))GO TO 20
17     COSINE = DSQRT((RAD + DIFF)/(2.*RAD))
       SINE = AV/(RAD*COSINE)
       GO TO 19
C   ON THIS PATH ROTATIONS WILL ALWAYS BE PERFORMED SINCE
```

```
C   ELEMENTS ON THE DIAGONAL NEED REORDERING.
18     SINE = DSQRT((RAD - DIFF)/(2.*RAD))
       IF(AV .LT. 0.0)SINE = -SINE
       COSINE = AV/(RAD*SINE)
C   CHECK TO SEE IF SIN(THETA) IS NEGLIGIBLE.  IF IT IS,
C   SKIP THE ROTATION.
19     IF(1. .LT. 1.+DABS(SINE))GO TO 21
20     NRSKIP = NRSKIP + 1
       GO TO 24
C   PERFORM THE ROTATION.
C   PREMULTIPLY BY THE ROTATION MATRIX.
21     DO 22 L1 = 1,N
       Q = A(I,L1)
       A(I,L1) = COSINE*Q + SINE*A(J,L1)
       A(J,L1) = -SINE*Q + COSINE*A(J,L1)
22     CONTINUE
C   POSTMULTIPLY BY THE TRANSFORM OF THE ROTATION MATRIX.
       DO 23 L1 = 1,N
       Q = A(L1,I)
       A(L1,I) = COSINE*Q + SINE*A(L1,J)
       A(L1,J) = -SINE*Q + COSINE*A(L1,J)
C   POSTMULTIPLY THE CURRENT PRODUCT OF ALL THE RT MATRICES
C   UP TO THIS POINT BY THE CURRENT RT MATRIX.  THIS PRODUCT
C   OF RT MATRICES WHICH HAS THE FORTRAN VARIABLE NAME RT
C   WILL EVENTUALLY BECOME THE MATRIX CONTAINING THE EIGENVECTORS
C   OF AX = (LAMBDA)X  FROM WHICH THE EIGENVECTORS OF
C   KX = (LAMBDA)MX  WILL BE OBTAINED.
       Q = RT(L1,I)
       RT(L1,I) = COSINE*Q + SINE*RT(L1,J)
       RT(L1,J) = -SINE*Q + COSINE*RT(L1,J)
23     CONTINUE
24     CONTINUE
25     CONTINUE
C   KEEP A TALLY OF THE NUMBER OF SWEEPS.
       NSWEEP = NSWEEP + 1
       IF(NSWEEP .GT. 100)GO TO 33
       WRITE(2,26)NRSKIP,NSWEEP
26     FORMAT(' ',5X,'THERE WERE ',I2,
      $' ROTATIONS SKIPPED ON SWEEP NUMBER ',I2)
C   SEE IF THE NUMBER OF ROTATIONS SKIPPED IS LESS THAN
C   OR EQUAL TO THE NUMBER OF ELEMENTS ABOVE THE MAIN DIAGONAL.
C   IF EQUAL TO, THEN CONVERGENCE HAS OCCURRED.
       IF(NRSKIP .LT. N*(N-1)/2)GO TO 15
C   AS AN ACCURACY CHECK SEE IF THE DOT PRODUCT OF THE FIRST
C   AND LAST EIGENVECTORS OF THE TRANSFORMED MATRIX
C   A = [LINV][M]LINVTR] IS NEAR ZERO.
       PROD1 = 0.0
       DO 27 J = 1,N
       PROD1 = PROD1 + RT(J,1)*RT(J,N)
```

```
27     CONTINUE
       WRITE(2,28)
28     FORMAT(/,' ',5X,'THE SCALAR PRODUCT OF THE FIRST AND LAST')
       WRITE(2,29)PROD1
29     FORMAT(' ',5X,'EIGENVECTORS OF THE TRANSFORMED MATRIX IS ',
      $F19.17/)
C   THE MATRIX RT(I,J) CURRENTLY CONTAINS THE EIGENVECTORS OF
C   A = [LINV][K][LINVTR] AS ITS COLUMNS.  EACH COLUMN OF RT(I,J)
C   IS EQUAL TO [LT][X] FOR A PARTICULAR X, WHERE THE X'S ARE
C   EIGENVECTORS OF K.  GET THE SQUARE MATRIX WHOSE COLUMNS ARE
C   THE EIGENVECTORS OF K AS [LINVTR][RT].
       CALL MATMPY(N,LINVTR,RT,PROD)
       DO 30 I = 1,N
       DO 30 J = 1,N
30     RT(I,J) = PROD(I,J)
C   NORMALIZE THE EIGENVECTORS X SO THAT THEIR FIRST COMPONENTS
C   ARE UNITY. IF THE MAGNITUDE OF THE FIRST COMPONENT IS VERY
C   SMALL COMPARED TO THE AVERAGE MAGNITUDE OF THE OTHER COMPONENTS,
C   NORMALIZATION WILL BE WITH RESPECT TO THE LARGEST COMPONENT.
       DO 42 J = 1,N
       SUM = 0.0
       DO 31 I = 1,N
31     SUM = SUM + DABS(RT(I,J))
       AV = SUM/N
       QUOT = DABS(RT(1,J))/AV
       IF(QUOT .LT. 0.000001)GO TO 40
       DO 32 I = 2,N
32     RT(I,J) = RT(I,J)/RT(1,J)
       RT(1,J) = 1.0D0
       GO TO 42
40     CALL SEARCH(RT,J,II,N)
       BIG = RT(II,J)
       DO 41 I = 1,N
41     RT(I,J) = RT(I,J)/BIG
42     CONTINUE
C   CALCULATE NATURAL CIRCULAR FREQUENCIES, RAD/SEC.
       DO 110 I = 1,N
       IF(A(I,I) .LE. 0.0)GO TO 43
       OMEGA(I) = DSQRT(A(I,I))
       GO TO 110
43     OMEGA(I) = 0.0
110    CONTINUE
C   WRITE OUT THE NUMBER OF SWEEPS USED.
33     WRITE(2,34)NSWEEP
34     FORMAT(/,' ',5X,'THERE WERE ',I3,' SWEEPS PERFORMED.',
      $/,5X,' THE EIGENVALUES AND EIGENVECTORS FOLLOW:')
       DO 39 JJ = 1,N
       J = N-JJ+1
       WRITE(2,35)JJ,A(J,J)
35     FORMAT(/,' ',5X,'LAMBDA (',I2,') = ',F20.4)
       WRITE(2,111)JJ,OMEGA(J)
```

```
111   FORMAT(' ',5X,'OMEGA(',I2,') = ',F20.4,' RAD/S')
      WRITE(2,36)
36    FORMAT(/,' ',5X,'THE ASSOCIATED EIGENVECTOR IS:')
      DO 37 I = 1,N
37    WRITE(2,38)RT(I,J)
38    FORMAT(' ',5X,D17.10)
39    CONTINUE
      STOP
      END

      SUBROUTINE DECOMP(A,N,L,LT)
      IMPLICIT REAL*8(A-H,O-Z)
      DIMENSION A(12,12)
      REAL*8 L(12,12),LT(12,12)
      DO 9 J = 1,N
      IF(J .EQ. 1)GO TO 7
      JM1 = J-1
      DO 6 I = J,N
      IF(I .NE. J)GO TO 4
      SUM = 0.0
      DO 3 K = 1,JM1
3     SUM = SUM + L(I,K)*L(J,K)
      L(J,J) = DSQRT(A(J,J) - SUM)
      GO TO 6
4     SUM = 0.0
      DO 5 K = 1,JM1
5     SUM = SUM + L(I,K)*L(J,K)
      L(I,J) = (A(I,J) - SUM)/L(J,J)
6     CONTINUE
      GO TO 9
7     L(1,1) = DSQRT(A(1,1))
      DO 8 I = 2,N
8     L(I,1) = A(I,1)/L(1,1)
9     CONTINUE
C  FILL IN ZERO VALUES OF MATRIX L
      DO 11 J = 2,N
      JM1 = J-1
      DO 11 I = 1,JM1
11    L(I,J) = 0.0
C  ASSIGN VALUES TO THE UPPER TRIANGULAR MATRIX LT
      DO 12 I = 1,N
      DO 12 J = 1,N
12    LT(I,J) = L(J,I)
      RETURN
      END

      SUBROUTINE MATINV(B,A,N)
C  MATRIX INVERSION USING GAUSS-JORDAN REDUCTION AND PARTIAL
C  PIVOTING.  MATRIX B IS THE MATRIX TO BE INVERTED AND A IS
```

```
C   THE INVERTED MATRIX.
      IMPLICIT REAL*8(A-H,O-Z)
      DIMENSION B(12,12),A(12,12),INTER(12,2)
      DO 2 I = 1,N
      DO 2 J = 1,N
2     A(I,J) = B(I,J)
C   CYCLE PIVOT ROW NUMBER FROM 1 TO N
      DO 12 K = 1,N
      JJ = K
      IF(K .EQ. N)GO TO 6
      KP1 = K + 1
      BIG = DABS(A(K,K))
C   SEARCH FOR LARGEST PIVOT ELEMENT
      DO 5 I = KP1,N
      AB = DABS(A(I,K))
      IF(BIG-AB)4,5,5
4     BIG = AB
      JJ = I
5     CONTINUE
C   MAKE DECISION ON NECESSITY OF ROW INTERCHANGE AND
C   STORE THE NUMBER OF THE TWO ROWS INTERCHANGED DURING KTH
C   REDUCTION.  IF NO INTERCHANGE, BOTH NUMBERS STORED EQUAL K
6     INTER(K,1) = K
      INTER(K,2) = JJ
      IF(JJ-K)7,9,7
7     DO 8 J = 1,N
      TEMP = A(JJ,J)
      A(JJ,J) = A(K,J)
8     A(K,J) = TEMP
C   CALCULATE ELEMENTS OF REDUCED MATRIX
C   FIRST CALCULATE NEW ELEMENTS OF PIVOT ROW
9     DO 10 J = 1,N
      IF(J .EQ. K)GO TO 10
      A(K,J) = A(K,J)/A(K,K)
10    CONTINUE
C   CALCULATE ELEMENT REPLACING PIVOT ELEMENT
      A(K,K) = 1./A(K,K)
C   CALCULATE NEW ELEMENTS NOT IN PIVOT ROW OR COLUMN
      DO 11 I = 1,N
      IF(I .EQ. K)GO TO 11
      DO 110 J = 1,N
      IF(J .EQ. K) GO TO 110
      A(I,J) = A(I,J) - A(K,J)*A(I,K)
110   CONTINUE
11    CONTINUE
C   CALCULATE NEW ELEMENTS FOR PIVOT COLUMN--EXCEPT PIVOT ELEMENT
      DO 120 I = 1,N
      IF(I .EQ. K) GO TO 120
      A(I,K) = -A(I,K)*A(K,K)
```

```
120     CONTINUE
12      CONTINUE
C   REARRANGE COLUMNS OF FINAL MATRIX OBTAINED
        DO 13 L = 1,N
        K = N-L+1
        KROW = INTER(K,1)
        IROW = INTER(K,2)
        IF(KROW .EQ. IROW)GO TO 13
        DO 130 I = 1,N
        TEMP = A(I,IROW)
        A(I,IROW) = A(I,KROW)
        A(I,KROW) = TEMP
130     CONTINUE
13      CONTINUE
        RETURN
        END

        SUBROUTINE MATMPY(N,A,B,C)
        IMPLICIT REAL*8(A-H,O-Z)
C   C IS THE PRODUCT MATRIX OF A AND B
        DIMENSION A(12,12),B(12,12),C(12,12)
        DO 2 I = 1,N
        DO 2 J = 1,N
        C(I,J) = 0.0
        DO 2 K = 1,N
2       C(I,J) = C(I,J) + A(I,K)*B(K,J)
        RETURN
        END

        SUBROUTINE SEARCH(RT,J,II,N)
C   THIS SUBROUTINE SEARCHES THE JTH COLUMN OF THE MATRIX RT
C   FOR THE LARGEST EIGENVECTOR COMPONENT.  ITS ROW NUMBER IS
C   ASSIGNED TO THE NAME II.
        IMPLICIT REAL*8(A-H,O-Z)
        DIMENSION RT(12,12)
        II = 1
        BIG = DABS(RT(1,J))
        DO 3 I = 2,N
        AB = DABS(RT(I,J))
        IF(BIG-AB)2,3,3
2       BIG = AB
        II = I
3       CONTINUE
        RETURN
        END
```

The subroutine subprograms DECOMP, MATINV, MATMPY, and SEARCH are used, respectively, for determining the Cholesky decomposition of the mass matrix **M**, determining the inverse of the lower triangular matrix **L**, multiplying *square* matrices, and searching for the largest eigenvector component in those instances in which normalization of the eigenvector is with respect to the largest component.

The printout that follows this paragraph is the solution of Example 5-20 using the JACOBI.FOR program with the mass and stiffness matrices shown in the example as data to the program (see Eqs. 5-126 and 5-127). The λ values shown are in terms of $EI/m_1 l^3$, and the ω values are in terms of $\sqrt{EI/m_1 l^3}$. Note as mentioned earlier that this program also yields the rigid-body mode for $\omega_1 = 0$.

```
THE STIFFNESS MATRIX K IS:

K( 1, 1) =    .3000000E+01
K( 1, 2) =   -.3000000E+01
K( 1, 3) =    .0000000E+00
K( 2, 1) =   -.3000000E+01
K( 2, 2) =    .6000000E+01
K( 2, 3) =   -.3000000E+01
K( 3, 1) =    .0000000E+00
K( 3, 2) =   -.3000000E+01
K( 3, 3) =    .3000000E+01
THE MASS MATRIX M IS:

M( 1, 1) =    .1000000E+01
M( 1, 2) =    .0000000E+00
M( 1, 3) =    .0000000E+00
M( 2, 1) =    .0000000E+00
M( 2, 2) =    .4000000E+01
M( 2, 3) =    .0000000E+00
M( 3, 1) =    .0000000E+00
M( 3, 2) =    .0000000E+00
M( 3, 3) =    .1000000E+01
THERE WERE  0 ROTATIONS SKIPPED ON SWEEP NUMBER  1
THERE WERE  0 ROTATIONS SKIPPED ON SWEEP NUMBER  2
THERE WERE  0 ROTATIONS SKIPPED ON SWEEP NUMBER  3
THERE WERE  1 ROTATIONS SKIPPED ON SWEEP NUMBER  4
THERE WERE  3 ROTATIONS SKIPPED ON SWEEP NUMBER  5

THE SCALAR PRODUCT OF THE FIRST AND LAST
EIGENVECTORS OF THE TRANSFORMED MATRIX IS  .0000000000000001


THERE WERE   5 SWEEPS PERFORMED.
THE EIGENVALUES AND EIGENVECTORS FOLLOW:

LAMBDA ( 1) =                .0000
OMEGA( 1) =                 .0000 RAD/S

THE ASSOCIATED EIGENVECTOR IS:
  .1000000000D+01
  .1000000000D+01
  .1000000000D+01
```

```
LAMBDA ( 2) =              3.0000
OMEGA( 2) =               1.7321 RAD/S

THE ASSOCIATED EIGENVECTOR IS:
   .1000000000D+01
   .1023688286D-16
  -.1000000000D+01

LAMBDA ( 3) =              4.5000
OMEGA( 3) =               2.1213 RAD/S

THE ASSOCIATED EIGENVECTOR IS:
   .1000000000D+01
  -.5000000000D+00
   .1000000000D+01
```

Jacobi's Method with the Flexibility Matrix

It was shown in Sec. 5-9 that the matrix equation for obtaining the natural circular frequencies and mode shapes of a system using the flexibility matrix is

$$\mathbf{AMX} = \lambda\mathbf{X} \tag{5-121}$$

Although $\mathbf{A}$ and $\mathbf{M}$ are both symmetric matrices, the matrix $\mathbf{AM}$ is usually not. However, Eq. 5-121 can be reduced rather readily to the standard form required for solution by Jacobi's method, as we shall see in the next few paragraphs.

It was pointed out in the preceding subsection that the mass matrix $\mathbf{M}$ is generally both symmetric and positive definite and can therefore be decomposed by the Cholesky decomposition into the product of a lower triangular matrix $\mathbf{L}$ and an upper triangular matrix $\mathbf{L}^T$ which is the transpose of $\mathbf{L}$. That is,

$$\mathbf{M} = \mathbf{LL}^T$$

Replacing $\mathbf{M}$ in Eq. 5-121 by its decomposition, we find that

$$\mathbf{ALL}^T\mathbf{X} = \lambda\mathbf{X}$$

Premultiplying both sides of the above by $\mathbf{L}^T$,

$$\mathbf{L}^T\mathbf{ALL}^T\mathbf{X} = \lambda\mathbf{L}^T\mathbf{X}$$

which can also be written as

$$[\mathbf{L}^T\mathbf{AL}][\mathbf{L}^T\mathbf{X}] = \lambda[\mathbf{L}^T\mathbf{X}]$$

or as

$$\mathbf{BY} = \lambda\mathbf{Y} \tag{5-151}$$

in which

$$\mathbf{B} = [\mathbf{L}^T\mathbf{AL}]$$

and

$$\mathbf{Y} = [\mathbf{L}^T\mathbf{X}]$$

It can be verified that $[\mathbf{L}^T\mathbf{A}\mathbf{L}]$ will be symmetric, which means that Eq. 5-151 can be solved using Jacobi's method. The eigenvalues λ_i that satisfy Eq. 5-151 are the same eigenvalues that satisfy Eq. 5-121 since no change was made in λ in developing Eq. 5-151 from Eq. 5-121. However, it should be apparent from the expression for **Y** above that the eigenvectors **X** of the matrix **AM** (the eigenvectors satisfying Eq. 5-121) must be obtained from

$$\mathbf{X} = (\mathbf{L}^T)^{-1}\mathbf{Y} \tag{5-152}$$

The Jacobi program on page 369 can be modified to solve Eq. 5-121 by applying it to Eq. 5-151 and then utilizing Eq. 5-152 to determine the eigenvectors of **AM**.

5-12 LUMPED-MASS MODELING OF DISTRIBUTED MASSES (RODS AND BEAMS)

Bodies with distributed mass such as cables, rods, and beams can be modeled as lumped-mass systems for purposes of analyzing their vibration characteristics. As we shall see in Chap. 6, lumped-mass models are employed in the modal-analysis method used to obtain the vibration response of systems subjected to various types of excitation.

There are two general procedures for modeling the distributed masses of systems. The procedure presented here is characterized by a diagonal mass matrix. The other is associated with the finite-element method discussed in Chap. 7, which discretizes the distributed-mass elements of a system and results in a symmetric mass matrix containing nonzero off-diagonal elements.

In Example 5-21 two different lumped-mass models were used to represent the distributed mass of a tightly stretched cable. Each model used five lumped masses, but the magnitudes of the masses were different. By comparing the natural circular frequencies obtained from each of the models with those given by the exact solution, it was found that the fundamental natural frequency obtained using one model was within 1 percent of the exact value, while that obtained from the other model was in error by 10 percent. This example is recalled to stress the importance of determining lumped-mass models that represent the distributed-mass systems they are modeling as accurately as possible.

Let us now investigate a procedure for determining reasonably accurate lumped-mass models for use in analyzing the axial (longitudinal) vibration of rods and the lateral vibration of beams. The number of lumped masses selected to represent a distributed mass having a theoretically infinite number of degrees of freedom depends upon the number of natural frequencies and mode shapes necessary to obtain results to some desired degree of accuracy. In general, the lowest natural frequency and associated mode shape of an n lumped-mass model will be the most accurate obtained, with the second natural frequency and mode shape the next most accurate, and so on. It is also generally true that as the number of properly selected lumped masses used to represent the system increases, the accuracy of the frequencies and mode shapes increases also in the hierarchy mentioned in the preceding sentence. Thus, for example, if the use of only three or four lumped masses gave a reasonably accurate value of the fundamental natural frequency, the use of perhaps five or six lumped masses might provide sufficiently accurate values for the two or three lowest natural frequencies and mode shapes.

Lumped-Mass Modeling of Rods (Axial Vibration)

To illustrate the preceding discussion, let us consider the uniform rod of length l and cross-sectional area A shown in Fig. 5-34. The rod is fixed at the left end and free at the right

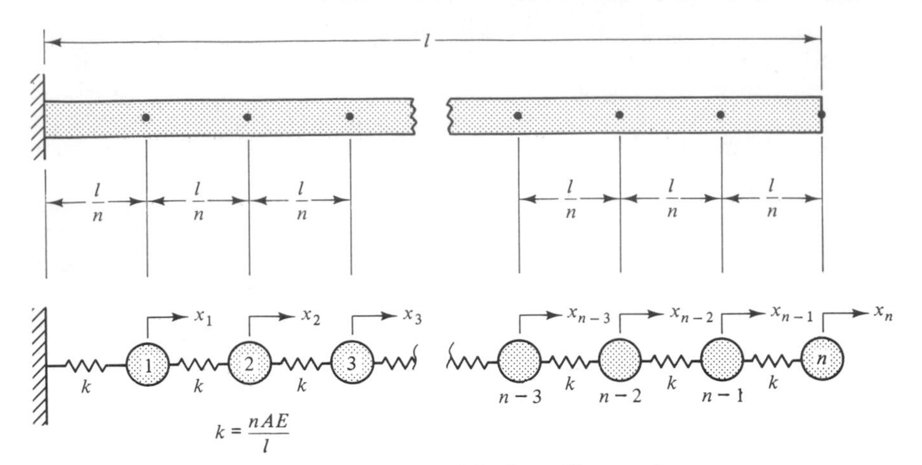

Figure 5-34 Typical n-lumped-mass model of a uniform rod.

end as shown. If the distributed mass of the rod is replaced by n equally spaced, concentrated masses, they will be l/n apart as shown in the figure. The total mass of the rod could simply be distributed equally among the n lumped masses so that

$$m_i = \frac{A\gamma l}{n} \qquad i = 1, 2, \ldots, n \tag{5-153}$$

in which γ is the mass density of the rod material. However, as we shall see in Example 5-22, such a distribution does not provide an accurate lumped-mass model of the rod.

A better lumped-mass model can be obtained using the following procedure:

1. Referring to Fig. 5-34, a lumped mass that is adjacent to a fixed or pinned support, such as m_1, is given a magnitude of one-half of the distributed mass lying between the fixed or pinned support and m_1, plus one-half of the distributed mass lying between m_1 and the adjacent mass m_2. Using this procedure,

$$m_1 = \frac{A\gamma l}{2n} + \frac{A\gamma l}{2n} = \frac{A\gamma l}{n} \tag{5-154}$$

in which l/n is the length of a rod segment between the lumped masses.

2. A mass m_i, which lies between the adjacent masses m_{i-1} and m_{i+1}, is given a magnitude of one-half of the distributed mass lying between m_{i-1} and m_i, plus one-half of the distributed mass lying between m_i and m_{i+1}. Thus,

$$m_2 = m_3 = \cdots = m_{n-1} = \frac{A\gamma l}{n} \tag{5-155}$$

3. The mass m_n at the free end of the rod is given a magnitude of one-half of the distributed mass lying between m_{n-1} and m_n only. Thus,

$$m_n = \frac{A\gamma l}{2n} \tag{5-156}$$

It should be noted that the total mass of the lumped masses determined by the above procedure is less than the total mass of the rod since one-half of the mass lying between the fixed or pinned support and m_1 is not included in the model.

Each of the spring constants k shown represents the stiffness of a segment of rod material between adjacent lumped masses. Thus, from part a of Table 2-1 on page 78, k is

$$k = \frac{AE}{l/n} \tag{5-157}$$

EXAMPLE 5-22

The uniform rod of length l and cross-sectional area A is modeled by the four lumped masses shown in Fig. 5-35. The material of the rod has a modulus of elasticity of E and a mass density of γ. The magnitudes of the four lumped masses are to be determined in two different ways as indicated in the following:

Model 1: Four lumped masses of equal magnitude as given by Eq. 5-153, so that

$$m_1 = m_2 = m_3 = m_4 = \frac{A\gamma l}{4}$$

Model 2: Four lumped masses having magnitudes determined from Eqs. 5-154, 5-155, and 5-156, so that

$$m_1 = \frac{A\gamma l}{4}$$

$$m_2 = m_3 = \frac{A\gamma l}{4}$$

$$m_4 = \frac{A\gamma l}{8}$$

We wish to determine the four lowest natural circular frequencies of the rod using each of the two models, and then compare these values with similar values obtained from the exact solution of the wave equation. The exact values can be determined from (see Sec. 8-2)

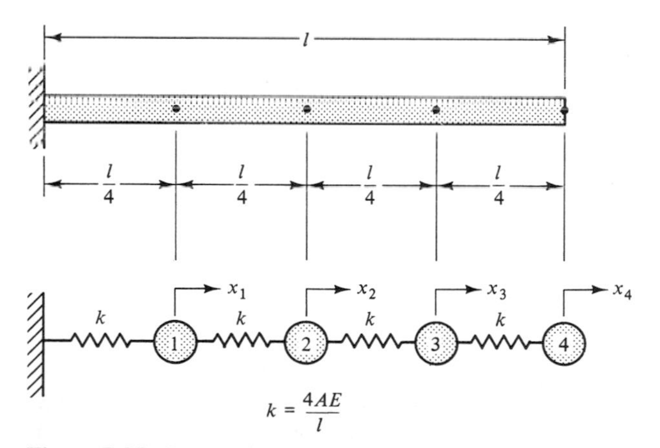

Figure 5-35 Lumped-mass model of Example 5-22.

$$\omega_i = \frac{i\pi}{2l}\sqrt{\frac{E}{\gamma}} = i\omega_1 \qquad i = 1, 3, 5, 7, \ldots \tag{5-158}$$

in which $i = 1, 3, 5,$ and 7 correspond to the first four natural circular frequencies of the rod.

Solution. The stiffness matrix **K** is the same for both models. Referring to Eq. 5-157 and recalling the definition of a stiffness coefficient while observing Fig. 5-35 we obtain

$$\mathbf{K} = \frac{AE}{l}\begin{bmatrix} 8 & -4 & 0 & 0 \\ -4 & 8 & -4 & 0 \\ 0 & -4 & 8 & -4 \\ 0 & 0 & -4 & 4 \end{bmatrix}$$

The mass matrices for the two models are:

Model 1:

$$\mathbf{M} = A\gamma l\begin{bmatrix} 0.25 & 0 & 0 & 0 \\ 0 & 0.25 & 0 & 0 \\ 0 & 0 & 0.25 & 0 \\ 0 & 0 & 0 & 0.25 \end{bmatrix}$$

Model 2:

$$\mathbf{M} = A\gamma l\begin{bmatrix} 0.25 & 0 & 0 & 0 \\ 0 & 0.25 & 0 & 0 \\ 0 & 0 & 0.25 & 0 \\ 0 & 0 & 0 & 0.125 \end{bmatrix}$$

Since the stiffness coefficients k_{ij} are divided by the mass elements m_i (see Eq. 5-36), the natural circular frequencies obtained from the numerical data values shown for **K** and **M** above will be in terms of $(1/l)\sqrt{E/\gamma}$ for comparison with the exact values determined from Eq. 5-158.

Referring to $(1/l)\sqrt{E/\gamma}$ as α, the natural circular frequencies ω_i obtained from the JACOBI.FOR program for both models are tabulated below for comparison with the exact values. The approximate percentage errors between exact and model values for the first four modes are shown in parentheses.

Mode number	Exact	Model 1	Model 2
1	1.571α	1.389α (16%)	1.561α (1%)
2	4.712α	4.000α (15%)	4.445α (6%)
3	7.854α	6.128α (22%)	6.652α (15%)
4	10.996α	7.518α (32%)	7.846α (29%)

It is apparent from the above values that Model 1 with the total mass of the rod divided into four equal lumped masses gives poor results for all four modes. However, Model 2 with the masses determined according to the procedure outlined by Eqs. 5-154 through 5-156 yields a first-mode value that is within 1 percent of the exact value, and a second-mode value that varies from the exact by only 6 percent.

The results also show that the first-mode eigenvector components obtained using Model 2 agree to four digits with the theoretical first-mode shape function

$$u = \sin\left(\frac{\pi x}{2l}\right)$$

(see Sec. 8-2) in which u is the axial displacement of the rod as a function of the location x along the rod.

A five-mass model in which the masses were lumped by the procedure used for Model 2 gives the first natural frequency within 0.4 percent of the exact value, and the second natural frequency within 4 percent. However, lumping five masses by the method used for Model 1 yields results for the first and second modes that are in error by 10 and 12 percent, respectively.

This example shows that the general procedure given by Eqs. 5-154, 5-155, and 5-156 provides reasonably accurate lumped-mass models for analyzing the axial vibration of rods.

EXAMPLE 5-23

The nonuniform rod shown in Fig. 5-36 has a length of $l = l_1 + l_2$, a cross-sectional area of A_1 for l_1 and A_2 for l_2, and is fixed at its right end. It has a modulus of elasticity of E and a mass density of γ. A rigid mass m_0 is attached to the left end of the rod and to a spring of stiffness k_0 as shown.

Assuming that the mass of the spring is negligible, determine (a) the elements m_i composing the mass matrix $\mathbf{M}$ for a six-lumped-mass model of the system, using the procedure outlined on page 380; (b) the stiffness elements k_{ij} of the stiffness matrix $\mathbf{K}$ for the same system.

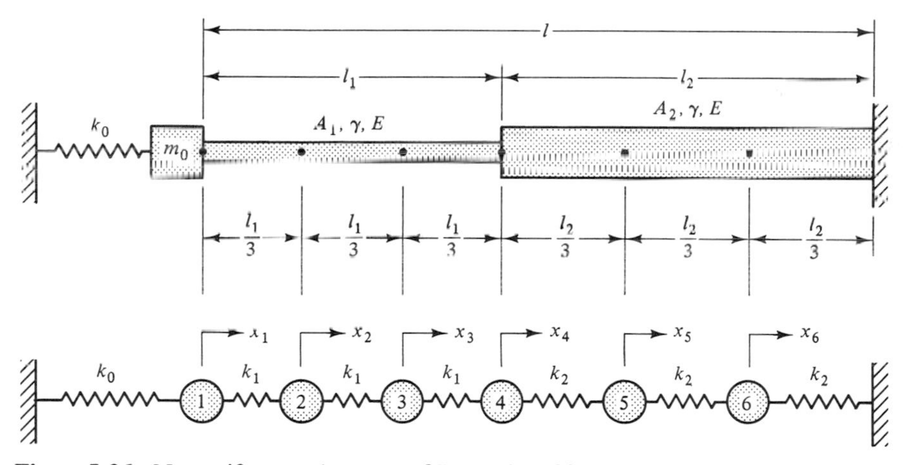

Figure 5-36 Nonuniform rod system of Example 5-23.

Solution. **a.** Dividing each of the rod portions l_1 and l_2 into three equal segments, the locations of the six lumped masses are as indicated by the black dots along the rod in the figure. The mass m_1 consists of the mass m_0, plus one-half of the mass of the rod lying between the masses m_1 and m_2, and is thus of magnitude

$$m_1 = m_0 + \frac{A_1 \gamma l_1}{6}$$

The lumped masses m_2 and m_3 are simply

$$m_2 = m_3 = \frac{A_1 \gamma l_1}{6} + \frac{A_1 \gamma l_1}{6} = \frac{A_1 \gamma l_1}{3}$$

The mass of m_4 consists of one-half of the mass of the smaller rod that lies between the masses m_3 and m_4, plus one-half of the mass of the larger rod that lies between the masses m_4 and m_5. Thus

$$m_4 = A_1 \gamma \left(\frac{l_1}{6} \right) + A_2 \gamma \left(\frac{l_2}{6} \right)$$

Finally,

$$m_5 = m_6 = \frac{A_2 \gamma l_2}{6} + \frac{A_2 \gamma l_2}{6} = \frac{A_2 \gamma l_2}{3}$$

It should be noted that the right one-half of the rod's mass lying between the fixed support and the mass m_6 is not included in the lumped-mass system.

b. Referring to Eq. 5-157, we see that

$$k_1 = \frac{A_1 E}{l_1/3} \qquad \text{(small section of rod)}$$

and that

$$k_2 = \frac{A_2 E}{l_2/3} \qquad \text{(large section of rod)}$$

Then, using the definitions of k_{ij} while inspecting Fig. 5-36, we can determine that

$$k_{11} = k_0 + k_1 = k_0 + \frac{A_1 E}{l_1/3}$$

$$k_{21} = k_{12} = -k_1 = \frac{-A_1 E}{l_1/3}$$

$$k_{31} = k_{13} = k_{41} = k_{14} = k_{51} = k_{15} = k_{61} = k_{16} = 0$$

$$k_{22} = 2k_1 = \frac{2A_1 E}{l_1/3}$$

$$k_{32} = k_{23} = -k_1 = \frac{-A_1 E}{l_1/3}$$

$$k_{42} = k_{24} = k_{52} = k_{25} = k_{62} = k_{26} = 0$$

$$k_{33} = 2k_1 = \frac{2A_1 E}{l_1/3}$$

$$k_{43} = k_{34} = -k_1 = \frac{-A_1 E}{l_1/3}$$

$$k_{53} = k_{35} = k_{63} = k_{36} = 0$$

$$k_{44} = k_1 + k_2 = \frac{A_1E}{l_1/3} + \frac{A_2E}{l_2/3}$$

$$k_{54} = k_{45} = -k_2 = \frac{-A_2E}{l_2/3}$$

$$k_{64} = k_{46} = 0$$

$$k_{55} = 2k_2 = \frac{2A_2E}{l_2/3}$$

$$k_{65} = k_{56} = -k_2 = \frac{-A_2E}{l_2/3}$$

$$k_{66} = 2k_2 = \frac{2A_2E}{l_2/3}$$

which are the elements of the stiffness matrix **K**.

Lumped-Mass Modeling of Beams

The general procedure outlined on page 380 for the lumped-mass modeling of rods also works well for modeling the distributed masses of various types of beams, and we shall use several examples to illustrate this procedure as applied to the lumped-mass modeling of uniform beams.

EXAMPLE 5-24

The overhanging beam shown in Fig. 5-37 has a mass per unit length of γ and is modeled by the four lumped masses indicated. Three of these lumped masses represent the distributed mass of the beam lying between the fixed and pinned supports while the fourth one represents the distributed mass lying between the pinned support and the free end of the beam. With this mass distribution, the lumped masses are located in regions of the beam's maximum amplitudes of vibration, which is a general criterion in modeling beams by lumped-mass systems. We wish to determine the mass matrix **M** for the lumped-mass model of the beam.

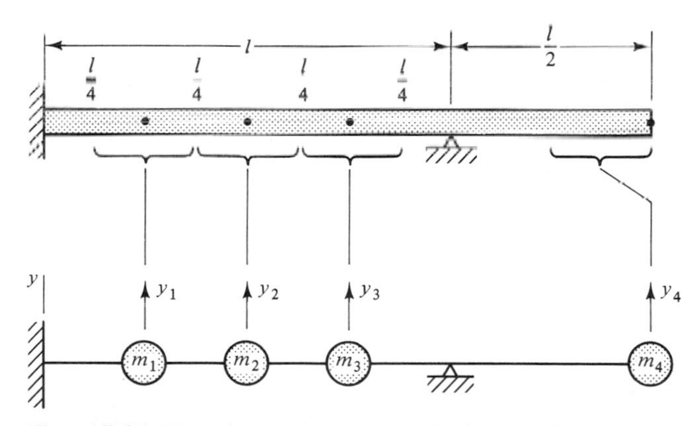

Figure 5-37 Four-lumped-mass model of beam with overhang of Example 5-24.

Solution. The mass m_1 has a magnitude of one-half of the mass of the beam lying between the fixed end and m_1, plus one-half of the mass of the beam lying between m_1 and m_2. The mass m_2 has a magnitude of one-half of the mass of the beam lying between m_1 and m_2, plus one-half of the mass of the beam lying between m_2 and m_3. The mass m_3 has a magnitude of one-half of the mass of the beam lying between m_2 and m_3, plus one-half of the mass of the beam lying between m_3 and the pinned support. The mass m_4 has a magnitude of one-half of the mass of the beam lying between the pinned support and the free end of the beam.

The mass matrix is thus

$$\mathbf{M} = \frac{\gamma l}{4}\begin{bmatrix} 1 & 0 & 0 & 0 \\ 0 & 1 & 0 & 0 \\ 0 & 0 & 1 & 0 \\ 0 & 0 & 0 & 1 \end{bmatrix}$$

EXAMPLE 5-25

The simply supported steel beam shown in Fig. 5-38 is a wide-flange beam, with an 18-in.-deep section and weighs 96 lb/ft. It is 40 ft long and has an area moment of inertia of $I = 1674.7$ in.^{4} We model the beam as three lumped masses and determine (a) the natural frequencies and mode shapes of the three-lumped-mass model; (b) the accuracy of the model, by comparing the natural frequencies and mode shapes determined from the model in part a with the exact values determined from the equations

$$\omega_i = \frac{i^2\pi^2}{l^2}\sqrt{\frac{EI}{\gamma}} = i^2\omega_1 \tag{5-159}$$

and

$$X_i = \sin\frac{i\pi x}{l} \qquad i = 1, 2, 3, \ldots \tag{5-160}$$

which evolve from Euler's beam equation (see Chap. 8). In these equations

ω_i = natural circular frequency of ith mode
X_i = normal-mode function of ith mode (function of x along the beam)
E = modulus of elasticity

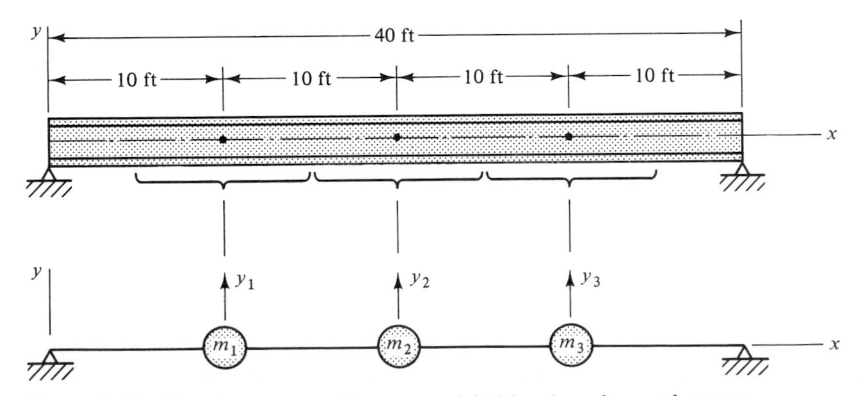

Figure 5-38 Simply supported beam modeled by three lumped masses.

I = area moment of inertia about neutral axis of bending
γ = mass of beam/unit length
l = length of beam

Solution. If we locate the three lumped masses at the quarter points of the beam, the beam will be divided into the four equal segments shown in Fig. 5-38, which is a logical choice because of the symmetrical distribution of the mass of the beam (uniform density and cross section). Following the procedure given on page 380 for modeling rods, and noting that the beam weighs 96 lb/ft, the magnitude of each of the lumped masses is

$$m_1 = m_2 = m_3 = \frac{10(96)}{386} = 2.487 \text{ lb} \cdot \text{s}^2/\text{in.}$$

in which $g = 386$ in./s^2. Note that the mass of both the left 5 ft and the right 5 ft of the beam is not included in the masses of m_1 and m_3, since the displacements are zero at the supports. Thus, the total mass of the model is only three fourths of the total mass of the beam. However, the lumped masses are concentrated in the region of the maximum amplitude of vibration of the beam, which, as mentioned before, is a criterion when modeling beams by lumped masses. The mass matrix is thus

$$\mathbf{M} = \begin{bmatrix} 2.487 & 0 & 0 \\ 0 & 2.487 & 0 \\ 0 & 0 & 2.487 \end{bmatrix}$$

It was pointed out in Sec. 5-3 that it is much easier to determine the flexibility coefficients a_{ij} for beams than it is to determine the stiffness coefficients k_{ij}. With this in mind, and referring to case e of Table 2-1 on page 78, the flexibility coefficients are calculated using the equation

$$y = \frac{Pbx}{6EIl}(l^2 - x^2 - b^2) \qquad x \le a \tag{5-161}$$

in which P is applied at $x = a$ as shown in Fig. 5-39. With a unit force $P = 1$ applied in turn at the locations of m_1, m_2, and m_3 (b = 360, 240, and 120 in.), Eq. 5-161 is used to obtain the flexibility matrix

$$\mathbf{A} = \frac{l^3}{1536EI}\begin{bmatrix} 18 & 22 & 14 \\ 22 & 32 & 22 \\ 14 & 22 & 18 \end{bmatrix} = 10^{-5}\begin{bmatrix} 2.580 & 3.153 & 2.006 \\ 3.153 & 4.586 & 3.153 \\ 2.006 & 3.153 & 2.580 \end{bmatrix}$$

in which $l = 480$ in., $E = 30(10)^6$ psi, and $I = 1674.7$ in.^{4}

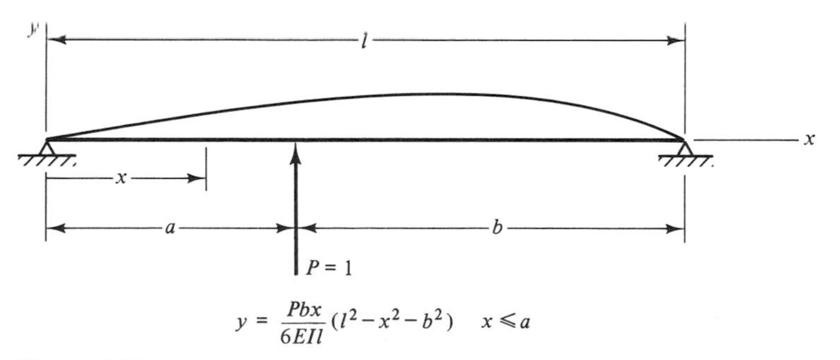

Figure 5-39

With the elements of the matrices **M** and **A** as data input to the computer program DEFLATEA.FOR, it provides us with the natural circular frequencies and mode shapes of the lumped-mass model. These results are tabulated below for comparison with the exact values (in parentheses) calculated from Eqs. 5-159 and 5-160.

Natural circular frequencies ω_i (rad/s)

Mode	Exact (Eq. 5-159)	Three-lumped-mass model
1	(66.70)	66.67 (0.1% error)
2	(266.80)	264.67 (0.8% error)
3	(600.30)	562.71 (6% error)

Eigenvectors—mode shapes

Lumped mass	First mode		Second mode		Third mode	
1	(0.707)	0.707	(1.000)	1.000	(0.707)	0.707
2	(1.000)	1.000	(0.000)	0.000	(−1.000)	−1.000
3	(0.707)	0.707	(−1.000)	−1.000	(0.707)	0.707

The tabulated results reveal that all the natural circular frequencies and mode shapes of the three-lumped-mass model are in excellent agreement with the exact values, with the exception of ω_3, which shows an error of 6 percent. The general shapes of the three modes of the model are shown in Fig. 5-40.

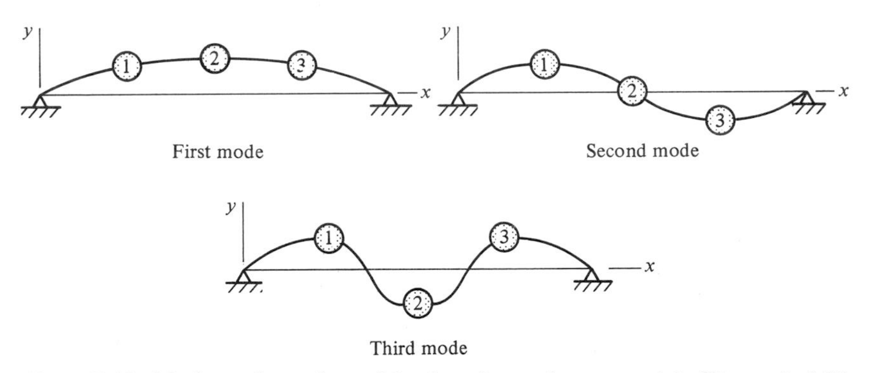

Figure 5-40 Mode configurations of the three-lumped-mass model of Example 5-25.

Beams of Nonuniform Cross Section

The general procedure discussed earlier for the lumped-mass modeling of rods and uniform beams also works well for modeling beams of nonuniform cross section, such as the tapered cantilever beam shown in Fig. 5-41. To illustrate some of the added considerations introduced by a nonuniform cross section, let us discuss the modeling of the tapered cantilever beam as we would in preparing to analyze its vibration in the x-y plane.

We use four lumped masses, spaced so that they divide the length of the beam into the four equal segments shown in Fig. 5-42. With lines b, c, and d located midway between the respective lumped masses, and line e located midway between m_4 and the fixed end of the beam, the lumped masses are given the following magnitudes:

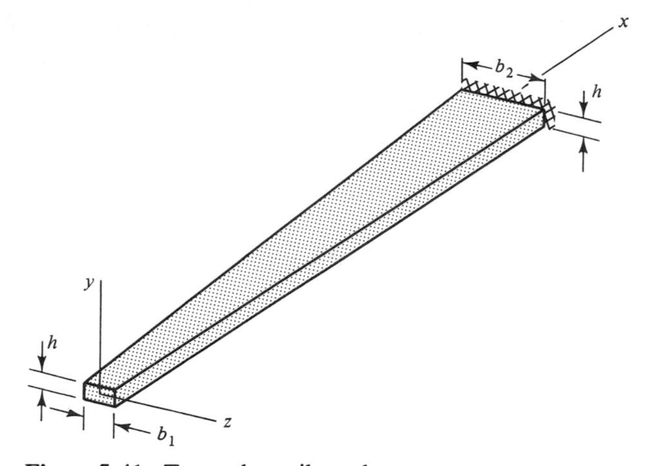

Figure 5-41 Tapered cantilever beam.

$$m_1 = m_{ab} = \text{mass of beam between } a \text{ and } b$$
$$m_2 = m_{bc} = \text{mass of beam between } b \text{ and } c$$
$$m_3 = m_{cd} = \text{mass of beam between } c \text{ and } d$$
$$m_4 = m_{de} = \text{mass of beam between } d \text{ and } e$$

It is generally quite difficult to determine the flexibility coefficients a_{ij} for nonuniform beams to enable us to use the DEFLATEA.FOR program as was done in Example 5-25. However, as we shall see in Chap. 7, the finite-element method does not require the de-

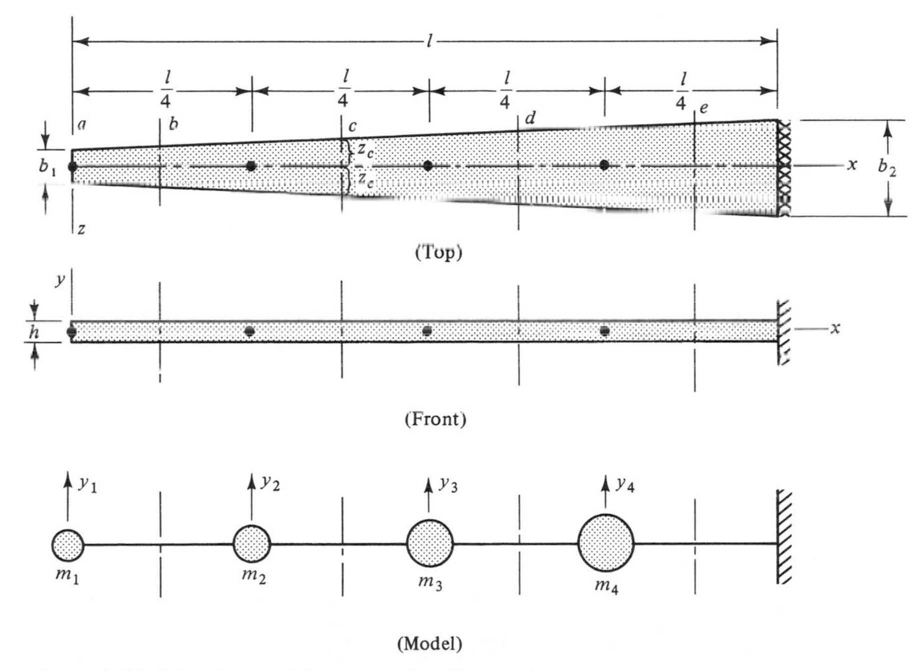

Figure 5-42 Four-lumped-mass model of tapered beam.

termination of either the flexibility coefficients or the lumped masses of nonuniform beams in determining their natural frequencies and mode shapes (see the tapered beam of Prob. 7-29).

PROBLEMS

Problems 5-1 through 5-17 (Sections 5-1 through 5-4)

5-1. Referring to the system shown in the accompanying figure, determine (a) the differential equations of motion of the system using Newton's second law, (b) the matrix form of the equations determined in part a, and (c) the stiffness matrix **K** for the system by just inspecting the system and applying the definition of k_{ij}.

Partial ans: $$\begin{bmatrix} m & 0 \\ 0 & 2m \end{bmatrix}\begin{Bmatrix} \ddot{x}_1 \\ \ddot{x}_2 \end{Bmatrix} + \begin{bmatrix} 3k & -k \\ -k & 3k \end{bmatrix}\begin{Bmatrix} x_1 \\ x_2 \end{Bmatrix} = \mathbf{0}$$

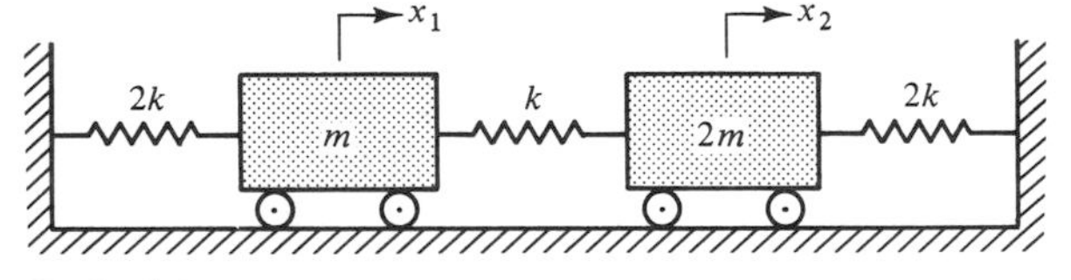

Prob. 5-1

5-2. Referring to the system shown in the accompanying figure, determine (a) the differential equations of motion of the system using Newton's second law, (b) the matrix form of the equations determined in part a, and (c) the stiffness matrix **K** for the system by just inspecting the system and applying the definition of a stiffness coefficient to obtain the elements k_{ij}.

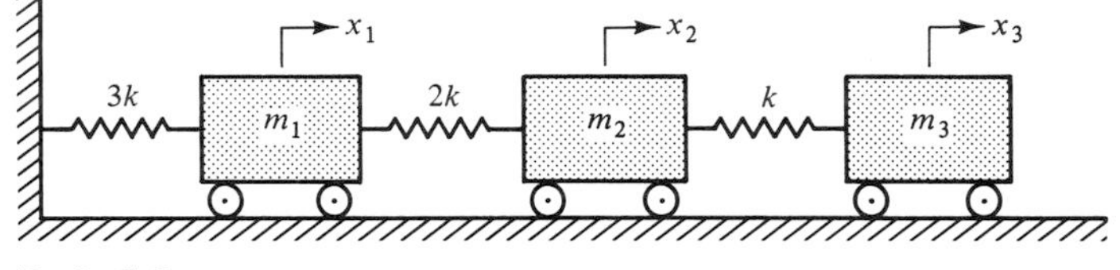

Prob. 5-2

5-3. In Example 5-6 the fourth column of the stiffness matrix **K** was determined by simply inspecting the six-degree-of-freedom system shown in Fig. 5-6 and applying the definition of a stiffness coefficient to obtain the elements k_{ij}. Determine the other columns of **K** in the same manner.

Partial ans: (elements of fifth column)

$$k_{15} = k_{25} = k_{45} = 0$$
$$k_{35} = -k_5$$
$$k_{55} = k_5 + k_6$$
$$k_{65} = -k_6$$

5-4. The nonuniform bar of length l, mass m, and centroidal mass moment of inertia $\bar{I}$ shown in the accompanying figure is supported by two springs, each of which has a stiffness of k. Assuming that the oscillation of the system is small in magnitude, and using the generalized coordinates q_1 and q_2 shown in the figure, determine (a) the differential equations of motion of the undamped free vibration of the system using Newton's second law, and (b) the stiffness matrix **K** using the definition of a stiffness coefficient to obtain the elements k_{ij}.

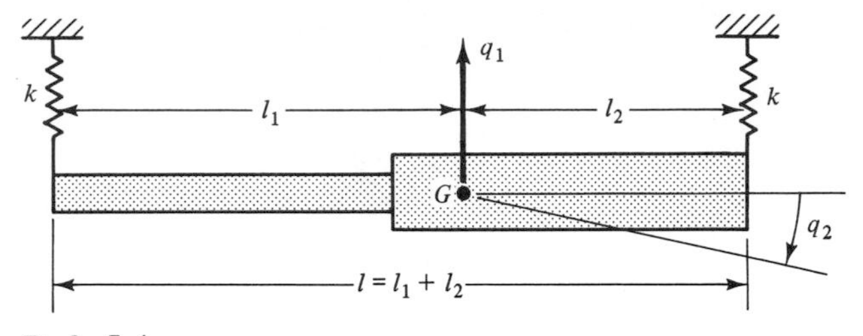

Prob. 5-4

5-5. In Example 5-7 the first column of the stiffness matrix **K** was determined for the system shown in Fig. 5-7 of that example. Determine the second and third columns of the stiffness matrix **K** for that system using the procedure discussed in the example.

Partial ans: $k_{13} = -kl/2$
$k_{23} = -3kl^2/4$
$k_{33} = 3kl^2/4$

5-6. The uniform bar shown in the accompanying figure has a mass of m_1, a centroidal mass moment of inertia of $\bar{I}$, and is supported by identical springs of stiffness k at each end. The mass m_3 is attached at the center of gravity G of the bar by means of another spring also of stiffness k as shown. Assuming small amplitudes of oscillation for the system, and using the generalized coordinates q_1, q_2, and q_3 shown in the figure, show that the differential equations of motion in matrix form are given by

$$\begin{bmatrix} m_1 & 0 & 0 \\ 0 & \bar{I} & 0 \\ 0 & 0 & m_3 \end{bmatrix} \begin{Bmatrix} \ddot{q}_1 \\ \ddot{q}_2 \\ \ddot{q}_3 \end{Bmatrix} + \begin{bmatrix} 3k & 0 & -k \\ 0 & 2kl^2 & 0 \\ -k & 0 & k \end{bmatrix} \begin{Bmatrix} q_1 \\ q_2 \\ q_3 \end{Bmatrix} = \mathbf{0}$$

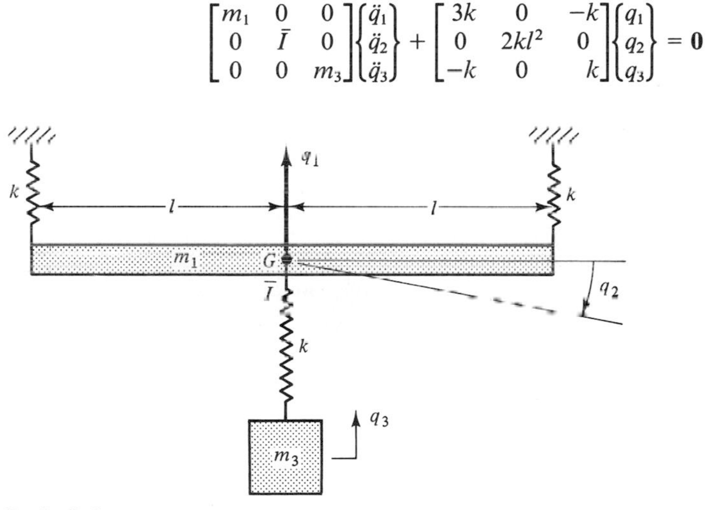

Prob. 5-6

5-7. The masses m_1 and m_2 are attached to a tightly stretched wire as shown in part a of the accompanying figure, with the wire having a tension T. Assuming small amplitudes of oscillation for the system and that the tension in the wire does not change when the masses are displaced y_1 and y_2 as shown in part b of the figure, (a) obtain the differential equations of motion of the system by determining its stiffness matrix **K**, and (b) verify your answer by deriving the differential equations of motion using Newton's second law.

Ans: $\begin{bmatrix} m_1 & 0 \\ 0 & m_2 \end{bmatrix} \begin{Bmatrix} \ddot{y}_1 \\ \ddot{y}_2 \end{Bmatrix} + \frac{T}{2l} \begin{bmatrix} 3 & -1 \\ -1 & 3 \end{bmatrix} \begin{Bmatrix} y_1 \\ y_2 \end{Bmatrix} = \mathbf{0}$

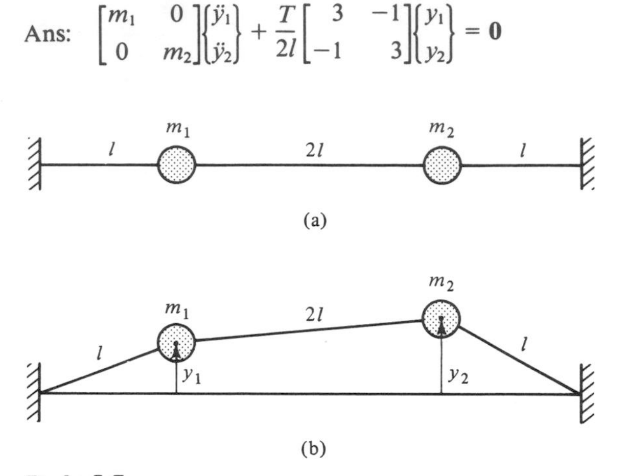

Prob. 5-7

5-8. Use Newton's second law to determine the differential equations of motion given by Eq. 5-12 that are for the system of Example 5-5.

5-9. Use the concept of dynamic equilibrium to determine the differential equations of motion of the two-degree-of-freedom system of Example 5-8.

5-10. An automobile is modeled as a simple four-degree-of-freedom system as shown in the accompanying figure. The parameters of the model are as follows:

c = equivalent damping coefficient for two shock absorbers (both front and back)
k_1 = equivalent stiffness of two springs (both front and back)
k_2 = equivalent stiffness of two tires (both front and back)
m_1 = combined mass of automobile body and motor
$\bar{I}$ = mass moment of inertia of combined mass of body and motor about its mass center G
m_2 = equivalent mass of two wheels (both front and back)

Assuming small amplitudes of oscillation for the system, verify that the stiffness matrix **K** and damping matrix **C** are, respectively,

$$\mathbf{K} = \begin{bmatrix} 2k_1 & k_1(l_2 - l_1) & -k_1 & -k_1 \\ k_1(l_2 - l_1) & k_1(l_1^2 + l_2^2) & k_1 l_1 & -k_1 l_2 \\ -k_1 & k_1 l_1 & (k_1 + k_2) & 0 \\ -k_1 & -k_1 l_2 & 0 & (k_1 + k_2) \end{bmatrix}$$

and

$$\mathbf{C} = \begin{bmatrix} 2c & c(l_2 - l_1) & -c & -c \\ c(l_2 - l_1) & c(l_1^2 + l_2^2) & cl_1 & -cl_2 \\ -c & cl_1 & c & 0 \\ -c & -cl_2 & 0 & c \end{bmatrix}$$

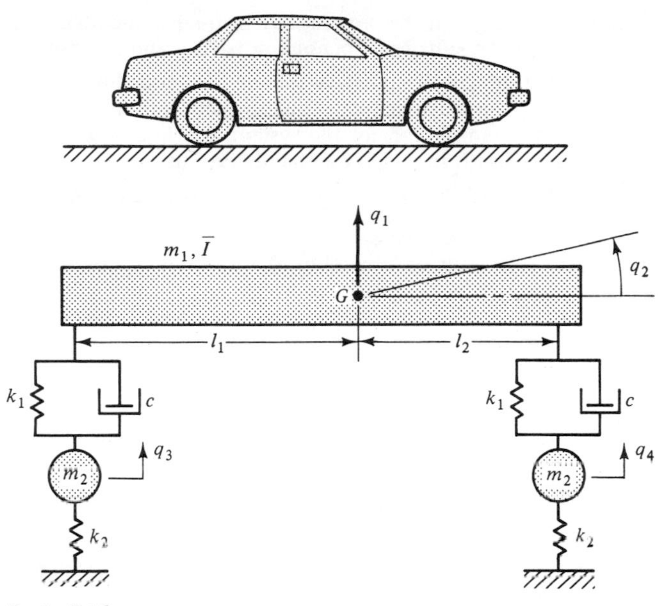

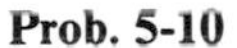

Prob. 5-10

5-11. The system shown in the accompanying figure consists of four gears having mass moments of inertia about their centroidal axes parallel to the shafts of $\bar{I}_1$, $\bar{I}_2$, $\bar{I}_3$, and $\bar{I}_4$ as indicated on the figure. The pitch radii of the meshing gears are r_2 and r_3 and are related by $r_3 = 2r_2$. Because of this relationship $\theta_2 = 2\theta_3$, and the system has three degrees of freedom that can be defined by the generalized coordinates θ_1, θ_2, and θ_4 or by θ_1, θ_3, and θ_4. Using the former set of coordinates, show that the differential equations of motion in matrix form are

$$\begin{bmatrix} I_1 & 0 & 0 \\ 0 & \left(I_2 + \dfrac{I_3}{4}\right) & 0 \\ 0 & 0 & I_4 \end{bmatrix} \begin{Bmatrix} \ddot{\theta}_1 \\ \ddot{\theta}_2 \\ \ddot{\theta}_4 \end{Bmatrix} + k \begin{bmatrix} 1 & -1 & 0 \\ -1 & \frac{5}{4} & \frac{1}{2} \\ 0 & \frac{1}{2} & 1 \end{bmatrix} \begin{Bmatrix} \theta_1 \\ \theta_2 \\ \theta_4 \end{Bmatrix} = \mathbf{0}$$

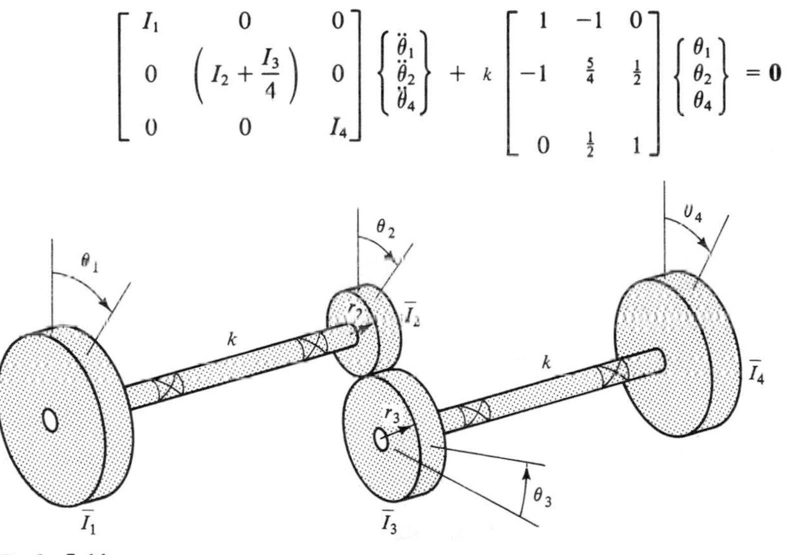

Prob. 5-11

5-12. Referring to the three-degree-of-freedom system shown in Fig. 5-14a, and letting $k_1 = k_2 = k_3 = k_4 = k$, determine (a) the flexibility matrix **A** for the system and (b) its stiffness matrix **K**. Recalling that the product **AK** is equal to the identity matrix **I** since $\mathbf{A}^{-1} = \mathbf{K}$, check to see if the results you obtained in parts a and b satisfy $\mathbf{AK} = \mathbf{I}$.

5-13. Determine the flexibility matrix **A** for the disk-and-shaft system shown in Fig. 5-5 of Example 5-5. Write the differential equations of motion in terms of **A**. Using the **K** matrix determined in Example 5-5 for the system (see Eq. 5-12), check to see if $\mathbf{A}^{-1} = \mathbf{K}$ by showing that $\mathbf{AK} = \mathbf{I}$, in which **I** is the identity matrix.

5-14. Determine the flexibility matrix **A** of the six-degree-of-freedom system shown in Fig. 5-6 on page 289.

Partial ans: (elements of fifth column)

$$a_{15} = \frac{1}{k_1}$$

$$a_{25} = \frac{1}{k_1} + \frac{1}{k_2}$$

$$a_{35} = a_{45} = \frac{1}{k_1} + \frac{1}{k_2} + \frac{1}{k_3}$$

$$a_{55} = a_{65} = \frac{1}{k_1} + \frac{1}{k_2} + \frac{1}{k_3} + \frac{1}{k_5}$$

5-15. Assuming small amplitudes of oscillation for the system shown in the accompanying figure, show that the damping matrix **C** is

$$\mathbf{C} = \begin{bmatrix} c_1 & 0 & 0 \\ 0 & 0 & 0 \\ 0 & 0 & c_2 l^2 \end{bmatrix}$$

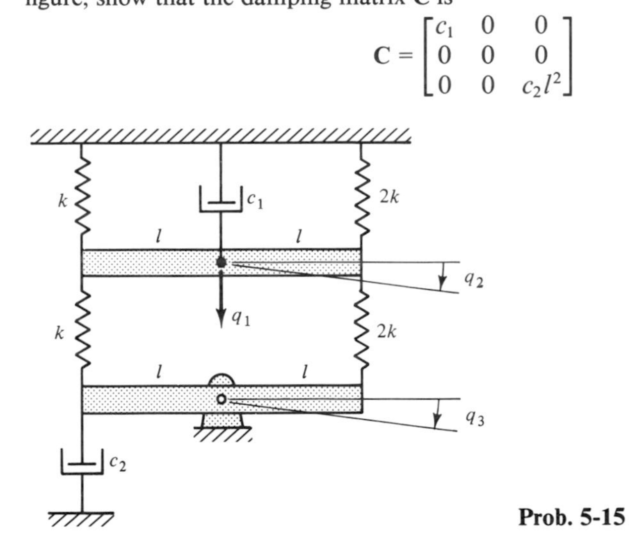

Prob. 5-15

5-16. Determine the flexibility matrix **A** of the disk-and-shaft system shown in Fig. 5-5 of Example 5-5 by inspecting the system and using the definition of a flexibility coefficient to determine the elements a_{ij}. Write the differential equations of motion in terms of **A**.

5-17. Determine the flexibility matrix **A** of the six-degree-of-freedom system shown in Fig. 5-6 of Example 5-6 by inspecting the system and using the definition of a flexibility coefficient to determine the elements a_{ij}.

Problems 5-18 through 5-26 (Sections 5-5 through 5-6) (Two-Degree-of-Freedom Systems)

5-18. Referring to the system shown in the accompanying figure, determine (a) the mass and stiffness matrices of the system, and (b) the system's natural circular frequencies and modal matrix $[u]$ if $k = 60$ N/m and $m = 1.5$ kg.

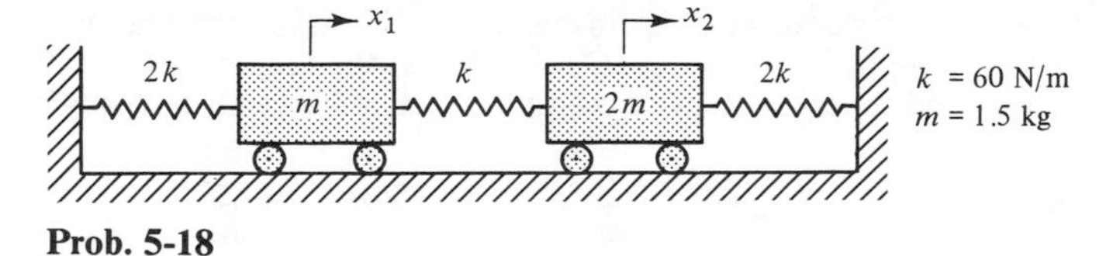

Prob. 5-18

5-19. The sphere of mass m is attached to the end of a cantilever beam that is fixed to a carriage of mass $2m$ as shown in the accompanying figure. The generalized coordinates of the system are the absolute displacements x_1 and x_2 of the carriage and sphere, respectively. Determine (a) the mass and stiffness matrices of the system, and (b) the system's natural circular frequencies and modal matrix $[u]$ if $k = 200$ lb/in. and $m = 2$ lb·s²/in. Does your solution satisfy the orthogonality relationship pertaining to the mass matrix?

Ans: $\omega_1 = 8.48$ rad/s, $\omega_2 = 16.68$ rad/s

$$[u] = \begin{bmatrix} 1.00 & 1.00 \\ 3.562 & -0.562 \end{bmatrix}$$

x_2 m k x_1 $2k$ $2m$ $2k$

k = 200 lb/in.
m = 2 lb·s²/in.

Prob. 5-19

5-20. The two identical cylinders of mass m, radius r, and centroidal mass moment of inertia $\bar{I} = mr^2/2$ shown in the accompanying figure are attached to each other and to rigid supports by springs having the stiffnesses shown. Assuming that the cylinders roll without slipping, determine (a) the differential equations of motion of the system in terms of the generalized coordinates θ_1 and θ_2 shown in the figure, and (b) the system's natural frequencies and modal matrix $[u]$ if $m = 2$ lb·s²/in., $r = 6$ in., and $k = 300$ lb/in.

Partial ans: $$\begin{bmatrix} 3m/2 & 0 \\ 0 & 3m/2 \end{bmatrix}\begin{Bmatrix} \ddot{\theta}_1 \\ \ddot{\theta}_2 \end{Bmatrix} + k\begin{bmatrix} 3 & -2 \\ -2 & 5 \end{bmatrix}\begin{Bmatrix} \theta_1 \\ \theta_2 \end{Bmatrix} = \mathbf{0}$$

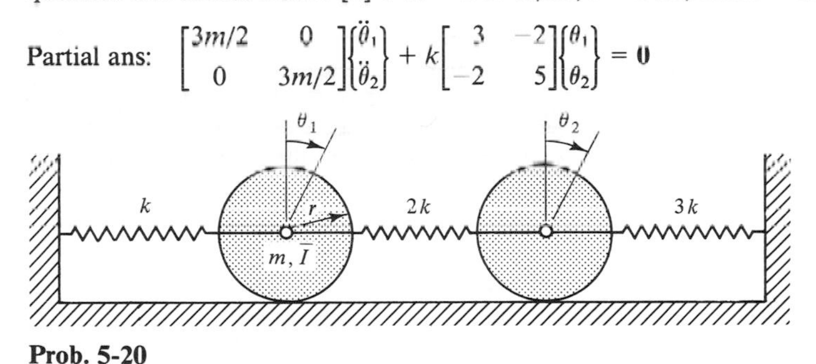

Prob. 5-20

5-21. The natural circular frequencies of a two-degree-of-freedom system are found to be $\omega_1 =$ 5.385 rad/s and $\omega_2 = 13.00$ rad/s. Show that the frequency equation of the system is

$$(\omega^2)^2 - 18.385\omega^2 + 70.000 = 0$$

5-22. The two carriages of mass m and $2m$, shown in the accompanying figure are connected to each other and to the rigid support at the left by identical springs of stiffness k as shown. The generalized coordinates of the system are the absolute displacements x_1 and x_2 of the respective carriages. Determine (a) the differential equations of motion of the system, and (b) the system's natural circular frequencies and modal matrix $[u]$.

Partial ans: $\omega_1 = 0.560 \sqrt{\dfrac{k}{m}}, \qquad \omega_2 = 1.785 \sqrt{\dfrac{k}{m}}$

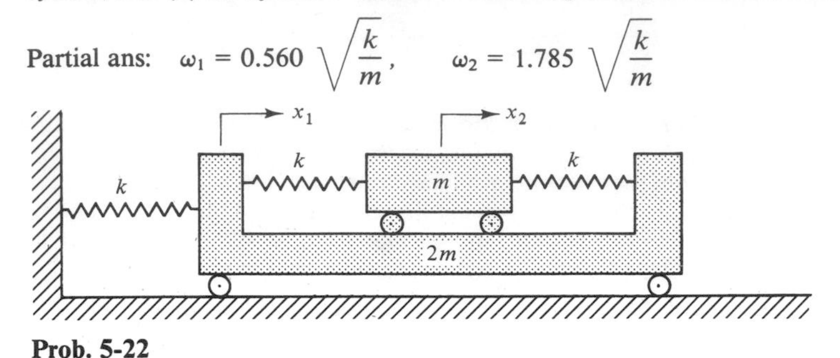

Prob. 5-22

5-23. The cylinder of mass m, radius r, and centroidal mass moment of inertia $\bar{I} = mr^2/2$ rolls without slipping on the platform of mass $2m$ as shown in the accompanying figure. The generalized coordinates x_1 and x_2 of the system are the absolute displacements of the platform and the mass center of the cylinder, respectively. Note that the absolute angular displacement of the cylinder is $(x_2 - x_1)/r$. Write the kinetic energy T of the system and determine (a) if there is dynamic coupling of the coordinates, (b) the differential equations of motion of the system, and (c) the system's natural circular frequencies and modal matrix $[u]$.

Partial ans: $\omega_1 = 0.559 \sqrt{\dfrac{k}{m}}, \qquad \omega_2 = 1.353 \sqrt{\dfrac{k}{m}}$

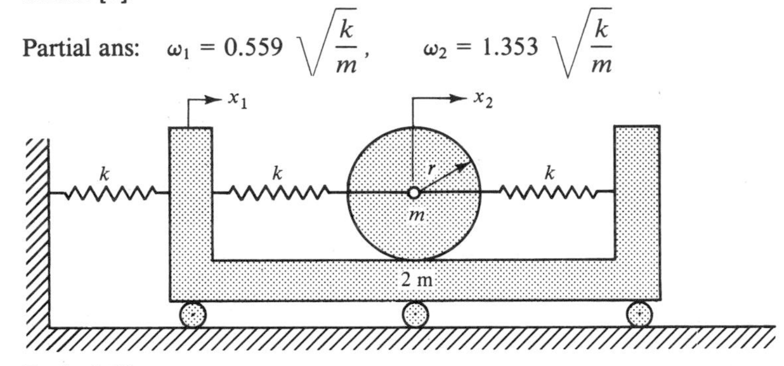

Prob. 5-23

5-24. A bar of circular cross section having a torsional spring constant of k is supported by the bearings at A and B as shown in the accompanying figure. Two identical light rods of length l are fixed to the ends of the bar as shown. Two spheres of mass m_1 and m_2 are attached to the lower ends of the rods as shown in the figure. Determine the natural circular frequencies and mode shapes of the system if its parameters are as shown in the figure. Neglect the mass of the rods and the bar.

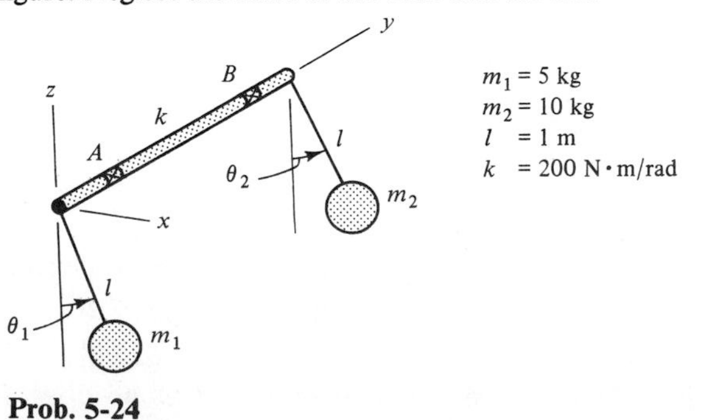

Prob. 5-24

5-25. A boat is pulling a skier as shown in the accompanying figure. The spring constant of the ski rope is $k = 100$ lb/in., and the weights of the skier and boat are 150 and 1500 lb, respectively. Determine (a) the natural circular frequencies and modal matrix $[u]$ of the system composed of the boat and skier, and (b) the values of the elements for the inverse of the stiffness matrix $\mathbf{K}$.

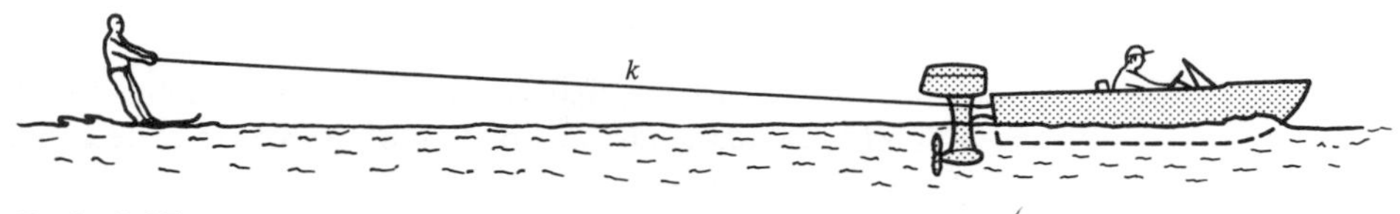

Prob. 5-25

5-26. Two small gears are attached to a slender shaft that is supported by the self-aligning bearings A and B as shown in part a of the accompanying figure. The engineer designing the system would like to determine the natural frequencies of the system experimentally but does not have the equipment necessary to do so. Therefore, rather than obtaining vibration measurements, he applies a 20-lb *static* load at gear 1 and measures the static deflections of the shaft at the location of the center of each gear as shown in part b of the figure. He then applies a 20-lb force statically at gear 2 and again measures the static deflections of the shaft at the location of the center of each gear as shown in part c of the figure. The natural frequencies and mode shapes are then calculated by using these static-load test data with the assumption that the mass of the shaft is negligible. Determine the frequencies and mode shapes calculated by the engineer.

Partial ans: $f_1 = 20.47$ Hz, $f_2 = 41.59$ Hz

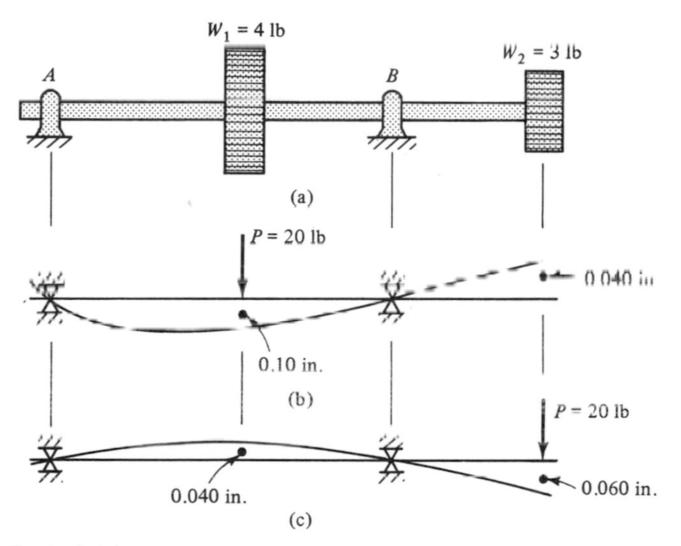

Prob. 5-26

Problems 5-27 through 5-45 (Sections 5-7 through 5-11)

5-27. Referring to Example 5-18, show that if the system shown there is released from rest in a second-mode configuration, its undamped free vibratory motion is

$$x_1 = \cos(25.55\,t), \qquad x_2 = 0.54\cos(25.55\,t), \qquad x_3 = -1.78\cos(25.55\,t)$$

5-28. An n-degree-of-freedom system is released from rest with the initial displacements of the masses corresponding to the mode shape of an rth mode. That is, the initial conditions are

$$\begin{Bmatrix} x_1 \\ x_2 \\ \vdots \\ x_n \end{Bmatrix}_{t=0} = \begin{Bmatrix} x_1 \\ x_2 \\ \vdots \\ x_n \end{Bmatrix}_r \qquad \begin{Bmatrix} \dot{x}_1 \\ \dot{x}_2 \\ \vdots \\ \dot{x}_n \end{Bmatrix}_{t=0} = \begin{Bmatrix} 0 \\ 0 \\ \vdots \\ 0 \end{Bmatrix}$$

Show that the undamped free vibratory motion consists only of the rth mode circular frequency ω_r with the amplitudes of vibration of the masses corresponding to the components of the eigenvector $\mathbf{X}_r$.

5-29. Referring to the building shown in Example 5-14, consider that it is released from rest with the initial displacement of each floor being

$$\begin{Bmatrix} x_1 \\ x_2 \\ x_3 \end{Bmatrix}_{t=0} = \begin{Bmatrix} 1.0 \text{ in.} \\ 2.0 \text{ in.} \\ 3.0 \text{ in.} \end{Bmatrix}$$

and use the modal-analysis approach to determine the undamped free vibratory motion of the building.

Partial ans: $x_3 = 2.93 \cos \omega_1 t + 0.0472 \cos \omega_2 t + 0.0229 \cos \omega_3 t$

5-30. Referring to Example 5-13, consider that the system shown there is released from rest with the initial displacements:

$$\begin{Bmatrix} \theta_1 \\ \theta_2 \\ \theta_3 \end{Bmatrix}_{t=0} = \begin{Bmatrix} 0.1 \text{ rad} \\ 0.2 \text{ rad} \\ 0.1 \text{ rad} \end{Bmatrix}$$

The parameter values for the system are $k = 5000$ lb-in./rad, and $I = 5$ lb $\cdot$ s^2 $\cdot$ in. Using the modal-analysis approach, determine the undamped free vibratory motion of the system.

Partial ans: $\theta_1 = 0.125 + 0.01481 \cos(26.83t) - 0.03981 \cos(52.73t)$

5-31. The three railroad cars of mass m each are connected by identical couplers represented as springs of stiffness k in the accompanying figure, and are traveling at a constant speed of 30 ft/s. Assuming that the displacements of the cars at $t = 0$ are $x_1 = x_2 = x_3 = 0$, use the concept of modal analysis to show that the displacements of the cars at any time t are $x_1 = x_2 = x_3 = 30t$.

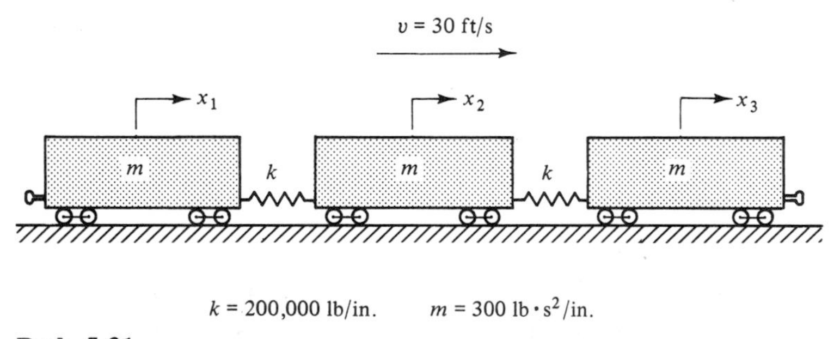

Prob. 5-31

5-32. The natural circular frequencies of a three-degree-of-freedom system are found to be

$\omega_1 = 1$, $\omega_2 = 3$, and $\omega_3 = 6$ rad/s. Show that the frequency equation of the system is

$$\lambda^3 - 46\lambda^2 + 369\lambda - 324 = 0$$

in which $\lambda = \omega^2$.

5-33. Use the iteration procedure discussed in Sec. 5-9 to determine the natural circular frequencies and mode shapes of the spring-and-mass system shown in the accompanying figure. Obtain the natural circular frequencies in terms of k and m.

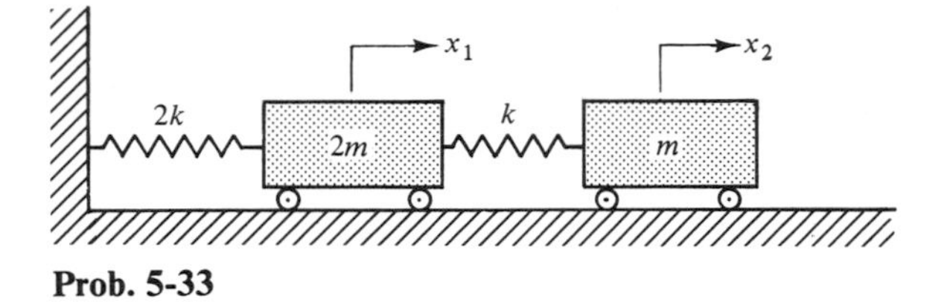

Prob. 5-33

5-34. Referring to the three-story building shown in Fig. 5-22 of Example 5-14, use the iteration procedure discussed in Sec. 5-9 to verify the values determined in the example for ω_1 and the corresponding mode shape $\mathbf{X}_1$. The values of k and m are as shown in Fig. 5-22.

5-35. A 40-in. steel shaft fixed at both ends has three disks attached to it as shown in the accompanying figure. Determine the natural circular frequencies and the modal matrix $[u]$ of the system if the diameter of the shaft is 1.5 in. Sketch the mode shapes and show the node locations. The modulus of rigidity (the shear modulus) of steel is $G = 12(10)^6$ psi, and the centroidal mass moment of inertia of each disk is as shown in the figure.

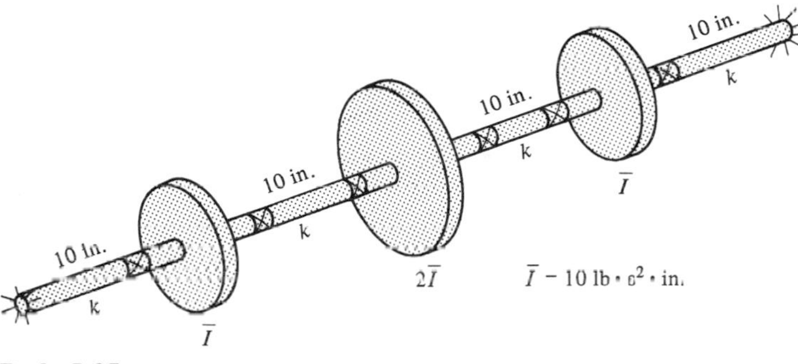

Prob. 5-35

5-36. The system (semidefinite) shown in the accompanying figure consists of a flexible cable that passes over a pulley of radius r and centroidal mass moment of inertia $\bar{I}$, two identical springs of stiffness k, and two identical blocks of mass m. Determine the stiffness matrix $\mathbf{K}$ of the system, and show that the natural circular frequencies of the system are

$$\omega_1 = 0 \qquad \omega_2 = 15.81 \text{ rad/s} \qquad \omega_3 = 17.50 \text{ rad/s}$$

and that the modal matrix is

$$[u] = \begin{bmatrix} 1.00 & 0.0 & 1.00 \\ 0.15 & -1.0 & -0.667 \\ 0.15 & 1.0 & -0.667 \end{bmatrix}$$

Check the orthogonality relationship pertaining to the mass matrix.

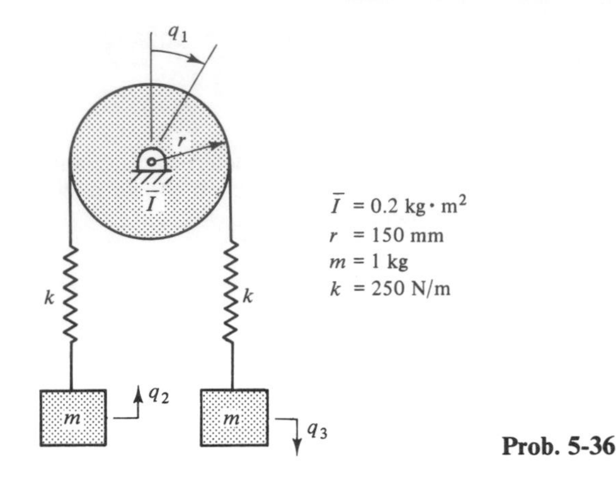

Prob. 5-36

5-37. The two pulleys having centroidal mass moments of inertia of $\bar{I}_1$ and $\bar{I}_2$ are connected to each other and to the block of mass m by the belt-and-spring arrangement shown in the accompanying figure. There is sufficient friction between the belt and the pulleys to prevent slipping. Determine the natural circular frequencies and normal-mode shapes for the system.

Partial ans: $\omega_1 = 24.16$ rad/s, $\omega_2 = 67.56$ rad/s, $\omega_3 = 130.78$ rad/s

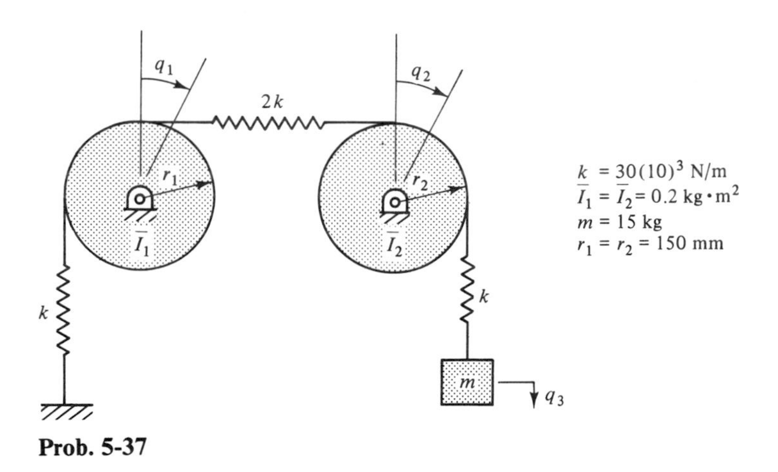

Prob. 5-37

5-38. A motor drives two fans in a small closed-circuit wind tunnel as shown in part a of the accompanying figure. The parameters of the two-fans-and-motor-armature system are shown with this system in part b of the figure, and have the values shown there. The speed of the motor is a constant 1800 rpm (30 Hz), and since each fan has four blades, resonance can occur if the system has a torsional natural frequency of either 30 or 120 Hz. Thus, to avoid resonance occurring during operation of the wind tunnel, the fan-and-motor system should be designed so that neither of its natural frequencies approaches these values. Other considerations dictate that in avoiding these resonant conditions it should be accomplished by making both of the natural frequencies of the system lie between the possible resonant frequencies. Assuming that the torsional stiffness of both the 24-in. and the 36-in. shaft sections is k, determine the diameters d_1 and d_2 of the shafts such that the lowest torsional natural frequency of the system f_2 ($f_1 = 0$ for rigid-body motion) is 35 Hz. Is f_3 then less than 120 Hz?

Partial ans: $d_1 = 0.59$ in., $d_2 = 0.653$ in.

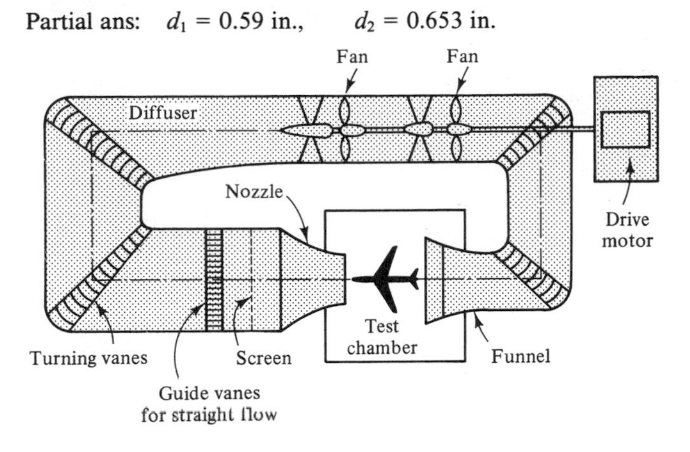

(a)

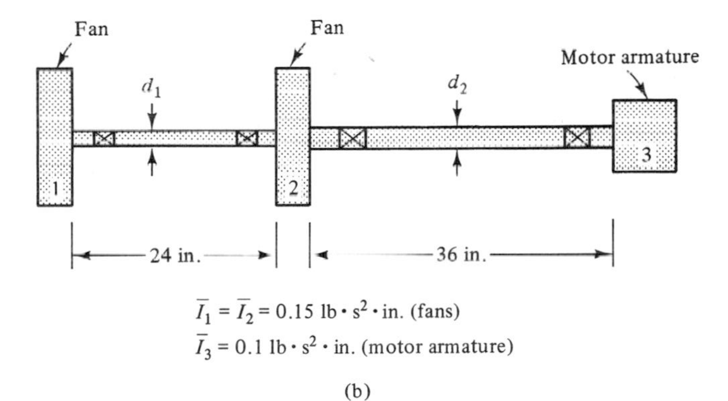

(b)

Prob. 5-38

5-39. The turbine shown in the accompanying figure has an operating speed of 3600 rpm and drives the generator at 1800 rpm through the gear arrangement shown. The shear modulus of elasticity for the steel in both shafts is $G = 80(10)^9$ Pa (80 gigapascals). Other parameter values of the system are as follows:

$$\begin{aligned}
\bar{I}_1 &= 3600 \text{ kg}\cdot\text{m}^2 \text{ (turbine)}\\
\bar{I}_2 &= 200 \text{ kg}\cdot\text{m}^2 \text{ (gear)}\\
\bar{I}_3 &= 800 \text{ kg}\cdot\text{m}^2 \text{ (gear)}\\
\bar{I}_4 &= 4800 \text{ kg}\cdot\text{m}^2 \text{ (generator armature)}\\
d_1 &= 150 \text{ mm (diameter of turbine shaft)}\\
l_1 &= 3.5 \text{ m (length of turbine shaft)}\\
d_2 &= 200 \text{ mm (diameter of generator shaft)}\\
l_2 &= 3 \text{ m (length of generator shaft)}
\end{aligned}$$

Using the values above, determine the natural frequencies of the turbine-generator system.

Ans: $f_1 = 0$ (rigid-body mode), $f_2 = 3.87$ Hz,
$f_3 = 12.38$ Hz

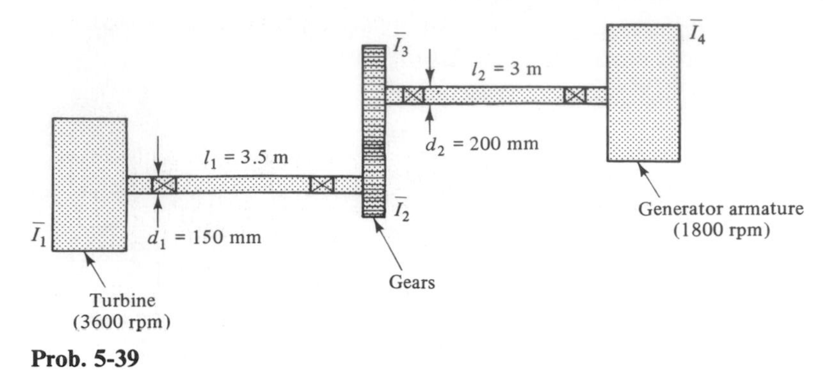

Prob. 5-39

5-40. It has been recommended that the turbine-generator system of Prob. 5-39, which was found to have a fundamental natural frequency of $f_2 = 3.87$ Hz in that analysis, be changed so that its lowest natural frequency will be 5 Hz. One possible way of increasing this frequency is by making the diameter of the turbine shaft larger than its original diameter of 150 mm. Adopt this approach, and determine the new diameter d_1 of the turbine shaft that will make $f_2 = 5$ Hz.

Ans: $d_1 = 227$ mm

5-41. A commercial feed mixer is shown in part a of the accompanying figure, and the gear-and-shaft system for driving the two mixer paddles is shown in part b of the figure. Each of the two bevel gears shown has 20 teeth (20T), two of the spur gears have 36 teeth (36T), and the common spur gear has 12 teeth (12T) as shown. With this arrangement, the angular displacement of each of the bevel gears is the same, and the angular displace-

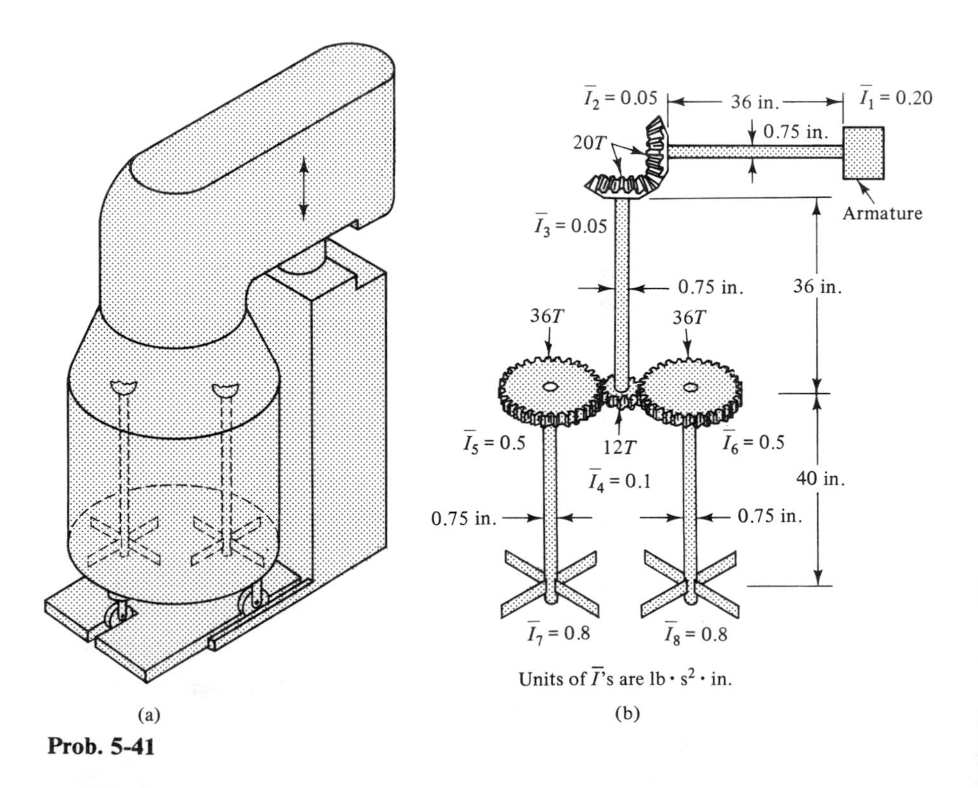

Prob. 5-41

ment of each of the two spur gears attached to the paddles is one-third that of the common spur gear. Determine the natural frequencies of the system.

Ans: $f_1 = 0$, $f_2 = 17.18$ Hz, $f_3 = 18.71$ Hz,
$f_4 = 38.00$ Hz, $f_5 = 80.92$ Hz

5-42. Vibration tests have been conducted on the rocket shown in the accompanying figure to determine among other characteristics its longitudinal natural frequencies and mode shapes. It is desired to develop a mathematical model of the rocket incorporating its physical parameters which will yield data conforming fairly accurately to the experimental data so that analytical analyses of the rocket can be made. The five-lumped-mass model of the rocket shown in the figure is selected as a preliminary model. Use the JACOBI.FOR program on page 369 to determine the longitudinal natural frequencies and mode shapes of the rocket as given by the model, and compare the frequencies obtained with the experimentally determined value of $f_2 = 14.1$ Hz, $f_3 = 25.9$ Hz, $f_4 = 39.2$ Hz, and $f_5 = 48.6$ Hz ($f_1 = 0$ for rigid-body mode). Make a sketch of the first elastic mode shape.

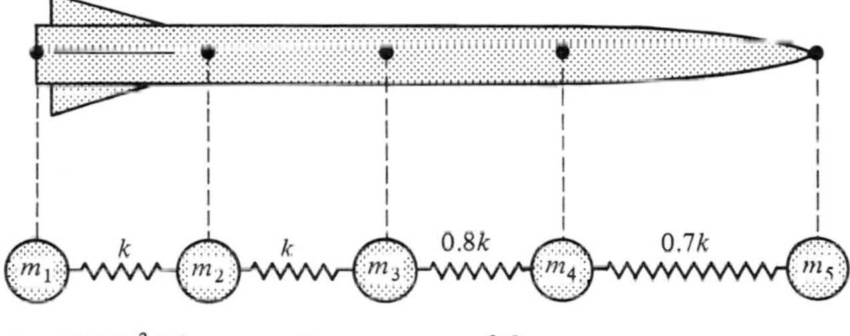

$k = 50(10)^3$ N/m, $m_1 = 4$ kg, $m_2 = m_3 = 0.8m_1$
$m_4 = 0.6m_1$, $m_5 = 0.4m_1$

Prob. 5-42

5-43. A light flexible cable-and-spring assembly is fastened to two blocks, passes over a pulley of radius r and centroidal mass moment of inertia of $\bar{I}$, and has both ends fastened to a rigid support as shown in the accompanying figure. Assuming that the cable does not slip on the pulley, and using the generalized coordinates q_1, q_2, and q_3 shown in the figure, determine (a) the mass and stiffness matrices **M** and **K**, respectively, and (b) the natural frequencies and modal matrix $[u]$ of the system using the system parameter values shown in the figure.

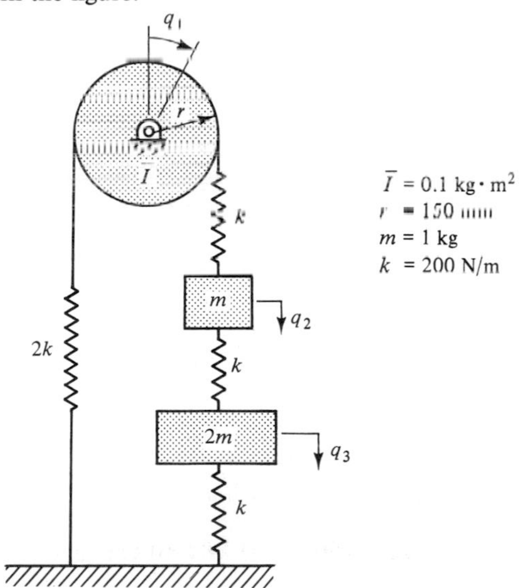

$\bar{I} = 0.1$ kg·m²
$r = 150$ mm
$m = 1$ kg
$k = 200$ N/m

Prob. 5-43

5-44. Two identical bars of length l, mass m, and centroidal mass moment of inertia $\bar{I}$ are pinned together at A and supported on springs as shown in the accompanying figure. Using the generalized coordinates q_1, q_2, and q_3 shown in the figure, determine (a) the differential equations of motion for the system assuming small oscillations and (b) the natural circular frequencies and modal matrix $[u]$ of the system. Sketch the mode shapes. (Hint: Use concept of dynamic equilibrium to derive the differential equations of motion, and normalize the angular displacements with respect to $q_1 = 0.01$ m to be consistent with small amplitudes of oscillation assumed.)

Partial ans: (third mode)

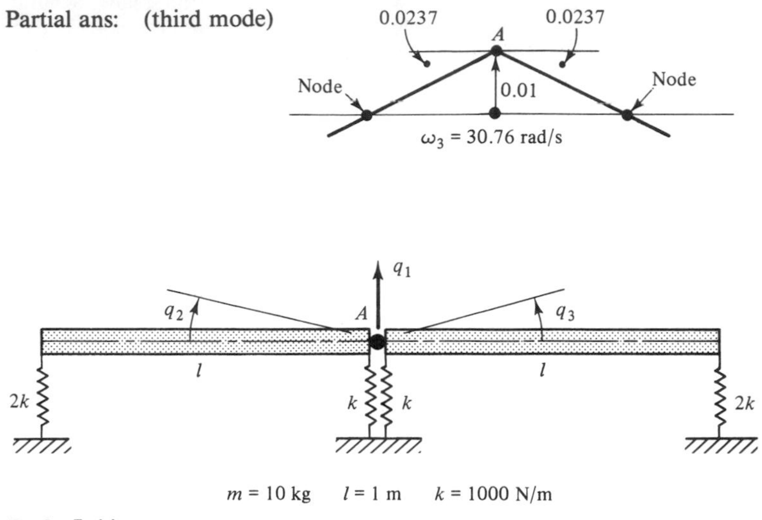

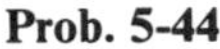
Prob. 5-44

5-45. Two tanks of mass m_2 and m_4 are suspended from very stiff girders of mass m_1 and m_3 as shown in the accompanying figure. Determine (a) the flexibility matrix **A** of the system, (b) its lowest natural circular frequency and corresponding mode shape using a pocket calculator and the iteration procedure discussed in Sec. 5-9, and (c) the four natural frequencies and modal matrix $[u]$ of the system using the DEFLATEA.FOR program.

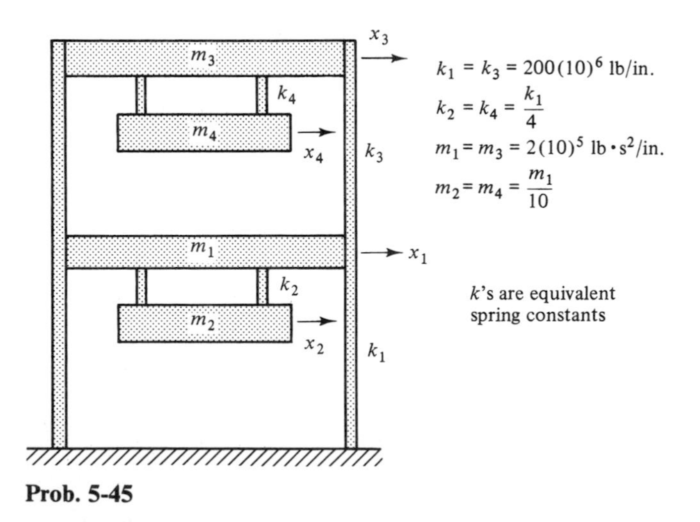

Prob. 5-45

Problems 5-46 through 5-53 (Section 5-12)

5-46. The uniform rod shown in Fig. 5-35 of Example 5-22 was modeled by two different four-lumped-mass systems determined by the two procedures described under Model 1 and Model 2 in that example. As a means of obtaining a comparative analysis, model the rod by two different five-lumped-mass systems using the same two procedures, and then using the JACOBI.FOR program on page 369 to determine the five natural circular frequencies for each system. Compare the four lowest frequencies obtained for each system with the four lowest exact values shown in Example 5-22.

5-47. The 40-ft-long simply supported steel beam shown in the accompanying figure was modeled as a three-lumped-mass system in Example 5-25. A motor that weighs 2000 lb is now attached to the center of this beam, and it is desired to model the resulting system again by a three-lumped-mass model and determine its natural frequencies and mode shapes.

Partial ans: $f_1 = 7.41$ Hz, $f_2 = 42.12$ Hz, $f_3 = 73.01$ Hz

$E = 30(10)^6$ psi $I = 1674.7$ in^4. $W = 2000$ lb

Prob. 5-47

5-48. The distributed mass of the overhanging beam shown in part a of the accompanying figure is to be modeled by the two lumped masses m_1 and m_2 shown in part b of the figure. Determine the natural circular frequencies of the two-lumped-mass model.

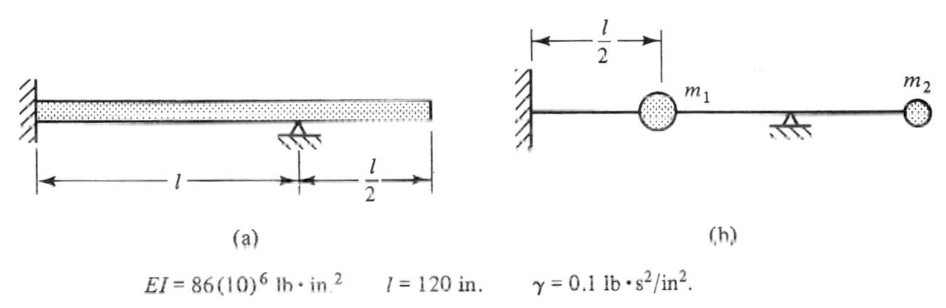

$EI = 86(10)^6$ lb · in.^{2} $l = 120$ in. $\gamma = 0.1$ lb · s^2/in^2.

Prob. 5-48

5-49. In Example 5-25 the flexibility matrix **A** was determined for the three-lumped-mass model of the beam shown in Fig. 5-38, which accompanies that example. This matrix was then used as input to the DEFLATEA.FOR program to obtain the natural circular frequencies and mode shapes of the beam. Use the program INVERT.FOR on page 303 to invert the **A** matrix of Example 5-25 to obtain the stiffness matrix **K** of the model, so that the latter can be used as input to the JACOBI.FOR program (see page 369) to again determine the natural circular frequencies and mode shapes of the beam. Compare the values determined in this problem with the corresponding values found in Example 5-25.

5-50. Model the 40-ft-long steel beam shown in Fig. 5-38 of Example 5-25 using three identical lumped masses, each of which has a mass equal to one-third the total mass of the beam.

Show that this is a poor lumped-mass model of the beam by comparing the natural circular frequencies it yields for the beam with those obtained in Example 5-25 in which the model was formulated using the recommended procedure.

Ans: $\omega_1 = 57.74$ rad/s (13.4 percent error)
$\omega_2 = 229.21$ rad/s (14 percent error)

5-51. In Example 5-24 the mass matrix **M** of the four-lumped-mass model of the overhanging beam shown in Fig. 5-37 accompanying that example was determined. Determine the flexibility matrix **A** of the same model in terms of the beam parameters EI and l, and use it in finding the first two natural circular frequencies and mode shapes of the beam. Express the natural circular frequencies in terms of the beam parameters EI, l, and γ. Sketch the first two mode shapes.

Partial ans: $$\omega_1 = 6.018 \sqrt{\frac{EI}{\gamma l^4}} \text{ rad/s}$$

$$\omega_2 = 18.77 \sqrt{\frac{EI}{\gamma l^4}} \text{ rad/s}$$

5-52. A 4-ft-long steel bar of circular cross section of 1.5 in. diameter is bent into an L shape and fixed to a rigid support at A as shown in part a of the accompanying figure. Model the bent bar as a two-lumped-mass system as shown in part b of the figure, and determine its natural frequencies and mode shapes. The specific weight of steel is 490 lb/ft³, the modulus of elasticity is $E = 30(10)^6$ psi, and the shear modulus is $G = 12(10)^6$ psi.

Partial ans: $f_1 = 20.56$ Hz, $f_2 = 41.51$ Hz

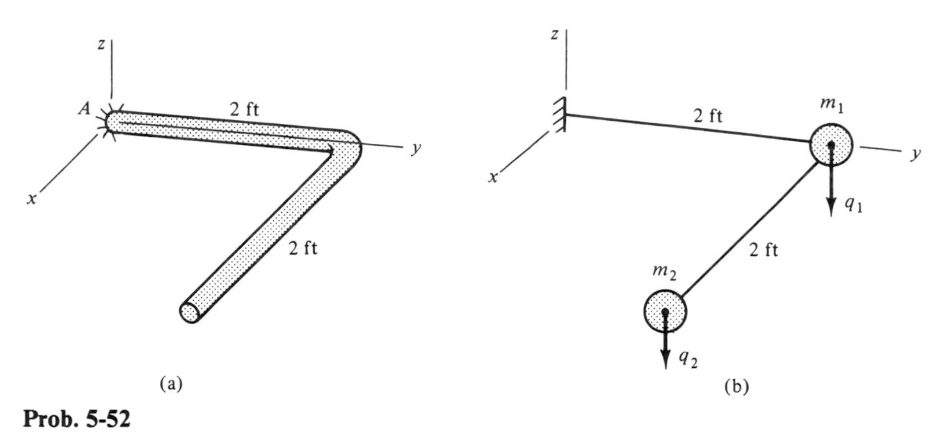

Prob. 5-52

5-53. In Example 2-12 the exact fundamental natural frequency of the composite wood-and-steel beam repeated in part a of the accompanying figure was found as 16.22 Hz using Rayleigh's method. The equivalent steel cross section of the beam determined in that example is repeated in part b of the accompanying figure. The properties of the beam materials are as follows:

	Specific weight, lb/in.³	Modulus of elasticity, psi
Wood	$w_w = 0.022$	$E_w = 1.5(10^6)$
Steel	$w_s = 0.284$	$E_s = 30(10)^6$

Model the beam as a three-lumped-mass system and determine its natural frequencies and mode shapes. Compare the fundamental frequency f_1 obtained using this model with the exact value of 16.22 Hz referred to above.

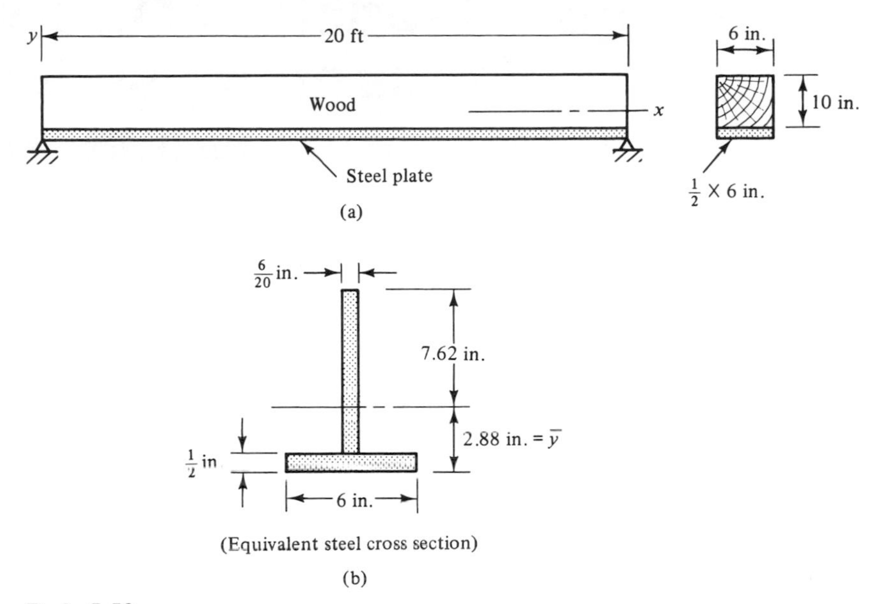

Prob. 5-53

Chapter 6

Forced Vibration of Discrete and Lumped-Mass Systems/ Modal Analysis

6-1 INTRODUCTION

In this chapter we discuss methods of determining the response of both discrete and lumped-mass systems subjected to either periodic or nonperiodic excitations, since both are analyzed by considering a finite number of degrees of freedom.

Although most real systems have many degrees of freedom, with *continuous* (distributed-mass) systems having a theoretically infinite number, many can be modeled as either discrete or lumped-mass systems having a relatively small number of degrees of freedom.[1] As a general rule, if the response of a model is to accurately correspond to that of the real system, it should have natural frequencies and mode shapes that are quite close to the fundamental and several successively higher frequencies and mode shapes of the real system.

The ultimate accuracy of the correspondence between a real system and the model depends upon the modeler's experience, the availability of experimental data from studies of similar systems, and the physical characteristics of the real system being analyzed.

The response of a forced system depends primarily upon the system's natural frequencies, the damping present, and the characteristics of the excitation to which it is subjected. If a periodic excitation contains a frequency component that has the same frequency as one of the system's natural frequencies, a condition referred to as resonance occurs that can result in large amplitudes of vibration. Excitations that induce forced vibrations are generally dynamic forces $F(t)$ and/or moments $M(t)$ or moving support motion $y(t)$. Examples of such excitations are shown in Fig. 6-1.

Considerable emphasis is given in this chapter to the method of *modal analysis* as

[1] As noted in Chap. 5, a discrete system is one in which the inertial, elastic, and damping properties are all clearly described by distinct masses, springs, and damping mechanisms, respectively. When the distributed properties of beams, rods, plates, and so on are discretized by modeling them as systems composed of lumped masses, lumped elastic elements, and modal damping, such systems are referred to in this text as lumped-mass systems.

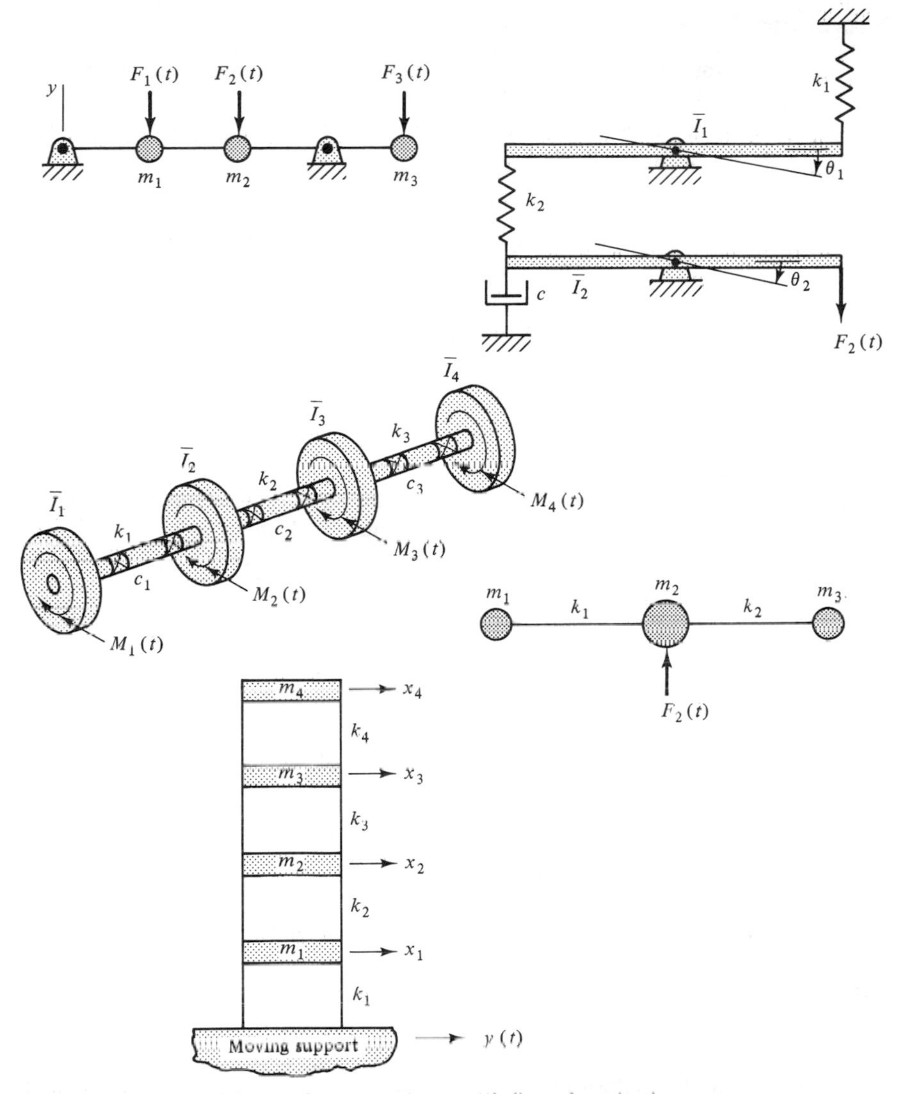

Figure 6-1 Typical lumped-mass systems with forced excitations.

a means of obtaining the response of multiple-degree-of-freedom systems because of its wide use at present in both analytical and experimental studies of vibration. As we shall see in subsequent discussions, this approach is a powerful tool for use in vibration analysis.

6-2 EQUATIONS OF MOTION—FORCED VIBRATION

In this section we develop the differential equations of motion of multiple-degree-of-freedom systems subjected to either force excitation or support excitation.

Force Excitation

The general form of the matrix equation of an n-degree-of-freedom system subjected to excitation forces is

$$\mathbf{M}\ddot{\mathbf{X}} + \mathbf{C}\dot{\mathbf{X}} + \mathbf{K}\mathbf{X} = \mathbf{F} = \begin{Bmatrix} F_1(t) \\ F_2(t) \\ F_3(t) \\ \vdots \\ F_n(t) \end{Bmatrix} \tag{6-1}$$

where $\mathbf{X}$ = a column matrix of linear or angular displacements
$\mathbf{F}$ = a column matrix of the excitation forces or moments as functions of time
$\mathbf{M}$ = mass matrix
$\mathbf{C}$ = damping matrix
$\mathbf{K}$ = stiffness matrix

The elements of **M, C,** and **K** can be determined as discussed in Chap. 5. The elements of **F** can be determined by considering the virtual work done by the excitation forces or moments acting on the system.[2] The virtual work done by an excitation force $F_i(t)$ for a virtual displacement of δx_i is

$$\delta W_i = F_i(t)\,\delta x_i \tag{6-2}$$

The virtual work done by an excitation moment $M_i(t)$ for a virtual displacement $\delta\theta_i$ is

$$\delta W_i = M_i(t)\,\delta\theta_i \tag{6-3}$$

If we let all but one of the coordinates of a system be fixed and then consider a virtual displacement δx_i of the mass associated with the unfixed one, $F_i(t)$ in matrix **F** will be the *sum* of all the excitation forces that do virtual work during that virtual displacement. Elements of the **F** matrix that are moments are determined in a similar manner using virtual displacements $\delta\theta_i$.

EXAMPLE 6-1

The automobile in Fig. 6-2a is modeled by the four-lumped-mass system shown in Fig. 6-2b. One cylinder of the engine is misfiring because of a fouled spark plug, which causes an excitation force $F(t)$ a distance l_3 from the mass center G of the auto. If we use the four generalized coordinates x_1, x_2, x_3, and θ shown in the figure, the differential equations of motion can be written in matrix form as

$$\begin{bmatrix} m_1 & 0 & 0 & 0 \\ 0 & m_2 & 0 & 0 \\ 0 & 0 & m_3 & 0 \\ 0 & 0 & 0 & \bar{I} \end{bmatrix} \begin{Bmatrix} \ddot{x}_1 \\ \ddot{x}_2 \\ \ddot{x}_3 \\ \ddot{\theta} \end{Bmatrix} + \mathbf{C} \begin{Bmatrix} \dot{x}_1 \\ \dot{x}_2 \\ \dot{x}_3 \\ \dot{\theta} \end{Bmatrix} + \mathbf{K} \begin{Bmatrix} x_1 \\ x_2 \\ x_3 \\ \theta \end{Bmatrix} = \begin{Bmatrix} F_1(t) \\ F_2(t) \\ F_3(t) \\ F_4(t) \end{Bmatrix} = \mathbf{F}$$

We wish to determine the elements of the column matrix **F** which represents the excitation forces and moments caused by the misfiring of the fouled plug.

Solution. Selecting x_1 as the displacement of the mass center G of m_1 and observing the rotation θ from a set of axes translating with the mass center, we obtain a set

[2] A discussion of virtual work can be found in most elementary statics texts. A small and arbitrary displacement δx is called a virtual displacement.

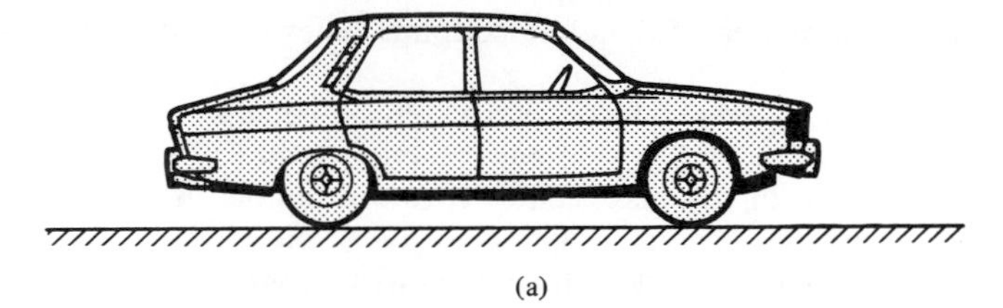

(a)

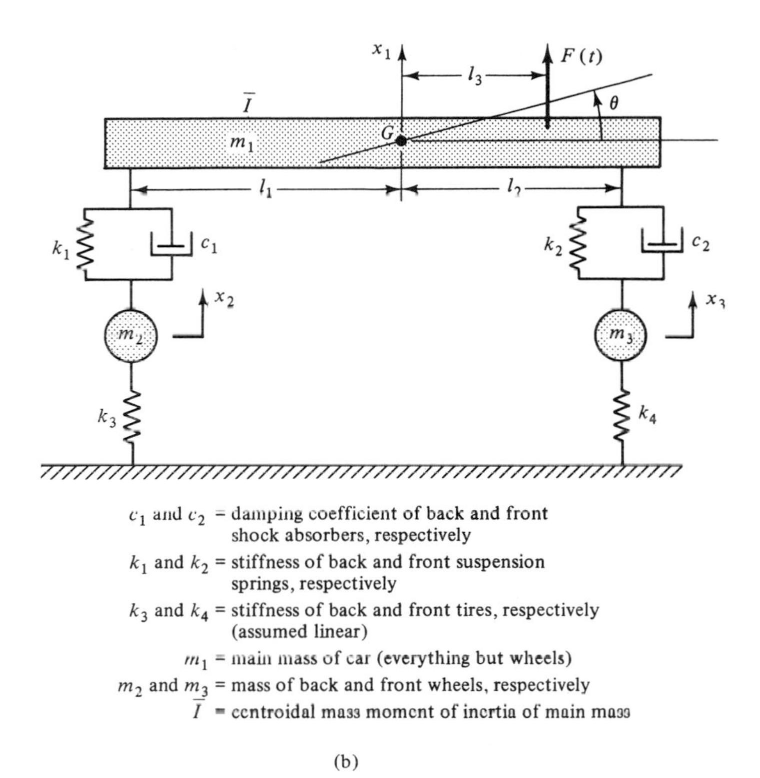

(b)

Figure 6-2 Car modeled as four-degree-of-freedom system.

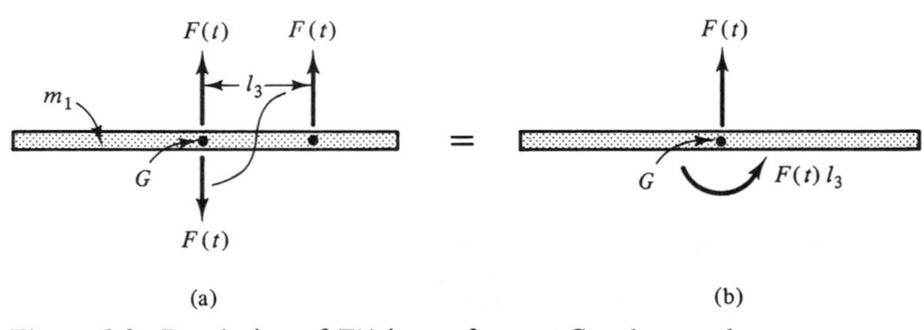

(a) (b)

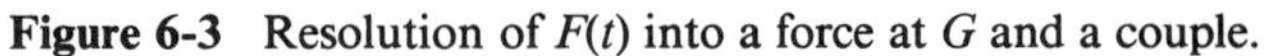

Figure 6-3 Resolution of $F(t)$ into a force at G and a couple.

of equations free of dynamic coupling as evidenced by the diagonal mass matrix shown above.

Adding the two forces $F(t)$ which are equal in magnitude but opposite in sense as shown in Fig. 6-3a, we obtain an equivalent force system that consists of a force acting through the mass center G and a couple of magnitude $F(t)l_3$ as shown in Fig. 6-3b.

The virtual work of the force due to a virtual displacement δx_1 of m_1 is

$$\delta W_1 = F(t)\,\delta x_1$$

which reveals that the first element of $\mathbf{F}$ is simply the force

$$F_1(t) = F(t)$$

Since no external forces are acting on the masses m_2 and m_3, it follows that

$$\delta W_2 = 0 \quad \therefore \quad F_2(t) = 0$$
$$\delta W_3 = 0 \quad \therefore \quad F_3(t) = 0$$

The virtual work of the couple $F(t)l_3$ due to an angular displacement $\delta\theta$ of the main mass is

$$\delta W_4 = F(t)l_3\,\delta\theta$$

so that the fourth element of $\mathbf{F}$ is

$$F_4(t) = F(t)l_3$$

which is a moment.

The column matrix of the forcing functions is thus

$$\mathbf{F} = \begin{Bmatrix} F(t) \\ 0 \\ 0 \\ F(t)l_3 \end{Bmatrix}$$

It should be apparent that the elements of $\mathbf{F}$ could also have been obtained by a simple inspection of Figs. 6-2 and 6-3.

Support Excitation

To derive the equations of motion of a multiple-degree-of-freedom system excited by support motion, let us consider a typical system, such as the three-degree-of-freedom system shown in Fig. 6-4 in which

x_i = absolute displacement of mass m_i
y = absolute displacement of moving support
z_i = displacement of m_i relative to moving support

The *absolute* displacements of m_1, m_2, and m_3 with respect to the *inertial* frame of reference shown are x_1, x_2, and x_3, respectively. The displacements of these masses *relative to the moving support* are z_1, z_2, and z_3, respectively. The *absolute* displacement of the moving support (with respect to the inertial frame of reference) is y. Thus, for the ith mass, we can write that

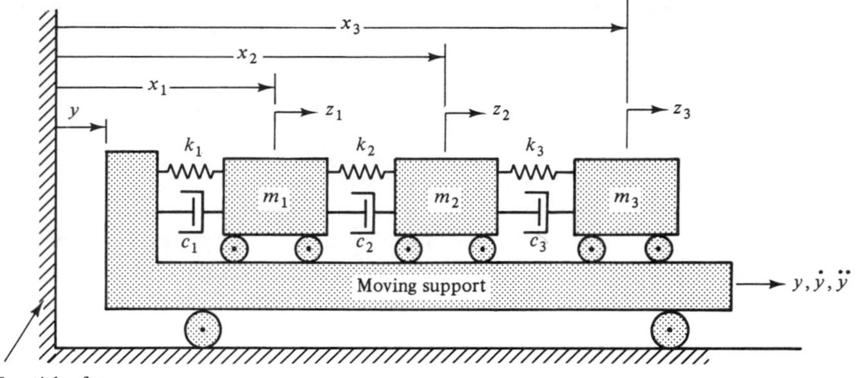

Figure 6-4 Typical system subjected to support excitation.

$$\left.\begin{aligned} x_i &= y + z_i \\ \dot{x}_i &= \dot{y} + \dot{z}_i \\ \ddot{x}_i &= \ddot{y} + \ddot{z}_i \end{aligned}\right\} \tag{6-4}$$

Referring again to Fig. 6-4, it can be seen that the only *forces* acting on the masses are those from the springs (the elastic elements) and the damping forces exerted by the dashpots. It should be apparent that the spring forces vary with the relative displacements z_i, while the damping forces are functions of the relative velocities $\dot{z}_i$. Therefore, since Newton's second law applies to the *absolute* accelerations of the masses $\ddot{x}_1$, $\ddot{x}_2$, and $\ddot{x}_3$, we can write that

$$\mathbf{M}\begin{Bmatrix} \ddot{x}_1 \\ \ddot{x}_2 \\ \ddot{x}_3 \end{Bmatrix} + \mathbf{C}\begin{Bmatrix} \dot{z}_1 \\ \dot{z}_2 \\ \dot{z}_3 \end{Bmatrix} + \mathbf{K}\begin{Bmatrix} z_1 \\ z_2 \\ z_3 \end{Bmatrix} = \mathbf{0} \tag{6-5}$$

Then noting from Eq. 6-4 that $\ddot{x}_i = \ddot{y} + \ddot{z}_i$, Eq. 6-5 can be expressed as

$$\mathbf{M}\begin{Bmatrix} \ddot{y} + \ddot{z}_1 \\ \ddot{y} + \ddot{z}_2 \\ \ddot{y} + \ddot{z}_3 \end{Bmatrix} + \mathbf{C}\begin{Bmatrix} \dot{z}_1 \\ \dot{z}_2 \\ \dot{z}_3 \end{Bmatrix} + \mathbf{K}\begin{Bmatrix} z_1 \\ z_2 \\ z_3 \end{Bmatrix} = \mathbf{0}$$

or

$$\mathbf{M}\begin{Bmatrix} \ddot{z}_1 \\ \ddot{z}_2 \\ \ddot{z}_3 \end{Bmatrix} + \mathbf{C}\begin{Bmatrix} \dot{z}_1 \\ \dot{z}_2 \\ \dot{z}_3 \end{Bmatrix} + \mathbf{K}\begin{Bmatrix} z_1 \\ z_2 \\ z_3 \end{Bmatrix} = -\mathbf{M}\begin{Bmatrix} \ddot{y} \\ \ddot{y} \\ \ddot{y} \end{Bmatrix} \tag{6-6}$$

Equation 6-6 expresses the differential equations of motion of the system in terms of its relative motion and an excitation source proportional to $\ddot{y}$, and can be written in the more compact matrix form as

$$\mathbf{M\ddot{Z}} + \mathbf{C\dot{Z}} + \mathbf{KZ} = -\mathbf{M\ddot{Y}} \tag{6-7}$$

In using Eq. 6-7, the damping matrix **C** and the stiffness matrix **K** are determined as though the moving support were stationary. For example, the first column of the stiffness matrix would be determined by finding the forces associated with the *relative* displacements $z_1 = 1$, $z_2 = z_3 = 0$, using the procedures discussed in Chap. 5.

The differential equations of motion of a system subjected to support motion can also be expressed in terms of the absolute motion of the system. Using Eqs. 6-4 and 6-5, it can be shown that they are given by

$$\mathbf{M\ddot{X}} + \mathbf{C\dot{X}} + \mathbf{KX} = \mathbf{C\dot{Y}} + \mathbf{KY} \tag{6-8}$$

in which the excitation terms include both the absolute displacement y and the absolute velocity $\dot{y}$ of the moving support as shown.

A disadvantage of the use of Eq. 6-8 to obtain the system response is that both y and $\dot{y}$ must be used for the excitation. In the case of a complicated support motion requiring the use of a computer to obtain a solution, the generation of both y and $\dot{y}$ may not be feasible, while the solution of Eq. 6-7 requires only the generation of $\ddot{y}$, which is often available as the output of an accelerometer. Therefore, in the vibration analysis of systems subjected to support excitation, the equations of motion in terms of the relative variable z are usually easier to use than those involving the absolute variable x.

6-3 GENERAL DISCUSSION OF DETERMINING FORCED-VIBRATION RESPONSE

In Sec. 6-2 we discussed the development of the differential equations of motion of multiple-degree-of-freedom systems subjected to either force or support excitation. In this section we discuss two basic approaches to solving such equations. They are

1. The simultaneous solution of *coupled* differential equations of motion. The steady-state response of two-degree-of-freedom systems subjected to *harmonic* excitations can generally be obtained quite readily by this method. However, if the excitations are not harmonic, it is usually necessary to use some numerical method such as the fourth-order Runge-Kutta method to obtain the total response of the system.
2. The modal-analysis (modal superposition) method, which involves solving a number of *uncoupled* differential equations (equal to the degrees of freedom). The uncoupled equations are expressed in terms of the system's principal coordinates, modal damping factors, and natural circular frequencies (eigenvalues). Each uncoupled equation has the general form of the differential equation of a single-degree-of-freedom system and is thus readily solved as such. The superposition of the solutions of these uncoupled equations, which yields the response of the system, is accomplished by means of the modal matrix $[u]$ (the eigenvectors) of the system.

Many special-purpose programs are available for the simultaneous solution of n coupled differential equations. One widely used program is CSMP (continuous system modeling program), which is a program written for the special purpose of simulating continuous systems in time on IBM computers.

Since there is seldom any logical method of determining physically realistic values for the c_{ij} elements of the damping matrix $\mathbf{C}$ which specifies the system damping, the simultaneous solution of coupled differential equations of motion is not a very practical means of obtaining the response of forced systems in which damping is an important parameter. However, the effect of damping on such systems is readily included when using modal analysis, since physically realistic values of the damping factor ζ can usually be determined from experimentally obtained data in the literature for various materials and types of structural systems.[3] For example, typical damping in reinforced-concrete structures might range from 1 to 5 percent ($\zeta = 0.01$ to 0.05), while for welded steel structures it could be from 0.1 to 0.3 percent. As we shall see later in this chapter, modal analysis is a powerful tool for obtaining the response of multiple-degree-of-freedom systems subjected to complicated excitations such as ground motion caused by earthquakes, and considerable emphasis will be placed on its use as a means of obtaining the forced response of such systems.

6-4 VIBRATION ABSORBER

Vibration absorbers can be used to virtually eliminate vibration in systems in which it is particularly undesirable, and to reduce excessive amplitudes of vibration in others. A common type of vibration absorber consists of a spring-and-mass system constructed such that its natural frequency is easily varied. This system is then attached to the principal system that is to have its amplitude of vibration reduced, and the frequency of the absorber system is then adjusted until the desired result is achieved.

For example, if the circular frequency ω of the disturbing force $F_0 \sin \omega t$ shown acting on the system in Fig. 6-5a is very near or equal to the natural circular frequency $\omega_n = \sqrt{k_1/m_1}$ of the system, the amplitude of vibration of the system could become very large owing to this resonant condition. However, if the *auxiliary* spring-and-mass system consisting of k_2 and m_2 is attached to the principal system as shown in Fig. 6-5b, the amplitude of the mass m_1 can be reduced essentially to zero if the natural circular frequency of the absorber is adjusted until it is equal to that of the disturbing force, that is, until $\sqrt{k_2/m_2} = \omega$.

Absorbers of this type are usually designed to have little damping and are "tuned" by varying either m_2 or k_2 or both. It is important to realize that the addition of an absorber of this type to a principal system results in a *combined* system having an added degree of freedom. The single-degree-of-freedom system shown in Fig. 6-5a, for example, becomes a two-degree-of-freedom system with the absorber added as shown in Fig. 6-5b.

To determine in more detail how such an absorber works, let us refer to the system shown in Fig. 6-5b. The stiffness matrix of this two-degree-of-freedom system is easily determined by inspecting the figure and recalling the definitions of stiffness coefficients as discussed in Chap. 5. Then, noting that the virtual work done by the excitation force $F_0 \sin \omega t$ owing to the virtual displacement δx_1 is

$$\delta W_1 = F_0 \sin \omega t \, (\delta x_1)$$

and that

$$\delta W_2 = 0$$

since there is no excitation force acting on m_2, the coupled differential equations of motion are found to be

$$\begin{bmatrix} m_1 & 0 \\ 0 & m_2 \end{bmatrix} \begin{Bmatrix} \ddot{x}_1 \\ \ddot{x}_2 \end{Bmatrix} + \begin{bmatrix} (k_1 + k_2) & -k_2 \\ -k_2 & k_2 \end{bmatrix} \begin{Bmatrix} x_1 \\ x_2 \end{Bmatrix} = \begin{Bmatrix} F_0 \sin \omega t \\ 0 \end{Bmatrix} \tag{6-9}$$

[3] T. H. H. Pian and F. C. Hallowell, "Structural Damping in a Simple Built-Up Beam," *Proceedings of the First U.S. National Congress on Applied Mechanics,* 1951, pp. 97–102. M. L. James, L. D. Lutes, and G. M. Smith, "Dynamic Properties of Reinforced and Prestressed Concrete Structural Elements," American Concrete Institute, vol. 61, no. 11, November 1964. Ai-Ting Yu, "Vibration Damping of Stranded Cable," *Proceedings of the Society for Experimental Stress Analysis,* vol. 9, no. 2, pp. 141–158, 1952.

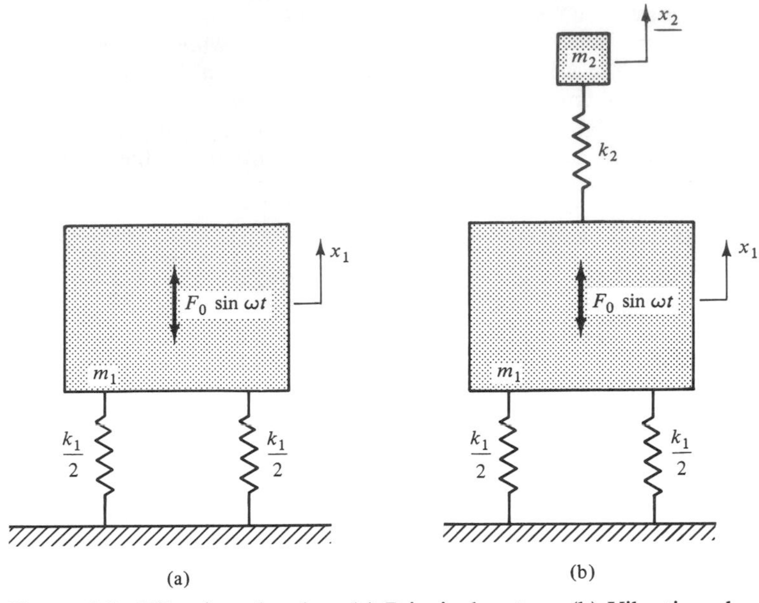

Figure 6-5 Vibration absorber. (a) Principal system. (b) Vibration absorber attached to principal system.

Premultiplying Eq. 6-9 by M^{-1}, we obtain

$$\left.\begin{aligned} \ddot{x}_1 + \frac{k_1 + k_2}{m_1} x_1 - \frac{k_2}{m_1} x_2 &= \frac{F_0}{m_1} \sin \omega t \\ \ddot{x}_2 - \frac{k_2}{m_2} x_1 + \frac{k_2}{m_2} x_2 &= 0 \end{aligned}\right\} \tag{6-10}$$

To determine the steady-state solution of these coupled equations, we let the imaginary part of $(F_0/m_1)e^{j\omega t}$ represent $(F_0/m_1) \sin \omega t$ and assume solutions of the form

$$x_1 = X_1 e^{j\omega t} \tag{6-11}$$

and

$$x_2 = X_2 e^{j\omega t} \tag{6-12}$$

Substituting Eqs. 6-11 and 6-12 and their appropriate derivatives into Eq. 6-10 yields the algebraic equations

$$\left.\begin{aligned} \left(\frac{k_1 + k_2}{m_1} - \omega^2\right) X_1 - \frac{k_2}{m_1} X_2 &= \frac{F_0}{m_1} \\ \frac{-k_2}{m_2} X_1 + \left(\frac{k_2}{m_2} - \omega^2\right) X_2 &= 0 \end{aligned}\right\} \tag{6-13}$$

from which we find that

$$X_1 = \frac{\dfrac{F_0}{m_1}\left(\dfrac{k_2}{m_2} - \omega^2\right)}{\left(\dfrac{k_1 + k_2}{m_1} - \omega^2\right)\left(\dfrac{k_2}{m_2} - \omega^2\right) - \dfrac{k_2^2}{m_1 m_2}} \tag{6-14}$$

and

$$X_2 = \frac{k_2/m_2}{k_2/m_2 - \omega^2} X_1 \tag{6-15}$$

It should be apparent from Eq. 6-14 that the amplitude X_1 is zero when k_2/m_2 of the vibration absorber is equal to ω^2. If the purpose of the absorber is to perform this reduction when the principal system is in resonance with the exciting force (when $k_1/m_1 = \omega^2$), it follows that

$$\frac{k_2}{m_2} = \frac{k_1}{m_1} \tag{6-16}$$

when $X_1 = 0$.

The two natural frequencies of the combined system depend upon the ratio of the absorber mass m_2 to the primary mass m_1. Therefore, this *mass ratio* of m_2/m_1 is an important parameter in the design of this type of vibration absorber. To see its effect on the response of the system, Eq. 6-14 is first transformed into nondimensional form by introducing the following notation:

$$\omega_{22}^2 = \frac{k_2}{m_2} = \frac{k_1}{m_1} \tag{6-17}$$

$$\mu = \frac{m_2}{m_1} = \frac{k_2}{k_1} \tag{6-18}$$

Using this notation, Eq. 6-14 can be written in the nondimensional form

$$\frac{X_1}{F_0/k_1} = \frac{1 - (\omega/\omega_{22})^2}{[(\omega/\omega_{22})^2]^2 - (2 + \mu)(\omega/\omega_{22})^2 + 1} \tag{6-19}$$

in which ω is the circular frequency of the disturbing force.

A plot of the *absolute* values of $X_1/(F_0/k_1)$ as a function of ω/ω_{22} for the mass ratio $\mu = 0.2$ is shown in Fig. 6-6. The plot shows that $X_1 = 0$ when $\omega/\omega_{22} = 1$ and becomes infinite when $\omega/\omega_{22} = 0.801$ or 1.248. The latter two values of ω/ω_{22} were obtained as the positive roots of the denominator of Eq. 6-19 when it was set equal to zero. Since these values define the two states of system resonance shown in Fig. 6-6, they can be used to determine the two natural circular frequencies of the system, ω_1 and ω_2. That is, for $\mu = 0.2$, the solution of

$$\left[\left(\frac{\omega}{\omega_{22}}\right)^2\right]^2 - 2.2\left(\frac{\omega}{\omega_{22}}\right)^2 + 1 = 0 \tag{6-20}$$

yields

$$\frac{\omega}{\omega_{22}} = 0.801, \quad 1.248$$

from which the two pertinent values of ω are

$$\omega_1 = 0.801\omega_{22} = 0.801\sqrt{\frac{k_1}{m_1}}$$

and

$$\omega_2 = 1.248\omega_{22} = 1.248\sqrt{\frac{k_1}{m_1}}$$

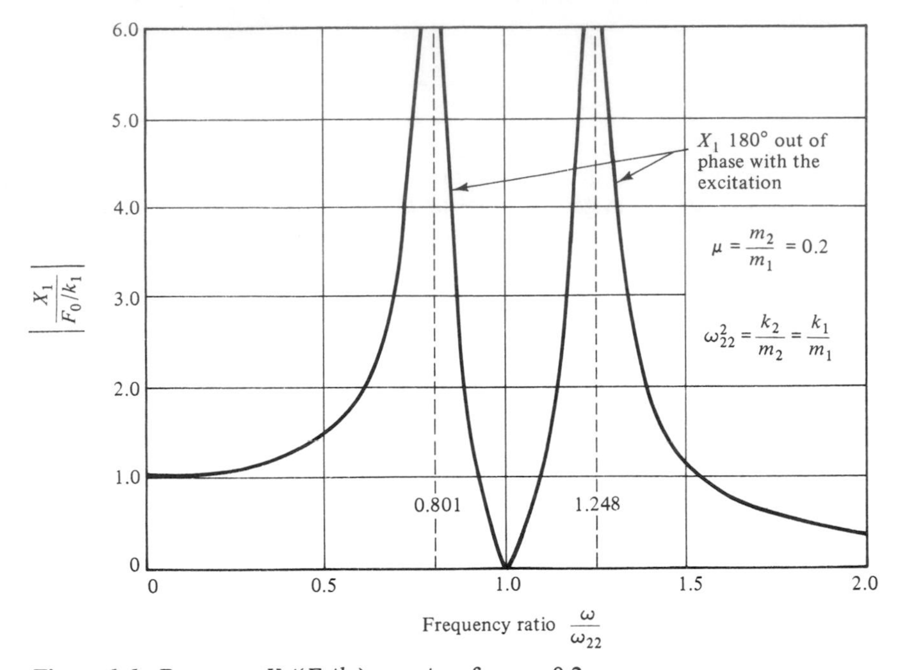

Figure 6-6 Response $X_1/(F_0/k_1)$ vs. ω/ω_{22} for $\mu = 0.2$.

in which $k_1/m_1 = k_2/m_2$. Therefore, we see that the two natural circular frequencies of the composite system are 0.801 and 1.248 times the natural circular frequency of the principal system.

Summarizing, Fig. 6-6 shows that with the absorber "tuned" to prevent vibration when the disturbing frequency is equal to the natural frequency of the principal system ($\omega/\omega_{22} = 1$), the principal system will have a zero amplitude of vibration at that particular disturbing frequency. However, if the disturbing frequency becomes 0.801 or 1.248 times the natural circular frequency of the principal system, the amplitude of the composite system response can become very large, since these frequencies are the resonant frequencies of the composite system.

Vibration Absorber Design

Since $\sqrt{k_2/m_2}$ of the absorber is equal to the excitation frequency ω when $X_1 = 0$, there are theoretically an infinite number of values of k_2 and m_2 that will satisfy the condition that $\sqrt{k_2/m_2} = \omega$. However, as we shall soon see, the important factors in determining the size of a vibration absorber are

1. The effect of the mass ratio $\mu = m_2/m_1$ on the natural frequencies
2. The effect of k_2 on the *allowable* amplitude X_2 of the absorber

To determine the effect of the mass ratio μ on the natural frequencies, we again set the denominator of Eq. 6-19 equal to zero, so that

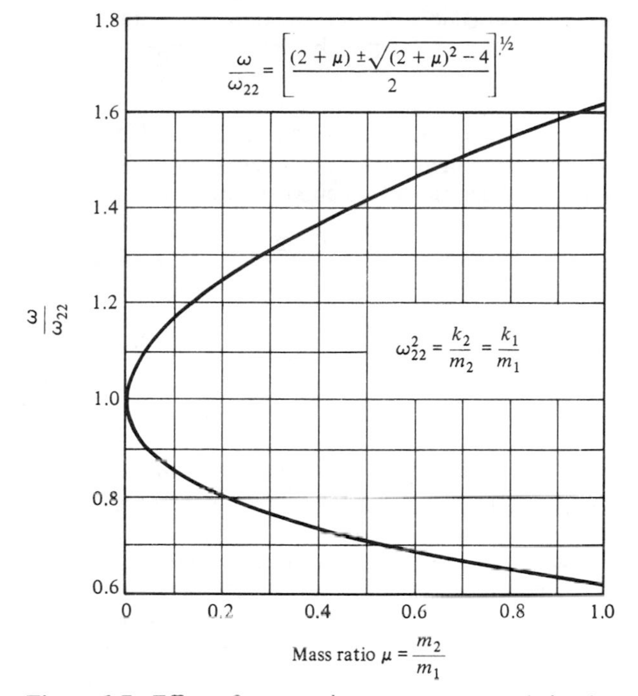

Figure 6-7 Effect of mass ratio m_2/m_1 on natural circular frequencies.

$$\left[\left(\frac{\omega}{\omega_{22}}\right)^2\right]^2 - (2+\mu)\left(\frac{\omega}{\omega_{22}}\right)^2 + 1 = 0 \tag{6-21}$$

in which the two values of ω are now the two natural circular frequencies of the combined system ω_1 and ω_2. The positive roots of this quadratic equation in $(\omega/\omega_{22})^2$ are shown plotted versus the mass ratio μ in Fig. 6-7. It should be apparent from this plot that the two natural frequencies of the composite system, which are above and below the natural frequency of the principal system, become farther apart as the size of the absorber increases. For example, when $\mu = 0.1$, resonance occurs at a frequency 0.85 and 1.17 times that of the principal system, but when $\mu = 0.2$, resonance occurs at a frequency 0.8 and 1.25 times that of the principal system. Therefore, if the frequency of the disturbing force fluctuates slightly above or below the frequency for which the absorber is tuned, a condition of resonance can occur if the mass m_2 of the absorber is too small.

When $k_2/m_2 = \omega_{22}^2$ and $\omega/\omega_{22} = 1$, we find from Eqs. 6-15, 6-18, and 6-19 that the amplitude X_2 of the absorber is

$$X_2 = \frac{-F_0}{k_2} \tag{6-22}$$

which is 180° out of phase with F_0. Therefore, the principal mass m_1 is subjected to both the excitation force $F_0 \sin \omega t$ and the force $-k_2X_2 \sin \omega t$ from the absorber. The combination of these forces corresponds to a condition of static equilibrium at any instant of time with $X_1 = 0$.

Vibration absorbers are susceptible to fatigue failure if the amplitude X_2 and accompanying stresses in the elastic element of the absorber are large. It can be seen from Eq. 6-22 that increasing k_2 reduces the amplitude X_2 for a given value of F_0. Since $\sqrt{k_2/m_2} = \omega$, an increase in k_2 requires a corresponding increase in m_2 for a given value of ω, which results in the separation of the natural frequencies as shown in Fig. 6-7.

A more thorough discussion of vibration absorbers and auxiliary-mass dampers used for controlling the amplitude of vibration of systems can be found in the *Shock and Vibration Handbook.*[4]

EXAMPLE 6-2

The pump shown in Fig. 6-8 is driven by a variable speed dc motor so that its speed can be varied to control the flow rate of the fluid being pumped. At a speed of 600 rpm the horizontal vibration of the pump-and-motor system is found to be so excessive that the speed-control system malfunctions. To eliminate this problem, a vibration absorber of stiffness k_2 and mass m_2 is to be attached to the pump housing as shown in the figure. The rod of the vibration absorber is threaded into a hole tapped in the pump housing so that it corresponds to a cantilever beam loaded at the free end having a spring constant of $k_2 = 3EI/l^3$.

As a preliminary step, a *temporary* vibration absorber consisting of a mass m_2 weighing 2 lb attached to the end of a $\frac{1}{4}$-in.-diameter steel rod of length l is tuned to 600 rpm (10 Hz) to obtain pertinent data for the design of a *permanent* absorber. By varying the speed of the dc motor with the temporary absorber attached, test data are obtained that reveal that the system has two natural frequencies of 8.9 and 11.2 Hz and that the amplitude X_2 of the absorber at 10 Hz is 0.095 in. It is desired to determine the following: (a) The length l of the rod of the temporary absorber when it is tuned to 10 Hz. (b) The apparent effective magnitude F_0 of the exciting force acting on the principal system (the pump and motor). (c) The apparent effective weight W_1 of the principal system. (d) The weight W_2 that the permanent absorber must have so that the two natural frequencies of the total system fall outside the 480 rpm (8 Hz) to 720 rpm (12 Hz) operating frequency range of the system. (e) The length l of the permanent absorber rod, and the magnitude of its spring constant k_2 if a $\frac{1}{4}$-in. steel rod is used. (f) The resulting stress σ in the permanent absorber if a $\frac{1}{4}$-in. rod is used. After determining σ, compare it with an allowable stress of 12,500 psi for the threaded portion of the rod that connects to the pump housing. The allowable stress is based upon an endurance limit of 60,000 psi for the steel used, a stress-concentration factor of 3.85,[5] and a factor of safety of 1.25.

Solution. **a.** Since the vibration absorber is tuned to 10 Hz, which is the natural frequency of the principal system, we note that

$$\omega^2 = \omega_{22}^2 = [10(2\pi)]^2 = \frac{k_2}{m_2} = \frac{k_2}{\frac{2}{386}}$$

from which

$$k_2 = 20.5 \text{ lb/in.}$$

[4] F. Everett Reed, "Dynamic Vibration Absorbers and Auxiliary Mass Dampers," chap. 6, *Shock and Vibration Handbook,* ed. by C. M. Harris and C. E. Crede, vol. 1, McGraw-Hill Book Co., New York, 1961.

[5] M. Hetenyi, "A Photoelastic Study of Bolt and Nut Fastenings," *Trans. ASME,* vol. 65, p. A-93, 1943.

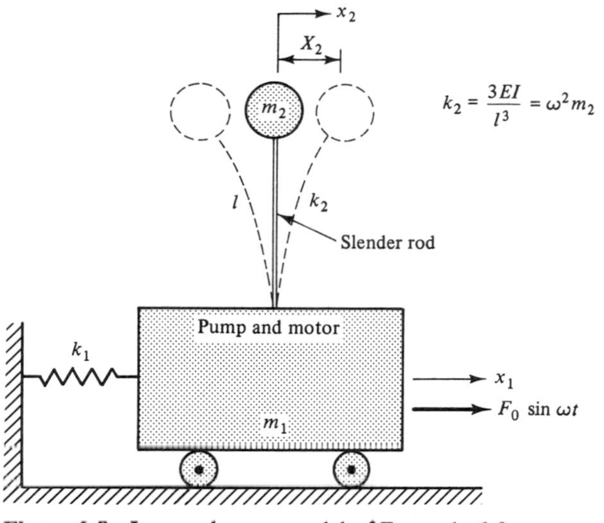

Figure 6-8 Lumped-mass model of Example 6-2.

Then, with $k_2 = 3EI/l^3$ for a cantilever beam, and the beam having a modulus of elasticity of $E = 30(10)^6$ psi and a diametral area moment of inertia of $I = \pi(0.25)^4/64$ in.^{4}, the length of the temporary absorber rod is found as

$$l = \left(\frac{3EI}{k_2}\right)^{1/3} = \left[\frac{3(30)10^6\pi(0.25)^4}{20.5(64)}\right]^{1/3} = 9.44 \text{ in.}$$

b. Using the test-data value of $X_2 = 0.095$ in., we find from Eq. 6-22 that the apparent magnitude of F_0 is

$$F_0 = k_2 X_2 = 20.5(0.095) = 1.95 \text{ lb}$$

c. Using the natural frequencies determined experimentally with the temporary absorber tuned to 10 Hz, we can write that

$$\frac{\omega_1}{\omega_{22}} = \frac{8.9}{10} = 0.89$$

and

$$\frac{\omega_2}{\omega_{22}} = \frac{11.2}{10} = 1.12$$

The mass ratio μ can now be determined from Fig. 6-7 or by substituting either of the above frequency ratios into Eq. 6-21. Substituting $\omega_1/\omega_{22} = 0.89$ into Eq. 6-21,

$$(0.89)^4 - (2 + \mu)(0.89)^2 + 1 = 0$$

from which $\mu = 0.055$. Thus, the apparent weight W_1 of the principal system is found as

$$W_1 = \frac{W_2}{\mu} = \frac{2}{0.055} = 36.4 \text{ lb}$$

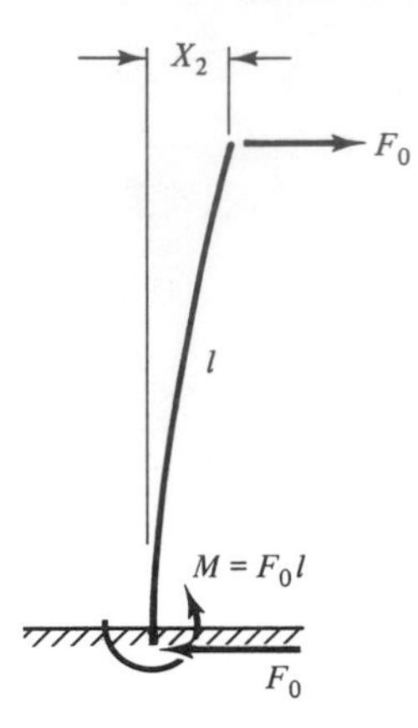

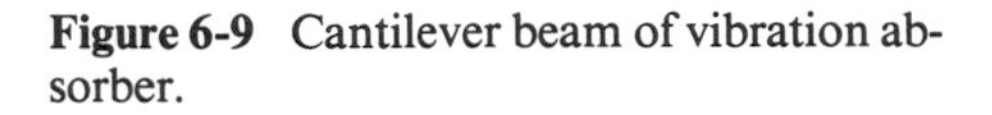

Figure 6-9 Cantilever beam of vibration absorber.

d. Using the upper and lower operating frequency limits of 12 and 8 Hz, we can determine the ratios

$$\frac{\omega}{\omega_{22}} = \frac{8}{10} = 0.8$$

and

$$\frac{\omega}{\omega_{22}} = \frac{12}{10} = 1.2$$

Figure 6-7 reveals that $\mu = 0.2$ when $\omega/\omega_{22} = 0.8$ and that $\mu = 0.14$ when $\omega/\omega_{22} = 1.2$. Therefore, we consider the largest value of μ to ensure that resonance will not occur within the frequency range of 8 to 12 Hz. Then, using the value of $W_1 = 36.4$ lb found in part c, the weight of the permanent vibration absorber is found to be

$$W_2 = 36.4(0.2) = 7.28 \text{ lb}$$

e. The magnitude of the spring constant of the absorber is

$$k_2 = m_2\omega^2 = \frac{7.28}{386}[10(2\pi)]^2 = 74.5 \text{ lb/in.}$$

and the length of the permanent vibration absorber is then

$$l = \left(\frac{3EI}{k_2}\right)^{1/3} = \left[\frac{3(30)10^6(1.92)10^{-4}}{74.5}\right]^{1/3} = 6.14 \text{ in.}$$

f. Figure 6-9 shows that the maximum moment in the rod is $F_0 l$. Therefore, the maximum flexural stress in the rod is

$$\sigma = \frac{Mc}{I} = \frac{F_0 l(d/2)}{I} \tag{6-23}$$

where $F_0 = 1.95$ lb (from part b)
$d = 0.215$ in. (diameter of rod[6])

[6] This diameter is based on the stress area of the $\frac{1}{4}$-in. threaded rod.

$I = \pi(0.215)^4/64 = 1.05(10)^{-4}$ in.^{4} (area moment of inertia)
$l = 6.14$ in. (from part e)

Substituting the above values into Eq. 6-23, we find that

$$\sigma = \frac{1.95(6.14)(0.215/2)}{1.05(10)^{-4}} = 12{,}258 \text{ psi}$$

This stress is less than the allowable stress of 12,500 psi; so the permanent vibration absorber should not be susceptible to fatigue failure.

6-5 COMPUTER PROGRAM FOR THE SIMULTANEOUS SOLUTION OF COUPLED DIFFERENTIAL EQUATIONS OF MOTION

The matrix equation of an n-degree-of-freedom system subjected to force or moment excitation has been given previously by Eq. 6-1 as

$$\mathbf{M}\ddot{\mathbf{X}} + \mathbf{C}\dot{\mathbf{X}} + \mathbf{K}\mathbf{X} = \mathbf{F} = \begin{Bmatrix} F_1(t) \\ F_2(t) \\ F_3(t) \\ \vdots \\ F_n(t) \end{Bmatrix} \tag{6-1}$$

Premultiplying this equation by the inverse of $\mathbf{M}$ gives

$$\ddot{\mathbf{X}} + \mathbf{M}^{-1}\mathbf{C}\dot{\mathbf{X}} + \mathbf{M}^{-1}\mathbf{K}\mathbf{X} = \mathbf{M}^{-1}\mathbf{F} = \begin{Bmatrix} F_1(t)/m_1 \\ F_2(t)/m_2 \\ F_3(t)/m_3 \\ \vdots \\ F_n(t)/m_n \end{Bmatrix} \tag{6-24}$$

in which $\mathbf{M}$ is considered as having only elements on the diagonal.

Either of the above equations represents n simultaneous (coupled) *second-order* differential equations, or $2n$ *first-order* simultaneous differential equations, with the dependent variables $x_1, \dot{x}_1, x_2, \dot{x}_2, \ldots, x_n, \dot{x}_n$. Either the n second-order or the $2n$ first-order differential equations can be solved numerically to obtain the time-domain response of a system. The classic fourth-order Runge-Kutta method with fixed step size is well suited for the numerical solution of the $2n$ first-order equations, which yields the forced-vibration response of an n-degree-of-freedom system.[7] We use this Runge-Kutta method in the FORTRAN computer program SIM2.FOR which is presented next.

The main program of SIM2.FOR calls a subroutine subprogram RKSFX that uses the classic fourth-order Runge-Kutta equations to solve any number of simultaneous first-order differential equations.[8] This subprogram solves the $2n$ first-order simultaneous differential equations represented by Eq. 6-24.

In using RKSFX, the equations of motion are expressed in the main program by the following first-order differential equations:

[7] M. L. James, G. M. Smith, and J. C. Wolford, *Applied Numerical Methods for Digital Computations,* 3d ed., p. 461, Harper & Row, New York, 1985.

[8] The interested reader will find a complete derivation in *Mathematical Methods for Digital Computers,* ed. by Anthony Ralston and Herbert S. Wilf, John Wiley & Sons, New York, 1960.

$$\left.\begin{aligned}
\frac{d\dot{x}_1}{dt} &= \frac{F_1(t)}{m_1} - \sum_{j=1}^{n} \frac{c_{1j}\dot{x}_j + k_{1j}x_j}{m_1} \\
\frac{dx_1}{dt} &= \dot{x}_1 \\
\frac{d\dot{x}_2}{dt} &= \frac{F_2(t)}{m_2} - \sum_{j=1}^{n} \frac{c_{2j}\dot{x}_j + k_{2j}x_j}{m_2} \\
\frac{dx_2}{dt} &= \dot{x}_2 \\
&\;\vdots \\
\frac{d\dot{x}_n}{dt} &= \frac{F_n(t)}{m_n} - \sum_{j=1}^{n} \frac{c_{nj}\dot{x}_j + k_{nj}x_j}{m_n} \\
\frac{dx_n}{dt} &= \dot{x}_n
\end{aligned}\right\} \tag{6-25}$$

In the computer program, the values of the dependent variables $\dot{x}_1, x_1, \dot{x}_2, x_2, \ldots, \dot{x}_n, x_n$ are stored in an array Y in which

y_1	corresponds to	$\dot{x}_1$
y_2	corresponds to	x_1
y_3	corresponds to	$\dot{x}_2$
y_4	corresponds to	x_2
$\vdots$	$\vdots$ $\vdots$	$\vdots$
y_{2n-1}	corresponds to	$\dot{x}_n$
y_{2n}	corresponds to	x_n

Thus, when ENTER INITIAL VALUE OF Y(1), FREE FORMAT appears on the console screen, the initial value of $\dot{x}_1$ should be entered. When the initial value of $Y(2)$ is requested, the initial value of x_1 is entered, and so forth.

The values of the expressions on the right-hand side of Eq. 6-25 are stored in an array FUNC. The elements of FUNC are defined by statements in the main program. As can be seen from Eq. 6-25, these defining expressions include the excitation functions $F_1(t)$, $F_2(t), \ldots, F_n(t)$, which are applied to the masses $m_1, m_2, \ldots, m_n$, respectively. The values of the exciting functions are obtained by referencing the function subprogram F(T, I), which appears immediately following the main program. In general, the excitation functions are functions of time, so that the current value of T is sent to the function subprogram when it is referenced, along with the value of I specifying which excitation function is required. That is, when I is equal to 1, the value of the function $F_1(t)$ acting on mass m_1 is required, and so on.

The function subprogram F(T, I) will vary for different problems, since it must specify the type of excitation function acting on each mass in a particular problem. Several examples of such subprograms for various types of excitation will be shown later.

The principal FORTRAN variables and array names, and the quantities that they represent in the main program, are as follows:

Variable or array name (main program)	Quantity
N	Number of simultaneous second-order differential equations, or degrees of freedom describing the system
L	Number of first-order differential equations. $L = 2N$
Y(I)	An array containing the values of the dependent variables (velocities and displacements of the L first-order equations)
FUNC(I)	An array containing the values of the right-hand sides of the L first-order equations
M(I)	An array containing the values of the N masses of the system
K(I,J)	Elements k_{ij} of the stiffness matrix $\mathbf{K}$
C(I,J)	Elements c_{ij} of the damping matrix $\mathbf{C}$
SUM	$\sum_{J=1}^{N} (c_{ij}\dot{x}_j + k_{ij}x_j)/m_i$ where i has one value from 1 to N for each value of SUM
T	Independent variable time
DELT	The time increment for the numerical integration
TMAX	The maximum time for which the integration is to run
IA	An integer variable assigned a value 1 if damping is to be considered and 2 if damping is to be neglected
IMAIN	An integer variable initialized at zero in the main program. It is sent to the subprogram RKSFX as the value of M in RKSFX where its value is incremented by 1 and used as the variable of a computed GO TO statement. Its value increases by 1 each time RKSFX is called within a particular time increment. It is reset to zero by RKSFX at the end of each time step of DELT
ISUB	An integer variable having a value 1 or 2 and used in a computed GO TO statement. RKSFX is called 4 times for each increment of time in the integration. ISUB receives a value of 1 from RKSFX as a result of the first call and retains that value until the fourth call, when it receives a value of 2

The main program and the subroutine subprogram contain DIMENSION statements capable of handling systems with up to 10 degrees of freedom. These DIMENSION statements must be changed for problems involving more than 10 degrees of freedom. The function subprogram F(T, I), which is shown following the main program, is for a sine pulse acting on mass m_4. This function subprogram is used in Example 6-3 to illustrate the use of the SIM2.FOR program on a four-degree-of-freedom system.

It should be noted that the program as written requires double-precision computations. The data input required for the program is entered by means of the console keyboard according to the messages appearing on the console screen prescribed by the WRITE(*,n) statements in the program. The program is as follows:

```
C   SIM2.FOR
C   THIS IS THE MAIN PROGRAM FOR SOLVING SETS OF SECOND-ORDER
C   DIFFERENTIAL EQUATIONS EXPRESSED IN MATRIX FORM AS SHOWN IN
C   EQ. 6-24.  THE RUNGE-KUTTA SUBROUTINE SUBPROGRAM RKSFX FOR
C   SOLVING SETS OF SIMULTANEOUS FIRST-ORDER EQUATIONS IS CALLED
C   BY THIS MAIN PROGRAM.  THE FUNCTION F(T,I) MUST BE EXPRESSED
C   IN A FUNCTION SUBPROGRAM WHICH IS REFERENCED BY THIS MAIN
```

```
C   PROGRAM.  THE INPUT DATA REQUIRED FOR A COMPUTER RUN ARE:
C   THE NUMBER OF SECOND-ORDER EQUATIONS N, TIME INCREMENT DELT,
C   MAXIMUM TIME OF INTEGRATION TMAX, VALUES OF THE N MASSES,
C   VALUES OF THE ELEMENTS OF THE SYMMETRIC K MATRIX ALONG AND
C   ABOVE THE MAIN DIAGONAL, A VALUE OF 1 FOR IA IF DAMPING IS
C   CONSIDERED AND 2 IF DAMPING IS TO BE NEGLECTED, VALUES OF
C   THE ELEMENTS OF THE SYMMETRIC DAMPING MATRIX C ALONG AND ABOVE
C   THE MAIN DIAGONAL WHEN IA = 1.
C   DEVICE * IN READ AND WRITE STATEMENTS IS THE CONSOLE.
C   DEVICE 2 IN WRITE STATEMENTS IS THE PRINTER.
C   * IN THE PLACE OF A FORMAT STATEMENT NUMBER MEANS FREE FORMAT.
      IMPLICIT DOUBLE PRECISION(A-H,O-Z)
      DOUBLE PRECISION M,K
      DIMENSION Y(20),FUNC(20),M(10),K(10,10),C(10,10)
      OPEN (2,FILE='PRN')
C ENTER NUMBER OF SIMULTANEOUS 2ND ORDER EQUATIONS.
      WRITE(*,2)
2     FORMAT(' ENTER NUMBER OF EQUATIONS, I2')
      READ(*,3)N
3     FORMAT(I2)
      L=2*N
C ENTER TIME INCREMENT AND MAXIMUM TIME.
      WRITE(*,4)
4     FORMAT(' ENTER DELT AND TMAX, 2F10.0')
      READ(*,*)DELT,TMAX
C READ IN THE MAGNITUDES OF THE N MASSES
      DO 8 I = 1,N
      WRITE(*,6)I
6     FORMAT(' ENTER VALUE FOR MASS M(',I2,'),F10.0')
      READ(*,7)M(I)
7     FORMAT(F10.0)
8     CONTINUE
C READ IN BY ROWS THE ELEMENTS OF THE SYMMETRIC K MATRIX
C ALONG AND ABOVE THE MAIN DIAGONAL.
      DO 12 I = 1,N
      DO 12 J = I,N
      WRITE(*,10)I,J
10    FORMAT(' ENTER K(',I2,',',I2,'), F10.0')
      READ(*,11)K(I,J)
11    FORMAT(F20.0)
12    CONTINUE
C ASSIGN VALUE TO SYMMETRIC K ELEMENTS BELOW MAIN DIAGONAL.
      DO 26 I = 2,N
      IMIN1 = I-1
      DO 26 J = 1,IMIN1
26    K(I,J) = K(J,I)
C WRITE OUT ELEMENTS OF K MATRIX BY ROWS.
      DO 27 I = 1,N
      DO 27 J = 1,N
```

```
27     WRITE(2,28)I,J,K(I,J)
28     FORMAT(' ','K(',I2,',',I2,') = ',E14.7)
       WRITE(*,29)
29     FORMAT(/,' ','IS DAMPING TO BE CONSIDERED ?  ENTER  1')
       WRITE(*,30)
30     FORMAT(' ','FOR YES OR 2 FOR NO USING FIELD SPEC. OF I1')
       READ(*,31)IA
31     FORMAT(I1)
       IF(IA .EQ. 1)GO TO 32
       GO TO 39
C READ IN BY ROWS THE ELEMENTS OF THE SYMMETRIC C MATRIX ALONG
C AND ABOVE THE MAIN DIAGONAL.
32     DO 35 I = 1,N
       DO 35 J = I,N
       WRITE(*,33)I,J
33     FORMAT(' ENTER C(',I2,',',I2,'), F10.0')
       READ(*,34)C(I,J)
34     FORMAT(F10.0)
35     CONTINUE
C ASSIGN VALUES TO SYMMETRIC C ELEMENTS BELOW MAIN DIAGONAL
       DO 36 I = 2,N
       IMIN1 = I-1
       DO 36 J = 1,IMIN1
36     C(I,J) = C(J,I)
C WRITE OUT THE ELEMENTS OF THE C MATRIX BY ROWS.
       DO 37 I = 1,N
       DO 37 J = 1,N
37     WRITE(2,38)I,J,C(I,J)
38     FORMAT(' ','C(',I2,',',I2,') = ',E14.7)
       GO TO 41
C SINCE DAMPING IS BEING NEGLECTED, ASSIGN ALL ELEMENTS OF
C THE C MATRIX ZERO VALUES.
39     DO 40 I = 1,N
       DO 40 J = 1,N
40     C(I,J) = 0.
41     CONTINUE
C INITIAL IMAIN AND T.
       IMAIN = 0.
       T = 0.0D0
C ENTER INITIAL VALUES OF THE VELOCITIES AND DISPLACEMENTS.
       DO 15 I = 1,L
       WRITE(*,13)I
13     FORMAT(' ENTER INITIAL VALUE OF Y(',I2,'),F10.0')
       READ(*,14)Y(I)
14     FORMAT(F10.0)
15     CONTINUE
C WRITE COLUMN HEADINGS AND INITIAL VALUES.  CHANGE
C FORMAT STATEMENT 16 AND 17 AS REQUIRED BY PROBLEM.
       WRITE(2,16)
```

```
16      FORMAT(' ','TIME',7X,'Y(2)',7X,'Y(4)',7X,'Y(6)',7X,'Y(8)'//)
        WRITE(2,17)T,(Y(I),I=2,L,2)
17      FORMAT(' ',F6.4,4F11.4)
18      IF(T .GE. TMAX-0.0001)GO TO 100
        GO TO 20
19      CALL RKSFX(L,IMAIN,ISUB,Y,FUNC,T,DELT)
        GO TO (20,25),ISUB
20      DO 24 NUM = 1,L
        IF(MOD(NUM,2))21,23,21
21      I = (NUM+1)/2
        SUM = 0.0
        DO 22 J = 1,N
        JJ = J+J
        SUM = SUM + C(I,J)*Y(JJ-1)/M(I) + K(I,J)*Y(JJ)/M(I)
22      CONTINUE
        FUNC(NUM) = F(T,I)/M(I)-SUM
        GO TO 24
23      FUNC(NUM) = Y(NUM-1)
24      CONTINUE
        GO TO 19
C WRITE OUT TIME AND DISPLACEMENTS
25      WRITE(2,17)T,(Y(I), I=2,L,2)
        GO TO 18
100     STOP
        END

        FUNCTION F(T,I)
        IMPLICIT DOUBLE PRECISION(A-H,O-Z)
        GO TO (10,20,30,40),I
10      F = 0.0D0
        GO TO 50
20      F = 0.0D0
        GO TO 50
30      F = 0.0D0
        GO TO 50
40      T1 = 0.1D0
        IF(T .GE. T1)GO TO 48
        IF(T .LE. T1)GO TO 45
45      F=2.D6*DSIN(3.141593*T/T1)
        GO TO 50
48      F = 0.0D0
50      CONTINUE
        RETURN
        END

        SUBROUTINE RKSFX(N,M,J,Y,F,T,DELT)
C  THIS DOUBLE-PRECISION SUBROUTINE SOLVES N SIMULTANEOUS
C  FIRST ORDER DIFFERENTIAL EQUATIONS USING THE CLASSIC
C  4TH-ORDER RUNGE-KUTTA METHOD
C
```

```
C   N     = NUMBER OF FIRST-ORDER DIFF. EQS. TO BE SOLVED
C   M     = THE INTEGER VARIABLE OF A COMPUTED GO TO STATEMENT
C           IN RKSFX.  IT MUST BE SET EQUAL TO ZERO BEFORE
C           RKSFX IS CALLED THE FIRST TIME
C   J     = AN INTEGER VARIABLE WHICH IS GIVEN A VALUE IN
C           SUBPROGRAM RKSFX AND USED IN THE CALLING PROGRAM AS
C           THE INTEGER VARIABLE OF A COMPUTED GO TO STATEMENT
C   Y     = ARRAY OF N DEPENDENT VARIABLES.  INITIAL VALUES ARE
C           ASSIGNED IN CALLING PROGRAM
C   F     = ARRAY OF DY/DT VALUES FOR THE N Y'S EXPRESSIONS
C           FOR THE N F(I)'S MUST BE GIVEN IN THE CALLING PROGRAM
C   T     = THE INDEPENDENT VARIABLE.  THE INITIAL VALUE IS
C           ASSIGNED IN THE CALLING PROGRAM
C   DELT = THE STEP SIZE OF THE INDEPENDENT VARIABLE. DELT
C           IS ASSIGNED IN THE CALLING PROGRAM
C
      IMPLICIT DOUBLE PRECISION(A-H,O-Z)
      DIMENSION Y(1),F(1),AK1(20),AK2(20),AK3(20),AK4(20)
      M=M+1
      GO TO(10,20,30,40),M
C   CALCULATE THE AK1(I)'S
10    DO 2 I = 1,N
      AK1(I)=DELT*F(I)
      Y(I)=Y(I)+AK1(I)/2.
2     CONTINUE
      T=T+DELT/2.
C   GO BACK TO CALLING PROGRAM TO CALCULATE NEW F(I)'S
      J=1
      GO TO 7
C   CALCULATE THE AK2(I)'S
20    DO 3 I - 1,N
      AK2(I)=DELT*F(I)
      Y(I)=Y(I)-AK1(I)/2.+AK2(I)/2.
3     CONTINUE
C   GO BACK TO CALLING PROGRAM TO CALCULATE NEW F(I)'S
      GO TO 7
C   CALCULATE THE AK3(I)'S
30    DO 4 I = 1,N
      AK3(I)=DELT*F(I)
      Y(I)=Y(I)-AK2(I)/2.+AK3(I)
4     CONTINUE
      T=T+DELT/2.
C   GO BACK TO CALLING PROGRAM TO CALCULATE NEW F(I)'S
      GO TO 7
C   CALCULATE THE AK4(I)'S
40    DO 5 I = 1,N
      AK4(I)=DELT*F(I)
5     CONTINUE
C   CALCULATE THE NEW Y(I) VALUES AT END OF STEP
      DO 6 I = 1,N
```

```
      Y(I)=Y(I)-AK3(I)+(AK1(I)+2.*AK2(I)+2.*AK3(I)+AK4(I))/6.
6     CONTINUE
      J=2
      M=0
7     RETURN
      END
```

EXAMPLE 6-3

The mass m_4 of the four-degree-of-freedom system shown in Fig. 6-10 is subjected to the sine pulse

$$F_4(t) = F_0 \sin \frac{\pi t}{t_1}$$

in which the duration of the pulse is $t_1 = 0.1$ s and its amplitude is $F_0 = 2(10)^6$ lb. (a) Use the values of k, c, and m shown in Fig. 6-10 to determine the data necessary as input to the program SIM2.FOR (see page 425) for obtaining the time response of the system. (b) Determine the time response of the system in time increments of DELT = 0.005 s up to a time of TMAX = 1.0 s.

Solution. **a.** From Fig. 6-10 we find the mass elements as

$$m_1 = m_2 = 2m = 2(6)10^3 = 12(10)^3 \text{ lb} \cdot \text{s}^2/\text{in.}$$
$$m_3 = m_4 = m = 6(10)^3 \text{ lb} \cdot \text{s}^2/\text{in.}$$

Using the definition of stiffness coefficients k_{ij} in conjunction with an inspection of Fig. 6-10, we obtain the following input data:

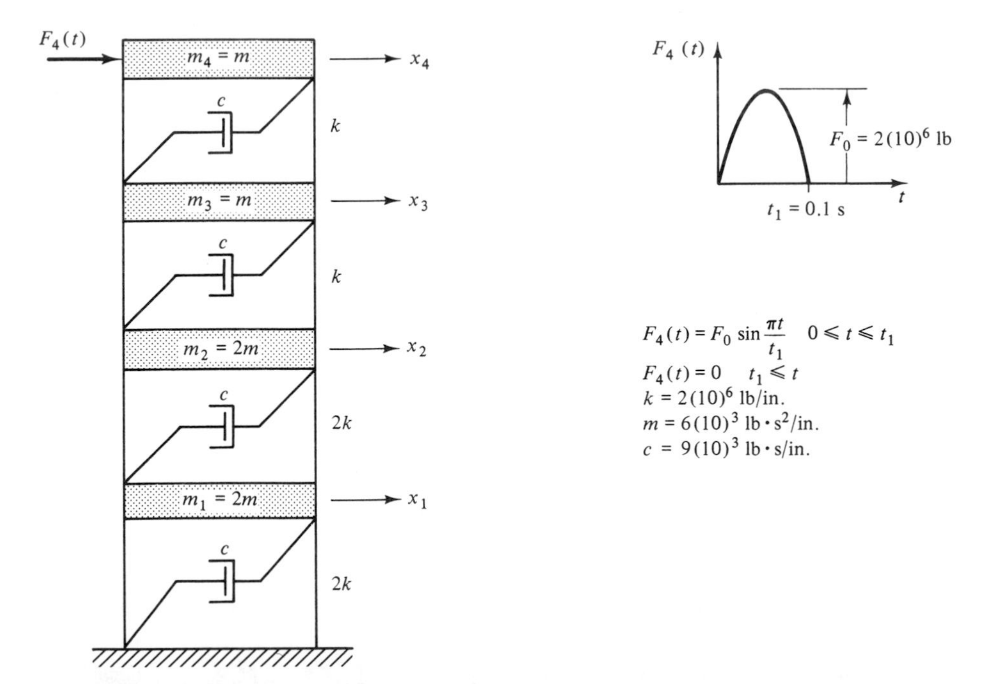

Figure 6-10 A four-degree-of-freedom system subjected to a sine pulse, Example 6-3.

$$\mathbf{K} = \begin{bmatrix} 4k & -2k & 0 & 0 \\ -2k & 3k & -k & 0 \\ 0 & -k & 2k & -k \\ 0 & 0 & -k & k \end{bmatrix} = 10^6 \begin{bmatrix} 8 & -4 & 0 & 0 \\ -4 & 6 & -2 & 0 \\ 0 & -2 & 4 & -2 \\ 0 & 0 & -2 & 2 \end{bmatrix}$$

Using the definition of damping coefficients c_{ij} on page 305 in conjunction with an inspection of Fig. 6-10, the following input data are obtained:

$$\mathbf{C} = \begin{bmatrix} 2c & -c & 0 & 0 \\ -c & 2c & -c & 0 \\ 0 & -c & 2c & -c \\ 0 & 0 & -c & c \end{bmatrix} = 10^3 \begin{bmatrix} 18 & -9 & 0 & 0 \\ -9 & 18 & -9 & 0 \\ 0 & -9 & 18 & -9 \\ 0 & 0 & -9 & 9 \end{bmatrix}$$

b. Since the highest natural circular frequency ω_4 of the system was found to be 32.32 rad/s from the program JACOBI.FOR (see page 369), a suitable value for the integration time increment Δt is determined as

$$\Delta t \leq \frac{2\pi}{32.32}\left(\frac{1}{20}\right) = 0.0097 \text{ s}$$

This shows that the value of DELT = 0.005 specified in the problem statement should yield very accurate results.

The function subprogram shown at the end of the SIM2.FOR program assigns values of zero to $F_1(t)$, $F_2(t)$, and $F_3(t)$, since the forces acting on the masses m_1, m_2, and m_3 are zero. That is, when I = 1, statement 10 assigns a value of 0.000 to $F_1(t)$, and so on. The sine pulse $F_4(t)$, acting on mass m_4, is generated by statements 40 through 48.

When a value is requested for IA on the console screen, it is assigned an integer value of 1 since damping is being considered. Because the initial conditions of x_i and $\dot{x}_i$ are zero, Y(1), Y(2), . . . , Y(8) are assigned a value of zero when requested on the console screen.

Initial and final portions of the printout obtained from the execution of the program are as follows:

TIME	Y(2)	Y(4)	Y(6)	Y(8)
0.0000	0.0000	0.0000	0.0000	0.0000
.0050	.0000	.0000	.0000	.0002
.0100	.0000	.0000	.0000	.0017
.0150	.0000	.0000	.0001	.0058
.0200	.0000	.0000	.0002	.0135
.0250	.0000	.0000	.0005	.0259
.0300	.0000	.0000	.0011	.0439
. .	. .	. .	. .	. .
.9600	.0531	.0994	.4562	.7052
.9650	.0770	.1272	.4770	.7287
.9700	.1006	.1560	.4972	.7501
.9750	.1238	.1857	.5166	.7694
.9800	.1464	.2162	.5355	.7866
.9850	.1683	.2473	.5539	.8017
.9900	.1894	.2790	.5719	.8148
.9950	.2097	.3111	.5895	.8259
1.0000	.2290	.3434	.6068	.8351

As explained previously, Y(2), Y(4), Y(6), and Y(8) in the printout correspond to the displacements x_1, x_2, x_3, and x_4, respectively.

With the use of computer graphics software, the appropriate statements for graphics can be added to the main program to obtain the plots of the response shown in Fig. 6-11. These plots obviously provide a much better overview of the time-response characteristics of the system than does just the printout of the response.

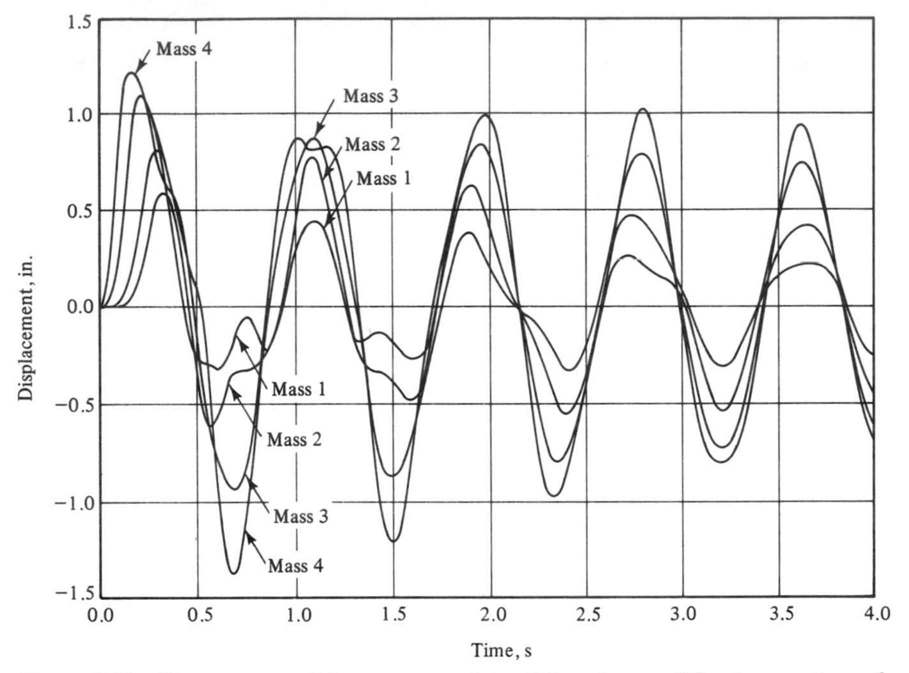

Figure 6-11 Computer graphics response plots of four-degree-of-freedom system of Example 6-3.

EXAMPLE 6-4

Referring to the discussion in Example 6-3, modify the function subprogram used in SIM2.FOR (see page 425) so that m_2 is subjected to the sine pulse

$$F_2(t) = F_0 \sin \frac{\pi t}{t_1}$$

with $F_1(t) = F_3(t) = F_4(t) = 0$. Use a pulse duration of $t_1 = 0.1$ s and pulse magnitude of $F_0 = 2(10)^6$ lb.

Solution

```
      FUNCTION F(T,I)
      IMPLICIT DOUBLE PRECISION(A-H,O-Z)
      GO TO (10,20,30,40),I
10    F = 0.0D0
      GO TO 50
20    T1 = 0.1D0
```

```
      IF(T .GE. T1)GO TO 25
      IF(T .LE. T1)GO TO 22
22    F = 2.D6*DSIN(3.141593*T/T1)
      GO TO 50
25    F = 0.0D0
      GO TO 50
30    F = 0.0D0
      GO TO 50
40    F = 0.0D0
50    CONTINUE
      RETURN
      END
```

EXAMPLE 6-5

Write a function subprogram for use in the SIM2.FOR program (see page 425) for the solution of a five-degree-of-freedom system, in which $F_2(t) = 10{,}000$ lb and $F_3(t) = 20{,}000$ lb are suddenly applied to the masses m_2 and m_3, respectively. $F_1(t) = F_4(t) = F_5(t) = 0$.

Solution

```
      FUNCTION F(T,I)
      IMPLICIT DOUBLE PRECISION(A-H,O-Z)
      GO TO (10,20,30,40,50),I
10    F = 0.0D0
      GO TO 60
20    F = 10.D3
      GO TO 60
30    F = 20.D3
      GO TO 60
40    F = 0.0D0
      GO TO 60
50    F = 0.0D0
60    CONTINUE
      RETURN
```

Free-Vibration Solutions

The program SIM2.FOR can also be used to obtain the solutions of systems vibrating freely as a result of initial conditions. To obtain such solutions, all of the forces $F_i(t)$ are assigned zero values in the function subprogram.

One procedure for checking the execution of the program is to obtain the *undamped* free-vibration solution of a three- or four-degree-of-freedom system, in which the *initial* values of the displacements correspond to the eigenvector components of one of the normal modes. If the program is executing properly, the displacement ratios of the undamped free vibration at *any* instant of time will correspond to the eigenvector components used for the initial conditions (see Prob. 6-11).

6-6 MODAL ANALYSIS (MODAL SUPERPOSITION METHOD)

Section 5-7 discussed the uncoupling of differential equations of motion and the basic concepts of using modal analysis to obtain the free vibration of multiple-degree-of-freedom systems resulting from initial conditions. The modal-analysis approach works equally well in obtaining the forced-vibration response of multiple-degree-of-freedom systems using uncoupled equations of motion expressed in terms of the system's natural circular frequencies and modal-damping factors.

Since the modal-analysis method makes use of uncoupled differential equations of motion that are similar to those of single-degree-of-freedom systems, the computations necessary to determine the forced-vibration response of systems by this method are generally simpler and more straightforward than the computations in methods requiring the simultaneous solution of coupled differential equations of motion.

As stated previously, it is usually difficult to determine physically significant values for the damping coefficients c_{ij} when the effect of damping is included in the coupled differential equations of a system. However, in the modal-analysis method the effect of damping on the response of a system subjected to some type of excitation can be given physical significance by using typical modal damping factors ζ obtained from experimental studies of various types of structural systems and materials as reported in the literature.

Uncoupled Equations—Force-Type Excitation

The general form of the matrix equation for an n-degree-of-freedom system was given by Eq. 6-1. To uncouple the equations of motion represented by this equation, we start with the linear transformation

$$\{x_i\} = [u]\{\delta_i\} \tag{6-26}$$

from which

$$\{\dot{x}_i\} = [u]\{\dot{\delta}_i\} \tag{6-27}$$

and

$$\{\ddot{x}_i\} = [u]\{\ddot{\delta}_i\} \tag{6-28}$$

in which $[u]$ is the *modal matrix.* In this matrix, the columns are the eigenvectors (mode shapes) and the δ_i's are the *principal* coordinates discussed in Sec. 5-7. Substituting Eqs. 6-26, 6-27, and 6-28 into Eq. 6-1 yields

$$\mathbf{M}[u]\{\ddot{\delta}_i\} + \mathbf{C}[u]\{\dot{\delta}_i\} + \mathbf{K}[u]\{\delta_i\} = \{F_i\} \tag{6-29}$$

in which $i = 1, 2, \ldots, n$ for an n-degree-of-freedom system. Premultiplying Eq. 6-29 by the transpose of $[u]$ gives

$$[u]^T\mathbf{M}[u]\{\ddot{\delta}_i\} + [u]^T\mathbf{C}[u]\{\dot{\delta}_i\} + [u]^T\mathbf{K}[u]\{\delta_i\} = [u]^T\{F_i\} \tag{6-30}$$

In Chap. 5, Eq. 5-111 was reduced to the uncoupled set of homogeneous equations given by Eq. 5-115 by considering the concept of proportional damping discussed in Sec. 5-7. In similar manner, since the left side of Eq. 6-30 is identical to Eq. 5-111, we can consider the damping matrix $\mathbf{C}$ to be proportional to either the mass matrix $\mathbf{M}$ or the stiffness matrix $\mathbf{K}$ and reduce Eq. 6-30 to a set of uncoupled equations having the form of

$$\begin{bmatrix} \nwarrow & & \\ & M_r & \\ & & \searrow \end{bmatrix}\begin{Bmatrix} \ddot{\delta}_1 \\ \ddot{\delta}_2 \\ \vdots \\ \ddot{\delta}_n \end{Bmatrix} + \begin{bmatrix} \nwarrow & & \\ & 2\zeta_r\omega_r M_r & \\ & & \searrow \end{bmatrix}\begin{Bmatrix} \dot{\delta}_1 \\ \dot{\delta}_2 \\ \vdots \\ \dot{\delta}_n \end{Bmatrix} + \begin{bmatrix} \nwarrow & & \\ & \omega_r^2 M_r & \\ & & \searrow \end{bmatrix}\begin{Bmatrix} \delta_1 \\ \delta_2 \\ \vdots \\ \delta_n \end{Bmatrix} = [u]^T\begin{Bmatrix} F_1 \\ F_2 \\ \vdots \\ F_n \end{Bmatrix} \tag{6-31}$$

where M_r = generalized mass of rth mode (see Eq. 5-84)

ω_r = undamped natural circular frequency of rth mode
ζ_r = modal damping factor of the rth mode

To determine the expanded form of the right-hand side of Eq. 6-31 for an rth mode equation, let us write the modal matrix $[u]$ for an n-degree-of-freedom system as

$$[u] = \begin{bmatrix} u_{11} & u_{12} & \cdots & u_{1r} & \cdots & u_{1n} \\ u_{21} & u_{22} & \cdots & u_{2r} & \cdots & u_{2n} \\ \cdot & \cdot & \cdot & \cdot & \cdot & \cdot \\ u_{n1} & u_{n2} & \cdots & u_{nr} & \cdots & u_{nn} \end{bmatrix} \tag{6-32}$$

Mode

in which the value of the second subscript denotes the column, or mode, number. For example, the second column is the eigenvector (mode shape) for the second mode. Since the jth column of $[u]$ is the jth row of $[u]^T$, the right-hand side of Eq. 6-31 can be written as

$$[u]^T\{F_i\} = \begin{bmatrix} u_{11} & u_{21} & u_{31} & \cdots & u_{n1} \\ u_{12} & u_{22} & u_{32} & \cdots & u_{n2} \\ \cdot & \cdot & \cdot & \cdot & \cdot \\ u_{1r} & u_{2r} & u_{3r} & \cdots & u_{nr} \\ \cdot & \cdot & \cdot & \cdot & \cdot \\ u_{1n} & u_{2n} & u_{3n} & \cdots & u_{nn} \end{bmatrix} \begin{Bmatrix} F_1 \\ F_2 \\ \cdot \\ F_r \\ \cdot \\ F_n \end{Bmatrix} \tag{6-33}$$

Thus, Eqs. 6-31 and 6-33 reveal that there are n uncoupled equations of the form

$$\ddot{\delta}_r + 2\zeta_r\omega_r\dot{\delta}_r + \omega_r^2\delta_r = \frac{u_{1r}F_1 + u_{2r}F_2 + u_{3r}F_3 + \cdots + u_{nr}F_n}{M_r}$$

or

$$\ddot{\delta}_r + 2\zeta_r\omega_r\dot{\delta}_r + \omega_r^2\delta_r = \frac{\sum_{i=1}^n u_{ir}F_i}{M_r} = E_r(t) \qquad \begin{aligned} r &= 1, 2, \ldots, n \\ i &= 1, 2, \ldots, n \text{ for each } r \end{aligned} \tag{6-34}$$

where
δ_r = principal coordinate of rth mode
ζ_r = modal damping factor of rth mode
ω_r = undamped natural circular frequency of rth mode
u_{ir} = eigenvector components of rth mode
$M_r = \sum_{i=1}^n m_i u_{ir}^2$ = generalized mass of rth mode (see Eq. 5-84)
F_i = excitation forces that are functions of time
$E_r(t) = \dfrac{\sum_{i=1}^n u_{ir}F_i}{M_r}$ = excitation function of rth mode

If each force in Eq. 6-34 can be expressed in terms of the same function of time $g(t)$, so that $F_i = f_i g(t)$, then $E_r(t)$ can be written as

$$E_r(t) = g(t)\,\frac{\sum_{i=1}^n u_{ir}f_i}{M_r}$$

in which

$$\frac{\sum_{i=1}^n u_{ir}f_i}{M_r}$$

is frequently referred to as the *mode participation factor*.

We shall see in the example problems that follow this discussion that the response of a system subjected to a force excitation can usually be obtained with reasonable accuracy by the superposition of only a few of the normal modes. For example, if a 40-degree-of-freedom system is subjected to a periodic excitation having frequency components that are in the neighborhood of the lower mode frequencies of the system, the response contributions of the higher modes can be neglected. With transient or shock-type excitations, the largest contribution to the response of a system comes from the fundamental mode response δ_1, followed by the second mode response δ_2, and so on.

Thus, if the response from the fourth mode were negligible, for example, the total response of an n-degree-of-freedom system as given by Eq. 6-26 would consist essentially of only the first three modes, so that

$$\left.\begin{aligned} x_1 &= u_{11}\delta_1 + u_{12}\delta_2 + u_{13}\delta_3 \\ x_2 &= u_{21}\delta_1 + u_{22}\delta_2 + u_{23}\delta_3 \\ &\cdot\ \cdot\ \cdot\ \cdot\ \cdot\ \cdot\ \cdot \\ x_n &= u_{n1}\delta_1 + u_{n2}\delta_2 + u_{n3}\delta_3 \end{aligned}\right\} \tag{6-35}$$

Equation 6-35 can also be written as

$$\begin{Bmatrix} x_1 \\ x_2 \\ \vdots \\ x_n \end{Bmatrix} = \delta_1 \begin{Bmatrix} u_1 \\ u_2 \\ \vdots \\ u_n \end{Bmatrix}_1 + \delta_2 \begin{Bmatrix} u_1 \\ u_2 \\ \vdots \\ u_n \end{Bmatrix}_2 + \delta_3 \begin{Bmatrix} u_1 \\ u_2 \\ \vdots \\ u_n \end{Bmatrix}_3 \tag{6-36}$$

First mode — Second mode — Third mode

in which the column matrices on the right-hand side of Eq. 6-36 correspond to the eigenvectors of the first three modes as indicated.

General Modal-Analysis Procedure for Obtaining the Response of a System Subjected to Force Excitation

The following steps summarize the general procedure:

1. Determine the mass elements m_i of the mass matrix **M**, and either the stiffness elements k_{ij} of the stiffness matrix **K** or the flexibility elements a_{ij} of the flexibility matrix **A**, of the system. As explained earlier in Sec. 5-3, it is easier in some problems to determine the flexibility coefficients than it is to determine the stiffness coefficients, and vice versa; so use the most convenient.
2. Determine the eigenvalues and eigenvectors from the values obtained for m_i, k_{ij}, or a_{ij} in the first step above. The computer program DEFLATEA.FOR given in Sec. 5-10 can be used with the flexibility coefficients a_{ij} as input data, or the program JACOBI.FOR given in Sec. 5-11 can be used with the stiffness coefficients k_{ij} as input data. It is suggested that DEFLATEA.FOR be used with the a_{ij} values when only the first few modes are necessary to obtain the response of a system that has a fairly large number of degrees of freedom. The program INVERT.FOR given in Sec. 5-3 can be used if it is desirable to obtain the a_{ij}'s from known k_{ij}'s, or vice versa.
3. Use the m_i values determined in step 1 and the eigenvectors obtained in step 2 to calculate the generalized mass terms from

$$M_r = \sum_{i=1}^{n} m_i u_{ir}^2 \qquad \begin{array}{l} r = 1, 2, \ldots, s \\ i = 1, 2, \ldots, n \text{ for each } r \end{array}$$

in which s is the number of modes to be used in determining the response. In most problems it is not necessary to calculate all the generalized mass terms, since only a few of the normal modes are necessary to obtain the response of the system. After some experience with modal analysis, the reader should acquire some intuitive sense of the number of modes required to obtain a sufficiently accurate solution of the response of a particular system.

4. Determine the forcing function terms $E_r(t)$, shown on the right-hand side of Eq. 6-34, from

$$E_r(t) = \frac{\sum_{i=1}^{n} u_{ir} F_i}{M_r} \qquad \begin{array}{l} r = 1, 2, \ldots, s \\ i = 1, 2, \ldots, n \text{ for each } r \end{array}$$

5. Determine the solutions for the normal-mode equations

$$\ddot{\delta}_r + 2\zeta_r \omega_r \dot{\delta}_r + \omega_r^2 \delta_r = E_r(t) \tag{6-34}$$

containing the $E_r(t)$ functions determined in step 4. The procedure for obtaining these solutions will generally depend upon the characteristics of the forces F_i. For example, if each $E_r(t)$ involves a single harmonic excitation frequency ω, the particular solutions of $\delta_1(t), \delta_2(t), \ldots, \delta_n(t)$ will correspond to steady-state responses of the normal modes, which will result in a steady-state vibration response of the system with a circular frequency ω. In the case of transient or shock-type forces, we shall see later that the maximum values of $\delta_1, \delta_2, \ldots, \delta_n$ can be determined from an appropriate *response spectrum* (see Chap. 4) of a single-degree-of-freedom system.

6. Substitute the solutions determined in step 5 into Eq. 6-26, to obtain the response of the forced system,

$$\{x_i\} = [u]\{\delta_i\} \tag{6-26}$$

To facilitate the computations required, and to observe the effect of the individual modes on the total response of the system, Eq. 6-26 can be written in the form of Eq. 6-36. Thus, with the participation of s modes in obtaining the total response of the system,

$$\begin{Bmatrix} x_1 \\ x_2 \\ \vdots \\ x_n \end{Bmatrix} = \delta_1 \begin{Bmatrix} u_1 \\ u_2 \\ \vdots \\ u_n \end{Bmatrix}_1 + \delta_2 \begin{Bmatrix} u_1 \\ u_2 \\ \vdots \\ u_n \end{Bmatrix}_2 + \cdots + \delta_s \begin{Bmatrix} u_1 \\ u_2 \\ \vdots \\ u_n \end{Bmatrix}_s \tag{6-37}$$

EXAMPLE 6-6

Figure 6-12a repeats the simply supported beam of Example 5-25, which is 40 ft long and subjected to the force $F_2(t)$ at its center as shown. It was determined in Example 5-25 that a three-lumped-mass model such as the one shown in Fig. 6-12b represented the distributed mass of the beam quite well, with the lumped masses obtained by using the procedure explained in Sec. 5-12.

Using the three-lumped-mass model shown in the figure, and the steps beginning on page 436, determine (a) the time response of each of the three lumped masses y_1, y_2, and y_3 if $F_2(t)$ is a suddenly applied force of 30,000 lb as shown in Fig. 6-12c; (b) the approximate

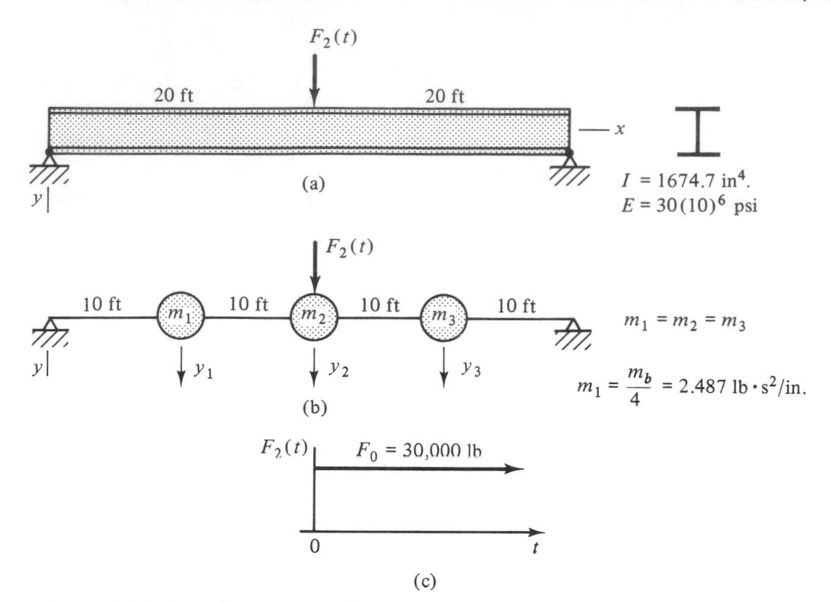

Figure 6-12 Simply supported beam subjected to force $F_2(t)$.

maximum response of the mass m_2 (the maximum value of y_2) and the time t_m at which this maximum occurs; (c) how close the value of y_2 obtained in part b is to being twice the static deflection of the beam for a 30,000-lb load (theoretically the deflection due to a suddenly applied load is twice that caused by the same magnitude of load applied statically).

Solution. **a.** Steps 1 and 2 of the general procedure beginning on page 436 were carried out in Example 5-25; so utilizing these results we note the following:

$$m_1 = m_2 = m_3 = 2.487 \text{ lb} \cdot \text{s}^2/\text{in.}$$

$$\omega_1 = 66.67 \text{ rad/s} \qquad \omega_2 = 264.67 \text{ rad/s} \qquad \omega_3 = 562.71 \text{ rad/s}$$

$$[u] = \left[\begin{Bmatrix} 0.707 \\ 1.000 \\ 0.707 \end{Bmatrix}_1 \begin{Bmatrix} 1.000 \\ 0.000 \\ -1.000 \end{Bmatrix}_2 \begin{Bmatrix} 0.707 \\ -1.000 \\ 0.707 \end{Bmatrix}_3 \right] \tag{6-38}$$

or

$$[u] = \left[\begin{Bmatrix} 1.000 \\ 1.414 \\ 1.000 \end{Bmatrix}_1 \begin{Bmatrix} 1.000 \\ 0.000 \\ -1.000 \end{Bmatrix}_2 \begin{Bmatrix} 1.000 \\ -1.414 \\ 1.000 \end{Bmatrix}_3 \right] \tag{6-39}$$

The modal matrix of Eq. 6-38, normalized with respect to the second eigenvector component, will yield the same response for the system as the modal matrix of Eq. 6-39 normalized with respect to the first eigenvector component (see Prob. 6-16). We now proceed with the third step.

Step 3. Using the modal matrix of Eq. 6-39, we obtain the generalized mass terms as

$$M_1 = \sum_{i=1}^{3} m_i u_{i1}^2 = 2.487[(1)^2 + (1.414)^2 + (1)^2] = 9.947$$

$$M_2 = \sum m_i u_{i2}^2 = 2.487[(1)^2 + (1)^2] = 4.974$$
$$M_3 = \sum m_i u_{i3}^2 = 2.487[(1)^2 + (-1.414)^2 + (1)^2] = 9.947$$

Step 4. Assuming that the displacements are positive downward, the forcing functions $E_r(t)$ are found to be

$$E_1(t) = \frac{\sum_{i=1}^{3} u_{i1}F_i}{M_1} = \frac{1.414F_2(t)}{9.947} = 0.142F_0$$

$$E_2(t) = \frac{\sum_{i=1}^{3} u_{i2}F_i}{M_2} = 0$$

$$E_3(t) = \frac{\sum_{i=1}^{3} u_{i3}F_i}{M_3} = \frac{-1.414F_2(t)}{9.947} = -0.142F_0$$

Step 5. Neglecting damping, the uncoupled equations are as follows:

$$\left.\begin{aligned} \ddot{\delta}_1 + \omega_1^2\delta_1 &= 0.142F_0 \\ \ddot{\delta}_2 + \omega_2^2\delta_2 &= 0 \\ \ddot{\delta}_3 + \omega_3^2\delta_3 &= -0.142F_0 \end{aligned}\right\} \tag{6-40}$$

The solutions of these equations are found to be

$$\left.\begin{aligned} \delta_1 &= A_1 \cos \omega_1 t + B_1 \sin \omega_1 t + \frac{0.142F_0}{\omega_1^2} \\ \delta_2 &= A_2 \cos \omega_2 t + B_2 \sin \omega_2 t \\ \delta_3 &= A_3 \cos \omega_3 t + B_3 \sin \omega_3 t - \frac{0.142F_0}{\omega_3^2} \end{aligned}\right\} \tag{6-41}$$

in which the A_i's and B_i's depend upon the *initial conditions* for each of the three lumped masses. Assuming the initial displacements and velocities of the three masses as all being zero, we can write from Eq. 6-37 that

$$\begin{Bmatrix} y_1 \\ y_2 \\ y_3 \end{Bmatrix}_{t=0} = \begin{Bmatrix} 0 \\ 0 \\ 0 \end{Bmatrix} = \delta_1\{u_i\}_1 + \delta_2\{u_i\}_2 + \delta_3\{u_i\}_3 \tag{6-42}$$

and

$$\begin{Bmatrix} \dot{y}_1 \\ \dot{y}_2 \\ \dot{y}_3 \end{Bmatrix}_{t=0} = \begin{Bmatrix} 0 \\ 0 \\ 0 \end{Bmatrix} = \dot{\delta}_1\{u_i\} + \dot{\delta}_2\{u_i\} + \dot{\delta}_3\{u_i\} \tag{6-43}$$

It should be apparent from Eqs. 6-42 and 6-43, that when $t = 0$,

$$\delta_1 = \delta_2 = \delta_3 = \dot{\delta}_1 = \dot{\delta}_2 = \dot{\delta}_3 = 0$$

Thus, for $t = 0$, we obtain from Eq. 6-41 that

$$A_1 = \frac{-0.142F_0}{\omega_1^2} \qquad B_1 = 0$$
$$A_2 = 0 \qquad B_2 = 0$$
$$A_3 = \frac{0.142F_0}{\omega_3^2} \qquad B_3 = 0$$

Substituting the expressions for the constants back into Eq. 6-41 gives

$$\left.\begin{aligned} \delta_1 &= \frac{0.142F_0}{\omega_1^2}(1 - \cos \omega_1 t) \\ \delta_2 &= 0 \\ \delta_3 &= \frac{0.142F_0}{\omega_3^2}(\cos \omega_3 t - 1) \end{aligned}\right\} \qquad (6\text{-}44)$$

Step 6. Substituting the values $F_0 = 30{,}000$ lb, $\omega_1 = 66.67$ rad/s, and $\omega_3 = 562.71$ rad/s into Eq. 6-44, and referring to the modal matrix of Eq. 6-39, we find that Eq. 6-37 gives the total response of the three-lumped-mass model of the beam as

$$\begin{Bmatrix} y_1 \\ y_2 \\ y_3 \end{Bmatrix} = 0.958(1 - \cos 66.67t)\begin{Bmatrix} 1.000 \\ 1.414 \\ 1.000 \end{Bmatrix} + 0.014(\cos 562.71t - 1)\begin{Bmatrix} 1.000 \\ -1.414 \\ 1.000 \end{Bmatrix} \qquad (6\text{-}45)$$

It should be noted at this point that the response of the beam depends almost entirely upon the contribution of the first mode, with the third mode adding little and the second mode adding nothing since $F_2(t)$ acts at the nodal point of the second mode of the beam (see Fig. 5-40).[9]

b. A good approximation of the *maximum* response of m_2 can be obtained by considering only the contribution of the first mode in Eq. 6-45. Thus, taking $66.67t_m = \pi$ so that $(1 - \cos 66.67t_m) = 2$, Eq. 6-45 yields

$$y_2)_{\max} \cong 0.958(2)(1.414) = 2.71 \text{ in.}$$

The time at which this occurs is found as

$$t_m = \frac{\pi}{66.67} = 0.047 \text{ s}$$

c. The static deflection Δ_s at the center of the beam is determined as

$$\Delta_s = \frac{F_0 l^3}{48\,EI} = \frac{30{,}000(480)^3}{48(30)(10)^6 1674.7} = 1.38 \text{ in.}$$

Since the theoretical maximum value of y_2 due to a suddenly applied force is twice the static deflection,

$$y_2)_{\max} = 2\Delta_s = 2(1.38) = 2.76 \text{ in.}$$

and we see that the value of $y_2 = 2.71$ obtained by using only the contribution of the first mode in part b is within 2 percent of the theoretical value. The reader should pause here to reflect that this result obtained using the modal-analysis method is remarkably accurate when it is recalled that the response of the beam was determined by modeling the distributed mass of the beam by only three lumped masses.

EXAMPLE 6-7

In this example we show how the response spectrum of a single-degree-of-freedom system can be used to obtain the absolute-maximum response of a multiple-degree-of-freedom system using the modal-analysis method.

[9] A node is a point of zero displacement of a vibrating element.

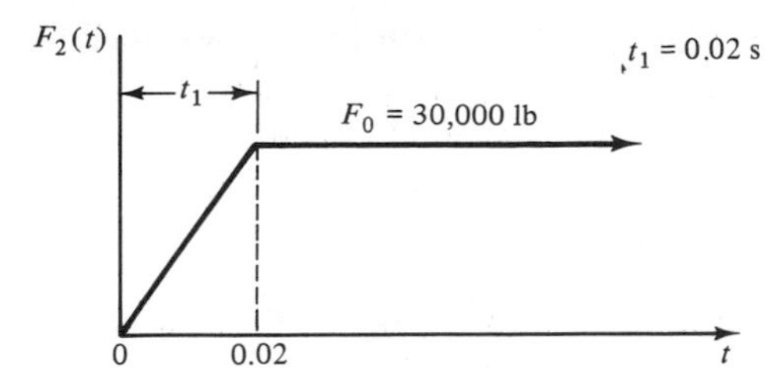

Figure 6-13 Step-function force $F_2(t)$ for Example 6-7.

The simply supported beam of Example 6-6 (see Fig. 6-12) is subjected to the step-function force $F_2(t)$ shown in Fig. 6-13. The rise time t_1 of this step function is 0.02 s. For ease of reference, let us recall the following quantities from Example 6-6 that are pertinent to the solution of this example:

$$m_1 = m_2 = m_3 = 2.487 \text{ lb} \cdot \text{s}^2/\text{in.}$$
$$\omega_1 = 66.67 \text{ rad/s} \qquad \omega_2 = 264.67 \text{ rad/s} \qquad \omega_3 = 562.71 \text{ rad/s}$$

$$[u] = \begin{bmatrix} 1.000 & 1.000 & 1.000 \\ 1.414 & 0.000 & -1.414 \\ 1.000 & -1.000 & 1.000 \end{bmatrix} \tag{6-39}$$

$$M_1 = 9.947 \qquad M_2 = 4.974 \qquad M_3 = 9.947$$

Use the response spectrum shown before in Fig. 4-15 to determine the absolute-maximum response of the beam modeled by the three lumped masses shown in Fig. 6-12b. Assume that the modal damping for each of the three modes is 5 percent. That is, $\zeta_1 = \zeta_2 = \zeta_3 = 0.05$.

Solution. Since each normal mode of Eq. 6-34 behaves like a single-degree-of-freedom system, the maximum response of each mode can be obtained from the response spectrum of Fig. 4-15. Then, from Eq. 6-37 the absolute-maximum response of the system becomes

$$\begin{Bmatrix} y_1 \\ y_2 \\ y_3 \end{Bmatrix}_{\text{abs max}} = (\delta_1)_m\{u_i\}_1 + (\delta_2)_m\{u_i\}_2 + (\delta_3)_m\{u_i\}_3 \tag{6-46}$$

in which $(\delta_1)_m$, $(\delta_2)_m$, and $(\delta_3)_m$ refer to the maximum values of the principal coordinates determined for modes 1, 2, and 3, respectively. While these maximum values occur at different values of time t owing to phase shifts in the time domain, the absolute-maximum response values of y_1, y_2, and y_3 determined from Eq. 6-46 are the upper bounds of the actual response values.

In using the response spectrum to obtain the maximum values of the principal coordinates $(\delta_r)_m$ for the various modes it is also necessary to determine the static response values $(\delta_r)_s$ of the principal coordinates from Eq. 6-34. To accomplish this, we assume that the forces F_i go from zero to their maximum values very slowly so that we can let $\ddot{\delta}_r = \dot{\delta}_r = 0$ in Eq. 6-34, and thus write that

$$(\delta_r)_s = \frac{(E_r)_s}{\omega_r^2} \tag{6-47}$$

where $(\delta_r)_s$ = static response of rth mode corresponding to very slowly applied forces

$(E_r)_s$ = static excitation function of rth mode corresponding to maximum values of forces F_i

ω_r = natural circular frequency of rth mode

Since the maximum value of the step function $F_2(t)$ is $F_0 = 30{,}000$ lb, Eq. 6-47 gives the *static-mode* response values as

$$(\delta_1)_s = \frac{(E_1)_s}{\omega_1^2} = \frac{u_{21}F_0}{\omega_1^2 M_1} = \frac{1.414(30{,}000)}{(66.67)^2(9.947)} = 0.959$$

$$(\delta_2)_s = \frac{(E_2)_s}{\omega_2^2} = \frac{u_{22}F_0}{\omega_2^2 M_2} = 0$$

$$(\delta_3)_s = \frac{(E_3)_s}{\omega_3^2} = \frac{u_{23}F_0}{\omega_3^2 M_3} = \frac{-1.414(30{,}000)}{(562.71)^2(9.947)} = -0.014$$

Considering that the natural period of a normal mode is $\tau_r = 2\pi/\omega_r$, and that the rise time of the step function is $t_1 = 0.02$ *s*, we obtain

$$\frac{t_1}{\tau_1} = \frac{0.02(66.67)}{2\pi} = 0.21$$

$$\frac{t_1}{\tau_2} = \frac{0.02(264.67)}{2\pi} = 0.84$$

$$\frac{t_1}{\tau_3} = \frac{0.02(562.71)}{2\pi} = 1.79$$

Thus, with $\zeta_1 = \zeta_2 = \zeta_3 = 0.05$, and using the values of t_1/τ_r just obtained, we can determine the following values of $(\delta_r)_m/(\delta_r)_s$ from an inspection of Fig. 4-15:

Mode r	t_1/τ_r	$x_{max}/(F_0/k) = (\delta_r)_m/(\delta_r)_s$
1	0.21	1.79
2	0.84	1.12
3	1.79	1.07

Then using these values of $(\delta_r)_m/(\delta_r)_s$, and recalling the values of $(\delta_r)_s$ previously calculated, we find that

$$(\delta_1)_m = 1.79(0.959) = 1.72$$
$$(\delta_2)_m = 1.12(0) = 0$$
$$(\delta_3)_m = 1.07(-0.014) = -0.015$$

Substituting these values into Eq. 6-46, we obtain the absolute-maximum response of the system as

$$\begin{Bmatrix} y_1 \\ y_2 \\ y_3 \end{Bmatrix}_{\text{abs max}} = 1.72\begin{Bmatrix} 1.000 \\ 1.414 \\ 1.000 \end{Bmatrix}_1 + (-0.015)\begin{Bmatrix} 1.000 \\ -1.414 \\ 1.000 \end{Bmatrix}_3$$

which yields

$$\begin{Bmatrix} y_1 \\ y_2 \\ y_3 \end{Bmatrix}_{\text{abs max}} = \begin{Bmatrix} 1.74\text{ in.} \\ 2.45\text{ in.} \\ 1.74\text{ in.} \end{Bmatrix}$$

when the absolute values of the third mode are used.

As in Example 6-6, the magnitude of the total response comes almost entirely from the contribution of the first mode, since the contribution of the second mode is zero and the contribution of the third mode is very small in comparison with that of the first mode.

EXAMPLE 6-8

The disk at each end of the damped disk-and-shaft system is subjected to a sinusoidally varying torque of $T_0 \sin \omega t$ as shown in Fig. 6-14. The value of the mass moment of inertia of each disk, and the torsional spring constant of each shaft segment, are also given in the figure.

We wish to determine the steady-state response of the system as a function of time t, modal damping factors ζ_r, and natural circular frequencies ω_r, using the steps beginning on page 436.

Solution. *Step 1.* Since the mass moments of inertia of the disks of this torsional-vibration system are the elements of the mass matrix,

$$\mathbf{M} = \begin{bmatrix} \bar{I}_1 & 0 & 0 & 0 \\ 0 & \bar{I}_2 & 0 & 0 \\ 0 & 0 & \bar{I}_3 & 0 \\ 0 & 0 & 0 & \bar{I}_4 \end{bmatrix} = \begin{bmatrix} 20 & 0 & 0 & 0 \\ 0 & 10 & 0 & 0 \\ 0 & 0 & 10 & 0 \\ 0 & 0 & 0 & 20 \end{bmatrix}$$

Because this is a semidefinite system (free-free), in which the flexibility coefficients a_{ij} are infinite, we must obviously determine the stiffness coefficients k_{ij}. Recalling the definition of stiffness coefficients while inspecting Fig. 6-14, we obtain

$$\mathbf{K} = \begin{bmatrix} k & -k & 0 & 0 \\ -k & 2k & -k & 0 \\ 0 & -k & 2k & -k \\ 0 & 0 & -k & k \end{bmatrix} = (10)^5 \begin{bmatrix} 1 & -1 & 0 & 0 \\ -1 & 2 & -1 & 0 \\ 0 & -1 & 2 & -1 \\ 0 & 0 & -1 & 1 \end{bmatrix}$$

Step 2. The eigenvalues and eigenvectors obtained from the computer program JACOBI.FOR using the values of the mass elements and stiffness coefficients determined in step 1 as input data are shown rounded off as

$$\omega_1 = 0.0 \text{ (rigid-body motion)}$$
$$\omega_2 = 56.02 \text{ rad/s}$$

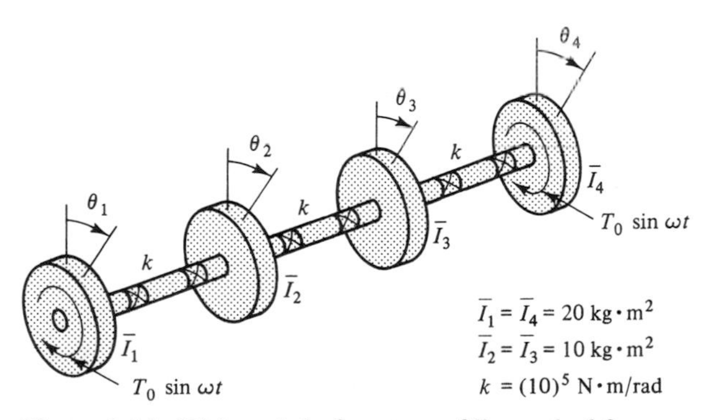

Figure 6-14 Disk-and-shaft system of Example 6-8.

$$\omega_3 = 122.47 \text{ rad/s}$$
$$\omega_4 = 178.50 \text{ rad/s}$$

$$[u] = \begin{bmatrix} 1.000 & 1.000 & 1.000 & 1.000 \\ 1.000 & 0.372 & -2.000 & -5.372 \\ 1.000 & -0.372 & -2.000 & 5.372 \\ 1.000 & -1.000 & 1.000 & -1.000 \end{bmatrix}$$

Step 3. Recalling that the mass moments of inertia $\bar{I}_i$ correspond to the mass elements m_i, the generalized mass terms are determined from

$$M_r = \sum_{i=1}^{4} \bar{I}_i u_{ir}^2$$

and are found to have values of

$$\begin{aligned} M_1 &= 10[2(1)^2 + (1)^2 + (1)^2 + 2(1)^2] &&= 60.0 \text{ kg}\cdot\text{m}^2 \\ M_2 &= 10[2(1)^2 + (0.372)^2 + (-0.372)^2 + 2(-1)^2] &&= 42.77 \text{ kg}\cdot\text{m}^2 \\ M_3 &= 10[2(1)^2 + (-2)^2 + (-2)^2 + 2(1)^2] &&= 120.0 \text{ kg}\cdot\text{m}^2 \\ M_4 &= 10[2(1)^2 + (-5.372)^2 + (5.372)^2 + 2(-1)^2] &&= 617.2 \text{ kg}\cdot\text{m}^2 \end{aligned}$$

Step 4. The forcing functions $E_r(t)$ are determined from

$$E_r(t) = \frac{\sum_{i=1}^{4} u_{ir} T_i}{M_r}$$

in which the T_i's are the excitation torques acting on the disks. With $T_1 = T_4 = T_0 \sin \omega t$, and $T_2 = T_3 = 0$, and replacing $\sin \omega t$ by $e^{j\omega t}$ for a particular solution, we obtain the forcing functions

$$E_1(t) = \frac{[(1)T_0 + (1)T_0]}{60.0} e^{j\omega t} = \frac{T_0}{30} e^{j\omega t} \quad \text{rad/s}^2$$

$$E_2(t) = \frac{[(1)T_0 + (-1)T_0]}{42.77} e^{j\omega t} = 0$$

$$E_3(t) = \frac{[(1)T_0 + (1)T_0]}{120.0} e^{j\omega t} = \frac{T_0}{60} e^{j\omega t} \quad \text{rad/s}^2$$

$$E_4(t) = \frac{[(1)T_0 + (-1)T_0]}{617.2} e^{j\omega t} = 0$$

Step 5. Using the excitation functions determined in step 4, and noting that $\omega_1 = 0$ for the rigid-body motion, the uncoupled modal equations are found to be

$$\ddot{\delta}_1 = \frac{T_0}{30} e^{j\omega t} \qquad \text{(rigid-body motion)}$$
$$\ddot{\delta}_2 + 2\zeta_2\omega_2\dot{\delta}_2 + \omega_2^2\delta_2 = 0$$
$$\ddot{\delta}_3 + 2\zeta_3\omega_3\dot{\delta}_3 + \omega_3^2\delta_3 = \frac{T_0}{60} e^{j\omega t}$$
$$\ddot{\delta}_4 + 2\zeta_4\omega_4\dot{\delta}_4 + \omega_4^2\delta_4 = 0$$

Assuming that the particular solutions of the above equations have the form of

$$\delta_r = A_r e^{j\omega t}$$

we obtain the particular solutions as

$$\delta_1 = \frac{-T_0}{30\omega^2} \sin \omega t$$
$$\delta_2 = \delta_4 = 0$$
$$\delta_3 = |A_3| \sin(\omega t - \phi_3)$$

where
$$|A_3| = \frac{T_0/(60\ \omega_3^2)}{\sqrt{[1 - (\omega/\omega_3)^2]^2 + [2\zeta_3(\omega/\omega_3)]^2}}$$

$$\tan \phi_3 = \frac{2\zeta_3(\omega/\omega_3)}{1 - (\omega/\omega_3)^2}$$

$\omega_3 = 122.47$ rad/s
ω = excitation frequency, rad/s
ζ_3 = modal damping factor for third mode

Step 6. Substituting the solutions determined in step 5 into Eq. 6-37, we obtain the steady-state response of the system as

$$\begin{Bmatrix} \theta_1 \\ \theta_2 \\ \theta_3 \\ \theta_4 \end{Bmatrix} = \frac{-T_0}{30\omega^2} \sin \omega t \begin{Bmatrix} 1.0 \\ 1.0 \\ 1.0 \\ 1.0 \end{Bmatrix}_{1\ \text{(rigid-body motion)}} + |A_3| \sin(\omega t - \phi_3) \begin{Bmatrix} 1.0 \\ -2.0 \\ -2.0 \\ 1.0 \end{Bmatrix}_3$$

This equation reveals that the oscillation of each disk consists of a rigid-body motion of amplitude $T_0/(30\omega^2)$ and circular frequency ω that is common to all the disks, plus an out-of-phase contribution of the same frequency from the third mode. Thus, the steady-state response of each disk consists of the superposition of two sinusoidal motions of frequency ω which are out of phase with each other by the phase angle ϕ_3.

Uncoupled Equations—Support Excitation

Let us begin our discussion by recalling from Sec. 6-2 that the matrix equation giving the coupled differential equations of motion in terms of the motion of the mass elements relative to the moving support is

$$\mathbf{M\ddot{Z}} + \mathbf{C\dot{Z}} + \mathbf{KZ} = -\mathbf{M\ddot{Y}} = -\begin{Bmatrix} m_1\ddot{y} \\ m_2\ddot{y} \\ \vdots \\ m_n\ddot{y} \end{Bmatrix} \tag{6-7}$$

In this equation the column matrices $\mathbf{Z}$, $\mathbf{\dot{Z}}$, and $\mathbf{\ddot{Z}}$ are the displacements, velocities, and accelerations, respectively, of the mass elements *relative* to the moving support. It should be noted that every element of the column matrix $\mathbf{\ddot{Y}}$ is the acceleration $\ddot{y}$ of the moving support.

As we shall soon see, the left side of the uncoupled normal-mode equations obtained from Eq. 6-7 will be identical to the left side of Eq. 6-34. Using the same procedure discussed at the beginning of this section to uncouple the equations of motion containing force-type excitation terms (Eqs. 6-26, 6-27, and 6-28), we use the linear transformations

$$\{z_i\} = [u]\{\delta_i\} \tag{6-48}$$
$$\{\dot{z}_i\} = [u]\{\dot{\delta}_i\} \tag{6-49}$$
$$\{\ddot{z}_i\} = [u]\{\ddot{\delta}_i\} \tag{6-50}$$

Substituting these transformations into Eq. 6-7 and premultiplying the result by $[u]^T$ gives

$$[u]^T\mathbf{M}[u]\{\ddot{\delta}_i\} + [u]^T\mathbf{C}[u]\{\dot{\delta}_i\} + [u]^T\mathbf{K}[u]\{\delta_i\} = -[u]^T\begin{Bmatrix} m_1\ddot{y} \\ m_2\ddot{y} \\ \vdots \\ m_n\ddot{y} \end{Bmatrix} \tag{6-51}$$

the left side of which is identical to the left side of Eq. 6-30.

Then, reviewing the discussion accompanying Eqs. 6-30 through 6-34, it should be apparent that the n uncoupled equations of Eq. 6-51 have the form of

$$\ddot{\delta}_r + 2\zeta_r\omega_r\dot{\delta}_r + \omega_r^2\delta_r = \frac{-[u_{1r}m_1 + u_{2r}m_2 + \cdots + u_{nr}m_n]}{M_r}\ddot{y}$$

or

$$\ddot{\delta}_r + 2\zeta_r\omega_r\dot{\delta}_r + \omega_r^2\delta_r = \frac{-\sum_{i=1}^{n} u_{ir}m_i}{M_r}\ddot{y} = -S_r\ddot{y} \qquad \begin{matrix} r = 1, 2, \ldots, n \\ i = 1, 2, \ldots, n \text{ for each } r \end{matrix} \tag{6-52}$$

where
δ_r = principal coordinate of rth mode
ζ_r = modal damping factor of rth mode
ω_r = undamped natural circular frequency of rth mode
u_{ir} = eigenvector components of rth mode
$M_r = \sum_{i=1}^{n} m_i u_{ir}^2$ = generalized mass of rth mode (see Eq. 5-84)
m_i = mass elements of diagonal mass matrix $\mathbf{M}$
$S_r = \dfrac{\sum_{i=1}^{n} u_{ir}m_i}{M_r}$ = mode-participation factor of rth mode
$\ddot{y}$ = acceleration of moving support

The six steps beginning on page 436, which were given previously for obtaining the response of a system subjected to force excitation, are also applicable for obtaining the response of a system subjected to support excitation $\ddot{y}$. However, steps 4, 5, and 6 require some added discussion to make sure that the appropriate equations are used in these steps for the latter case.

Step 4. The mode-participation factor of the rth mode is

$$S_r = \frac{\sum_{i=1}^{n} u_{ir}m_i}{M_r} \qquad \begin{matrix} r = 1, 2, \ldots, s \\ i = 1, 2, \ldots, n \text{ for each } r \end{matrix}$$

in which s indicates the number of modes selected for determining the total response of an n-degree-of-freedom system.

Step 5. The normal-mode equations are given by Eq. 6-52 as

$$\ddot{\delta}_r + 2\zeta_r\omega_r\dot{\delta}_r + \omega_r^2\delta_r = -S_r\ddot{y}$$

and the solutions obtained from these equations are used in step 6 to obtain the total response of the system caused by the support excitations.

Step 6. The system response z_i of the mass elements relative to the moving support is obtained from Eq. 6-48 repeated here as

$$\{z_i\} = [u]\{\delta_i\} \tag{6-48}$$

in which the δ_i's are the solutions of the normal-mode equations determined in step 5. To facilitate the computations, and to observe the effect of the individual modes, Eq. 6-48 can be written as

$$\begin{Bmatrix} z_1 \\ z_2 \\ \vdots \\ z_n \end{Bmatrix} = \delta_1 \begin{Bmatrix} u_1 \\ u_2 \\ \vdots \\ u_n \end{Bmatrix}_1 + \delta_2 \begin{Bmatrix} u_1 \\ u_2 \\ \vdots \\ u_n \end{Bmatrix}_2 + \cdots + \delta_s \begin{Bmatrix} u_1 \\ u_2 \\ \vdots \\ u_n \end{Bmatrix}_s \tag{6-53}$$

which yields the total response with the participation of s modes.

EXAMPLE 6-9

The small airplane shown in Fig. 6-15a has a loaded weight of 3200 lb and a normal landing speed of approximately 75 mph. The runway has a contour that can be modeled fairly accurately by a sine wave with a wavelength of 80 ft and an amplitude of Y_0, as shown in the figure.

Vibration tests of the loaded airplane indicate that it can be modeled reasonably well by four lumped masses distributed as shown in Fig. 6-15b. The lumped masses and the spring constants of the four-lumped-mass model have the following values:

$$m_1 = m_3 = \frac{500}{386} = 1.30 \text{ lb}\cdot\text{s}^2/\text{in.}$$

$$m_2 = \frac{2000}{386} = 5.18 \text{ lb}\cdot\text{s}^2/\text{in.}$$

$$m_4 = \frac{200}{386} = 0.52 \text{ lb}\cdot\text{s}^2/\text{in.}$$

$k_1 = k_2 = 3400$ lb/in. (wings)

$k_3 = 850$ lb/in. (landing-gear system)

$k_4 = 2000$ lb/in. (tires)

Assuming that the plane lands at a speed of 75 mph and that it remains in contact with the runway after touchdown for a sufficient length of time to reach a steady-state condition, it is desired to determine its steady-state response *relative to the runway* (z_i). Although a damping mechanism is not shown in Fig. 6-15, assume that the damping factor for all four modes is 0.05.

Solution. Figure 6-15b shows that the runway contour given by

$$y = Y_0 \sin \frac{2\pi vt}{l}$$

provides the support excitation of the four-lumped-mass model of the plane. With a landing speed of 75 mph, which is 110.0 ft/s, the circular excitation frequency ω for a wavelength of $l = 80$ ft is

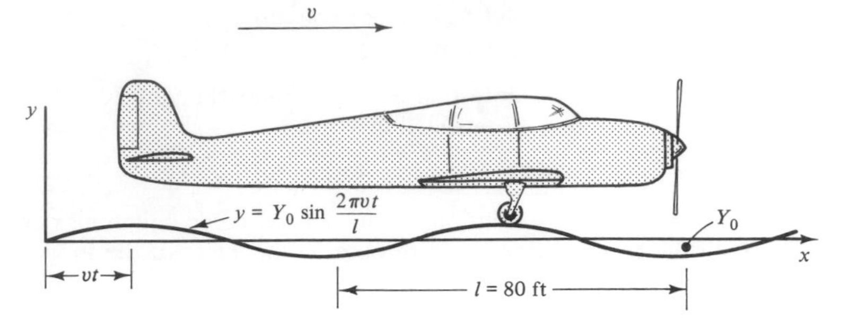

(a)

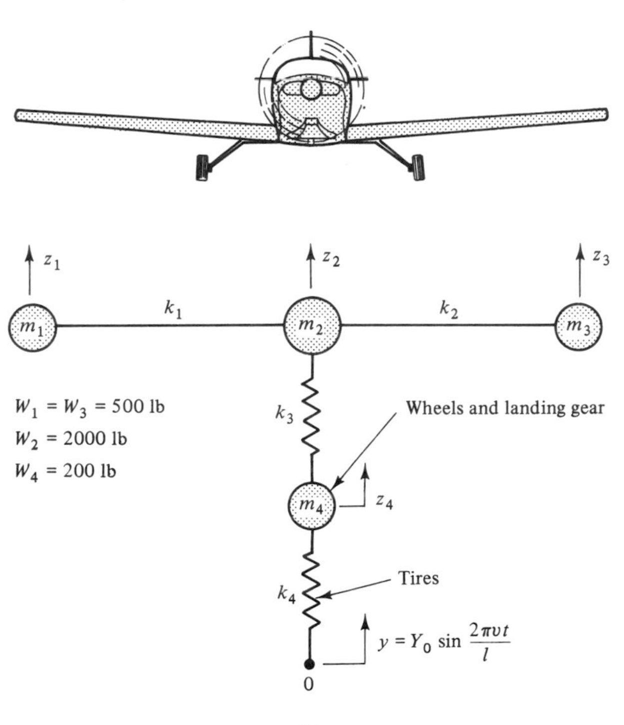

(b)

Figure 6-15 Airplane of Example 6-9 modeled as four lumped masses.

$$\omega = \frac{2\pi v}{l} = \frac{2\pi(110.0)}{80} = 8.64 \text{ rad/s}$$

The acceleration of the support excitation can then be obtained as

$$\ddot{y} = -Y_0\omega^2 \sin \omega t = -(8.64)^2 Y_0 \sin(8.64\ t)$$

Step 1. The stiffness coefficients k_{ij} are determined as though the moving support were stationary. To do this, we consider the point 0 shown in Fig. 6-15b as fixed with $z_1 = 1$ and $z_2 = z_3 = z_4 = 0$, to obtain the first column of the stiffness matrix as

$$\left.\begin{aligned} k_{11} &= k_1 = 3400 \\ k_{21} &= -k_1 = -3400 \\ k_{31} &= 0 \\ k_{41} &= 0 \end{aligned}\right\} \text{(first column of } K\text{)}$$

Proceeding in a similar manner with $z_2 = 1$, $z_1 = z_3 = z_4 = 0$ for the second column of **K**, and so on, the stiffness matrix is obtained as

$$\mathbf{K} = \begin{bmatrix} 3400 & -3400 & 0 & 0 \\ -3400 & 7650 & -3400 & -850 \\ 0 & -3400 & 3400 & 0 \\ 0 & -850 & 0 & 2850 \end{bmatrix}$$

It is suggested that the reader verify the last three columns of this matrix for practice, considering the definition of stiffness coefficients in conjunction with Fig. 6-15b.

Step 2. Using the values of the masses given previously in the statement of the problem and the elements of the stiffness matrix **K** as input data to the program JACOBI.FOR on page 369, the following natural circular frequencies and normal-mode shapes are obtained (rounded to two places after the decimal point) from the printout:

$$\begin{aligned} \omega_1 &= 8.69 \text{ rad/s} \\ \omega_2 &= 51.14 \text{ rad/s} \\ \omega_3 &= 62.65 \text{ rad/s} \\ \omega_4 &= 74.65 \text{ rad/s} \end{aligned}$$

$$[u] = \begin{bmatrix} 1.00 & 1.00 & 1.00 & 1.00 \\ 0.97 & 0.00 & -0.50 & -1.13 \\ 1.00 & -1.00 & 1.00 & 1.00 \\ 0.29 & 0.00 & -0.53 & 20.22 \end{bmatrix}$$

Step 3. The natural circular frequencies determined in step 2 reveal that the lowest one, $\omega_1 = 8.69$ rad/s, is the only one near the support excitation circular frequency of $\omega = 8.64$ rad/s. Therefore, the response contributions of the second, third, and fourth modes will be negligible compared with that of the first mode. Thus, in considering only the first mode, we determine only the generalized mass M_1 for that mode as

$$M_1 = \sum_{i=1}^{4} m_i u_{i1}^2 = [1.30(1)^2 + 5.18(0.97)^2 + 1.30(1)^2 + 0.52(0.29)^2]$$

$$= 7.52$$

Step 4. Recalling that $\ddot{y} = -Y_0\omega^2 \sin \omega t$, the support excitation function for the first mode is

$$-S_1\ddot{y} = \frac{-\sum_{i=1}^{4} u_{i1} m_i}{M_1} \ddot{y}$$

or

$$-S_1\ddot{y} = \frac{[(1)(1.30) + 0.97(5.18) + (1)(1.30) + 0.29(0.52)]}{7.52} Y_0\omega^2 \sin \omega t$$

which reduces to

$$-S_1\ddot{y} = 1.03Y_0\omega^2 \sin \omega t$$

Step 5. The normal-mode equation for the first mode is

$$\ddot{\delta}_1 + 2\zeta_1\omega_1\dot{\delta}_1 + \omega_1^2\delta_1 = 1.03Y_0\omega^2 \sin \omega t \tag{6-54}$$

Replacing sin ωt by the imaginary part of $e^{j\omega t}$, and assuming a particular solution of Eq. 6-54 to have the form of

$$\delta_1 = A_1 e^{j\omega t}$$

we obtain the steady-state solution as

$$\delta_1 = |A_1| \sin(\omega t - \phi_1) \tag{6-55}$$

$$|A_1| = \frac{1.03Y_0(\omega/\omega_1)^2}{\sqrt{[1 - (\omega/\omega_1)^2]^2 + [2\zeta(\omega/\omega_1)]^2}} \tag{6-56}$$

where

$$\tan\phi_1 = \frac{2\zeta(\omega/\omega_1)}{1 - (\omega/\omega_1)^2} \tag{6-57}$$

$$\omega = 8.64 \text{ rad/s}$$

$$\omega_1 = 8.69 \text{ rad/s}$$

$$\zeta_1 = 0.05$$

$$Y_0 = \text{amplitude of sine wave}$$

Substituting the appropriate values into Eqs. 6-56 and 6-57 yields

$$|A_1| = 10.18\ Y_0$$

and

$$\phi_1 = 1.45 \text{ rad} = 83.1°$$

Equation 6-55 then becomes

$$\delta_1 = 10.18\ Y_0 \sin(8.64t - \phi_1) \tag{6-58}$$

Step 6. Substituting Eq. 6-58 into Eq. 6-53, we obtain the steady-state response of the four lumped masses (the airplane) relative to the point 0 (the runway) as

$$\begin{Bmatrix} z_1 \\ z_2 \\ z_3 \\ z_4 \end{Bmatrix} = 10.18\ Y_0 \sin(8.64t - 1.45) \begin{Bmatrix} 1.00 \\ 0.97 \\ 1.00 \\ 0.29 \end{Bmatrix} \tag{6-59}$$

6-7 DETERMINING STRUCTURAL RESPONSE TO EARTHQUAKES BY MODAL ANALYSIS

Let us now consider the use of modal analysis to obtain the response of linear multiple-degree-of-freedom systems subjected to earthquake excitation. This method involves the use of data obtained from earthquake response spectrums, such as those discussed in Chap. 4 and shown in Figs. 4-19 and 4-21. This earlier discussion is mentioned at this point since it is important to the discussion that follows that the reader remember that a response spectrum for a particular excitation is a plot of the maximum responses of a single-degree-of-freedom system subjected to that excitation for a range of natural frequencies and damping values.

In Chap. 4 we derived the differential equation of motion of a single-degree-of-freedom system subjected to support motion as

$$\ddot{z} + 2\zeta\omega_n\dot{z} + \omega_n^2 z = -\ddot{y} \tag{4-35}$$

where ζ = damping factor
ω_n = undamped natural circular frequency
z = displacement of mass m relative to moving support
$\ddot{y}$ = absolute acceleration of moving support

The modal equation for the rth mode of a multiple-degree-of-freedom system subjected to support excitation was developed in the preceding section and is repeated here as

$$\ddot{\delta}_r + 2\zeta_r\omega_r\dot{\delta}_r + \omega_r^2\delta_r = -S_r\ddot{y} \tag{6-52}$$

This equation is identical to Eq. 4-35, except for the factor S_r. Therefore, if $\zeta = \zeta_r$, $\omega_n = \omega_r$, and $\ddot{y}$ is the same support acceleration in both equations, it follows that the maximum response $(\delta_r)_{max}$ of Eq. 6-52 should be S_r times that of z_{max} of the single-degree-of-freedom system. That is,

$$(\delta_r)_{max} = S_r z_{max} = S_r A_r \tag{6-60}$$

in which A_r is used to denote z_{max} in the equations that follow.

With this relationship established, and looking back at Eq. 6-53, it follows that the maximum response of an n-degree-of-freedom system relative to a moving support can be determined by the superposition of s modal responses as

$$\begin{Bmatrix} z_1 \\ z_2 \\ \cdot \\ \cdot \\ z_n \end{Bmatrix}_{max} = S_1A_1\begin{Bmatrix} u_1 \\ u_2 \\ \vdots \\ u_n \end{Bmatrix}_1 + S_2A_2\begin{Bmatrix} u_1 \\ u_2 \\ \vdots \\ u_n \end{Bmatrix}_2 + \cdots + S_sA_s\begin{Bmatrix} u_1 \\ u_2 \\ \vdots \\ u_n \end{Bmatrix}_s \tag{6-61}$$

where $S_r = \dfrac{\sum_{i=1}^n u_{ir}m_i}{M_r}$ = mode-participation factor of the rth mode due to support excitation
u_{ir} = eigenvector components of rth mode
$M_r = \sum_{i=1}^n m_i u_{ir}^2$ = generalized mass of rth mode
z_i = displacement of mass m_i relative to moving support
A_r = maximum relative response z_{max} of a support-excited single-degree-of-freedom system having an ω_r and ζ_r corresponding to those of the rth mode of a multiple-degree-of-freedom system subjected to the same excitation

Since the maximum responses of the s modes in Eq. 6-61 generally occur at different times, this equation will not yield the actual maximum relative displacements of the masses. Therefore, various means are employed to determine good approximations of these values for design purposes. If the terms in Eq. 6-61 are added algebraically, for example, the values obtained for the maximum relative displacements could be smaller than the actual values. If the same terms are added using their absolute values, we obtain what are referred to as the *absolute-maximum* relative displacements. These are the largest relative displacements that the masses could possibly have, thus providing an upper bound for design purposes.

Another practice is to compute the *root sum square* of the same terms, to obtain "average" maximum relative displacement values, using

$$(z_i)_{\max} = \sqrt{(S_1 A_1 u_{i1})^2 + (S_2 A_2 u_{i2})^2 + \cdots + (S_s A_s u_{is})^2} \qquad (6\text{-}62)$$

Example 6-10 illustrates the determination of both the absolute-maximum relative displacements and the maximum relative displacements obtained using Eq. 6-62.

EXAMPLE 6-10

To illustrate the use of the modal-analysis method, let us consider the five-story, reinforced-concrete building shown in Fig. 6-16a, which is modeled as a five-degree-of-freedom system. The lumped masses at floor and roof levels are considered as rigid bodies, having a horizontal motion caused by the shear deformations of the columns, as shown in Fig. 6-16b. The spring constants k_i shown in the figure are the equivalent spring constants of the columns that act as springs in parallel.

As a means of evaluating the potential for damage to the building when subjected to a strong earthquake, it is desired to determine the following: (a) the root-sum-square response $\{z_i\}_{\max}$ of the building, using the response spectrum for the 1940 El Centro earth-

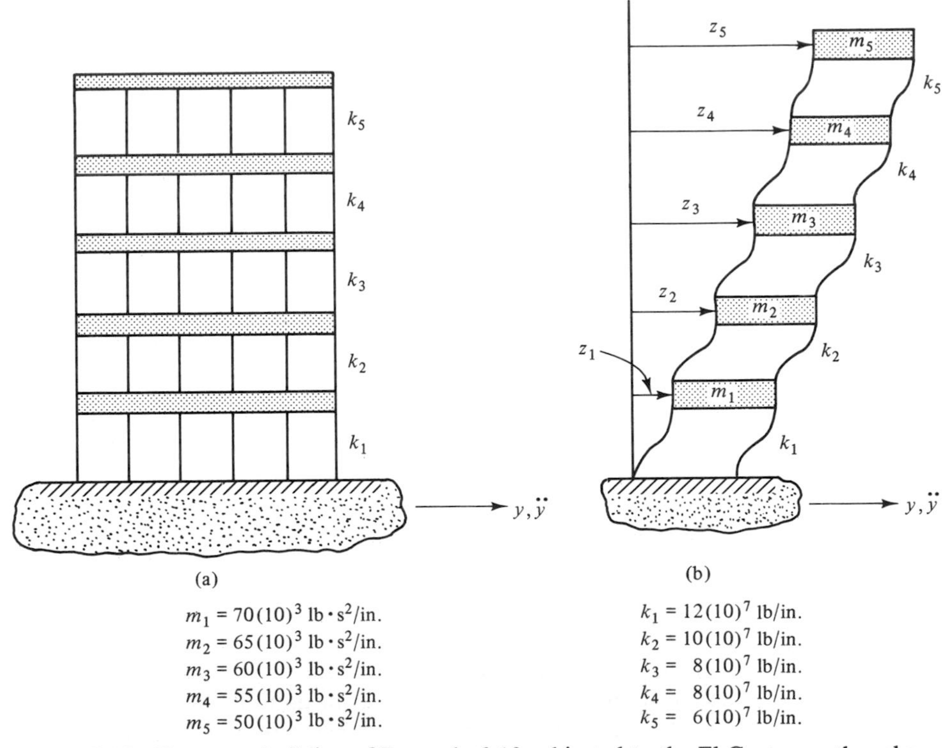

Figure 6-16 Five-story building of Example 6-10 subjected to the El Centro earthquake.

quake shown in Fig. 4-21 on page 262; (b) the absolute-maximum response of the building; (c) the shear deformations of the columns, as determined using the root-sum-square relative displacements determined in part a; (d) the resultant shear forces V_i in the columns corresponding to the shear deformations determined in part c; (e) the possibility of the building's suffering appreciable structural damage if each of the columns has the capability of deforming elastically up to a maximum of 0.75 in.

Solution. **a.** Let us begin by assuming that the superposition of the first three modes is sufficient to adequately determine the total maximum response of the building, and that all three modes have the same modal damping factor of $\zeta_r = 0.02$ (2 percent damping in the reinforced-concrete building). If the contribution of the third mode is found to be appreciable, we then add the contribution of the fourth and fifth modes. Our analysis begins with the use of the steps outlined on page 436.

Step 1. The lumped masses used are shown in Fig. 6-16. Using the definition of stiffness coefficients k_{ij} while inspecting Fig. 6-16b, the stiffness matrix is determined as

$$\mathbf{K} = 10^7 \begin{bmatrix} 22 & -10 & 0 & 0 & 0 \\ -10 & 18 & -8 & 0 & 0 \\ 0 & -8 & 16 & -8 & 0 \\ 0 & 0 & -8 & 14 & -6 \\ 0 & 0 & 0 & -6 & 6 \end{bmatrix}$$

Step 2. Using the lumped-mass and stiffness-coefficient values as input data to the computer program JACOBI.FOR (see page 369), the natural circular frequencies (the eigenvalues) are found to be[10]

$$\omega_1 = 11.86 \text{ rad/s} \qquad \omega_2 = 31.32 \text{ rad/s} \qquad \omega_3 = 47.44 \text{ rad/s}$$
$$\omega_4 = 62.54 \text{ rad/s} \qquad \omega_5 = 71.00 \text{ rad/s}$$

and the normal-mode shapes (the eigenvectors) are as shown in the modal matrix,

$$[u] = \begin{bmatrix} 1.00 & 1.00 & 1.00 & 1.00 & 1.00 \\ 2.10 & 1.51 & 0.62 & -0.54 & -1.33 \\ 3.24 & 0.95 & -0.99 & -0.75 & 1.20 \\ 4.03 & -0.31 & -0.93 & 1.24 & -0.81 \\ 4.57 & -1.71 & 1.06 & -0.55 & 0.25 \end{bmatrix}$$

Step 3. Using

$$M_r = \sum_{i=1}^{5} m_i u_{ir}^2$$

the generalized mass terms for the first three modes are found to be

$$\begin{aligned} M_1 &= 10^3[70(1)^2 + 65(2.10)^2 + 60(3.24)^2 + 55(4.03)^2 + 50(4.57)^2] \\ &= 29.24(10)^5 \\ M_2 &= 10^3[70(1)^2 + 65(1.51)^2 + 60(0.95)^2 + 55(-0.31)^2 + 50(-1.71)^2] \\ &= 42.38(10)^4 \\ M_3 &= 10^3[70(1)^2 + 65(0.62)^2 + 60(-0.99)^2 + 55(0.93)^2 + 50(1.06)^2] \\ &= 25.75(10)^4 \end{aligned}$$

[10] Computer printout values rounded to two digits following the decimal point.

Step 4. The mode-participation factors due to support excitation are determined from

$$S_r = \frac{\sum_{i=1}^{5} u_{ir} m_i}{M_r}$$

which yields the following values for the first three modes:

$$S_1 = 10^3 \frac{[(1)70 + (2.10)65 + (3.24)60 + (4.03)55 + (4.57)50]}{29.24(10)^5}$$
$$= 0.29$$
$$S_2 = 10^3 \frac{[(1)70 + (1.51)65 + (0.95)60 + (-0.31)55 + (-1.71)50]}{42.38(10)^4}$$
$$= 0.29$$
$$S_3 = 10^3 \frac{[(1)70 + (0.62)65 + (-0.99)60 + (0.93)55 + (1.06)50]}{25.75(10)^4}$$
$$= 0.60$$

Step 5. To determine the maximum response values A_r from the response spectrum of Fig. 4-21, we must determine the natural periods of the normal modes. Thus, using the ω_r's determined in step 2, we calculate the periods as

$$\tau_1 = \frac{2\pi}{11.86} = 0.53 \text{ s}$$

$$\tau_2 = \frac{2\pi}{31.32} = 0.20 \text{ s}$$

$$\tau_3 = \frac{2\pi}{47.44} = 0.13 \text{ s}$$

Then using the 2 percent damping curve of the response spectrum of Fig. 4-21, we obtain the desired $A_r(z_{\max})$ values corresponding to the natural periods just determined. These values are as shown in the following table:

Mode	Period	A_r
1	0.53 s	0.15 ft = 1.80 in.
2	0.20 s	0.02 ft = 0.24 in.
3	0.13 s	0.006 ft = 0.07 in.

Substituting the S_r and A_r values determined into Eq. 6-61, the superposition of the first three modes yields the maximum relative response equations as

$$\begin{Bmatrix} z_1 \\ z_2 \\ z_3 \\ z_4 \\ z_5 \end{Bmatrix}_{\max} = 0.29(1.8)\begin{Bmatrix} 1.00 \\ 2.10 \\ 3.24 \\ 4.03 \\ 4.57 \end{Bmatrix}_1 + 0.29(0.24)\begin{Bmatrix} 1.00 \\ 1.51 \\ 0.95 \\ -0.31 \\ -1.71 \end{Bmatrix}_2 + 0.60(0.07)\begin{Bmatrix} 1.00 \\ 0.62 \\ -0.99 \\ -0.93 \\ 1.06 \end{Bmatrix}_3 \qquad (6\text{-}63)$$

The terms necessary for determining the root-sum-square response of the building can now be determined using the values shown in Eq. 6-63. Substituting these values into Eq. 6-62, we obtain the following:

$$\begin{Bmatrix} z_1 \\ z_2 \\ z_3 \\ z_4 \\ z_5 \end{Bmatrix} = \begin{Bmatrix} 0.53 \text{ in.} \\ 1.10 \text{ in.} \\ 1.69 \text{ in.} \\ 2.10 \text{ in.} \\ 2.39 \text{ in.} \end{Bmatrix} \qquad \text{(root-sum-square)}$$

b. The absolute-maximum response of the building is determined from Eq. 6-63 by using the absolute values of the numbers in the column matrices, and is found to be

$$\begin{Bmatrix} z_1 \\ z_2 \\ z_3 \\ z_4 \\ z_5 \end{Bmatrix} = \begin{Bmatrix} 0.63 \text{ in.} \\ 1.23 \text{ in.} \\ 1.80 \text{ in.} \\ 2.16 \text{ in.} \\ 2.55 \text{ in.} \end{Bmatrix} \qquad \text{(absolute-maximum)}$$

Comparing the two sets of response values, it can be seen that the root-sum-square values are 3 to 16 percent lower than the absolute-maximum values, depending upon which floor (mass) is considered.

c. The shear deformations of the columns are determined from

$$\Delta_{i+1} = z_{i+1} - z_i \qquad (i = 0, 1, 2, 3, 4)$$

in which z_0 corresponds to the displacement of the support with respect to itself, which is obviously zero. Using the maximum displacement of each floor relative to the ground displacement, as obtained using the root-sum-square averaging process, the shear deformations in the columns are found to be

$$\begin{aligned} \Delta_1 &= 0.53 \text{ in.} \\ \Delta_2 &= 0.57 \text{ in.} \\ \Delta_3 &= 0.59 \text{ in.} \\ \Delta_4 &= 0.41 \text{ in.} \\ \Delta_5 &= 0.29 \text{ in.} \end{aligned}$$

It is left as an exercise for the reader to calculate the shear deformations using the absolute-maximum relative displacements from part b.

d. The resultant shear forces V_i in the columns are obtained by multiplying the shear deformations determined in part c by the respective column spring constants k_i shown in Fig. 6-16, and are determined as

$$\begin{Bmatrix} V_1 \\ V_2 \\ V_3 \\ V_4 \\ V_5 \end{Bmatrix} = \begin{Bmatrix} (0.53)12(10)^7 \\ (0.57)10(10)^7 \\ (0.59)8(10)^7 \\ (0.41)8(10)^7 \\ (0.29)6(10)^7 \end{Bmatrix} = (10)^6 \begin{Bmatrix} 63.6 \text{ lb} \\ 57.0 \text{ lb} \\ 47.2 \text{ lb} \\ 32.8 \text{ lb} \\ 17.4 \text{ lb} \end{Bmatrix}$$

e. Since none of the column shear deformations determined in part c exceed the limiting elastic shear deformation of 0.75 in., no major structural damage to the building would be anticipated. The reader should note that the contributions of the second and third modes to the total response are quite small in comparison with the contribution of the first mode. Therefore, there is no need to investigate the contributions of even higher modes as suggested earlier in the example.

Although the modal-analysis procedure just illustrated for determining the response of a structure to support excitation such as an earthquake is relatively straightforward, the correspondence of the results obtained to the actual physical performance of the structure depends upon how well the model represents the structure. The ability to model structural systems efficiently depends upon the modeler's experience and knowledge of the results of experimental tests performed on typical full-sized structures.

One simple procedure for lumping the masses of a building such as the one shown in Fig. 6-16 is to lump one-half of the mass of the walls and columns *above* and *below* the floor with that of the floor to determine the mass for that floor level. This approach for lumping masses is analogous to that explained in Sec. 5-12 for the lumped-mass modeling of beams.

6-8 COMMENTS CONCERNING COMPUTER SOLUTIONS

The numerical computations required to obtain the vibration response of discrete or lumped-mass systems are easily accomplished by today's microcomputers.

As discussed previously, there are two general approaches that can be used to obtain the time response of a system. The FORTRAN program SIM2.FOR presented in Sec. 6-5 obtains the response of an n-degree-of-freedom system by the simultaneous solution of n coupled differential equations. The modal superposition method, or modal analysis, discussed in Sec. 6-6 generally utilizes more straightforward numerical calculations than those involved in the program SIM2.FOR.

The two methods both yield accurate results, and it has been shown in several examples that the superposition of a large number of modes in a modal analysis is usually not necessary to obtain reasonably accurate results.

Computer Program for Simultaneous Integration of Modal Equations—MODALEQ.FOR

Since the differential equation

$$\ddot{\delta}_r + 2\zeta_r\omega_r\dot{\delta}_r + \omega_r^2\delta_r = E_r(t) = \frac{\sum_{i=1}^{n} u_{ir}F_i}{M_r} \tag{6-34}$$

for the rth mode is analogous to a single-degree-of-freedom system, the FORTRAN programs given in Chap. 4 (such as STEP.FOR on page 251) can be modified to obtain the numerical integrations of several modal differential equations corresponding to Eq. 6-34. In doing this at each Δt step during the integration of each of the separate modal differential equations being used, the program would include the following computation for, say, the first two modes of an n-degree-of-freedom system:

$$\begin{Bmatrix} x_1 \\ x_2 \\ \vdots \\ x_n \end{Bmatrix} = \delta_1 \begin{Bmatrix} u_1 \\ u_2 \\ \vdots \\ u_n \end{Bmatrix}_1 + \delta_2 \begin{Bmatrix} u_1 \\ u_2 \\ \vdots \\ u_n \end{Bmatrix}_2 \tag{6-64}$$

in which δ_1 and δ_2 would be the values from the numerical integrations at time t, with $\{u_i\}_1$ and $\{u_i\}_2$ being the modal vectors for modes one and two, respectively.

The FORTRAN program MODALEQ.FOR simultaneously integrates the desired number of modal equations (Eq. 6-34) by means of the fourth-order Runge-Kutta equations. If we use the program to obtain the time response of a system, it is necessary to first determine the natural circular frequencies (eigenvalues) and the mode shapes (eigenvectors) of the system that are to be included in the superposition of the modes (Eq. 6-64).

The function subprogram FUNCT(T) shown following the end of the main program is the *arbitrary-function* generator discussed on page 261, in which the time and force data (TDAT and FDAT) are stored in a data file such as FORT11.DAT. The function subprogram F(T, I) with a computed GO TO statement defines the force excitation acting on each of the masses M(I).

The variable names and the quantities they represent, as well as two function subprogram names and their descriptions, are as follows:

Name	Quantity
M(I)	Lumped masses, m_i
MM(R)	Generalized masses, $\Sigma\, m_i u_{ir}^2$
R	The mode number (r = 1, 2, etc.)
DEL(R)	Principal coordinates δ_r of the r modes
DDEL(R)	$\dot{\delta}_r$, time derivative of δ_r
RMAX	The number of modes to be summed
Y(I)	Displacements of masses
N	The number of degrees of freedom of the system
ZETA(R)	The modal damping factors ζ_r for the r modes
OMEGA(R)	The undamped natural circular frequencies ω_r for the r modes
DELT	The time step increment for integration
T	Time t
TMAX	Maximum time for integration (run time)
U(I,J)	The modal matrix array
SUM	$\Sigma_{i=1}^{n} u_{ir}F_i$ for a particular rth mode
E	Excitation function for a particular rth mode, $(\Sigma u_{ir}F_i)/M_r$
F(T, I)	Name of function subprogram for determining the excitation forces acting on masses, m_i
FUNCT(T)	Function subprogram name for the arbitrary function generator
TDAT(I)	Array storing the values of the independent variable time t for the arbitrary function generator
FDAT(I)	Array storing the dependent variable values for the force excitation of the arbitrary function generator

The MODALEQ.FOR program is as follows:

```
C   MODALEQ.FOR
C   MAIN PROGRAM FOR DETERMINING SYSTEM RESPONSE BY SUPERPOSITION
C   OF SOLUTIONS TO THE NORMAL MODE EQUATIONS.  ANY NUMBER OF
C   MODES MAY BE SELECTED FOR SUPERPOSITION.  THE DECOUPLED MODAL
C   SECOND-ORDER DIFFERENTIAL EQUATIONS (EQ. 6-34) ARE SOLVED
C   USING THE FOURTH-ORDER RUNGE-KUTTA EQUATIONS.  THE FUNCTION
C   SUBPROGRAM F(T,I) DESCRIBES THE FORCING FUNCTIONS ACTING ON
C   THE DIFFERENT MASSES.  THE FUNCTION SUBPROGRAM FUNCT(T) IS AN
C   ARBITRARY FUNCTION GENERATOR WHICH GETS THE DATA FOR THE
C   FORCING FUNCTION FROM A DATA FILE.  THE FUNCTION SUBPROGRAM
```

```
C   F(T,I) AND DATA FILE WILL HAVE TO BE REVISED FOR DIFFERENT
C   PROBLEMS.
C   DEVICE * IN READ AND WRITE STATEMENTS IS THE CONSOLE.
C   DEVICE 2 IN WRITE STATEMENTS IS THE PRINTER.
C   DEVICE 10 IN READ STATEMENTS IS THE DATA FILE 'FORT11.DAT'
C   * IN THE PLACE OF A FORMAT STATEMENT NUMBER MEANS FREE FORMAT.
      REAL M,MM
      INTEGER R,RMAX
      DIMENSION DEL(10),DDEL(10),AK1(10),AK2(10),AK3(10),AK4(10)
      DIMENSION M(10),Y(10)
      COMMON U(10,10),MM(10),ZETA(10),OMEGA(10),TDAT(129),
     *FDAT(129),N
      OPEN (2,FILE='PRN')
      OPEN (10,FILE='FORT11.DAT',STATUS='OLD')
C   ENTER NUMBER OF DATA POINTS FOR ARBITRARY FUNCTION GENERATOR
      WRITE(*,1)
1     FORMAT(' ENTER NO. OF DATA POINTS FOR FUNCTION GEN., I3',/)
      READ(*,3)N1
C   READ DATA FROM DATA FILE (FORT11.DAT)
C
      DO 110 I = 1,N1
      READ(10,*)TDAT(I),FDAT(I)
110   CONTINUE
C   FILL TDAT ARRAY WITH ARBITRARY INCREASING VALUES AND FILL
C   FDAT ARRAY WITH ZERO VALUES
      N1P1 = N1+1
      DO 115 I = N1P1,129
      TDAT(I) = TDAT(I-1) + 1.
      FDAT(I) = 0.0
115   CONTINUE
      DO 120 I = 1,N1
      WRITE(*,105)TDAT(I),FDAT(I)
120   CONTINUE
105   FORMAT(' ',5X,F10.3,5X,F10.0)
C
C   ENTER THE NUMBER OF DEGREES OF FREEDOM OF THE SYSTEM
      WRITE(*,2)
2     FORMAT(' ENTER NUMBER OF DEGREES OF FREEDOM, I2',/)
      READ(*,3)N
3     FORMAT(I3)
C   ENTER TIME INCREMENT AND PROGRAM RUN TIME
      WRITE(*,4)
4     FORMAT(' ENTER DELT AND TMAX, 2F10.0',/)
      READ(*,*)DELT,TMAX
C   READ IN THE MAGNITUDES OF THE N MASSES
      DO 8 I = 1,N
      WRITE(*,6)I
6     FORMAT(' ENTER VALUE OF MASS M(',I2,'),F10.0',/)
      READ(*,7)M(I)
```

```
7       FORMAT(F10.0)
8       CONTINUE
C   READ IN THE NUMBER OF MODES TO BE SUMMED IN DETERMINING THE
C   TIME RESPONSE
        WRITE(*,9)
9       FORMAT(' ENTER NUMBER OF MODES TO BE SUPERIMPOSED, I2',/)
        READ(*,10)RMAX
10      FORMAT(I2)
C   READ IN THE REQUIRED PART OF THE MODAL MATRIX U BY COLUMNS
        DO 14 J = 1,RMAX
        DO 14 I = 1,N
        WRITE(*,12)I,J
12      FORMAT(' ENTER U(',I2,',',I2,'),F10.0',/)
        READ(*,13)U(I,J)
13      FORMAT(F20.0)
14      CONTINUE
C   WRITE OUT THE PARTIAL (OR FULL) MODAL MATRIX BY COLUMNS
        DO 15 J = 1,RMAX
        DO 15 I = 1,N
15      WRITE(2,16)I,J,U(I,J)
16      FORMAT(' ','U(',I2,',',I2,') = ',E14.7)
C   ENTER THE MODAL DAMPING FACTORS
        DO 19 R = 1,RMAX
        WRITE(*,17)R
17      FORMAT(' ENTER VALUE FOR ZETA (',I2,'),F10.0',/)
        READ(*,18)ZETA(R)
18      FORMAT(F10.0)
19      CONTINUE
C   ENTER THE NATURAL FREQUENCIES
        DO 22 R = 1,RMAX
        WRITE(*,20)R
20      FORMAT(' ENTER VALUE FOR OMEGA (',I2,'), F10.0',/)
        READ(*,21)OMEGA(R)
21      FORMAT(F10.0)
22      CONTINUE
C   CALCULATE THE GENERALIZED MASSES
        DO 24 R = 1,RMAX
        MM(R) = 0.0
        DO 23 I = 1,N
23      MM(R) = MM(R) + M(I)*U(I,R)**2
24      CONTINUE
C   ENTER THE INITIAL VALUES
        DO 27 R = 1,RMAX
        WRITE(*,25)R
25      FORMAT(' ENTER INITIAL VALUE OF DEL (',I2,'), F10.0',/)
        READ(*,26)DEL(R)
26      FORMAT(F10.0)
27      CONTINUE
        DO 30 R = 1,RMAX
        WRITE(*,28)R
```

```
28    FORMAT(' ENTER INITIAL VALUE OF DDEL (',I2,'), F10.0',/)
      READ(*,29)DDEL(R)
29    FORMAT(F10.0)
30    CONTINUE
      T = 0.0
C  WRITE COLUMN HEADINGS.  CHANGE FORMAT STATEMENT 31 AS
C  REQUIRED BY PROBLEM
      WRITE(2,31)
31    FORMAT(' ','TIME',7X,'Y(1)',7X,'Y(2)',7X,'Y(3)',
     *7X,'Y(4)',7X,'Y(5)'//)
C  CALCULATE THE DISPLACEMENT VALUES
40    DO 33 I = 1,N
      Y(I) = 0.0
      DO 32 R = 1,RMAX
32    Y(I) = Y(I) + DEL(R)*U(I,R)
33    CONTINUE
C  WRITE OUT TIME AND DISPLACEMENT VALUES
      WRITE(2,34)T,(Y(I),I=1,N)
C  FORMAT STATEMENT NUMBER 34 WHICH FOLLOWS MUST BE CHANGED AS
C  REQUIRED BY THE PROBLEM
34    FORMAT(' ',F6.4,5F11.4)
      IF(T .GE. TMAX-0.00001)GO TO 100
      DO 35 R = 1,RMAX
      AK1(R) = DELT*FUNC(T,R,DEL(R),DDEL(R))
      AK2(R) = DELT*FUNC(T+DELT/2.,R,DEL(R) + DELT*DDEL(R)/2.,
     *DDEL(R)+AK1(R)/2.)
      AK3(R) = DELT*FUNC(T+DELT/2.,R,DEL(R)+DELT*(DDEL(R)/2. +
     *AK1(R)/4.),DDEL(R)+AK2(R)/2.)
      AK4(R)=DELT*FUNC(T+DELT,R,DEL(R)+DELT*(DDEL(R)+
     *AK2(R)/2.),DDEL(R)+AK3(R))
      DEL(R) = DEL(R)+DELT*(DDEL(R)+(AK1(R)+AK2(R)+AK3(R))/6.)
      DDEL(R)=DDEL(R)+(AK1(R)+2.*AK2(R)+2.*AK3(R)+AK4(R))/6.
35    CONTINUE
      T = T + DELT
      GO TO 40
100   STOP
      END
C
      FUNCTION FUNC(T,R,DISP,VEL)
      REAL MM
      INTEGER R
      COMMON U(10,10),MM(10),ZETA(10),OMEGA(10),TDAT(129),
     *FDAT(129),N
      SUM = 0.0
      DO 5 I = 1,N
5     SUM = SUM + U(I,R)*F(T,I)
      E = SUM/MM(R)
      FUNC = -2.*ZETA(R)*OMEGA(R)*VEL-OMEGA(R)**2*DISP + E
```

```
      RETURN
      END
C
C
      FUNCTION F(T,I)
      GO TO (10,20,30,40,50),I
10    F = FUNCT(T)
      GO TO 70
20    F = FUNCT(T)
      GO TO 70
30    F = FUNCT(T)
      GO TO 70
40    F = FUNCT(T)
      GO TO 70
50    F = FUNCT(T)
70    CONTINUE
      RETURN
      END
C
C
      FUNCTION FUNCT(T)
      COMMON U(10,10),MM(10),ZETA(10),OMEGA(10),TDAT(129),
     *FDAT(129),N
      IMAX=129
      IMIN=1
      DO 3 L = 1,7
      IMED=(IMAX+IMIN)/2
      IF(T .LE. TDAT(IMED))GO TO 2
      IMIN=IMED
      GO TO 3
2     IMAX=IMED
3     CONTINUE
      SLOPE=(FDAT(IMAX)-FDAT(IMIN))/(TDAT(IMAX)-TDAT(IMIN))
      FUNCT=FDAT(IMIN)+SLOPE*(T-TDAT(IMIN))
      RETURN
      END
```

EXAMPLE 6-11

The five-story building shown in Fig. 6-17 is located in a region in which tornadoes occur quite frequently. To determine the ability of the building to withstand tornadic winds, it is proposed to determine the time response of a lumped-mass model of the building when such a wind force $F(t)$ acts on each mass as shown in Fig. 6-17. It is assumed that the wind force shown in Fig. 6-18 represents a typical tornadic wind, in which the negative values result from negative pressures that occur on the windward side of the building. The 19 data values shown in Fig. 6-18 are to be stored in a data file FORT11.DAT, which is read into the main program MODALEQ.FOR for use by the arbitrary-function generator FUNCT(T).

It is desired to obtain the time response of the building using the program MODALEQ.FOR. It is assumed that the superposition of the first three modes is adequate to

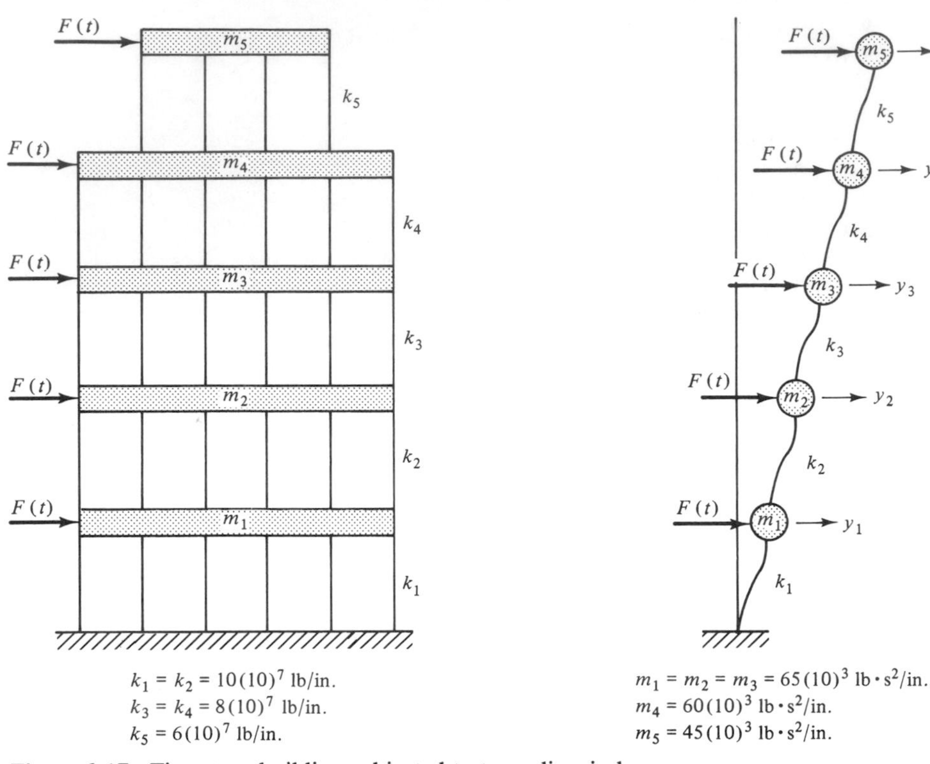

Figure 6-17 Five-story building subjected to tornadic winds.

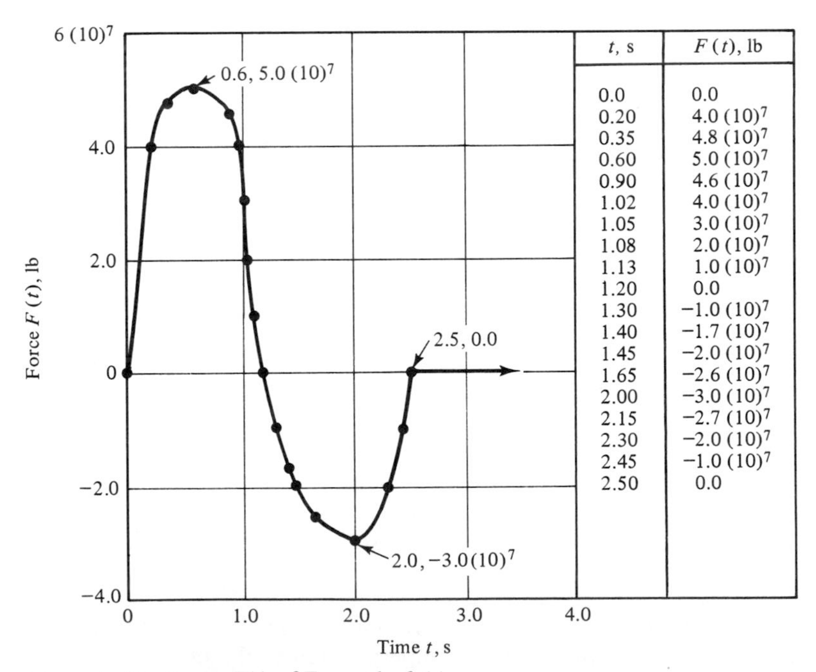

t, s	$F(t)$, lb
0.0	0.0
0.20	$4.0(10)^7$
0.35	$4.8(10)^7$
0.60	$5.0(10)^7$
0.90	$4.6(10)^7$
1.02	$4.0(10)^7$
1.05	$3.0(10)^7$
1.08	$2.0(10)^7$
1.13	$1.0(10)^7$
1.20	0.0
1.30	$-1.0(10)^7$
1.40	$-1.7(10)^7$
1.45	$-2.0(10)^7$
1.65	$-2.6(10)^7$
2.00	$-3.0(10)^7$
2.15	$-2.7(10)^7$
2.30	$-2.0(10)^7$
2.45	$-1.0(10)^7$
2.50	0.0

Figure 6-18 Force $F(t)$ of Example 6-11.

determine the response with the required accuracy and that the modal damping factor ζ_r for all modes is 0.01.

Solution. The first step is to determine the first three natural circular frequencies (eigenvalues) and corresponding mode shapes (eigenvectors) for use as input data to the computer program. Inspecting Fig. 6-17 while recalling the definition of the stiffness coefficients k_{ij}, the stiffness matrix is obtained as

$$\mathbf{K} = 10^7 \begin{bmatrix} 20 & -10 & 0 & 0 & 0 \\ -10 & 18 & -8 & 0 & 0 \\ 0 & -8 & 16 & -8 & 0 \\ 0 & 0 & -8 & 14 & -6 \\ 0 & 0 & 0 & -6 & 6 \end{bmatrix}$$

Using the data from the stiffness matrix and the values of the masses m_i shown in Fig. 6-17 as input to the program JACOBI.FOR (see page 369), the natural circular frequencies and corresponding mode shapes for the first three modes are obtained as

$$\omega_1 = 11.4876 \qquad \omega_2 = 31.0801 \qquad \omega_3 = 46.8806$$

$$\begin{Bmatrix} 1.0 \\ 1.9142228 \\ 2.8517555 \\ 3.4835188 \\ 3.8661676 \end{Bmatrix}_1 \quad \begin{Bmatrix} 1.0 \\ 1.3721190 \\ 0.7603585 \\ -0.4481703 \\ -1.6266227 \end{Bmatrix}_2 \quad \begin{Bmatrix} 1.0 \\ 0.5714377 \\ -0.9846310 \\ -0.7824525 \\ 1.2068531 \end{Bmatrix}_3$$

The other data for input to the program MODALEQ.FOR are as follows:

```
     N1 = 19 (number of data points in data file
             FORT11.DAT for arbitrary function generator)
      N = 5 (number of degrees of freedom)
   DELT = 0.01 (time step increment)
   TMAX = 3.0 (run time)
   M(1) = M(2) = M(3) = 65.E3 }
   M(4) = 60.E3               } (values of masses m_i)
   M(5) = 45.E3               }
   RMAX = 3 (number of modes to be superimposed)
ZETA(R) = 0.01 (modal damping factors for R = 1, 2, and 3)
 DEL(R) = 0.0 (initial values of δ1, δ2, and δ3)
DDEL(R) = 0.0 (initial values of δ̇1, δ̇2, and δ̇3)
```

A partial printout from the program of the time response of the displacements is as follows:

TIME	Y(1)	Y(2)	Y(3)	Y(4)	Y(5)
0.0000	0.0000	0.0000	0.0000	0.0000	0.0000
.0100	.0004	.0006	.0005	.0005	.0008
.0200	.0034	.0046	.0040	.0043	.0060
.0300	.0111	.0154	.0137	.0146	.0201
.0400	.0253	.0358	.0327	.0349	.0470
.0500	.0475	.0681	.0644	.0689	.0903
.0600	.0782	.1142	.1120	.1204	.1534
.0700	.1176	.1754	.1789	.1931	.2391

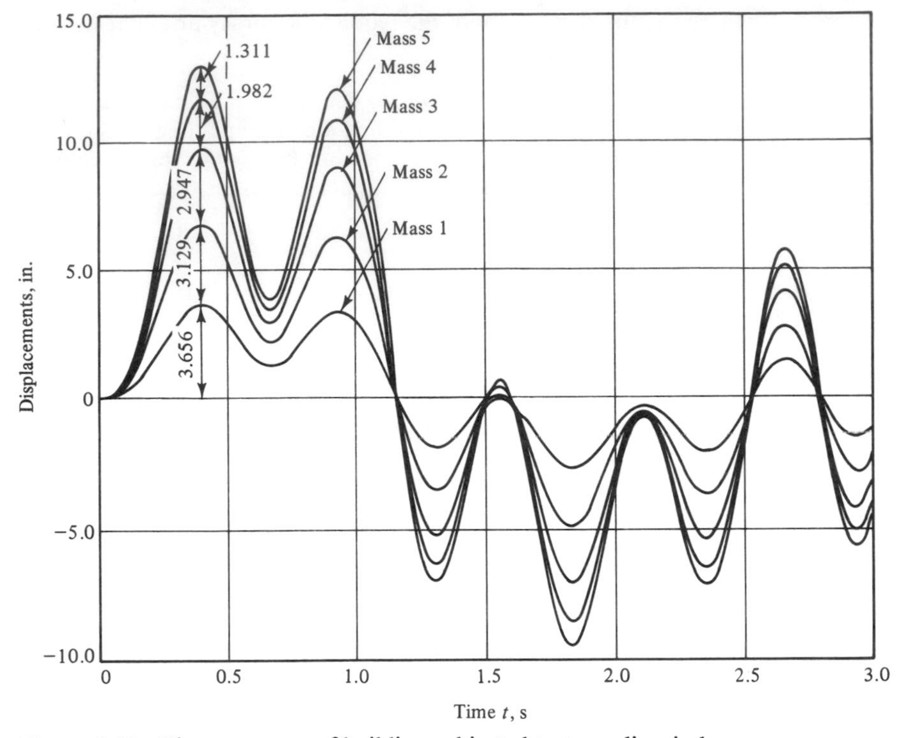

Figure 6-19 Time response of building subjected to tornadic winds.

```
.0800   .1657   .2526   .2680    .2908    .3503
.0900   .2220   .3462   .3818    .4167    .4897
.1000   .2862   .4564   .5223    .5737    .6600
  .   .   .   .   .   .   .   .   .   .   .
.3800 3.6453 6.7593 9.6868 11.6611 12.9744
.3900 3.6558 6.7845 9.7310 11.7134 13.0247  ← Max
.4000 3.6475 6.7761 9.7292 11.7097 13.0101
.4100 3.6218 6.7354 9.6807 11.6495 12.9323
.4200 3.5804 6.6637 9.5851 11.5328 12.7941
```

The computer-graphics plot of the time-response data shown in Fig. 6-19 is obtained by adding the appropriate plot statements to the program MODALEQ.FOR. The maximum *distortions* in the columns are shown on the plots, and the maximum distortion of 3.656 in. indicated for the bottom columns indicates that considerable structural damage would probably occur to the building.

PROBLEMS

Problems 6-1 through 6-10 (Sections 6-1 through 6-4)

6-1. The system shown in the accompanying figure is subjected to a sinusoidal excitation force $F_0 \sin \omega t$. Determine (a) the differential equations of motion of the system, (b) the steady-state response amplitudes X_1 and X_2, and (c) the natural circular frequencies of the system. For parts b and c let $m_1 = m_2 = m$ and $k_1 = k_2 = k$.

Partial ans: $$X_1 = \frac{F_0 k/m^2}{(\omega^2)^2 - (3k/m)\omega^2 + k^2/m^2},$$

$$\omega_1 = 0.618\sqrt{\frac{k}{m}}$$

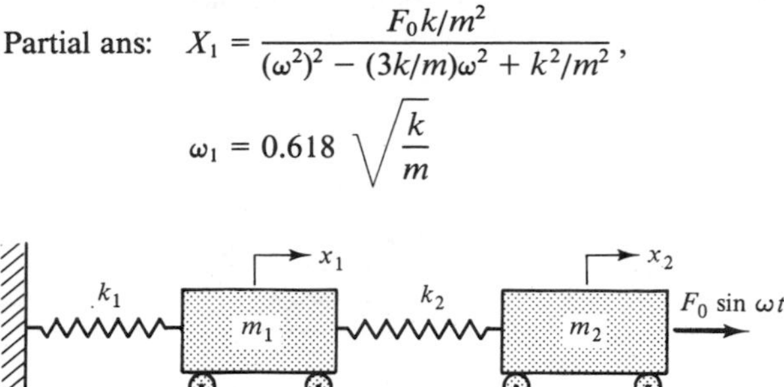

Prob. 6-1

6-2. For the system shown in the accompanying figure the absolute displacements of the masses m and $2m$ are x_1 and x_2, respectively. When the system is subjected to the support motion $y = Y \sin \omega t$, $x_1 = z_1 + y$ and $x_2 = z_2 + y$, in which z_1 and z_2 are the respective displacements of the masses m and $2m$ relative to the moving support. Determine (a) the relative steady-state response amplitudes Z_1 and Z_2 and (b) the natural circular frequencies of the system if $k = 100$ N/m and $m = 0.1$ kg.

Partial ans: (b) $f_1 = 4.06$ Hz, $\quad f_2 = 9.86$ Hz

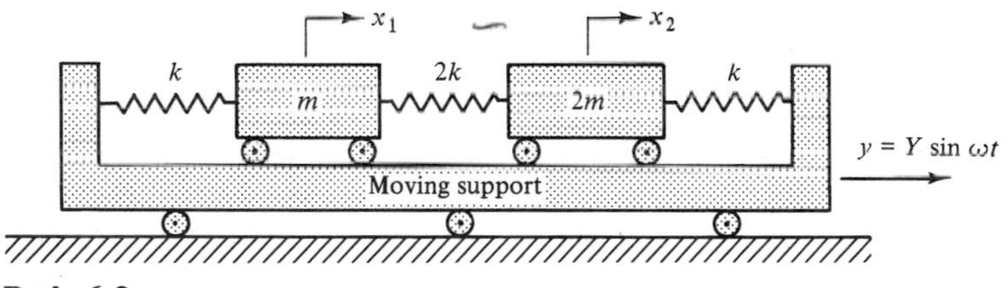

Prob. 6-2

6-3. The gear-and-shaft system shown in the accompanying figure is subjected to the moment $M = M_0 \sin \omega t$ which acts on the center one of the three gears. The centroidal mass moment of inertia of each gear and the stiffness of each shaft is as shown in the figure. The mass of the shafts is negligible with respect to that of the gears. Determine (a) the differential equations of motion of the system and (b) the amplitudes of the steady-state response of the system.

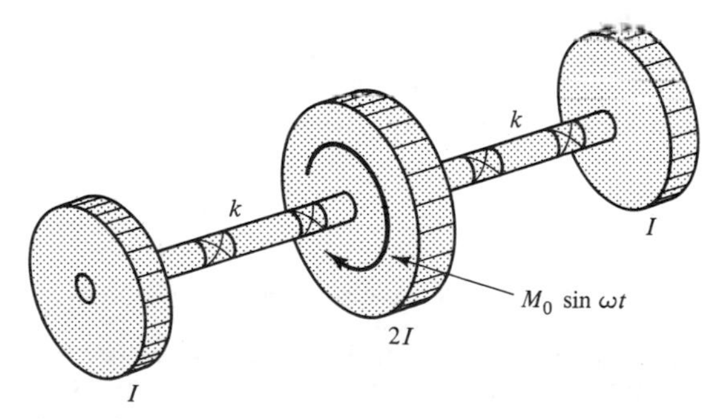

Prob. 6-3

6-4. The system shown in the accompanying figure consists of three identical masses and three identical springs, and is subjected to the excitation force $F = F_0 \sin \omega t$. (a) Determine the differential equations of motion of the system, and (b) show that if $\omega = \sqrt{k/m}$, the amplitudes of the steady-state response are $X_1 = F_0/k$, $X_2 = 0$, and $X_3 = -F_0/k$.

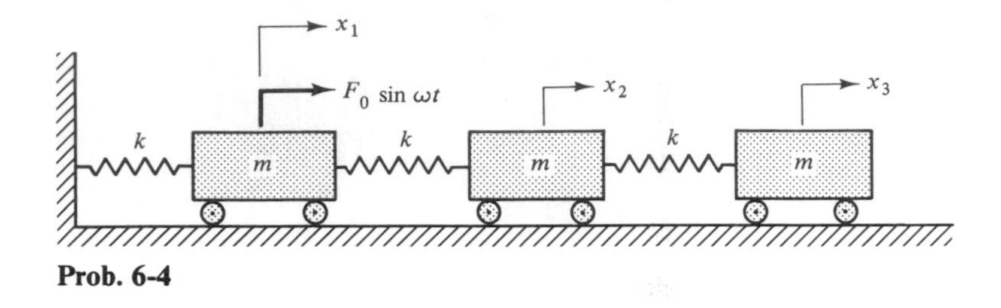

Prob. 6-4

6-5. The gear-and-shaft system shown in the accompanying figure is subjected to the moment $M = M_0 \sin \omega t$ as shown. Using the parameter values given below, and considering the mass of the shafts as negligible when compared with that of the gears, determine (a) the differential equations of motion of the system and (b) the nondimensional amplitude ratios $\theta_1/(M_0/k)$, $\theta_2/(M_0/k)$, and $\theta_4/(M_0/k)$ for the steady-state response of the system if the excitation frequency is 2.5 Hz. The parameter values are: $I_1 = 30 \text{ kg}\cdot\text{m}^2$, $I_2 = 5 \text{ kg}\cdot\text{m}^2$, $I_3 = 20 \text{ kg}\cdot\text{m}^2$, $I_4 = 25 \text{ kg}\cdot\text{m}^2$, $k = 6000 \text{ N}\cdot\text{m/rad}$, and $r_3 = 2r_2$.

Ans: $\theta_1/(M_0/k) = -2.97$, $\theta_2/(M_0/k) = -0.31$,
$\theta_4/(M_0/k) = -5.43$

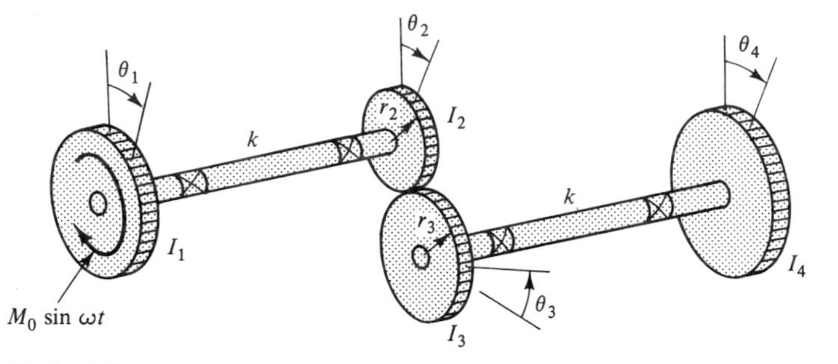

Prob. 6-5

6-6. The low-frequency oscillations constituting the roll of a ship depend upon the location of the metacenter M of the ship relative to its mass center G. This metacenter, which is located a distance h from G as shown in the accompanying figure, is determined by the intersection of the vertical centerline of the ship and the line of action of the buoyant force F_b shown. The bed in a typical cabin is located a distance d_2 from the centerline of the ship as shown in part b of the figure, is mounted on four identical springs of stiffness k, and has the additional parameter values:

$W = 50$ lb (weight)
$d_1 = 72$ in.
$\bar{I} = 110 \text{ lb}\cdot\text{s}^2\cdot\text{in.}$ (centroidal mass moment of inertia)

If x and θ are the vertical and angular displacements, respectively, of the bed, and the rolling motion of the ship is characterized by $\psi = \psi_0 \sin \omega t$, determine (a) the differential equations of motion of the bed and (b) the practicality of attempting to eliminate 80 percent of the rolling motion of the bed if the frequency of the rolling motion is 0.2 Hz (see Sec. 3-8).

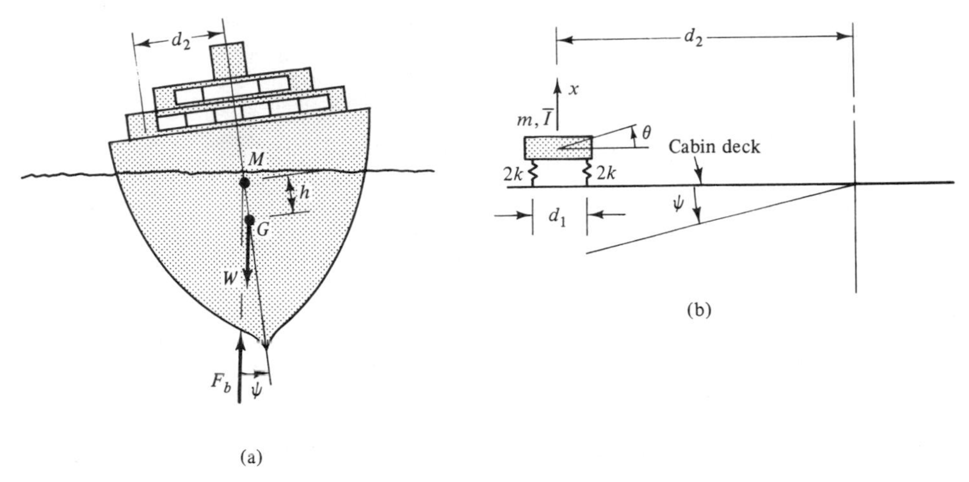

Prob. 6-6

6-7. The cylinder, which has a mass m, radius r, and centroidal mass moment of inertia $\bar{I} = mr^2/2$, rolls without slipping on a platform of mass $2m$ as shown in the accompanying figure. The generalized coordinates x_1 and x_2 shown are the absolute displacements of the mass centers of the platform and cylinder, respectively. When the platform is subjected to the sinusoidal force $F = F_0 \sin \omega t$, determine (a) the differential equations of motion of the system, (b) the steady-state response amplitudes X_1 and X_2 of the system, and (c) the condition under which the cylinder will act as a vibration absorber for the platform (such that $X_1 = 0$).

Partial ans: (c) $\dfrac{k}{m} = \dfrac{3}{4}\omega^2 \qquad (X_1 = 0)$

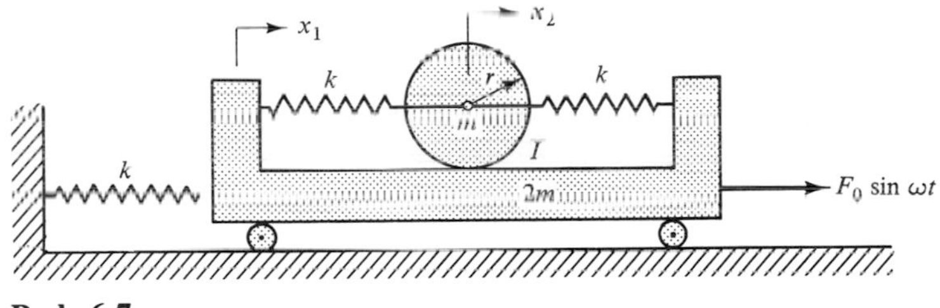

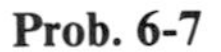
Prob. 6-7

6-8. A uniform bar of length l and centroidal mass moment of inertia $\bar{I}$ is attached to a torsional spring of stiffness k_1 as shown in the accompanying figure, and then is subjected to the force $F = F_0 \sin \omega t$ at its left end as shown. The spring of stiffness k_2 and the mass m_2 are attached to the right end of the bar to act as a vibration absorber in eliminating large angular displacements θ of the bar when $\omega = \sqrt{k_1/\bar{I}}$. Determine a relationship between the values of m_2 and k_2 of the absorber that will make $\theta = 0$.

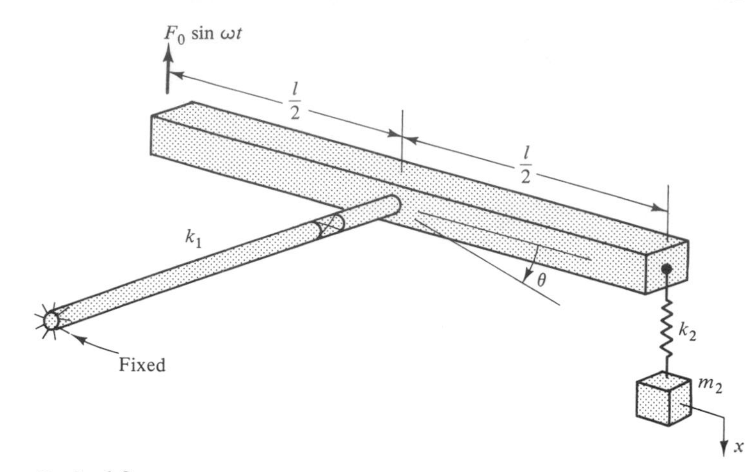

Prob. 6-8

6-9. The fluid-carrying pipe shown in the accompanying figure is supported on a floor that vibrates with the motion $Y \sin \omega t$ because of an eccentricity in a nearby machine. A section of the pipe is observed to vibrate with excessive amplitudes at a frequency of 5.5 Hz, and it is proposed to eliminate the excessive vibration by attaching a vibration absorber to the pipe. The absorber consists of two identical spheres each of mass $m_2/2$ attached to the ends of two identical slender rods each of stiffness $k_2/2$ (and negligible mass) which are in turn fastened to the pipe as shown in part b of the figure to eliminate any torsional loading of the pipe. The pipe and absorber can then be modeled by the spring-and-mass system shown in part c of the figure. As a preliminary step in the design of the absorber, a temporary absorber with two identical spheres of total weight $W_2 = 6$ lb is tuned to 5.5 Hz, and the system composed of the pipe and absorber is observed to have two natural frequencies of 4.70 and 6.44 Hz. Determine (a) the differential equations of motion of the lumped-mass model shown, (b) the apparent effective weight W_1 of the pipe and fluid, and (c) what the total weight W_2 of the two identical spheres and the effective spring constant k_2 of the two identical rods should be for a permanent absorber if the natural frequencies of the system are to lie outside the region of 4.5 to 6.0 Hz.

Partial ans: (b) $W_1 = 60$ lb; (c) $W_2 = 10.8$ lb, $k_2 = 33.4$ lb/in.

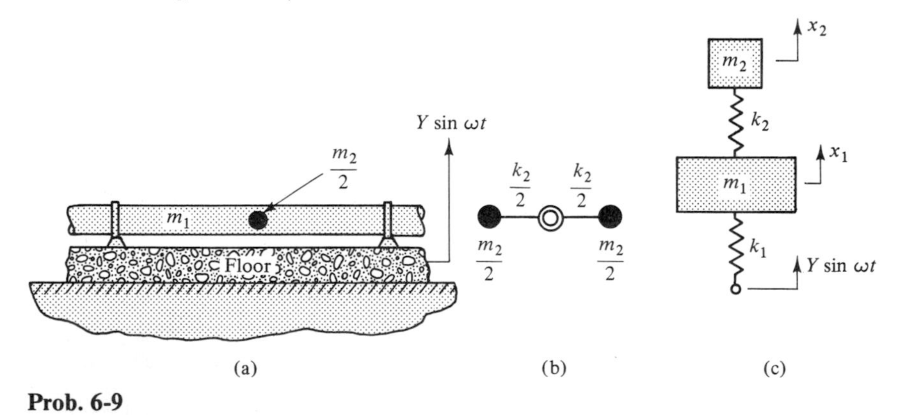

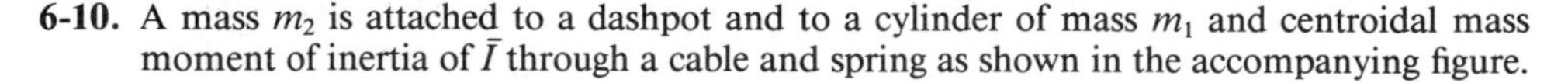

(a) (b) (c)

Prob. 6-9

6-10. A mass m_2 is attached to a dashpot and to a cylinder of mass m_1 and centroidal mass moment of inertia of $\bar{I}$ through a cable and spring as shown in the accompanying figure.

The two springs shown each have a stiffness of k, and the cylinder rolls without slipping when the mass m_2 is subjected to the excitation force $F_0 \sin \omega t$. Determine (a) the differential equations of motion of the system and (b) the conditions under which the cylinder will act as a vibration absorber for the mass m_2 (so that $q_2 = 0$).

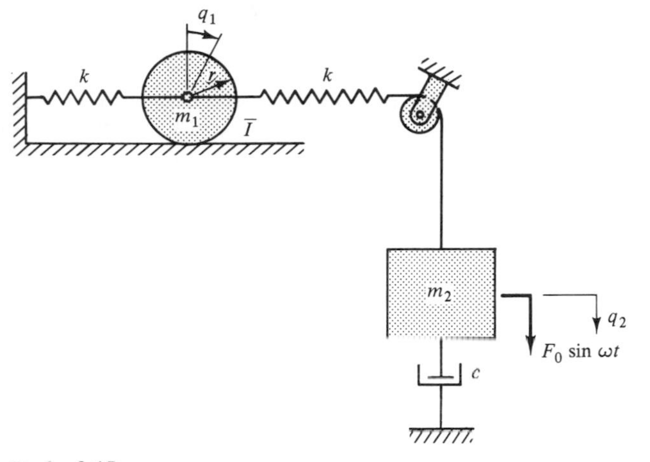

Prob. 6-10

Problems 6-11 through 6-15 (Section 6-5)

6-11. Use the JACOBI.FOR program on page 369 to obtain the undamped natural circular frequencies and eigenvectors for the four-degree-of-freedom system shown in Fig. 6-10 of Example 6-3. Use the eigenvector components of one of the normal modes found as initial conditions in obtaining an undamped free vibration solution of the system using the SIM2.FOR program. This process will provide a check to determine if the SIM2.FOR program is executing properly, since the ratios of the displacements found at any instant in time should correspond to the eigenvector components used as the initial conditions.

6-12. Modify the function subprogram in the main program SIM2.FOR so that each of the masses of the four-degree-of-freedom system of Example 6-3 (see Fig. 6-10) is subjected to the sine pulse $F(t) = F_0 \sin \pi t/t_1$ in which $t_1 = 0.1$ s and $F_0 = 2(10)^6$ lb.

6-13. The 36-ft-long beam shown in the accompanying figure weighs 1800 lb and has a stiffness

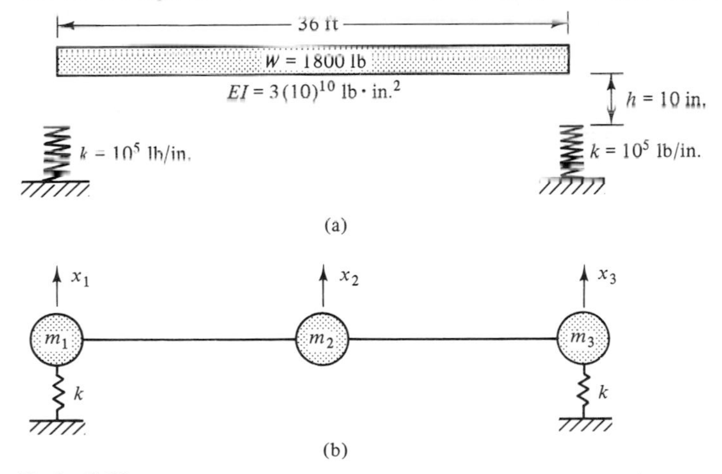

Prob. 6-13

factor of $EI = 3(10)^{10}$ lb · in.^{2} It is dropped from a height of $h = 10$ in. onto coil springs, each of which has a stiffness of $k = (10)^5$ lb/in. Model the beam by using three lumped masses as shown in part b of the figure, assuming that the masses of the coil springs are negligible when compared with the mass of the beam. Use the SIM2.FOR program to determine the time at which the beam begins its rebound from the coil springs.

Ans: 11 ms $< t <$ 11.5 ms

6-14. A steel pipe 36 ft long has an outside diameter of $d_o = 10$ in. and an inside diameter of $d_i = 9$ in. The left end of the pipe is pinned and the right end is dropped from a height of 10 ft onto the support B as shown in part a of the accompanying figure. Assume that the right end of the beam does not rebound after striking the support at B. Model the beam by three lumped masses spaced as shown in the figure, and determine the maximum absolute value of the displacement of mass m_2 that occurs during the period corresponding to the lowest natural frequency of the system ($\tau_1 = 2\pi/\omega_1$). Also note the time at which this maximum occurs. It is suggested that the natural frequencies be determined so that a suitable time increment DELT can be selected for integration using the SIM2.FOR program.

Ans: $|x_2|(\text{max}) = 6.3839$ in. at $t = 0.1300$ s

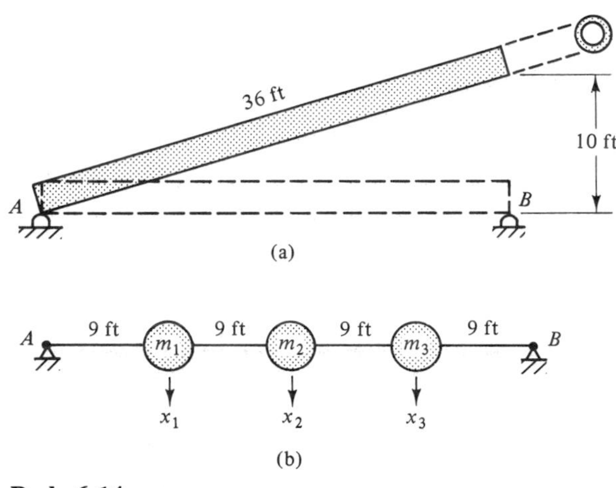

Prob. 6-14

6-15. Three identical bars pinned at their centers are connected to each other and to three supports by five identical springs of stiffness k and two dashpots with damping coefficients of c as shown in the accompanying figure. Each bar is $2l$ in length and has a centroidal mass moment of inertia of $\bar{I}$. The values of these parameters are shown in the figure. With the system at rest, the left end of the top bar is subjected to the sinusoidal pulse force

$$F(t) = 80 \sin \frac{\pi t}{t_1} \quad \text{N}$$

shown graphically in part b of the figure. Determine (a) the stiffness matrix **K** and the damping matrix **C**, (b) the undamped natural frequencies and mode shapes of the system, and (c) the time response of the system for a time interval up to $t = 0.2$ s using the SIM2.FOR program on page 425 (the time increment DELT selected for an accurate integration is to have a value no greater than $\tau_3/20$).

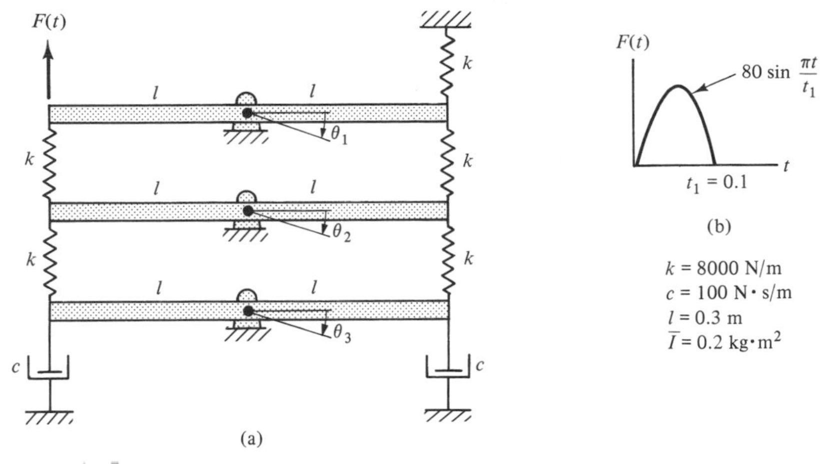

Prob. 6-15

Problems 6-16 through 6-32 (Sections 6-6 through 6-7)

6-16. In Example 6-6 the modal matrix of Eq. 6-39 was normalized with respect to the first eigenvector component and used to obtain the response of the three-lumped-mass model of a beam. Show that the modal matrix of Eq. 6-38 when normalized with respect to the second eigenvalue component shown in the example will yield the same response as shown by Eq. 6-45.

6-17. Refer to Prob. 6-14, and use modal analysis to obtain the time response of the 36-ft-long pipe that has its right end dropped from a height of 10 ft. Compare the response value obtained for $t = 0.13$ s by this method with the corresponding value obtained by the numerical integration in the SIM2.FOR program (the answer shown for Prob. 6-14).

6-18. The damping of the four-degree-of-freedom system shown in Fig. 6-10 of Example 6-3 was represented schematically by four dashpots having damping coefficients of $c = 9(10)^3$ lb·s/in. It is desired to obtain the damping factors ζ_1 and ζ_2 for the first- and second-modes, respectively, which correspond to the damping provided by the dashpots. This is done by using the SIM2.FOR program on page 425 with initial conditions corresponding to the first- and second-mode configurations. These are obtained by first using the appropriate system parameter values from Example 6-3 in the JACOBI.FOR program (see page 369) to determine the first two natural circular frequencies and mode shapes of the system with damping neglected. The damping factors can then be determined from the damped free time response data obtained from SIM2.FOR for the two separate modes used as initial conditions.

Ans: $\zeta_1 = 0.011, \quad \zeta_2 = 0.029$

6-19. Use the results obtained in the solution of Example 6-9 for the relative response z_i of the four-lumped-mass model of the airplane to determine the steady-state absolute acceleration $\ddot{x}_i$ of the four lumped masses as functions of time.

6-20. A machine on the second floor of the three-story building shown in the accompanying figure subjects the structure to the horizontal force $F(t) = F_0 \cos \omega t$. Use modal analysis, and show that if $\omega = \sqrt{k/m}$ the steady-state response is

$$x_1 = \frac{F_0}{k} \cos \omega t$$

$$x_2 = 0$$

$$x_3 = -\frac{F_0}{k} \cos \omega t$$

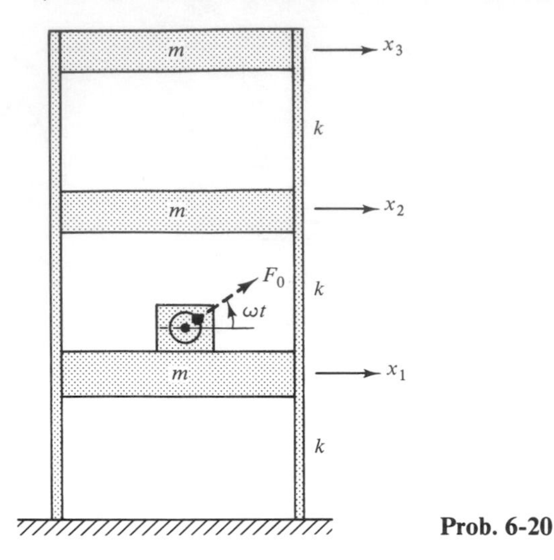

Prob. 6-20

Assume that the mass of the machine is small compared with the mass m of each of the floors.

6-21. The four-degree-of-freedom system shown in the accompanying figure is subjected to the sinusoidal force $F_0 \sin \omega t$ shown. Considering that the damping factor $\zeta_r = 0.05$ for all modes, and that the excitation frequency is $\omega = \sqrt{k/m}$, show that the steady-state response is

$$\begin{Bmatrix} x_1 \\ x_2 \\ x_3 \\ x_4 \end{Bmatrix} = \frac{33.33F_0}{k} \sin(\omega_2 t - \phi_2) \begin{Bmatrix} 1.000 \\ 1.000 \\ 0.000 \\ -1.000 \end{Bmatrix}$$

in which $\omega_2 = \sqrt{k/m}$.

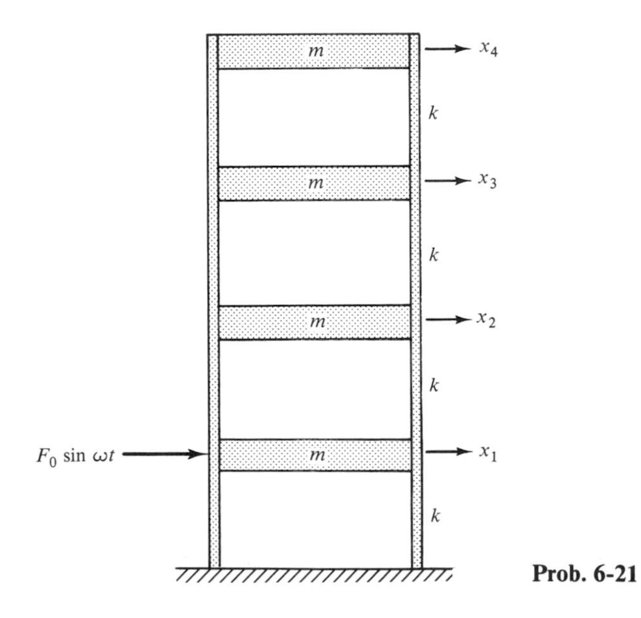

Prob. 6-21

6-22. The mass m_1 shown in part a of the accompanying figure is subjected to a pulse of 600 N for 0.1 s, which is shown graphically in part b of the figure. Using the parameter values shown in the figure, determine (a) the time response of the system for a time interval of $0 \leq t \leq 0.1$ by modal analysis and (b) the forces F_1 and F_2 that the springs shown exert on the masses at the end of the pulse ($t = 0.1$ s).

Partial ans: (b) $F_1 = -650.5$ N, $F_2 = -543.5$ N

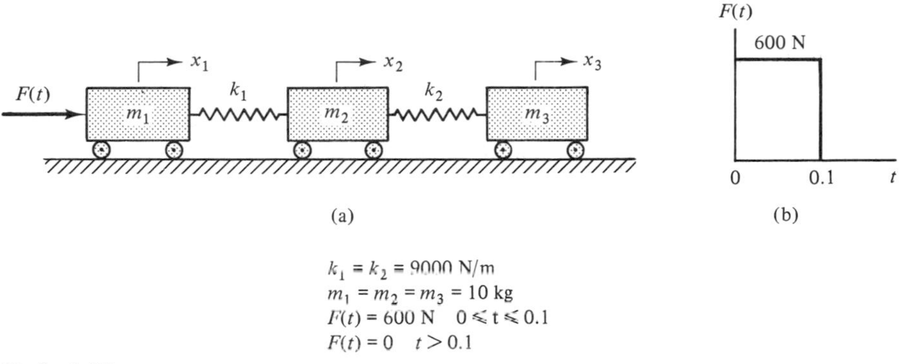

Prob. 6-22

6-23. The uniform rod of length L and cross-sectional area A shown in the accompanying figure is to be modeled by four lumped masses using the procedure explained in Sec. 5-12 and illustrated in Example 5-22. Determine the time response of the rod as given by the four-lumped-mass model when the right end of the rod is subjected to the suddenly applied force shown in part b of the figure with damping neglected. The material of the rod has a modulus of elasticity of E and a mass density of γ.

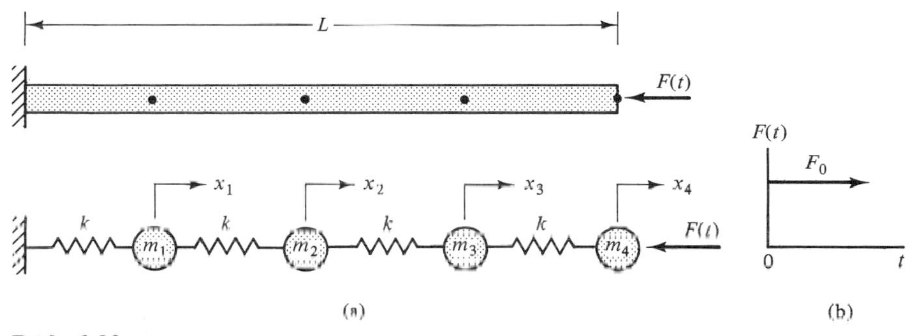

Prob. 6-23

6-24. The 20-ft-long wooden beam shown in part a of the accompanying figure is fabricated by gluing two rough dimension 2- by 6-in. planks to two rough dimension 2- by 8-in. planks to form the cross section shown in part b of the figure. The modulus of elasticity and specific weight of the wood are $E = 1.5(10)^6$ psi and $w = 0.022$ lb/in.^{3}, respectively. Knowing the allowable stresses for both the wood and the glue used, the design load for a *static* load of P_0 at the center of the simply supported beam has been found to depend upon the allowable shear stress of the glue rather than that of the wood, and it is reached when $P_0 = 6000$ lb. To evaluate the behavior of the glued joints under dynamic loading conditions, it is proposed to subject the beam to a step-function force with a rise time of t_1 as shown in part c of the figure. The force $F(t)$ is to be applied by a servohydraulic testing machine that is programmed to control the rise time t_1 and the magnitude F_0. With the beam modeled by the five lumped masses shown in part d of the figure, the natural frequencies, natural periods, and eigenvectors for the first three modes have been found to be as follows:

$$f_1 = 14.42 \text{ Hz} \qquad f_2 = 57.62 \text{ Hz} \qquad f_3 = 128.84 \text{ Hz}$$
$$\tau_1 = 0.06934 \text{ s} \qquad \tau_2 = 0.01735 \text{ s} \qquad \tau_3 = 0.00776 \text{ s}$$
$$\begin{Bmatrix} 1.000 \\ 1.732 \\ 2.000 \\ 1.732 \\ 1.000 \end{Bmatrix}_1 \qquad \begin{Bmatrix} 1.000 \\ 1.000 \\ 0.000 \\ -1.000 \\ -1.000 \end{Bmatrix}_2 \qquad \begin{Bmatrix} 1.000 \\ 0.000 \\ -1.000 \\ 0.000 \\ 1.000 \end{Bmatrix}_3$$

If the rise time is selected as $t_1 = 30$ ms (0.03 s), what should be the magnitude of F_0 to make the dynamic deflection of the center of the beam essentially the same as the static deflection caused by the load of 6000 lb?

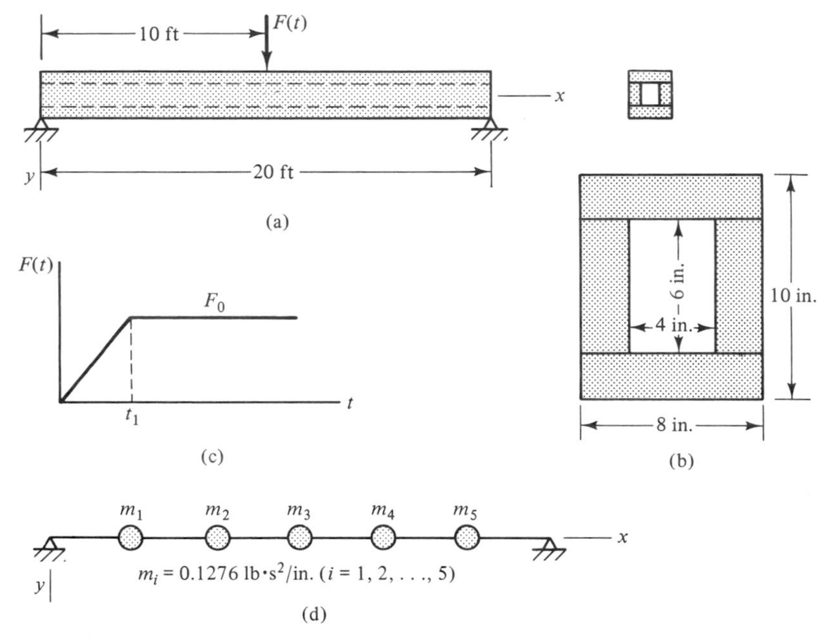

Prob. 6-24

6-25. The rocket shown in part a of the accompanying figure is in horizontal flight with a constant velocity of v_0 when it is subjected to the thrust $T(t)$ shown in part b of the figure by firing the rocket engine. The thrust builds up linearly with time and reaches a maximum value of $15(10)^4$ N in 2.0 s as shown. When the thrust reaches its maximum value, the rigid-body acceleration is 3 g's. A simplified model of the rocket consists of the five lumped masses and four elastic elements shown below the rocket. Neglect damping and the mass loss due to the firing of the engine, and determine an analytical solution for the time response of the simplified model of the rocket that is valid for the time interval $0 \leq t \leq 2$. The solution is to include the rigid-body motion and the first two elastic vibration modes. As a hint, in applying the initial conditions on $\{\dot{x}_i\}$ in the equation

$$\{\dot{x}_i\} = \dot{\delta}_1\{u_i\}_1 + \dot{\delta}_2\{u_i\}_2 + \dot{\delta}_3\{u_i\}_3 \qquad (i = 1, 2, \ldots, 5)$$

where $\dot{\delta}_1$, $\dot{\delta}_2$, and $\dot{\delta}_3$ are expressed in terms of B_1, B_2, and B_3, obtain B_2 and B_3 by setting $\dot{\delta}_2$ and $\dot{\delta}_3$ equal to zero (with $t = 0$). Then determine B_1. These solutions will satisfy all

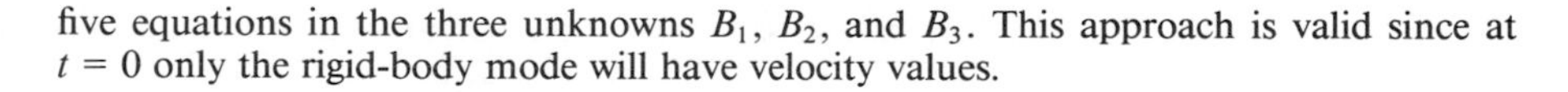
five equations in the three unknowns B_1, B_2, and B_3. This approach is valid since at $t = 0$ only the rigid-body mode will have velocity values.

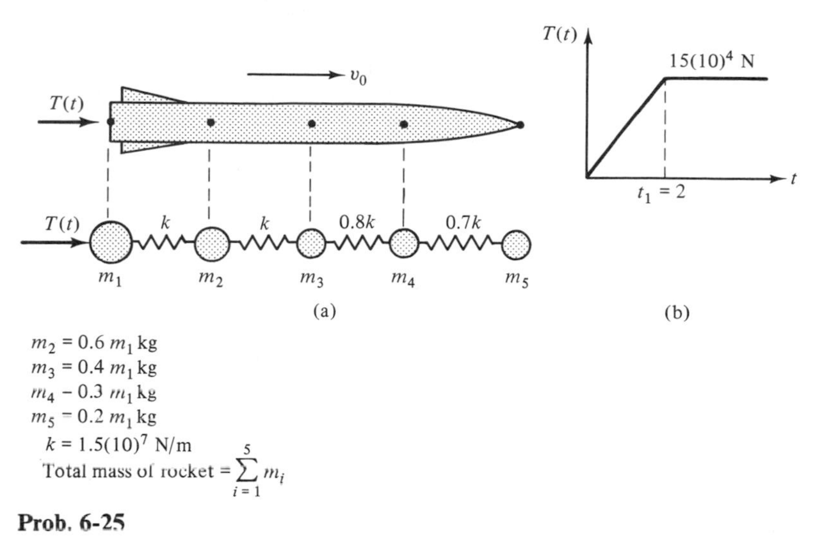

Prob. 6-25

6-26. The undamped differential equations of motion of the masses m_1 and m_2 in the system shown in the accompanying figure can be obtained in terms of the absolute displacements x_1 and x_2 from Eq. 6-8 as

$$\mathbf{M\ddot{X}} + \mathbf{KX} = \mathbf{KY}$$

in which the elements of the column matrix $\mathbf{Y}$ are the displacement $y(t)$ of the moving support. Show that the *decoupled* equations of motion are

$$\ddot{\delta}_1 + \omega_1^2 \delta_1 = \frac{(u_{11} + u_{21})ky(t)}{M_1}$$

and

$$\ddot{\delta}_2 + \omega_2^2 \delta_2 = \frac{(u_{12} + u_{22})ky(t)}{M_2}$$

in which M_1 and M_2 are the generalized mass terms and

$$[u] = \begin{bmatrix} u_{11} & u_{12} \\ u_{21} & u_{22} \end{bmatrix}$$

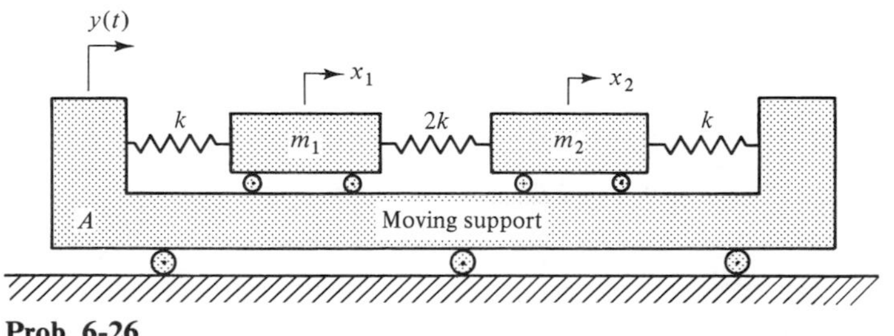

Prob. 6-26

6-27. The two identical masses m_1 and m_2 rest on the moving support A and are connected to each other and to the ends of the moving support by the springs shown in part a of the accompanying figure. The motion of the moving support is described by the velocity $\dot{y}(t)$ as indicated in part b of the figure. Neglect damping, and determine (a) the displacements of m_1 and m_2 as functions of time relative to the moving support for the time interval $0 \leq t \leq t_1$, (b) the forces in the elastic elements when $t_1 = 0.5$ s and $v_0 = 50$ in./s, and (c) whether any one of the elastic elements will fail at that time knowing that each of them can withstand a force of 125,000 lb.

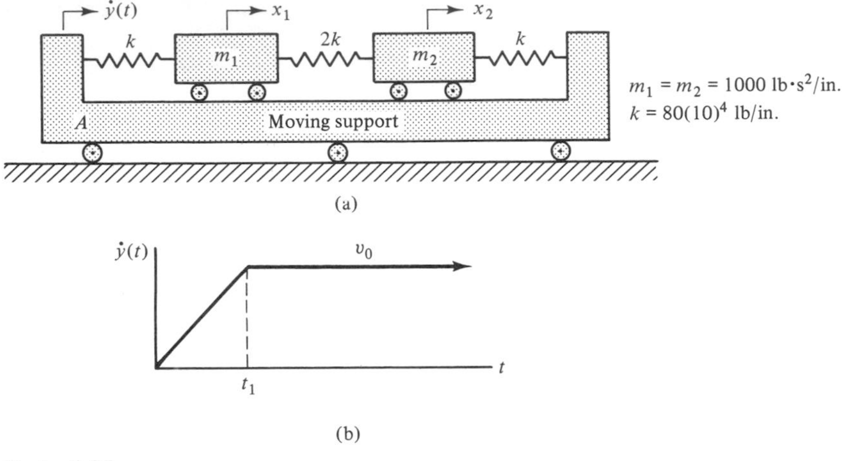

Prob. 6-27

6-28. A 1-in.-diameter aluminum rod 40 in. long is subjected to a sine pulse $F(t)$ of duration t_1 at its left end as shown in part a of the accompanying figure. The specific weight of the aluminum is 0.1 lb/in.^{3}, and its modulus of elasticity is $E = 10(10)^6$ psi. Model the rod by the four lumped masses shown in part b of the figure using the procedure explained in Sec. 5-12. Neglecting damping, determine (a) the maximum response of the system using the response spectrum shown in Fig. 4-16 on page 254 and (b) the stress σ in the portion of the rod lying between the mass m_4 and the fixed end.

Partial ans: (b) $\sigma = 38,200$ psi

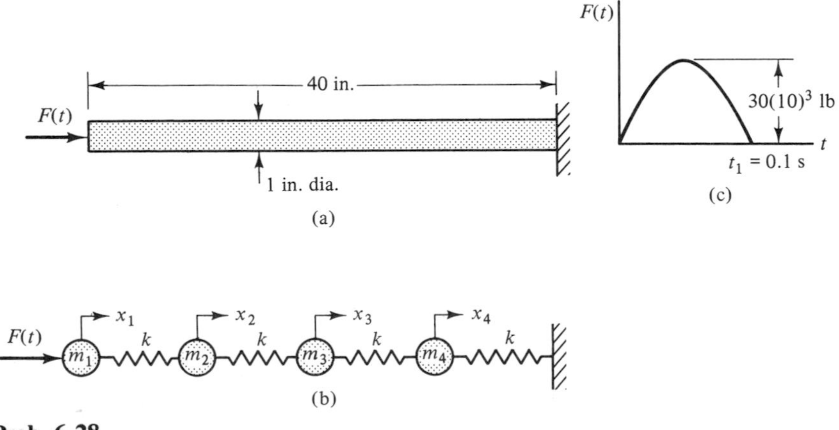

Prob. 6-28

6-29. Refer to the five-story building shown in Fig. 6-16 of Example 6-10, and assuming that the damping is given by $\zeta_r = 0.05$ for all modes, use the response spectrum of the 1971

San Fernando, California, earthquake shown in Fig. 4-19 on page 260 to determine the following: (a) the root-sum-square of the response $\{z_i\}_{max}$; (b) the absolute-maximum response; (c) the shear deformations of the columns using the root-sum-square relative displacements determined in part a; (d) the resultant shear forces V_i in the columns resulting from the shear deformations determined in part c; (e) whether the building will suffer appreciable structural damage if the columns can deform elastically up to a maximum value of 0.75 in.

6-30. It is proposed to add a sixth story to the five-story building shown in Fig. 6-16 of Example 6-10 and to include a swimming pool on the roof of this additional story. Assuming that $k_6 = 6(10)^7$ lb/in. and that $m_6 = 75(10)^3$ lb·s²/in. for this sixth floor, use the response spectrum of the 1940 El Centro earthquake shown in Fig. 4-21 on page 262 to determine (a) the root-sum-square response $\{z_i\}_{max}$ of the building for an assumed damping factor of $\zeta_r = 0.02$ for all modes, (b) the shear deformations of the columns using the root-sum-square values determined in part a, and (c) whether you would recommend the addition of the sixth floor knowing that the deformations of the columns in the building must be no more than 0.75 in. if they are to remain in the elastic range of the material they are made of.

6-31. The simply supported beam shown in part a of the accompanying figure is modeled by a lumped-mass system as shown in part b of the figure and is to be used in a modal analysis to obtain the response of the system when the excitation source is the acceleration $\ddot{y}_B$ of pin B as shown in the figure. Assuming that $\cos\theta \cong 1$, (a) determine the undamped differential equations of motion of the lumped-mass system in terms of z_i, x_i, $\ddot{y}_B$, and the stiffness matrix $\mathbf{K}$, and (b) show that the uncoupled differential equation of motion for an rth mode of an n-degree-of-freedom system is given by

$$\ddot{\delta}_r + 2\zeta_r\omega_r\dot{\delta}_r + \omega_r^2\delta_r = \frac{-[\sum_{i=1}^{n} u_{ir}m_ix_i]\ddot{y}_B}{lM_r}$$

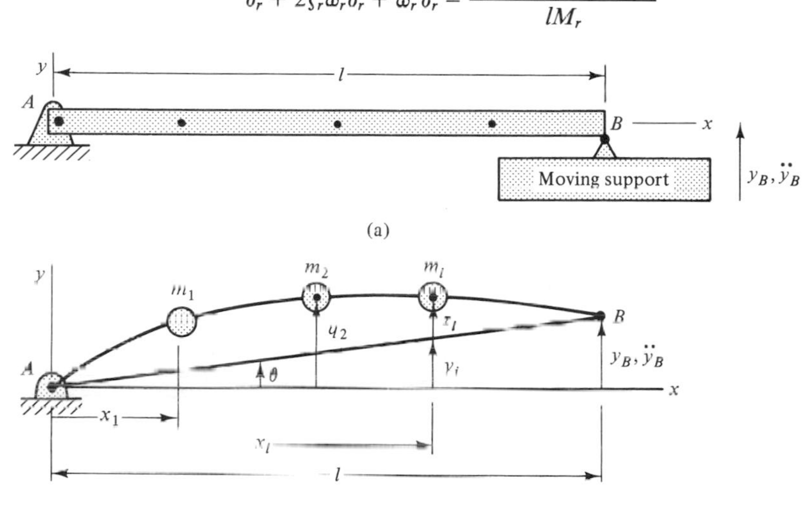

q_i = total absolute displacement of mass m_i
y_i = displacement of m_i due to rigid body motion of line AB
z_i = displacement of m_i relative to the rigid body motion
x_i = location of mass m_i along beam
$z_i = q_i - y_i$

Prob. 6-31

6-32. The overhanging steel beam [$E = 30(10)^6$ psi] shown in the accompanying figure has two weights W_1 and W_2 attached to it, each of which weighs 2000 lb. It is subjected to the suddenly applied force $F(t)$ shown in part c of the figure. Model the system by the four lumped masses shown in part b of the figure, and determine (a) the response $\{y_i\}$ of the lumped-mass model as a function of time in terms of the normal modes and (b) the maximum response of the mass m_4. It is suggested that a simple computer program be written to determine part b. Neglect damping.

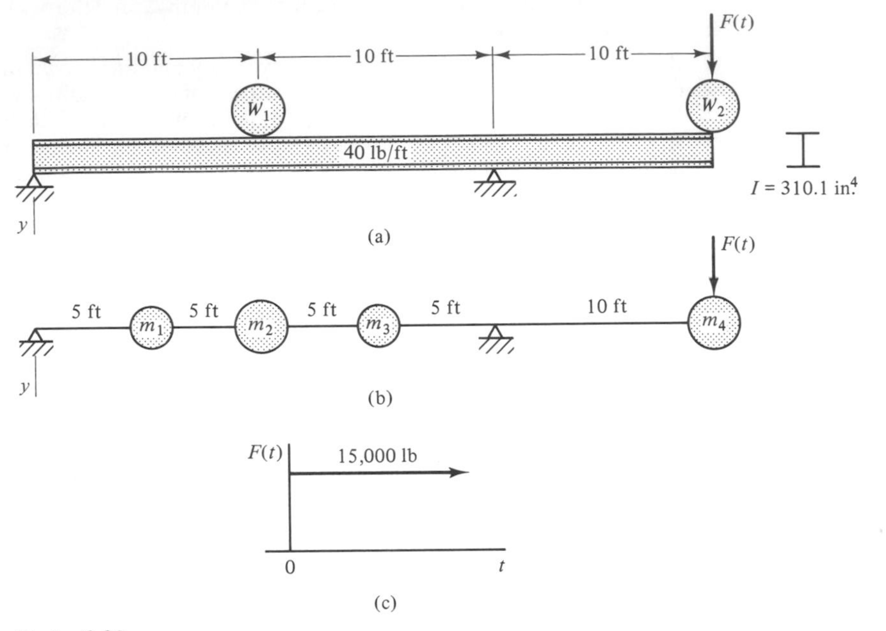

Prob. 6-32

Problems 6-33 through 6-36 (Section 6-8)

6-33. In Example 6-11, the first three modes were assumed as adequate to obtain the time response of the building when it was subjected to the tornadic wind forces $F(t)$ shown in Fig. 6-18 of that example. In that example a low damping factor of $\zeta_r = 0.01$ was assumed for all modes. Assume now that the damping factor is $\zeta_r = 0.05$ for the first three modes, and obtain the time response of the building using the program MODALEQ.FOR shown on page 457. Compare the maximum distortion found in the bottom columns with 5 percent damping assumed with the maximum distortion of 3.656 in. found in Example 6-11 with 1 percent damping. In your opinion, is it a serious matter if the *exact value* of the damping is not known for this type of problem?

6-34. As explained in Sec. 6-8, the MODALEQ.FOR program is written for the purpose of simultaneously integrating a desired number of modal equations having the form of

$$\ddot{\delta}_r + 2\zeta_r\omega_r\dot{\delta}_r + \omega_r^2\delta_r = \frac{\sum_{i=1}^{n} u_{ir}F_i}{M_r} \qquad \text{(see Eq. 6-34)}$$

in which F_i $(i = 1, 2, \ldots, n)$ are the force excitations acting on the corresponding masses m_i. Modify the subprogram F(T, I) shown in that program so that the program can be used to obtain the relative response $\{z_i\}$ of the building by simultaneously integrating the desired number of modal equations of the form

$$\ddot{\delta}_r + 2\zeta_r\omega_r\dot{\delta}_r + \omega_r^2\delta_r = \frac{-[\sum_{i=1}^{n} u_{ir}m_i]\ddot{y}}{M_r} \qquad \text{(see Eq. 6-52)}$$

in which the support acceleration $\ddot{y}$ is generated by the arbitrary function generator, the function subprogram FUNCT(T). As a hint, note that $-m_i\ddot{y}$ is a force to which each mass is subjected. Therefore, F_i in $\sum_{i=1}^{n} u_{ir}F_i/M_r$, which is determined by the function F(T, I), can be replaced by $-m_i\ddot{y}$ and determined by F(T, I) as modified.

6-35. A reinforced-concrete tower 200 ft high is to be constructed near a rock quarry in which considerable blasting occurs. The tower is to be supported on a rigid foundation that is to rest in turn on an elastic foundation (soil) as shown in part a of the accompanying figure. Pertinent parameter values are

$d_0 = 20$ ft (outside diameter of tower)
$d_i = 18$ ft (inside diameter of tower)
$E = 5(10)^6$ psi (modulus of elasticity of concrete)
$w = 150$ lb/ft³ (specific weight of concrete)
$k_t = 300(10)^{10}$ lb · in./rad

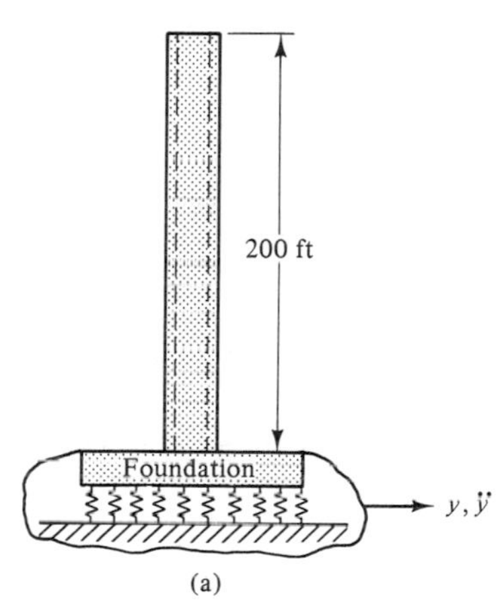

(a)

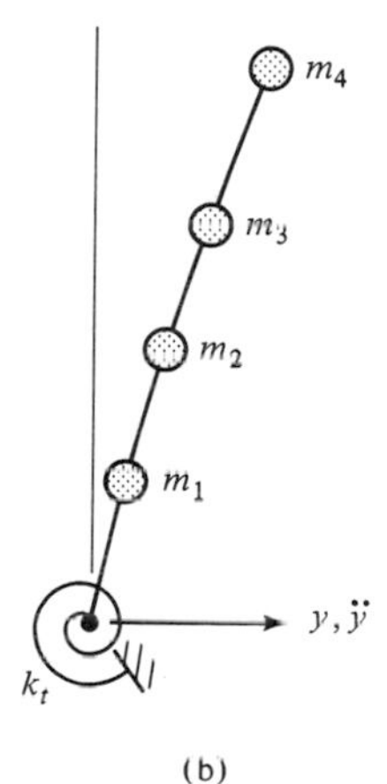

(b)

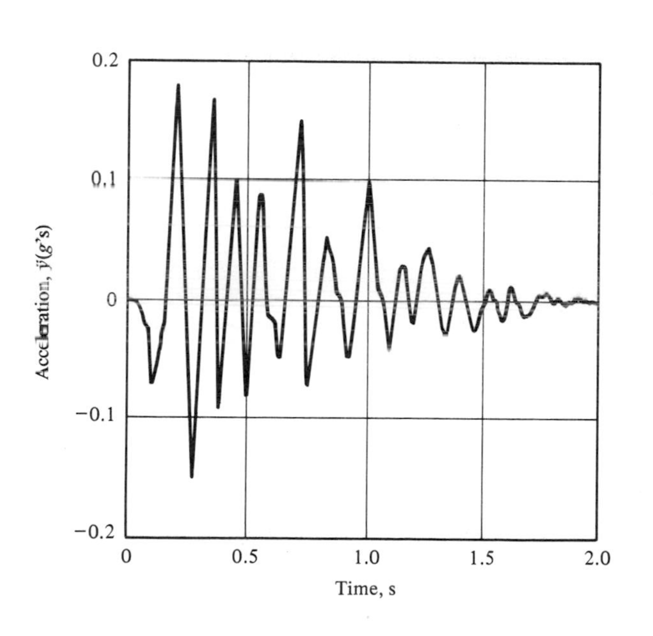

Time, s	Acceleration, in./s²
0.0	0.0
0.10	−27.02
0.20	69.48
0.28	−57.90
0.35	65.62
0.38	−34.74
0.45	38.60
0.50	−30.88
0.51	34.74
0.65	−19.30
0.72	57.90
0.75	−27.02
0.83	19.30
0.93	−19.30
1.00	34.74
1.10	−15.44
1.15	11.58
1.20	−7.72
1.26	15.44
1.45	−11.58
1.53	3.86
1.60	−7.72
1.63	3.86
1.70	−3.86
1.75	0.0

(c)

Prob. 6-35

As the vibration analyst, you are to evaluate the dynamic response of the tower to a typical ground acceleration $\ddot{y}(t)$ such as shown in both graphical and tabular form in part c of the figure. For this purpose the tower is to be modeled by the four-lumped-mass system shown in part b of the figure, in which k_t is the effective torsional spring constant representing the interaction of the soil and the rigid foundation. In the process of making your analysis, determine the following: (a) The flexibility matrix **A**. (b) The natural circular frequencies and mode shapes for the first two modes. (c) The time response of the displacements z_i relative to the ground of the four-lumped-mass model if the damping for all modes is assumed as $\zeta_r = 0.05$. Use the program MODALEQ.FOR on page 457 with modifications of the function subprogram F(T, I). (See hint in Prob. 6-34.) (d) The stiffness matrix **K** using the INVERT.FOR program on page 303. (e) The forces $\mathbf{K}\{z_i\}$ that the masses exert on the elastic elements when z_1 is a maximum. (Assuming that the damping forces are small in comparison with the elastic forces, it follows from $\mathbf{M\ddot{Z}} + \mathbf{KZ} = -\mathbf{M\ddot{Y}}$ and Newton's second law that $-\mathbf{KZ}$ corresponds to forces that the elastic elements exert on the lumped masses. Thus, $\{P_i\} = \mathbf{KZ}$ corresponds to forces that the lumped masses exert *on the structure*.) (f) Whether the concrete tower will crack from the forces $\mathbf{K}\{z_i\}$ determined in part e knowing that the concrete will crack if the bending stress $\sigma = Mc/I$ exceeds 400 psi.

6-36. Two disks having the weights and radii shown in the accompanying figure are attached to a 1.5-in-diameter steel shaft [$E = 30(10)^6$ psi] that is 60 in. long. The shaft is assumed to be simply supported by the self-aligning bearings A and B shown. As the system vibrates, the motion of each of the rigid disks is that of plane motion (translation and rotation) as

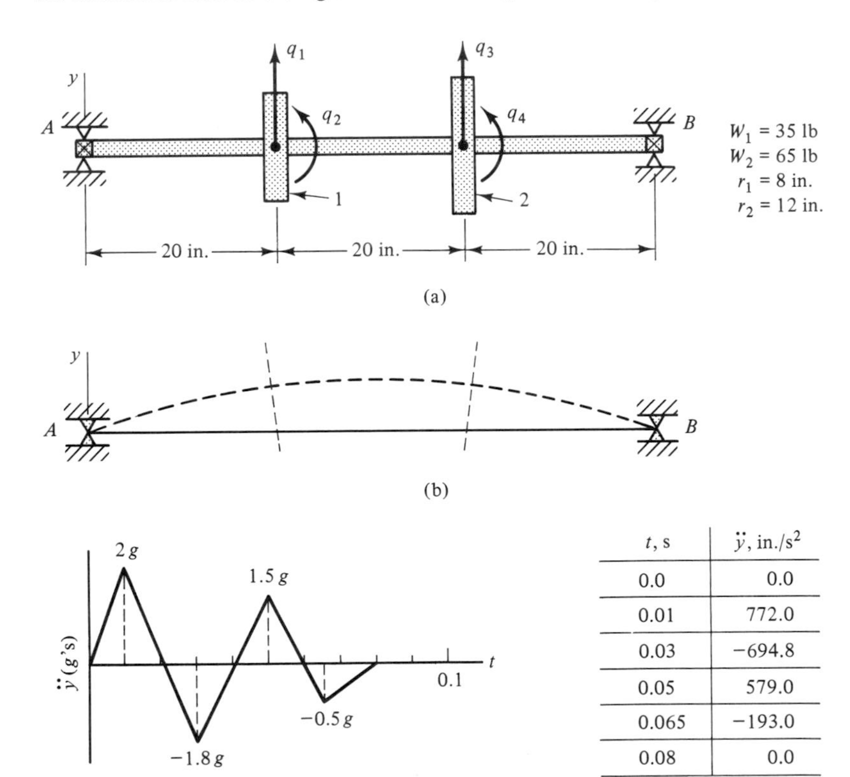

t, s	$\ddot{y}$, in./s^2
0.0	0.0
0.01	772.0
0.03	−694.8
0.05	579.0
0.065	−193.0
0.08	0.0

Prob. 6-36

shown in part b of the figure. The generalized coordinates accompanying this motion are thus q_1, q_2, q_3, and q_4 as shown. Determine (a) the flexibility matrix **A**, (b) the first two natural circular frequencies and mode shapes of the system, and (c) the maximum time response $\{z_i\}$ of the system relative to the bearing supports when a drop hammer located near the disk-and-shaft system subjects the bearing supports to the acceleration $\ddot{y}$ shown in part c of the figure (use the modified MODALEQ.FOR program of Prob. 6-34).

Partial ans: (c) Maximum response values at $t = 0.052$ s
$z_1 = 0.0989$ in. $z_2 = 0.0031$ rad
$z_3 = 0.1023$ in. $z_4 = -0.0030$ rad

Chapter 7

Lagrange's Equations and the Finite-Element Method in Vibration Analysis

7-1 INTRODUCTION

The availability and sophistication of modern digital computers has made possible the extensive use of the finite-element method for analyzing complex structures. This method consists of representing a complex structure by an assembly of simple discrete elements, such as rods, beams, plates, or two-force truss members. For example, the simple frame in Fig. 7-1 is modeled by the three beam elements shown. Elements 1 and 2 are common at point *B*, and elements 2 and 3 are common at point *C*. These common points are referred to as *joints* or *nodes*. Since the term "node" is so commonly used for indicating a point of zero displacement in vibrating systems, we shall refer to these connecting points as *joints* in the discussions that follow.

The distributed mass and stiffness properties of the elements modeling a system appear in mass and stiffness matrices associated with those elements. These matrices are then assembled into single mass and stiffness matrices **M** and **K** representing the mass and stiffness properties of the system. These matrices are then used to determine a set of dynamic equations for the system, equal in number to the number of generalized coordinates used to define the displacements of the system. For example, with the four generalized coordinates q_1, q_2, q_3, and q_4 used to represent the linear and angular displacements of the frame shown in Fig. 7-1, the mass and stiffness matrices **M** and **K** obtained by assembling the mass and stiffness matrices of the elements would be 4×4 matrices analogous to a four-degree-of-freedom system. In this manner, the frame, which theoretically has an infinite number of degrees of freedom, is reduced to a discrete four-degree-of-freedom system by using the finite elements shown.

Since an in-depth study of finite-element methods is beyond the scope of this text, we shall limit our discussions in this chapter to some of the more basic concepts pertinent to the solution of vibration problems involving structural systems that can be modeled

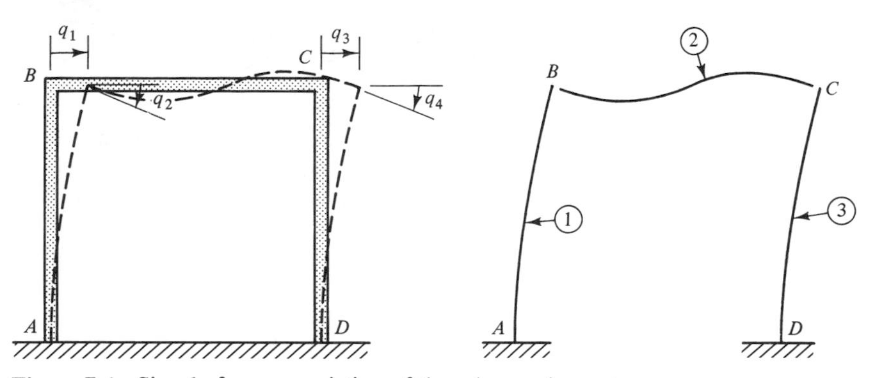

Figure 7-1 Simple frame consisting of three beam elements.

two-dimensionally by rod, beam, or truss elements. For a more comprehensive treatment of the finite-element method, there are a number of books devoted entirely to that subject.[1]

The first section after this introduction discusses the use of Lagrange's equations in determining the differential equations of motion of systems. Lagrange's equations are presented at this point since the authors feel that an understanding of the use of these equations facilitates the development of the basic concepts of the finite-element method for vibration analysis.

7-2 LAGRANGE'S EQUATIONS

In Chaps. 5 and 6 the equations of motion for discrete and lumped-mass systems were derived using Newton's second law or by using the definitions of influence coefficients. Lagrange's equations, which are widely used in engineering and physics, offer yet another approach to obtaining the equations of motion of such systems. As mentioned in the introductory remarks of this chapter, understanding the use of Lagrange's equations also aids in understanding the use of the finite-element method.

Lagrange's equations, which are given here without proof, are[2]

$$\frac{d}{dt}\left(\frac{\partial T}{\partial \dot{q}_i}\right) - \frac{\partial T}{\partial q_i} + \frac{\partial U}{\partial q_i} = Q_i \qquad i = 1, 2, \ldots, n \tag{7-1}$$

where q_i = generalized coordinates (linear or angular displacements)
T = total kinetic energy of a system (generally a function of the $\dot{q}_i$'s, but sometimes also a function of the q_i's)
U = change in the potential energy of a system with respect to its potential energy in the static-equilibrium position (a function of the q_i's)

[1] Y. C. Pao, *A First Course in Finite Element Analysis,* Allyn and Bacon, Boston, Mass., 1986. O. C. Zienkiewicz, *The Finite Element Method,* McGraw-Hill Book Co., New York, 1977. K. H. Huebner and E. A. Thornton, *The Finite Element Method for Engineers,* 2d ed., John Wiley & Sons, New York, 1983.

[2] For derivation of Lagrange's equations see M. J. Forray, *Variational Calculus in Science and Engineering,* pp. 75–81, McGraw-Hill Book Co., New York, 1968.

Q_i = generalized nonpotential forces or moments (nonconservative forces or moments) resulting from excitation forces or moments that add energy to the system and damping forces and moments that remove energy from it

The applications of Eq. 7-1 to an n-degree-of-freedom system will yield n equations of motion, which can then be written in matrix form. Lagrange's equations for an n-degree-of-freedom system having n independent (generalized) coordinates are

$$\begin{aligned}
\frac{d}{dt}\left(\frac{\partial T}{\partial \dot{q}_1}\right) - \frac{\partial T}{\partial q_1} + \frac{\partial U}{\partial q_1} &= Q_1 \\
\frac{d}{dt}\left(\frac{\partial T}{\partial \dot{q}_2}\right) - \frac{\partial T}{\partial q_2} + \frac{\partial U}{\partial q_2} &= Q_2 \\
\vdots \qquad\qquad \vdots \qquad\quad \vdots \quad &\quad \vdots \\
\frac{d}{dt}\left(\frac{\partial T}{\partial \dot{q}_n}\right) - \frac{\partial T}{\partial q_n} + \frac{\partial U}{\partial q_n} &= Q_n
\end{aligned}$$

It should be noted that since all the quantities T, U, and Q_i that appear in Lagrange's equations are scalar quantities, only an abbreviated free-body diagram of the system, showing just the external nonpotential forces and/or moments, is needed.

In general, the mass elements of a system are acted upon by

1. Potential forces and/or moments arising from springs and gravity
2. Nonpotential forces such as damping forces
3. Nonpotential excitation forces and/or moments

It should be recalled at this point that the work done by a potential force between any two points is independent of the path followed, while the work done by a nonpotential force is not.

The Q_i term of Eq. 7-1 can be damping force(s) or moment(s) that dissipate energy from a system, and/or excitation force(s) or moment(s) that add energy to the system, since both are nonpotential. They are determined by considering the differential work dW_i done by all of the nonpotential forces and moments due to a differential displacement dq_i of a *particular* q_i. In equation form this is expressed as

$$dW_i = Q_i \, dq_i \qquad i = 1, 2, \ldots, n \tag{7-2}$$

in which the Q_i terms are forces when q_i is a linear displacement and moments when q_i is an angular displacement.

Generalized Coordinates

Let us recall from Sec. 5-2 that the generalized coordinates that define the configurations of a vibrating system are coordinates that are independent of each other and that there is one generalized coordinate associated with each degree of freedom. Such coordinates are independent because a change in any one of them does not necessitate a change in any of the others.

As we shall see in Examples 7-2 and 7-3, it can be easier in certain types of problems to initially express the kinetic and potential energy of a system in terms of coordinates that are not necessarily all independent of each other. These coordinates are then related by *constraint* equations (usually geometric relationships) to ultimately yield the kinetic and potential energy in terms of the independent coordinates (the generalized coordinates) of the system. The examples that follow illustrate applications of the use of Lagrange's equations to determine the differential equations of motion of several types of systems.

EXAMPLE 7-1

The system in Fig. 7-2a consists of a pulley, a mass m, two springs of stiffness k_1 and k_2, and a dashpot with a damping coefficient of c, all connected as shown in the figure. The pulley has a mass moment of inertia $\bar{I}$ about its axis of rotation and is subjected to the moment $M(t)$ shown. We wish to determine the equations of motion of the system using Lagrange's equations.

Solution. The system has two degrees of freedom, and the generalized coordinates are $q_1 = \theta$ and $q_2 = x$. The kinetic energy of the system consists of the rotational energy of the pulley, and the translational energy of the mass m, and is thus

$$T = \tfrac{1}{2}\bar{I}\dot{\theta}^2 + \tfrac{1}{2}m\dot{x}^2 \tag{7-3}$$

Referring to Lagrange's equations, we obtain

$$\frac{d}{dt}\left(\frac{\partial T}{\partial \dot{\theta}}\right) = \bar{I}\ddot{\theta} \tag{7-4a}$$

and

$$\frac{d}{dt}\left(\frac{\partial T}{\partial \dot{x}}\right) = m\ddot{x} \tag{7-4b}$$

Since the kinetic energy is not a function of either θ or x,

$$\frac{\partial T}{\partial \theta} = 0$$

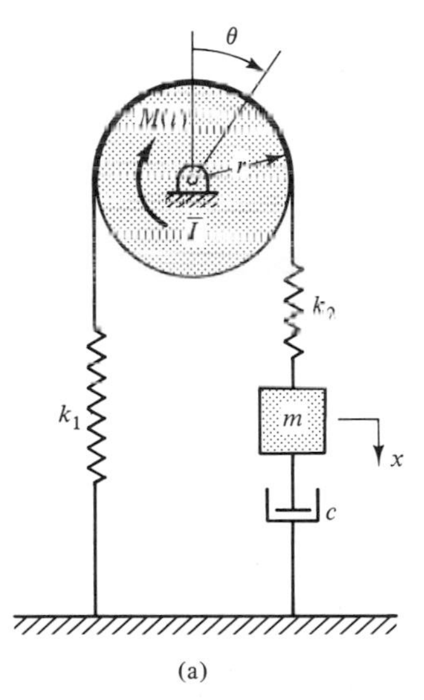

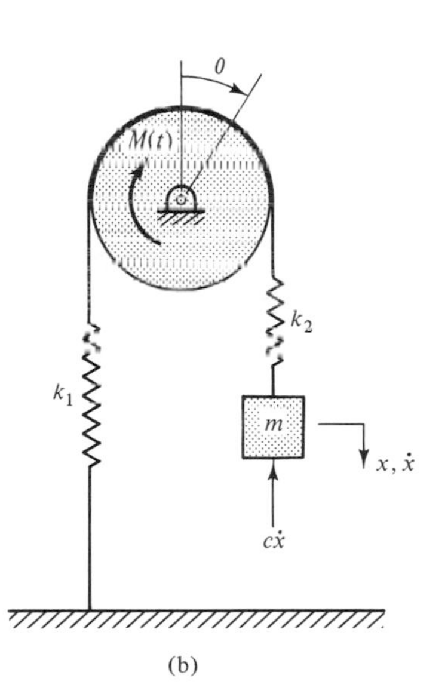

Figure 7-2 System of Example 7-1.

and

$$\frac{\partial T}{\partial x} = 0$$

Since the springs support the mass m in the static-equilibrium position ($\theta = x = 0$) of the system, the change in the potential energy of the system from its potential energy in the static-equilibrium position depends only upon the end displacements of the springs from their static-equilibrium positions. These displacements are $r\theta$ for k_1 and $(x - r\theta)$ for k_2. The change in the potential energy of the system is thus (see Sec. 2-8)

$$U = \tfrac{1}{2}k_1(r\theta)^2 + \tfrac{1}{2}k_2(x - r\theta)^2$$

Referring again to Lagrange's equations, we obtain

$$\frac{\partial U}{\partial \theta} = k_1 r^2\theta + k_2(x - r\theta)(-r) \tag{7-5a}$$

and

$$\frac{\partial U}{\partial x} = k_2(x - r\theta) \tag{7-5b}$$

An abbreviated free-body diagram showing just the external nonpotential forces and/or moments acting on a system is usually helpful in determining the differential work dW_i done by the forces and/or moments due to a differential displacement dq_i of a particular coordinate q_i. The moment $M(t)$ and damping force $c\dot{x}$ acting on the system in this example are shown on such a partial free-body diagram in Fig. 7-2b. With x held fixed during a differential angular displacement $d\theta$, the differential work done by the moment moving through $d\theta$ (see Eq. 7-2) is

$$dW_\theta = Q_\theta\, d\theta = M(t)\, d\theta$$

which shows that

$$Q_\theta = M(t) \tag{7-6}$$

Then with θ held fixed during a differential translation dx, the work done by the damping force moving through dx is

$$dW_x = Q_x\, dx = -c\dot{x}\, dx$$

which shows that

$$Q_x = -c\dot{x} \tag{7-7}$$

The latter term is negative since damping forces always oppose motion.

Substituting the preceding pertinent equations into Eq. 7-1 and simplifying, we obtain the differential equations of motion of the system as

$$\left.\begin{aligned} \bar{I}\ddot{\theta} + (k_1 + k_2)r^2\theta - k_2 r x &= M(t) \\ \text{and} \qquad m\ddot{x} + k_2 x + c\dot{x} - k_2 r\theta &= 0 \end{aligned}\right\} \tag{7-8}$$

These equations can be expressed in matrix form as

$$\begin{bmatrix} \bar{I} & 0 \\ 0 & m \end{bmatrix} \begin{Bmatrix} \ddot{\theta} \\ \ddot{x} \end{Bmatrix} + \begin{bmatrix} 0 & 0 \\ 0 & c \end{bmatrix} \begin{Bmatrix} \dot{\theta} \\ \dot{x} \end{Bmatrix} + \begin{bmatrix} (k_1 + k_2)r^2 & -k_2 r \\ -k_2 r & k_2 \end{bmatrix} \begin{Bmatrix} \theta \\ x \end{Bmatrix} = \begin{Bmatrix} M(t) \\ 0 \end{Bmatrix} \tag{7-9}$$

EXAMPLE 7-2

The system shown in Fig. 7-3 consists of two identical springs attached to a trolley of mass m_1 that supports a mass m_2 by means of a rod of length l and negligible mass pinned at point O.

We wish to determine the equations of motion of the system in terms of its generalized coordinates, and then linearize these equations by assuming that the oscillations of the rod supporting m_2 are small.

Solution. Using the x-y axes shown, the displacement of the mass m_1 is defined by the single coordinate x_1. However, the displacement of the mass m_2 can be defined by the coordinates x_2 and y_2, or x_2 and θ. This indicates that all of the four coordinates shown in the figure are not independent coordinates of the system. Figure 7-3a shows that these four coordinates can be related by the following two expressions:

$$x_2 = x_1 + l\sin(\theta)$$

and

$$y_2 = l\cos(\theta) \tag{7-10}$$

which are known as *constraint equations*. Since x_2 and y_2 can be expressed in terms of x_1 and θ, these equations show that only two independent coordinates are required to define the configurations of the system, and the system thus has only two degrees of freedom. Of the various combinations of two coordinates available, let us select x_1 and θ for our use.

Taking the derivatives of the expressions in Eq. 7-10 yields the velocity components of m_2 as

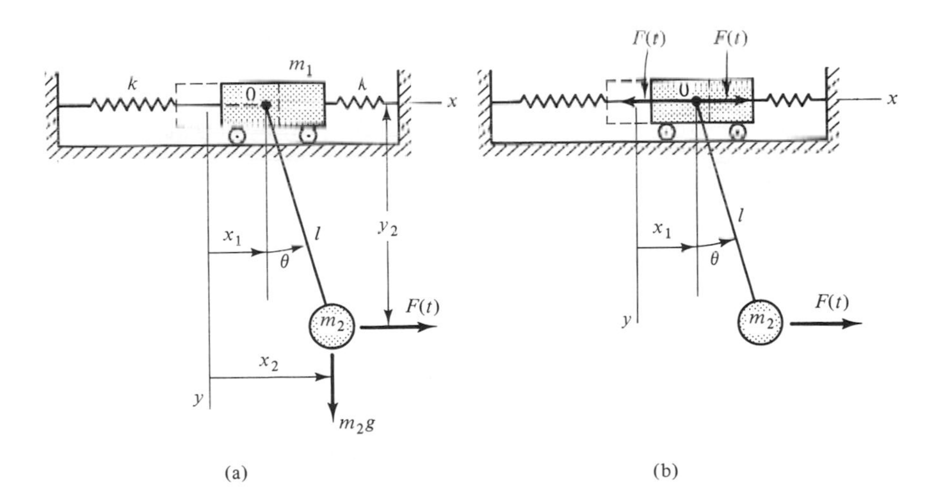

Figure 7-3 System of Example 7-2.

and

$$\left.\begin{aligned} \dot{x}_2 &= \dot{x}_1 + l\cos(\theta)\dot{\theta} \\ \dot{y}_2 &= -l\sin(\theta)\dot{\theta} \end{aligned}\right\} \tag{7-11}$$

The kinetic energy of the system in terms of $\dot{x}_1$, $\dot{x}_2$, $\dot{y}_2$ is given by

$$T = \tfrac{1}{2}m_1\dot{x}_1^2 + \tfrac{1}{2}m_2(\dot{x}_2^2 + \dot{y}_2^2) \tag{7-12}$$

To determine the kinetic energy of the system in terms of the generalized coordinates x_1 and θ that we selected, the expressions for $\dot{x}_2$ and $\dot{y}_2$ from Eq. 7-11 are substituted into Eq. 7-12 to obtain

$$T = \tfrac{1}{2}m_1\dot{x}_1^2 + \tfrac{1}{2}m_2(\dot{x}_1^2 + 2\dot{x}_1 l\cos(\theta)\dot{\theta} + l^2\dot{\theta}^2)$$

or

$$T = \tfrac{1}{2}(m_1 + m_2)\dot{x}_1^2 + m_2\dot{x}_1 l\cos(\theta)\dot{\theta} + \tfrac{1}{2}m_2 l^2\dot{\theta}^2 \tag{7-13}$$

Since the mass m_2 is not supported by any elastic elements, there is no strain energy in the system in the static-equilibrium position. Therefore, the change in potential energy associated with the vertical displacement of m_2 must be included as part of the total change in the potential energy of the system (see Sec. 2-8). Using the x axis shown in the figure as the datum, the change in the potential energy of the system is given by

$$U = kx_1^2 + m_2 gl(1 - \cos\theta) \tag{7-14}$$

which consists of the strain energy in the springs and the potential energy of position of m_2.

To facilitate the determination of the differential work done during differential changes in x_1 and θ, two equal and opposite forces of the same magnitude $F(t)$ as the applied force are added at point O to form the *equivalent* system shown in Fig. 7-3b. Inspection of the system shows that it consists of a force $F(t)$ acting through point O on the trolley and a couple of magnitude $F(t)l\cos\theta$. The differential work done is thus

$$dW_{x_1} = Q_{x_1}\,dx_1 = F(t)\,dx_1$$

and

$$dW_\theta = Q_\theta\,d\theta = F(t)l\cos(\theta)\,d\theta$$

which shows that

$$Q_{x_1} = F(t) \tag{7-15}$$

and

$$Q_\theta = F(t)l\cos\theta \tag{7-16}$$

Referring to Lagrange's equations with $q_1 = x_1$, Eqs. 7-13 and 7-14 yield the following expressions:

$$\frac{\partial T}{\partial \dot{x}_1} = (m_1 + m_2)\dot{x}_1 + m_2 l\cos(\theta)\dot{\theta}$$

$$\frac{d}{dt}\left(\frac{\partial T}{\partial \dot{x}_1}\right) = (m_1 + m_2)\ddot{x}_1 - m_2 l\sin(\theta)(\dot{\theta})^2 + m_2 l\cos(\theta)\ddot{\theta} \tag{7-17}$$

$$\frac{\partial T}{\partial x_1} = 0 \tag{7-18}$$

$$\frac{\partial U}{\partial x_1} = 2kx_1 \tag{7-19}$$

Substituting Eqs. 7-15, 7-17, 7-18, and 7-19 into Eq. 7-1 yields the *nonlinear* differential equation

$$(m_1 + m_2)\ddot{x}_1 + 2kx_1 - m_2 l \sin(\theta)(\dot{\theta})^2 + m_2 l \cos(\theta)\ddot{\theta} = F(t) \tag{7-20}$$

Referring to Lagrange's equations again with $q_2 = \theta$, Eqs. 7-13 and 7-14 yield the following expressions:

$$\frac{\partial T}{\partial \dot{\theta}} = m_2 \dot{x}_1 l \cos(\theta) + m_2 l^2 \dot{\theta}$$

$$\frac{d}{dt}\left(\frac{\partial T}{\partial \dot{\theta}}\right) = m_2 \ddot{x}_1 l \cos(\theta) - m_2 \dot{x}_1 l \sin(\theta)\dot{\theta} + m_2 l^2 \ddot{\theta} \tag{7-21}$$

$$\frac{\partial T}{\partial \theta} = -m_2 \dot{x}_1 l \sin(\theta)\dot{\theta} \tag{7-22}$$

$$\frac{\partial U}{\partial \theta} = m_2 g l \sin(\theta) \tag{7-23}$$

Substituting Eqs. 7-16, 7-21, 7-22, and 7-23 into Eq. 7-1 yields another *nonlinear* differential equation,

$$m_2 l \cos(\theta)\ddot{x}_1 + m_2 l^2 \ddot{\theta} + m_2 g l \sin(\theta) = F(t) l \cos(\theta) \tag{7-24}$$

Assuming small oscillations of the rod, Eqs. 7-20 and 7-24 can be linearized by assuming that $\sin\theta \cong \theta$, $\cos\theta \cong 1$, and $\theta\dot{\theta}^2 \cong 0$. The linearized equations of motion are then

$$(m_1 + m_2)\ddot{x}_1 + 2kx_1 + m_2 l \ddot{\theta} = F(t)$$

and

$$m_2 l^2 \ddot{\theta} + m_2 g l \theta + m_2 l \ddot{x}_1 = F(t) l$$

which appear in matrix form as

$$\begin{bmatrix} (m_1 + m_2) & m_2 l \\ m_2 l & m_2 l^2 \end{bmatrix} \begin{Bmatrix} \ddot{x}_1 \\ \ddot{\theta} \end{Bmatrix} + \begin{bmatrix} 2k & 0 \\ 0 & m_2 g l \end{bmatrix} \begin{Bmatrix} x_1 \\ \theta \end{Bmatrix} = \begin{Bmatrix} F(t) \\ F(t) l \end{Bmatrix} \tag{7-25}$$

EXAMPLE 7-3

The gear system shown in Fig. 7-4 consists of four gears and two shafts connected as shown. Gears 2 and 4 have a gear ratio of $n = r_2/r_4$, and gear 1 is subjected to the torque $M(t)$ shown.

We wish to determine the differential equations of motion of the system using Lagrange's equations.

Solution. The angular displacement θ_4 of gear 4 can be expressed in terms of the angular displacement θ_2 of gear 2 as

$$\theta_4 = \frac{r_2}{r_4}\theta_2 = n\theta_2 \tag{7-26}$$

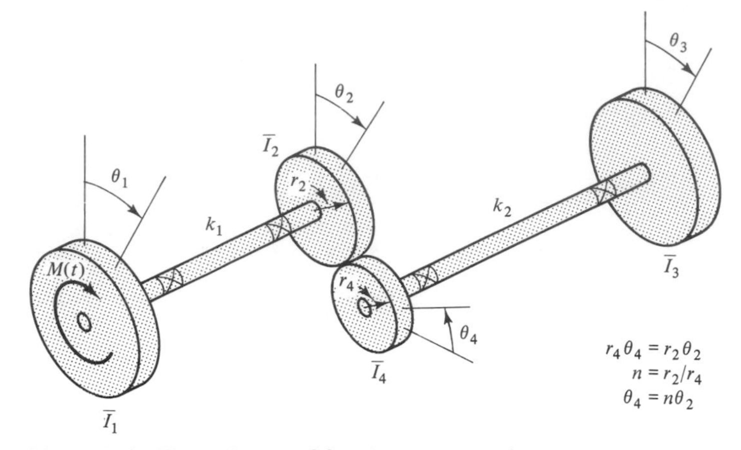

Figure 7-4 Three-degree-of-freedom system of Example 7-3.

Equation 7-26 is a constraint equation and, in geometrically relating θ_2 and θ_4, indicates that only three of the four coordinates shown in Fig. 7-4 are independent coordinates. Since the system configurations can be defined by three generalized coordinates, the system has just three degrees of freedom. There are two sets of independent coordinates that we can choose from: (θ_1, θ_2, and θ_3) or (θ_1, θ_3, and θ_4). Let us choose the first of the two for use in the ensuing discussion.

The kinetic and potential energy of the system can be most easily determined by expressing them initially in terms of all four of the system coordinates shown in Fig. 7-4 and then using the constraint equation to eliminate θ_4. Using all four coordinates, the kinetic energy of the system is given by

$$T = \tfrac{1}{2}\bar{I}_1\dot{\theta}_1^2 + \tfrac{1}{2}\bar{I}_2\dot{\theta}_2^2 + \tfrac{1}{2}\bar{I}_3\dot{\theta}_3^2 + \tfrac{1}{2}\bar{I}_4\dot{\theta}_4^2$$

Substituting Eq. 7-26 into the above, we obtain the kinetic energy in terms of the generalized coordinates θ_1, θ_2, and θ_3 as

$$T = \tfrac{1}{2}\bar{I}_1\dot{\theta}_1^2 + \tfrac{1}{2}(\bar{I}_2 + n^2\bar{I}_4)\dot{\theta}_2^2 + \tfrac{1}{2}\bar{I}_3\dot{\theta}_3^2 \tag{7-27}$$

With the angular displacements shown in Fig. 7-4 representing positive displacements, the angle of twist of the shaft supporting disks 1 and 2 is $(\theta_1 - \theta_2)$, while that of the other shaft is $(\theta_3 + \theta_4)$. The potential (strain) energy of the system in terms of all four coordinates is thus

$$U = \tfrac{1}{2}k_1(\theta_1 - \theta_2)^2 + \tfrac{1}{2}k_2(\theta_4 + \theta_3)^2$$

Substituting Eq. 7-26 into the above, the potential energy of the system in terms of the generalized coordinates is

$$U = \tfrac{1}{2}k_1(\theta_1 - \theta_2)^2 + \tfrac{1}{2}k_2(n\theta_2 + \theta_3)^2 \tag{7-28}$$

Referring to Fig. 7-4, we see that the moment $M(t)$ acting on disk 1 is the only external nonpotential force or moment acting on the system. Therefore,

$$\left.\begin{aligned} dW_{\theta_1} &= M(t)\,d\theta_1 & \therefore\quad Q_{\theta_1} &= M(t) \\ dW_{\theta_2} &= 0 & \therefore\quad Q_{\theta_2} &= 0 \\ dW_{\theta_3} &= 0 & \therefore\quad Q_{\theta_3} &= 0 \end{aligned}\right\} \tag{7-29}$$

Referring to Lagrange's equations, Eqs. 7-27 and 7-28 yield

$$\frac{d}{dt}\left(\frac{\partial T}{\partial \dot{\theta}_1}\right) = \bar{I}_1\ddot{\theta}_1 \qquad \frac{\partial U}{\partial \theta_1} = k_1(\theta_1 - \theta_2)$$

$$\frac{d}{dt}\left(\frac{\partial T}{\partial \dot{\theta}_2}\right) = (\bar{I}_2 + n^2\bar{I}_4)\ddot{\theta}_2 \qquad \frac{\partial U}{\partial \theta_2} = k_1(\theta_1 - \theta_2)(-1) + k_2(n\theta_2 + \theta_3)(n)$$

$$\frac{d}{dt}\left(\frac{\partial T}{\partial \dot{\theta}_3}\right) = \bar{I}_3\ddot{\theta}_3 \qquad \frac{\partial U}{\partial \theta_3} = k_2(n\theta_2 + \theta_3)$$

$$\frac{\partial T}{\partial \theta_1} = \frac{\partial T}{\partial \theta_2} = \frac{\partial T}{\partial \theta_3} = 0$$

We are now ready to substitute the above relationships and those shown in Eq. 7-29 into Eq. 7-1 to obtain the equations of motion as

$$\bar{I}_1\ddot{\theta}_1 + k_1\theta_1 - k_1\theta_2 = M(t)$$
$$(\bar{I}_2 + n^2\bar{I}_4)\ddot{\theta}_2 + (k_1 + n^2k_2)\theta_2 - k_1\theta_1 + nk_2\theta_3 = 0$$
$$\bar{I}_3\ddot{\theta}_3 + k_2\theta_3 + nk_2\theta_2 = 0$$

which in matrix form are given by

$$\begin{bmatrix} \bar{I}_1 & 0 & 0 \\ 0 & (\bar{I}_2 + n^2\bar{I}_4) & 0 \\ 0 & 0 & \bar{I}_3 \end{bmatrix} \begin{Bmatrix} \ddot{\theta}_1 \\ \ddot{\theta}_2 \\ \ddot{\theta}_3 \end{Bmatrix} + \begin{bmatrix} k_1 & -k_1 & 0 \\ -k_1 & (k_1 + n^2k_2) & nk_2 \\ 0 & nk_2 & k_2 \end{bmatrix} \begin{Bmatrix} \theta_1 \\ \theta_2 \\ \theta_3 \end{Bmatrix} = \begin{Bmatrix} M(t) \\ 0 \\ 0 \end{Bmatrix} \tag{7-30}$$

7-3 MASS AND STIFFNESS MATRICES FOR ROD AND BEAM ELEMENTS

Various types of elements are used in the finite-element analysis of different types of systems, and a general discussion of all these elements is beyond the intended scope of this text. Our discussions will be limited to rod elements experiencing axial or torsional deformations, beam elements having axial and/or bending deformations, and plane truss elements.

The procedure for lumping the distributed masses of rods and beams discussed in Sec. 5-12 led to diagonal mass matrices for the systems. As we shall see in this section, the procedure involving the distributed mass and stiffness properties of finite rod, beam, and truss elements leads to mass and stiffness matrices that are always symmetric, and generally full.

Rod Element—Axial Deformation

Let us first consider a finite rod element that is axially loaded. Figure 7-5a shows a rod of length l subjected to an axially distributed force $w(x, t)$. The axial (longitudinal) motion of any section a distance x from the left end of the rod can be expressed as

$$u(x, t) = \phi_1(x)u_1 + \phi_2(x)u_2 \tag{7-31}$$

in which $\phi_1(x)$ and $\phi_2(x)$ are appropriate shape functions and u_1 and u_2 are *functions of time* that give the axial displacements of the rod at $x = 0$ and $x = l$, respectively. In Fig. 7-5b the distributed force $w(x, t)$ has been replaced by the two forces f_1 and f_2 acting at the

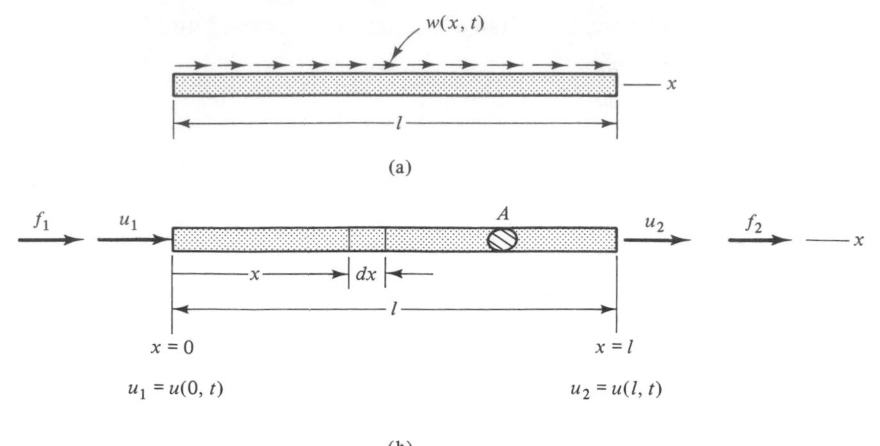

Figure 7-5 Uniform rod element.

ends of the rod as shown. These two *joint forces* are used to represent the effects of $w(x, t)$ on the rod element.

In using Eq. 7-31 to express the motion of the rod, the displacements u_1 and u_2, which we shall henceforth refer to as *joint displacements,* in effect become the generalized coordinates of the rod element. In this manner the rod element, with its distributed mass and theoretically infinite number of degrees of freedom, is modeled as a two-degree-of-freedom system.[3]

With u_1 and u_2 the joint displacements of the rod, we can write the boundary conditions

$$\left.\begin{aligned} u(0, t) &= u_1 \\ u(l, t) &= u_2 \end{aligned}\right\} \tag{7-32}$$

Referring to Eq. 7-31, we see that for this expression to satisfy the boundary conditions expressed by Eq. 7-32, the shape functions must have the following *boundary values:*

$$\left.\begin{aligned} \phi_1(0) &= 1 \\ \phi_2(0) &= 0 \\ \phi_1(l) &= 0 \\ \phi_2(l) &= 1 \end{aligned}\right\} \tag{7-33}$$

We are now ready to select shape functions that will satisfy the boundary conditions of Eq. 7-32 using the values shown in Eq. 7-33. Remembering from elementary mechanics of materials that the static displacement along a rod subjected to an axial load varies linearly, let us assume linear shape functions of the form of

$$\phi_1(x) = a_1 + b_1 x$$

and

$$\phi_2(x) = a_2 + b_2 x$$

[3] See Chap. 8 on Continuous Systems.

Substituting $\phi_1(0) = 1$ and $\phi_1(l) = 0$ into the first of the assumed shape functions to solve for a_1 and b_1, we find that

$$\phi_1(x) = 1 - \frac{x}{l} \tag{7-34}$$

In similar fashion, using $\phi_2(0) = 0$ and $\phi_2(l) = 1$ in the second shape function, we obtain

$$\phi_2(x) = \frac{x}{l} \tag{7-35}$$

These linear shape functions, which are shown in Fig. 7-6, are not the only shape functions that will satisfy the boundary conditions specified by Eqs. 7-32 and 7-33. For example, the trigonometric functions

$$\phi_1(x) = \cos\frac{\pi x}{2l} \tag{7-36}$$

$$\phi_2(x) = \sin\frac{\pi x}{2l} \tag{7-37}$$

also satisfy the boundary conditions. However, the use of these functions does not yield results as accurate as that obtained using the linear functions given by Eqs. 7-34 and 7-35. Therefore, the latter are commonly used in determining the mass and stiffness matrices of rod elements in the finite-element method (see Prob. 7-8).

If we write the differential equations of motion of the rod element in terms of the joint displacements u_1 and u_2, the equations will appear in matrix form as

$$\begin{bmatrix} m_{11} & m_{12} \\ m_{21} & m_{22} \end{bmatrix} \begin{Bmatrix} \ddot{u}_1 \\ \ddot{u}_2 \end{Bmatrix} + \begin{bmatrix} k_{11} & k_{12} \\ k_{21} & k_{22} \end{bmatrix} \begin{Bmatrix} u_1 \\ u_2 \end{Bmatrix} = \begin{Bmatrix} Q_1 \\ Q_2 \end{Bmatrix} \tag{7-38}$$

in which

$$\begin{bmatrix} m_{11} & m_{12} \\ m_{21} & m_{22} \end{bmatrix} = [m]_e \qquad \text{(mass matrix of rod element)}$$

$$\begin{bmatrix} k_{11} & k_{12} \\ k_{21} & k_{22} \end{bmatrix} = [k]_e \qquad \text{(stiffness matrix of rod element)}$$

The next step is to determine the m_{ij}'s and k_{ij}'s of the mass and stiffness matrices. We begin by expressing the kinetic energy of the rod in terms of the joint displacements u_1 and u_2. Since the velocity of any section of the rod a distance x from the left end of the rod is $\dot{u}(x, t)$, the kinetic energy of the rod element of mass per unit length γ is

$$T = \frac{\gamma}{2}\int_0^l [\dot{u}(x, t)]^2\, dx \tag{7-39}$$

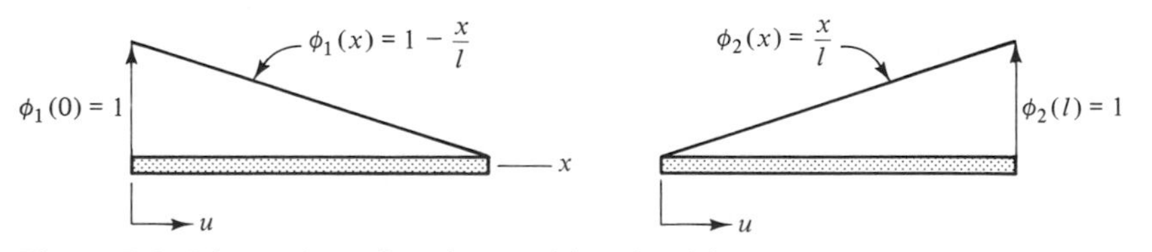

Figure 7-6 Linear shape functions $\phi_1(x)$ and $\phi_2(x)$.

in which

$$\dot{u}(x, t) = \phi_1(x)\dot{u}_1 + \phi_2(x)\dot{u}_2$$

is the time derivative of Eq. 7-31. Substituting the right side of the above equation into Eq. 7-39, the kinetic energy is

$$T = \frac{\gamma}{2}\int_0^l [\phi_1(x)\dot{u}_1 + \phi_2(x)\dot{u}_2]^2\,dx \tag{7-40}$$

To determine the potential energy, which is the strain energy of the rod element, we begin with the two strain relationships[4]

$$\left.\begin{aligned} \epsilon &= \frac{\sigma}{E} \\ \text{and} \qquad & \\ \epsilon &= \frac{du}{dx} \end{aligned}\right\} \tag{7-41}$$

where u = axial displacement of a section along the rod
σ = stress on the section
E = modulus of elasticity of rod material

The strain energy *per unit volume* of the rod is then

$$\Delta U = \tfrac{1}{2}\sigma\epsilon = \tfrac{1}{2}E\epsilon^2$$

and the strain energy in a differential volume $dA\,dx$ is

$$dU = \frac{E}{2}\left(\frac{du}{dx}\right)^2 dA\,dx \tag{7-42}$$

Integrating over the volume of the rod element gives us the total strain energy as

$$U = \frac{E}{2}\int_0^l \int_A \left(\frac{du}{dx}\right)^2 dA\,dx$$

from which

$$U = \frac{EA}{2}\int_0^l \left(\frac{du}{dx}\right)^2 dx$$

or, since u is a function of both x and t,

$$U = \frac{EA}{2}\int_0^l \left[\frac{\partial u}{\partial x}(x, t)\right]^2 dx \tag{7-43}$$

in which A is the cross-sectional area of the rod element. Referring to Eq. 7-31, we note that

$$\frac{\partial u}{\partial x}(x, t) = \phi_1'(x)u_1 + \phi_2'(x)u_2 \tag{7-44}$$

[4] From elementary mechanics of materials and the theory of elasticity, respectively.

Substituting Eq. 7-44 into Eq. 7-43,

$$U = \frac{EA}{2} \int_0^l [\phi_1'(x)u_1 + \phi_2'(x)u_2]^2 \, dx \tag{7-45}$$

With u_1 and u_2 the generalized coordinates, and recalling Lagrange's equations, we can use Eq. 7-40 to find that

$$\frac{\partial T}{\partial \dot{u}_1} = \gamma \int_0^l [\phi_1(x)\dot{u}_1 + \phi_2(x)\dot{u}_2]\phi_1(x) \, dx$$

and

$$\frac{\partial T}{\partial \dot{u}_2} = \gamma \int_0^l [\phi_1(x)\dot{u}_1 + \phi_2(x)\dot{u}_2]\phi_2(x) \, dx$$

from which

$$\frac{d}{dt}\left(\frac{\partial T}{\partial \dot{u}_1}\right) = \left[\gamma \int_0^l \phi_1(x)\phi_1(x) \, dx\right]\ddot{u}_1 + \left[\gamma \int_0^l \phi_1(x)\phi_2(x) \, dx\right]\ddot{u}_2 \tag{7-46}$$

and

$$\frac{d}{dt}\left(\frac{\partial T}{\partial \dot{u}_2}\right) = \left[\gamma \int_0^l \phi_2(x)\phi_1(x)dx\right]\ddot{u}_1 + \left[\gamma \int_0^l \phi_2(x)\phi_2(x) \, dx\right]\ddot{u}_2 \tag{7-47}$$

Comparing Eqs. 7-46 and 7-47 with Eq. 7-38, we see that the terms within the brackets of the former correspond to the m_{ij} elements of the latter. Therefore,

$$m_{ij} = \gamma \int_0^l \phi_i(x)\phi_j(x) \, dx \tag{7-48}$$

If we now substitute the linear shape functions,

$$\phi_1(x) = 1 - \frac{x}{l}$$

and

$$\phi_2(x) = \frac{x}{l}$$

(from Eqs. 7-34 and 7-35) into Eq. 7-48, we find that

$$m_{11} = \gamma \int_0^l \left(1 - \frac{x}{l}\right)^2 dx = \frac{\gamma l}{3}$$

$$m_{12} = \gamma \int_0^l \left(1 - \frac{x}{l}\right)\left(\frac{x}{l}\right) dx = \frac{\gamma l}{6}$$

$$m_{21} = m_{12} = \gamma \int_0^l \left(\frac{x}{l}\right)\left(1 - \frac{x}{l}\right) dx = \frac{\gamma l}{6}$$

$$m_{22} = \gamma \int_0^l \left(\frac{x}{l}\right)^2 dx = \frac{\gamma l}{3}$$

The mass matrix of the rod element is thus the symmetric matrix

$$[m]_e = \frac{\gamma l}{6}\begin{bmatrix} 2 & 1 \\ 1 & 2 \end{bmatrix} \tag{7-49}$$

Using Lagrange's equations with Eq. 7-45, which gives the potential energy of the rod element, we find that

$$\frac{\partial U}{\partial u_1} = EA \int_0^l [\phi_1'(x)u_1 + \phi_2'(x)u_2]\phi_1'(x)\, dx$$

and

$$\frac{\partial U}{\partial u_2} = EA \int_0^l [\phi_1'(x)u_1 + \phi_2'(x)u_2]\phi_2'(x)\, dx$$

which can also be expressed as

$$\frac{\partial U}{\partial u_1} = \left[EA \int_0^l \phi_1'(x)\phi_1'(x)\, dx\right]u_1 + \left[EA \int_0^l \phi_1'(x)\phi_2'(x)\, dx\right]u_2 \tag{7-50}$$

and

$$\frac{\partial U}{\partial u_2} = \left[EA \int_0^l \phi_2'(x)\phi_1'(x)\, dx\right]u_1 + \left[EA \int_0^l \phi_2'(x)\phi_2'(x)\, dx\right]u_2 \tag{7-51}$$

Comparing Eqs. 7-50 and 7-51 with Eq. 7-38, it can be seen that the terms within the brackets of the former correspond to the k_{ij} elements of the latter. Therefore,

$$k_{ij} = EA \int_0^l \phi_i'(x)\phi_j'(x)\, dx \tag{7-52}$$

Taking the derivatives of Eqs. 7-34 and 7-35, we obtain

$$\left.\begin{aligned} \phi_1'(x) &= -1/l \\ \phi_2'(x) &= 1/l \end{aligned}\right\} \tag{7-53}$$

Substituting these derivative expressions into Eq. 7-52, we find that the elements of the stiffness matrix are

$$k_{11} = EA \int_0^l \left(\frac{-1}{l}\right)^2 dx = \frac{EA}{l}$$

$$k_{12} = EA \int_0^l \left(\frac{-1}{l}\right)\left(\frac{1}{l}\right) dx = \frac{-EA}{l}$$

$$k_{21} = EA \int_0^l \left(\frac{1}{l}\right)\left(\frac{-1}{l}\right) dx = \frac{-EA}{l}$$

$$k_{22} = EA \int_0^l \left(\frac{1}{l}\right)^2 dx = \frac{EA}{l}$$

Thus, the stiffness matrix for the rod element is found to be the symmetric matrix

$$[k]_e = \frac{EA}{l}\begin{bmatrix} 1 & -1 \\ -1 & 1 \end{bmatrix} \tag{7-54}$$

To determine the nonpotential joint force f_1 and f_2 representing the effects of the distributed force $w(x, t)$ on the rod element, we refer to Eq. 7-2 in the discussion of Lagrange's equations in Sec. 7-2. Using this equation,

$$\left.\begin{aligned} dW_1 &= Q_1\, du_1 = f_1\, du_1 \\ dW_2 &= Q_2\, du_2 = f_2\, du_2 \end{aligned}\right\} \tag{7-55}$$

in which dW_1 is the differential work done by $w(x, t)$ accompanying a differential displacement du_1 of the left end with the right end fixed, while dW_2 is the differential work done by $w(x, t)$ accompanying a differential displacement du_2 of the right end with the left end fixed. Substituting the expressions for $\phi_1(x)$ and $\phi_2(x)$ given by Eqs. 7-34 and 7-35 into Eq. 7-31,

$$u(x, t) = \left(1 - \frac{x}{l}\right)u_1 + \left(\frac{x}{l}\right)u_2$$

Then, with the right end of the rod fixed, it follows that the *displacement function* due to a differential displacement of the left end is

$$d[u(x, t)]_1 = \left(1 - \frac{x}{l}\right) du_1$$

Similarly, with the left end fixed, the displacement function due to a differential displacement at the right end is

$$d[u(x, t)]_2 = \left(\frac{x}{l}\right) du_2$$

The differential work dW_1 is thus

$$dW_1 = \int_0^l \{w(x, t)\, d[u(x, t)]_1\}\, dx$$

or

$$dW_1 = \left[\int_0^l w(x, t)\left(1 - \frac{x}{l}\right) dx\right] du_1$$

from which

$$Q_1 = f_1 = \int_0^l w(x, t)\left(1 - \frac{x}{l}\right) dx \tag{7-56}$$

In similar fashion we can write that

$$dW_2 = \int_0^l \{w(x, t)\, d[u(x, t)]_2\}\, dx$$

or

$$dW_2 = \left[\int_0^l w(x, t)\left(\frac{x}{l}\right) dx\right] du_2$$

from which

$$Q_2 = f_2 = \int_0^l w(x, t)\left(\frac{x}{l}\right) dx \tag{7-57}$$

From Eq. 7-38 and the preceding discussion we see that the equations of motion for a rod element in terms of the joint displacements u_1 and u_2 can be expressed as

$$m_{11}\ddot{u}_1 + m_{12}\ddot{u}_2 + k_{11}u_1 + k_{12}u_2 = f_1$$
$$m_{21}\ddot{u}_1 + m_{22}\ddot{u}_2 + k_{21}u_2 + k_{22}u_2 = f_2$$

or written more compactly in matrix form as

$$[m]_e\begin{Bmatrix}\ddot{u}_1\\ \ddot{u}_2\end{Bmatrix} + [k]_e\begin{Bmatrix}u_1\\ u_2\end{Bmatrix} = \begin{Bmatrix}f_1\\ f_2\end{Bmatrix} \tag{7-58}$$

in which

$$[m]_e = \frac{\gamma l}{6}\begin{bmatrix}2 & 1\\ 1 & 2\end{bmatrix} \qquad [k]_e = \frac{EA}{l}\begin{bmatrix}1 & -1\\ -1 & 1\end{bmatrix}$$

$$f_1 = \int_0^l w(x, t)\left(1 - \frac{x}{l}\right) dx \qquad f_2 = \int_0^l w(x, t)\left(\frac{x}{l}\right) dx$$

and the following parameters pertain:

l = length of rod element
γ = mass of rod element per unit length
A = cross-sectional area of rod element
E = modulus of elasticity

The reader is reminded that the stiffness matrix $[k]_e$, which was determined from a consideration of the potential energy of the rod, can also be determined using the fundamental definition of the stiffness coefficients k_{ij} given in Sec. 5-3 (see Prob. 7-7).

Rod Element—Angular Deformation

The mass and stiffness matrices of a torsionally loaded rod element, such as the one shown in Fig. 7-7, are analogous to those determined for an axially loaded rod element in the preceding discussion. The angles of twist θ_1 and θ_2 shown in the figure are the angular joint displacements (the generalized coordinates) and are analogous to the joint displacements u_1 and u_2 of the axially loaded rod. The torques f_1 and f_2 acting at the ends of the rod element are *joint torques* representing the effects of the distributed torque $w(x, t)$ shown in the figure.

An expression similar to that of Eq. 7-31 can be written to express the angular motion of the rod at any section a distance x from the left end of the rod as

$$\theta(x, t) = \phi_1(x)\theta_1 + \phi_2(x)\theta_2 \tag{7-59}$$

in which $\phi_1(x)$ and $\phi_2(x)$ are appropriate shape functions and θ_1 and θ_2 are *functions of time* that give the angular displacements (joint displacements) at $x = 0$ and $x = l$, respectively. Considering θ_1 and θ_2 as the generalized coordinates of the rod element, we see that it can be considered as a two-degree-of-freedom system similar to the axially deformed rod.

The equations pertinent to determining the mass and stiffness elements m_{ij} and k_{ij} and the Q_i's of a torsional rod element are

$$T = \frac{\rho J}{2}\int_0^l [\dot{\theta}(x, t)]^2\, dx \tag{7-60}$$

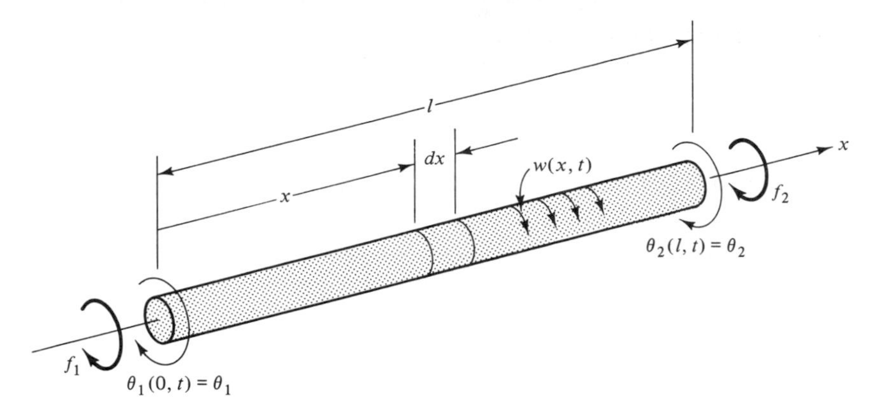

Figure 7-7 Torsional rod element.

$$U = \frac{GJ}{2} \int_0^l \left[\frac{\partial \theta}{\partial x}(x, t)\right]^2 dx \tag{7-61}$$

$$dW_i = \int_0^l [w(x, t) d[\theta(x, t)]_i \, dx \tag{7-62}$$

where
T = kinetic energy of element
$\dot{\theta}(x, t)$ = angular velocity
U = potential (strain) energy of element
ρ = mass density (mass per unit volume)
J = polar moment of inertia of cross-sectional area
G = shear modulus of elasticity
dW_i = differential work done by the distributed external torque due to a differential change $d\theta_i$ of the coordinate θ_i $(i = 1, 2)$
$d[\theta(x, t)]_i$ = angular displacement function for a differential angular displacement $d\theta_i$ at joint i $(i = 1, 2)$ with the other joint fixed

Using the above expressions, and following the procedure used with the axially loaded rod, we can obtain the mass and stiffness matrices and nonpotential moment expressions for a torsionally loaded rod element as

$$[m]_e = \frac{\rho J l}{6} \begin{bmatrix} 2 & 1 \\ 1 & 2 \end{bmatrix} \tag{7-63}$$

$$[k]_e = \frac{GJ}{l} \begin{bmatrix} 1 & -1 \\ -1 & 1 \end{bmatrix} \tag{7-64}$$

$$f_1 = \int_0^l w(x, t)\left(1 - \frac{x}{l}\right) dx \tag{7-65}$$

$$f_2 = \int_0^l w(x, t)\left(\frac{x}{l}\right) dx \tag{7-66}$$

Beam Element—Bending Deformation

Let us now turn our attention to determining the mass and stiffness matrices of a uniform beam element such as the one shown in Fig. 7-8, which is of length l and mass per unit

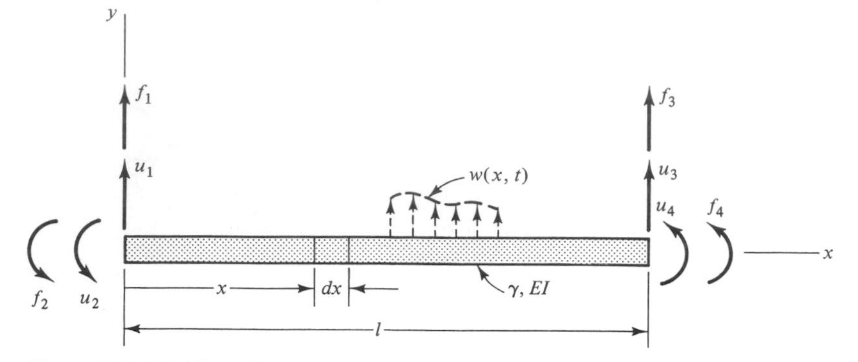

Figure 7-8 Uniform beam element.

length γ and has a bending stiffness of EI. The *lateral* displacement of any section a distance x along the beam element can be expressed as

$$y(x, t) = \phi_1(x)u_1 + \phi_2(x)u_2 + \phi_3(x)u_3 + \phi_4(x)u_4 \tag{7-67}$$

or more simply as

$$y(x, t) = \sum_{i=1}^{4} \phi_i(x)u_i \tag{7-68}$$

in which $\phi_1(x)$, $\phi_2(x)$, $\phi_3(x)$, and $\phi_4(x)$ are appropriate shape functions, and u_1, u_2, u_3, and u_4 are functions of time defining the displacements at the ends of the beam element (joint displacements). It can be seen in Fig. 7-8 that u_1 and u_3 are *lateral* joint displacements, while u_2 and u_4 are *angular* joint displacements. These four joint displacements can be considered as the generalized coordinates of the beam element, which reduces it from a distributed mass having a theoretically infinite number of degrees of freedom to an element with just four degrees of freedom.

The effects of the distributed transverse force $w(x, t)$ shown acting on the element in Fig. 7-8 are represented by the nonpotential joint forces f_1 and f_3 associated with the lateral joint displacements u_1 and u_3 and the nonpotential joint moments f_2 and f_4 associated with the angular joint displacements u_2 and u_4.

Referring once more to Fig. 7-8, and assuming small angular displacements, we can write the boundary conditions in terms of the joint displacements as

$$\left.\begin{aligned} y(0, t) &= u_1 \\ \frac{\partial y}{\partial x}(0, t) &= u_2 \\ y(l, t) &= u_3 \\ \frac{\partial y}{\partial x}(l, t) &= u_4 \end{aligned}\right\} \tag{7-69}$$

Referring to Eq. 7-67, it can be seen that for this expression to satisfy the boundary conditions of Eq. 7-69, the shape functions $\phi_i(x)$ and their derivatives with respect to x, $\phi_i'(x)$ must have the following boundary values:

$$\begin{aligned} &\phi_1(0) = 1, \quad \phi_1'(0) = 0, \quad \phi_1(l) = 0, \quad \phi_1'(l) = 0 \\ &\phi_2(0) = 0, \quad \phi_2'(0) = 1, \quad \phi_2(l) = 0, \quad \phi_2'(l) = 0 \end{aligned}$$

$$\phi_3(0) = 0, \qquad \phi_3'(0) = 0, \qquad \phi_3(l) = 1, \qquad \phi_3'(l) = 0$$
$$\phi_4(0) = 0, \qquad \phi_4'(0) = 0, \qquad \phi_4(l) = 0, \qquad \phi_4'(l) = 1$$

Noting that there are four boundary values known for each shape function and derivative, and recalling from elementary beam theory that the elastic curves of beams are commonly given by polynomials, we shall assume that the shape functions are third-degree polynomials having the form of[5]

$$\phi_i(x) = a_i + b_i x + c_i x^2 + d_i x^3 \qquad (i = 1, 2, 3, 4) \tag{7-70}$$

The four constants a_i, b_i, c_i, and d_i are then determined by substituting the above boundary values into Eq. 7-70 and its derivative. For example, to determine the polynomial representing $\phi_1(x)$, the first row of boundary values above are substituted into Eq. 7-70 and its derivative to find that

$$a_1 = 1 \qquad b_1 = 0$$
$$c_1 = -\frac{3}{l^2} \qquad d_1 = \frac{2}{l^3}$$

Then

$$\phi_1(x) = 1 - 3\left(\frac{x}{l}\right)^2 + 2\left(\frac{x}{l}\right)^3$$

The four shape functions determined in this manner are shown sketched in Fig. 7-9 for future reference, since they will be used subsequently in determining the mass matrix of the beam element. The second derivatives of the functions are also shown in the figure, since they will be used later in determining the stiffness matrix of the beam element.

Proceeding with our determination of the mass and stiffness matrices of the beam element, we note that the velocity of any section along the beam is $\dot{y}(x, t)$, so that the kinetic energy of the beam element can be expressed as

$$T = \frac{\gamma}{2}\int_0^l [\dot{y}(x, t)]^2\, dx$$

or as

$$T = \frac{\gamma}{2}\int_0^l [\phi_1(x)\dot{u}_1 + \phi_2(x)\dot{u}_2 + \phi_3(x)\dot{u}_3 + \phi_4(x)\dot{u}_4]^2\, dx \tag{7-71}$$

Since we can consider the joint displacements u_1, u_2, u_3, and u_4 as the generalized coordinates of the beam element, we first take the partial derivative of T with respect to a particular $\dot{u}_j$ and then the time derivative of that expression to obtain

$$\frac{d}{dt}\left(\frac{\partial T}{\partial \dot{u}_j}\right) = \gamma \int_0^l [\phi_1(x)\ddot{u}_1 + \phi_2(x)\ddot{u}_2 + \phi_3(x)\ddot{u}_3 + \phi_4(x)\ddot{u}_4]\phi_j(x)\, dx \tag{7-72}$$

in which $j = 1, 2, 3, 4$. Upon examining Eq. 7-72, it should be apparent that all the possible terms for a term-by-term integration can be expressed as

$$\left[\gamma \int_0^l \phi_i(x)\phi_j(x)\, dx\right]\ddot{u}_i \qquad \begin{pmatrix} i = 1, 2, 3, 4 \\ j = 1, 2, 3, 4 \end{pmatrix}$$

[5] With four boundary values, a third-degree polynomial with four constants is the highest-order polynomial we can use.

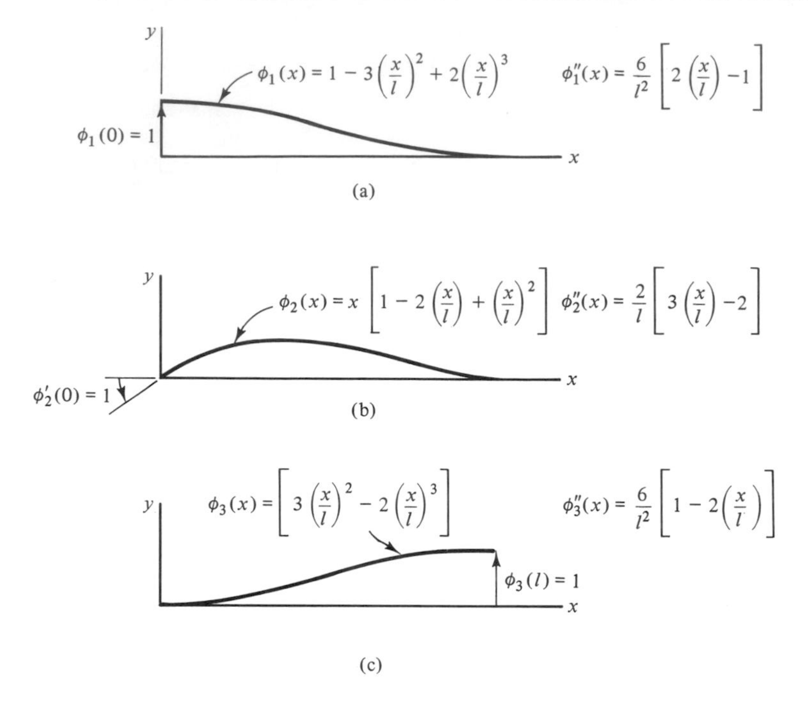

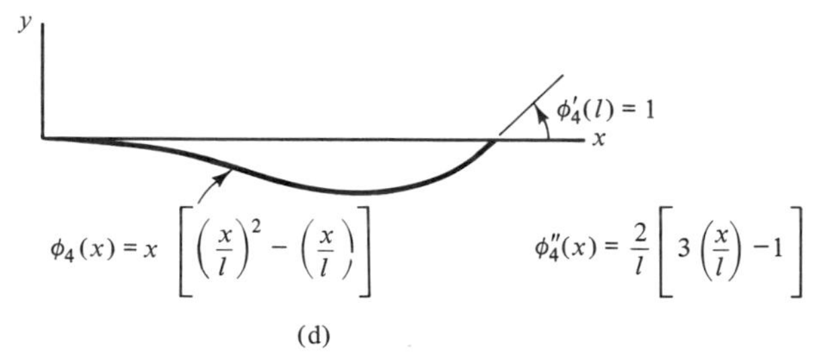

Figure 7-9 Shape functions $\phi_i(x)$ and derivatives $\phi_i''(x)$.

in which the expression within the bracket corresponds to the mass elements m_{ij}. Thus, the mass elements for the finite beam element can be determined from

$$m_{ij} = \gamma \int_0^l \phi_i(x)\phi_j(x)\,dx \qquad \begin{pmatrix} i = 1, 2, 3, 4 \\ j = 1, 2, 3, 4 \end{pmatrix} \tag{7-73}$$

using the shape functions shown in Fig. 7-9. For example, referring to Fig. 7-9a for $i = j = 1$, we can write Eq. 7-73 as

$$m_{11} = \gamma \int_0^l [\phi_1(x)]^2\,dx = \gamma \int_0^l \left[1 - 3\left(\frac{x}{l}\right)^2 + 2\left(\frac{x}{l}\right)^3\right]^2 dx$$

to find that

$$m_{11} = \tfrac{156}{420}\,\gamma l$$

Similarly, for $i = 1, j = 2$, reference to Fig. 7-9a and b enables us to write Eq. 7-73 as

$$m_{12} = m_{21} = \gamma \int_0^l \left[1 - 3\left(\frac{x}{l}\right)^2 + 2\left(\frac{x}{l}\right)^3\right](x)\left[1 - 2\left(\frac{x}{l}\right) + \left(\frac{x}{l}\right)^2\right] dx$$

from which
$$m_{12} = m_{21} = \tfrac{22}{420}\gamma l^2$$

Proceeding in a similar manner with the other combinations of i and j values, the symmetric mass matrix of the beam element is obtained as

$$[m]_e = \frac{\gamma l}{420}\begin{bmatrix} 156 & 22l & 54 & -13l \\ 22l & 4l^2 & 13l & -3l^2 \\ 54 & 13l & 156 & -22l \\ -13l & -3l^2 & -22l & 4l^2 \end{bmatrix} \tag{7-74}$$

The stiffness matrix for the beam element is determined by considering the potential energy (the strain energy) of the beam. Referring to page 92, we can write from Eq. 2-69 that

$$U = \frac{EI}{2}\int_0^l \left[\frac{\partial^2 y}{\partial x^2}(x, t)\right]^2 dx \tag{7-75}$$

If we take the second partial derivative of Eq. 7-67 with respect to x, which is

$$\frac{\partial^2 y}{\partial x^2}(x, t) = \phi_1''(x)u_1 + \phi_2''(x)u_2 + \phi_3''(x)u_3 + \phi_4''(x)u_4$$

and substitute it into Eq. 7-75, we find that

$$U = \frac{EI}{2}\int_0^l [\phi_1''(x)u_1 + \phi_2''(x)u_2 + \phi_3''(x)u_3 + \phi_4''(x)u_4]^2\, dx \tag{7-76}$$

Then taking the partial derivative of U with respect to a particular generalized coordinate u_j, we obtain

$$\frac{\partial U}{\partial u_j} = EI\int_0^l [\phi_1''(x)u_1 + \phi_2''(x)u_2 + \phi_3''(x)u_3 + \phi_4''(x)u_4]\phi_j''(x)\, dx \tag{7-77}$$

in which $j = 1, 2, 3, 4$. Examining this equation, it should be apparent that all the possible terms for a term-by-term integration have the form of

$$\left[EI\int_0^l \phi_i''(x)\phi_j''(x)\, dx\right]u_i \qquad \begin{pmatrix} i = 1, 2, 3, 4 \\ j = 1, 2, 3, 4 \end{pmatrix}$$

Referring to Eq. 7-38 and considering four degrees of freedom, it can be seen that the bracketed term above corresponds to the stiffness elements k_{ij}. Thus, the k_{ij}'s of the stiffness matrix for the finite beam element can be determined from

$$k_{ij} = EI\int_0^l \phi_i''(x)\phi_j''(x)\, dx \qquad \begin{pmatrix} i = 1, 2, 3, 4 \\ j = 1, 2, 3, 4 \end{pmatrix} \tag{7-78}$$

using the $\phi_i''(x)$'s shown in Fig. 7-9. For example, referring to Fig. 7-9a for $i = j = 1$, Eq. 7-78 can be written as

$$k_{11} = EI\int_0^l [\phi_1''(x)]^2\, dx = EI\int_0^l \left(\frac{6}{l^2}\right)^2\left[2\left(\frac{x}{l}\right) - 1\right]^2 dx$$

from which

$$k_{11} = \frac{12EI}{l^3}$$

Similarly, referring to Fig. 7-9a and b for $i = 1, j = 2$, Eq. 7-78 can be written as

$$k_{12} = k_{21} = EI \int_0^l \frac{6}{l^2}\left[2\left(\frac{x}{l}\right) - 1\right]\left(\frac{2}{l}\right)\left[3\left(\frac{x}{l}\right) - 2\right] dx$$

from which

$$k_{12} = k_{21} = \frac{6EI}{l^2}$$

Proceeding in similar fashion with the other combinations of i and j values, the symmetric stiffness matrix of the beam element is obtained as[6]

$$[k]_e = \frac{EI}{l^3}\begin{bmatrix} 12 & 6l & -12 & 6l \\ 6l & 4l^2 & -6l & 2l^2 \\ -12 & -6l & 12 & -6l \\ 6l & 2l^2 & -6l & 4l^2 \end{bmatrix} \tag{7-79}$$

The nonpotential joint forces f_1 and f_3 and joint moments f_2 and f_4 representing the effects of the external distributed force $w(x, t)$ acting on the beam element (see Fig. 7-8) are determined by considering the differential work dW_i done by $w(x, t)$ for a differential displacement du_i, which is

$$dW_i = \int_0^l \{w(x, t)d[y(x, t)]_i\} \, dx \qquad i = 1, 2, 3, 4 \tag{7-80}$$

For a differential displacement du_i, with all other coordinates held fixed, the displacement function is

$$d[y(x, t)]_i = \phi_i(x) \, du_i$$

and

$$dW_i = \left[\int_0^l [w(x, t)\phi_i(x)] \, dx\right] du_i$$

Then, since

$$dW_i = Q_i \, du_i = f_i \, du_i$$

we see that

$$f_i = \int_0^l [w(x, t)\phi_i(x)] \, dx \qquad i = 1, 2, 3, 4 \tag{7-81}$$

[6] This stiffness matrix can also be determined using the fundamental definition of the stiffness coefficients k_{ij} discussed in Sec. 5-3.

From Eq. 7-81,

$$\left.\begin{aligned} f_1 &= \int_0^l [w(x, t)\phi_1(x)]\,dx \\ f_2 &= \int_0^l [w(x, t)\phi_2(x)]\,dx \\ f_3 &= \int_0^l [w(x, t)\phi_3(x)]\,dx \\ f_4 &= \int_0^l [w(x, t)\phi_4(x)]\,dx \end{aligned}\right\} \tag{7-82}$$

in which $\phi_1(x)$, $\phi_2(x)$, $\phi_3(x)$, and $\phi_4(x)$ are found in Fig. 7-9.

Beam Element—Combined Bending and Axial Deformations

When beam elements are subjected to both bending and axial deformations, three joint displacements are required at each end of the element as shown in Fig. 7-10. The element thus has six generalized coordinates and six degrees of freedom, which means that its mass and stiffness matrices are 6×6 matrices.

It is common practice in linear finite-element analysis to assume that the axial forces associated with the axial joint displacements u_1 and u_4 shown in Fig. 7-10 have only a negligible effect on the shape functions associated with the joint displacements u_2, u_3, u_5, and u_6. With this assumption, the mass and stiffness matrices for the beam element in Fig. 7-10 can be derived from the mass and stiffness matrices previously determined for the rod element with axial deformation and the beam element with bending deformation shown in Fig. 7-11a and b. The mass and stiffness matrices of the beam element of Fig. 7-10 are shown in Fig. 7-11c.

To illustrate how the mass and stiffness matrices of the beam element shown in Fig. 7-11c were derived from the mass and stiffness matrices of the rod element in Fig. 7-11a and the beam element in Fig. 7-11b, let us consider first only the mass matrices of these elements. The subscripts of the m_{ij}'s of the mass matrix of each element correspond to the subscripts of the generalized coordinates of that element. Thus, the m_{ij}'s of the mass matrices of the three elements can be related by establishing a correspondence between the generalized coordinates of the elements. For example, Fig. 7-11 shows that the generalized coordinates u_1 and u_2 of the rod element correspond to the generalized coordinates u_1 and u_4 of the beam element in Fig. 7-11c since they describe respective axial displacements at the joints

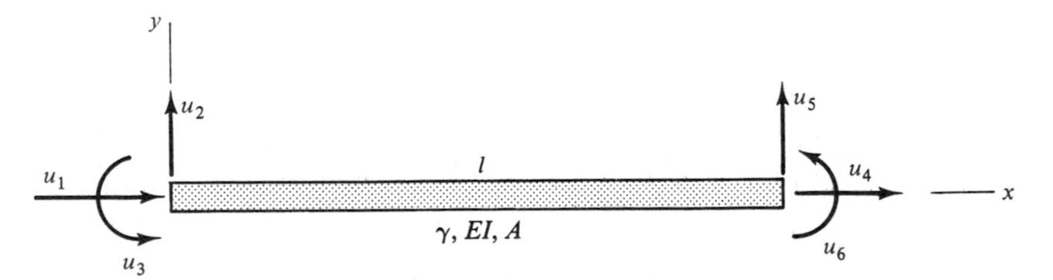

Figure 7-10 Beam element with bending and axial deformations.

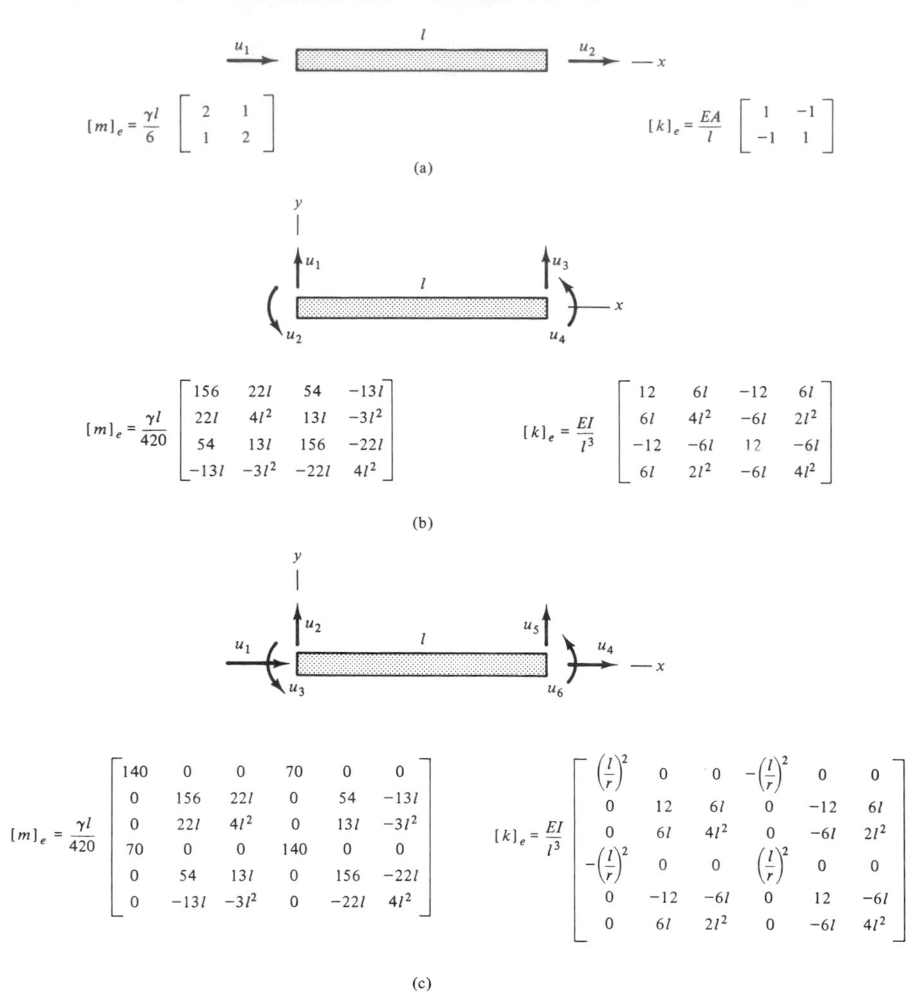

Figure 7-11 Mass and stiffness matrices of finite rod and beam elements (local matrices). (a) Rod element—axial deformation. (b) Beam element—bending deformation. (c) Beam element—axial and bending deformations.

of the two elements. With this correspondence, m_{12} of the mass matrix of the rod element is m_{14} of the mass matrix of the beam shown in Fig. 7-11c, and m_{11} and m_{22} in the former are m_{11} and m_{44} in the latter. Similarly, the generalized coordinates u_1 and u_3 of the beam element shown in Fig. 7-11b correspond to the generalized coordinates u_2 and u_5 of the beam element in Fig. 7-11c since they describe respective lateral displacements at the joints of the two elements. Thus, m_{11}, m_{13}, and m_{33} of the mass matrix of the beam in Fig. 7-11b are m_{22}, m_{25}, and m_{55} of the mass matrix of the beam in Fig. 7-11c. Similarly, u_2 and u_4 of the beam in Fig. 7-11b correspond to u_3 and u_6 of the beam in Fig. 7-11c, so that m_{22}, m_{24}, and m_{44} of the mass matrix in Fig. 7-11b are m_{33}, m_{36}, and m_{66}, respectively, of the mass matrix in Fig. 7-11c. Continuing in this manner, we find that m_{12}, m_{14}, m_{23}, and m_{34} of the mass matrix in Fig. 7-11b correspond to m_{23}, m_{26}, m_{35}, and m_{56}, respectively, of the mass matrix in Fig. 7-11c, and complete the nonzero elements of the upper triangular portion of the latter matrix. The m_{ij}'s shown in the lower half of the mass matrix in Fig. 7-11c obviously come from symmetry.

The stiffness matrix shown in Fig. 7-11c was derived from the stiffness matrices shown in Fig. 7-11a and b in the same manner, with the terms of the matrices in Fig. 7-11a and c related by

$$r = \sqrt{\frac{I}{A}}$$

In this expression, r is the radius of gyration of the beam cross section about the same centroidal bending axis for which I is determined.

In using the finite-element method for a tapered beam, it is only necessary to determine the cross-sectional areas, area moment of inertias, and masses per unit length at the mid-sections of the beam segments. For example, at section b midway between stations 1 and 2 in Fig. 5-42,

$$A_b = 2z_b h$$

$$I_b = \frac{2z_b h^3}{12}$$

$$\gamma_b = 2z_b h\rho$$

in which $2z_b$ is the width of the beam, and ρ is the mass density of the beam material at that section (see Prob. 7-29).

From the preceding discussion of rod and beam elements, it should be apparent that the equations of motion of these and other finite elements can be written in terms of their joint displacements as

$$[m]_e\{\ddot{u}_i\}_e + [k]_e\{u_i\}_e = \{f_i\}_e \tag{7-83}$$

where $\{u_i\}_e$ = axial, lateral, or angular joint displacements
$[m]_e$ = mass matrix of element
$[k]_e$ = stiffness matrix of element
$\{f_i\}_e$ = joint forces and moments

In the section that follows we are interested in the coordinate transformations that are used in assembling elements into systems, and in writing the equations of motion of these systems.

7-4 COORDINATE TRANSFORMATIONS—USE OF TRANSFORMATION MATRIX

The axes used to reference the joint displacements of finite elements are usually selected to simplify the determination of the mass and stiffness properties of the elements. For example, it was logical to identify the joint displacements $u_1, u_2, \ldots, u_6$ of the beam element in Fig. 7-10 with respect to the x, y, z axes in the figure since they are the axes commonly used with beams in elementary beam theory. Such axes associated with the element are referred to as *local axes*, and the joint displacements referenced with respect to them are known as *local joint displacements* and designated by u_i as shown in Fig. 7-10.

The axes associated with a system composed of finite elements, such as the $\bar{x}$, $\bar{y}$, $\bar{z}$ axes in Fig. 7-12a, are referred to as *global axes*, and the joint displacements of the system

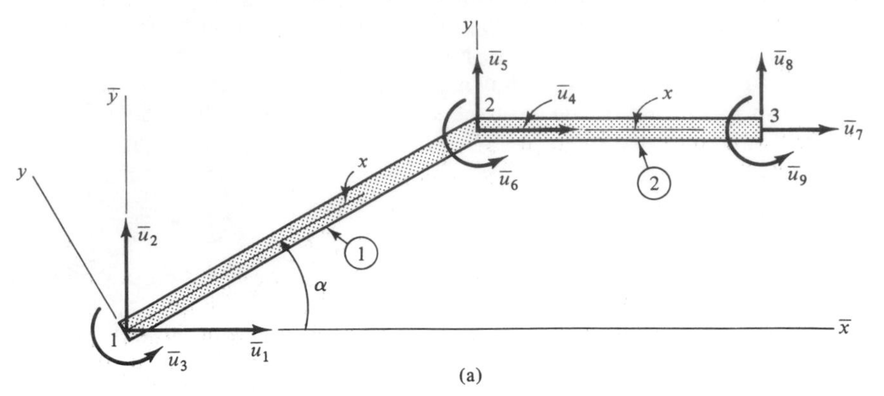

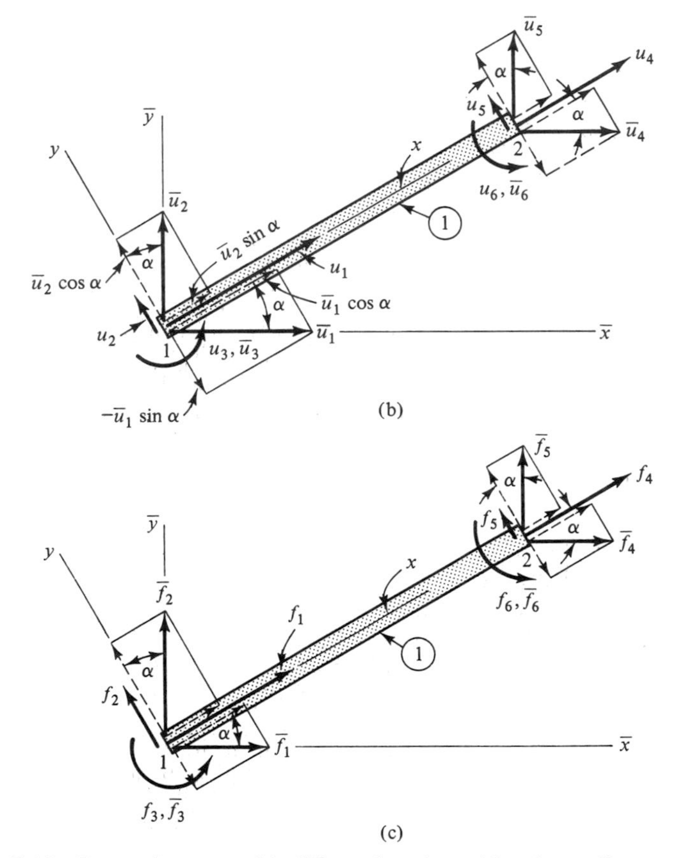

Figure 7-12 Beam elements with differently oriented local coordinates.

referenced with respect to them are known as global joint displacements and designated as $\bar{u}_i$. The orientation of the global axes is arbitrary but is generally selected to be parallel to as many as possible of the local axes of the system elements, the reasons for which will become clear in the discussions that follow.

The local axes of each element retain their orientation with respect to the element, regardless of the orientation the element takes in space as part of a system. For example, the system shown in Fig. 7-12a consists of two beam elements, each with its individual set of local x, y, z axes. However, as can be seen, the local axes of element 1 are at an angle α from the global $\bar{x}$, $\bar{y}$ axes, while the local axes of element 2 are parallel to the global axes.

Because the elements do vary in orientation with respect to the system axes, a method of relating the local joint displacements of the elements to the corresponding joint displacements of the system must be provided. This is accomplished by expressing the local joint displacements u_i of each element in terms of components of the global joint displacements $\bar{u}_i$, which process is referred to as a *coordinate transformation*. The relationships determined in the process form a *transformation matrix* that relates the local and global joint displacements of the elements and, as we shall see later, also relates their local and global mass and stiffness matrices and local and global force-moment matrices.

To derive the transformation matrix relating coordinate systems for which z and $\bar{z}$ are parallel, let us consider element 1 in Fig. 7-12b. The local joint displacements of the element are $u_1, u_2, \ldots, u_6$, while the global joint displacements of the system are $\bar{u}_1, \bar{u}_2, \ldots, \bar{u}_6$. From the figure it can be seen that the local and global joint displacements at joint 1 are related by

$$\begin{aligned} u_1 &= \bar{u}_1 \cos\alpha + \bar{u}_2 \sin\alpha \\ u_2 &= -\bar{u}_1 \sin\alpha + \bar{u}_2 \cos\alpha \\ u_3 &= \bar{u}_3 \end{aligned}$$

while at joint 2 they are related by

$$\begin{aligned} u_4 &= \bar{u}_4 \cos\alpha + \bar{u}_5 \sin\alpha \\ u_5 &= -\bar{u}_4 \sin\alpha + \bar{u}_5 \cos\alpha \\ u_6 &= \bar{u}_6 \end{aligned}$$

These equations can be expressed in matrix form as

$$\begin{Bmatrix} u_1 \\ u_2 \\ u_3 \\ u_4 \\ u_5 \\ u_6 \end{Bmatrix}_e = \begin{bmatrix} \cos\alpha & \sin\alpha & 0 & 0 & 0 & 0 \\ -\sin\alpha & \cos\alpha & 0 & 0 & 0 & 0 \\ 0 & 0 & 1 & 0 & 0 & 0 \\ 0 & 0 & 0 & \cos\alpha & \sin\alpha & 0 \\ 0 & 0 & 0 & -\sin\alpha & \cos\alpha & 0 \\ 0 & 0 & 0 & 0 & 0 & 1 \end{bmatrix} \begin{Bmatrix} \bar{u}_1 \\ \bar{u}_2 \\ \bar{u}_3 \\ \bar{u}_4 \\ \bar{u}_5 \\ \bar{u}_6 \end{Bmatrix}_e \tag{7-84}$$

or more simply as

$$\{u_i\}_e = \mathbf{R}\{\bar{u}_i\}_e \tag{7-85}$$

in which the *transformation matrix* is

$$\mathbf{R} = \begin{bmatrix} \cos\alpha & \sin\alpha & 0 & 0 & 0 & 0 \\ -\sin\alpha & \cos\alpha & 0 & 0 & 0 & 0 \\ 0 & 0 & 1 & 0 & 0 & 0 \\ 0 & 0 & 0 & \cos\alpha & \sin\alpha & 0 \\ 0 & 0 & 0 & -\sin\alpha & \cos\alpha & 0 \\ 0 & 0 & 0 & 0 & 0 & 1 \end{bmatrix} \tag{7-86}$$

Let us next consider the joint forces and moments shown in Fig. 7-12c, in which $f_1, f_2, \ldots, f_6$ are referenced with respect to the local axes of the element and $\bar{f}_1, \bar{f}_2, \ldots, \bar{f}_6$ are referenced with respect to the global axes of the system. Since the orientation of the joint forces and moments is identical to that of the joint displacements with respect to both sets of axes, it follows that the transformation matrix of Eq. 7-86 will also relate the local and global forces and moments, so that

$$\begin{Bmatrix} f_1 \\ f_2 \\ f_3 \\ f_4 \\ f_5 \\ f_6 \end{Bmatrix}_e = \mathbf{R} \begin{Bmatrix} \bar{f}_1 \\ \bar{f}_2 \\ \bar{f}_3 \\ \bar{f}_4 \\ \bar{f}_5 \\ \bar{f}_6 \end{Bmatrix}_e \tag{7-87}$$

or more simply,

$$\{f_i\}_e = \mathbf{R}\{\bar{f}_i\}_e \tag{7-88}$$

At this point an important property of the $\mathbf{R}$ matrix should be noted. It is easy to show that

$$\mathbf{R}^T\mathbf{R} = \mathbf{I} \qquad \text{(identity matrix)} \tag{7-89}$$

Then, since $\mathbf{R}^{-1}\mathbf{R} = \mathbf{I}$, it follows from Eq. 7-89 that

$$\mathbf{R}^T = \mathbf{R}^{-1} \tag{7-90}$$

which by definition shows that $\mathbf{R}$ is an *orthogonal* matrix. With this property noted, let us turn our attention to Eq. 7-83, which gives the equations of motion of a finite element in terms of its local joint displacements, and which is repeated here for the sake of convenience as

$$[m]_e\{\ddot{u}_i\}_e + [k]_e\{u_i\}_e = \{f_i\}_e \tag{7-83}$$

where $\{u_i\}_e$ = local joint displacements of element
$[m]_e$ = local mass matrix of element
$[k]_e$ = local stiffness matrix of element
$\{f_i\}_e$ = local joint forces and moments acting on element

If we now substitute the relationships obtained from Eqs. 7-85 and 7-88 into Eq. 7-83, we obtain

$$[m]_e\mathbf{R}\{\ddot{\bar{u}}_i\}_e + [k]_e\mathbf{R}\{\bar{u}_i\}_e = \mathbf{R}\{\bar{f}_i\}_e$$

Premultiplying this equation by $\mathbf{R}^{-1}$, and recalling that $\mathbf{R}^T = \mathbf{R}^{-1}$, we obtain

$$\mathbf{R}^T[m]_e\mathbf{R}\{\ddot{\bar{u}}_i\}_e + \mathbf{R}^T[k]_e\mathbf{R}\{\bar{u}_i\}_e = \{\bar{f}_i\}_e \tag{7-91}$$

which yields the equations of motion of an element in terms of the global joint displacements and forces and moments. If we let

$$\mathbf{R}^T[m]_e\mathbf{R} = [\bar{m}]_e \tag{7-92a}$$

and

$$\mathbf{R}^T[k]_e\mathbf{R} = [\bar{k}]_e \tag{7-92b}$$

Eq. 7-91 can be written as

$$[\bar{m}]_e\{\ddot{\bar{u}}_i\}_e + [\bar{k}]_e\{\bar{u}_i\}_e = \{\bar{f}_i\}_e \tag{7-93}$$

Comparing Eq. 7-93 with Eq. 7-83, we see that Eqs. 7-92a and 7-92b define the global mass and stiffness matrices, respectively, of an element.

Premultiplying Eq. 7-88 by $\mathbf{R}^{-1}$ so that

$$\mathbf{R}^{-1}\{f_i\}_e = \{\bar{f}_i\}_e$$

and again recalling that $\mathbf{R}^T = \mathbf{R}^{-1}$, we see that the global force and/or moment vector is defined by

$$\{\bar{f}_i\}_e = \mathbf{R}^T\{f_i\}_e \tag{7-94}$$

The relationships given in Eqs. 7-92 and 7-94 are of fundamental importance to the finite-element method since they provide the means of *assembling* the finite elements of a system and determining its equations of motion, as discussed in the next section.

EXAMPLE 7-4

The system shown in Fig. 7-12a is formed from two finite beam elements such as the one shown in both Figs. 7-10 and 7-11c. The local mass and stiffness matrices $[m]_e$ and $[k]_e$ for the beam element are as shown in Fig. 7-11c.

We wish to determine the global mass and stiffness matrices for the two elements by using the transformation relationships given by Eq. 7-92. For the sake of simplicity, we let $C = \cos\alpha$ and $S = \sin\alpha$.

Solution. Using the matrices shown in Fig. 7-11c, the transformation matrix $\mathbf{R}$ given by Eq. 7-86, and the transpose $\mathbf{R}^T$ of the latter, the matrix multiplication indicated by Eq. 7-92a yields the global mass matrix of element 1 as

$$[\bar{m}]_1 = \frac{\gamma l}{420}\begin{bmatrix} \begin{pmatrix}140C^2\\+156S^2\end{pmatrix} & -16CS & -22lS & \begin{pmatrix}70C^2\\+54S^2\end{pmatrix} & 16CS & 13lS \\ -16CS & \begin{pmatrix}140S^2\\+156C^2\end{pmatrix} & 22lC & 16CS & \begin{pmatrix}70S^2\\+54C^2\end{pmatrix} & -13lC \\ -22lS & 22lC & 4l^2 & -13lS & 13lC & -3l^2 \\ \begin{pmatrix}70C^2\\+54S^2\end{pmatrix} & 16CS & -13lS & \begin{pmatrix}140C^2\\+156S^2\end{pmatrix} & -16CS & 22lS \\ 16CS & \begin{pmatrix}70S^2\\+54C^2\end{pmatrix} & 13lC & -16CS & \begin{pmatrix}140S^2\\+156C^2\end{pmatrix} & -22lC \\ 13lS & -13lC & -3l^2 & 22lS & 22lC & 4l^2 \end{bmatrix} \tag{7-95}$$

and its global stiffness matrix from Eq. 7-92b as

$$[\bar{k}]_1 = \frac{EI}{l^3}\begin{bmatrix} \begin{pmatrix}(l/r)^2C^2\\+12S^2\end{pmatrix} & \begin{pmatrix}(l/r)^2CS\\-12CS\end{pmatrix} & -6lS & \begin{pmatrix}-(l/r)^2C^2\\-12S^2\end{pmatrix} & \begin{pmatrix}-(l/r)^2CS\\+12CS\end{pmatrix} & 6lS \\ \begin{pmatrix}(l/r)^2CS\\-12CS\end{pmatrix} & \begin{pmatrix}(l/r)^2S^2\\+12C^2\end{pmatrix} & 6lC & \begin{pmatrix}-(l/r)^2CS\\+12CS\end{pmatrix} & \begin{pmatrix}-(l/r)^2S^2\\-12C^2\end{pmatrix} & 6lC \\ -6lS & 6lC & 4l^2 & 6lS & -6lC & 2l^2 \\ \begin{pmatrix}-(l/r)^2C^2\\-12S^2\end{pmatrix} & \begin{pmatrix}-(l/r)^2CS\\+12CS\end{pmatrix} & 6lS & \begin{pmatrix}(l/r)^2C^2\\+12S^2\end{pmatrix} & \begin{pmatrix}(l/r)^2CS\\-12CS\end{pmatrix} & 6lS \\ \begin{pmatrix}-(l/r)^2CS\\+12CS\end{pmatrix} & \begin{pmatrix}-(l/r)^2S^2\\-12C^2\end{pmatrix} & -6lC & \begin{pmatrix}(l/r)^2CS\\-12CS\end{pmatrix} & \begin{pmatrix}(l/r)^2S^2\\+12C^2\end{pmatrix} & -6lC \\ -6lS & 6lC & 2l^2 & 6lS & -6lC & 4l^2 \end{bmatrix} \tag{7-96}$$

For element 2

$$[\bar{m}]_2 = [m]_2$$

and

$$[\bar{k}]_2 = [k]_2$$

since $\alpha = 0$. The reader is reminded that in using the global mass and stiffness matrices given by Eqs. 7-95 and 7-96, the parameters l, r, γ, E, and I are local parameters of the finite beam element.

7-5 EQUATIONS OF MOTION FOR SYSTEMS COMPOSED OF FINITE ELEMENTS

Let us now discuss some concepts that are basic to assembling the finite elements forming a system to obtain the equations of motion of the system. For our discussion we consider the system in Fig. 7-13, which consists of three beam elements joined as shown.

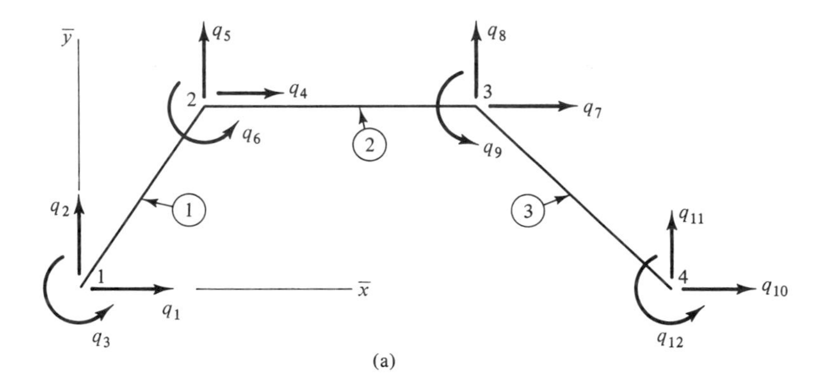

(a)

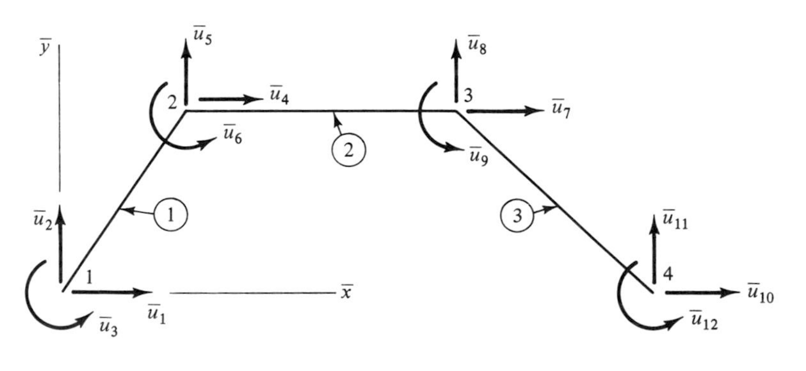

(b)

Figure 7-13 System formed by three finite beam elements. (a) Generalized coordinates of the complete system. (b) Joint displacements of elements in terms of global coordinates.

The generalized coordinates $q_1, q_2, \ldots, q_{12}$ shown in Fig. 7-13a represent the system's joint displacements with reference to the global axes $\bar{x}$, $\bar{y}$, and $\bar{z}$. The generalized coordinates correspond in direction, sense, and subscripting to the global joint displacements $\bar{u}_1$, $\bar{u}_2$, $\ldots$, $\bar{u}_{12}$ of the elements shown in Fig. 7-13b. It is initially assumed that there are no constraints on the system, leaving it as a free-free system capable of rigid-body motion. The imposition of constraints reduces the number of joint displacements (generalized coordinates), which reduces the number of degrees of freedom of the system. For example, if the lower joint of element 1 were to be fixed, the system would have only nine degrees of freedom instead of twelve, since the generalized coordinates q_1, q_2, and q_3 would be eliminated.

The global joint displacements associated with each of the finite elements are

Element 1	Element 2	Element 3
$\begin{Bmatrix} \bar{u}_1 \\ \bar{u}_2 \\ \bar{u}_3 \\ \bar{u}_4 \\ \bar{u}_5 \\ \bar{u}_6 \end{Bmatrix}_1$	$\begin{Bmatrix} \bar{u}_4 \\ \bar{u}_5 \\ \bar{u}_6 \\ \bar{u}_7 \\ \bar{u}_8 \\ \bar{u}_9 \end{Bmatrix}_2$	$\begin{Bmatrix} \bar{u}_7 \\ \bar{u}_8 \\ \bar{u}_9 \\ \bar{u}_{10} \\ \bar{u}_{11} \\ \bar{u}_{12} \end{Bmatrix}_3$

with displacements common to two elements indicating common joints for those elements.

The correspondence of the direction, sense, and subscripting between the global joint displacements and generalized coordinates mentioned above is part of a systematic scheme for relating the two in an n-degree-of-freedom system. The relationship can be expressed as

$$\{\bar{u}_i\}_e = \mathbf{A}_e \begin{Bmatrix} q_1 \\ q_2 \\ \vdots \\ q_n \end{Bmatrix} \tag{7-97}$$

in which $\mathbf{A}_e$ is a *rectangular* matrix consisting of zeros and ones, and $\{\bar{u}_i\}_e$ is a column matrix containing the global joint displacements of a particular element. For example, for element 1 in Fig. 7-13, Eq. 7-97 becomes

$$\begin{Bmatrix} \bar{u}_1 \\ \bar{u}_2 \\ \bar{u}_3 \\ \bar{u}_4 \\ \bar{u}_5 \\ \bar{u}_6 \end{Bmatrix}_1 = \begin{bmatrix} 1 & 0 & 0 & 0 & 0 & 0 & 0 & 0 & 0 & 0 & 0 & 0 \\ 0 & 1 & 0 & 0 & 0 & 0 & 0 & 0 & 0 & 0 & 0 & 0 \\ 0 & 0 & 1 & 0 & 0 & 0 & 0 & 0 & 0 & 0 & 0 & 0 \\ 0 & 0 & 0 & 1 & 0 & 0 & 0 & 0 & 0 & 0 & 0 & 0 \\ 0 & 0 & 0 & 0 & 1 & 0 & 0 & 0 & 0 & 0 & 0 & 0 \\ 0 & 0 & 0 & 0 & 0 & 1 & 0 & 0 & 0 & 0 & 0 & 0 \end{bmatrix} \begin{Bmatrix} q_1 \\ q_2 \\ q_3 \\ q_4 \\ q_5 \\ q_6 \\ \vdots \\ q_{12} \end{Bmatrix}$$

while for element 2 the relationship is

$$\begin{Bmatrix} \bar{u}_4 \\ \bar{u}_5 \\ \bar{u}_6 \\ \bar{u}_7 \\ \bar{u}_8 \\ \bar{u}_9 \end{Bmatrix}_2 = \begin{bmatrix} 0 & 0 & 0 & 1 & 0 & 0 & 0 & 0 & 0 & 0 & 0 & 0 \\ 0 & 0 & 0 & 0 & 1 & 0 & 0 & 0 & 0 & 0 & 0 & 0 \\ 0 & 0 & 0 & 0 & 0 & 1 & 0 & 0 & 0 & 0 & 0 & 0 \\ 0 & 0 & 0 & 0 & 0 & 0 & 1 & 0 & 0 & 0 & 0 & 0 \\ 0 & 0 & 0 & 0 & 0 & 0 & 0 & 1 & 0 & 0 & 0 & 0 \\ 0 & 0 & 0 & 0 & 0 & 0 & 0 & 0 & 1 & 0 & 0 & 0 \end{bmatrix} \begin{Bmatrix} q_1 \\ q_2 \\ q_3 \\ q_4 \\ q_5 \\ q_6 \\ \vdots \\ q_{12} \end{Bmatrix}$$

It can be seen that all the matrix elements in a row of $\mathbf{A}_e$ are zero except for the one that relates $\bar{u}_i$ to q_i and that the number of rows and columns of $\mathbf{A}_e$ correspond to the number of global joint displacements of a particular finite element and the number of generalized coordinates of the system, respectively.

The basic relationships required for assembling the elements to obtain the mass and stiffness matrices of the complete system are determined by considering the kinetic and strain energy of the system formed by the elements. The kinetic energy of an n-degree-of-freedom system can be expressed as[7]

$$T = \frac{1}{2} \sum_{i=1}^{n} \sum_{j=1}^{n} m_{ij} \dot{q}_i \dot{q}_j$$

or in matrix form as

$$T = \tfrac{1}{2}\{\dot{q}_i\}^T \mathbf{M} \{\dot{q}_i\} \tag{7-98}$$

in which $\mathbf{M}$ is the mass matrix of the system. The strain energy of the system can be expressed as

$$U = \frac{1}{2} \sum_{i=1}^{n} \sum_{j=1}^{n} k_{ij} q_i q_j$$

or in matrix form as

$$U = \tfrac{1}{2}\{q_i\}^T \mathbf{K} \{q_i\} \tag{7-99}$$

in which $\mathbf{K}$ is the stiffness matrix of the system.[7]

It is suggested that the reader verify Eqs. 7-98 and 7-99 using the mass and stiffness matrices and the q_i's of Example 7-1.

Recalling the direct correspondence mentioned earlier between the global joint displacements and the generalized coordinates of a system as evidenced in Fig. 7-13, we can express the kinetic and strain energy of a particular element of the system by analogy to Eqs. 7-98 and 7-99 as

$$T_e = \tfrac{1}{2}\{\dot{\bar{u}}_i\}_e^T [\bar{m}]_e \{\dot{\bar{u}}_i\}_e \tag{7-100}$$

and

$$U_e = \tfrac{1}{2}\{\bar{u}_i\}_e^T [\bar{k}]_e \{\bar{u}_i\}_e \tag{7-101}$$

Since the total energy of the system can be obtained by adding the kinetic and strain energies of the finite elements of the system, we can write that

$$T = \frac{1}{2} \sum_{e=1}^{p} \{\dot{\bar{u}}_i\}_e^T [\bar{m}]_e \{\dot{\bar{u}}_i\}_e \tag{7-102}$$

and

$$U = \frac{1}{2} \sum_{e=1}^{p} \{\bar{u}_i\}_e^T [\bar{k}]_e \{\bar{u}_i\}_e \tag{7-103}$$

in which p is the number of elements in the system.

[7] M. J. Forray, *Variational Calculus in Science and Engineering,* pp. 75–81, McGraw-Hill Book Co., New York, 1968.

Referring to Eq. 7-97, we can write that

$$\{\bar{u}_i\}_e^T = \{q_i\}^T \mathbf{A}_e^T \tag{7-104a}$$

$$\{\dot{\bar{u}}_i\}_e = \mathbf{A}_e\{\dot{q}_i\} \tag{7-104b}$$

$$\{\dot{\bar{u}}_i\}_e^T = \{\dot{q}_i\}^T \mathbf{A}_e^T \tag{7-104c}$$

since the transpose of the product of two matrices is the product in reverse order of their transposes. Substituting Eqs. 7-104b and 7-104c into Eq. 7-102 yields

$$T = \frac{1}{2}\sum_{e=1}^{p} \{\dot{q}_i\}^T \mathbf{A}_e^T [\bar{m}]_e \mathbf{A}_e \{\dot{q}_i\} \tag{7-105}$$

Comparing the above equation with Eq. 7-98 reveals that the mass matrix *of the system* is

$$\mathbf{M} = \sum_{e=1}^{p} \mathbf{A}_e^T [\bar{m}]_e \mathbf{A}_e \tag{7-106}$$

in which $[\bar{m}]_e$ is the global mass matrix of an individual element (see Eq. 7-92a).

Similarly, substituting Eqs. 7-97 and 7-104a into Eq. 7-103 gives

$$U = \frac{1}{2}\sum_{e=1}^{p} \{q_i\}^T \mathbf{A}_e^T [\bar{k}]_e \mathbf{A}_e \{q_i\} \tag{7-107}$$

Comparing this equation with Eq. 7-99 shows that the stiffness matrix *of the system* is

$$\mathbf{K} = \sum_{e=1}^{p} \mathbf{A}_e^T [\bar{k}]_e \mathbf{A}_e \tag{7-108}$$

in which $[\bar{k}]_e$ is the global stiffness matrix of an individual element (see Eq. 7-92b). Equations 7-106 and 7-108 provide the means of calculating the mass and stiffness matrices of a system from the global mass and stiffness matrices of the elements, which have been obtained by transformations from the local mass and stiffness matrices of the elements, which have been obtained in turn from the mass and stiffness properties of the finite elements using local coordinates.

To determine the generalized nonpotential forces Q_i we recall that in Lagrange's equations the differential work $dW_i = Q_i\, dq_i$ is due to the differential change dq_i of a particular coordinate q_i. Thus, we can write for an n-degree-of-freedom system that

$$dW_j = \{dq_i\}^T\{Q_i\} \qquad i = 1, 2, \ldots, n \tag{7-109}$$

in which the value of j is equal to the i value of the one nonzero dq. Recalling that the global joint forces $\{f_i\}_e$ result from the distributed nonpotential force $w(x, t)$ acting on an element and that each element at a joint contributes only one nonzero term to $\{d\bar{u}\}_e^T$ for each dW_j, we can write that

$$dW_j = \sum_{e=1}^{p} \{d\bar{u}_i\}_e^T \{\bar{f}_i\}_e \tag{7-110}$$

From Eq. 7-104a we see that

$$\{d\bar{u}_i\}_e^T = \{dq_i\}^T \mathbf{A}_e^T \tag{7-111}$$

which when substituted into Eq. 7-110 gives

$$dW_j = \{dq_i\}^T \sum_{e=1}^{p} \mathbf{A}_e^T \{\bar{f}_i\}_e \tag{7-112}$$

Comparing Eqs. 7-109 and 7-112, we see that

$$\{Q_i\} = \sum_{e=1}^{p} \mathbf{A}_e^T \{\bar{f}_i\}_e \tag{7-113}$$

in which

$$\{\bar{f}_i\}_e = \mathbf{R}^T \{f_i\}_e$$

as expressed in Eq. 7-94.

With the undamped, coupled differential equations of motion given by

$$\mathbf{M}\{\ddot{q}_i\} + \mathbf{K}\{q_i\} = \{Q_i\} \qquad i = 1, 2, \ldots, n \tag{7-114}$$

it should be apparent that the purpose of the preceding discussion has been to determine the following expressions for use in Eq. 7-114:

$$\mathbf{M} = \sum_{e=1}^{p} \mathbf{A}_e^T [\bar{m}]_e \mathbf{A}_e \qquad \text{(see Eq. 7-106)}$$

$$\mathbf{K} = \sum_{e=1}^{p} \mathbf{A}_e^T [\bar{k}]_e \mathbf{A}_e \qquad \text{(see Eq. 7-108)}$$

$$\{Q_i\} = \sum_{e=1}^{p} \mathbf{A}_e^T \{\bar{f}_i\}_e \qquad \text{(see Eq. 7-113)}$$

with

$$\left.\begin{aligned} [\bar{m}]_e &= \mathbf{R}^T [m]_e \mathbf{R} \\ [\bar{k}]_e &= \mathbf{R}^T [k]_e \mathbf{R} \end{aligned}\right\} \qquad \text{(see Eq. 7-92)}$$

$$\{\bar{f}_i\}_e = \mathbf{R}^T \{f_i\}_e \qquad \text{(see Eq. 7-94)}$$

After some additional comments on the formulation of $\{Q_i\}$, introducing damping, and a discussion of boundary (support) conditions, we shall be ready to apply the finite-element method in a vibration analysis.

If a system is subjected to concentrated forces $P_i(t)$ and/or concentrated moments $M_i(t)$ *at the joints* of the system in addition to distributed forces and/or moments acting on the elements, they will become part of the appropriate ith components of the generalized nonpotential force and/or moment vector $\{Q_i\}$. If they are the only forces and/or moments acting on the system, they will be the only components of $\{Q_i\}$. For example, the frame in Fig. 7-14 has only the joint force $P(t)$ and moment $M(t)$ acting on the joints shown. They give rise to the vector components Q_4 and Q_9 since they are directly associated with the generalized coordinates q_4 and q_9, as can be seen by comparing Fig. 7-14a and b, and $\{Q_i\}$ in Eq. 7-114 would appear as

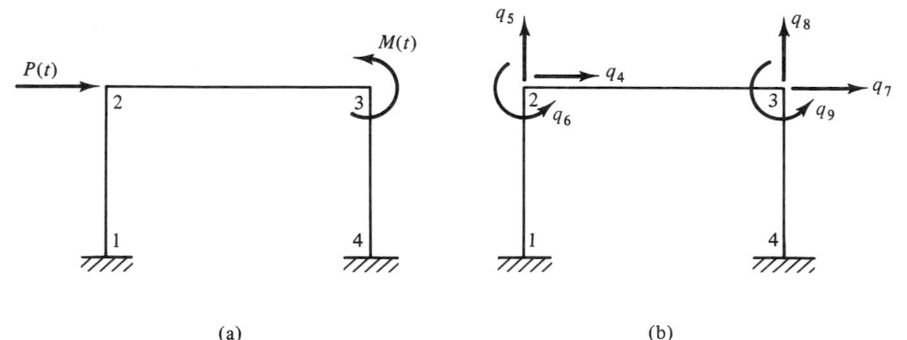

Figure 7-14 Concentrated force $P(t)$ and moment $M(t)$ acting at joints of a frame system.

$$\{Q_i\} = \begin{Bmatrix} Q_4 \\ Q_5 \\ Q_6 \\ Q_7 \\ Q_8 \\ Q_9 \end{Bmatrix} = \begin{Bmatrix} P(t) \\ 0 \\ 0 \\ 0 \\ 0 \\ M(t) \end{Bmatrix}$$

Damping can be introduced into the system equations (Eq. 7-114) by way of a damping matrix **C**. The matrix **C** is obtained from damping coefficient matrices $[c]_e$ of the elements by the same procedure used to obtain **M** and **K**. The drawback of this approach is the difficulty in formulating realistic $[c]_e$ matrices for the elements.

An alternate, and more practical, approach is to determine the response of a damped system by using the modal-analysis method discussed in Sec. 6-6 in conjunction with the finite-element method. This procedure will be illustrated in Example 7-14.

Boundary (Support) Conditions

Up to now all the joints of a system have been assumed as unconstrained so that rigid-body motion of the system was possible. However, most systems have supports of some type that constrain the system and reduce the number of joint displacements (generalized coordinates) and thus the degrees of freedom of the system, and rigid-body motion is impossible.

A simple procedure for eliminating the generalized coordinates associated with the global joint displacements $\bar{u}_i$ that go to zero because of constraints imposed on the system is to eliminate the corresponding ith rows and columns from the **M** and **K** matrices of the equations of motion determined for the system with no constraints. The corresponding rows of $\{\ddot{q}_i\}$, $\{q_i\}$, and $\{Q_i\}$ are also eliminated.

To illustrate this procedure, let us consider the nonuniform beam of Fig. 7-15a, which is modeled by the two beam elements shown and subjected to the force $P(t)$. The global joint displacements $\bar{u}_1, \bar{u}_2, \ldots, \bar{u}_6$ and the generalized coordinates $q_1, q_2, \ldots, q_6$ for the unconstrained beam system are shown in Fig. 7-15a. The actual beam system is fixed at the left end, pinned at the right end, and subjected to the force $P(t)$ as shown in Fig. 7-15b. Referring to Eq. 7-114, the equations of motion of the free-free beam are given by

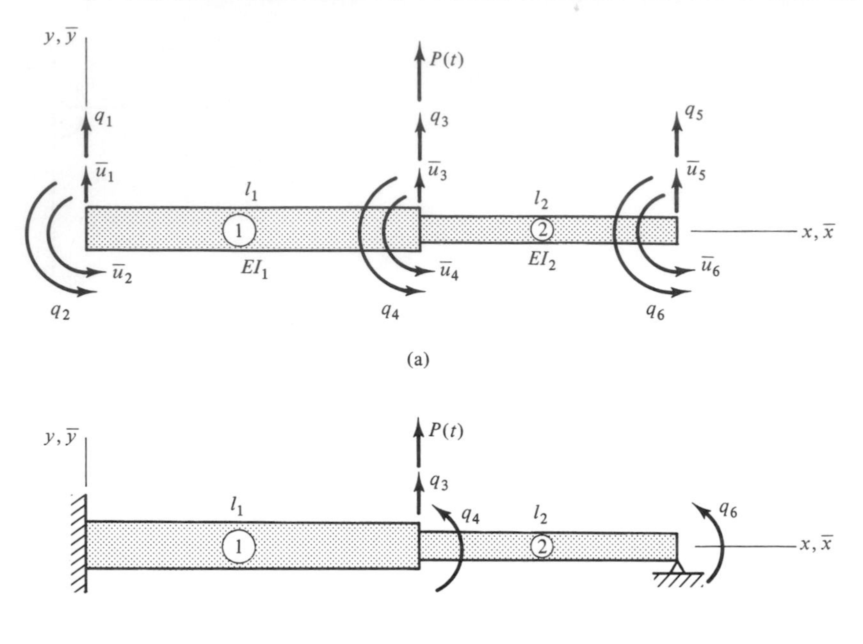

Figure 7-15 Nonuniform beam modeled as two beam elements.

$$\mathbf{M}\begin{Bmatrix} \ddot{q}_1 \\ \ddot{q}_2 \\ \ddot{q}_3 \\ \ddot{q}_4 \\ \ddot{q}_5 \\ \ddot{q}_6 \end{Bmatrix} + \mathbf{K}\begin{Bmatrix} q_1 \\ q_2 \\ q_3 \\ q_4 \\ q_5 \\ q_6 \end{Bmatrix} = \begin{Bmatrix} 0 \\ 0 \\ P(t) \\ 0 \\ 0 \\ 0 \end{Bmatrix} \tag{7-115}$$

in which **M** and **K** are 6×6 matrices determined by assembling the global matrices $[\bar{m}]_e$ and $[\bar{k}]_e$, respectively, of the two beam elements. Since $q_1 = q_2 = q_5 = 0$ for the support conditions shown, the *first, second,* and *fifth* rows and columns of **M** and **K** are eliminated. The equations of motion for the resulting fixed-pinned beam modeled by the two beam elements then become

$$\begin{bmatrix} m_{33} & m_{34} & m_{36} \\ m_{43} & m_{44} & m_{46} \\ m_{63} & m_{64} & m_{66} \end{bmatrix}\begin{Bmatrix} \ddot{q}_3 \\ \ddot{q}_4 \\ \ddot{q}_6 \end{Bmatrix} + \begin{bmatrix} k_{33} & k_{34} & k_{36} \\ k_{43} & k_{44} & k_{46} \\ k_{63} & k_{64} & k_{66} \end{bmatrix}\begin{Bmatrix} q_3 \\ q_4 \\ q_6 \end{Bmatrix} = \begin{Bmatrix} P(t) \\ 0 \\ 0 \end{Bmatrix}$$

7-6 OUTLINE FOR USE OF FINITE-ELEMENT METHOD IN VIBRATION ANALYSIS

Having discussed the basic concepts of the finite-element method as it pertains to vibration analysis, let us outline a general procedure for applying this method, in which the finite elements and corresponding mass and stiffness matrices shown in Fig. 7-11 are considered.

1. Select the type of element from Fig. 7-11 that is to be used to model the system. Sketch the system, showing the number of elements used, and indicating the global axes selected for the system. Determine the angle α between the local axes of each element used and the global axes selected for the system. As a general rule, the number of elements used will depend upon the accuracy desired, and consequently upon the experience and judgment of the analyst. Since the finite-element method discretizes the distributed mass and stiffness properties of a system, increasing the number of elements generally improves the accuracy of the results but also increases the number of computations required.
2. Assume initially that all the system joints are unconstrained, so that all possible joint displacements of the system exist for the type of elements used to form the system. Show the global joint displacements as $\bar{u}_1, \bar{u}_2, \ldots, \bar{u}_n$ on the sketch of the system, and subscript them in correct numerical sequence at each joint. The correct sequencing for each type of element shown in Fig. 7-11 will be given in the examples that follow and must be observed if the transformation matrix given by Eq. 7-86 and the mass and stiffness matrices shown in Fig. 7-11 are to be used, since such sequencing was used in deriving each of the latter. Next show the generalized coordinates $q_1, q_2, \ldots, q_n$ representing the joint displacements of the system on the sketch, observing the same numerical sequence for their subscripts as used for the $\bar{u}_i$'s. An appropriate $\mathbf{A}_e$ matrix is then written for each element of the system such that the relationship

$$\{\bar{u}_i\}_e = \mathbf{A}_e \begin{Bmatrix} q_1 \\ q_2 \\ \vdots \\ q_n \end{Bmatrix} \tag{7-97}$$

is satisfied for each element of the unconstrained system.
3. Write the local mass and stiffness matrices $[m]_e$ and $[k]_e$ for each element of the system as obtained from Fig. 7-11.
4. Use the local mass and stiffness matrices written for each element in step 3 to determine the global mass and stiffness matrices of each element from

$$[\bar{m}]_e = \mathbf{R}^T[m]_e\mathbf{R} \tag{7-92a}$$

$$[\bar{k}]_e = \mathbf{R}^T[k]_e\mathbf{R} \tag{7-92b}$$

in which $\mathbf{R}$ is the transformation matrix given by Eq. 7-86. If distributed forces $w(x, t)$ act on any of the elements, determine the global joint forces and/or moments from

$$\{\bar{f}_i\}_e = \mathbf{R}^T\{f_i\}_e \tag{7-94}$$

The reader is reminded at this point that when the local axes of an element are parallel to the global axes of the system

$$[\bar{m}]_e = [m]_e \qquad [\bar{k}]_e = [\bar{k}]_e \qquad \{\bar{f}_i\}_e = \{f_i\}_e$$

5. Using the $\mathbf{A}_e$ matrices determined in step 2 and the $[\bar{m}]_e$, $[\bar{k}]_e$, $[\bar{f}_i]_e$ matrices found in step 4, determine the assembled mass matrix $\mathbf{M}$, stiffness matrix $\mathbf{K}$, and non-potential force and/or moment vector $\{Q_i\}$ *for the system* from

$$\mathbf{M} = \sum_{e=1}^{p} \mathbf{A}_e^T[\bar{m}]_e\mathbf{A}_e \tag{7-106}$$

$$\mathbf{K} = \sum_{e=1}^{p} \mathbf{A}_e^T[\bar{k}]_e\mathbf{A}_e \tag{7-108}$$

$$\{Q_i\} = \sum_{e=1}^{p} \mathbf{A}_e^T\{\bar{f}_i\}_e \tag{7-113}$$

in which p is the number of elements composing the system. The reader is reminded that in this step **M** and **K** are for an unconstrained (free-free) system capable of rigid-body motion. As explained previously with the discussion of the frame of Fig. 7-14, if any concentrated forces $P_i(t)$ and/or moments $M_i(t)$ act at any of the system joints, they must be added to the ith component of $\{Q_i\}$ as determined from Eq. 7-113 above.

6. Consider the generalized coordinates (system joint displacements) that became zero when the system support constraints are imposed. Referring to the subscript of each of these coordinates in turn, delete the corresponding *row and column* of both **M** and **K** determined in step 5, and the corresponding row of $\{Q_i\}$ also determined in step 5. This procedure eliminates the zero coordinates. With the reduced **M, K,** and $\{Q_i\}$ matrices determined, Eq. 7-114 establishes the undamped equations of motion of the system.

Having determined the equations of motion by the finite-element method as outlined in the preceding steps, we can use the values of m_{ij} and k_{ij} remaining in **M** and **K,** respectively, after step 6 to determine the natural undamped circular frequencies ω_r and corresponding mode shapes $\{X_i\}_r$ (r is the normal mode number) of the system.

The JACOBI.FOR program given in Chap. 5 on page 369 provides an excellent means of obtaining the natural circular frequencies and mode shapes since it also yields the rigid-body mode shapes for zero frequencies of free-free systems. However, in the computer programs presented in Secs. 7-7 and 7-8, which utilize the steps just outlined in analyzing plane frames and trusses, the program JACOBI.FOR is rewritten as a subprogram that is CALLed to determine the natural circular frequencies and normal-mode shapes.

Forced-Vibration Response

With the natural circular frequencies and mode shapes of a system determined, the response of the forced system can be obtained by using the modal-analysis method discussed in Chap. 6. Recalling that this method involves the superposition of solutions of the uncoupled normal-mode equations of motion, we refer to Eq. 6-34 on page 435 to write the uncoupled equation for the rth mode in terms of the principal coordinates as[8]

$$\ddot{\delta}_r + 2\zeta_r\omega_r\dot{\delta}_r + \omega_r^2\delta_r = \frac{\{X_i\}_r^T\{Q_i\}}{M_r} \qquad r = 1, 2, \ldots, n \tag{7-116}$$

where δ_r = principal coordinate of the rth mode
ζ_r = modal damping factor of the rth mode
ω_r = undamped natural circular frequency of the rth mode

[8] In Chap. 6, $\{u_i\}_r$ was used to express the eigenvector for the rth mode. To avoid confusion with $\{u_i\}$, which is used in this chapter to indicate local joint displacements, we use $\{X_i\}_r$ here to denote the rth-mode eigenvectors.

$\{X_i\}_r$ = normal-mode shape (eigenvector) of rth mode
$M_r = \{X_i\}_r^T \mathbf{M}\{X_i\}_r$ = generalized mass of the rth mode
$\mathbf{M}$ = mass matrix of system from step 6
$\{Q_i\}$ = generalized nonpotential force vector from step 6

After the solutions of the normal-mode equations are obtained (see Eq. 7-116), the response of the system can be obtained by the superposition of these solutions. Thus, referring to Eq. 6-37 on page 437, we can express the superposition of s modes as

$$\begin{Bmatrix} q_1 \\ q_2 \\ \vdots \\ q_n \end{Bmatrix} = \delta_1 \begin{Bmatrix} X_1 \\ X_2 \\ \vdots \\ X_n \end{Bmatrix}_1 + \delta_2 \begin{Bmatrix} X_1 \\ X_2 \\ \vdots \\ X_n \end{Bmatrix}_2 + \cdots + \delta_s \begin{Bmatrix} X_1 \\ X_2 \\ \vdots \\ X_n \end{Bmatrix}_s \tag{7-117}$$

which yields the total response of the forced system.

It was shown in Chap. 6 that the modal-analysis method yields reasonably accurate results with the superposition of only a few of the lower modes since the contributions of the higher modes to the total response are usually negligible.

In the case of transient or shock-type excitation forces, maximum values of δ_1, δ_2, $\ldots$, δ_s for use in Eq. 7-117 can be determined from appropriate response spectrums of single-degree-of-freedom systems (see Chap. 4). The use of such spectrums to obtain the absolute-maximum response of a system was illustrated in Example 6-7.

EXAMPLE 7-5

A uniform rod of length L, cross-sectional area A, modulus of elasticity E, and mass per unit length γ is fixed at the right end as shown in Fig. 7-16a and is modeled by the three finite elements shown in Fig. 7-16b. We wish to determine (a) the equations of motion of the rod using the steps beginning on page 519, (b) the natural circular frequencies and corresponding mode shapes of the axial vibration of the rod as modeled by the three finite elements shown in the figure using the JACOBI.FOR program on page 369. The natural circular frequencies are to be determined in terms of A, E, γ, and L.

Solution. a. *Step 1.* Three finite elements such as the one shown in Fig. 7-11a, each of length $L/3$, are used to model the system. The global axis $\bar{x}$ is as shown in Fig. 7-16c, and it is obvious that the local x axes of all three elements are parallel to the global axis so that $\alpha_1 = \alpha_2 = \alpha_3 = 0$.

Step 2. Although the rod is fixed at its right end, we assume initially that all the joints are free and show the global axial joint displacements as $\bar{u}_1$, $\bar{u}_2$, $\bar{u}_3$, and $\bar{u}_4$ in Fig. 7-16c. Since only axial displacements of the system are being considered, the sequencing of the subscripts of the $\bar{u}_i$'s is evident. The generalized coordinates q_1, q_2, q_3, and q_4 are then shown in the same figure as the joint displacements of the system (the rod of length L) and are subscripted in the same order as the $\bar{u}_i$'s. Equation 7-97 is then written in turn for each element of the *assumed* free-free system using an appropriate $\mathbf{A}_e$ matrix as follows:

$$\begin{Bmatrix} \bar{u}_1 \\ \bar{u}_2 \end{Bmatrix}_1 = \begin{bmatrix} 1 & 0 & 0 & 0 \\ 0 & 1 & 0 & 0 \end{bmatrix}_1 \begin{Bmatrix} q_1 \\ q_2 \\ q_3 \\ q_4 \end{Bmatrix} \quad \therefore \quad \mathbf{A}_1^T = \begin{bmatrix} 1 & 0 \\ 0 & 1 \\ 0 & 0 \\ 0 & 0 \end{bmatrix}$$

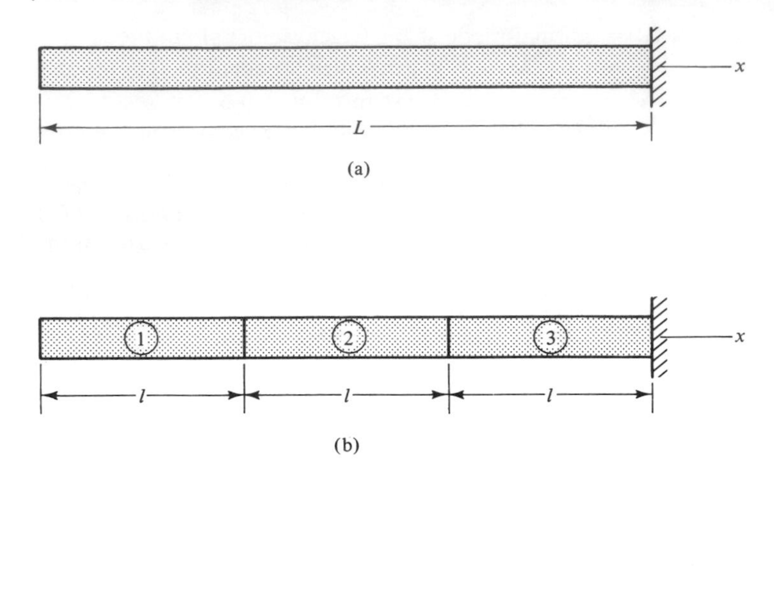

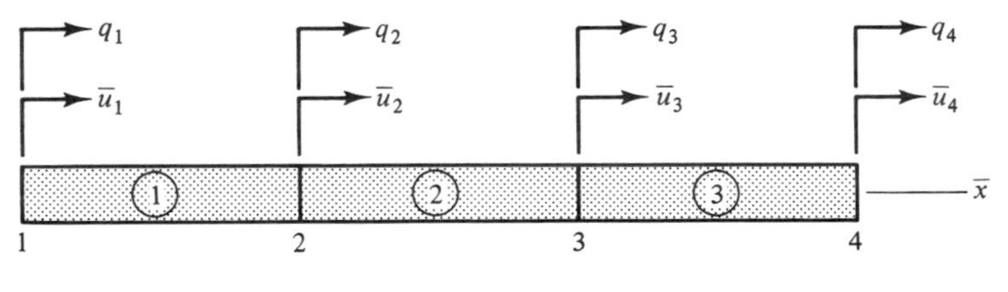

(c)

Figure 7-16 Uniform rod of Example 7-5.

$$\begin{Bmatrix} \bar{u}_2 \\ \bar{u}_3 \end{Bmatrix}_2 = \begin{bmatrix} 0 & 1 & 0 & 0 \\ 0 & 0 & 1 & 0 \end{bmatrix}_2 \begin{Bmatrix} q_1 \\ q_2 \\ q_3 \\ q_4 \end{Bmatrix} \qquad \therefore \quad \mathbf{A}_2^T = \begin{bmatrix} 0 & 0 \\ 1 & 0 \\ 0 & 1 \\ 0 & 0 \end{bmatrix}$$

$$\begin{Bmatrix} \bar{u}_3 \\ \bar{u}_4 \end{Bmatrix}_3 = \begin{bmatrix} 0 & 0 & 1 & 0 \\ 0 & 0 & 0 & 1 \end{bmatrix}_3 \begin{Bmatrix} q_1 \\ q_2 \\ q_3 \\ q_4 \end{Bmatrix} \qquad \therefore \quad \mathbf{A}_3^T = \begin{bmatrix} 0 & 0 \\ 0 & 0 \\ 1 & 0 \\ 0 & 1 \end{bmatrix}$$

Step 3. Since the three elements are identical, we can write their local mass and stiffness matrices from Fig. 7-11a as

$$[m]_e = \frac{\gamma L}{18} \begin{bmatrix} 2 & 1 \\ 1 & 2 \end{bmatrix} \qquad e = 1, 2, 3$$

$$[k]_e = \frac{3EA}{L} \begin{bmatrix} 1 & -1 \\ -1 & 1 \end{bmatrix} \qquad e = 1, 2, 3$$

Step 4. Since the local axes of each element are parallel to the global axes of the system ($\alpha_i = 0$), we can express their mass and stiffness matrices very simply as

$$[\bar{m}]_e = [m]_e = \frac{\gamma L}{18}\begin{bmatrix} 2 & 1 \\ 1 & 2 \end{bmatrix} \qquad e = 1, 2, 3$$

$$[\bar{k}]_e = [k]_e = \frac{3EA}{L}\begin{bmatrix} 1 & -1 \\ -1 & 1 \end{bmatrix} \qquad e = 1, 2, 3$$

Since there are no external forces acting on any of the elements, the joint forces are all zero. That is,

$$\{\bar{f}_i\}_e = \{f_i\}_e = 0 \qquad e = 1, 2, 3$$

Step 5. Using the results obtained in steps 2 and 4, and referring to Eq. 7-106 on page 515, we can write that

$$\mathbf{A}_1^T[\bar{m}]_1\mathbf{A}_1 = \frac{\gamma L}{18}\begin{bmatrix} 1 & 0 \\ 0 & 1 \\ 0 & 0 \\ 0 & 0 \end{bmatrix}\begin{bmatrix} 2 & 1 \\ 1 & 2 \end{bmatrix}\begin{bmatrix} 1 & 0 & 0 & 0 \\ 0 & 1 & 0 & 0 \end{bmatrix}$$

$$= \frac{\gamma L}{18}\begin{bmatrix} 2 & 1 & 0 & 0 \\ 1 & 2 & 0 & 0 \\ 0 & 0 & 0 & 0 \\ 0 & 0 & 0 & 0 \end{bmatrix}$$

$$\mathbf{A}_2^T[\bar{m}]_2\mathbf{A}_2 = \frac{\gamma L}{18}\begin{bmatrix} 0 & 0 \\ 1 & 0 \\ 0 & 1 \\ 0 & 0 \end{bmatrix}\begin{bmatrix} 2 & 1 \\ 1 & 2 \end{bmatrix}\begin{bmatrix} 0 & 1 & 0 & 0 \\ 0 & 0 & 1 & 0 \end{bmatrix}$$

$$= \frac{\gamma L}{18}\begin{bmatrix} 0 & 0 & 0 & 0 \\ 0 & 2 & 1 & 0 \\ 0 & 1 & 2 & 0 \\ 0 & 0 & 0 & 0 \end{bmatrix}$$

$$\mathbf{A}_3^T[\bar{m}]_3\mathbf{A}_3 = \frac{\gamma L}{18}\begin{bmatrix} 0 & 0 \\ 0 & 0 \\ 1 & 0 \\ 0 & 1 \end{bmatrix}\begin{bmatrix} 2 & 1 \\ 1 & 2 \end{bmatrix}\begin{bmatrix} 0 & 0 & 1 & 0 \\ 0 & 0 & 0 & 1 \end{bmatrix}$$

$$= \frac{\gamma L}{18}\begin{bmatrix} 0 & 0 & 0 & 0 \\ 0 & 0 & 0 & 0 \\ 0 & 0 & 2 & 1 \\ 0 & 0 & 1 & 2 \end{bmatrix}$$

The summation of the above three matrices yields the mass matrix $\mathbf{M}$ for the system (the rod of length L) as

$$\mathbf{M} = \sum_{e=1}^{3} \mathbf{A}_e^T[\bar{m}]_e\mathbf{A}_e = \frac{\gamma L}{18}\begin{bmatrix} 2 & 1 & 0 & 0 \\ 1 & 4 & 1 & 0 \\ 0 & 1 & 4 & 1 \\ 0 & 0 & 1 & 2 \end{bmatrix} \tag{7-118}$$

In similar manner, Eq. 7-108 is used to obtain the stiffness matrix $\mathbf{K}$ of the system as

$$\mathbf{K} = \sum_{e=1}^{3} \mathbf{A}_e^T[\bar{k}]_e\mathbf{A}_e = \frac{3AE}{L}\begin{bmatrix} 1 & -1 & 0 & 0 \\ -1 & 2 & -1 & 0 \\ 0 & -1 & 2 & -1 \\ 0 & 0 & -1 & 1 \end{bmatrix} \tag{7-119}$$

The reader should note that both **M** and **K** are symmetric matrices as anticipated. Since no external forces act at the joints of the system, the generalized nonpotential forces Q_i are zero, so

$$\{Q_i\} = \{0\} \tag{7-120}$$

The reader is again reminded that the **M** and **K** matrices just determined are for a free-free rod of length L. Therefore, if these matrices were used as input data to the JACOBI.FOR program, the output from the program would yield one rigid-body mode shape ($X_1 = X_2 = X_3 = X_4 = 1.0$) of zero frequency, and three elastic mode frequencies and mode shapes.

Step 6. If we wish to determine the natural circular frequencies and mode shapes of the system when it is fixed at the right end, the generalized coordinate q_4 must be eliminated since the joint displacement it is associated with is zero. This is accomplished by deleting the fourth row and fourth column of the matrices in Eqs. 7-118 and 7-119. Doing this, we find that the reduced mass and stiffness matrices for the constrained rod are given by

$$\mathbf{M} = \frac{\gamma L}{18}\begin{bmatrix} 2 & 1 & 0 \\ 1 & 4 & 1 \\ 0 & 1 & 4 \end{bmatrix} \tag{7-121}$$

and

$$\mathbf{K} = \frac{3AE}{L}\begin{bmatrix} 1 & -1 & 0 \\ -1 & 2 & -1 \\ 0 & -1 & 2 \end{bmatrix} \tag{7-122}$$

Referring to Eqs. 7-114, 7-120, 7-121, and 7-122, the equations of motion of the free-fixed rod of Fig. 7-16b are given by

$$\mathbf{M}\begin{Bmatrix} \ddot{q}_1 \\ \ddot{q}_2 \\ \ddot{q}_3 \end{Bmatrix} + \mathbf{K}\begin{Bmatrix} q_1 \\ q_2 \\ q_3 \end{Bmatrix} = \{0\}$$

or

$$\begin{Bmatrix} \ddot{q}_1 \\ \ddot{q}_2 \\ \ddot{q}_3 \end{Bmatrix} + \mathbf{M}^{-1}\mathbf{K}\begin{Bmatrix} q_1 \\ q_2 \\ q_3 \end{Bmatrix} = \{0\} \tag{7-123}$$

b. Using the numerical values shown in Eqs. 7-121 and 7-122 as input data to the JACOBI.FOR program beginning on page 369, and noting that $\mathbf{M}^{-1}\mathbf{K}$ of Eq. 7-123 contains the factor $AE/\gamma L^2$, we obtain the following natural circular frequencies and mode shapes:

$$\omega_1 = \frac{1.589}{L}\sqrt{\frac{AE}{\gamma}} \qquad \omega_2 = \frac{5.196}{L}\sqrt{\frac{AE}{\gamma}} \qquad \omega_3 = \frac{9.426}{L}\sqrt{\frac{AE}{\gamma}}$$

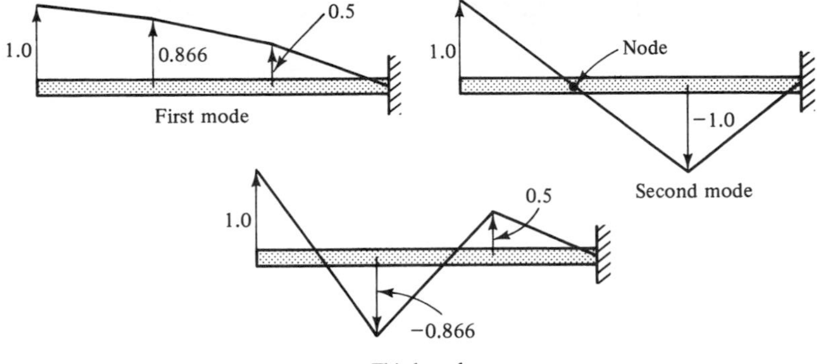

Figure 7-17 Mode configurations for free-fixed rod of Example 7-5.

$$\{X_i\}_1 = \begin{Bmatrix} 1.000 \\ 0.866 \\ 0.500 \end{Bmatrix} \qquad \{X_i\}_2 = \begin{Bmatrix} 1.000 \\ 0.000 \\ -1.000 \end{Bmatrix} \qquad \{X_i\}_3 = \begin{Bmatrix} 1.000 \\ -0.866 \\ 0.500 \end{Bmatrix}$$

Plots of the mode shapes are shown in Fig. 7-17.

It is interesting to note that the fundamental natural circular frequency given by $\omega_1 = 1.589\sqrt{AE/\gamma}/L$ is within 1 percent of the exact value given by $\omega_1 = \pi\sqrt{AE/\gamma}/2L$ on page 594 of Chap. 8. However, the ω_2 and ω_3 values shown vary from their exact values by 9 and 17 percent, respectively. As mentioned earlier, more accurate values can be obtained for ω_2 and ω_3 by using more elements to model the rod, which will result in an even more accurate value of ω_1 also.

EXAMPLE 7-6

The uniform rod of length L, for which the natural circular frequencies and eigenvectors were found in Example 7-5, is subjected to a suddenly applied force in this example. The force is uniformly distributed over the region of $0 \le x \le L/6$ as shown in Fig. 7-18, so that $w(x, t)$ acting on element 1 is given by

$$w(x, t) = \begin{Bmatrix} w_0 & 0 \le x \le L/6 \\ 0 & x > L/6 \end{Bmatrix}$$

We wish to determine the response of the forced vibration of the rod as a function of time using the modal-analysis method. Neglecting damping, determine the time response as approximated by the superposition of the first two normal modes.

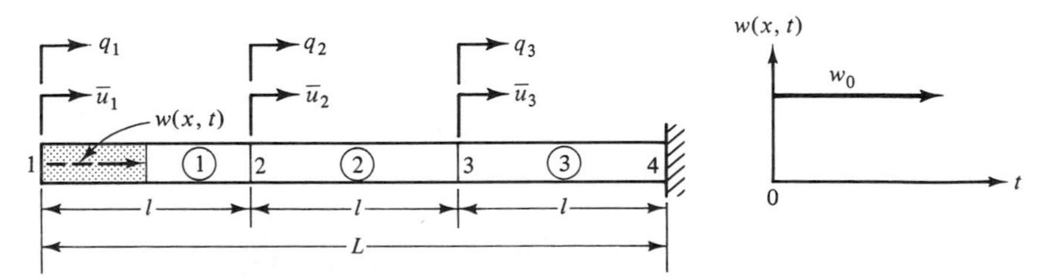

Figure 7-18 Rod of Example 7-6 subjected to force $w(x, t)$ acting on element 1.

Solution. With damping neglected, Eq. 7-116 for the first two normal modes yields

$$\ddot{\delta}_1 + \omega_1^2 \delta_1 = \frac{\{X_i\}_1^T \{Q_i\}}{M_1} \tag{7-124}$$

and

$$\ddot{\delta}_2 + \omega_2^2 \delta_2 = \frac{\{X_i\}_2^T \{Q_i\}}{M_2} \tag{7-125}$$

in which ω_1, ω_2, $\{X_i\}_1$, and $\{X_i\}_2$ are the first- and second-mode frequencies and eigenvectors, respectively, which were found for the rod in Example 7-5.

The generalized nonpotential force vector $\{Q_i\}$ and the generalized mass terms M_1 and M_2 must next be determined, and we begin by referring to Eqs. 7-56 and 7-57 to obtain the local joint forces for element 1 as

$$\left.\begin{aligned} f_1 &= w_0 \int_0^{l/2} \left(1 - \frac{x}{l}\right) dx = \tfrac{3}{8} w_0 l \\ f_2 &= w_0 \int_0^{l/2} \left(\frac{x}{l}\right) dx = \tfrac{1}{8} w_0 l \end{aligned}\right\} \tag{7-126}$$

Since the local axes of the elements are parallel to the global axes of the system,

$$\begin{Bmatrix} \bar{f}_1 \\ \bar{f}_2 \end{Bmatrix}_1 = \begin{Bmatrix} f_1 \\ f_2 \end{Bmatrix}_1 = \begin{Bmatrix} \dfrac{3w_0 l}{8} \\ \dfrac{w_0 l}{8} \end{Bmatrix}$$

and

$$\{\bar{f}_i\}_e = \{0\} \qquad e = 2, 3$$

Referring to Eq. 7-113 and the elements of $\mathbf{A}_1$ shown in Example 7-5, we can write that

$$\{Q_i\} = \sum_{e=1}^{3} \mathbf{A}_e^T \{\bar{f}_i\}_e = \mathbf{A}_1^T \{\bar{f}_i\}_1$$

which becomes

$$\{Q_i\} = \begin{bmatrix} 1 & 0 \\ 0 & 1 \\ 0 & 0 \\ 0 & 0 \end{bmatrix} \begin{Bmatrix} \dfrac{3w_0 l}{8} \\ \dfrac{w_0 l}{8} \end{Bmatrix} = \begin{Bmatrix} \dfrac{3w_0 l}{8} \\ \dfrac{w_0 l}{8} \\ 0 \\ 0 \end{Bmatrix} \tag{7-127}$$

Recalling from Example 7-5 that $q_4 = 0$ since the right end of the rod is fixed, we delete the fourth row of $\{Q_i\}$ in Eq. 7-127 and premultiply the resulting equation by the transpose of the mode shapes determined in Example 7-5 to obtain

$$\{X_i\}_1^T\{Q_i\} = [1.000 \quad 0.866 \quad 0.500]\begin{Bmatrix} \dfrac{3w_0 l}{8} \\ \dfrac{w_0 l}{8} \\ 0 \end{Bmatrix} = 0.483\ w_0 l \qquad (7\text{-}128)$$

and

$$\{X_i\}_2^T\{Q_i\} = [1.000 \quad 0.000 \quad -1.000]\begin{Bmatrix} \dfrac{3w_0 l}{8} \\ \dfrac{w_0 l}{8} \\ 0 \end{Bmatrix} = 0.375\ w_0 l \qquad (7\text{-}129)$$

Using the mass matrix **M** determined in Example 7-5 for the free-fixed rod (see Eq. 7-121), we find the generalized mass terms to be

$$M_1 = \{X_i\}_1^T\mathbf{M}\{X_i\}_1 = [1.000 \quad 0.866 \quad 0.500]\left(\frac{\gamma l}{18}\right)\begin{bmatrix} 2 & 1 & 0 \\ 1 & 4 & 1 \\ 0 & 1 & 4 \end{bmatrix}\begin{Bmatrix} 1.000 \\ 0.866 \\ 0.500 \end{Bmatrix}$$

yielding

$$M_1 = 0.478\ \gamma l \qquad (7\text{-}130)$$

and

$$M_2 = \{X_i\}_2^T\mathbf{M}\{X_i\}_2 = [1.000 \quad 0.000 \quad -1.000]\left(\frac{\gamma l}{18}\right)\begin{bmatrix} 2 & 1 & 0 \\ 1 & 4 & 1 \\ 0 & 1 & 4 \end{bmatrix}\begin{Bmatrix} 1.000 \\ 0.000 \\ -1.000 \end{Bmatrix}$$

yielding

$$M_2 = 0.333\gamma l \qquad (7\text{-}131)$$

Substituting Eqs. 7-128, 7-129, 7-130, and 7-131 into Eqs. 7-124 and 7-125, we obtain

$$\ddot{\delta}_1 + \omega_1^2\delta_1 = 1.01\frac{w_0}{\gamma} \qquad (7\text{-}132)$$

and

$$\ddot{\delta}_2 + \omega_2^2\delta_2 = 1.13\frac{w_0}{\gamma} \qquad (7\text{-}133)$$

in which

$$\omega_1 = \frac{1.589}{L}\sqrt{\frac{AE}{\gamma}}$$

and

$$\omega_2 = \frac{5.196}{L}\sqrt{\frac{AE}{\gamma}}$$

Designating $(1.01w_0/\gamma) = F_1$ and $(1.13w_0/\gamma) = F_2$, and noting from Eq. 7-117 that $\delta_1 = \dot{\delta}_1 = \delta_2 = \dot{\delta}_2 = 0$ to satisfy the *initial conditions* of $q_1 = \dot{q}_1 = q_2 = \dot{q}_2 = q_3 = \dot{q}_3 = 0$, the total solutions of Eqs. 7-132 and 7-133 are found to be

$$\delta_1 = \frac{F_1}{\omega_1^2}(1 - \cos\omega_1 t)$$

and

$$\delta_2 = \frac{F_2}{\omega_2^2}(1 - \cos\omega_2 t)$$

Substituting these solutions and the eigenvectors $\{X_i\}_1$ and $\{X_i\}_2$ determined in Example 7-5 into Eq. 7-117 yields

$$\begin{Bmatrix} q_1 \\ q_2 \\ q_3 \end{Bmatrix} = \frac{F_1}{\omega_1^2}(1 - \cos\omega_1 t)\begin{Bmatrix} 1.000 \\ 0.866 \\ 0.500 \end{Bmatrix} + \frac{F_2}{\omega_2^2}(1 - \cos\omega_2 t)\begin{Bmatrix} 1.000 \\ 0.000 \\ -1.000 \end{Bmatrix}$$

which is the time response of the rod as determined by the superposition of the first two normal modes. Noting from the results obtained in Example 7-5 that $\omega_2^2 \gg \omega_1^2$, we see that the contribution of the second mode to the response is small compared with that of the first mode.

EXAMPLE 7-7

Owing to a slight eccentricity in the thrust alignment of its motor, the rocket shown in Fig. 7-19a has developed lateral vibrations in its free flight after burnout. The rocket is to be analyzed by considering it as a uniform beam modeled by the two finite elements shown in Fig. 7-19c, and having the following parameters:

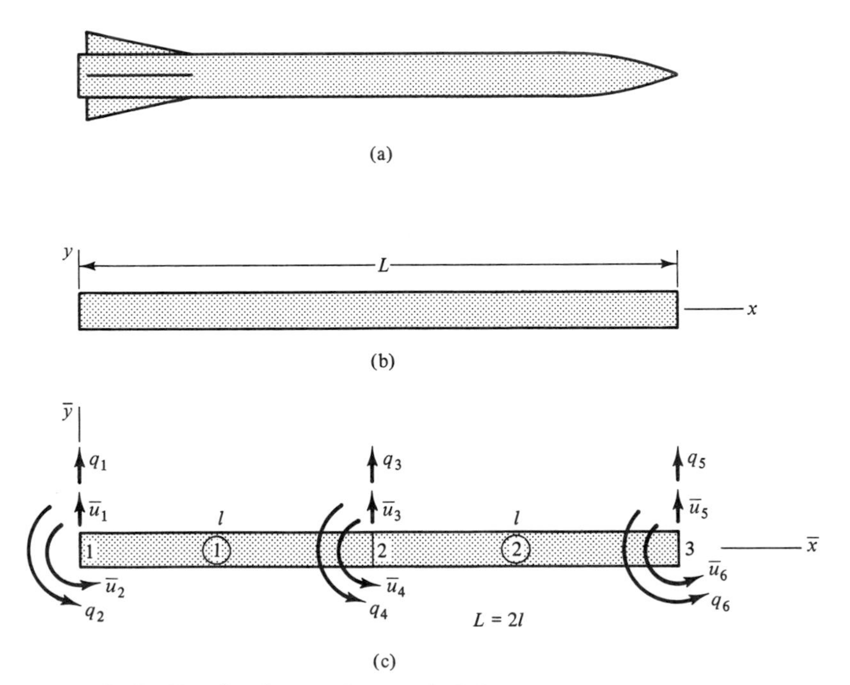

Figure 7-19 Free-free beam of Example 7-7.

$$L = 200 \text{ in.} \quad \text{(overall length)}$$
$$EI = (10)^{10} \text{ lb} \cdot \text{in.}^2 \quad \text{(bending stiffness)}$$
$$\gamma = 0.042 \text{ lb} \cdot \text{s}^2/\text{in.}^2 \quad \text{(mass per unit length)}$$

Since the rocket is in free flight with no external constraints, it provides an example of a free-free beam. Neglect axial deformations of the rocket.

Using this rather crude model of the rocket, but one that we consider satisfactory for a preliminary analysis, we wish to (a) determine the natural circular frequencies and corresponding mode shapes for the lateral vibration of the rocket modeled as a free-free beam composed of two finite elements; (b) compare the lowest elastic mode frequency obtained with the exact value of $\omega_1 = 22.3733 \sqrt{EI/\gamma L^4}$ given in Chap. 8.

Solution. **a.** *Step 1.* Since we are neglecting axial deformations, we can use two identical elements such as the element in Fig. 7-11b. The global axes $\bar{x}$, $\bar{y}$ are as shown in Fig. 7-19c, and the local x, y axes of the finite elements are obviously parallel to the global axes so that $\alpha_1 = \alpha_2 = 0$.

Step 2. The global joint displacements $\bar{u}_1, \bar{u}_2, \ldots, \bar{u}_6$ are shown in Fig. 7-19c with the same numerical sequence of subscripts at each system joint as shown for the element in Fig. 7-11b. The generalized coordinates $q_1, q_2, \ldots, q_6$ that represent the joint displacements of the system are added next as shown in Fig. 7-19c, with the same subscripting sequence as used for the $\bar{u}_i$'s. Referring again to Fig. 7-19c, we write the following relationships between the $\bar{u}_i$'s and q_i's of each element using Eq. 7-97 and appropriate $\mathbf{A}_e$ matrices:

$$\begin{Bmatrix} \bar{u}_1 \\ u_2 \\ \bar{u}_3 \\ \bar{u}_4 \end{Bmatrix}_1 = \begin{bmatrix} 1 & 0 & 0 & 0 & 0 & 0 \\ 0 & 1 & 0 & 0 & 0 & 0 \\ 0 & 0 & 1 & 0 & 0 & 0 \\ 0 & 0 & 0 & 1 & 0 & 0 \end{bmatrix}_1 \begin{Bmatrix} q_1 \\ q_2 \\ q_3 \\ q_4 \\ q_5 \\ q_6 \end{Bmatrix} \qquad \text{(element 1)}$$

$$\begin{Bmatrix} \bar{u}_3 \\ \bar{u}_4 \\ \bar{u}_5 \\ \bar{u}_6 \end{Bmatrix}_2 = \begin{bmatrix} 0 & 0 & 1 & 0 & 0 & 0 \\ 0 & 0 & 0 & 1 & 0 & 0 \\ 0 & 0 & 0 & 0 & 1 & 0 \\ 0 & 0 & 0 & 0 & 0 & 1 \end{bmatrix}_2 \begin{Bmatrix} q_1 \\ q_2 \\ q_3 \\ q_4 \\ q_5 \\ q_6 \end{Bmatrix} \qquad \text{(element 2)}$$

The transposes of the respective $\mathbf{A}_e$ matrices determined here for use later in step 5 are

$$\mathbf{A}_1^T = \begin{bmatrix} 1 & 0 & 0 & 0 \\ 0 & 1 & 0 & 0 \\ 0 & 0 & 1 & 0 \\ 0 & 0 & 0 & 1 \\ 0 & 0 & 0 & 0 \\ 0 & 0 & 0 & 0 \end{bmatrix}$$

and

$$\mathbf{A}_2^T = \begin{bmatrix} 0 & 0 & 0 & 0 \\ 0 & 0 & 0 & 0 \\ 1 & 0 & 0 & 0 \\ 0 & 1 & 0 & 0 \\ 0 & 0 & 1 & 0 \\ 0 & 0 & 0 & 1 \end{bmatrix}$$

Step 3. Since the finite elements are identical, we can write the local mass and stiffness matrices of both from Fig. 7-11b as

$$[m]_e = \frac{\gamma l}{420}\begin{bmatrix} 156 & 22l & 54 & -13l \\ 22l & 4l^2 & 13l & -3l^2 \\ 54 & 13l & 156 & -22l \\ -13l & -3l^2 & -22l & 4l^2 \end{bmatrix} \qquad e = 1, 2$$

and

$$[k]_e = \frac{EI}{l^3}\begin{bmatrix} 12 & 6l & -12 & 6l \\ 6l & 4l^2 & -6l & 2l^2 \\ -12 & -6l & 12 & -6l \\ 6l & 2l^2 & -6l & 4l^2 \end{bmatrix} \qquad e = 1, 2$$

Step 4. Since the local axes of each element are parallel to the global axes of the rocket system, we can write the global matrices as

$$[\bar{m}]_e = [m]_e \qquad e = 1, 2$$
$$[\bar{k}]_e = [k]_e \qquad e = 1, 2$$

Since there are no external forces acting on the system, all the joint forces are zero, so that

$$\{\bar{f}_i\}_e = \{f_i\}_e = \mathbf{0} \qquad e = 1, 2$$

Step 5. Using $[\bar{m}]_e$, $[\bar{k}]_e$, $\mathbf{A}_1$, $\mathbf{A}_1^T$, $\mathbf{A}_2$, $\mathbf{A}_2^T$, and $\{\bar{f}_i\}_e$ as determined in the preceding steps, we obtain the mass and stiffness matrices and nonpotential force vector of the system as

$$\mathbf{M} = \sum_{e=1}^{2} \mathbf{A}_e^T[\bar{m}]_e\mathbf{A}_e = \frac{\gamma l}{420}\begin{bmatrix} 156 & 22l & 54 & -13l & 0 & 0 \\ 22 & 4l^2 & 13l & -3l^2 & 0 & 0 \\ 54 & 13l & 312 & 0 & 54 & -13l \\ -13l & -3l^2 & 0 & 8l^2 & 13l & -3l^2 \\ 0 & 0 & 54 & 13l & 156 & -22l \\ 0 & 0 & -13l & -3l^2 & -22l & 4l^2 \end{bmatrix} \tag{7-134}$$

$$\mathbf{K} = \sum_{e=1}^{2} \mathbf{A}_e^T[\bar{k}]_e\mathbf{A}_e = \frac{EI}{l^3}\begin{bmatrix} 12 & 6l & -12 & 6l & 0 & 0 \\ 6l & 4l^2 & -6l & 2l^2 & 0 & 0 \\ -12 & -6l & 24 & 0 & -12 & 6l \\ 6l & 2l^2 & 0 & 8l^2 & -6l & 2l^2 \\ 0 & 0 & -12 & -6l & 12 & -6l \\ 0 & 0 & 6l & 2l^2 & -6l & 4l^2 \end{bmatrix} \tag{7-135}$$

$$\{Q_i\} = \sum_{e=1}^{2} \mathbf{A}_e^T\{\bar{f}_i\}_e = \mathbf{0} \tag{7-136}$$

Step 6. Substituting the values of $l = 100$, $EI = (10)^{10}$, and $\gamma = 0.042$ into Eqs. 7-134 and 7-135, we find that for the free-free system

$$\mathbf{M} = \begin{bmatrix} 1.56 & 22.00 & 0.54 & -13.00 & 0.00 & 0.00 \\ 22.00 & 400.00 & 13.00 & -300.00 & 0.00 & 0.00 \\ 0.54 & 13.00 & 3.12 & 0.00 & 0.54 & -13.00 \\ -13.00 & -300.00 & 0.00 & 800.00 & 13.00 & -300.00 \\ 0.00 & 0.00 & 0.54 & 13.00 & 1.56 & -22.00 \\ 0.00 & 0.00 & -13.00 & -300.00 & -22.00 & 400.00 \end{bmatrix} \tag{7-137}$$

and

$$\mathbf{K} = (10)^4 \begin{bmatrix} 12 & 600 & -12 & 600 & 0 & 0 \\ 600 & 40{,}000 & -600 & 20{,}000 & 0 & 0 \\ -12 & -600 & 24 & 0 & -12 & 600 \\ 600 & 20{,}000 & 0 & 80{,}000 & -600 & 20{,}000 \\ 0 & 0 & -12 & -600 & 12 & -600 \\ 0 & 0 & 600 & 20{,}000 & -600 & 40{,}000 \end{bmatrix} \tag{7-138}$$

We can now use the numerical values in Eqs. 7-137 and 7-138 as input data to the JACOBI.FOR program beginning on page 369, to obtain the six natural circular frequencies and corresponding mode shapes (eigenvectors). These will include two rigid-body modes of zero frequency ($\omega_0 = 0$), and four elastic-mode frequencies ω_1, ω_2, ω_3, and ω_4. The two rigid-body mode shapes for $\omega_0 = 0$ and the first elastic-mode frequency ω_1 and mode shape are

$$\omega_0 = 0 \qquad \omega_0 = 0 \qquad \omega_1 = 273.54 \text{ rad/s}$$

$$\{X_i\}_0 = \begin{Bmatrix} 1.0000 \\ 0.2283 \\ 23.8314 \\ 0.2283 \\ 46.6663 \\ 0.2283 \end{Bmatrix} \qquad \{X_i\}_0 = \begin{Bmatrix} 1.0000 \\ -0.0076 \\ 0.2420 \\ -0.0076 \\ -0.5159 \\ -0.0076 \end{Bmatrix} \qquad \{X_i\}_1 = \begin{Bmatrix} 1.0000 \\ -0.0233 \\ -0.6118 \\ 0.0000 \\ 1.0000 \\ 0.0233 \end{Bmatrix}$$

The mode configurations are shown in Fig. 7-20.

The reader should pause at this point to associate the eigenvector components shown above with the generalized coordinates shown in Fig. 7-19c. For example, the eigenvector components $X_2 = X_4 = X_6 = 0.2283$ for the first rigid-body mode shown in Fig. 7-20a correspond to the *positive angular* displacements of the system joints designated by the generalized coordinates q_2, q_4, and q_6. It should be apparent from Fig. 7-20a and b that each rigid-body mode configuration is made up of a combination of translation and rotation with zero bending. These modes thus represent *positions* of the beam in space rather than bent shapes of the rocket as given by the elastic modes such as the one shown in Fig. 7-20c.

The two zero frequencies of the rigid-body modes represent a case of equal eigenvalues as discussed in Sec. 5-9. Thus, the eigenvectors shown for the rigid-body modes are not unique, since any linear combination of the two obtained from the JACOBI.FOR program will also satisfy the matrix equation $\mathbf{M}^{-1}\mathbf{K}\mathbf{X} = \lambda\mathbf{X}$.

b. With the exact value of the first elastic-mode frequency being

$$\omega_1 = 22.3733 \sqrt{\frac{(10)^{10}}{(0.042)(200)^4}} = 272.93 \text{ rad/s}$$

the value of $\omega_1 = 273.54$ obtained by the finite-element method and using just two elements varies by only 0.2 percent from the exact value.

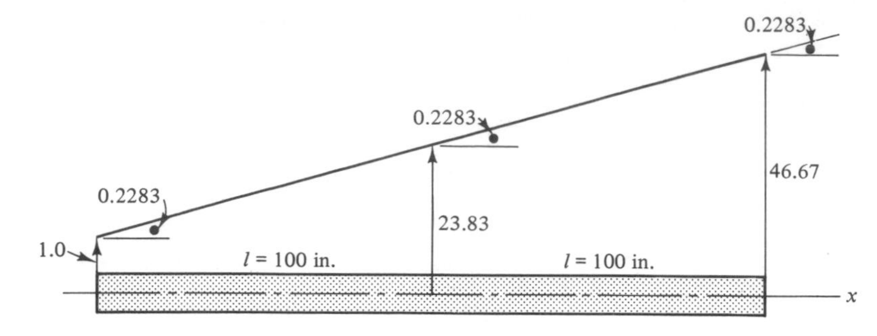

($\omega_0 = 0$) Rigid body mode

(a)

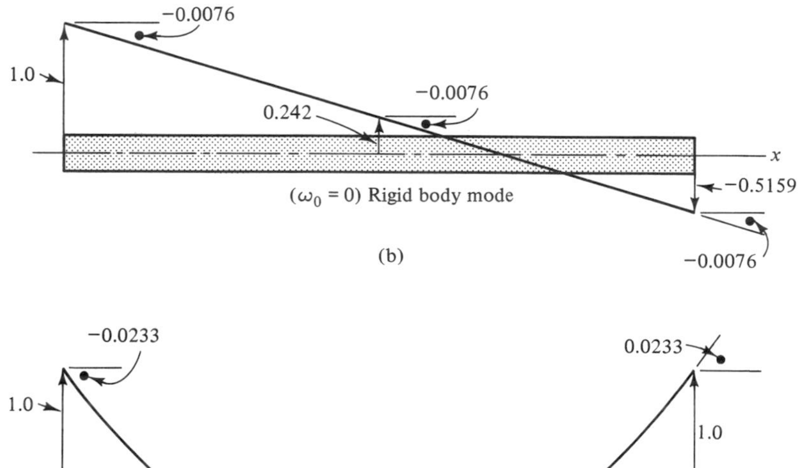

($\omega_0 = 0$) Rigid body mode

(b)

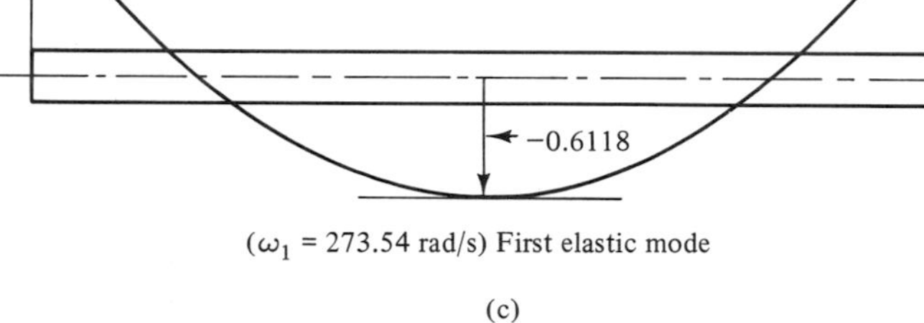

($\omega_1 = 273.54$ rad/s) First elastic mode

(c)

Figure 7-20 Mode shapes for Example 7-7.

EXAMPLE 7-8

The center and both ends of the beam shown in Fig. 7-21a are simply supported to form a two-span continuous beam. The beam parameters are

$$
\begin{aligned}
L &= 200 \text{ in.} \\
l &= 100 \text{ in.} \\
EI &= (10)^{10} \text{ lb} \cdot \text{in.}^2 \\
\gamma &= 0.042 \text{ lb} \cdot \text{s}^2/\text{in.}^2
\end{aligned}
$$

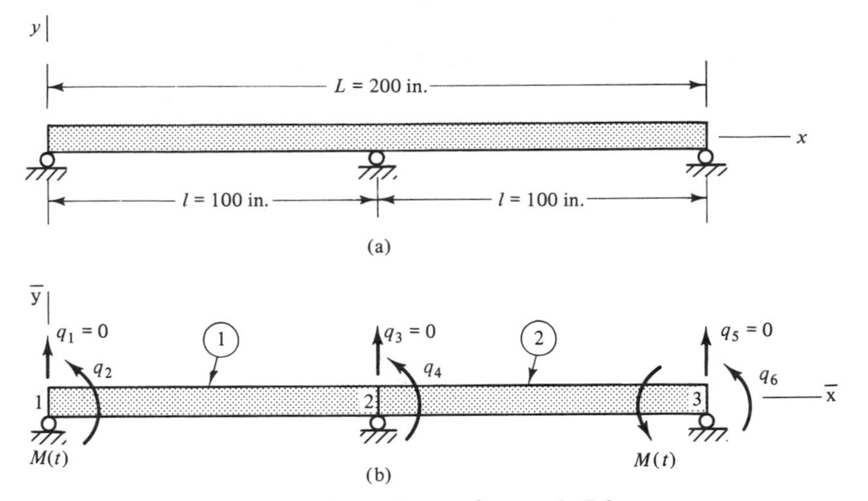

Figure 7-21 Two-span continuous beam of Example 7-8.

and it is subjected to a moment $M(t)$ at its right end as shown. We wish to determine (a) the equations of motion of the continuous beam; (b) the natural circular frequencies and corresponding mode shapes of the continuous beam.

Solution. **a.** Comparing the beam of this example with the beam of Example 7-7 which was used to provide a rough model of a rocket, it can be seen that they are identical in their free-free states, with identical parameter values. Therefore, the first five steps outlined in Example 7-7 can be used verbatim as written, with the exception of step 5, which must be modified to include the concentrated moment at the right end of the beam.

Beginning with this modification, we note that since $M(t)$ is an external moment applied at a system joint (the right end of the beam) as shown in Fig. 7-21b, it is associated with the generalized coordinate q_6 representing the angular displacement of that joint, and thus becomes Q_6 in $\{Q_i\}$ (see Eq. 7-139 below).

With the first five steps now completed, let us move to step 6 and begin by comparing Figs. 7-19c and 7-21b. We note that only the generalized coordinates q_2, q_4, and q_6 are working coordinates since the vertical displacements of the supports are constrained to zero. Therefore, we refer to the mass and stiffness matrices of the system for its free-free state given by Eqs. 7-137 and 7-138 in Example 7-7 and delete the first, third, and fifth *rows and columns* from each of them to obtain the equations of motion of the continuous beam subjected to the concentrated moment $M(t)$ as

$$(10)^2\begin{bmatrix} 4 & -3 & 0 \\ -3 & 8 & -3 \\ 0 & -3 & 4 \end{bmatrix}\begin{Bmatrix} \ddot{q}_2 \\ \ddot{q}_4 \\ \ddot{q}_6 \end{Bmatrix} + (10)^8\begin{bmatrix} 4 & 2 & 0 \\ 2 & 8 & 2 \\ 0 & 2 & 4 \end{bmatrix}\begin{Bmatrix} q_2 \\ q_4 \\ q_6 \end{Bmatrix} = \begin{Bmatrix} 0 \\ 0 \\ M(t) \end{Bmatrix} \tag{7-139}$$

b. Using the values shown in Eq. 7-139 for the elements of the system mass and stiffness matrices as input data to the JACOBI.FOR program beginning on page 369, the following natural circular frequencies and normal-mode shapes are obtained from the program:

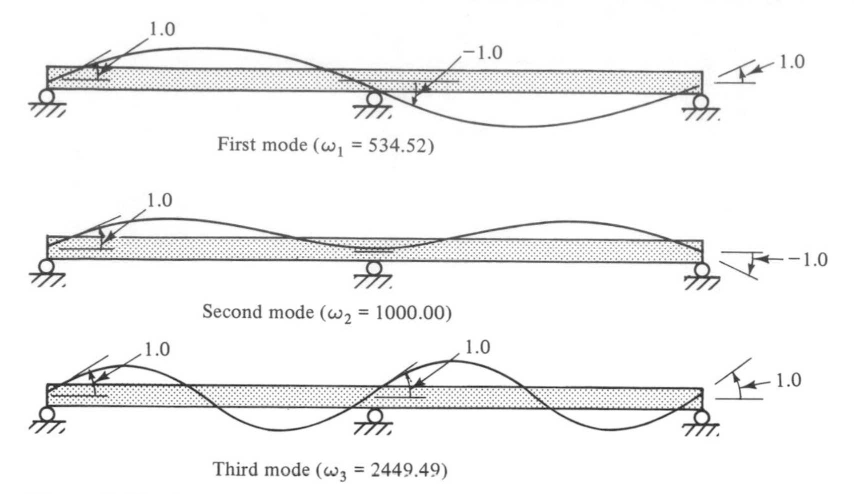

Figure 7-22 Normal-mode shapes of two-span continuous beam of Example 7-8.

$$\omega_1 = 534.52 \text{ rad/s} \qquad \omega_2 = 1000.00 \text{ rad/s} \qquad \omega_3 = 2449.49 \text{ rad/s}$$

$$\begin{Bmatrix} X_2 \\ X_4 \\ X_6 \end{Bmatrix}_1 = \begin{Bmatrix} 1.00 \\ -1.00 \\ 1.00 \end{Bmatrix} \qquad \begin{Bmatrix} X_2 \\ X_4 \\ X_6 \end{Bmatrix}_2 = \begin{Bmatrix} 1.00 \\ 0.00 \\ -1.00 \end{Bmatrix} \qquad \begin{Bmatrix} X_2 \\ X_4 \\ X_6 \end{Bmatrix}_3 = \begin{Bmatrix} 1.00 \\ 1.00 \\ 1.00 \end{Bmatrix}$$

The mode shapes are shown in Fig. 7-22. Since only the three generalized coordinates q_2, q_4, and q_6 were used to represent the angular displacements of the continuous beam, the natural circular frequencies and normal-mode shapes cannot be expected to be of great accuracy. As a matter of interest, the exact solution for the two-span continuous beam obtained by considering a simply supported beam of half the length of the continuous beam gives the fundamental natural circular frequency as $\omega_1 = 481.58$ rad/s. Thus, the fundamental frequency obtained in this example of $\omega_1 = 534.52$ is in error from the exact by about 11 percent. However, the first-mode shape shown in Fig. 7-22 is in close agreement with that given by the two half-sine waves of the exact solution (see Chap. 8). Fairly accurate results for the first two modes of the two-span continuous beam can be obtained by the finite-element method if the beam is modeled by four finite elements instead of just two (see Prob. 7-20).

EXAMPLE 7-9

The frame shown in Fig. 7-23a is fixed at joints 1 and 4, and it is assumed that the welded corners at joints 2 and 3 are rigid enough to maintain the frame members at right angles to each other at these joints during sidesway of the frame. The length of each vertical column is l, and the length of the horizontal member is $2l$. All the frame members are W18 × 60 steel beams, and the frame has the following parameter values:

$$l = 240 \text{ in.}$$
$$EI = 30(10)^6 984 = 2.952(10)^{10} \text{ lb} \cdot \text{in.}^2$$
$$\gamma = \frac{60}{12(386)} = 0.013 \text{ lb} \cdot \text{s}^2/\text{in.}^2$$
$$r = 7.47 \text{ in.} \quad \text{(radius of gyration)}$$

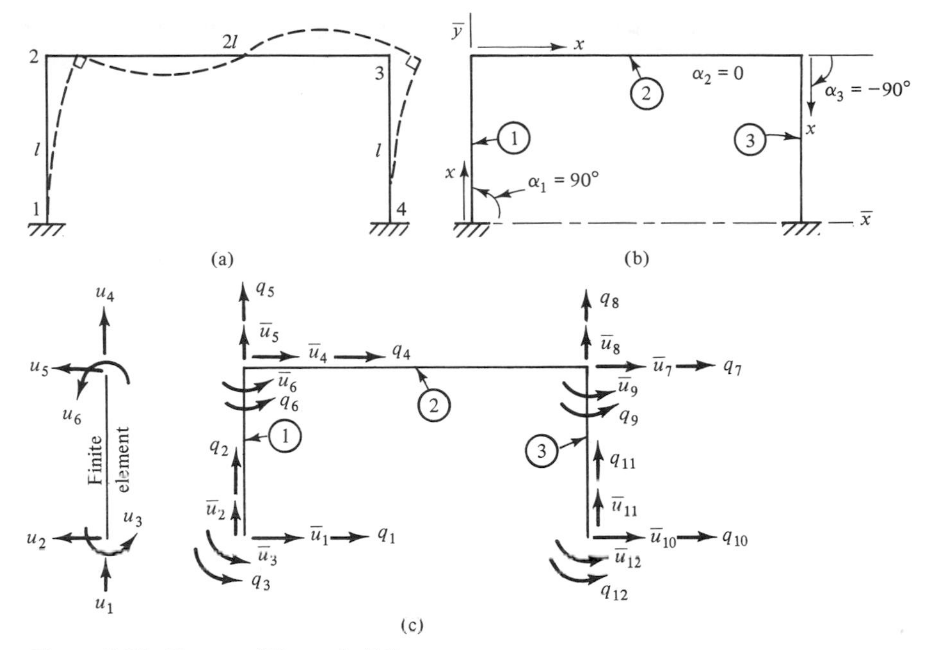

Figure 7-23 Frame of Example 7-9.

The frame is to be modeled by the three finite elements shown in Fig. 7-23b, with the assumption that the axial deformations of the vertical beam elements have only negligible effect on the lateral vibrations of the frame.

The purpose of our analysis is to determine the natural circular frequencies and corresponding mode shapes of the frame as modeled by the three finite beam elements.

Solution. Step 1. We select three elements such as the one shown in Fig. 7-11c to model the frame. The global $\bar{x}$, $\bar{y}$ axes are as shown in Fig. 7-23b, and it is obvious that the angles from the global axes to the local axes of the elements are $\alpha_1 = 90°$, $\alpha_2 = 0°$, and $\alpha_3 = -90°$ for the elements 1, 2, and 3, respectively. The most general element, that of Fig. 7-11c, is selected for all three of the frame elements to ensure that all the joint displacements of the frame pertinent to its lateral vibration are included. The joint displacements that are fixed at zero or assumed to be negligible are then removed by eliminating the appropriate rows and columns from the **M** and **K** matrices determined for the frame in its initially assumed free-free state.

Step 2. The global joint displacements $\bar{u}_1, \bar{u}_2, \ldots, \bar{u}_{12}$ are added as shown in Fig. 7-23c. Note that beginning with joint 1 the numerical *sequence* of the subscripts of the $\bar{u}_i$'s is the same at every joint of the system, with the $\bar{u}$ having the smallest subscript placed parallel to the global $\bar{x}$ axis. This corresponds to the sequencing shown in Fig. 7-12 from which the transformation matrix was derived, and to the sequencing of the subscripts of the local joint displacements u of the finite element in Fig. 7-11c. Such sequencing must be observed for Eq. 7-84 to relate the local joint displacements u_i of a finite element to the corresponding global joint displacements $\bar{u}_i$ for any orientation of the element in a system. The generalized coordinates q_1, $q_2, \ldots, q_{12}$ representing the joint displacements of the unconstrained system are next added as shown in Fig. 7-23c, with their subscripts sequenced the same as the corresponding global joint displacements. Equation 7-97 is then used to relate the global joint displacements and generalized coordinates of the three elements as follows:

$$\begin{Bmatrix} \bar{u}_1 \\ \bar{u}_2 \\ \bar{u}_3 \\ \bar{u}_4 \\ \bar{u}_5 \\ \bar{u}_6 \end{Bmatrix}_1 = \mathbf{A}_1 \begin{Bmatrix} q_1 \\ q_2 \\ q_3 \\ \vdots \\ q_{12} \end{Bmatrix} \qquad \begin{Bmatrix} \bar{u}_4 \\ \bar{u}_5 \\ \bar{u}_6 \\ \bar{u}_7 \\ \bar{u}_8 \\ \bar{u}_9 \end{Bmatrix}_2 = \mathbf{A}_2 \begin{Bmatrix} q_1 \\ q_2 \\ q_3 \\ \vdots \\ q_{12} \end{Bmatrix} \qquad \begin{Bmatrix} \bar{u}_7 \\ \bar{u}_8 \\ \bar{u}_9 \\ \bar{u}_{10} \\ \bar{u}_{11} \\ \bar{u}_{12} \end{Bmatrix}_3 = \mathbf{A}_3 \begin{Bmatrix} q_1 \\ q_2 \\ q_3 \\ \vdots \\ q_{12} \end{Bmatrix}$$

in which the appropriate $\mathbf{A}_e$ matrices are, respectively,

$$\mathbf{A}_1 = \begin{bmatrix} 1 & 0 & 0 & 0 & 0 & 0 & 0 & 0 & 0 & 0 & 0 & 0 \\ 0 & 1 & 0 & 0 & 0 & 0 & 0 & 0 & 0 & 0 & 0 & 0 \\ 0 & 0 & 1 & 0 & 0 & 0 & 0 & 0 & 0 & 0 & 0 & 0 \\ 0 & 0 & 0 & 1 & 0 & 0 & 0 & 0 & 0 & 0 & 0 & 0 \\ 0 & 0 & 0 & 0 & 1 & 0 & 0 & 0 & 0 & 0 & 0 & 0 \\ 0 & 0 & 0 & 0 & 0 & 1 & 0 & 0 & 0 & 0 & 0 & 0 \end{bmatrix}$$

$$\mathbf{A}_2 = \begin{bmatrix} 0 & 0 & 0 & 1 & 0 & 0 & 0 & 0 & 0 & 0 & 0 & 0 \\ 0 & 0 & 0 & 0 & 1 & 0 & 0 & 0 & 0 & 0 & 0 & 0 \\ 0 & 0 & 0 & 0 & 0 & 1 & 0 & 0 & 0 & 0 & 0 & 0 \\ 0 & 0 & 0 & 0 & 0 & 0 & 1 & 0 & 0 & 0 & 0 & 0 \\ 0 & 0 & 0 & 0 & 0 & 0 & 0 & 1 & 0 & 0 & 0 & 0 \\ 0 & 0 & 0 & 0 & 0 & 0 & 0 & 0 & 1 & 0 & 0 & 0 \end{bmatrix}$$

$$\mathbf{A}_3 = \begin{bmatrix} 0 & 0 & 0 & 0 & 0 & 0 & 1 & 0 & 0 & 0 & 0 & 0 \\ 0 & 0 & 0 & 0 & 0 & 0 & 0 & 1 & 0 & 0 & 0 & 0 \\ 0 & 0 & 0 & 0 & 0 & 0 & 0 & 0 & 1 & 0 & 0 & 0 \\ 0 & 0 & 0 & 0 & 0 & 0 & 0 & 0 & 0 & 1 & 0 & 0 \\ 0 & 0 & 0 & 0 & 0 & 0 & 0 & 0 & 0 & 0 & 1 & 0 \\ 0 & 0 & 0 & 0 & 0 & 0 & 0 & 0 & 0 & 0 & 0 & 1 \end{bmatrix}$$

Step 3. In step 3 we would usually write the local mass and stiffness matrices of the elements used to form the system. Then in step 4 we would use the transformation matrix of Eq. 7-86 and its transpose to determine the global mass and stiffness matrices of each element as indicated by Eq. 7-92. However, in Example 7-4, we performed this step for a finite element such as the one shown in Fig. 7-11c and obtained its global mass and stiffness matrices as shown by Eqs. 7-95 and 7-96, respectively, which are in terms of the angle α which is measured from the global axes of the system to the local axes of the element ($C = \cos\alpha_i$ and $S = \sin\alpha_i$). Therefore, we can now proceed to step 4.

Step 4. Substituting the appropriate values for C and S into Eq. 7-95 for each element in turn, we obtain the three global mass matrices:

$$[\bar{m}]_1 = \frac{\gamma l}{420} \begin{bmatrix} 156 & 0 & -22l & 54 & 0 & 13l \\ 0 & 140 & 0 & 0 & 70 & 0 \\ -22l & 0 & 4l^2 & -13l & 0 & -3l^2 \\ 54 & 0 & -13l & 156 & 0 & 22l \\ 0 & 70 & 0 & 0 & 140 & 0 \\ 13l & 0 & -3l^2 & 22l & 0 & 4l^2 \end{bmatrix} \qquad \begin{matrix} \text{(element 1)} \\ (\alpha_1 = 90^\circ) \end{matrix}$$

$$[\bar{m}]_2 = \frac{\gamma l}{420}\begin{bmatrix} 280 & 0 & 0 & 140 & 0 & 0 \\ 0 & 312 & 88l & 0 & 108 & -52l \\ 0 & 88l & 32l^2 & 0 & 52l & -24l^2 \\ 140 & 0 & 0 & 280 & 0 & 0 \\ 0 & 108 & 52l & 0 & 312 & -88l \\ 0 & -52l & -24l^2 & 0 & -88l & 32l^2 \end{bmatrix} \qquad \begin{matrix}\text{(element 2)} \\ (\alpha_2 = 0)\end{matrix}$$

$$[\bar{m}]_3 = \frac{\gamma l}{420}\begin{bmatrix} 156 & 0 & 22l & 54 & 0 & -13l \\ 0 & 140 & 0 & 0 & 70 & 0 \\ 22l & 0 & 4l^2 & 13l & 0 & -3l^2 \\ 54 & 0 & 13l & 156 & 0 & -22l \\ 0 & 70 & 0 & 0 & 140 & 0 \\ -13l & 0 & -3l^2 & -22l & 0 & 4l^2 \end{bmatrix} \qquad \begin{matrix}\text{(element 3)} \\ (\alpha_3 = -90^\circ)\end{matrix}$$

Proceeding in similar fashion with Eq. 7-96, we obtain the global stiffness matrices as

$$[\bar{k}]_1 = \frac{EI}{l^3}\begin{bmatrix} 12 & 0 & -6l & -12 & 0 & -6l \\ 0 & (l/r)^2 & 0 & 0 & -(l/r)^2 & 0 \\ -6l & 0 & 4l^2 & 6l & 0 & 2l^2 \\ -12 & 0 & 6l & 12 & 0 & 6l \\ 0 & -(l/r)^2 & 0 & 0 & (l/r)^2 & 0 \\ -6l & 0 & 2l^2 & 6l & 0 & 4l^2 \end{bmatrix} \qquad \begin{matrix}\text{(element 1)} \\ (\alpha_1 = 90^\circ)\end{matrix}$$

$$[\bar{k}]_2 = \frac{EI}{8l^3}\begin{bmatrix} 4(l/r)^2 & 0 & 0 & -4(l/r)^2 & 0 & 0 \\ 0 & 12 & 12l & 0 & -12 & 12l \\ 0 & 12l & 16l^2 & 0 & -12l & 8l^2 \\ -4(l/r)^2 & 0 & 0 & 4(l/r)^2 & 0 & 0 \\ 0 & -12 & -12l & 0 & 12 & -12l \\ 0 & 12l & 8l^2 & 0 & -12l & 16l^2 \end{bmatrix} \qquad \begin{matrix}\text{(element 2)} \\ (\alpha_2 = 0)\end{matrix}$$

$$[\bar{k}]_3 = \frac{EI}{l^3}\begin{bmatrix} 12 & 0 & 6l & -12 & 0 & 6l \\ 0 & (l/r)^2 & 0 & 0 & -(l/r)^2 & 0 \\ 6l & 0 & 4l^2 & -6l & 0 & 2l^2 \\ -12 & 0 & -6l & 12 & 0 & -6l \\ 0 & -(l/r)^2 & 0 & 0 & (l/r)^2 & 0 \\ 6l & 0 & 2l^2 & -6l & 0 & 4l^2 \end{bmatrix} \qquad \begin{matrix}\text{(element 3)} \\ (\alpha_3 = -90^\circ)\end{matrix}$$

Step 5. Using the results obtained from steps 2 and 4 in Eqs. 7-106 and 7-108, which are repeated above each of the following mass and stiffness matrices of the system, respectively, we obtain the latter as

$$\mathbf{M} = \sum_{e=1}^{3} \mathbf{A}_e^T [\bar{m}]_e \mathbf{A}_e \tag{7-106}$$

$$\mathbf{M} = \frac{\gamma l}{420}\begin{bmatrix} 156 & 0 & -22l & 54 & 0 & 13l & 0 & 0 & 0 & 0 & 0 & 0 \\ 0 & 140 & 0 & 0 & 70 & 0 & 0 & 0 & 0 & 0 & 0 & 0 \\ -22l & 0 & 4l^2 & -13l & 0 & -3l^2 & 0 & 0 & 0 & 0 & 0 & 0 \\ 54 & 0 & -13l & 436 & 0 & 22l & 140 & 0 & 0 & 0 & 0 & 0 \\ 0 & 70 & 0 & 0 & 452 & 88l & 0 & 108 & -52l & 0 & 0 & 0 \\ 13l & 0 & -3l^2 & 22l & 88l & 36l^2 & 0 & 52l & -24l^2 & 0 & 0 & 0 \\ 0 & 0 & 0 & 140 & 0 & 0 & 436 & 0 & 22l & 54 & 0 & -13l \\ 0 & 0 & 0 & 0 & 108 & 52l & 0 & 452 & -88l & 0 & 70 & 0 \\ 0 & 0 & 0 & 0 & -52l & -24l^2 & 22l & -88l & 36l^2 & 13l & 0 & -3l^2 \\ 0 & 0 & 0 & 0 & 0 & 0 & 54 & 0 & 13l & 156 & 0 & -22l \\ 0 & 0 & 0 & 0 & 0 & 0 & 0 & 70 & 0 & 0 & 140 & 0 \\ 0 & 0 & 0 & 0 & 0 & 0 & -13l & 0 & -3l^2 & -22l & 0 & 4l^2 \end{bmatrix}$$

$$\mathbf{K} = \sum_{e=1}^{3} \mathbf{A}_e^T [\bar{k}]_e \mathbf{A}_e \tag{7-108}$$

$$\mathbf{K} = \frac{EI}{l^3} \begin{bmatrix} 12 & 0 & -6l & -12 & 0 & -6l & 0 & 0 & 0 & 0 & 0 & 0 \\ 0 & \left(\frac{l}{r}\right)^2 & 0 & 0 & -\left(\frac{l}{r}\right)^2 & 0 & 0 & 0 & 0 & 0 & 0 & 0 \\ -6l & 0 & 4l^2 & 6l & 0 & 2l^2 & 0 & 0 & 0 & 0 & 0 & 0 \\ -12 & 0 & 6l & \left[12 + (l/r)^2/2\right] & 0 & 6l & -\frac{1}{2}\left(\frac{l}{r}\right)^2 & 0 & 0 & 0 & 0 & 0 \\ 0 & -\left(\frac{l}{r}\right)^2 & 0 & 0 & \left[\frac{3}{2} + (l/r)^2\right] & \frac{3}{2}l & 0 & -\frac{3}{2} & \frac{3}{2}l & 0 & 0 & 0 \\ -6l & 0 & 2l^2 & 6l & \frac{3}{2}l & 6l^2 & 0 & -\frac{3}{2}l & l^2 & 0 & 0 & 0 \\ 0 & 0 & 0 & -\frac{1}{2}\left(\frac{l}{r}\right)^2 & 0 & 0 & \left[12 + (l/r)^2/2\right] & 0 & 6l & -12 & 0 & 6l \\ 0 & 0 & 0 & 0 & -\frac{3}{2} & -\frac{3}{2}l & 0 & \left[\frac{3}{2} + (l/r)^2\right] & -\frac{3}{2}l & 0 & -\left(\frac{l}{r}\right)^2 & 0 \\ 0 & 0 & 0 & 0 & \frac{3}{2}l & l^2 & 6l & -\frac{3}{2}l & 6l^2 & -6l & 0 & 2l^2 \\ 0 & 0 & 0 & 0 & 0 & 0 & -12 & 0 & -6l & 12 & 0 & -6l \\ 0 & 0 & 0 & 0 & 0 & 0 & 0 & -\left(\frac{l}{r}\right)^2 & 0 & 0 & \left(\frac{l}{r}\right)^2 & 0 \\ 0 & 0 & 0 & 0 & 0 & 0 & 6l & 0 & 2l^2 & -6l & 0 & 4l^2 \end{bmatrix}$$

Step 6. Since the frame is fixed at joints 1 and 4 as shown in Fig. 7-23a, the system joint displacements represented by q_1, q_2, q_3, q_{10}, q_{11}, and q_{12} in Fig. 7-23c are zero. Therefore, we delete both rows and columns 1, 2, 3, 10, 11, and 12 from the **M** and **K** matrices determined in step 5. To eliminate the axial deformations of the columns, which Fig. 7-23c shows involves the generalized coordinates q_5 and q_8, we also delete the fifth and eighth rows and columns of **M** and **K**. With these deletions made, we obtain the following reduced mass and stiffness matrices for the system constituted by the constrained frame:

$$\mathbf{M} = \frac{\gamma l}{420} \begin{bmatrix} 436 & 22l & 140 & 0 \\ 22l & 36l^2 & 0 & -24l^2 \\ 140 & 0 & 436 & 22l \\ 0 & -24l^2 & 22l & 36l^2 \end{bmatrix} \tag{7-140}$$

and

$$\mathbf{K} = \frac{EI}{l^3} \begin{bmatrix} \left[12 + (l/r)^2/2\right] & 6l & -(l/r)^2/2 & 0 \\ 6l & 6l^2 & 0 & l^2 \\ -(l/r)^2/2 & 0 & \left[12 + (l/r)^2/2\right] & 6l \\ 0 & l^2 & 6l & 6l^2 \end{bmatrix} \tag{7-141}$$

Thus, we see that q_4, q_6, q_7, and q_9 are the remaining generalized coordinates, so that the equations of motion of the constrained frame are given by

$$\mathbf{M}\begin{Bmatrix} \ddot{q}_4 \\ \ddot{q}_6 \\ \ddot{q}_7 \\ \ddot{q}_9 \end{Bmatrix} + \mathbf{K}\begin{Bmatrix} q_4 \\ q_6 \\ q_7 \\ q_9 \end{Bmatrix} = \mathbf{0}$$

in which **M** and **K** have the matrix elements shown in Eqs. 7-140 and 7-141.

Substituting the values

$$\begin{aligned} l &= 240 \text{ in.} \\ EI &= 2.952(10)^{10} \text{ lb} \cdot \text{in.}^2 \\ \gamma &= 0.0130 \text{ lb} \cdot \text{s}^2/\text{in.}^2 \\ r &= 7.47 \text{ in.} \end{aligned}$$

into Eqs. 7-140 and 7-141 yields the following matrices for use as data input to the JACOBI.FOR program:

$$\mathbf{M} = \begin{bmatrix} 3.24 & 39.22 & 1.04 & 0.00 \\ 39.22 & 15{,}403.89 & 0.00 & 10{,}269.26 \\ 1.04 & 0.00 & 3.24 & 39.22 \\ 0.00 & -10{,}269.26 & 39.22 & 15{,}403.89 \end{bmatrix}$$

$$\mathbf{K} = (10)^6 \begin{bmatrix} 1.13 & 3.08 & -1.10 & 0.00 \\ 3.08 & 738.00 & 0.00 & 123.00 \\ -1.10 & 0.00 & 1.13 & 3.08 \\ 0.00 & 123.00 & 3.08 & 738.00 \end{bmatrix}$$

The natural circular frequencies and normal-mode shapes of the frame as obtained from the program are shown in Fig. 7-24.

The preceding examples have illustrated the use of each of the three finite elements shown in Fig. 7-11. In each example the steps discussed on page 519 were used to obtain the system mass and stiffness matrices as data input to the JACOBI.FOR program. In the next section we write a general program FINITEL.FOR for analyzing systems formed of rod elements, either of the beam elements of Fig. 7-11, or combinations of the beam elements as used in plane frames, for example. As we shall see, the program performs all the steps necessary to obtain the system mass and stiffness matrices of a system composed of p elements, and then computes the natural circular frequencies and mode shapes desired for the system. This is accomplished by utilizing the more general element of Fig. 7-11c in the program, obtaining the mass and stiffness matrices of a system composed of p elements of this type, and then reducing these general matrices to ones pertinent to the other types of systems mentioned above by simply eliminating the rows and columns in each matrix that correspond to the subscripts of the generalized coordinates having zero values (due to system constraints) in the particular system.

A similar program TRUSS.FOR for plane-truss systems is developed in Sec. 7-8 following a discussion of the truss element.

It should be noted here that occasionally there are systems for which the natural frequency values obtained are found to be very sensitive to very small differences in the numerical values of the elements of the mass or stiffness matrices. The first mode is generally the one that is significantly affected, with the higher modes and all the mode shapes affected but slightly. As an example, the frame of Example 7-9 was analyzed in that example using values in the system mass and stiffness matrices that were calculated on a pocket calculator and rounded off to two digits after the decimal point, which resulted in many of the values

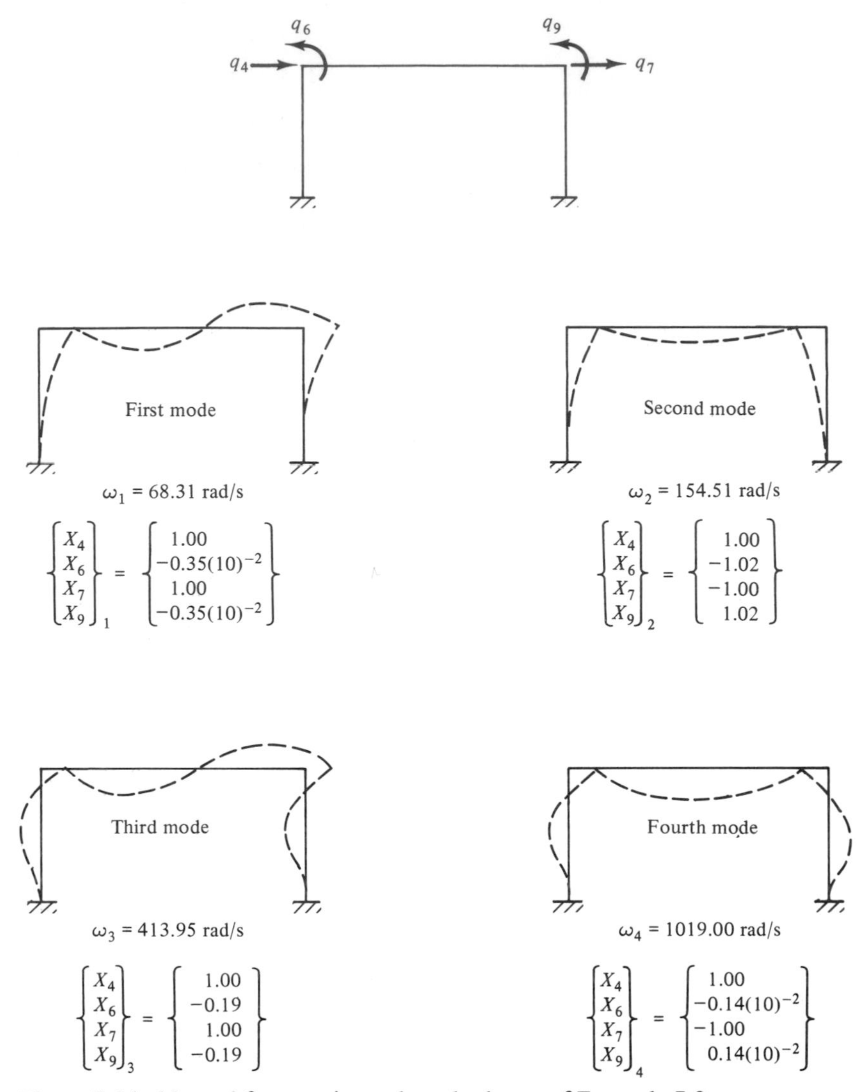

$$\begin{Bmatrix} X_4 \\ X_6 \\ X_7 \\ X_9 \end{Bmatrix}_1 = \begin{Bmatrix} 1.00 \\ -0.35(10)^{-2} \\ 1.00 \\ -0.35(10)^{-2} \end{Bmatrix} \qquad \begin{Bmatrix} X_4 \\ X_6 \\ X_7 \\ X_9 \end{Bmatrix}_2 = \begin{Bmatrix} 1.00 \\ -1.02 \\ -1.00 \\ 1.02 \end{Bmatrix}$$

$$\begin{Bmatrix} X_4 \\ X_6 \\ X_7 \\ X_9 \end{Bmatrix}_3 = \begin{Bmatrix} 1.00 \\ -0.19 \\ 1.00 \\ -0.19 \end{Bmatrix} \qquad \begin{Bmatrix} X_4 \\ X_6 \\ X_7 \\ X_9 \end{Bmatrix}_4 = \begin{Bmatrix} 1.00 \\ -0.14(10)^{-2} \\ -1.00 \\ 0.14(10)^{-2} \end{Bmatrix}$$

Figure 7-24 Natural frequencies and mode shapes of Example 7-9.

having only three significant digits.[9] The same frame was then analyzed using the program FINITEL.FOR mentioned earlier, in which all the calculations, including those leading to the values obtained for the system mass and stiffness matrices, were done in double precision. The fundamental frequencies obtained by the two procedures varied by approximately 13 percent, with the value determined by FINITEL.FOR being the lower of the two. The values for the higher-mode frequencies and all the mode shapes were essentially the same.

Upon observing the discrepancy between the two solutions, small changes were made

[9] Perhaps only three significant digits were justified, since the measurements from which the values were calculated were of limited accuracy. For example, the mass per unit length of beams is often given with just three-digit accuracy, as is the modulus of elasticity E of materials.

in turn in the values of the elements of the mass and stiffness matrices, and it was found that the system was sensitive to very small differences in the values of the elements of the stiffness matrix but tolerant of similar differences in the values of the elements of the mass matrix. This seemed to indicate that the analysis of the frame of Example 7-9 provided an example of an *ill-conditioned eigenvalue problem,* and while a discussion of such problems is beyond the scope of this text, the reader is advised here that such problems do exist, and is referred to a discussion of this subject by Gourlay and Watson.[10] They warn that the eigenvalue problem $\mathbf{AX} = \lambda\mathbf{BX}$ ($\mathbf{KX} = \lambda\mathbf{MX}$ in Example 7-9) can be ill-conditioned if $\mathbf{B}$ is ill-conditioned with respect to inversion. It is interesting to note that in Example 7-9 the matrix $\mathbf{A}$ (the stiffness matrix) was the one found to be ill-conditioned with respect to inversion, while $\mathbf{B}$ (the mass matrix) was not. Gourlay and Watson further state that an analysis of the sensitivity of the eigenvectors of a matrix with respect to perturbations of the elements of the matrix is extremely complex.

7-7 COMPUTER PROGRAM FOR THE VIBRATION ANALYSIS OF PLANE FRAMES USING BEAM ELEMENTS (FINITEL.FOR)

The program FINITEL.FOR given in this section for determining the natural circular frequencies and mode shapes of plane frames follows the steps given in Sec. 7-6 for the most part; so a flowchart for the program will not be shown. Comment statements throughout the program should aid the reader in correlating the program with the steps given in Sec. 7-6 beginning on page 519. The major differences between these steps and those of the computer program appear in the procedure for assembling the mass and stiffness matrices of the elements into the system matrices.

It was shown in Sec. 7-5 that the system mass and stiffness matrices are given by

$$\mathbf{M} = \sum_{e=1}^{p} \mathbf{A}_e^T [\bar{m}]_e \mathbf{A}_e \tag{7-106}$$

and

$$\mathbf{K} = \sum_{e=1}^{p} \mathbf{A}_e^T [\bar{k}]_e \mathbf{A}_e \tag{7-108}$$

respectively. The $\mathbf{A}$ matrices are those which relate the global joint displacements of the elements to the generalized coordinates of the system (see Eq. 7-97). So that all the $\mathbf{A}$ matrices need not be entered into the program or generated as part of it, an equivalent assembly procedure given by Pao[11] for the assembly of stiffness matrices is used in this program in assembling both the stiffness and mass matrices of the system. This scheme, which is illustrated here in assembling the stiffness matrices, utilizes the following relationships:

[10] A. R. Gourlay and G. A. Watson, *Computational Methods for Matrix Eigenproblems,* pp. 23–25, 125, John Wiley & Sons, London, 1973.

[11] Y. C. Pao, *A First Course in Finite Element Analysis,* p. 49, Allyn and Bacon, Boston, Mass., 1986.

For $e = 1, 2, 3, \ldots, N_e$

$$k_{pq}^{(e)} \rightarrow K_{r_p c_q} \qquad (\text{for } p, q = 1, 2, \ldots, n) \tag{7-142}$$

and

$$r_p = r_{n_f j-m+1} = c_{n_f j-m+1} = n_f N_{ej} - m + 1 \tag{7-143}$$

for $j = 1, 2, \ldots, n_n$ and $m = 1, 2, \ldots, n_f$. In these equations the symbols are defined as follows:

$k_{pq}^{(e)}$ = stiffness coefficients of the eth element
N_e = total number of elements
$K_{r_p c_q}$ = stiffness coefficients of the system stiffness matrix
r_p, c_q = subscripts of the system stiffness coefficients
n_f = number of degrees of freedom of each joint (three for plane frames using beam elements with axial deformations)
n_n = number of joints per element (two for beam elements)
n = order of element stiffness matrices ($n = n_n \times n_f = 6$ for beam elements)
m = variable $1, 2, \ldots, n_f$
j = variable 1 to n_n ($n_n = 2$ for beam elements)
N_{ej} = jth joint number of the eth element

The symbol $\rightarrow$ is interpreted as meaning "contributing to the cumulative sum" of the element stiffness coefficients that make up the particular system stiffness coefficient given on the right of the symbol.

The quantities N_{ej} are elements of what is called the *node-number matrix.* We shall refer to it as the *joint-number matrix* to avoid confusion with the use of node as a point of zero displacement in a vibrating body. This matrix has one row for each element of the system, the first row corresponding to element 1, the second row to element 2, and so forth. The quantities that make up the row are the joint numbers of that particular element. Since beam elements have only two joints, the joint-number matrix for our analysis will consist of just two columns.

To verify that this assembly scheme gives the same results as that obtained by using Eqs. 7-106 and 7-108, let us determine the coefficients of the system stiffness matrix of Example 7-9 to which the coefficients of the stiffness matrix for element 2 contribute. The joint-number matrix for the frame in Example 7-9 is

$$\mathbf{N} = \begin{bmatrix} 1 & 2 \\ 2 & 3 \\ 3 & 4 \end{bmatrix}$$

Then with $e = 2$, since we are considering element 2,

$$r_p = r_{n_f j-m+1} \qquad (\text{or } c_q) = n_f N_{ej} - m + 1$$

$j = 1$, $m = 1$: $$r_{(3)(1)-1+1} = \underset{\uparrow p}{r_3} = \underset{\uparrow q}{c_3} = (3)(2) - 1 + 1 = 6$$

$j = 2$, $m = 1$: $$r_{(3)(2)-1+1} = \underset{\uparrow p}{r_6} = \underset{\uparrow q}{c_6} = (3)(3) - 1 + 1 = 9$$

$j = 1$, $m = 2$: $$r_{(3)(1)-2+1} = r_2 = c_2 = (3)(2) - 2 + 1 = 5$$

$$\begin{matrix} j = 2 \\ m = 2 \end{matrix} \qquad r_{(3)(2)-2+1} = r_5 = c_5 = (3)(3) - 2 + 1 = 8$$

$$\begin{matrix} j = 1 \\ m = 3 \end{matrix} \qquad r_{(3)(1)-3+1} = r_1 = c_1 = (3)(2) - 3 + 1 = 4$$

$$\begin{matrix} j = 2 \\ m = 3 \end{matrix} \qquad r_{(3)(2)-3+1} = r_4 = c_4 = (3)(3) - 3 + 1 = 7$$

We see that when subscript p of $k_{pq}^{(2)}$ is equal to 3, subscript r_p of $K_{r_p c_q}$ is equal to 6 (or when subscript q of $k_{pq}^{(2)}$ is equal to 3, subscript c_q of $K_{r_p c_q}$ is equal to 6). Similarly, when subscript p is equal to 6, subscript r_p is equal to 9 (or when subscript q is equal to 6, subscript c_q is equal to 9), and so forth. Although values for p and q are calculated from the same formula, it should be noted that the subscript values of p and q of $k_{pq}^{(2)}$ are not necessarily the same. Thus, it should be apparent that

$$\begin{matrix} k_{11}^{(2)} \rightarrow K_{44} & k_{22}^{(2)} \rightarrow K_{55} \\ k_{12}^{(2)} \rightarrow K_{45} & \\ k_{13}^{(2)} \rightarrow K_{46} & \vdots \quad \vdots \quad \vdots \\ k_{14}^{(2)} \rightarrow K_{47} & k_{63}^{(2)} \rightarrow K_{96} \\ k_{15}^{(2)} \rightarrow K_{48} & k_{64}^{(2)} \rightarrow K_{97} \\ k_{16}^{(2)} \rightarrow K_{49} & k_{65}^{(2)} \rightarrow K_{98} \\ k_{21}^{(2)} \rightarrow K_{54} & k_{66}^{(2)} \rightarrow K_{99} \end{matrix}$$

Examination of the system stiffness matrix determined in step 5 of Example 7-9 shows that the coefficients of the stiffness matrix for element 2 have indeed contributed to the coefficients of the system matrix as indicated above. This assembly procedure is well suited to programming on the computer, and we shall therefore use it in our computer program.

The computer program FINITEL.FOR is written for analyzing beams and plane frames made up of beam elements such as the one shown in Fig. 7-11c. However, it can also be used for systems composed of beam elements in which the axial displacements are zero or assumed to be so, such as the beam shown in Fig. 7-11b. This is accomplished by applying boundary conditions that fix at zero the generalized coordinate values corresponding to the axial displacements of the beam elements. This will be illustrated in Examples 7-10 and 7-11. The program can also be used for rods composed of rod elements such as the one shown in Fig. 7-11a by applying boundary conditions that fix at zero the generalized coordinate values corresponding to both the lateral and rotational displacements of the rod elements. This will be illustrated in Example 7-12.

With the array dimensions shown in the program listing, plane frames consisting of up to six beam elements and having up to six joints (18 degrees of freedom before boundary constraints are applied) can be analyzed. The program can be easily modified to analyze frames with more elements than six by increasing the array sizes in the DIMENSION statements.

If the program is used for problems in which axial displacements are not considered (such as the rocket problem of Example 7-7), values of 1.0 can be read in for the cross-sectional areas of all such elements. If the program is used for rods, in which bending is not considered (such as the rod in Example 7-5), a value of 1.0 can be read in for the area moment of inertia of each rod element. Although values of 1.0 are read in for the cross-sectional areas or area moments of inertia of the elements, these values will have no effect on the solution, since the coordinates that are fixed will in effect delete the rows and columns of the system matrices that involve cross-sectional areas or area moments of inertia.

Rather than obtaining the reduced system mass and stiffness matrices from the finite-element computer program and entering them into the JACOBI.FOR program given in Chap. 5, the JACOBI.FOR program is used now as a *subprogram* (called JCBI) of the main finite-element program FINITEL.FOR. The subprogram JCBI and its subprograms DECOMP, MATINV, MATMPY, and SEARCH are listed in a subroutine file called LIBRARY.FOR. The main program FINITEL.FOR and the subprograms that are in LIBRARY.FOR are then compiled separately prior to linking to obtain the execution program. The FINITEL.FOR program as written can be run on a microcomputer having as little as 64K of random access memory. Readers having computers with additional memory who wish to use the program to analyze systems larger than those discussed in the examples will generally find that they can accomplish this by simply changing the DIMENSION statements to allow for larger arrays.

The most important program variables and array names and the quantities that they represent are as follows:

Variable or array name	Quantity
NUMEL	The number of beam elements
A(I)	The cross-sectional areas of the beam elements
E(I)	Moduli of elasticity of beam elements
IA(I)	Area moments of inertia of the beam elements
GAMMA(I)	Mass per unit length of the beam elements
NJTS	The number of joints in the system
JNM(I,J)	The joint-number matrix
X(I), Y(I)	The x and y coordinates of the system joints in the global coordinate system. Used to calculate the values for sin α and cos α of the transformation matrix **R**
L	Length of element (calculated from x and y coordinates of joints of element)
NB	The number of boundary conditions (or the number of generalized coordinates to be fixed at zero)
N	The number of degrees of freedom of the system with no boundary conditions. The full mass and stiffness matrices will be N $\times$ N. N = NJTS*3
EN	A DO variable representing element numbers
KEL(I,J,EN)	A three-dimensional array containing the element stiffness matrices expressed in the local coordinate systems. Each "plane" of KEL contains the stiffness matrix of element EN (1 $\leq$ EN $\leq$ NUMEL)
MEL(I,J,EN)	A three-dimensional array containing the element mass matrices expressed in the local coordinate systems
KEG(I,J,EN)	A three-dimensional array containing the element stiffness matrices expressed in the global coordinate system. Each "plane" of KEG contains the stiffness matrix of element EN (1 $\leq$ EN $\leq$ NUMEL)
MEG(I,J,EN)	A three-dimensional array containing the element mass matrices expressed in the global coordinate system
TK(I,J)	Initially a 6 $\times$ 6 null matrix that becomes the matrix $[k]_e\mathbf{R}$
TM(I,J)	Initially a 6 $\times$ 6 null matrix that becomes the matrix $[m]_e\mathbf{R}$
SINA	The sine of angle α (α is the angle from the global x axis to the local x axis of an element)
COSA	The cosine of angle α
R(I,J)	The transformation matrix for transforming global joint displacements to local joint displacements (see Eq. 7-84)
RT(I,J)	The transpose of the transformation matrix

Variable or array name	Quantity
SUB(J1) J1 = J*3−M+1	An array that contains the values of subscripts of system mass and stiffness coefficients. The subscripts of SUB are the subscripts of element stiffness or mass coefficients
ROWSUB	Takes on the various values of SUB and is the first subscript of a particular system mass or stiffness coefficient
COLSUB	Takes on the various values of SUB and is the second subscript of a particular mass or stiffness coefficient
SK(I,J)	The system stiffness matrix
SM(I,J)	The system mass matrix
NF	The number of degrees of freedom of the system after applying boundary conditions
RSK(I,J)	The reduced system stiffness matrix obtained after applying boundary conditions
RSM(I,J)	The reduced system mass matrix obtained after applying boundary conditions
CFIX(I)	An array containing the generalized coordinate numbers whose values are to be fixed at zero

The computer program FINITEL.FOR is as follows:

```
C   FINITEL.FOR
C   FINITE-ELEMENT PROGRAM FOR VIBRATION ANALYSIS OF BEAMS,
C   RODS, AND PLANE FRAMES USING BEAM ELEMENTS.  SUBROUTINES
C   JCBI, DECOMP, MATINV, MATMPY, AND SEARCH ARE IN LIBRARY.FOR
C   AND SHOULD BE COMPILED SEPARATELY.  AFTER FINITEL.FOR HAS
C   BEEN COMPILED, FINITEL.OBJ MUST BE LINKED WITH LIBRARY.OBJ.
C   DEVICE * IN READ AND WRITE STATEMENTS IS THE CONSOLE.
C   DEVICE 2 IN WRITE STATEMENTS IS THE PRINTER.
C   * IN THE PLACE OF A FORMAT STATEMENT NUMBER MEANS FREE FORMAT.
      IMPLICIT REAL*8 (A-H,O-Z)
      REAL*8 L,IA,KEL,MEL,KEG,MEG
      INTEGER SUB,ROWSUB,COLSUB,B,Z,EN,CFIX
      DIMENSION KEL(6,6,6),MEL(6,6,6),KEG(6,6,6),MEG(6,6,6),
     $RT(6,6),
     $R(6,6),TK(6,6),TM(6,6),SK(18,18),SM(18,18),RSK(12,12),
     $RSM(12,12),E(6),A(6),X(6),Y(6),GAMMA(6),IA(6)
      DIMENSION JNM(6,2),CFIX(12),SUB(6)
      OPEN (2,FILE='PRN')
C   READ IN PROBLEM DATA AS INDICATED BY MESSAGES ON CONSOLE. DATA
C   READ IN IS PRINTED OUT. (PROGRAM STATEMENTS 2 THROUGH 40)
      WRITE(*,2)
2     FORMAT(' ', 'ENTER THE NUMBER OF BEAM ELEMENTS (I1)',/)
      READ(*,3)NUMEL
3     FORMAT(I1)
      DO 4 I = 1,NUMEL
      WRITE(*,5)I
4     READ(*,6)A(I)
5     FORMAT(/,' ENTER A(',I1,') (F20.0)',/)
6     FORMAT(F20.0)
      WRITE(2,7)
```

```
7       FORMAT(6X,'THE AREA ARRAY  A  IS:',/)
        DO 8 I = 1,NUMEL
8       WRITE(2,9)I,A(I)
9       FORMAT(6X,'A(',I1,') = ',E14.7)
        DO 10 I = 1,NUMEL
        WRITE(*,11)I
10      READ(*,12)E(I)
11      FORMAT(/,' ENTER E(',I1,')  (F20.0)',/)
12      FORMAT(F20.0)
        WRITE(2,13)
13      FORMAT(/,6X,'THE ELASTICITY ARRAY  E  IS;',/)
        DO 14 I = 1,NUMEL
14      WRITE(2,15)I,E(I)
15      FORMAT(6X,'E(',I1,') = ',E14.7)
        DO 16 I = 1,NUMEL
        WRITE(*,17)I
16      READ(*,18)IA(I)
17      FORMAT(/,' ENTER IA(',I1,')  (F20.0)',/)
18      FORMAT(F20.0)
        WRITE(2,19)
19      FORMAT(/,6X,'THE MOMENT OF INERTIA ARRAY  IA  IS;',/)
        DO 20 I = 1,NUMEL
20      WRITE(2,21)I,IA(I)
21      FORMAT(6X,'IA(',I1,') = ',E14.7)
        DO 82 I = 1,NUMEL
        WRITE(*,83)I
82      READ(*,84)GAMMA(I)
83      FORMAT(/,' ENTER GAMMA(',I1,')   (F20.0)',/)
84      FORMAT(F20.0)
        WRITE(2,85)
85      FORMAT(/,6X, 'THE GAMMA ARRAY IS;',/)
        DO 86 I = 1,NUMEL
86      WRITE(2,87)I,GAMMA(I)
87      FORMAT(6X,'GAMMA(',I1,') = ',E14.7)
        WRITE(*,22)
22      FORMAT(/,' ENTER THE NUMBER OF JOINTS, NJTS  (I1)',/)
        READ(*,23)NJTS
23      FORMAT(I1)
        DO 24 I = 1,NUMEL
        DO 24 J = 1,2
        WRITE(*,25)I,J
24      READ(*,26)JNM(I,J)
25      FORMAT(/,' ENTER JNM(',I1,',',I1,')   (I1)',/)
26      FORMAT(I1)
        WRITE(2,27)
27      FORMAT(/,6X,'THE JOINT-NUMBER MATRIX IS;',/)
        DO 28 I = 1,NUMEL
28      WRITE(2,29)JNM(I,1),JNM(I,2)
29      FORMAT(10X,I5,I4)
```

```
      DO 30 I = 1,NJTS
      WRITE(*,31)I,I
30    READ(*,*)X(I),Y(I)
31    FORMAT(/' ENTER JOINT COORD. X(',I1,'),Y(',I1,')(2F20.0)',/)
      WRITE(2,33)
33    FORMAT(/,6X,'THE JOINT COORDINATES ARE;',/)
      DO 34 I = 1,NJTS
34    WRITE(2,35)I,X(I),I,Y(I)
35    FORMAT(6X,'X(',I1,') = ',E14.7,5X,'Y(',I1,') = ',E14.7)
      WRITE(*,36)
36    FORMAT(/,' ENTER THE NUMBER OF FIXED COORDINATES (I2)',/)
      READ(*,37)NB
37    FORMAT(I2)
      IF(NB .EQ. 0)GO TO 94
      DO 38 I = 1,NB
      WRITE(*,39)I
38    READ(*,40)CFIX(I)
39    FORMAT(/,' ENTER CFIX(',I2,')  (I2)',/)
40    FORMAT(I2)
      WRITE(2,41)
41    FORMAT(/,6X,'ARRAY CFIX IS;',/)
      DO 42 I = 1,NB
42    WRITE(2,43)I,CFIX(I)
43    FORMAT(6X,'CFIX(',I2,') = ',I2)
C  GENERATE NULL 3-DIMENSIONAL ARRAYS KEL AND MEL.  PLANES OF KEL
C  AND MEL WILL LATER CONTAIN THE LOCAL ELEMENT STIFFNESS AND MASS
C  MATRICES, RESPECTIVELY.
94    DO 44 I = 1,6
      DO 44 J = 1,6
      DO 44 M = 1,NUMEL
      KEL(I,J,M) = 0.
44    MEL(I,J,M)= 0.
C  GENERATE NULL MATRICES  R  AND  RT  WHICH WILL LATER BECOME THE
C  TRANSFORMATION MATRIX AND ITS TRANSPOSE, RESPECTIVELY.
      DO 45 I = 1,6
      DO 45 J = 1,6
      R(I,J)=0.0
45    RT(I,J)=0.0
C  GENERATE THE LOCAL ELEMENT STIFFNESS MATRICES AND STORE IN THE 3
C  3-DIMENSIONAL STIFFNESS ARRAY KEL (SEE FIG. 7-11 FOR THE
C  EQUATION USED).  EACH PLANE IN THE 3-DIM. ARRAY IS ONE ELEMENT
C  STIFFNESS MATRIX.
      DO 100 EN=1,NUMEL
      IC=JNM(EN,1)
      ID=JNM(EN,2)
      L=DSQRT((X(ID)-X(IC))**2+(Y(ID)-Y(IC))**2)
      QUOT=IA(EN)/A(EN)
      R1=DSQRT(QUOT)
      F=E(EN)*IA(EN)/L
```

```
      P=F/R1**2
      Q=4.*P*R1**2
      S=3.*Q/(2.*L)
      T=S*2./L
      SINA=(Y(ID)-Y(IC))/L
      COSA=(X(ID)-X(IC))/L
      KEL(1,1,EN)=P
      KEL(1,4,EN)=-P
      KEL(2,2,EN)=T
      KEL(2,3,EN)=S
      KEL(2,5,EN)=-T
      KEL(2,6,EN)=S
      KEL(3,3,EN)=Q
      KEL(3,5,EN)=-S
      KEL(3,6,EN)=Q/2.
      KEL(4,4,EN)=P
      KEL(5,5,EN)=T
      KEL(5,6,EN)=-S
      KEL(6,6,EN)=Q
      DO 46 I=2,6
      IM1 = I-1
      DO 46 J=1,IM1
46    KEL(I,J,EN)=KEL(J,I,EN)
C  GENERATE THE LOCAL ELEMENT MASS MATRICES AND STORE THEM IN THE
C  3-DIMENSIONAL MASS ARRAY MEL (SEE FIG. 7-11 FOR THE EQUATION
C  USED).  EACH PLANE OF THE 3-DIM. ARRAY MEL CONTAINS ONE LOCAL
C  ELEMENT MASS MATRIX.
      F=GAMMA(EN)*L/420.
      P=70.*F
      P2=2.*P
      Q=156.*F
      S=22.*L*F
      T=54.*F
      U=4.*L*L*F
      V=13.*L*F
      W=3.*L*L*F
      MEL(1,1,EN)=P2
      MEL(1,4,EN)=P
      MEL(2,2,EN)=Q
      MEL(2,3,EN)=S
      MEL(2,5,EN)=T
      MEL(2,6,EN)=-V
      MEL(3,3,EN)=U
      MEL(3,5,EN)=V
      MEL(3,6,EN)=-W
      MEL(4,4,EN)=P2
      MEL(5,5,EN)=Q
      MEL(5,6,EN)=-S
      MEL(6,6,EN)=U
```

```
      DO 47 I = 2,6
      IM1=I-1
      DO 47 J=1,IM1
47    MEL(I,J,EN)=MEL(J,I,EN)
C  GENERATE THE TRANSFORMATION MATRIX  R  AND ITS TRANSPOSE RT.
      R(1,1)=COSA
      R(1,2)=SINA
      R(2,1)=-SINA
      R(2,2)=COSA
      R(3,3)=1.
      R(4,4)=COSA
      R(4,5)=SINA
      R(5,4)=-SINA
      R(5,5)=COSA
      R(6,6)=1.
      DO 48 I=1,3
      DO 48 J=1,3
      RT(I,J)=R(J,I)
48    RT(I+3,J+3)=R(J+3,I+3)
C  DETERMINE THE ELEMENT STIFFNESS MATRICES IN THE GLOBAL
C  COORDINATE SYSTEM (EQ. 7-92B) AND STORE THEM IN THE 3-DIM.
C  STIFFNESS ARRAY KEG.  EACH PLANE OF THE 3-DIM. ARRAY CONTAINS
C  ONE GLOBAL ELEMENT STIFFNESS MATRIX.
      DO 95 I=1,6
      DO 95 J=1,6
      TK(I,J)=0.0
      DO 95 K=1,6
95    TK(I,J)=TK(I,J)+KEL(I,K,EN)*R(K,J)
      DO 96 I=1,6
      DO 96 J=1,6
      KEG(I,J,EN)=0.0
      DO 96 K=1,6
96    KEG(I,J,EN)=KEG(I,J,EN)+RT(I,K)*TK(K,J)
C  DETERMINE THE ELEMENT MASS MATRICES IN THE GLOBAL SYSTEM (EQ.
C  7-92A) AND STORE THEM IN THE 3-DIM. MASS ARRAY MEG.  EACH PLANE
C  OF THE 3-DIM. ARRAY CONTAINS ONE GLOBAL ELEMENT MASS MATRIX.
      DO 97 I=1,6
      DO 97 J=1,6
      TM(I,J)=0.0
      DO 97 K=1,6
97    TM(I,J)=TM(I,J)+MEL(I,K,EN)*R(K,J)
      DO 98 I=1,6
      DO 98 J=1,6
      MEG(I,J,EN)=0.0
      DO 98 K=1,6
98    MEG(I,J,EN)=MEG(I,J,EN)+RT(I,K)*TM(K,J)
100   CONTINUE
C  GENERATE NULL MATRICES SK AND SM WHICH WILL BECOME THE
C  SYSTEM STIFFNESS AND MASS MATRICES, RESPECTIVELY.
      N=NJTS*3
```

```
      DO 49 I=1,N
      DO 49 J=1,N
      SK(I,J)=0.
49    SM(I,J)=0.
C  ASSEMBLE THE SYSTEM STIFFNESS AND MASS MATRICES.
      DO 51 I=1,NUMEL
      DO 50 J=1,2
      DO 50 M=1,3
      J1=J*3-M+1
50    SUB(J1)=3*JNM(I,J)-M+1
      DO 51 B=1,6
      DO 51 Z=1,6
      ROWSUB=SUB(B)
      COLSUB=SUB(Z)
      SK(ROWSUB,COLSUB)=SK(ROWSUB,COLSUB)+KEG(B,Z,I)
51    SM(ROWSUB,COLSUB)=SM(ROWSUB,COLSUB)+MEG(B,Z,I)
C  CALCULATE THE NUMBER OF DEGREES OF FREEDOM AND REMOVE ROWS AND
C  COLUMNS FROM THE SYSTEM STIFFNESS AND MASS MATRICES.
      NF=N-NB
      IF(NB .EQ. 0)GO TO 69
      NA=1
      KL=N-1
62    JC=1
63    IF(JC .EQ. CFIX(NA))GO TO 64
      JC=JC+1
      IF(JC .EQ. N)GO TO 68
      GO TO 63
64    DO 65 I=1,N
      DO 65 J=JC,KL
      SK(I,J)=SK(I,J+1)
      SM(I,J)=SM(I,J+1)
65    CONTINUE
      DO 66 J=1,N
      DO 66 I=JC,KL
      SK(I,J)=SK(I+1,J)
      SM(I,J)=SM(I+1,J)
66    CONTINUE
      IF(NA .EQ. NB)GO TO 68
      NA=NA+1
      DO 67 I=NA,NB
67    CFIX(I)=CFIX(I)-1
      GO TO 62
68    CONTINUE
C  ASSIGN REDUCED STIFFNESS AND MASS MATRIX ELEMENTS TO ARRAY
C  NAMES RSK AND RSM, RESPECTIVELY.
69    DO 70 I=1,NF
      DO 70 J=1,NF
      RSK(I,J)=SK(I,J)
70    RSM(I,J)=SM(I,J)
C  WRITE OUT THE REDUCED STIFFNESS AND MASS MATRICES OBTAINED
```

```
C   FROM THE BOUNDARY CONDITIONS.
        WRITE(2,71)
71      FORMAT(/,' THE REDUCED SYSTEM STIFFNESS MATRIX IS:',/)
        WRITE(2,72) ((RSK(I,J),J=1,NF),I=1,NF)
72      FORMAT(' ',6E11.4/)
        WRITE(2,73)
73      FORMAT(/,' THE REDUCED SYSTEM MASS MATRIX IS:',/)
        WRITE(2,74) ((RSM(I,J),J=1,NF),I=1,NF)
74      FORMAT(' ',6E11.4/)
        WRITE(2,200)
200     FORMAT(//,' ')
C   CALL SUBPROGRAM JCBI TO CALCULATE FREQUENCIES AND MODE SHAPES
        CALL JCBI(NF,RSK,RSM)
        STOP
        END
C   LIBRARY.FOR
C   SUBROUTINES JCBI, DECOMP, MATINV, MATMPY, AND SEARCH AS REQUIRED
C   BY FINITEL.FOR AND TRUSS.FOR
        SUBROUTINE JCBI(N,K,M)
        IMPLICIT REAL*8 (A-H,O-Z)
        REAL*8 K,M,L,LT,LINV,LINVTR
        DIMENSION K(12,12),RT(12,12),A(12,12)
        DIMENSION OMEGA(12),M(12,12),L(12,12),LT(12,12)
        DIMENSION LINV(12,12),LINVTR(12,12),PROD(12,12)
        CALL DECOMP(M,N,L,LT)
        CALL MATINV(L,LINV,N)
        DO 204 I = 1,N
        DO 204 J = 1,N
204     LINVTR(I,J) = LINV(J,I)
        CALL MATMPY(N,K,LINVTR,PROD)
        CALL MATMPY(N,LINV,PROD,A)
        DO 14 I = 1,N
        DO 13 J = 1,N
        RT(I,J) = 0.0
13      CONTINUE
        RT(I,I) = 1.0
14      CONTINUE
        NSWEEP = 0
15      NRSKIP = 0
        NMIN1 = N-1
        DO 25 I = 1,NMIN1
        IP1 = I+1
        DO 24 J = IP1,N
        AV = 0.5*(A(I,J) + A(J,I))
        DIFF = A(I,I) - A(J,J)
        RAD = DSQRT(DIFF*DIFF + 4.*AV*AV)
        IF(RAD .EQ. 0.0)GO TO 20
        IF(DIFF .LT. 0.0)GO TO 18
        IF(DABS(A(I,I)) .EQ. DABS(A(I,I))+100.*DABS(AV))GO TO 16
        GO TO 17
```

```
16    IF(DABS(A(J,J)) .EQ. DABS(A(J,J))+100.*DABS(AV))GO TO 20
17    COSINE = DSQRT((RAD + DIFF)/(2.*RAD))
      SINE = AV/(RAD*COSINE)
      GO TO 19
18    SINE = DSQRT((RAD - DIFF)/(2.*RAD))
      IF(AV .LT. 0.0)SINE = -SINE
      COSINE = AV/(RAD*SINE)
19    IF(1. .LT. 1.+DABS(SINE))GO TO 21
20    NRSKIP = NRSKIP + 1
      GO TO 24
21    DO 22 L1 = 1,N
      Q = A(I,L1)
      A(I,L1) = COSINE*Q + SINE*A(J,L1)
      A(J,L1) = -SINE*Q + COSINE*A(J,L1)
22    CONTINUE
      DO 23 L1 = 1,N
      Q = A(L1,I)
      A(L1,I) = COSINE*Q + SINE*A(L1,J)
      A(L1,J) = -SINE*Q + COSINE*A(L1,J)
      Q = RT(L1,I)
      RT(L1,I) = COSINE*Q + SINE*RT(L1,J)
      RT(L1,J) = -SINE*Q + COSINE*RT(L1,J)
23    CONTINUE
24    CONTINUE
25    CONTINUE
C  KEEP A TALLY OF THE NUMBER OF SWEEPS.
      NSWEEP = NSWEEP + 1
      IF(NSWEEP .GT. 100)GO TO 33
      WRITE(2,26)NRSKIP,NSWEEP
26    FORMAT(' ',5X,'THERE WERE ',I2,
     $' ROTATIONS SKIPPED ON SWEEP NUMBER ',I2)
      IF(NRSKIP .LT. N*(N-1)/2)GO TO 15
      PROD1 = 0.0
      DO 27 J = 1,N
      PROD1 = PROD1 + RT(J,1)*RT(J,N)
27    CONTINUE
      WRITE(2,28)
28    FORMAT(/,' ',5X,'THE SCALAR PRODUCT OF THE FIRST AND LAST')
      WRITE(2,29)PROD1
29    FORMAT(' ',5X,'EIGENVECTORS OF THE TRANSFORMED MATRIX IS ',
     $F19.17/)
      CALL MATMPY(N,LINVTR,RT,PROD)
      DO 30 I = 1,N
      DO 30 J = 1,N
30    RT(I,J) = PROD(I,J)
      DO 42 J = 1,N
      SUM = 0.0
      DO 31 I = 1,N
```

```
31      SUM = SUM + DABS(RT(I,J))
        AV = SUM/N
        QUOT = DABS(RT(1,J))/AV
        IF(QUOT .LT. 0.000001)GO TO 40
        DO 32 I = 2,N
32      RT(I,J) = RT(I,J)/RT(1,J)
        RT(1,J) = 1.0D0
        GO TO 42
40      CALL SEARCH(RT,J,II,N)
        BIG = RT(II,J)
        DO 41 I = 1,N
41      RT(I,J) = RT(I,J)/BIG
42      CONTINUE
        DO 110 I = 1,N
        IF(A(I,I) .LE. 0.0)GO TO 43
        OMEGA(I) = DSQRT(A(I,I))
        GO TO 110
43      OMEGA(I) = 0.0
110     CONTINUE
33      WRITE(2,34)NSWEEP
34      FORMAT(/,' ',5X,'THERE WERE ',I3,' SWEEPS PERFORMED.',
       $/,5X,' THE EIGENVALUES AND EIGENVECTORS FOLLOW:')
        DO 39 JJ = 1,N
        J = N-JJ+1
        WRITE(2,35)JJ,A(J,J)
35      FORMAT(/,' ',5X,'LAMBDA (',I2,') = ',F20.4)
        WRITE(2,111)JJ,OMEGA(J)
111     FORMAT(' ',5X,'OMEGA(',I2,') = ',F20.4,' RAD/S')
        WRITE(2,36)
36      FORMAT(/,' ',5X,'THE ASSOCIATED EIGENVECTOR IS:')
        DO 37 I = 1,N
37      WRITE(2,38)RT(I,J)
38      FORMAT(' ',5X,D17.10)
39      CONTINUE
        RETURN
        END

        SUBROUTINE DECOMP(A,N,L,LT)
        IMPLICIT REAL*8(A-H,O-Z)
        DIMENSION A(12,12)
        REAL*8 L(12,12),LT(12,12)
        DO 9 J = 1,N
        IF(J .EQ. 1)GO TO 7
        JM1 = J-1
        DO 6 I = J,N
        IF(I .NE. J)GO TO 4
        SUM = 0.0
        DO 3 K = 1,JM1
3       SUM = SUM + L(I,K)*L(J,K)
```

```
      L(J,J) = DSQRT(A(J,J) - SUM)
      GO TO 6
4     SUM = 0.0
      DO 5 K = 1,JM1
5     SUM = SUM + L(I,K)*L(J,K)
      L(I,J) = (A(I,J) - SUM)/L(J,J)
6     CONTINUE
      GO TO 9
7     L(1,1) = DSQRT(A(1,1))
      DO 8 I = 2,N
8     L(I,1) = A(I,1)/L(1,1)
9     CONTINUE
C  FILL IN ZERO VALUES OF MATRIX L
      DO 11 J = 2,N
      JM1 = J-1
      DO 11 I = 1,JM1
11    L(I,J) = 0.0
C  ASSIGN VALUES TO THE UPPER TRIANGULAR MATRIX LT
      DO 12 I = 1,N
      DO 12 J = 1,N
12    LT(I,J) = L(J,I)
      RETURN
      END

      SUBROUTINE MATINV(B,A,N)
C  MATRIX INVERSION USING GAUSS-JORDAN REDUCTION AND PARTIAL
C  PIVOTING.  MATRIX B IS THE MATRIX TO BE INVERTED AND A IS
C  THE INVERTED MATRIX.
      IMPLICIT REAL*8(A-H,O-Z)
      DIMENSION B(12,12),A(12,12),INTER(12,2)
      DO 2 I = 1,N
      DO 2 J = 1,N
2     A(I,J) = B(I,J)
C  CYCLE PIVOT ROW NUMBER FROM 1 TO N
      DO 12 K = 1,N
      JJ = K
      IF(K .EQ. N)GO TO 6
      KP1 = K + 1
      BIG = DABS(A(K,K))
C  SEARCH FOR LARGEST PIVOT ELEMENT
      DO 5 I = KP1,N
      AB = DABS(A(I,K))
      IF(BIG-AB)4,5,5
4     BIG = AB
      JJ = I
5     CONTINUE
C  MAKE DECISION ON NECESSITY OF ROW INTERCHANGE AND
C  STORE THE NUMBER OF THE TWO ROWS INTERCHANGED DURING KTH
C  REDUCTION.  IF NO INTERCHANGE, BOTH NUMBERS STORED EQUAL K
6     INTER(K,1) = K
```

```
      INTER(K,2) = JJ
      IF(JJ-K)7,9,7
7     DO 8 J = 1,N
      TEMP = A(JJ,J)
      A(JJ,J) = A(K,J)
8     A(K,J) = TEMP
C  CALCULATE ELEMENTS OF REDUCED MATRIX
C  FIRST CALCULATE NEW ELEMENTS OF PIVOT ROW
9     DO 10 J = 1,N
      IF(J .EQ. K)GO TO 10
      A(K,J) = A(K,J)/A(K,K)
10    CONTINUE
C  CALCULATE ELEMENT REPLACING PIVOT ELEMENT
      A(K,K) = 1./A(K,K)
C  CALCULATE NEW ELEMENTS NOT IN PIVOT ROW OR COLUMN
      DO 11 I = 1,N
      IF(I .EQ. K)GO TO 11
      DO 110 J - 1,N
      IF(J .EQ. K) GO TO 110
      A(I,J) = A(I,J) - A(K,J)*A(I,K)
110   CONTINUE
11    CONTINUE
C  CALCULATE NEW ELEMENTS FOR PIVOT COLUMN--EXCEPT PIVOT ELEMENT
      DO 120 I = 1,N
      IF(I .EQ. K) GO TO 120
      A(I,K) = -A(I,K)*A(K,K)
120   CONTINUE
12    CONTINUE
C  REARRANGE COLUMNS OF FINAL MATRIX OBTAINED
      DO 13 L = 1,N
      K = N-L+1
      KROW = INTER(K,1)
      IROW = INTER(K,2)
      IF(KROW .EQ. IROW)GO TO 13
      DO 130 I = 1,N
      TEMP = A(I,IROW)
      A(I,IROW) = A(I,KROW)
      A(I,KROW) = TEMP
130   CONTINUE
13    CONTINUE
      RETURN
      END

      SUBROUTINE MATMPY(N,A,B,C)
      IMPLICIT REAL*8(A-H,O-Z)
C  C IS THE PRODUCT MATRIX OF A AND B
      DIMENSION A(12,12),B(12,12),C(12,12)
      DO 2 I = 1,N
      DO 2 J = 1,N
```

```
      C(I,J) = 0.0
      DO 2 K = 1,N
2     C(I,J) = C(I,J) + A(I,K)*B(K,J)
      RETURN
      END

      SUBROUTINE SEARCH(RT,J,II,N)
C  THIS SUBROUTINE SEARCHES THE JTH COLUMN OF THE MATRIX RT
C  FOR THE LARGEST EIGENVECTOR COMPONENT.  ITS ROW NUMBER IS
C  ASSIGNED TO THE NAME II.
      IMPLICIT REAL*8(A-H,O-Z)
      DIMENSION RT(12,12)
      II = 1
      BIG = DABS(RT(1,J))
      DO 3 I = 2,N
      AB = DABS(RT(I,J))
      IF(BIG-AB)2,3,3
2     BIG = AB
      II = I
3     CONTINUE
      RETURN
      END
```

The three examples that follow illustrate the use of this program.

EXAMPLE 7-10

The frame shown in Fig. 7-25a is fixed at joints 1 and 4, and it is assumed that the welded corners at joints 2, 3, 5, and 6 are rigid enough to maintain the frame members at right angles during sidesway of the frame. The elements composing the frame are numbered 1 through 6 as shown by the numbers in circles, and the joints are numbered 1 through 6 as shown.

All the frame members are W18 $\times$ 60 steel beams, with the four vertical ones each 240 in. long and the horizontal ones 480 in. long. Other pertinent parameter values that are applicable to all six beam elements are

$$E = 30(10)^6 \text{ psi} \quad \text{(modulus of elasticity)}$$
$$I = 984 \text{ in.}^4 \quad \text{(area moment of inertia)}$$
$$A = 17.634 \text{ in.}^2 \quad \text{(cross-sectional area)}$$
$$\gamma = 0.013 \text{ lb}\cdot\text{s}^2/\text{in.}^2 \quad \text{(mass per unit length)}$$

We wish to use the program FINITEL.FOR to determine the natural circular frequencies and mode shapes of the frame when it is modeled by the six finite beam elements shown.

Solution. The generalized coordinates are given subscript values ranging from 1 to 18 as shown in Fig. 7-25b. Note that the coordinate with the lowest subscript value is parallel to the global axis $\bar{x}$ at each joint and that the numerical sequence of the subscript values is counterclockwise at each joint, with the rotational coordinate having the largest subscript value in each case. Progression is from joint to joint according to increasing joint number, and the order of the joints is selected as shown in the figure.

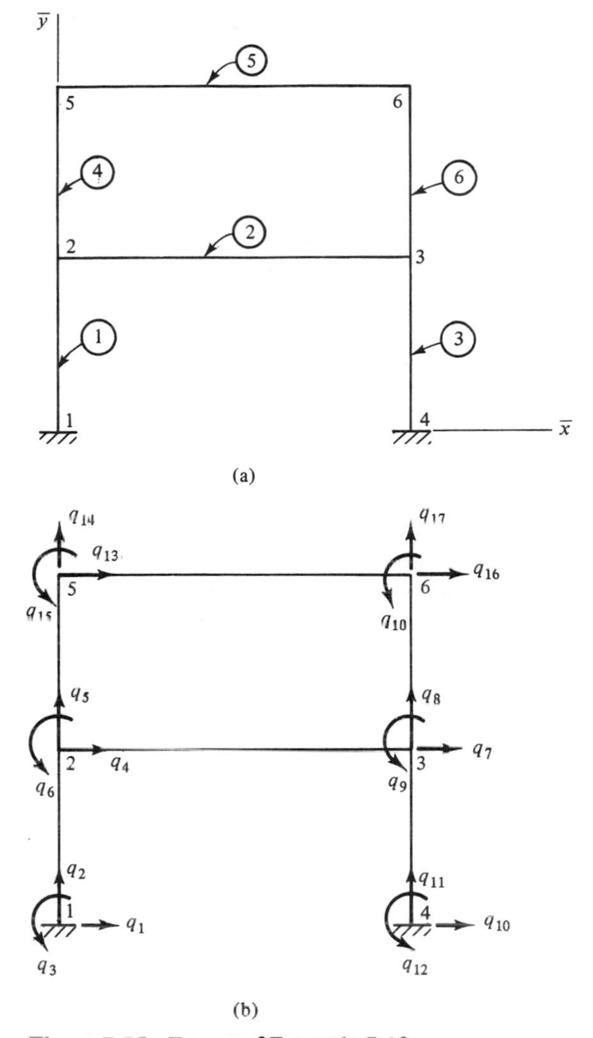

Figure 7-25 Frame of Example 7-10.

The axial deformations of the vertical frame elements are to be neglected; so the vertical displacements at joints 2, 3, 5, and 6 will be assumed to be zero. Consequently the generalized coordinates q_5, q_8, q_{14}, and q_{17} are given values of zero as boundary conditions. Since the lower ends of elements 1 and 3 are fixed, the coordinates q_1, q_2, q_3, q_{10}, q_{11}, and q_{12} are also given values of zero.

The inputs to FINITEL.FOR for this example are

```
NUMEL = 6
A(I) = 17.634 (I = 1, 2, 3, 4, 5, 6)
E(I) = 30.E6 (I = 1, 2, 3, 4, 5, 6)
IA(I) = 984.0 (I = 1, 2, 3, 4, 5, 6)
GAMMA(I) = 0.013 (I = 1, 2, 3, 4, 5, 6)
```

$$\text{NJTS} = 6, \quad \text{JNM(I,J)} = \begin{bmatrix} 1 & 2 \\ 2 & 3 \\ 3 & 4 \\ 2 & 5 \\ 5 & 6 \\ 6 & 3 \end{bmatrix}$$

$$\text{X(I), Y(I)} = \begin{matrix} 0.0, & 0.0 \\ 0.0, & 240. \\ 480., & 240. \\ 480., & 0.0 \\ 0.0, & 480. \\ 480., & 480. \end{matrix}$$

$$\text{NB} = 10, \quad \text{CFIX(I)} = [1\ 2\ 3\ 5\ 8\ 10\ 11\ 12\ 14\ 17]^T$$

It should be noted that the order of the matrix elements in any row of JNM(I, J) can be reversed without affecting the solution. For example, the order of 6 and 3 shown in the last row above could be changed to 3 and 6.

The first three natural circular frequencies and normal-mode shape values obtained from the computer program are as follows:

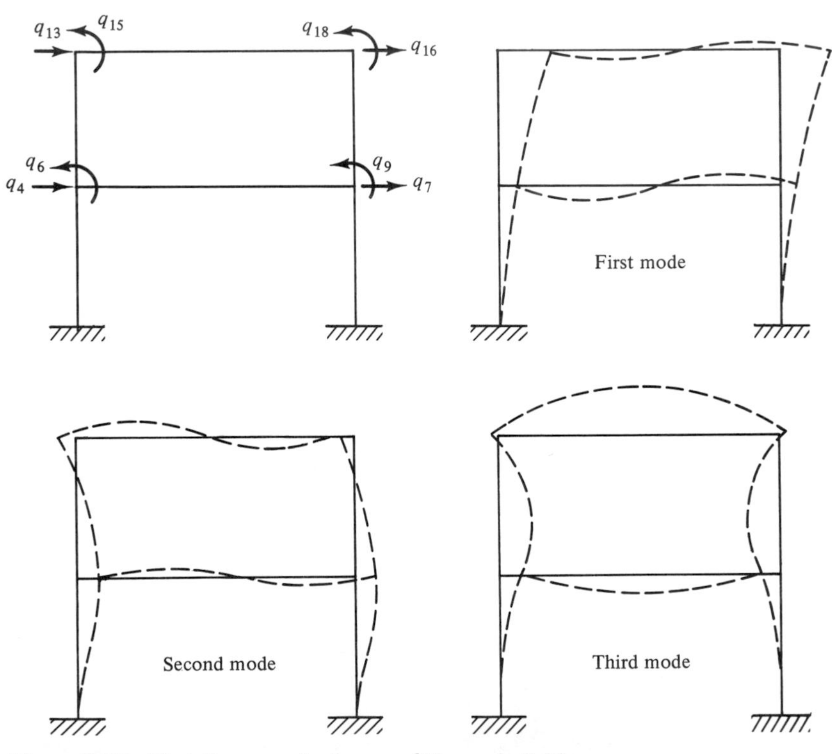

Figure 7-26 First three mode shapes of Example 7-10.

$$\omega_1 = 27.62 \text{ rad/s} \qquad\qquad \omega_2 = 97.44 \text{ rad/s}$$

$$\begin{Bmatrix} X_4 \\ X_6 \\ X_7 \\ X_9 \\ X_{13} \\ X_{15} \\ X_{16} \\ X_{18} \end{Bmatrix}_1 = \begin{Bmatrix} 1.000 \\ -0.441(10)^{-2} \\ 1.000 \\ -0.441(10)^{-2} \\ 2.142 \\ -0.273(10)^{-2} \\ 2.142 \\ -0.273(10)^{-2} \end{Bmatrix} \qquad \begin{Bmatrix} X_4 \\ X_6 \\ X_7 \\ X_9 \\ X_{13} \\ X_{15} \\ X_{16} \\ X_{18} \end{Bmatrix}_2 = \begin{Bmatrix} 1.000 \\ 0.057(10)^{-2} \\ 1.000 \\ 0.057(10)^{-2} \\ -0.712 \\ 0.624(10)^{-2} \\ -0.712 \\ 0.624(10)^{-2} \end{Bmatrix}$$

$$\omega_3 = 137.48 \text{ rad/s}$$

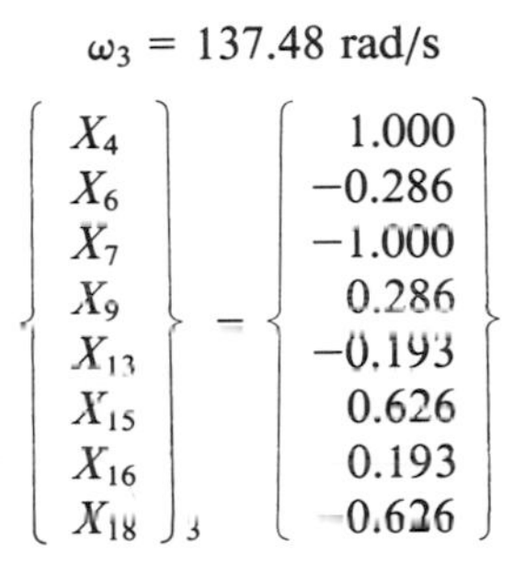

$$\begin{Bmatrix} X_4 \\ X_6 \\ X_7 \\ X_9 \\ X_{13} \\ X_{15} \\ X_{16} \\ X_{18} \end{Bmatrix}_3 = \begin{Bmatrix} 1.000 \\ -0.286 \\ -1.000 \\ 0.286 \\ -0.193 \\ 0.626 \\ 0.193 \\ -0.626 \end{Bmatrix}$$

The first three normal-mode shapes are shown graphically in Fig. 7-26.

EXAMPLE 7-11

In Example 7-7, the rocket shown in Fig. 7-19 accompanying that example was modeled by two beam elements having no axial deformations (see Fig. 7-11b), and its natural circular frequencies and normal-mode shapes were determined by following the steps given on page 519.

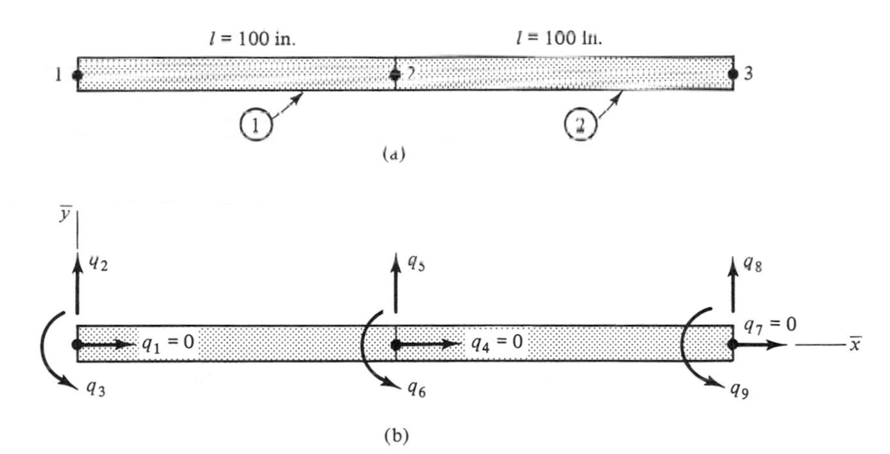

Figure 7-27 Rocket model for FINITEL.FOR program.

The computer program FINITEL.FOR can also be used to determine the natural circular frequencies and mode shapes of the rocket, and in this example we are interested in determining the necessary data input to that program to accomplish that end.

Solution. With the joints numbered in the order shown in Fig. 7-27a, the generalized coordinates are subscripted as shown in Fig. 7-27b. Since axial deformations are to be neglected, the coordinates q_1, q_4, and q_7 will be fixed at zero as boundary conditions. We shall read in values of 1.0 for the cross-sectional areas to satisfy the command for those input data. Actually, any value other than zero can be used for that purpose. The values used will have no effect on the solution since the zero values given to the coordinates q_1, q_4, and q_7 will in effect delete the rows and columns of the system mass and stiffness matrices involving the cross-sectional areas.

With the value of $EI = (10)^{10}$ lb·in.^{2} given in Example 7-7 and with $A(I) = 1.0$ as mentioned above, the input data to the program are as follows:

```
NUMEL = 2
A(I) = 1.0 (I = 1, 2)
E(I) = 1.0E10 (I = 1, 2)
IA(I) = 1.0 (I = 1, 2)
GAMMA(I) = 0.042 (I = 1, 2)
```

$$\text{NJTS} = 3\ , \quad \text{JNM(I,J)} = \begin{bmatrix} 1 & 2 \\ 2 & 3 \end{bmatrix}$$

$$\text{X(I), Y(I)} = \begin{matrix} 0.0, & 0.0 \\ 100.0, & 0.0 \\ 200.0, & 0.0 \end{matrix}$$

$$\text{NB} = 3, \quad \text{CFIX(I)} = [1\ \ 4\ \ 7]^T$$

Using the above in the program FINITEL.FOR yields the two rigid-body modes associated with zero frequencies, and the bending-mode frequency and mode shape determined previously in Example 7-7 as shown on page 531.

EXAMPLE 7-12

In this example we illustrate the use of the FINITEL.FOR program in obtaining the natural circular frequencies and mode shapes of a rod modeled by three finite rod elements as shown in Fig. 7-28. The four joints are numbered as shown in Fig. 7-28a, and the number of each element is shown circled. The pertinent parameter values are

$$A = 1.20 \text{ in.} \quad \text{(cross-sectional area)}$$
$$E = 30(10)^6 \text{ psi} \quad \text{(modulus of elasticity)}$$
$$\gamma = \frac{490(1.2)}{1728(386)} = 8.815(10)^{-4} \text{ lb}\cdot\text{s}^2/\text{in.}^2 \quad \text{(mass per in.)}$$

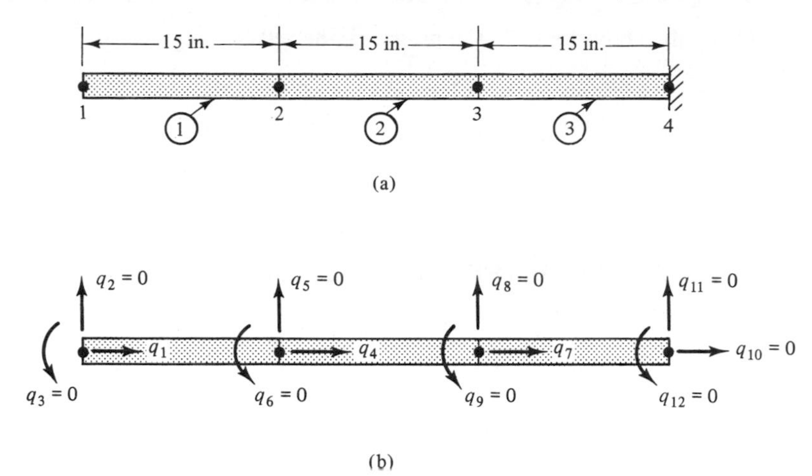

Figure 7-28 Rod model for FINITEL.FOR program.

We wish to determine the necessary input data to the program to obtain the natural circular frequencies and mode shapes of the rod when modeled by the three rod elements shown.

Solution. Since the beam elements in the computer program are to be reduced to rod elements by constraining their lateral and rotational displacements, the generalized coordinates $q_2, q_3, q_5, q_6, q_8, q_9, q_{11}$, and q_{12} are fixed at zero owing to these constraints, and q_{10} is also fixed at zero since the rod is fixed at its right end. The rod model is thus as shown in Fig. 7-28b. Since bending is not considered, a value of 1.0 will be assigned to IA(I) for the area moment of inertia of each finite element to satisfy the command for those input data, just as was done with the cross-sectional areas in Example 7-11. With the above in mind, the inputs to the program are as follows:

```
NUMEL = 3
A(I) = 1.2 (I = 1, 2, 3)
E(I) = 30.E6 (I = 1, 2, 3)
IA(I) = 1.0 (I = 1, 2, 3)
GAMMA(I) = 8.815E-4 (I = 1, 2, 3)
```

$$\text{NJTS} = 4, \quad \text{JNM(I,J)} = \begin{bmatrix} 1 & 2 \\ 2 & 3 \\ 3 & 4 \end{bmatrix}$$

```
                    0.0, 0.0
                   15.0, 0.0
     X(I), Y(I) =  30.0, 0.0
                   45.0, 0.0
```

$$\text{NB} = 9, \quad \text{CFIX(I)} = [2 \ 3 \ 5 \ 6 \ 8 \ 9 \ 10 \ 11 \ 12]^T$$

Using the above input to the program, the results obtained are as follows:

$$\omega_1 = 7135.0 \text{ rad/s} \qquad \omega_2 = 23{,}335.1 \text{ rad/s} \qquad \omega_3 = 42{,}333.2 \text{ rad/s}$$

$$\begin{Bmatrix} X_1 \\ X_4 \\ X_7 \end{Bmatrix}_1 = \begin{Bmatrix} 1.000 \\ 0.866 \\ 0.500 \end{Bmatrix} \qquad \begin{Bmatrix} X_1 \\ X_4 \\ X_7 \end{Bmatrix}_2 = \begin{Bmatrix} 1.000 \\ 0.000 \\ -1.000 \end{Bmatrix} \qquad \begin{Bmatrix} X_1 \\ X_4 \\ X_7 \end{Bmatrix}_3 = \begin{Bmatrix} 1.000 \\ -0.866 \\ 0.500 \end{Bmatrix}$$

7-8 COMPUTER PROGRAM FOR THE VIBRATION ANALYSIS OF PLANE TRUSSES (TRUSS.FOR)

The computer program FINITEL.FOR developed in Sec. 7-7 was written for use in the analysis of plane frames and utilized the general finite beam element shown in Fig. 7-11c. Example 7-10 illustrates that this program can also be used to analyze frames composed of beam elements having negligible axial deformations by applying boundary conditions in the program that fix at zero the values of the generalized coordinates associated with the axial displacements. Example 7-12 shows that FINITEL.FOR can also be used for the analysis of rods composed of finite rod elements by applying boundary conditions in the program that fix at zero the values of the generalized coordinates associated with the lateral and rotational displacements of the system joints.

However, this program cannot be used for the analysis of plane trusses, since the general system mass and stiffness matrices developed in the program utilizing the general beam element of Fig. 7-11c cannot be reduced to those of a truss system by simply removing the rotational joint displacements and eliminating the corresponding rows and columns in the matrices. This follows from the fact that the local mass and stiffness matrices of a truss element cannot be obtained from those of the beam element of Fig. 7-11c by simply removing the rotational joint displacements and eliminating the corresponding rows and columns of the matrices shown for the beam element. The reader can confirm this by performing such a procedure and comparing the result obtained with the correct local mass and stiffness matrices shown for a truss element in Fig. 7-29. Therefore, we must develop another program for analyzing plane trusses.

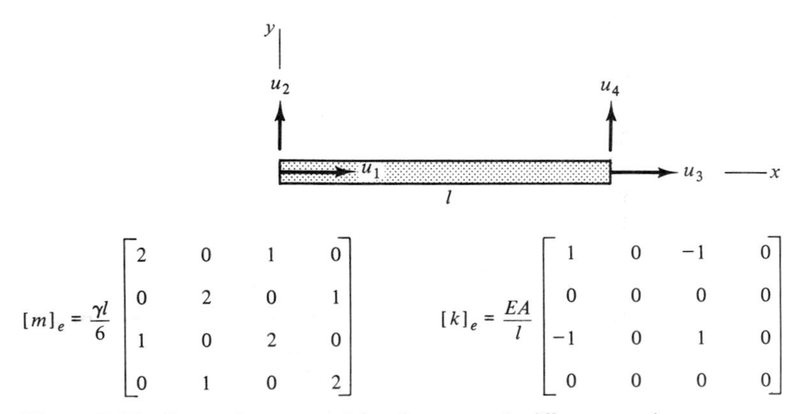

Figure 7-29 Truss element and local mass and stiffness matrices.

The local mass and stiffness matrices shown in Fig. 7-29 can be derived by the use of Lagrange's equations in the same manner discussed previously for determining the local mass and stiffness matrices of rod elements or beam elements with no axial deformations (see Prob. 7-25). (The local stiffness matrix can also be determined directly by applying the definition of stiffness coefficients.)

To appreciate the fundamental differences between truss and beam elements, consider the truss element shown in Fig. 7-29. Note that no rotational joint displacements are shown at the joints of this element. This is because, although a truss element can rotate as a rigid body (u_2 positive and u_4 negative, for example), it cannot have joint rotations due to bending deformations, since truss elements in a truss system are not subjected to bending when the system is correctly designed (the forces in the members at each joint are concurrent and the loads on the system act at the joints).

Since our discussion has shown that we cannot use the program FINITEL.FOR in analyzing trusses, we use a program TRUSS.FOR that utilizes the mass and stiffness matrices shown in Fig. 7-29. This program has many similarities to FINITEL.FOR, and the variable and array names used in the two programs represent the same quantities.

The transformation matrix **R** for truss elements can be obtained from the transformation matrix of Eq. 7-86 by simply eliminating the third and sixth rows and columns of the latter since there are no joint rotations to relate between the local and global coordinate systems; it is given by

$$\mathbf{R} = \begin{bmatrix} \cos\alpha & \sin\alpha & 0 & 0 \\ -\sin\alpha & \cos\alpha & 0 & 0 \\ 0 & 0 & \cos\alpha & \sin\alpha \\ 0 & 0 & -\sin\alpha & \cos\alpha \end{bmatrix} \tag{7-144}$$

Only the main program TRUSS.FOR is given here. The subprograms required are the same as the ones used with FINITEL.FOR. As written, TRUSS.FOR can be used for the analysis of trusses having up to seven truss elements and nine joints. It can be readily expanded to handle trusses with more than these numbers of elements and joints if sufficient computer memory is provided.

The inputs to the program that are necessary in analyzing a truss are illustrated in Example 7-13.

```
C   TRUSS.FOR
C   FINITE-ELEMENT PROGRAM FOR VIBRATION ANALYSIS OF TRUSSES
C   USING TRUSS ELEMENTS.  SUBROUTINES JCBI, DECOMP, MATINV,
C   MATMPY, AND SEARCH ARE IN LIBRARY.FOR AND SHOULD BE
C   COMPILED SEPARATELY.  AFTER TRUSS.FOR HAS BEEN COMPILED,
C   TRUSS.OBJ MUST BE LINKED WITH LIBRARY.OBJ.
C   DEVICE * IN READ AND WRITE STATEMENTS IS THE CONSOLE.
C   DEVICE 2 IN WRITE STATEMENTS IS THE PRINTER.
C   * IN THE PLACE OF A FORMAT STATEMENT NUMBER MEANS FREE FORMAT.
      IMPLICIT REAL*8 (A-H,O-Z)
```

```
      REAL*8 L,IA,KEL,MEL,KEG,MEG
      INTEGER SUB,ROWSUB,COLSUB,B,Z,EN,CFIX
      DIMENSION KEL(4,4,7),MEL(4,4,7),KEG(4,4,7),MEG(4,4,7),
     $RT(4,4),
     $R(4,4),TK(4,4),TM(4,4),SK(18,18),SM(18,18),RSK(12,12),
     $RSM(12,12),E(7),A(7),X(7),Y(7),GAMMA(7)
      DIMENSION JNM(7,2),CFIX(12),SUB(4)
      OPEN (2,FILE='PRN')
C   READ IN PROBLEM DATA AS INDICATED BY MESSAGES ON CONSOLE. DATA
C   READ IN IS PRINTED OUT. (PROGRAM STATEMENTS 2 THROUGH 40)
      WRITE(*,2)
2     FORMAT(' ', 'ENTER THE NUMBER OF TRUSS ELEMENTS (I1)',/)
      READ(*,3)NUMEL
3     FORMAT(I1)
      DO 4 I = 1,NUMEL
      WRITE(*,5)I
4     READ(*,6)A(I)
5     FORMAT(/,' ENTER A(',I1,') (F20.0)',/)
6     FORMAT(F20.0)
      WRITE(2,7)
7     FORMAT(6X,'THE AREA ARRAY  A  IS:',/)
      DO 8 I = 1,NUMEL
8     WRITE(2,9)I,A(I)
9     FORMAT(6X,'A(',I1,') = ',E14.7)
      DO 10 I = 1,NUMEL
      WRITE(*,11)I
10    READ(*,12)E(I)
11    FORMAT(/,' ENTER E(',I1,')  (F20.0)',/)
12    FORMAT(F20.0)
      WRITE(2,13)
13    FORMAT(/,6X,'THE ELASTICITY ARRAY  E  IS;',/)
      DO 14 I = 1,NUMEL
14    WRITE(2,15)I,E(I)
15    FORMAT(6X,'E(',I1,') = ',E14.7)
      DO 82 I = 1,NUMEL
      WRITE(*,83)I
82    READ(*,84)GAMMA(I)
83    FORMAT(/,' ENTER GAMMA(',I1,')   (F20.0)',/)
84    FORMAT(F20.0)
      WRITE(2,85)
85    FORMAT(/,6X, 'THE GAMMA ARRAY IS;',/)
      DO 86 I = 1,NUMEL
86    WRITE(2,87)I,GAMMA(I)
87    FORMAT(6X,'GAMMA(',I1,') = ',E14.7)
      WRITE(*,22)
22    FORMAT(/,' ENTER THE NUMBER OF JOINTS, NJTS  (I1)',/)
      READ(*,23)NJTS
```

```
23      FORMAT(I1)
        DO 24 I = 1,NUMEL
        DO 24 J = 1,2
        WRITE(*,25)I,J
24      READ(*,26)JNM(I,J)
25      FORMAT(/,' ENTER JNM(',I1,',',I1,')    (I1)',/)
26      FORMAT(I1)
        WRITE(2,27)
27      FORMAT(/,6X,'THE JOINT-NUMBER MATRIX IS;',/)
        DO 28 I = 1,NUMEL
28      WRITE(2,29)JNM(I,1),JNM(I,2)
29      FORMAT(10X,I5,I4)
        DO 30 I = 1,NJTS
        WRITE(*,31)I,I
30      READ(*,*)X(I),Y(I)
31      FORMAT(/' ENTER JOINT COORD. X(',I1,'),Y(',I1,')(2F20.0)',/)
        WRITE(2,33)
33      FORMAT(/,6X,'THE JOINT COORDINATES ARE;',/)
        DO 34 I = 1,NJTS
34      WRITE(2,35)I,X(I),I,Y(I)
35      FORMAT(6X,'X(',I1,') = ',E11.7,6X,'Y(',I1,') = ',E14.7)
        WRITE(*,36)
36      FORMAT(/,' ENTER THE NUMBER OF FIXED COORDINATES (I2)',/)
        READ(*,37)NB
37      FORMAT(I2)
        IF(NB .EQ. 0)GO TO 94
        DO 38 I = 1,NB
        WRITE(*,39)I
38      READ(*,40)CFIX(I)
39      FORMAT(/,' ENTER CFIX(',I2,')  (I2)',/)
40      FORMAT(I2)
        WRITE(2,41)
41      FORMAT(/,6X,'ARRAY CFIX IS;',/)
        DO 42 I = 1,NB
42      WRITE(2,43)I,CFIX(I)
43      FORMAT(6X,'CFIX(',I2,') = ',I2)
C   GENERATE NULL 3-DIMENSIONAL ARRAYS KEL AND MEL.  PLANES OF KEL
C   AND MEL WILL LATER CONTAIN THE LOCAL ELEMENT STIFFNESS AND MASS
C   MATRICES, RESPECTIVELY.
94      DO 44 I = 1,4
        DO 44 J = 1,4
        DO 44 M = 1,NUMEL
        KEL(I,J,M) = 0.
44      MEL(I,J,M)= 0.
C   GENERATE NULL MATRICES  R  AND  RT  WHICH WILL LATER BECOME THE
C   TRANSFORMATION MATRIX AND ITS TRANSPOSE, RESPECTIVELY.
```

```
      DO 45 I = 1,4
      DO 45 J = 1,4
      R(I,J)=0.0
45    RT(I,J)=0.0
C  GENERATE THE LOCAL ELEMENT STIFFNESS MATRICES AND STORE IN THE
C  3-DIMENSIONAL STIFFNESS ARRAY KEL (SEE FIG. 7-11 FOR THE
C  EQUATION USED).  EACH PLANE IN THE 3-DIM. ARRAY IS ONE ELEMENT
C  STIFFNESS MATRIX.
      DO 100 EN=1,NUMEL
      IC=JNM(EN,1)
      ID=JNM(EN,2)
      L=DSQRT((X(ID)-X(IC))**2+(Y(ID)-Y(IC))**2)
      F=E(EN)*A(EN)/L
      SINA=(Y(ID)-Y(IC))/L
      COSA=(X(ID)-X(IC))/L
      KEL(1,1,EN)=F
      KEL(1,3,EN)=-F
      KEL(3,1,EN)=-F
      KEL(3,3,EN)=F
C  GENERATE THE LOCAL ELEMENT MASS MATRICES AND STORE THEM IN THE
C  3-DIMENSIONAL MASS ARRAY MEL (SEE FIG. 7-11 FOR THE EQUATION
C  USED).  EACH PLANE OF THE 3-DIM. ARRAY MEL CONTAINS ONE LOCAL
C  ELEMENT MASS MATRIX.
      F=GAMMA(EN)*L/6.
      P=2.*F
      MEL(1,1,EN)=P
      MEL(1,3,EN)=F
      MEL(2,2,EN)=P
      MEL(2,4,EN)=F
      MEL(3,1,EN)=F
      MEL(3,3,EN)=P
      MEL(4,2,EN)=F
      MEL(4,4,EN)=P
C  GENERATE THE TRANSFORMATION MATRIX  R  AND ITS TRANSPOSE RT.
      R(1,1)=COSA
      R(1,2)=SINA
      R(2,1)=-SINA
      R(2,2)=COSA
      R(3,3)=COSA
      R(3,4)=SINA
      R(4,3)=-SINA
      R(4,4)=COSA
      DO 48 I=1,2
      DO 48 J=1,2
      RT(I,J)=R(J,I)
48    RT(I+2,J+2)=R(J+2,I+2)
```

```
C   DETERMINE THE ELEMENT STIFFNESS MATRICES IN THE GLOBAL
C   COORDINATE SYSTEM (EQ. 7-92B) AND STORE THEM IN THE 3-DIM.
C   STIFFNESS ARRAY KEG.  EACH PLANE OF THE 3-DIM. ARRAY CONTAINS
C   ONE GLOBAL ELEMENT STIFFNESS MATRIX.
        DO 95 I=1,4
        DO 95 J=1,4
        TK(I,J)=0.0
        DO 95 K=1,4
95      TK(I,J)=TK(I,J)+KEL(I,K,EN)*R(K,J)
        DO 96 I=1,4
        DO 96 J=1,4
        KEG(I,J,EN)=0.0
        DO 96 K=1,4
96      KEG(I,J,EN)=KEG(I,J,EN)+RT(I,K)*TK(K,J)
C   DETERMINE THE ELEMENT MASS MATRICES IN THE GLOBAL SYSTEM (EQ.
C   7-92A) AND STORE THEM IN THE 3-DIM. MASS ARRAY MEG.  EACH PLANE
C   OF THE 3-DIM. ARRAY CONTAINS ONE GLOBAL ELEMENT MASS MATRIX.
        DO 97 I=1,4
        DO 97 J=1,4
        TM(I,J)=0.0
        DO 97 K=1,4
97      TM(I,J)=TM(I,J)+MEL(I,K,EN)*R(K,J)
        DO 98 I=1,4
        DO 98 J=1,4
        MEG(I,J,EN)=0.0
        DO 98 K=1,4
98      MEG(I,J,EN)=MEG(I,J,EN)+RT(I,K)*TM(K,J)
100     CONTINUE
C   GENERATE NULL MATRICES SK AND SM WHICH WILL BECOME THE
C   SYSTEM STIFFNESS AND MASS MATRICES, RESPECTIVELY.
        N=NJTS*2
        DO 49 I=1,N
        DO 49 J=1,N
        SK(I,J)=0.
49      SM(I,J)=0.
C   ASSEMBLE THE SYSTEM STIFFNESS AND MASS MATRICES.
        DO 51 I=1,NUMEL
        DO 50 J=1,2
        DO 50 M=1,2
        J1=J*2-M+1
50      SUB(J1)=2*JNM(I,J)-M+1
        DO 51 B=1,4
        DO 51 Z=1,4
        ROWSUB=SUB(B)
        COLSUB=SUB(Z)
        SK(ROWSUB,COLSUB)=SK(ROWSUB,COLSUB)+KEG(B,Z,I)
51      SM(ROWSUB,COLSUB)=SM(ROWSUB,COLSUB)+MEG(B,Z,I)
```

```
C   CALCULATE THE NUMBER OF DEGREES OF FREEDOM AND REMOVE ROWS AND
C   COLUMNS FROM THE SYSTEM STIFFNESS AND MASS MATRICES.
      NF=N-NB
      IF(NB .EQ. 0)GO TO 69
      NA=1
      KL=N-1
62    JC=1
63    IF(JC .EQ. CFIX(NA))GO TO 64
      JC=JC+1
      IF(JC .EQ. N)GO TO 68
      GO TO 63
64    DO 65 I=1,N
      DO 65 J=JC,KL
      SK(I,J)=SK(I,J+1)
      SM(I,J)=SM(I,J+1)
65    CONTINUE
      DO 66 J=1,N
      DO 66 I=JC,KL
      SK(I,J)=SK(I+1,J)
      SM(I,J)=SM(I+1,J)
66    CONTINUE
      IF(NA .EQ. NB)GO TO 68
      NA=NA+1
      DO 67 I=NA,NB
67    CFIX(I)=CFIX(I)-1
      GO TO 62
68    CONTINUE
C   ASSIGN REDUCED STIFFNESS AND MASS MATRIX ELEMENTS TO ARRAY
C   NAMES RSK AND RSM, RESPECTIVELY.
69    DO 70 I=1,NF
      DO 70 J=1,NF
      RSK(I,J)=SK(I,J)
70    RSM(I,J)=SM(I,J)
C   WRITE OUT THE REDUCED STIFFNESS AND MASS MATRICES OBTAINED
C   FROM THE BOUNDARY CONDITIONS.
      WRITE(2,71)
71    FORMAT(/,' THE REDUCED SYSTEM STIFFNESS MATRIX IS:',/)
      WRITE(2,72) ((RSK(I,J),J=1,NF),I=1,NF)
72    FORMAT(' ',6E11.4/)
      WRITE(2,73)
73    FORMAT(/,' THE REDUCED SYSTEM MASS MATRIX IS:',/)
      WRITE(2,74) ((RSM(I,J),J=1,NF),I=1,NF)
74    FORMAT(' ',6E11.4/)
      WRITE(2,200)
200   FORMAT(//,' ')
C   CALL SUBPROGRAM JCBI TO CALCULATE FREQUENCIES AND MODE SHAPES
      CALL JCBI(NF,RSK,RSM)
      STOP
      END
```

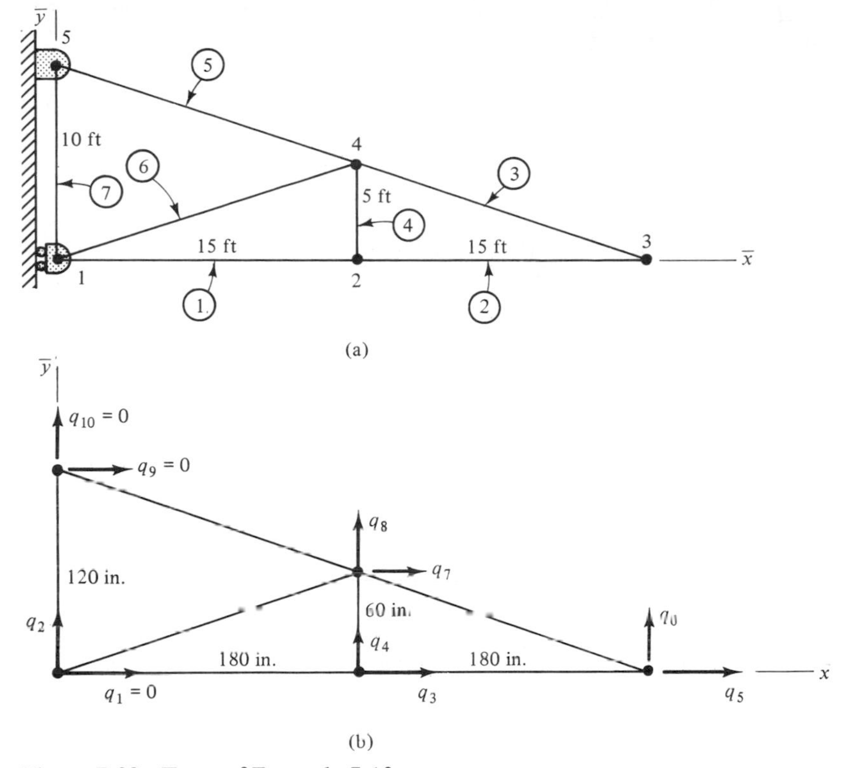

Figure 7-30 Truss of Example 7-13.

EXAMPLE 7-13

All the members of the truss shown in Fig. 7-30 are L4 × 4 × $\frac{3}{8}$ angles having the following properties:

$$
\begin{aligned}
A &= 2.86 \text{ in.}^2 \quad \text{(cross-sectional area)} \\
E &= 30(10)^6 \text{ psi} \quad \text{(modulus of elasticity)} \\
w &= 9.8 \text{ lb/ft} \quad \text{(weight per unit length)} \\
\gamma &= \frac{9.8}{12(386)} = 2.12(10)^{-3} \text{ lb} \cdot \text{s}^2/\text{in.}^2 \quad \text{(mass per in.)}
\end{aligned}
$$

We use the program TRUSS.FOR to determine the natural circular frequencies and mode shapes of the truss.

Solution. There are five joints and seven truss elements in the truss system. The elements and joints are numbered as shown in Fig. 7-30a, and the generalized coordinates are subscripted as shown in Fig. 7-30b using the sequencing at each joint shown in Fig. 7-29. It should be apparent from Fig. 7-30 that the boundary conditions at joints 1 and 5 require that the coordinates q_1, q_9, and q_{10} be fixed at zero. Thus, the inputs to the program are

```
NUMEL = 7
A(I) = 2.86 (I = 1, 2, 3, 4, 5, 6, 7)
E(I) = 30.E6 (I = 1, 2, 3, 4, 5, 6, 7)
GAMMA(I) = 2.12E-3 (I = 1, 2, 3, 4, 5, 6, 7)
```

$$\text{NJTS} = 5 \;, \quad \text{JNM(I,J)} = \begin{bmatrix} 1 & 2 \\ 2 & 3 \\ 3 & 4 \\ 4 & 2 \\ 4 & 5 \\ 4 & 1 \\ 5 & 1 \end{bmatrix}$$

```
                     0.0,   0.0
                   180.0,   0.0
X(I), Y(I)   =     360.0,   0.0
                   180.0,  60.0
                     0.0, 120.0
```

$$\text{NB} = 3 \;, \quad \text{CFIX(I)} = [1\ 9\ 10]^T$$

Using the above input to the TRUSS.FOR program, the results obtained for the first two modes are as follows:

$$\omega_1 = 187.7 \text{ rad/s} \qquad\qquad \omega_2 = 403.2 \text{ rad/s}$$

$$\begin{Bmatrix} X_2 \\ X_3 \\ X_4 \\ X_5 \\ X_6 \\ X_7 \\ X_8 \end{Bmatrix}_1 = \begin{Bmatrix} 1.000 \\ 7.181 \\ 25.616 \\ 14.145 \\ 102.443 \\ -3.767 \\ 25.253 \end{Bmatrix} \qquad \begin{Bmatrix} X_2 \\ X_3 \\ X_4 \\ X_5 \\ X_6 \\ X_7 \\ X_8 \end{Bmatrix}_2 = \begin{Bmatrix} 1.000 \\ -1.869 \\ 8.880 \\ -3.480 \\ -9.171 \\ 0.947 \\ 8.619 \end{Bmatrix}$$

EXAMPLE 7-14

Figure 7-31 shows the truss for which the first two natural circular frequencies and mode shapes were determined in Example 7-13. The suddenly applied force $F(t)$ shown in Fig. 7-31c acts at joint 3 of the truss as shown. We assume that the damping in the truss system is on the order of 5 percent ($\zeta = 0.05$) for all modes and that the superposition of the first two modes provides a reasonably accurate dynamic response analysis.

With the preceding in mind, and utilizing the results of Example 7-13, we wish to determine the maximum allowable value that F_0 can have before the absolute maximum displacement q_6 exceeds 3 in.

The first two natural circular frequencies and corresponding eigenvectors (mode shapes) are shown in the solution of Example 7-13, and the reduced mass matrix **M** for the system that was obtained from the printout in the solution of Example 7-13 (but which is not shown there) is as follows:

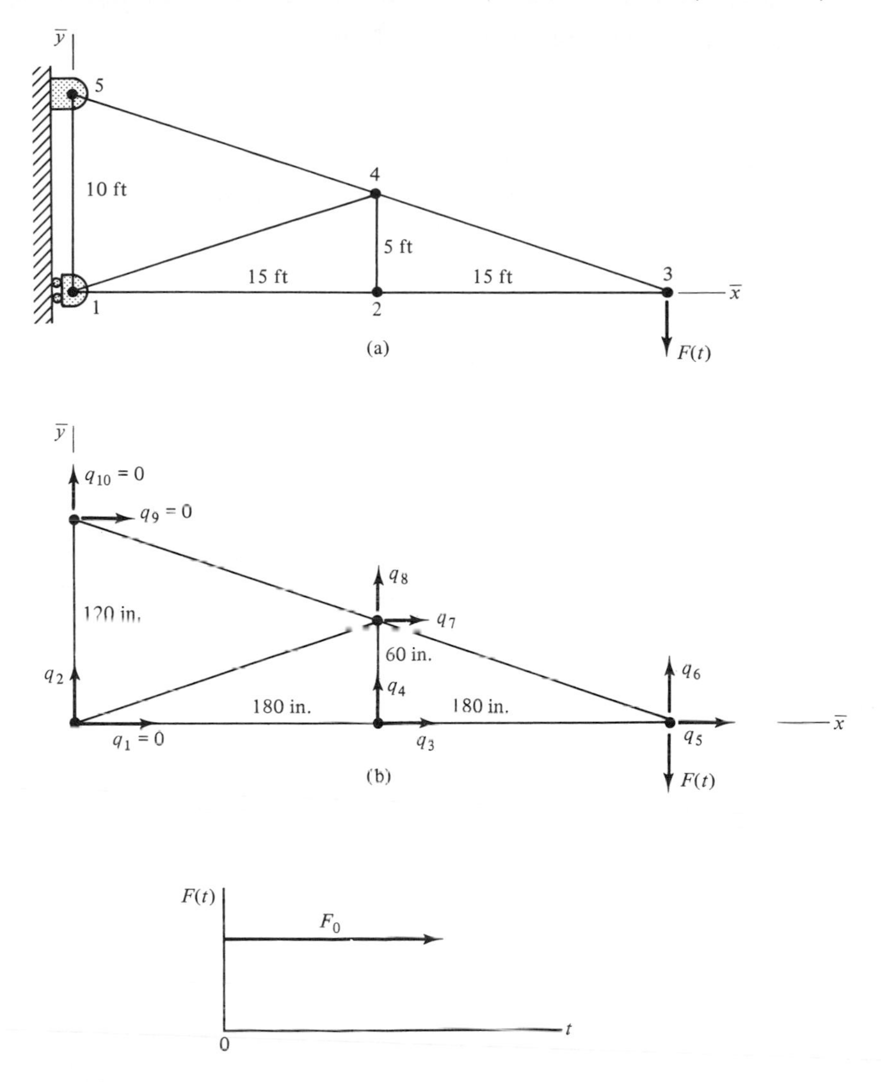

Figure 7-31 Truss subjected to suddenly applied force $F(t)$.

$$
\mathbf{M} = \begin{bmatrix}
0.3461 & 0.0000 & 0.0636 & 0.0000 & 0.0000 & 0.0000 & 0.0670 \\
 & 0.2968 & 0.0000 & 0.0636 & 0.0000 & 0.0212 & 0.0000 \\
 & & 0.2968 & 0.0000 & 0.0636 & 0.0000 & 0.0212 \\
 & & & 0.2613 & 0.0000 & 0.0670 & 0.0000 \\
 & \text{(symmetric)} & & & 0.2613 & 0.0000 & 0.0670 \\
 & & & & & 0.4446 & 0.0000 \\
 & & & & & & 0.4446
\end{bmatrix}
$$

Solution. Equation 6-34 is repeated here as

$$\ddot{\delta}_r + 2\zeta_r\omega_r\dot{\delta}_r + \omega_r^2\delta_r = \frac{\{X_i\}_r^T\{Q_i\}}{M_r} \tag{6-34}$$

where $\{X_i\}_r$ = eigenvector of the rth mode
$M_r = \{X_i\}_r^T\mathbf{M}\{X_i\}_r$ = generalized mass of rth mode
ω_r = natural frequency of rth mode
ζ_r = damping factor of rth mode
$\{Q_i\}$ = nonpotential excitation vector

Noting that $F(t)$ acts in a sense opposite to that of the generalized coordinate q_6 as shown in Fig. 7-31, we write that

$$\begin{Bmatrix} Q_2 \\ Q_3 \\ Q_4 \\ Q_5 \\ Q_6 \\ Q_7 \\ Q_8 \end{Bmatrix} = \begin{Bmatrix} 0 \\ 0 \\ 0 \\ 0 \\ -F_0 \\ 0 \\ 0 \end{Bmatrix}$$

Using Eq. 6-37, the maximum response of the truss from the superposition of the first two modes is

$$\begin{Bmatrix} q_2 \\ q_3 \\ q_4 \\ q_5 \\ q_6 \\ q_7 \\ q_8 \end{Bmatrix} = (\delta_1)_m \begin{Bmatrix} X_2 \\ X_3 \\ X_4 \\ X_5 \\ X_6 \\ X_7 \\ X_8 \end{Bmatrix}_1 + (\delta_2)_m \begin{Bmatrix} X_2 \\ X_3 \\ X_4 \\ X_5 \\ X_6 \\ X_7 \\ X_8 \end{Bmatrix}_2$$

which shows that

$$q_6 = (\delta_1)_m(X_6)_1 + (\delta_2)_m(X_6)_2 \tag{7-145}$$

The maximum-response values $(\delta_1)_m$ and $(\delta_2)_m$ in the above equation are readily obtained using the response spectrum shown in Fig. 4-15. This figure shows that for $t_1/\gamma_n = 0$ and $\zeta = 0.05$,

$$\frac{x_{\max}}{F_0/k} = 1.85$$

so that

$$\frac{(\delta_r)_m}{(\delta_r)_s} = 1.85$$

in which $(\delta_r)_s$ is associated with static forces (slowly applied forces). Then, since $\ddot{\delta}_r = \dot{\delta}_r = 0$ for slowly applied forces, Eq. 6-34 shows that

$$(\delta_r)_s = \frac{\{X_i\}_r^T\{Q_i\}}{\omega_r^2 M_r} \tag{7-146}$$

Using the values of the elements of the reduced system mass matrix **M** shown earlier in this example and the eigenvectors shown in the solution portion of Example 7-13, the following values are found for the generalized mass terms for the first two modes:

$$M_1 = \{X_i\}_1^T \mathbf{M} \{X_i\}_1 = 4013.9 \text{ lb} \cdot \text{s}^2/\text{in.}$$

and

$$M_2 = \{X_i\}_2^T \mathbf{M} \{X_i\}_2 = 68.24 \text{ lb} \cdot \text{s}^2/\text{in.}$$

Substituting the appropriate values into Eq. 7-146,

$$(\delta_1)_s = \frac{102.443(-F_0)}{(187.7)^2(4013.9)} = -7.24(10)^{-7}F_0$$

and

$$(\delta_2)_s = \frac{-9.171(-F_0)}{(403.2)^2(68.24)} = 8.27(10)^{-7}F_0$$

Then, since $(\delta_r)_m/(\delta_r)_s = 1.85$ as determined earlier,

$$(\delta_1)_m = 1.85(-7.24)(10)^{-7}F_0 = -13.39(10)^{-7}F_0$$

and

$$(\delta_2)_m = 1.85(8.27)(10)^{-7}F_0 = 15.30(10)^{-7}F_0$$

Substituting the latter two values above and the first- and second-mode values of the eigenvector component X_6 from Example 7-13 into Eq. 7-145, and with $q_6 = -3.0$ in. being the maximum allowable deflection of coordinate q_6, we can write that

$$q_6 = -3.0 = -13.39(10)^{-7}F_0(102.443) + 15.30(10)^{-7}F_0(-9.171)$$

from which the maximum allowable value of the suddenly applied force is found to be

$$F_0 = 19{,}841 \text{ lb}$$

PROBLEMS

Problems 7-1 through 7-6 (Sections 7-1 and 7-2)

7-1. Determine the differential equations of motion of the system shown in the accompanying figure by (a) using Lagrange's equations and (b) using the definition of the influence

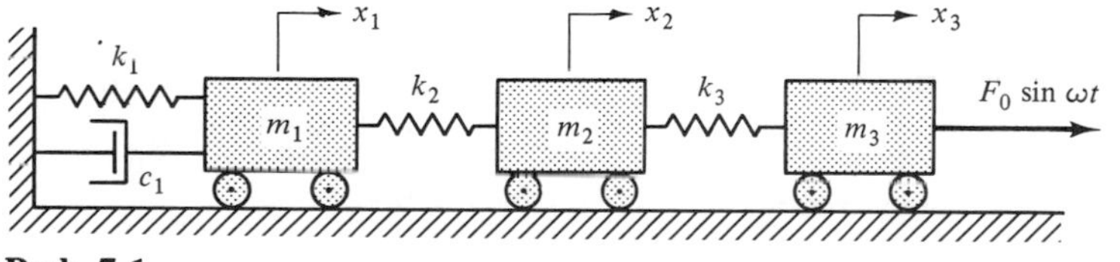

Prob. 7-1

coefficients k_{ij} and c_{ij} given in Sec. 5-3 and an inspection of the system to write the stiffness matrix **K** and damping matrix **C**. Compare the time and effort required by the two methods.

7-2. A cylinder of mass m_2, radius r, and centroidal mass moment of inertia $\bar{I} = m_2 r^2/2$ rolls without slipping on a platform of mass m_1 that is subjected to a force $F_0 \sin \omega t$ as shown in the accompanying figure. Using the generalized coordinates x_1 and x_2 as the respective absolute displacements of m_1 and m_2, determine the differential equations of motion of the system using Lagrange's equations.

Ans: $$\begin{bmatrix} (m_1 + m_2/2) & -m_2/2 \\ -m_2/2 & (3/2)m_2 \end{bmatrix}\begin{Bmatrix} \ddot{x}_1 \\ \ddot{x}_2 \end{Bmatrix} + \begin{bmatrix} (k_1 + k_2) & -(k_1 + k_2) \\ -(k_1 + k_2) & (k_1 + k_2) \end{bmatrix}\begin{Bmatrix} x_1 \\ x_2 \end{Bmatrix} = \begin{Bmatrix} F_0 \sin \omega t \\ 0 \end{Bmatrix}$$

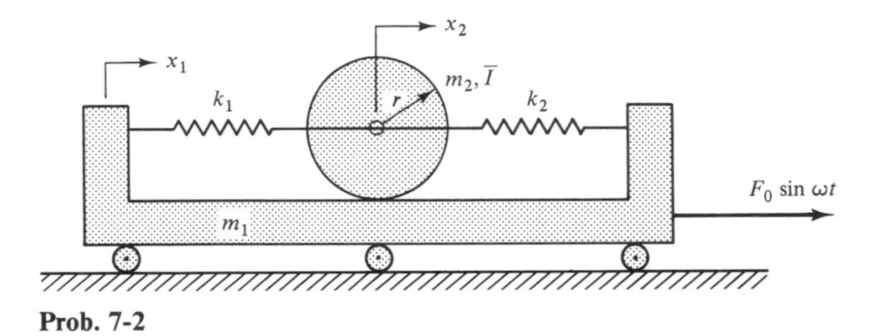

Prob. 7-2

7-3. The uniform bar that is part of the system shown in the accompanying figure has a mass of m_1 and a centroidal mass moment of inertia $\bar{I}$ and is supported by the two springs of stiffness k as shown. The mass m_3 is hung from the bar by means of another spring of stiffness k and is subjected to the force $F(t)$ shown. Assuming small amplitudes of oscillation, determine the differential equations of motion of the system using Lagrange's equations.

Partial ans: $m_3\ddot{q}_3 + k(q_3 - q_1) = F(t)$

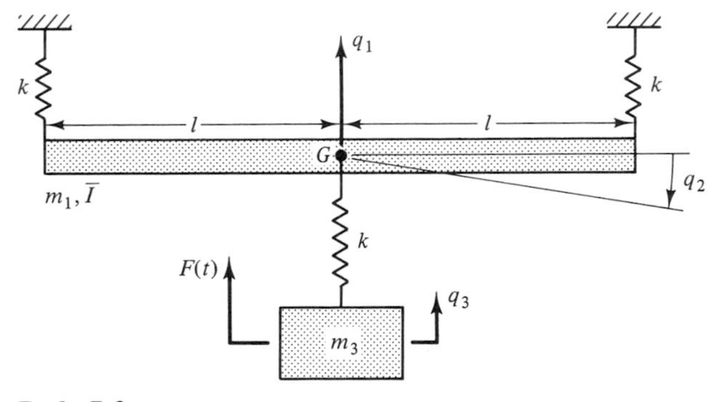

Prob. 7-3

7-4. Use Lagrange's equations to determine the differential equations of motion of the system shown in the accompanying figure. Check the equations determined by writing the stiffness matrix **K** and the damping matrix **C** from an inspection of the system using the definitions of influence coefficients as explained in Sec. 5-3.

Partial ans: $m_1\ddot{x}_1 + c_2(\dot{x}_1 - \dot{x}_2) + (k_1 + k_2 + k_3)x_1 - k_2(x_2 + x_3) = F_1(t)$

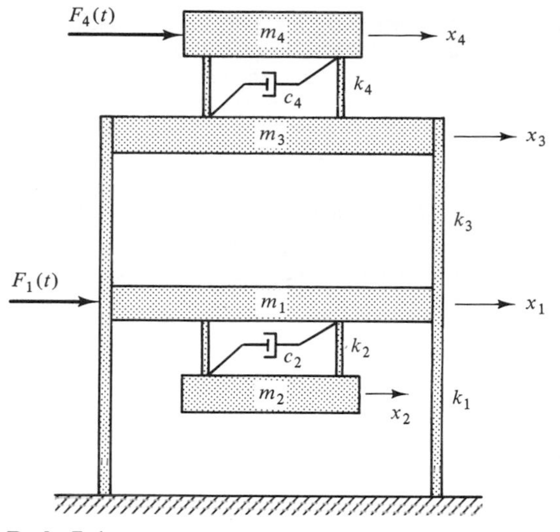

Prob. 7-4

7-5. A motor drives two mixer paddles of a commercial mixer through a gear-shaft system as shown in the accompanying figure. Each of the bevel gears has 20 teeth (20T), the central

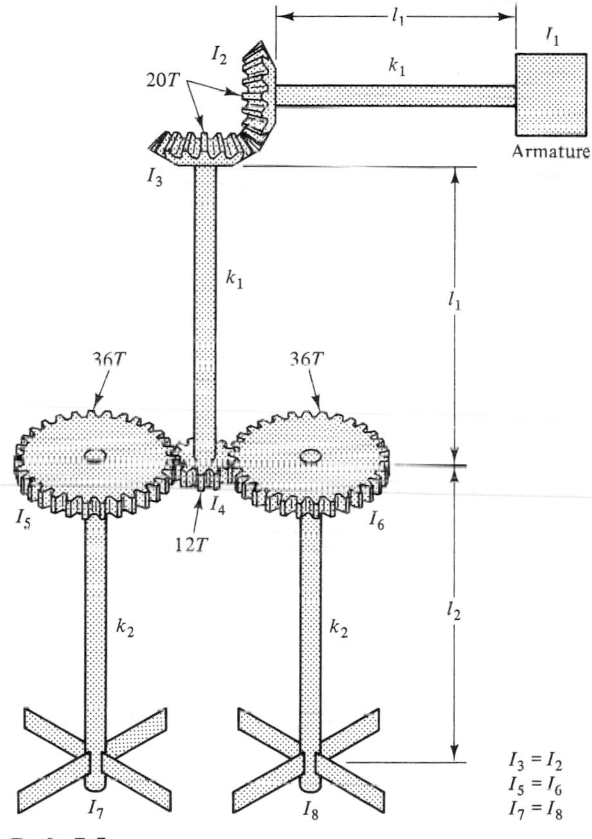

Prob. 7-5

spur gear has 12, and the other two spur gears have 36 each. With this arrangement, the angular displacement of each of the bevel gears is the same, and the angular displacement of each of the two spur gears attached to the paddles is one-third that of the central spur gear. Write the kinetic energy T of the system in terms of the generalized coordinates q_1, q_2, q_4, q_7, and q_8, with their subscripts corresponding to the subscripts of the mass moments of inertia I_1, I_2, I_4, I_7, and I_8, respectively, of the motor armature and the gears. Also write the potential energy U of the system in terms of the torsional spring constants k_1 and k_2 and the generalized coordinates. Assume that $I_5 = I_6 = 5I_4$.

7-6. Referring to Prob. 5-10, determine the differential equations of motion of the four-lumped-mass model of the automobile using Lagrange's equations. Assume small amplitudes of oscillation.

Problems 7-7 through 7-16 (Sections 7-3 through 7-6)

7-7. The stiffness matrix **K** of a rod element was determined in Sec. 7-3 by writing the potential energy U of the rod element and then using $\partial U/\partial u_1$ and $\partial U/\partial u_2$ of Lagrange's equations. Now determine **K** of the rod element by considering the definition of stiffness coefficients k_{ij} given in Sec. 5-3.

7-8. The accompanying figure shows a rod of length L modeled by two rod elements of length l and fixed to a support at its left end. It has a modulus of elasticity of E, a cross-sectional area A, and a mass per unit length γ. Determine (a) the mass and stiffness matrices of the rod elements using the trigonometric functions given by Eqs. 7-36 and 7-37, (b) the lowest natural circular frequency ω_1 of the rod as modeled by the two finite elements using the matrices determined in part a, and (c) the lowest natural circular frequency of the rod using the mass and stiffness matrices of a rod element shown in Fig. 7-11. Compare the results obtained in parts b and c with the exact value of $\omega_1 = (\pi/2L)\sqrt{EA/\gamma}$. What conclusion can you draw from a comparison of the two approximate solutions with the exact value?

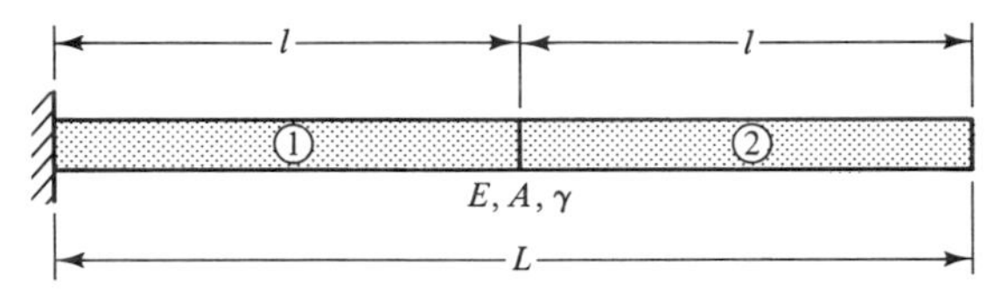

Prob. 7-8

7-9. Determine the natural frequencies and mode shapes of the rod of Example 7-5 if it is free-free instead of fixed at $x = L$. It is suggested that the JACOBI.FOR program (see p. 369) be used to obtain the solution.

7-10. Element 2 of the rod of Example 7-6 is subjected to a suddenly applied force that is uniformly distributed over the element as shown in the accompanying figure. That is,

$$w(x, t) = \begin{Bmatrix} 0 & 0 \le x < l \\ w_0 & l \le x \le 2l \\ 0 & x > 2l \end{Bmatrix}$$

Determine (a) the joint forces $\{f_i\}$ and $\{Q_i\}$ for the free-fixed rod due to $w(x, t)$, and (b) the time response of the rod using the modal-analysis method. Utilize results from Examples 7-5 and 7-6 wherever possible.

Partial ans: $\{Q_i\} = w_0 \dfrac{l}{2} \begin{Bmatrix} 0 \\ 1 \\ 1 \end{Bmatrix}$

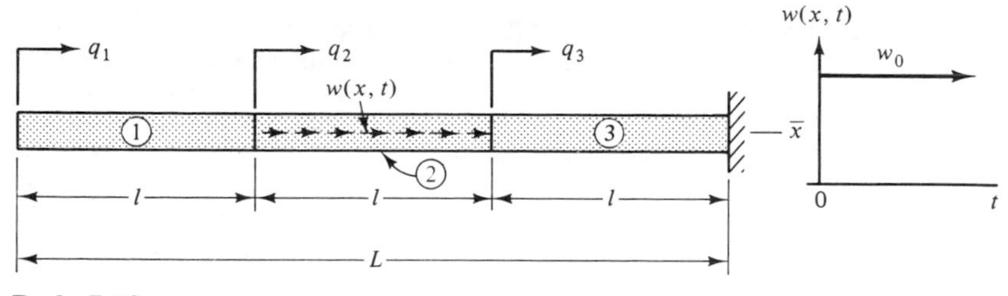

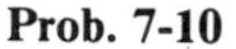
Prob. 7-10

7-11. The beam element shown in the accompanying figure is subjected to a concentrated force $P(t)$ at the position shown. Determine the joint forces and moments f_1, f_2, f_3, and f_4 due to this force.

Ans: $\{f_i\} = \dfrac{P(t)}{64}\begin{Bmatrix} 10 \\ 3l \\ 54 \\ -9l \end{Bmatrix}$

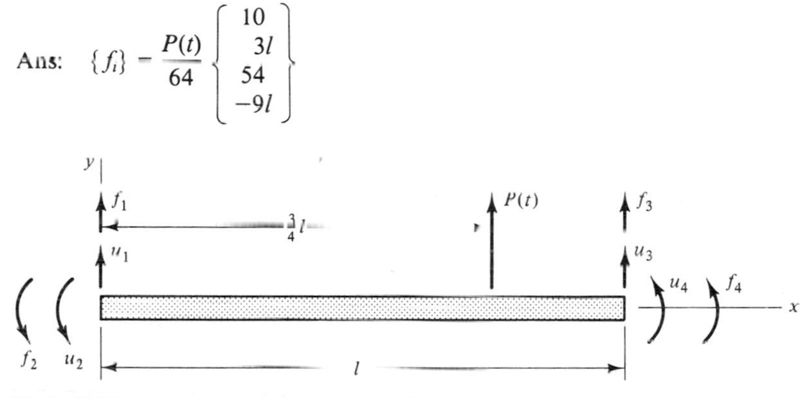

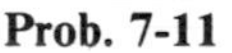
Prob. 7-11

7-12. Model the nonuniform beam shown in the accompanying figure by the two elements shown there, and using the finite-element method determine the natural circular frequencies and mode shapes of the beam if it is fixed-fixed.

Ans: $\omega_1 = 43.04$ rad/s

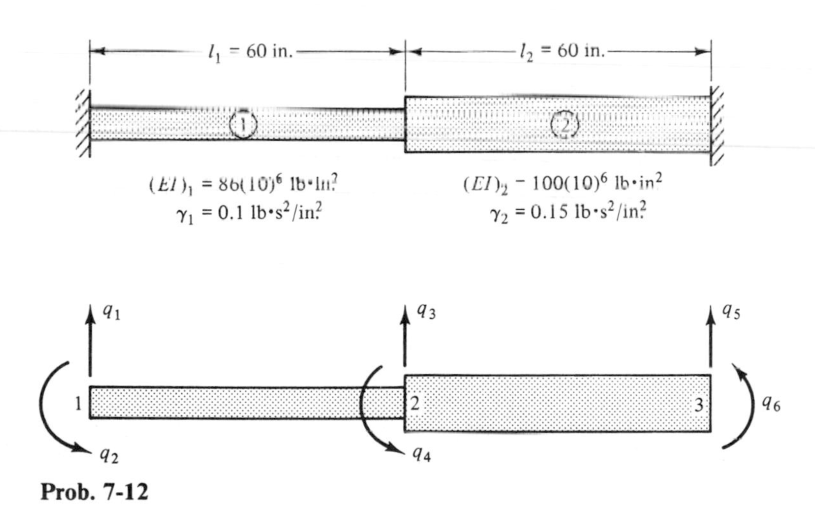

Prob. 7-12

7-13. The right end of the two-span continuous beam of Example 7-8 is subjected to a moment $M(t)$ as shown in Fig. 7-21. Use the results obtained in Example 7-8 and determine the time response of the beam if $M(t)$ is suddenly applied as shown in the accompanying figure. Neglect damping and consider that the superposition of the first two modes is adequate for the time-response solution.

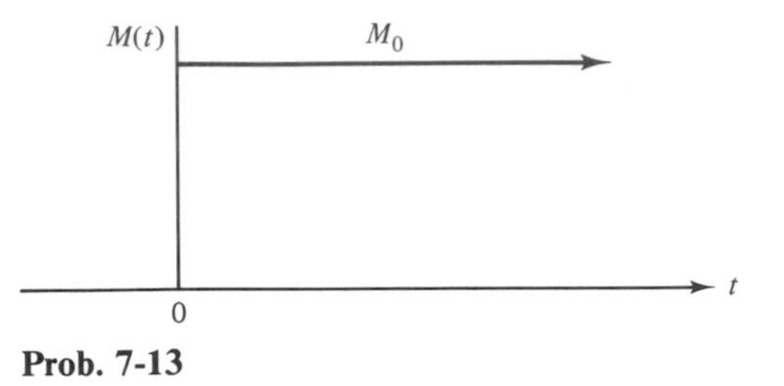

Prob. 7-13

7-14. The two-span continuous beam in Example 7-8 was subjected to the moment $M(t)$ as shown in Fig. 7-21. If instead of this moment, the beam is subjected to a uniformly distributed force that varies sinusoidally in magnitude as shown in the accompanying figure, determine the generalized nonpotential force vector $\{Q_i\}$ and the equations of motion of the beam for this excitation.

Partial ans: $$\begin{Bmatrix} Q_2 \\ Q_4 \\ Q_6 \end{Bmatrix} = w_0 \sin \omega t \begin{Bmatrix} (100)^2/12 \\ 0 \\ -(100)^2/12 \end{Bmatrix}$$

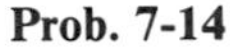

Prob. 7-14

7-15. The simply supported beam shown in the accompanying figure is subjected to a force that is given by the step function $P(t)$ shown which has a rise time of 0.5 s and reaches a magnitude of 100,000 lb. Noting that the physical properties of the beam are identical to those of the free-free beam of Example 7-7, use the results of that example to determine (a) the natural circular frequencies and mode shapes of the simply supported beam and (b) the maximum response of the beam due to the force $P(t)$ using the modal-analysis method with only the first mode considered. It is suggested that the response spectrum shown in Fig. 4-15 on page 253 be used to determine the maximum response. Is $P(t)$ a dynamic type of excitation for this beam?

Ans: $\omega_1 = 120.87$ rad/s

$$\begin{Bmatrix} X_2 \\ X_3 \\ X_4 \\ X_6 \end{Bmatrix}_1 = \begin{Bmatrix} 1.00 \\ 63.68 \\ 0.00 \\ -1.00 \end{Bmatrix} \qquad \begin{Bmatrix} q_2 \\ q_3 \\ q_4 \\ q_6 \end{Bmatrix}_{max} = \begin{Bmatrix} -0.026 \text{ in.} \\ -1.66 \text{ in.} \\ 0.0 \text{ in.} \\ 0.026 \text{ in.} \end{Bmatrix}$$

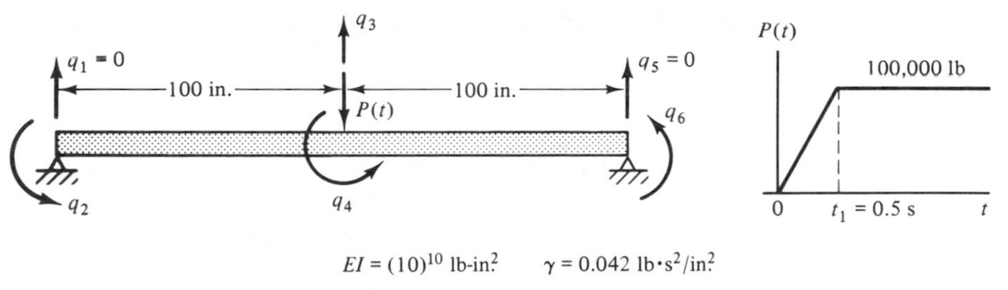

Prob. 7-15

7-16. The simple frame shown in the accompanying figure is fixed at A and free at B and is modeled by the two beam elements shown. Both elements have the same physical property values shown in the figure. Determine (a) the differential equations of motion of the frame if it is subjected to the force $P(t)$ shown and (b) the natural circular frequencies and mode shapes of the frame. Sketch the first three mode shapes. (Neglect axial extension of the horizontal element.)

Partial ans: $\omega_1 = 75.16$ rad/s

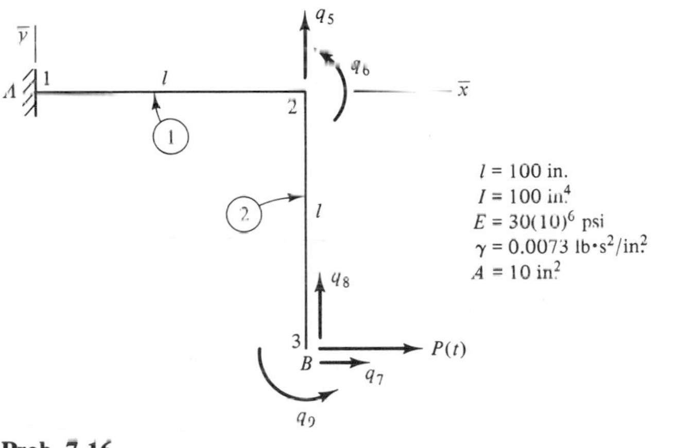

Prob. 7-16

Problems 7-17 through 7-29 (Sections 7-7 and 7-8)

7-17. Use the FINITEL.FOR program to obtain the natural frequencies and mode shapes for the nonuniform beam of Prob. 7-12.

7-18. Use the FINITEL.FOR program to obtain the natural frequencies and mode shapes for the simply supported beam of Prob. 7-15.

7-19. Use the FINITEL.FOR program to obtain the natural frequencies and mode shapes for the frame of Prob. 7-16.

7-20. In Example 7-8 the two-span continuous beam shown in Fig. 7-21 was modeled by just two finite beam elements, so that only three generalized coordinates corresponding to the respective angular displacements of the three joints were used. Since the values obtained for the natural circular frequencies and mode shapes for this two-element model cannot be expected to be very accurate, model the same beam using four beam elements that are each 50 in. in length. Use the FINITEL.FOR program to obtain the natural circular frequencies and mode shapes of this model, and compare the ω_1 obtained with both the value obtained in Example 7-8 and the exact value of $\omega_1 = 481.58$ rad/s. Sketch the first three mode shapes.

Ans: $\omega_1 = 483.49$ rad/s

7-21. In Example 7-9 the frame shown in Fig. 7-23 has joints 1 and 4 fixed. Determine the natural circular frequencies and mode shapes of the same frame, but with joints 1 and 4 pinned instead of fixed, using the program FINITEL.FOR.

7-22. The frame shown in the accompanying figure is fixed at both A and B and is subjected to the sinusoidal excitation $P(t) = P_0 \sin \omega t$ shown. Determine the amplitude of the steady-state response of the frame in terms of P_0 when the excitation frequency ω is equal to the lowest natural circular frequency ω_1 of the frame. Assume that the damping factor ζ_1 for the first mode is 0.001.

Ans: $$\begin{Bmatrix} q_4 \\ q_5 \\ q_6 \end{Bmatrix} = 5.05(10)^{-7} P_0 \begin{Bmatrix} 1.000 \\ 0.411 \\ 1.285 \end{Bmatrix}$$

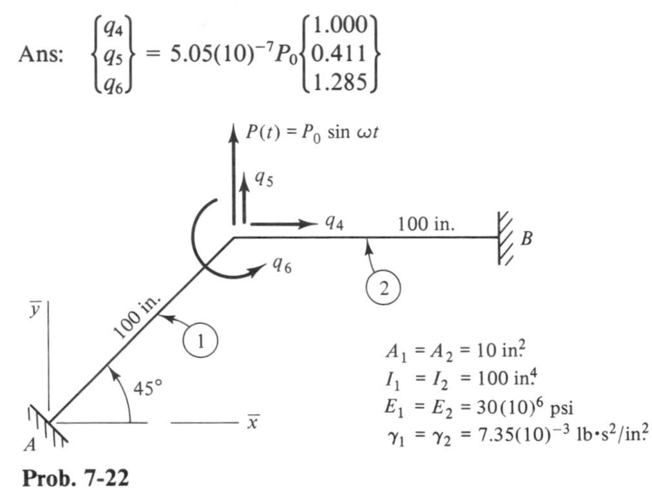

Prob. 7-22

7-23. The beam shown in the accompanying figure is fixed at A and pinned at B. Model this overhanging beam by three beam elements, and use the FINITEL.FOR program to obtain its natural circular frequencies and modes shapes. Sketch the first two mode shapes.

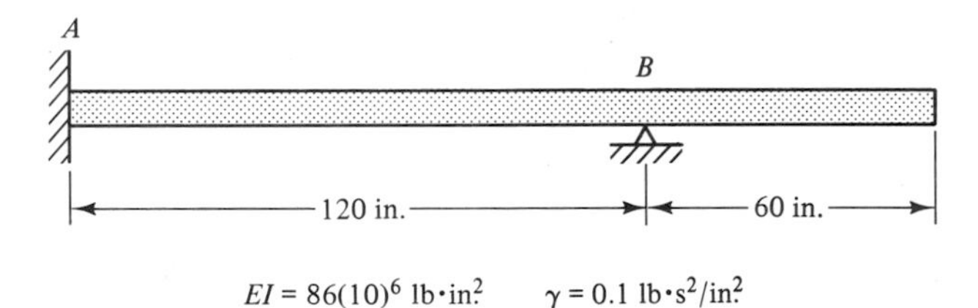

$EI = 86(10)^6$ lb·in.^{2} $\gamma = 0.1$ lb·s^2/in.^{2}

Prob. 7-23

7-24. In Example 7-7 the rocket shown in Fig. 7-19 was modeled by two beam elements not subject to axial deformations as shown in Fig. 7-19c. For this problem use two beam elements for which the generalized coordinates associated with axial displacements of the system are included, and use the FINITEL.FOR program to obtain the natural circular frequencies and modes shapes of the rocket model. Use an effective cross-sectional area of 40 in.^{2} for each element. Sketch the three rigid-body mode configurations and the first elastic-bending mode.

7-25. Derive the local mass and stiffness matrices for the truss element shown in Fig. 7-29. *Hint:* Consider that the displacement of a point on the element is described by the following functions:

$$u(x, t) = \phi_1(x)u_1 + \phi_3(x)u_3$$
$$y(x, t) = \phi_2(x)u_2 + \phi_4(x)u_4$$

7-26. The overhanging truss shown in the accompanying figure is pinned at A and supported on rollers at B. All members of the truss have the following properties:

$$A = 2.10(10)^{-3}\ \text{m}^2 \quad \text{(cross-sectional area)}$$
$$E = 2.068(10)^{11}\ \text{Pa} \quad \text{(modulus of elasticity)}$$
$$w = 185.0\ \text{N/m} \quad \text{(weight per m)}$$
$$\gamma = \frac{185.0}{9.81} = 18.86\ \text{kg/m} \quad \text{(mass per m)}$$

Use the TRUSS.FOR program to obtain the natural frequencies and mode shapes of the truss.

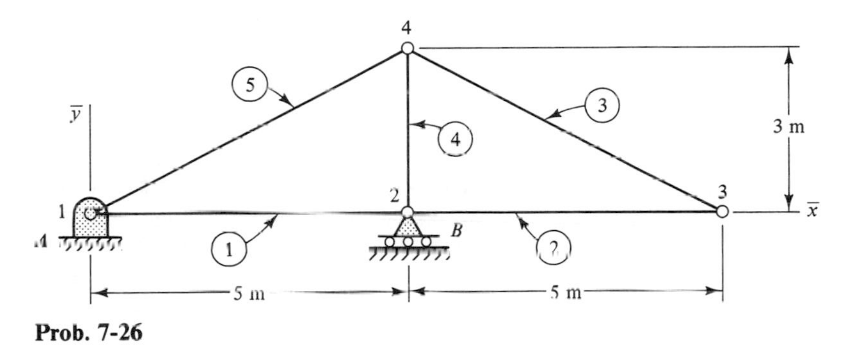

Prob. 7-26

7-27. The truss shown in the accompanying figure is pinned at A and supported on rollers at B. All the members of the truss are fabricated from L5 × 3 × $\frac{1}{2}$ angle stock and have the following properties:

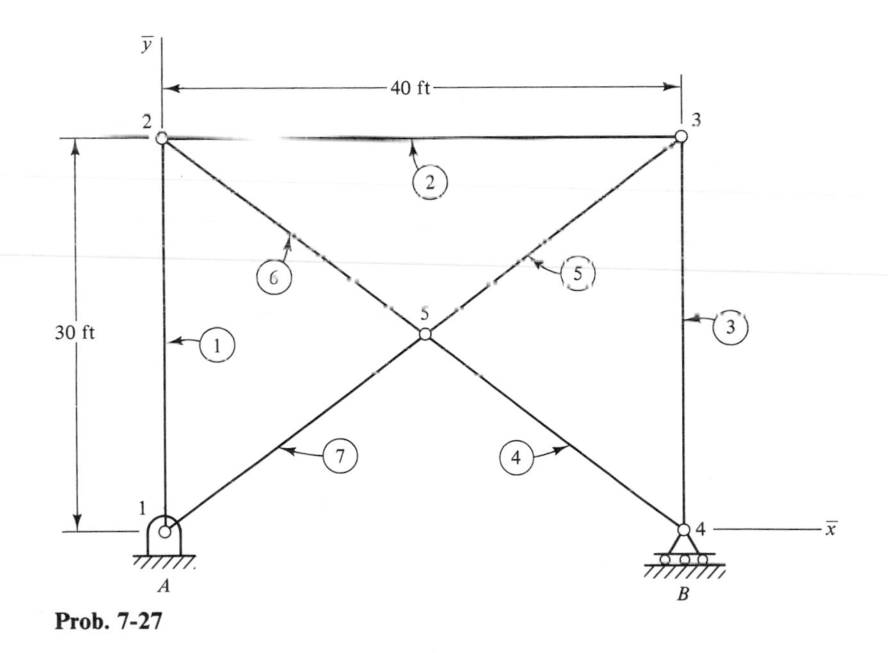

Prob. 7-27

$$A = 3.75 \text{ in.}^2 \quad \text{(cross-sectional area)}$$
$$E = 30(10)^6 \text{ psi} \quad \text{(modulus of elasticity)}$$
$$w = 12.8 \text{ lb/ft} \quad \text{(weight per ft)}$$
$$\gamma = \frac{12.8}{12(386)} = 2.76(10)^{-3} \text{ lb} \cdot \text{s}^2/\text{in.}^2 \quad \text{(mass per in.)}$$

Use the TRUSS.FOR program to obtain the natural circular frequencies and mode shapes of the truss.

Partial ans: $\omega_1 = 124.73$ rad/s

7-28. The frame shown in the accompanying figure is pinned at supports A and B. All the members of the frame are W18 × 60 beams that have the following properties:

$$A = 17.64 \text{ in.}^2 \quad \text{(cross-sectional area)}$$
$$E = 30(10)^6 \text{ psi} \quad \text{(modulus of elasticity)}$$
$$I = 984 \text{ in.}^4 \quad \text{(area moment of inertia)}$$
$$\gamma = 0.013 \text{ lb} \cdot \text{s}^2/\text{in.}^2 \quad \text{(mass per in.)}$$

The length l of the vertical members is 240 in., and the length of the horizontal member is 480 in. ($2l$). As mentioned above, the frame is pinned at joints 1 and 4, and zero rotation is assumed at joints 2 and 3 so that the horizontal member remains straight and at an angle of 90° to each of the vertical members as shown in part b of the figure. Determine the following: (a) the lowest natural circular frequency ω_1 of the frame found by considering it as a simple spring-and-mass system for which the distributed mass of the frame is lumped by the procedure explained in Sec. 5-12 and for which the equivalent spring constant of the two vertical members together is

$$k_e = \frac{2(3EI)}{l^3}$$

(b) the natural circular frequency ω_1 and mode shape of the first mode using the FINITEL.FOR program with the assumed zero rotation of joints 2 and 3 as constraints. Compare the value of ω_1 determined in part a with that obtained using the FINITEL.FOR program.

Ans: (a) $\omega_1 = 37.00$ rad/s
(b) $\omega_1 = 37.16$ rad/s

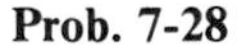

Prob. 7-28

7-29. The dimensions of a small tapered aluminum beam are shown in part a of the accompanying figure. The first natural frequency f_1 of this beam vibrating in the x-y plane was determined experimentally to be 88.2 Hz. Model the tapered beam as four elements as shown in b. Calculate the cross-sectional areas, the area moments of inertia, and the masses per unit length of the elements by the procedure for tapered beams explained in Sec. 7-3 on page 507, and determine the first natural frequency f_1 by using the FINITEL.FOR program. Neglect axial extension of the elements. Compare the value of f_1

obtained from the finite-element model with the experimental value. The mass density ρ and modulus of elasticity E are $2.59(10)^{-4}$ lb · s²/in.⁴ and $10(10)^6$ psi, respectively.

Ans: $f_1 = 86.89$ Hz

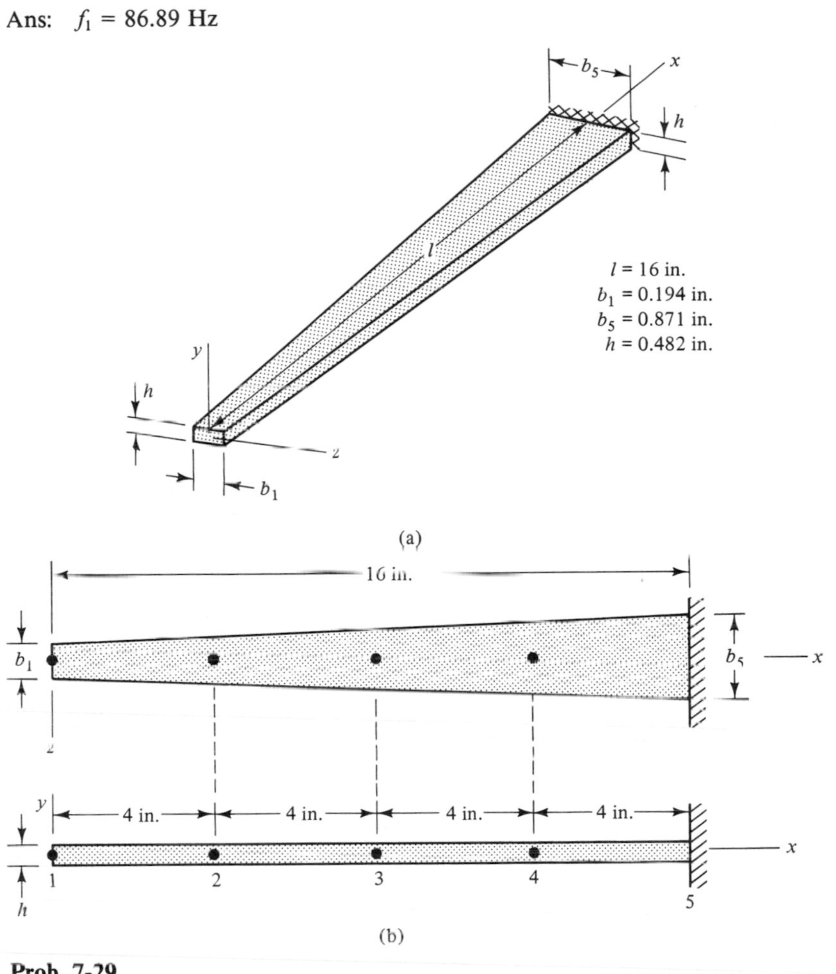

Prob. 7-29

Chapter 8

Continuous Systems

8-1 INTRODUCTION

In previous chapters the distributed mass and stiffness properties of various types of systems were lumped to form discrete models characterized by finite numbers of degrees of freedom, so that they could be modeled mathematically by ordinary differential equations.

In this chapter we consider the mass and stiffness properties of systems such as rods and beams in their continuous (distributed) form. Since the motion of every point of a continuous body is considered, such a system has a theoretically infinite number of degrees of freedom. Partial differential equations of motion are used to describe continuous systems since both space and time functions are required to describe the vibratory motion of the systems.

Since the partial differential equations of motion of continuous systems involve derivatives with respect to spatial coordinates as well as with respect to time, their solutions involve constants that must be determined from boundary conditions as well as initial conditions. They are thus referred to as *boundary-value* problems. Obtaining analytical solutions of partial differential equations modeling bodies having nonuniform cross sections is generally quite difficult. In such cases it is necessary to use some numerical technique to solve the equations. Such systems are generally better analyzed using some method that involves discretizing the system as discussed in Chaps. 5, 6, and 7. In this chapter we consider only bodies of uniform cross section and consisting of materials that are completely elastic, isotropic, and homogeneous.

Since the vibration analyses of complex continuous systems can be made by using appropriate discrete physical models that lead to mathematical models consisting of ordinary differential equations, the reader might wonder why it is important to study continuous systems as such. There are several reasons. First, the study of simple continuous systems leads us to an understanding of the necessity of using discrete models for some types of systems just mentioned. Another is that the exact solutions that can be obtained for some systems can be used as references in modeling the distributed masses of other systems as discrete-mass systems. Finally, the concepts of normal modes and modal superposition encountered in analyzing continuous systems as such are important in understanding the vibration characteristics of systems that are discrete in nature.

8-2 ONE-DIMENSIONAL WAVE EQUATION

The one-dimensional wave equation in conjunction with appropriate physical constants describes the motion of various types of systems, including the transverse vibration of taut strings and long flexible cables, the longitudinal vibration of rods, and the torsional vibration of cylindrical rods.

Taut Strings

Let us consider the transverse vibration of the tightly stretched string shown in Fig. 8-1a which is stretched by the force T_0 shown in Fig. 8-1b. As we shall soon see, the transverse motion y of any section along the string can be expressed as

$$y = \phi(x)q(t) \tag{8-1}$$

in which $\phi(x)$ is a function of the spatial coordinate x and $q(t)$ is a function of time t.

It is assumed that the amplitude of vibration of the string is small so that the tension T_0 in the string remains essentially constant during vibration and that the string offers no resistance to bending. Considering a differential length dx of the string, which has a mass per unit length of γ, we obtain the free body shown in Fig. 8-1b which includes the inertia effect shown as a dashed vector. Then, using the concept of dynamic equilibrium, we sum the forces and inertia effect in the y direction equal to zero to obtain

$$\sum F_y = 0 = T_0 \sin \theta_1 - T_0 \sin \theta_2 + \gamma \frac{\partial^2 y}{\partial t^2} dx \tag{8-2}$$

For small displacements, we let

$$\sin \theta_1 = \frac{\partial y}{\partial x}$$

and

$$\sin \theta_2 = \frac{\partial y}{\partial x} + \frac{\partial}{\partial x}\left(\frac{\partial y}{\partial x}\right) dx$$

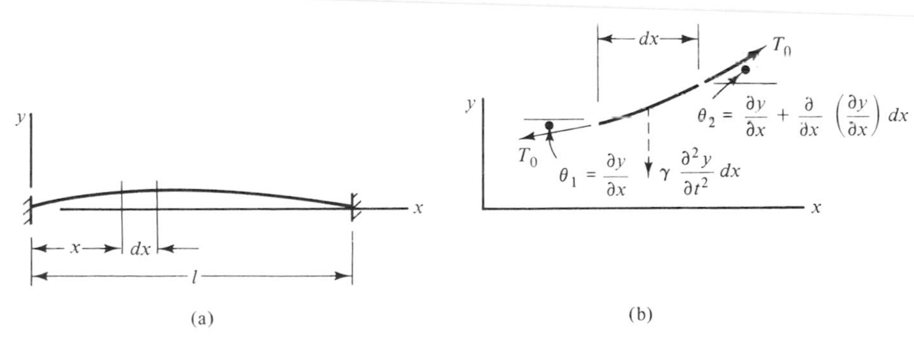

Figure 8-1 Tightly stretched string.

so that Eq. 8-2 reduces to the one-dimensional wave equation

$$c^2 \frac{\partial^2 y}{\partial x^2} = \frac{\partial^2 y}{\partial t^2} \tag{8-3}$$

in which $c^2 = T_0/\gamma$.

Solutions of the partial differential equations encountered in this chapter are generally obtained by means of the *separation-of-variables* method. Using this method, the partial differential equation is reduced to a set of ordinary differential equations by assuming that its solution is the product of functions of the independent variables appearing in the partial differential equation. Therefore, we assume a solution of the wave equation having the form of Eq. 8-1. Taking successive partial derivatives of Eq. 8-1 with respect to both x and t, we obtain

$$\frac{\partial^2 y}{\partial x^2} = \frac{d^2\phi}{dx^2}(x)q(t)$$

and

$$\frac{\partial^2 y}{\partial t^2} = \frac{d^2 q}{dt^2}(t)\phi(x)$$

Substituting these expressions into Eq. 8-3 gives

$$c^2 \frac{d^2\phi}{dx^2}(x)q(t) = \frac{d^2 q}{dt^2}(t)\phi(x)$$

Dividing both sides of this equation by $q(t)$ and $\phi(x)$ yields

$$c^2 \frac{d^2\phi}{dx^2}(x)\frac{1}{\phi(x)} = \frac{d^2 q}{dt^2}(t)\frac{1}{q(t)} \tag{8-4}$$

Since the left side of Eq. 8-4 is a function of x only and the right side is a function of t only, each side must be equal to a constant. If we let this constant be $-\omega^2$, we can obtain the following two ordinary differential equations from Eq. 8-4:

$$\frac{d^2\phi}{dx^2}(x) + \left(\frac{\omega}{c}\right)^2 \phi(x) = 0 \tag{8-5}$$

and

$$\frac{d^2 q}{dt^2}(t) + \omega^2 q(t) = 0 \tag{8-6}$$

Recognizing the familiar form of Eqs. 8-5 and 8-6, their solutions can be readily written as

$$\phi(x) = A \cos\frac{\omega}{c}x + B \sin\frac{\omega}{c}x \tag{8-7}$$

and

$$q(t) = C \cos\omega t + D \sin\omega t \tag{8-8}$$

Substituting Eqs. 8-7 and 8-8 into Eq. 8-1, the solution of the one-dimensional wave equation given by Eq. 8-3 is found to be

$$y = \left[A \cos\frac{\omega}{c}x + B \sin\frac{\omega}{c}x\right][C \cos\omega t + D \sin\omega t] \tag{8-9}$$

It should be explained at this point that a negative ω^2 was selected as the constant for use with Eq. 8-4 instead of a positive ω^2 so that the solution of Eq. 8-6 would be in terms of trigonometric functions of time which are periodic rather than hyperbolic functions that go to infinity with time. Solutions of the latter type are typical for *unstable* systems. In the vibrating-string problem it is apparent from the physical requirements of the problem that the system is *stable.* Since every position of the string is reproduced over and over with time, it is evident that a solution in terms of periodic functions of time is required. If $\omega = 0$, from Eqs. 8-1, 8-5, and 8-6, a solution of the type

$$y = (Ax + B)(Ct + D)$$

results. This solution is not periodic, which is consistent with the physical characteristics of a system for which $\omega = 0$.

To determine the natural circular frequencies and mode shapes of the string in terms of its physical parameters, it is necessary to consider the boundary conditions associated with the constraints shown in Fig. 8-1a. Considering the string as fixed at $x = 0$ and $x = l$ as shown, the two boundary conditions that must be satisfied for all values of time are

$$(1) \quad y(0, t) = 0$$
$$(2) \quad y(l, t) = 0$$

Substituting the first of these conditions into Eq. 8-9 gives

$$y(0, t) = 0 = A(C \cos \omega t + D \sin \omega t)$$

which reveals that $A = 0$. Substituting the second of the conditions into Eq. 8-9, with $A = 0$, we obtain

$$y(l, t) = 0 = B \sin \frac{\omega l}{c}(C \cos \omega t + D \sin \omega t) \qquad (8\text{-}10)$$

For a nontrivial solution ($B \neq 0$), it should be apparent that Eq. 8-10 can be satisfied for all values of time if

$$\sin \frac{\omega l}{c} = 0 \qquad (8\text{-}11)$$

Equation 8-11 is referred to as a *frequency equation,* and it should be apparent that this equation is satisfied when

$$\frac{\omega_i l}{c} = i\pi \qquad i = 1, 2, 3, \ldots$$

Thus, the natural frequencies of a tightly stretched string are given by

$$\omega_i = \frac{i\pi c}{l} = \frac{i\pi}{l}\sqrt{\frac{T_0}{\gamma}} \quad \text{rad/s} \qquad i = 1, 2, 3, \ldots \qquad (8\text{-}12)$$

or

$$f_i = \frac{\omega_i}{2\pi} = \frac{i}{2l}\sqrt{\frac{T_0}{\gamma}} \quad \text{Hz} \qquad (8\text{-}13)$$

Referring to Eqs. 8-9 and 8-12 and conveniently letting $B = 1$, it should be apparent that the normal-mode shape for the ith mode corresponding to ω_i is

$$\phi_i(x) = \sin \frac{i\pi x}{l} \qquad i = 1, 2, 3, \ldots \tag{8-14}$$

Since the sum of the solutions of a linear system is also a solution, Eqs. 8-9 and 8-14 yields a solution as

$$y = \sum_{i=1,2,3,\ldots}^{\infty} \sin \frac{i\pi x}{l} (C_i \cos \omega_i t + D_i \sin \omega_i t) \tag{8-15}$$

in which the constants C_i and D_i for each normal mode depend upon the initial conditions and the natural circular frequency

$$\omega_i = \frac{i\pi}{l} \sqrt{\frac{T_0}{\gamma}}$$

The reader should note that the solution given by Eq. 8-15 is the superposition of the normal (principal) modes, which is analogous to the manner in which solutions are obtained for discrete systems using the modal-analysis method.

Let us now consider a general procedure for obtaining the constants C_i and D_i when the initial displacement and velocity of the string are given, respectively, by

$$y(x, 0) = f(x)$$
$$\dot{y}(x, 0) = g(x)$$

With these initial conditions, Eq. 8-15 yields

$$f(x) = \sum_{i=1,2,\ldots}^{\infty} C_i \sin \frac{i\pi x}{l}$$

and its partial derivative with respect to t gives

$$g(x) = \sum_{i=1,2,\ldots}^{\infty} D_i \omega_i \sin \frac{i\pi x}{l}$$

Multiplying each of the last two equations by

$$\sin \frac{j\pi x}{l} \qquad (j = 1, 2, 3, \ldots)$$

and integrating from $x = 0$ to $x = l$, we obtain

$$\int_0^l f(x) \sin \frac{j\pi x}{l}\, dx = \int_0^l \left(\sum C_i \sin \frac{i\pi x}{l} \right) \sin \frac{j\pi x}{l}\, dx \tag{8-16}$$

and

$$\int_0^l g(x) \sin \frac{j\pi x}{l}\, dx = \int_0^l \left(\sum D_i \omega_i \sin \frac{i\pi x}{l} \right) \sin \frac{j\pi x}{l}\, dx \tag{8-17}$$

If we now recall from Sec. 1-5 that $\sin i\pi x/l$ and $\sin j\pi x/l$ are *orthogonal functions* over the interval $0 < x < l$, we realize that a term-by-term integration of either Eq. 8-16 or Eq. 8-17 yields all zeros except for the one term in each when $i = j$. Thus, the integration of these two equations yields the following expressions for the constants:

$$C_i = \frac{2}{l}\int_0^l f(x)\sin\frac{i\pi x}{l}\,dx \tag{8-18}$$

and

$$D_i = \frac{2}{l\omega_i}\int_0^l g(x)\sin\frac{i\pi x}{l}\,dx \tag{8-19}$$

EXAMPLE 8-1

A force P holds a string tightly stretched in the configuration shown in Fig. 8-2. We wish to determine the motion of the string after the force P is suddenly removed.

Solution. Noting the initial configuration of the string shown in Fig. 8-2, and recalling the slope-intercept form of the equation of a straight line, we can write that

$$y(x, 0) = f(x) = \begin{Bmatrix} \dfrac{2\Delta x}{l} & 0 \le x \le \dfrac{l}{2} \\ \dfrac{2\Delta}{l}(l - x) & \dfrac{l}{2} \le x \le l \end{Bmatrix}$$

Since the string is initially at rest,

$$\dot{y}(x, 0) = g(x) = 0$$

Referring to Eq. 8-18, we see that

$$C_i = \frac{4\Delta}{l^2}\int_0^{l/2} x\sin\frac{i\pi x}{l}\,dx + \frac{4\Delta}{l^2}\int_{l/2}^{l}(l - x)\sin\frac{i\pi x}{l}\,dx$$

which integrates to

$$C_i = \frac{8\Delta}{\pi^2 i^2}\sin\frac{i\pi}{2} = \frac{8\Delta}{\pi^2 i^2}(-1)^{(i-1)/2} \qquad i = 1, 3, 5, \ldots$$

Equation 8-19, with $g(x) = 0$, reveals that

$$D_i = 0$$

Substituting C_i and D_i into Eq. 8-15, the motion of the string is

$$y = \frac{8\Delta}{\pi^2}\sum_{i=1,3,\ldots}^{\infty}\frac{(-1)^{(i-1)/2}}{i^2}\sin\frac{i\pi x}{l}\cos\omega_i t \tag{8-20}$$

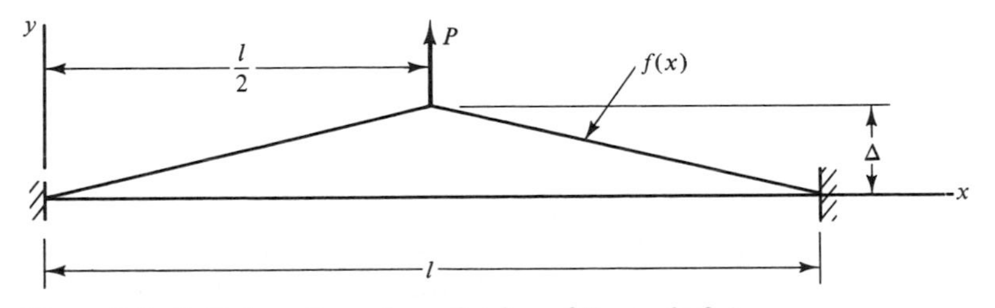

Figure 8-2 Initial configuration of string of Example 8-1.

in which

$$\omega_i = \frac{i\pi}{l}\sqrt{\frac{T_0}{\gamma}}$$

Since the initial configuration $f(x)$ of the string is symmetrical about $x = l/2$, the motion of the string consists of the superposition of all the *odd* modes. That is, the sine waves

$$\phi_i(x) = \sin\frac{i\pi x}{l}$$

are symmetric about $x = l/2$ for $i = 1, 3, 5, \ldots$ but are asymmetric about $x = l/2$ for $i = 2, 4, 6, \ldots$. It should also be noted that the contribution to the motion from the higher modes decreases fairly rapidly as the mode number increases because of the $1/i^2$ factor. The reader should note that this is analogous to what was found in analyzing the response of discrete systems by the modal-analysis method in Chaps. 5 and 6, in which the lower modes were found to dominate the response. As we shall see later, the *forced vibration* of continuous systems can be obtained in the form of

$$y = \sum_{i=1,2,\ldots}^{\infty} \phi_i(x)q_i \tag{8-21}$$

in which $\phi_i(x)$ are the functions for the normal modes and q_i are the *generalized coordinates* that are functions of time.

Wave Equation for Rods of Viscoelastic Materials

In preceding discussions in which we used the wave equation to analyze the vibration of a string, damping was neglected. Various types of damping models are used to simulate the damping characteristics of continuous systems. One that is very common that has been discussed in some detail in other sections of this text assumes that the damping is proportional to the velocity (*viscous damping*). Another employs a *complex modulus* of elasticity (see Sec. 3-11), and yet another assumes a viscoelastic material in which the stress depends not only upon the magnitude of the strain of the material but also upon the strain rate. The latter is referred to as *viscoelastic damping*, and we shall be concerned with this model here.

One of the simpler models used to simulate the damping of a viscoelastic material is the Kelvin model shown in Fig. 8-3. The spring is analogous to the modulus of elasticity

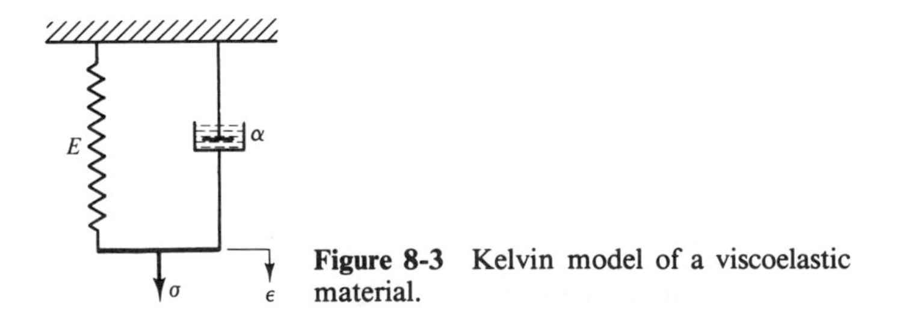

Figure 8-3 Kelvin model of a viscoelastic material.

E of the material and represents the perfectly elastic behavior of the material, while the dashpot is analogous to the resistance that is proportional to the strain rate $\dot{\epsilon}$. Thus, the stress σ depends not only upon the strain ϵ but also upon the strain rate $\dot{\epsilon}$. That is,

$$\sigma = E\epsilon + \alpha\dot{\epsilon} \tag{8-22}$$

in which α is a viscoelastic constant of the material.

Let us now use this model and derive the wave equation for the longitudinal vibration of the slender rod shown in Fig. 8-4, which is considered to consist of a viscoelastic material. Expressing the longitudinal displacement of a section of the rod a distance x from the left end as u, the strain can be written as

$$\epsilon = \frac{\partial u}{\partial x}$$

It then follows from Eq. 8-22 that

$$\sigma = E\frac{\partial u}{\partial x} + \alpha\frac{\partial^2 u}{\partial x\,\partial t} \tag{8-23}$$

Since the stress is essentially uniform across the cross-sectional area A_c of a slender rod, the force F shown in Fig. 8-4 can be expressed as

$$F = \sigma A_c = EA_c\frac{\partial u}{\partial x} + \alpha A_c\frac{\partial^2 u}{\partial x\,\partial t} \tag{8-24}$$

from which

$$\frac{\partial F}{\partial x} = EA_c\frac{\partial^2 u}{\partial x^2} + \alpha A_c\frac{\partial^3 u}{\partial x^2\,\partial t} \tag{8-25}$$

Designating the mass per unit volume of the material as ρ, and considering an element of length dx, the application of Newton's second law to the free body of the element shown in Fig. 8-4 yields

$$\frac{\partial F}{\partial x} = \rho A_c\frac{\partial^2 u}{\partial t^2} \tag{8-26}$$

Combining Eqs. 8-25 and 8-26, we find that

$$\frac{E}{\rho}\frac{\partial^2 u}{\partial x^2} + \frac{\alpha}{\rho}\frac{\partial^3 u}{\partial x^2\,\partial t} = \frac{\partial^2 u}{\partial t^2}$$

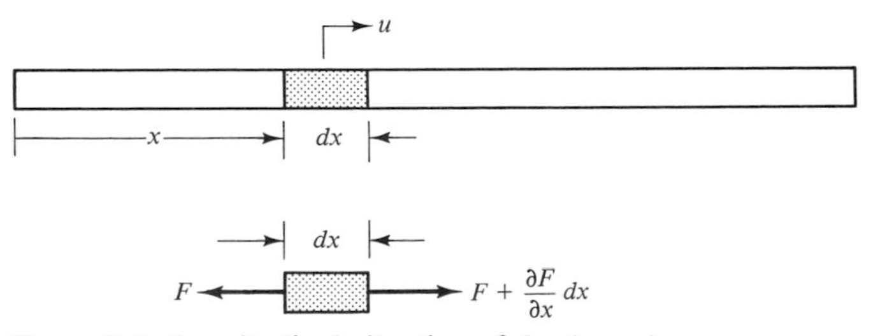

Figure 8-4 Longitudinal vibration of slender rod.

which becomes

$$c^2 \frac{\partial^2 u}{\partial x^2} + b^2 \frac{\partial^3 u}{\partial x^2 \partial t} = \frac{\partial^2 u}{\partial t^2} \tag{8-27}$$

with $c^2 = E/\rho$ and $b^2 = \alpha/\rho$.

Equation 8-27 is the wave equation for the longitudinal vibration of a slender rod of viscoelastic material. The reader should note that if $\alpha = 0$, Eq. 8-27 reduces to the same form as that of Eq. 8-3 derived for the undamped vibrating string. It can be shown (see Prob. 8-2) that Eq. 8-27 is also applicable to the *torsional* vibration of a circular bar when the following relationships are used:

$$\theta = \text{angle of twist} \quad (\text{dependent variable replacing } u)$$

$$c^2 = \frac{G}{\rho} = \frac{\text{shear modulus of elasticity}}{\text{mass density of material}}$$

$$b^2 = \frac{\alpha}{\rho} = \frac{\text{viscoelastic constant of material in shear}}{\text{mass density of material}}$$

To obtain a solution of Eq. 8-27 using the separation-of-variables method, we assume a solution of the form

$$u(x, t) = \phi(x)q(t)$$

or more simply

$$u(x, t) = \phi q \tag{8-28}$$

in which it is understood that $\phi = \phi(x)$ is a function of x and $q = q(t)$ is a function of time. It follows then that

$$\frac{\partial^2 u}{\partial x^2} = \frac{d^2\phi}{dx^2} q$$

$$\frac{\partial^3 u}{\partial x^2\, \partial t} = \frac{d^2\phi}{dx^2}\frac{dq}{dt} = \frac{d^2\phi}{dx^2} \dot{q}$$

$$\frac{\partial^2 u}{\partial t^2} = \phi \frac{d^2 q}{dt^2} = \phi \ddot{q}$$

Substituting these expressions into Eq. 8-27, we obtain

$$c^2 \frac{d^2\phi}{dx^2} q + b^2 \frac{d^2\phi}{dx^2} \dot{q} = \phi \ddot{q}$$

or

$$(c^2 q + b^2 \dot{q}) \frac{d^2\phi}{dx^2} = \phi \ddot{q}$$

Dividing both sides of the latter equation by $(c^2 q + b^2 \dot{q})$ and ϕ gives

$$\frac{d^2\phi}{dx^2}\frac{1}{\phi} = \frac{\ddot{q}}{c^2 q + b^2 \dot{q}} \tag{8-29}$$

Since the left side of Eq. 8-29 is a function of x only and the right side is a function of t

only, each side must be equal to a constant. If we select $-\lambda^2$ as this constant, we obtain the following two ordinary differential equations:

$$\frac{d^2\phi}{dx^2} + \lambda^2\phi = 0 \tag{8-30}$$

and

$$\ddot{q} + b^2\lambda^2\dot{q} + c^2\lambda^2 q = 0 \tag{8-31}$$

If we let $b^2\lambda^2 = 2\zeta\omega$ and $c^2\lambda^2 = \omega^2$, Eq. 8-31 can be written in the more familiar form of

$$\ddot{q} + 2\zeta\omega\dot{q} + \omega^2 q = 0 \tag{8-32}$$

The solution of Eq. 8-30 should be recognized as being

$$\phi(x) = A\cos\frac{\omega}{c}x + B\sin\frac{\omega}{c}x \tag{8-33}$$

in which $\omega/c = \lambda$ and $c = \sqrt{E/\rho}$. It can be shown that the expression for c yields the velocity of a stress wave (the velocity of sound) in the material.[1]

If we recall that Eq. 8-32 has the same form as the differential equation of motion determined earlier for a damped free system in Sec. 2-3, we can refer to Eq. 2-20 to see that the solution of Eq. 8-32 is

$$q = e^{-\zeta\omega t}(C\cos\omega\sqrt{1-\zeta^2}t + D\sin\omega\sqrt{1-\zeta^2}t) \tag{8-34}$$

Recalling from the discussion of the vibrating string that there are theoretically an infinite number of undamped natural circular frequencies ω_i for a continuous system, the solution of Eq. 8-27 can be written as the sum of the normal modes, so that

$$u(x,t) = \sum_{i=1,2,\ldots}^{\infty} \phi_i(x)e^{-\zeta_i\omega_i t}(C_i\cos\omega_i\sqrt{1-\zeta_i^2}t + D_i\sin\omega_i\sqrt{1-\zeta_i^2}t) \tag{8-35}$$

The normal-mode functions $\phi_i(x)$ and natural circular frequencies ω_i are determined from Eq. 8-33 with the use of the pertinent boundary conditions in Eq. 8-28. With the damping terms in Eqs. 8-31 and 8-32 equal to each other, the expression for ω_i is used to determine an expression for the damping factor ζ_i of each normal mode in terms of the physical parameters of the rod. This procedure is illustrated in Example 8-2.

EXAMPLE 8-2

The rod of length l shown in Fig. 8-5 is fixed at $x = 0$ and free at $x = l$.

Considering the rod material as viscoelastic, determine (a) the undamped natural circular frequencies ω_i; (b) the normal mode functions $\phi_i(x)$; (c) the damping factor ζ_i in terms of the physical parameters of the rod for each normal mode.

Solution. **a.** It was shown in the preceding discussion that the solution of the wave equation for a rod of viscoelastic material can be written as

$$u(x,t) = \left(A\cos\frac{\omega}{c}x + B\sin\frac{\omega}{c}x\right)q \tag{8-36}$$

[1] S. Timoshenko and D. H. Young, *Advanced Dynamics,* p. 110, McGraw-Hill Book Co., New York, 1948.

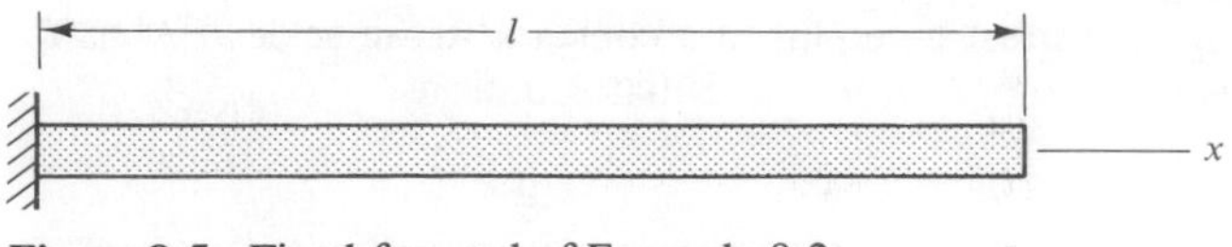

Figure 8-5 Fixed-free rod of Example 8-2.

in which q is the function of time given by Eq. 8-34. Since the displacement u is zero at $x = 0$, and the stress and strain are zero at $x = l$, the boundary conditions are

$$(1) \quad u(0, t) = 0 \qquad (2) \quad \frac{\partial u}{\partial x}(l, t) = \epsilon = 0$$

Applying the first boundary condition in Eq. 8-36, we find that

$$A = 0$$

Applying the second boundary condition in the partial derivative of Eq. 8-36 with respect to x, with $A = 0$, we find that

$$\frac{\partial u}{\partial x}(l, t) = 0 = \left(B \frac{\omega}{c} \cos \frac{\omega l}{c}\right) q \tag{8-37}$$

from which it should be apparent that the frequency equation is

$$\cos \frac{\omega l}{c} = 0$$

which is satisfied when

$$\frac{\omega_i l}{c} = \frac{i\pi}{2} \qquad i = 1, 3, 5, \ldots$$

The undamped natural circular frequencies are thus found to be

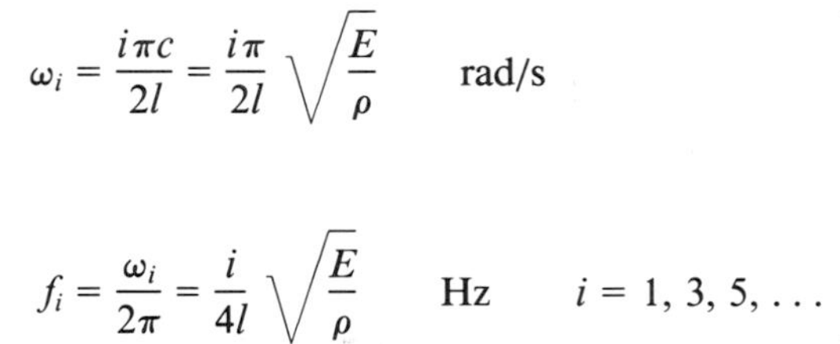

$$\omega_i = \frac{i\pi c}{2l} = \frac{i\pi}{2l}\sqrt{\frac{E}{\rho}} \qquad \text{rad/s}$$

or

$$f_i = \frac{\omega_i}{2\pi} = \frac{i}{4l}\sqrt{\frac{E}{\rho}} \qquad \text{Hz} \qquad i = 1, 3, 5, \ldots \tag{8-38}$$

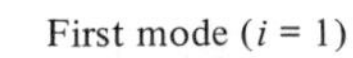

Figure 8-6 First two normal-mode shapes for rod of Example 8-2.

b. With $\omega_i/c = i\pi/2l$, $A = 0$, and $B = 1$ for convenience, the normal-mode functions are found from Eq. 8-33 as

$$\phi_i(x) = \sin \frac{i\pi x}{2l} \qquad i = 1, 3, 5, \ldots$$

and the first two mode shapes are shown in Fig. 8-6.

The reader should note that the normal-mode functions are not affected by the damping. This observation justifies the manner in which the eigenvalues and eigenvectors of *undamped* discrete and lumped-mass systems were determined in Chap. 5.

c. Recalling that

$$2\zeta_i\omega_i = b^2\lambda_i^2$$

and

$$c^2\lambda_i^2 = \omega_i^2$$

we can write that

$$\zeta_i = \left(\frac{b}{c}\right)^2 \frac{\omega_i}{2}$$

Combining the definitions of b^2 and c^2 given just below Eq. 8-27,

$$\left(\frac{b}{c}\right)^2 = \frac{\alpha}{E}$$

Using

$$\omega_i = \frac{i\pi}{2l}\sqrt{\frac{E}{\rho}}$$

from Eq. 8-38, the damping factor can be written as

$$\zeta_i = \frac{i\alpha\pi}{4l\sqrt{E\rho}} \qquad i = 1, 3, 5, \ldots \tag{8-39}$$

Equation 8-39 reveals that the higher the mode number, the larger the damping factor for the fixed-free rod of this example. This relationship is a general characteristic of viscoelastic damping and distinguishes it from viscous damping for which the damping factors decrease in magnitude as the mode numbers become higher (see Example 8-4).

8-3 TRANSCENDENTAL FREQUENCY EQUATIONS (WAVE EQUATION)

When masses and/or springs are attached at the ends of cables or rods as shown schematically in Fig. 8-7, the frequency equations obtained using the wave equation are generally transcendental equations, the roots of which must be determined by some numerical method. The analysis of such a system is illustrated in Example 8-3.

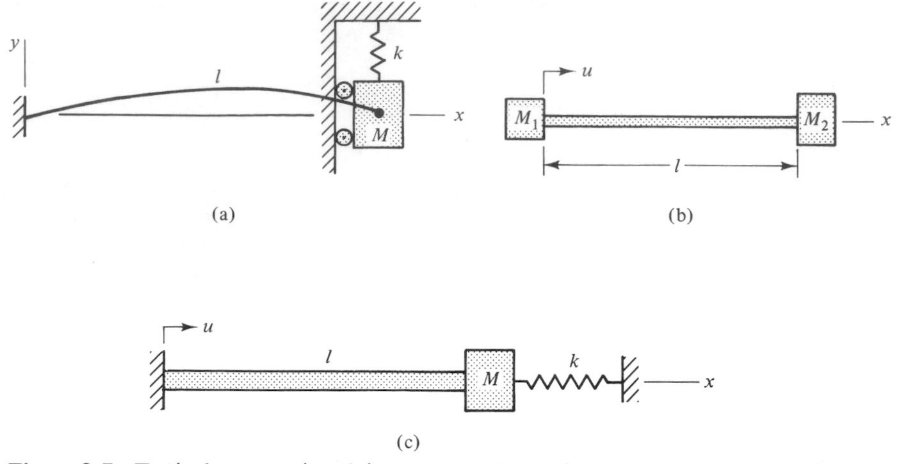

Figure 8-7 Typical systems involving transcendental frequency equations. (a) Tightly stretched cable with attached mass M and spring k. (b) Slender rod with attached masses M_1 and M_2. (c) Slender rod with attached mass M and spring k.

EXAMPLE 8-3

Let us consider a mass M that is attached at the end of a slender rod of mass m_r as shown in Fig. 8-8. The mass ratio is $m_r/M = 0.5$, $E = 30(10)^6$ psi, $\rho = 7.35(10)^{-4}$ lb·s²/in.⁴, and $l = 100$ in. We shall neglect damping in the system.

We wish to determine (a) the frequency equation; (b) the natural frequency of the first mode.

Solution. **a.** With damping neglected ($\zeta = 0$), it can be seen from Eqs. 8-28, 8-33, and 8-34 that the solution describing the undamped free vibration of a rod is

$$u(x, t) = \left(A \cos \frac{\omega x}{c} + B \sin \frac{\omega x}{c}\right)(C \cos \omega t + D \sin \omega t) \tag{8-40}$$

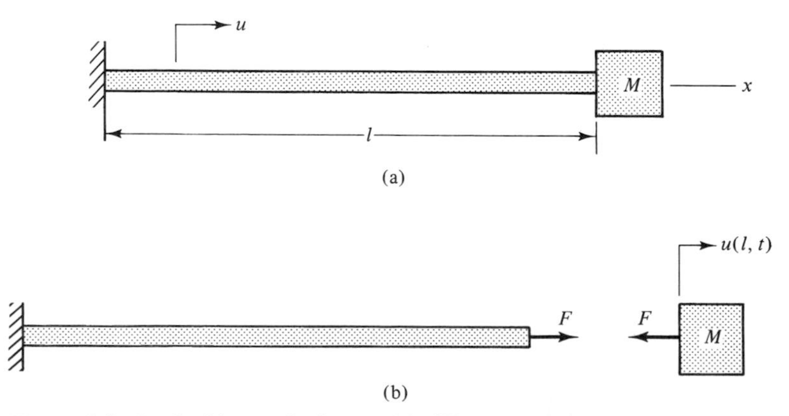

Figure 8-8 Rod with attached mass M of Example 8-3.

At $x = 0$, the boundary condition is given by

$$u(0, t) = 0$$

At $x = l$, the boundary condition is determined by considering the free-body diagram of the mass M shown in Fig. 8-8b, in which the force F between the rod and the mass M is equal to the stress σ times the cross-sectional area A_c of the rod. Since $\sigma = E\epsilon$, we can write that

$$F = \sigma A_c = EA_c \frac{\partial u}{\partial x}(l, t)$$

Then noting that F acts on M with a sense opposite to the positive sense assumed for both u and the acceleration $\partial^2 u/\partial t^2$, the application of Newton's second law yields

$$-EA_c \frac{\partial u}{\partial x}(l, t) = M \frac{\partial^2 u}{\partial t^2}(l, t)$$

Taking the second partial of Eq. 8-40 with respect to time with $x = l$, we find the second boundary condition as

$$-EA_c \frac{\partial u}{\partial x}(l, t) = -Mu(l, t)\omega^2$$

Using the first boundary condition with Eq. 8-40, we find that

$$A = 0$$

Equation 8-40 then reduces to

$$u(x, t) = \left(B \sin \frac{\omega x}{c}\right) q(t)$$

from which

$$\frac{\partial u}{\partial x}(x, t) = \left(B \frac{\omega}{c} \cos \frac{\omega x}{c}\right) q(t)$$

With $x = l$, the second boundary condition above gives

$$EA_c \frac{\omega}{c} \cos \frac{\omega l}{c} = M\omega^2 \sin \frac{\omega l}{c}$$

Next multiplying both sides of the last equation above by l/c, and recalling that $c^2 = E/\rho$ and $m_r = A_c l \rho$, we obtain the frequency equation as

$$\frac{\omega l}{c} \tan \frac{\omega l}{c} = \frac{m_r}{M}$$

or

$$\beta_i \tan \beta_i = \frac{m_r}{M} \qquad (8\text{-}41)$$

in which $\beta_i = \omega_i l/c$ are the roots of this transcendental equation for a specified value of m_r/M.

b. Since $\tan \beta_i$ goes to infinity at $\pi/2$, it should be apparent that β_i for the first mode with $m_r/M = 0.5$ lies between zero and $\pi/2$. In determining the roots, it is convenient to write Eq. 8-41 as

$$f(\beta_i) = \beta_i \tan \beta_i - 0.5 = 0 \tag{8-42}$$

for which $f(\beta_i)$ changes sign as a root is passed. Substituting values of β_i into Eq. 8-42, and looking for a sign change in $f(\beta_i)$ to indicate that a root has been passed, a root value can be determined to the desired accuracy quite quickly.

Following this procedure, the smallest root corresponding to the lowest natural circular frequency ω_1 is found to be

$$\beta_1 = \frac{\omega_1 l}{c} = 0.6533$$

in which $c = \sqrt{E/\rho}$. If we then substitute the parameter values given at the beginning of the example into the above equation, we find that the fundamental frequency of the rod-and-mass system is

$$\omega_1 = \frac{0.6533}{100}\sqrt{\frac{30(10)^{10}}{7.35}} = 1319.87 \text{ rad/s}$$

$$f_1 = \frac{\omega_1}{2\pi} = 210.06 \text{ Hz}$$

8-4 SURGE IN HELICAL SPRINGS

Another important application of the wave equation is its use in the analysis of the longitudinal motion of helical springs. If one end of a helical spring is fixed and the other end is suddenly compressed and then held fixed, a compression wave is formed that travels along the spring. This compression wave is reflected from the fixed ends of the spring and continues with this motion until it is damped out. This phenomenon is referred to as *surge,* and the natural frequencies of this wave are of interest in spring design since it is undesirable for the frequency of a disturbance to coincide with any of the spring's natural frequencies.

The wave equation for the longitudinal vibration of rods is easily modified to make it applicable to the analysis of helical springs. To illustrate this discussion, let us consider the helical spring and valve system shown in Fig. 8-9. The spring is always in compression so that the valve remains in contact with the rotating cam. Assuming that the motion of the valve is sinusoidal, the motion of the lower end of the spring is given by

$$u = u_0 \sin \omega t$$

The upper end of the spring, which has a length l as shown, is considered fixed. Measuring x from the upper end of the spring, the boundary (end) conditions of the spring are

$$\left.\begin{aligned} &(1) \quad u(0, t) = 0 \\ \text{and} \quad &(2) \quad u(l, t) = u_0 \sin \omega t \end{aligned}\right\} \tag{8-43}$$

in which the latter varies with time.

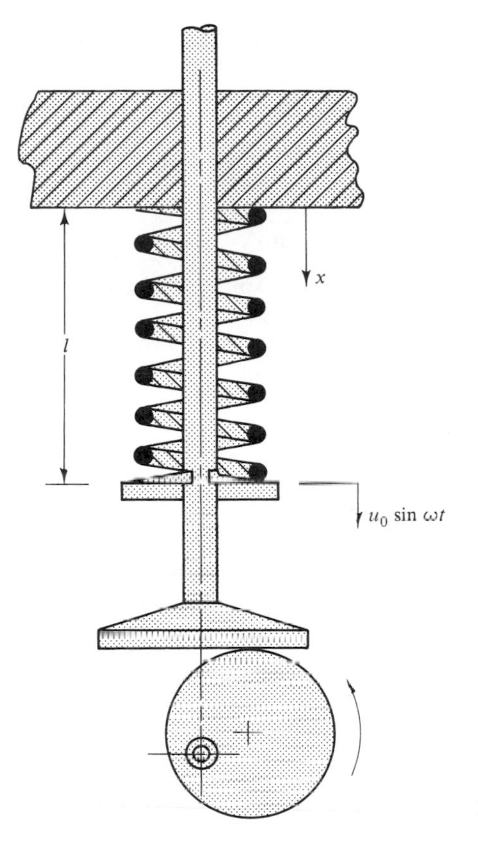

Figure 8-9 Helical spring and valve system.

Referring to Eq. 8-27, the wave equation for the longitudinal vibration of a theoretically undamped rod is

$$c^2 \frac{\partial^2 u}{\partial x^2} = \frac{\partial^2 u}{\partial t^2} \tag{8-44}$$

where $c^2 = E/\rho$
E = modulus of elasticity
ρ = mass of rod per unit volume

Equation 8-44 is also applicable to the longitudinal motion of a helical spring if $c^2 = E/\rho$ is replaced by

$$c^2 = \frac{kl^2}{m_s} \tag{8-45}$$

where k = spring constant of helical spring
l = length of active spring
m_s = total mass of spring

To justify the use of Eq. 8-45, we note that the spring constant k of a slender rod of length l and cross-sectional area A_c is given by

$$k = \frac{A_c E}{l} \tag{8-46}$$

from which

$$E = \frac{kl}{A_c}$$

Dividing the latter equation by ρ (the mass density of the rod material) gives

$$\frac{E}{\rho} = \frac{kl}{A_c\rho} = \frac{kl^2}{A_c\rho l} \tag{8-47}$$

Considering $A_c\rho l$ as the total mass m_s of the helical spring, and comparing Eqs. 8-47 and 8-45, it can be seen that E/ρ is analogous to c^2 in Eq. 8-45.

Since the motion of one end of the spring is sinusoidal with a circular frequency ω, we know that the solution of Eq. 8-44 has the form of

$$u(x, t) = \left(A \cos \frac{\omega x}{c} + B \sin \frac{\omega x}{c}\right) \sin \omega t \tag{8-48}$$

in which $c = l\sqrt{k/m_s}$. Using the boundary conditions given by Eq. 8-43 to determine the constants in Eq. 8-48, we find that

$$A = 0$$

$$B = \frac{u_0}{\sin(\omega l/c)}$$

The motion of the spring shown in Fig. 8-9 is thus given by

$$u(x, t) = u_0 \frac{\sin(\omega x/c)}{\sin(\omega l/c)} \sin \omega t \tag{8-49}$$

in which $c = l\sqrt{k/m_s}$. Equation 8-49 indicates that the displacement $u(x, t)$ goes to infinity for values of x other than $x = 0$ or $x = l$ when

$$\omega_i = \frac{i\pi c}{l} = i\pi \sqrt{\frac{k}{m_s}} \qquad \text{rad/s}$$

or

$$f_i = \frac{\omega_i}{2\pi} = \frac{i}{2} \sqrt{\frac{kg}{W_s}} \qquad \text{Hz} \qquad i = 1, 2, 3, \ldots \tag{8-50}$$

in which W_s is the total weight of the spring.

This reveals that a condition of resonance occurs for a helical spring, fixed at one end and driven at the other, when the driving frequency is given by Eq. 8-50. It is of interest to note that Eq. 8-50 is identical to the equation yielding the natural frequencies of a helical spring fixed at both ends (see Prob. 8-5).[2]

8-5 BEAM EQUATIONS

Partial differential equations that describe the vibration characteristics of beams appear in different forms, depending upon the damping model considered and whether or not rotary-

[2] J. C. Wolford and G. M. Smith, "Surge of Helical Springs," *Mechanical Engineering News,* vol. 13, no. 1, pp. 4–9, ASEE, February 1976.

inertia and/or shear effects are included. In general, the transverse vibration of a beam is described by

$$y(x, t) = \phi(x)q(t) \tag{8-51}$$

in which $\phi(x)$ is a function of the spatial coordinate x and $q(t)$ is a function of time t.

Euler's Beam Equation

One of the simpler equations describing the transverse vibration of beams is Euler's beam equation. It is easily derived by considering an elemental length dx of a beam and the internal shear forces and moments acting on it. An element a distance x from the left end of the beam is shown in Fig. 8-10a. The senses of the forces and moments shown acting on the free body of the element in Fig. 8-10b are positive in accordance with elementary beam theory. The same beam theory provides us with the following relations for beams of uniform cross section:

$$y = \text{deflection} \tag{8-52a}$$

$$\frac{\partial y}{\partial x} = \text{slope} \tag{8-52b}$$

$$EI\frac{\partial^2 y}{\partial x^2} = M \quad \text{(moment)} \tag{8-52c}$$

$$EI\frac{\partial^3 y}{\partial x^3} = \frac{\partial M}{\partial x} = V \quad \text{(shear)} \tag{8-52d}$$

$$EI\frac{\partial^4 y}{\partial x^4} = \frac{\partial V}{\partial x} = w(x) \quad \text{(load intensity)} \tag{8-52e}$$

Designating the mass of the beam per unit length as γ, and noting that the acceleration of the differential beam element is $\partial^2 y/\partial t^2$, Newton's second law states that

$$\sum F_y = \gamma \frac{\partial^2 y}{\partial t^2}\,dx$$

which when applied to the free body of the element in Fig. 8-10b yields

$$V - V - \frac{\partial V}{\partial x}\,dx = \gamma \frac{\partial^2 y}{\partial t^2}\,dx$$

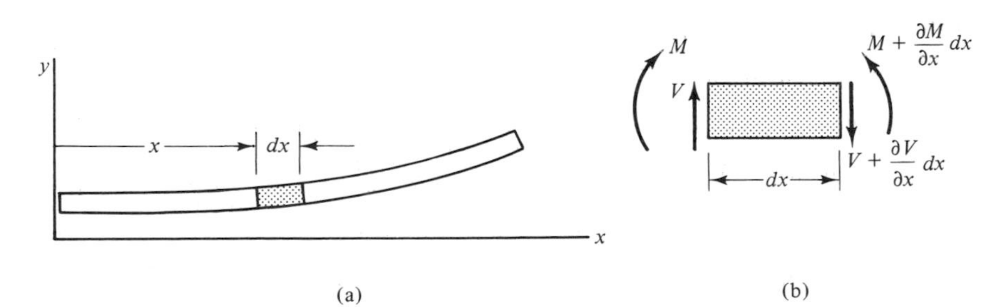

Figure 8-10 Uniform beam.

from which

$$\frac{\partial V}{\partial x} = -\gamma \frac{\partial^2 y}{\partial t^2} \tag{8-53}$$

Combining Eqs. 8-52e and 8-53, we obtain

$$\frac{EI}{\gamma}\frac{\partial^4 y}{\partial x^4} + \frac{\partial^2 y}{\partial t^2} = 0$$

or

$$a^2 \frac{\partial^4 y}{\partial x^4} + \frac{\partial^2 y}{\partial t^2} = 0 \tag{8-54}$$

in which $a^2 = EI/\gamma$. Equation 8-54 is Euler's beam equation.

To obtain a solution of this equation by the separation-of-variables method, we assume a solution having the form shown in Eq. 8-51 that leads to the two equations

$$\frac{\partial^4 y}{\partial x^4} = \frac{d^4 \phi}{dx^4} q$$

and

$$\frac{\partial^2 y}{\partial t^2} = \phi \frac{d^2 q}{dt^2} = \phi \ddot{q}$$

Substituting the above into Eq. 8-54 yields

$$a^2 \frac{d^4 \phi}{dx^4} q = -\phi \ddot{q}$$

Dividing both sides of this equation by ϕ and q, we obtain

$$a^2 \frac{d^4 \phi}{dx^4} \frac{1}{\phi} = -\frac{\ddot{q}}{q} \tag{8-55}$$

Since the left side of Eq. 8-55 is a function of x only and the right side is a function of t only, both sides must be equal to a constant. Looking at the right side of Eq. 8-55 and thinking in terms of sinusoidal functions and their derivatives, ω^2 would seem a logical choice for this constant. Equating both sides of Eq. 8-55 to ω^2, we obtain the following ordinary differential equations:

$$\ddot{q} + \omega^2 q = 0 \tag{8-56}$$

and

$$\frac{d^4 \phi}{dx^4} - \left(\frac{\omega}{a}\right)^2 \phi = 0$$

The second of the above equations can also be written as

$$\frac{d^4 \phi}{dx^4} - k^4 \phi = 0 \tag{8-57}$$

in which $k^4 = \omega^2/a^2$.

By now the solution of Eq. 8-56 should be recognized as being

$$q = A \cos \omega t + B \sin \omega t \tag{8-58}$$

To obtain a solution of Eq. 8-57, we assume a solution of the form of

$$\phi = e^{rx} \tag{8-59}$$

and determine expressions for r that will satisfy Eq. 8-57. To accomplish this, we first take the necessary derivatives to find that

$$\frac{d^4\phi}{dx^4} = r^4 e^{rx} \tag{8-60}$$

and then substitute this equation and Eq. 8-59 into Eq. 8-57 to obtain

$$r^4 - k^4 = 0$$

or

$$(r^2 - k^2)(r^2 + k^2) = 0 \tag{8-61}$$

The four roots of Eq. 8-61 for which Eq. 8-59 satisfies Eq. 8-57 are

$$r_{1,2} = \pm k$$

and

$$r_{3,4} = \pm jk \qquad (j = \sqrt{-1})$$

The solution of Eq. 8-57 can then be written as the sum of the solutions, giving

$$\phi = Ce^{kx} + De^{-kx} + Ee^{jkx} + Fe^{-jkx} \tag{8-62}$$

in which $k = \sqrt{\omega/a}$.

Assuming E and F as complex constants having the forms of

$$E = C_3 - jC_4$$
$$F = C_3 + jC_4$$

and using the identities

$$e^{kx} = \cosh kx + \sinh kx$$
$$e^{-kx} = \cosh kx - \sinh kx$$
$$e^{jkx} = \cos kx + j \sin kx$$
$$e^{-jkx} = \cos kx - j \sin kx$$

Eq. 8-62 can be transformed into

$$\phi = C_1 \cosh kx + C_2 \sinh kx + C_3 \cos kx + C_4 \sin kx \tag{8-63}$$

in which the constants C_1, C_2, C_3, and C_4 are real constants and related to the constants C, D, E, and F of Eq. 8-62 by

$$C_1 = C + D \qquad C_3 = E + F$$
$$C_2 = C - D \qquad C_4 = j(E - F)$$

We can thus write the solution of Euler's beam equation as

$$y(x, t) = (C_1 \cosh kx + C_2 \sinh kx + C_3 \cos kx + C_4 \sin kx)q(t) \tag{8-64}$$

in which $q(t)$ is the function of time given by Eq. 8-58.

Recalling that

$$\omega = ak^2 = k^2\sqrt{\frac{EI}{\gamma}} \tag{8-65}$$

it can be seen that the natural circular frequency ω depends upon the value of k. Since k appears in the shape function ϕ, the values of k for a particular beam obviously depend upon the support constraints at the ends of the beam. Since there are four constants in the parenthetical portion of Eq. 8-64, four boundary conditions are necessary to obtain the natural frequencies and normal-mode functions for a beam. Let us now illustrate what has been discussed in the preceding paragraphs by determining the natural frequencies and normal-mode shapes for both a pinned-pinned beam and a fixed-pinned beam.

Pinned-Pinned Beam

Looking first at the pinned-pinned beam shown in Fig. 8-11, it can be seen that the deflection y and moment M are zero at the ends of the beam for all values of time. The four boundary conditions are thus

$$(1)\quad y(0, t) = 0 \qquad\qquad (3)\quad y(l, t) = 0$$

$$(2)\quad \frac{\partial^2 y}{\partial x^2}(0, t) = \frac{M}{EI} = 0 \qquad (4)\quad \frac{\partial^2 y}{\partial x^2}(l, t) = \frac{M}{EI} = 0$$

Using the first and second boundary conditions with Eq. 8-64 and its second partial derivative with respect to x yields

$$C_1 + C_3 = 0$$
$$C_1 - C_3 = 0$$

Solving these two equations simultaneously, we find that

$$C_1 = C_3 = 0$$

Using the third and fourth boundary conditions in similar manner, we find that

$$C_2 \sinh kl + C_4 \sin kl = 0$$
$$C_2 \sinh kl - C_4 \sin kl = 0$$

Since $kl = 0$ represents a trivial solution and $\sinh kl = 0$ only at $kl = 0$, we see that $C_2 = 0$ and

$$\sin kl = 0$$

Thus,

$$k_i l = i\pi \qquad i = 1, 2, 3, \ldots$$

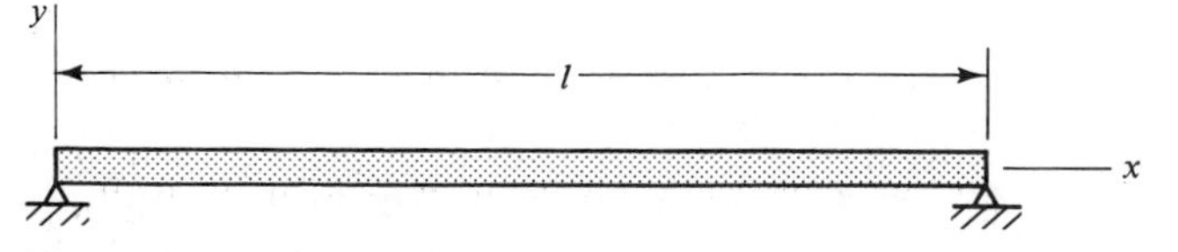

Figure 8-11 Pinned-pinned beam.

and the natural frequencies are determined from Eq. 8-65 as

$$\left.\begin{aligned} \omega_i &= a k_i^2 = \frac{i^2\pi^2}{l^2}\sqrt{\frac{EI}{\gamma}} \qquad \text{rad/s} \\ &\text{or} \\ f_i &= \frac{\omega_i}{2\pi} = \frac{i^2\pi}{2l^2}\sqrt{\frac{EI}{\gamma}} \qquad \text{Hz} \qquad i = 1, 2, 3, \ldots \end{aligned}\right\} \tag{8-66}$$

Since $C_1 = C_2 = C_3 = 0$, the natural-mode-shape functions determined from Eq. 8-63 are

$$\phi_i(x) = C_4 \sin \frac{i\pi x}{l} \qquad i = 1, 2, 3, \ldots \tag{8-67}$$

The configurations for the first three modes of the pinned-pinned beam are shown in Fig. 8-12.

As we shall see later in analyzing the forced vibration of beams, it is convenient to normalize the mode-shape functions so that

$$\int_0^l \phi_i^2(x)\, dx = l \tag{8-68}$$

Normalizing the mode-shape functions of Eq. 8-67 in this fashion, we see that

$$C_4^2 \int_0^l \sin^2 \frac{i\pi x}{l}\, dx = l$$

from which $C_4 = \sqrt{2}$. The normal-mode functions then become

$$\phi_i(x) = \sqrt{2} \sin \frac{i\pi x}{l} \qquad i = 1, 2, 3, \ldots \tag{8-69}$$

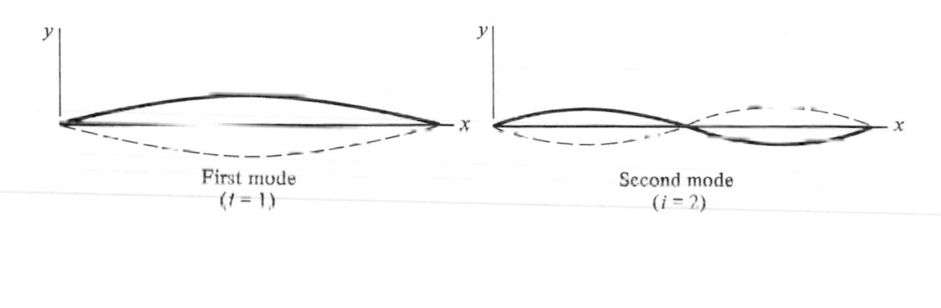

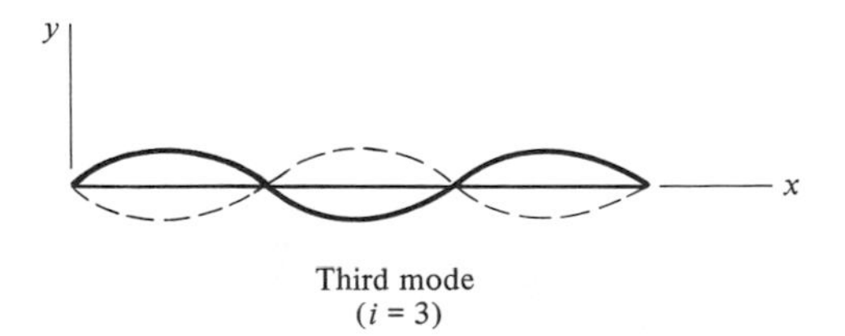

Figure 8-12 First three mode configurations of a pinned-pinned beam.

Fixed-Pinned Beam

We now look at the fixed-pinned beam shown in Fig. 8-13 to see that it has the boundary conditions:

$$(1)\quad y(0, t) = 0 \qquad (3)\quad y(l, t) = 0$$

$$(2)\quad \frac{\partial y}{\partial x}(0, t) = 0 \qquad (4)\quad \frac{\partial^2 y}{\partial x^2}(l, t) = \frac{M}{EI} = 0$$

Using the first and second of these conditions with Eq. 8-64, we obtain

$$C_1 + C_3 = 0$$
$$C_2 + C_4 = 0$$

from which

$$C_3 = -C_1$$
$$C_4 = -C_2$$

In similar manner, the use of the third and fourth boundary conditions with Eq. 8-64 yields

$$\left.\begin{aligned} C_1(\cosh kl - \cos kl) + C_2(\sinh kl - \sin kl) = 0 \\ C_1(\cosh kl + \cos kl) + C_2(\sinh kl + \sin kl) = 0 \end{aligned}\right\} \tag{8-70}$$

Since these algebraic equations are homogeneous, the frequency equation is obtained by setting the determinant $|D|$ of the coefficients of C_1 and C_2 equal to zero. Upon expanding this determinant and simplifying, we obtain the transcendental frequency equation

$$\frac{\sin kl}{\cos kl} = \frac{\sinh kl}{\cosh kl}$$

or

$$\tan kl = \tanh kl \tag{8-71}$$

One method of obtaining the roots of Eq. 8-71 is to sketch a graph of tan kl and tanh kl over a reasonable range of kl values as shown in Fig. 8-14 to obtain approximate root values, and then trial and error around these approximate values with a calculator to determine more accurate values. Using this procedure, the first three roots are found as

$$k_1 l = 3.926602 \qquad k_2 l = 7.068583 \qquad k_3 l = 10.210176$$

We then find from Eq. 8-65 that

$$\omega_1 = \frac{15.4182}{l^2}\sqrt{\frac{EI}{\gamma}} \text{ rad/s} \qquad f_1 = \frac{\omega_1}{2\pi} = \frac{2.4539}{l^2}\sqrt{\frac{EI}{\gamma}} \text{ Hz}$$

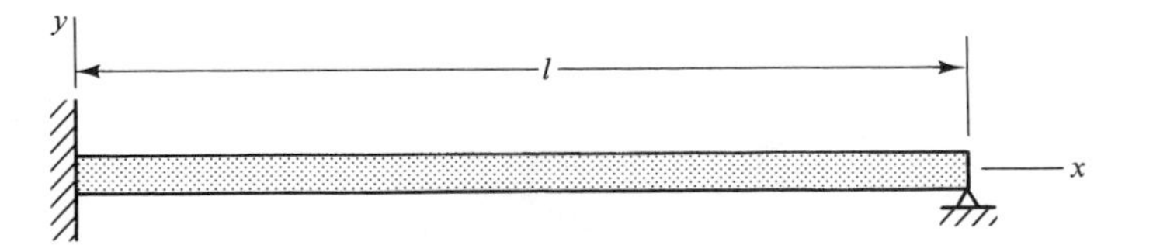

Figure 8-13 Fixed-pinned beam.

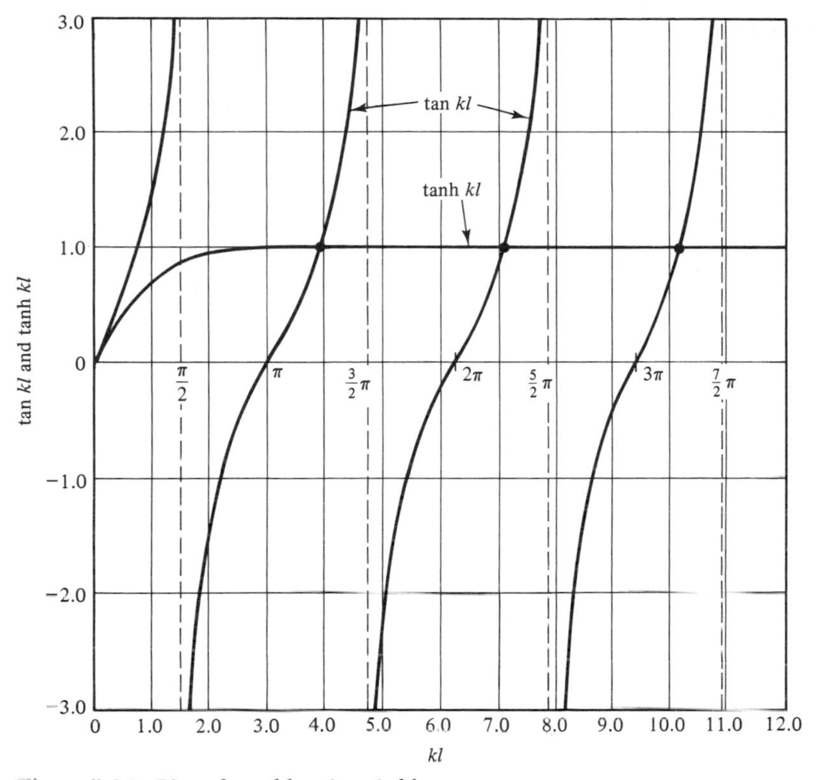

Figure 8-14 Plot of tan kl and tanh kl.

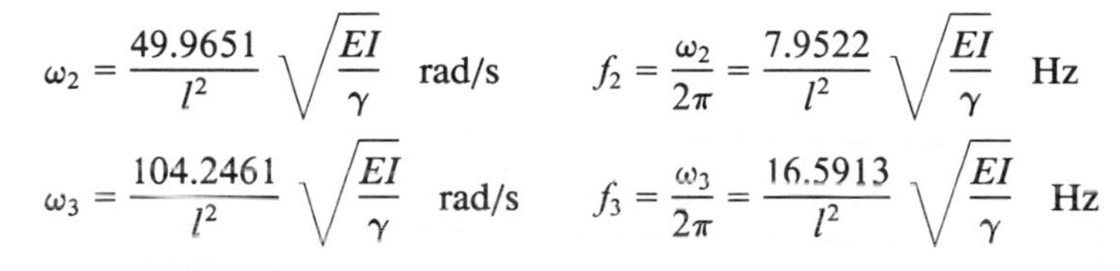

$$\omega_2 = \frac{49.9651}{l^2}\sqrt{\frac{EI}{\gamma}} \text{ rad/s} \qquad f_2 = \frac{\omega_2}{2\pi} = \frac{7.9522}{l^2}\sqrt{\frac{EI}{\gamma}} \text{ Hz}$$

$$\omega_3 = \frac{104.2461}{l^2}\sqrt{\frac{EI}{\gamma}} \text{ rad/s} \qquad f_3 = \frac{\omega_3}{2\pi} = \frac{16.5913}{l^2}\sqrt{\frac{EI}{\gamma}} \text{ Hz}$$

In determining the normal-mode functions, since there are no unique values for C_1

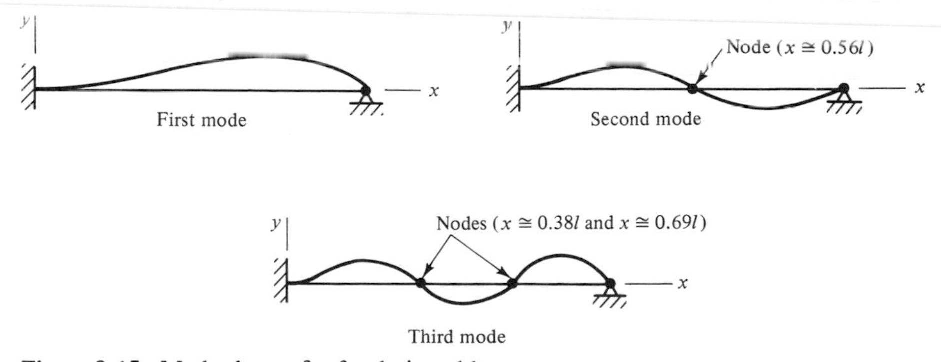

Figure 8-15 Mode shapes for fixed-pinned beam.

TABLE 8-1

For all beams below: $\omega_i = ak_i^2 = k_i^2\sqrt{\dfrac{EI}{\gamma}}$ $\quad \displaystyle\int_0^l \phi_i^2(x)\,dx = l$

Pinned-Pinned

$y(0, t) = y(l, t) = 0$

$\dfrac{\partial^2 y}{\partial x^2}(0, t) = \dfrac{\partial^2 y}{\partial x^2}(l, t) = 0$

$\phi_i(x) = \sqrt{2}\sin\dfrac{i\pi x}{l}$

$k_i l = i\pi \qquad i = 1, 2, 3, \ldots$

Fixed-Pinned $\quad \phi_i(x) = \cosh k_i x - \cos k_i x - \alpha_i(\sinh k_i x - \sin k_i x)$

$\alpha_i = \dfrac{\cosh k_i l - \cos k_i l}{\sinh k_i l - \sin k_i l}$

$y(0, t) = \dfrac{\partial y}{\partial x}(0, t) = 0$

$y(l, t) = \dfrac{\partial^2 y}{\partial x^2}(l, t) = 0$

$k_1 l = 3.926602$	$k_2 l = 7.068583$	$k_3 l = 10.210176$
$\alpha_1 = 1.000777$	$\alpha_2 = 1.000001$	$\alpha_3 = 1.000000$

Fixed-Free $\quad \phi_i(x) = \cosh k_i x - \cos k_i x - \alpha_i(\sinh k_i x - \sin k_i x)$

$\alpha_i = \dfrac{\cosh k_i l + \cos k_i l}{\sinh k_i l + \sin k_i l}$

$y(0, t) = \dfrac{\partial y}{\partial x}(0, t) = 0$

$\dfrac{\partial^2 y}{\partial x^2}(l, t) = \dfrac{\partial^3 y}{\partial x^3}(l, t) = 0$

$k_1 l = 1.875104$	$k_2 l = 4.694091$	$k_3 l = 7.854757$
$\alpha_1 = 0.734096$	$\alpha_2 = 1.018466$	$\alpha_3 = 0.999225$

Fixed-Fixed $\quad \phi_i(x) = \cosh k_i x - \cos k_i x - \alpha_i(\sinh k_i x - \sin k_i x)$

$\alpha_i = \dfrac{\cosh k_i l - \cos k_i l}{\sinh k_i l - \sin k_i l}$

$y(0, t) = \dfrac{\partial y}{\partial x}(0, t) = 0$

$y(l, t) = \dfrac{\partial y}{\partial x}(l, t) = 0$

$k_1 l = 4.730041$	$k_2 l = 7.853205$	$k_3 l = 10.995607$
$\alpha_1 = 0.982502$	$\alpha_2 = 1.000777$	$\alpha_3 = 0.999966$

Free-Free $\quad \phi_i(x) = \cosh k_i x + \cos k_i x - \alpha_i(\sinh k_i x + \sin k_i x)$

$\alpha_i = \dfrac{\cosh k_i l - \cos k_i l}{\sinh k_i l - \sin k_i l}$

$\dfrac{\partial^2 y}{\partial x^2}(0, t) = \dfrac{\partial^3 y}{\partial x^3}(0, t) = 0$

$\dfrac{\partial^2 y}{\partial x^2}(l, t) = \dfrac{\partial^3 y}{\partial x^3}(l, t) = 0$

$k_1 l = 4.730041$	$k_2 l = 7.853205$	$k_3 l = 10.995607$
$\alpha_1 = 0.982502$	$\alpha_2 = 1.000777$	$\alpha_3 = 0.999966$

and C_2, one of them must be expressed in terms of the other. This can be done using either of the expressions shown in Eq. 8-70. Using the first expression with $C_1 = 1$, we obtain

$$C_2 = \frac{-(\cosh kl - \cos kl)}{\sinh kl - \sin kl}$$

Recalling that $C_3 = -C_1$ and that $C_4 = -C_2$ and designating that

$$\alpha_i = -C_2 = \frac{\cosh k_i l - \cos k_i l}{\sinh k_i l - \sin k_i l} \tag{8-72}$$

for the roots $k_i l$, Eq. 8-63 can be written as

$$\phi_i(x) = \cosh k_i x - \cos k_i x - \alpha_i(\sinh k_i x - \sin k_i x)$$

The normal-mode functions can be obtained from this equation by substituting the $k_i l$ values determined earlier as the first three roots of Eq. 8-71 into Eq. 8-72 to obtain

$$\alpha_1 = 1.000777$$
$$\alpha_2 = 1.000001$$
$$\alpha_3 = 1.000000$$

The configurations for the first three modes of the fixed-pinned beam are shown in Fig. 8-15.

The natural frequencies and normal-mode functions for beams with other types of end conditions can be determined in the manner just discussed. Table 8-1 gives the end conditions, the normal-mode functions, and the first three roots of the frequency equations, the α_i equations, and the first three α_i values for five beams with the end conditions shown. At the top center of the table is an equation, applicable to all of the beams shown, for determining the natural circular frequencies ω_i of the beams. All the normal-mode functions have been normalized to satisfy the integral expression at the top right of the table. The information shown in this table has been tabulated by Young and Felgar.[3]

8-6 PROPERTIES OF NORMAL-MODE FUNCTIONS

Let us now determine the *orthogonality properties* of the normal-mode functions shown in Table 8-1. Since the differential equation shown by Eq. 8-57 must hold for each normal mode, we can write for modes i and j that

$$\frac{d^4\phi_i}{dx^4} = \lambda_i\phi_i \tag{8-73}$$

and

$$\frac{d^4\phi_j}{dx^4} = \lambda_j\phi_j \tag{8-74}$$

where $\lambda_i = k_i^4 = \omega_i^2/a^2$
$\lambda_j = k_j^4 = \omega_j^2/a^2$
$a^2 = EI/\gamma$

[3] D. Young and R. P. Felgar, Jr., *Tables of Characteristic Functions Representing Normal Modes of Vibration of a Beam,* The University of Texas Publication, no. 4913, July 1, 1949.

If we now multiply both sides of Eq. 8-73 by ϕ_j and both sides of Eq. 8-74 by ϕ_i, and put the resulting products in integral form, we obtain

$$\int_0^l \left(\frac{d^4\phi_i}{dx^4}\right)\phi_j\,dx = \lambda_i \int_0^l \phi_i\phi_j\,dx \tag{8-75}$$

and

$$\int_0^l \left(\frac{d^4\phi_j}{dx^4}\right)\phi_i\,dx = \lambda_j \int_0^l \phi_i\phi_j\,dx \tag{8-76}$$

with the limits shown.

Integrating the left-hand side of each of these equations by parts, and then integrating by parts the integrals obtained in those integrations yields

$$\left[\frac{d^3\phi_i}{dx^3}\phi_j\right]_0^l - \left[\frac{d^2\phi_i}{dx^2}\frac{d\phi_j}{dx}\right]_0^l + \int_0^l \left(\frac{d^2\phi_i}{dx^2}\right)\left(\frac{d^2\phi_j}{dx^2}\right)dx = \lambda_i \int_0^l \phi_i\phi_j\,dx \tag{8-77}$$

and

$$\left[\frac{d^3\phi_j}{dx^3}\phi_i\right]_0^l - \left[\frac{d^2\phi_j}{dx^2}\frac{d\phi_i}{dx}\right]_0^l + \int_0^l \left(\frac{d^2\phi_i}{dx^2}\right)\left(\frac{d^2\phi_j}{dx^2}\right)dx = \lambda_j \int_0^l \phi_i\phi_j\,dx \tag{8-78}$$

It is easy to show that the first two terms in each of the above equations always vanish at $x = 0$ and $x = l$ when the boundary conditions used to obtain the normal-mode functions for the different beams in Table 8-1 are applied. For example, the first two terms vanish for a fixed-free beam since the boundary conditions at $x = 0$ and $x = l$ require that

$$\phi_i(0) = \phi_j(0) = \frac{d\phi_i}{dx}(0) = \frac{d\phi_j}{dx}(0) = 0$$

and

$$\frac{d^2\phi_i}{dx^2}(l) = \frac{d^2\phi_j}{dx^2}(l) = \frac{d^3\phi_i}{dx^3}(l) = \frac{d^3\phi_j}{dx^3}(l) = 0$$

With the first two terms always equal to zero, subtracting Eq. 8-78 from Eq. 8-77 yields

$$(\lambda_i - \lambda_j)\int_0^l \phi_i\phi_j\,dx = 0 \tag{8-79}$$

Since $\lambda_i \neq \lambda_j$, it is apparent from Eq. 8-79 that

$$\int_0^l \phi_i\phi_j\,dx = 0 \qquad (i \neq j) \tag{8-80}$$

Substituting Eq. 8-80 into either of Eq. 8-77 or 8-78 with the first two terms equal to zero, we see that

$$\int_0^l \left(\frac{d^2\phi_i}{dx^2}\right)\left(\frac{d^2\phi_j}{dx^2}\right)dx = 0 \qquad (i \neq j) \tag{8-81}$$

Substituting Eq. 8-80 into Eq. 8-75, we also see that

$$\int_0^l \left(\frac{d^4\phi_i}{dx^4}\right)\phi_j\,dx = 0 \qquad (i \neq j) \tag{8-82}$$

Equations 8-80, 8-81, and 8-82 are the *orthogonality relationships* for the normal-mode functions of Euler's beam equation.

When $i = j$, the normal-mode functions shown in Table 8-1 have been normalized, so that

$$\int_0^l \phi_i^2\, dx = l \tag{8-83}$$

Then substituting Eq. 8-83 into Eq. 8-77, and recalling that $\lambda_i = \omega_i^2/a^2$, we find that

$$\int_0^l \left(\frac{d^2\phi_i}{dx^2}\right)^2 dx = \lambda_i l = \frac{\omega_i^2 l}{a^2} \tag{8-84}$$

As we shall see later, the use of Eqs. 8-83 and 8-84 facilitates the time-response analysis of beams when using the superposition of the normal modes.

It can also be shown that[4]

$$\int_0^l \phi_i^2\, dx = \frac{l}{4}\{[\phi_i(l)]^2 - 2\phi_i'(l)\phi_i'''(l) + [\phi_i''(l)]^2\} \tag{8-85}$$

where $\phi_i'(l) = \left.\dfrac{1}{k_i}\dfrac{d\phi_i}{dx}\right|_{x=l}$

$$\phi_i''(l) = \left.\frac{1}{k_i^2}\frac{d^2\phi_i}{dx^2}\right|_{x=l}$$

$$\phi_i'''(l) = \left.\frac{1}{k_i^3}\frac{d^3\phi_i}{dx^3}\right|_{x=l}$$

To show that the right-hand side of Eq. 8-85 is equal to l for all the normal-mode functions shown in Table 8-1, let us just arbitrarily consider the *second* mode of a fixed-free beam for which

$$k_2 l = 4.694091 \qquad \alpha_2 = 1.018466$$

as shown in the table. Since the *moment* and *shear* are zero at $x = l$ for the fixed-free beam, $\phi_i''(l)$ and $\phi_i'''(l)$ are zero for all modes and Eq. 8-85 becomes

$$\int_0^l \phi_i^2\, dx = \frac{l}{4}[\phi_i(l)]^2$$

Substituting the values shown above for $k_2 l$ and α_2 into the expression for $\phi_i(x)$ shown in Table 8-1 for the fixed-free beam yields

$$\phi_2(l) = -2.000$$

so that

$$\frac{l}{4}[\phi_2(l)]^2 = \frac{l}{4}(-2.000)^2 = l$$

This verifies that

$$\int_0^l \phi_2^2\, dx = l$$

[4] S. Timoshenko and D. H. Young, *Vibration Problems in Engineering,* 3d ed., p. 328, D. Van Nostrand Company, New York, 1955.

Using a similar procedure for other modes and other beams would lead each time to the conclusion that the right-hand side of Eq. 8-85 is equal to l for all the normal-mode functions shown in Table 8-1.

8-7 ANALYZING THE VIBRATION OF CONTINUOUS SYSTEMS BY THE MODAL-SUPERPOSITION METHOD

Both the free and forced vibration of continuous systems characterized by either the wave equation or Euler's beam equation can be analyzed by means of the superposition of the normal modes. The general procedure consists of assuming a solution of the form of

$$y = \sum_{i=1,2,\ldots}^{\infty} \phi_i q_i \tag{8-86}$$

in which q_i is a generalized coordinate that is a function of time and ϕ_i is a normal-mode function that is a function of the spatial coordinate x. The normal-mode function for a particular problem is determined from the pertinent boundary conditions, while the generalized coordinate is obtained from the solution of a second-order differential equation derived from the use of Lagrange's equations (see Sec. 7-2). To illustrate the general procedure, let us first consider the vibration of several systems characterized by the wave equation, and then several more characterized by Euler's beam equation.

Vibration of Tightly Stretched Strings and Cables

Since the use of Lagrange's equations is an integral part of our procedure, they are repeated here for convenience as

$$\frac{d}{dt}\left(\frac{\partial T}{\partial \dot{q}_i}\right) - \frac{\partial T}{\partial q_i} + \frac{\partial U}{\partial q_i} = Q_i \tag{7-1}$$

To utilize these equations we must first determine expressions for the kinetic energy T and the potential energy U of a system. For continuous systems these expressions can be determined by the use of Eq. 8-86. For systems subjected to a forcing function, the generalized nonpotential force Q_j is determined by considering the differential work dW_j done by the forcing function due to a differential change dq_j of a particular coordinate q_j. When the system is damped, the determination of Q_j will include the differential work dW_j done by the energy-dissipating force due to the same differential change dq_j of a particular coordinate q_j.

The normal-mode function of a tightly stretched string or cable fixed at both ends was determined in Sec. 8-2 as

$$\phi_i = \sin\frac{i\pi x}{l} \qquad i = 1, 2, 3, \ldots \tag{8-14}$$

The strain energy of a string that is stretched until it has an initial tension of T_0 and that is then displaced laterally y from that initial position (see Eq. 2-93) is given by

$$U = \frac{T_0}{2}\int_0^l \left(\frac{\partial y}{\partial x}\right)^2 dx = \frac{T_0}{2}\int_0^l \left(\sum \phi_i' q_i\right)^2 dx \tag{8-87}$$

in which

$$\phi_i' = \frac{d\phi_i}{dx} = \frac{i\pi}{l} \cos \frac{i\pi x}{l}$$

Then using Eq. 8-87 to obtain the partial derivative of U with respect to a particular q_i such as q_j, we obtain

$$\frac{\partial U}{\partial q_j} = T_0 \int_0^l \left(\sum \phi_i' q_i\right) \phi_j' \, dx \tag{8-88}$$

in which

$$\phi_i' = \frac{i\pi}{l} \cos \frac{i\pi x}{l}$$

and

$$\phi_j' = \frac{j\pi}{l} \cos \frac{j\pi x}{l}$$

Recalling from Sec. 1-4 that ϕ_i' and ϕ_j' are orthogonal functions over the interval $0 < x < l$, we can write that

$$\int_0^l \phi_i' \phi_j' \, dx = \frac{ij\pi^2}{l^2} \int_0^l \cos \frac{i\pi x}{l} \cos \frac{j\pi x}{l} \, dx = \begin{cases} 0 & i \neq j \\ \dfrac{i^2\pi^2}{2l} & i = j \end{cases}$$

Thus, for $i = j$, we can drop the summation in Eq. 8-88 and write that

$$\frac{\partial U}{\partial q_i} = \frac{T_0 i^2 \pi^2}{2l} q_i \tag{8-89}$$

The kinetic energy of a uniform string of mass γ per unit length is given by

$$T = \frac{\gamma}{2} \int_0^l \left(\frac{\partial y}{\partial t}\right)^2 dx = \frac{\gamma}{2} \int_0^l \left(\sum \phi_i \dot{q}_i\right)^2 dx \tag{8-90}$$

from which

$$\frac{\partial T}{\partial \dot{q}_j} = \gamma \int_0^l \left(\sum \phi_i \dot{q}_i\right) \phi_j \, dx \tag{8-91}$$

Then since ϕ_i and ϕ_j are orthogonal functions over the interval $0 < x < l$ and

$$\int_0^l \phi_i \phi_j \, dx = \int_0^l \sin \frac{i\pi x}{l} \sin \frac{j\pi x}{l} \, dx = \begin{cases} 0 & i \neq j \\ \dfrac{l}{2} & i = j \end{cases}$$

we can drop the summation in Eq. 8-91 for $i = j$ and write that

$$\frac{\partial T}{\partial \dot{q}_i} = \gamma \frac{l}{2} \dot{q}_i$$

from which

$$\frac{d}{dt}\left(\frac{\partial T}{\partial \dot{q}_i}\right) = \gamma \frac{l}{2} \ddot{q}_i \tag{8-92}$$

Since T is not a function of q_i,

$$\frac{\partial T}{\partial q_i} = 0 \tag{8-93}$$

Substituting Eqs. 8-89, 8-92, and 8-93 into Lagrange's equations, we see that the undamped free vibration of a tightly stretched string is given by

$$\gamma \frac{l}{2} \ddot{q}_i + T_0 \frac{i^2\pi^2}{2l} q_i = 0$$

or

$$\ddot{q}_i + \omega_i^2 q_i = 0 \qquad i = 1, 2, 3, \ldots \tag{8-94}$$

in which

$$\omega_i = \frac{i\pi}{l}\sqrt{\frac{T_0}{\gamma}}$$

gives the undamped natural circular frequencies as shown earlier by Eq. 8-12. The solution of Eq. 8-94 is

$$q_i = C_i \cos \omega_i t + D_i \sin \omega_i t \tag{8-95}$$

Then substituting the normal-mode function given by Eq. 8-14 and the expression for q_i given by Eq. 8-95 into Eq. 8-86, we obtain the undamped free vibration of the string as

$$y = \sum_{i=1,2,\ldots}^{\infty} \sin \frac{i\pi x}{l} (C_i \cos \omega_i t + D_i \sin \omega_i t) \tag{8-96}$$

in which C_i and D_i depend upon the initial conditions. The reader should note that Eq. 8-96 is identical to Eq. 8-15, which was obtained by the separation-of-variables method.

EXAMPLE 8-4

In the preceding paragraphs we discussed the procedure for determining the free response of a tightly stretched string by the modal-superposition method. In this example we wish to determine the steady-state response of a tightly stretched cable when it is subjected to a sinusoidal force $P_0 \sin \omega t$ at its third point as shown in Fig. 8-16. It is assumed that the damping of the cable is viscous in nature (proportional to the velocity) so that the damping force per unit length is $c(\partial y/\partial t)$, in which c is the damping coefficient.

Solution. We recall from the discussion at the beginning of this section that the generalized nonpotential forces Q_j are determined from

$$dW_j = Q_j \, dq_j \tag{8-97}$$

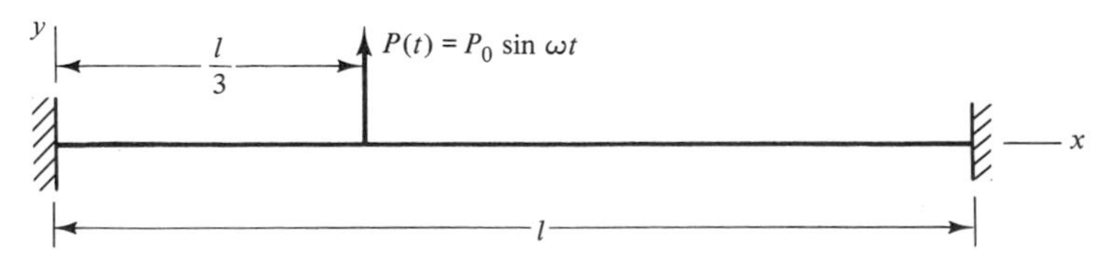

Figure 8-16 Tightly stretched cable of Example 8-4.

The differential work dW_j done by the excitation force $P(t)$ and by the damping force $c(\partial y/\partial t)$ due to a differential change dq_j of a particular coordinate q_j is given by

$$dW_j = P_0 \sin \omega t \, dy|_{x=l/3} - c \int_0^l \left[\frac{\partial y}{\partial t} dy\right] dx \tag{8-98}$$

in which the damping term is negative since it removes energy from the system. In this equation dy is the differential displacement due to a differential change dq_j of a *particular* coordinate q_j. Thus, referring to Eq. 8-86, we see that

$$dy = \phi_j \, dq_j = \sin \frac{j\pi x}{l} dq_j$$

and Eq. 8-98 becomes

$$dW_j = P_0 \sin \omega t \sin \frac{j\pi}{3} dq_j - \left[c \int_0^l (\sum \phi_i \dot{q}_i)\phi_j \, dx\right] dq_j \tag{8-99}$$

in which

$$\phi_i = \sin \frac{i\pi x}{l}$$

and

$$\phi_j = \sin \frac{j\pi x}{l}$$

Utilizing the orthogonality properties of ϕ_i and ϕ_j with $i = j$, Eq. 8-99 yields

$$dW_i = \left[P_0 \sin \omega t \sin \frac{i\pi}{3} - \frac{cl}{2} \dot{q}_i\right] dq_i$$

from which we find the generalized nonpotential forces as

$$Q_i = P_0 \sin \omega t \sin \frac{i\pi}{3} - \frac{cl}{2} \dot{q}_i \tag{8-100}$$

Upon substituting Eqs. 8-89, 8-92, 8-93, and 8-100 into Lagrange's equations and simplifying, we obtain

$$\ddot{q}_i + \frac{c}{\gamma} \dot{q}_i + \frac{i^2\pi^2 T_0}{l^2 \gamma} q_i = \frac{2P_0}{\gamma l} \sin \frac{i\pi}{3} \sin \omega t \qquad i = 1, 2, 3, \ldots$$

which can also be written as

$$\ddot{q}_i + 2\zeta_i \omega_i \dot{q}_i + \omega_i^2 q_i = F_0 \sin \frac{i\pi}{3} \sin \omega t \tag{8-101}$$

where $2\zeta_i \omega_i = \dfrac{c}{\gamma}$

$\omega_i = \dfrac{i\pi}{l} \sqrt{\dfrac{T_0}{\gamma}}$ = natural undamped frequency for the ith mode

ζ_i = viscous damping factor for the ith mode

$F_0 = \dfrac{2P_0}{\gamma l}$

Combining the first two of the latter expressions above, the viscous damping factor ζ_i can be expressed as

$$\zeta_i = \frac{c}{\gamma 2\omega_i} = \frac{cl}{2\gamma i\pi}\sqrt{\frac{\gamma}{T_0}} \qquad i = 1, 2, 3, \ldots$$

which reveals that it *decreases* as the mode number increases. This is a general characteristic of continuous systems when viscous damping is assumed. The reader might recall that in Example 8-2 the damping factor for a viscoelastic damping model was found to be proportional to the normal mode, indicating that it *increased* as the mode number increased.

To obtain the particular solution of Eq. 8-101 that yields the steady-state response of the cable, we assume a solution of the form of

$$q_i = A_i e^{j\omega t} \qquad j = \sqrt{-1}$$

and substitute it and the appropriate derivatives of it into Eq. 8-101 to obtain

$$\left.\begin{aligned} |A_i| &= \frac{F_0 \sin(i\pi/3)}{\omega_i^2\sqrt{[1-(\omega/\omega_i)^2]^2 + [2\zeta_i(\omega/\omega_i)]^2}} \\ &\text{with} \\ \tan\phi_i &= \frac{2\zeta_i(\omega/\omega_i)}{1-(\omega/\omega_i)^2} \end{aligned}\right\} \tag{8-102}$$

Thus,

$$q_i = |A_i|\sin(\omega t - \phi_i)$$

so that the steady-state response of the cable is found to be

$$y = \sum \phi_i q_i = \sum_{i=1,2,3,\ldots}^{\infty} |A_i| \sin\frac{i\pi x}{l}\sin(\omega t - \phi_i) \tag{8-103}$$

in which $|A_i|$ and ϕ_i are as given by Eq. 8-102. The reader should note that $|A_3| = 0$ since $P(t)$ acts at $x = l/3$, which is a node of the third normal mode. It should also be noted that the value of $|A_i|$ decreases fairly rapidly since ω_i^2 is proportional to i^2. Because of this, it is not usually necessary to sum a large number of terms of Eq. 8-103 to obtain the steady-state response fairly accurately. The reader might recall that this was also found to be true in determining the response of discrete and lumped-mass systems by the modal-analysis method in Chap. 6.

Longitudinal Vibration of Slender Rods

The longitudinal vibration of slender rods can be analyzed by the same procedure as described for tightly stretched strings and cables. The strain and kinetic energies of a slender rod are given, respectively, by

$$\left.\begin{aligned} U &= \frac{A_c E}{2}\int_0^l \left(\frac{\partial u}{\partial x}\right)^2 dx \\ &\text{and} \\ T &= \frac{A_c \rho}{2}\int_0^l \left(\frac{\partial u}{\partial t}\right)^2 dx \end{aligned}\right\} \tag{8-104}$$

where u = longitudinal displacement
A_c = cross-sectional area of rod
E = modulus of elasticity
ρ = mass density of rod material

Torsional Vibration of Shafts of Circular Cross Section

The procedure just discussed is also applicable to the analysis of torsional shafts of circular cross section. The strain and kinetic energies are given, respectively, by

$$U = \frac{GJ}{2} \int_0^l \left(\frac{\partial \theta}{\partial x}\right)^2 dx$$

and (8-105)

$$T = \frac{J\rho}{2} \int_0^l \left(\frac{\partial \theta}{\partial t}\right)^2 dx$$

where θ = angle of twist, radians
G = shear modulus, or modulus of rigidity
J = polar amount of inertia of cross-sectional area
ρ = mass density of shaft material

Flexural Vibration of Beams

The general procedure for analyzing the transverse vibration of beams characterized by Euler's beam equation is essentially the same as that discussed in the preceding examples of systems characterized by the wave equation. The strain energy in a beam due to bending (see Eq. 2-69) is given by

$$U = \frac{EI}{2} \int_0^l \left(\frac{\partial^2 y}{\partial x^2}\right)^2 dx = \frac{EI}{2} \int_0^l \left(\sum \phi_i'' q_i\right)^2 dx \tag{8-106}$$

from which

$$\frac{\partial U}{\partial q_j} = EI \int_0^l \left(\sum \phi_i'' q_i\right) \phi_j'' \, dx \tag{8-107}$$

where ϕ_i and ϕ_j = normal-mode functions of beam for the ith and jth modes, respectively, as shown in Table 8-1

ϕ_i'' and ϕ_j'' = $\dfrac{d^2\phi_i}{dx^2}$ and $\dfrac{d^2\phi_i}{dx^2}$

Recalling the orthogonality properties of ϕ_i'' and ϕ_j'' given by Eq. 8-81, and referring to Eq. 8-84, we see that when $i = j$, Eq. 8-107 yields

$$\frac{\partial U}{\partial q_i} = \frac{EI\omega_i^2 l}{EI/\gamma} q_i = \gamma \omega_i^2 l q_i \tag{8-108}$$

in which ω_i is the undamped natural circular frequency of the ith mode of the beam.

The kinetic energy of a uniform beam of mass per unit length γ is given by

$$T = \frac{\gamma}{2} \int_0^l \left(\frac{\partial y}{\partial t}\right)^2 dx = \frac{\gamma}{2} \int_0^l \left(\sum \phi_i \dot{q}_i\right)^2 dx \tag{8-109}$$

We can then determine that

$$\frac{d}{dt}\left(\frac{\partial T}{\partial \dot{q}_j}\right) = \gamma \int_0^l \left(\sum \phi_i \ddot{q}_i\right)\phi_j \, dx \tag{8-110}$$

Recalling the orthogonality properties of ϕ_i and ϕ_j given by Eq. 8-80, and considering that the normal-mode functions ϕ_i are normalized so that the integral of ϕ_i^2 is equal to l as indicated by Eq. 8-83, Eq. 8-110 yields

$$\frac{d}{dt}\left(\frac{\partial T}{\partial \dot{q}_i}\right) = \gamma l \ddot{q}_i \tag{8-111}$$

Since T is not a function of q_i,

$$\frac{\partial T}{\partial q_i} = 0 \tag{8-112}$$

Upon substituting Eqs. 8-108, 8-111, and 8-112 into Lagrange's equations we obtain

$$\ddot{q}_i + \omega_i^2 q_i = 0 \tag{8-113}$$

in which the undamped natural circular frequencies are found from

$$\omega_i = k_i^2 \sqrt{\frac{EI}{\gamma}}$$

and depend upon the boundary conditions of the beam (see the k_i values for the beams in Table 8-1 on page 608.

Since the solution of Eq. 8-113 is

$$q_i = C_i \cos \omega_i t + D_i \sin \omega_i t$$

we can refer to Eq. 8-86 to see that the undamped free vibration of the beam is given by

$$y = \sum_{i=1,2,\ldots}^{\infty} \phi_i (C_i \cos \omega_i t + D_i \sin \omega_i t) \tag{8-114}$$

in which ϕ_i are the normal-mode functions satisfying the boundary conditions of the beam being analyzed and the constants C_i and D_i are determined from the initial conditions.

EXAMPLE 8-5

A uniform steel beam of length l and weight W is pinned at A and held at rest initially in the position shown in Fig. 8-17a. We assume that damping in the beam is negligible and ignore the energy lost during the impact between the end of the beam and the support at B. We also assume that the beam does not rebound from B. Consider that the beam is a steel pipe and that the following parameter values pertain:

l = 180 in. (length of pipe)
r = 2 in. (outside radius of pipe)
I = 4.79 in.^{4} (area moment of inertia)
γ = 1.97(10)$^{-3}$ lb·s^2/in.^{2} (mass per unit length)
E = 30(10)^6 psi (modulus of elasticity)
h = 72 in. (initial height of right end of beam)

After the beam is released from its rest position, we wish to determine the following:

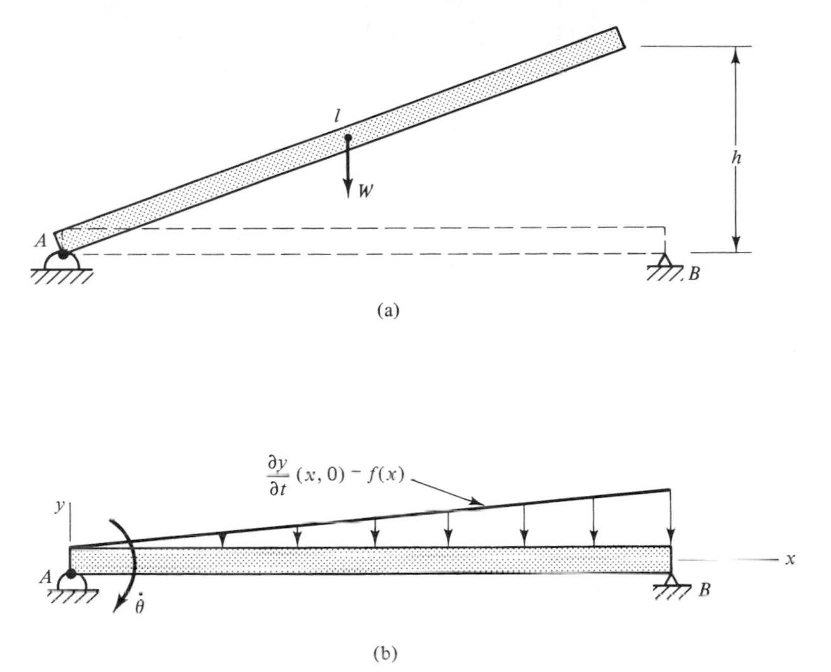

Figure 8-17 Beam of Example 8-5. (a) Position of beam at rest. (b) Velocity of beam cross sections just before impact with support B.

(a) the motion of the beam after the impact with support B; (b) the general equation of the bending stress σ in the fibers a distance c from the neutral axis of the beam section at $x = l/2$; (c) the maximum stress σ_m on a cross section of the beam at $x = l/2$.

Solution. **a.** Table 8-1 shows that the normal-mode function for a simply supported beam (pinned-pinned) is

$$\phi_i = \sqrt{2} \sin \frac{i\pi x}{l} \qquad i = 1, 2, 3, \ldots$$

Referring to Eq. 8-114, the undamped free vibration of the beam after impact is given by

$$y = \sqrt{2} \sum_{i=1,2,\ldots}^{\infty} \sin \frac{i\pi x}{l} (C_i \cos \omega_i t + D_i \sin \omega_i t) \tag{8-115}$$

in which

$$\omega_i = \left(\frac{i\pi}{l}\right)^2 \sqrt{\frac{EI}{\gamma}}$$

To determine the constants C_i and D_i, we note that at the instant after (or before) impact the initial conditions are

$$(1) \quad y(x, 0) = 0$$

$$(2) \quad \frac{\partial y}{\partial t}(x, 0) = f(x)$$

Inspecting Eq. 8-115, we see that $C_i = 0$ to satisfy the first initial condition, which reduces Eq. 8-115 to

$$y = \sqrt{2} \sum_{i=1,2,\ldots}^{\infty} D_i \sin \frac{i\pi x}{l} \sin \omega_i t \tag{8-116}$$

The constant D_i depends upon the second of the initial conditions shown, involving the velocity of the beam cross sections immediately before impact. Considering the beam as a rigid body with an angular velocity of $\dot{\theta}$ just before impact, the velocity of any section along the beam just before impact is $-x\dot{\theta}$. Equating $-x\dot{\theta}$ to the partial derivative of y with respect to t as obtained from Eq. 8-116, we obtain

$$\frac{\partial y}{\partial t}(x, 0) = -x\dot{\theta} = \sqrt{2} \sum_{i=1,2,\ldots}^{\infty} D_i \omega_i \sin \frac{i\pi x}{l} \tag{8-117}$$

Since the potential energy of position $Wh/2$ of the beam at the instant it is released from rest is equal to its kinetic energy $I_A\dot{\theta}^2/2$ at the instant before impact, we can write that

$$\frac{Wh}{2} = \frac{I_A\dot{\theta}^2}{2} \tag{8-118}$$

In this equation the mass moment of inertia of the beam about the pinned end A is $I_A = Wl^2/3g$. Equation 8-118 then yields the angular velocity $\dot{\theta}$ just prior to impact as

$$\dot{\theta} = \frac{1}{l}\sqrt{3gh} \tag{8-119}$$

Multiplying both sides of Eq. 8-117 by $\sin(j\pi x/l)$ and integrating over the length l of the beam gives

$$-\dot{\theta}\int_0^l x \sin \frac{j\pi x}{l}\, dx = \sqrt{2}\int_0^l \left(\sum D_i \omega_i \sin \frac{i\pi x}{l}\right) \sin \frac{j\pi x}{l}\, dx \tag{8-120}$$

Observing the orthogonality properties of $\sin(i\pi x/l)$ and $\sin(j\pi x/l)$, the integration of Eq. 8-120 with $i = j$ yields

$$\frac{\dot{\theta}l^2}{i\pi} \cos i\pi = \sqrt{2}\, D_i \omega_i \frac{l}{2} \qquad i = 1, 2, 3, \ldots$$

so that

$$D_i = \frac{\sqrt{2}\dot{\theta}l}{i\pi\omega_i}(-1)^i \tag{8-121}$$

where $\dot{\theta} = \frac{1}{l}\sqrt{3gh}$

$(-1)^i = \cos i\pi \qquad (i = 1, 2, 3, \ldots)$

$$\omega_i = \left(\frac{i\pi}{l}\right)^2 \sqrt{\frac{EI}{\gamma}}$$

Substituting Eq. 8-121 into Eq. 8-116, the undamped free vibration of the beam after impact with support B is given by

$$y = \frac{2l^2}{\pi^3}\sqrt{\frac{3gh\gamma}{EI}}\sum_{i=1,2,\ldots}^{\infty}\frac{(-1)^i}{i^3}\sin\frac{i\pi x}{l}\sin\omega_i t \tag{8-122}$$

b. Recalling from elementary mechanics of materials that

$$M = EI\frac{\partial^2 y}{\partial x^2}$$

we can express the stress on a fiber a distance c from the neutral axis of a beam as

$$\sigma = \frac{Mc}{I} = E\frac{\partial^2 y}{\partial x^2}c \tag{8-123}$$

Therefore, the stress on such a fiber in a section at $x = l/2$ is given by

$$\sigma\left(\frac{l}{2}, t\right) = \frac{2c}{\pi}\sqrt{\frac{3gh\gamma E}{I}}\sum_{i=1,3,5,\ldots}^{\infty}\frac{1}{i}\sin\frac{i\pi}{2}\sin\omega_i t \tag{8-124}$$

c. To determine the maximum stress σ_m in the steel pipe at $x = l/2$, we note that $\omega_i t$ can be expressed in terms of the fundamental natural circular frequency ω_1 as

$$\omega_i t = i^2\omega_1 t$$

or in terms of the fundamental period τ_1 as

$$\omega_i t = i^2 2\pi\frac{t}{\tau_1} \tag{8-125}$$

When $t = \tau_1/4$, it should be apparent from Eq. 8-125 that $\omega_i t = i^2\pi/2$ so that $\sin\omega_i t = 1$ for $i = 1, 3, 5, \ldots$. Therefore, at $t = \tau_1/4$ the maximum stress σ_m on a fiber a distance c from the neutral axis of a section at $x = l/2$ is given by

$$\sigma_m\left(\frac{l}{2}, \frac{\tau_1}{4}\right) = \frac{2c}{\pi}\sqrt{\frac{3gh\gamma E}{I}}\left(1 - \frac{1}{3} + \frac{1}{5} - \frac{1}{7} + \cdots\right) \tag{8-126}$$

Since it can be shown (see Prob. 1-16) that the series in Eq. 8-126

$$1 - \frac{1}{3} + \frac{1}{5} - \frac{1}{7} + \cdots = \frac{\pi}{4}$$

Eq. 8-126 becomes

$$\sigma_m\left(\frac{l}{2}, \frac{\tau_1}{4}\right) = \frac{c}{2}\sqrt{\frac{3gh\gamma E}{I}}$$

Substituting the parameter values given at the beginning of this example into the above equation yields

$$\sigma_m\left(\frac{l}{2}, \frac{\tau_1}{4}\right) = \frac{2}{2}\sqrt{\frac{3(386)(72)1.97(10)^{-3}30(10)^6}{4.79}} = 32{,}074 \text{ psi}$$

which is a stress of appreciable magnitude. The reader should realize that this computed stress is larger than the actual stress because of the simplifying assumptions that no energy was lost during impact and that the damping was negligible. However, it does show that rather large magnitudes of stress can be produced by dropping a beam in this fashion.

EXAMPLE 8-6

A pinned-pinned beam of length l and mass per unit length γ rests on an elastic foundation of stiffness per unit length of k_0 as shown in Fig. 8-18. The beam is subjected to a suddenly applied force $f(t)$ of intensity w_0 distributed uniformly along the beam as shown in Fig. 8-18a. Assume that the mass of the elastic foundation is negligible compared with that of the beam, and that damping is negligible.

We wish to determine the total response of the beam using the modal-superposition method and the information given in Table 8-1 on page 608.

Solution. From Table 8-1 the normal-mode functions for a pinned-pinned beam are found to be

$$\phi_i = \sqrt{2} \sin \frac{i\pi x}{l}$$

from which

$$\phi_i'' = -\sqrt{2} \left(\frac{i\pi}{l}\right)^2 \sin \frac{i\pi x}{l}$$

The strain energy of the system consists of the energy stored in the beam due to bending and that stored in the elastic foundation. Thus,

$$U = \frac{EI}{2} \int_0^l \left(\frac{\partial^2 y}{\partial x^2}\right)^2 dx + \frac{k_0}{2} \int_0^l y^2\, dx$$

or

$$U = \frac{EI}{2} \int_0^l \left(\sum \phi_i'' q_i\right)^2 dx + \frac{k_0}{2} \int_0^l \left(\sum \phi_i q_i\right)^2 dx \tag{8-127}$$

from which

$$\frac{\partial U}{\partial q_j} = EI \int_0^l \left(\sum \phi_i'' q_i\right)\phi_j''\, dx + k_0 \int_0^l \left(\sum \phi_i q_i\right)\phi_j\, dx \tag{8-128}$$

for a particular generalized coordinate q_j.

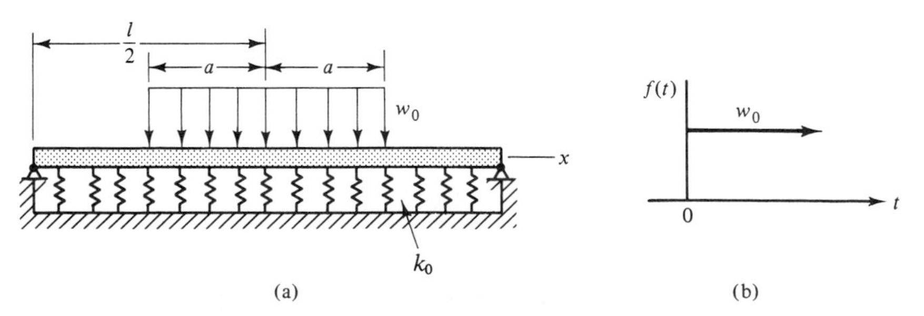

Figure 8-18 Pinned-pinned beam on elastic foundation.

Once again recalling the orthogonality properties of the normal-mode functions, and referring to Eqs. 8-83 and 8-84, it can be seen that with $i = j$ Eq. 8-128 yields

$$\frac{\partial U}{\partial q_i} = (\omega_i^2 \gamma l + k_0 l) q_i \tag{8-129}$$

Referring to Table 8-1 or Eq. 8-66, we see that

$$\omega_i^2 = \left(\frac{i\pi}{l}\right)^4 \frac{EI}{\gamma}$$

so that Eq. 8-129 can be written as

$$\frac{\partial U}{\partial q_i} = \left[\left(\frac{i\pi}{l}\right)^4 EI + k_0\right] l q_i \tag{8-130}$$

Referring to Eq. 8-111,

$$\frac{d}{dt}\left(\frac{\partial T}{\partial \dot{q}_i}\right) = \gamma l \ddot{q}_i$$

The differential work dW_i done by the excitation force $f(t)$ due to a differential change dq_i of a particular coordinate q_i is

$$dW_i = Q_i\, dq_i = -w_0 \int_{\frac{l}{2}-a}^{\frac{l}{2}+a} [dy]\, dx \tag{8-131}$$

in which $dy = \sqrt{2} \sin(i\pi x/l)\, dq_i$. We can then obtain the generalized nonpotential force Q_i from Eq. 8-131 as

$$Q_i = -w_0\sqrt{2} \int_{\frac{l}{2}-a}^{\frac{l}{2}+a} \sin \frac{i\pi x}{l}\, dx = \frac{-2\sqrt{2} w_0 l}{i\pi} \sin \frac{i\pi}{2} \sin \frac{i\pi a}{l} \tag{8-132}$$

Upon substituting Eqs. 8-111, 8-130, and 8-132 into Lagrange's equations and simplifying, we find that

$$\ddot{q}_i + \left[\left(\frac{i\pi}{l}\right)^4 \frac{EI}{\gamma} + \frac{k_0}{\gamma}\right] q_i = \frac{-2\sqrt{2} w_0}{\gamma i\pi} \sin \frac{i\pi}{2} \sin \frac{i\pi a}{l}$$

or that

$$\ddot{q}_i + \omega_i^2 q_i = \frac{-2\sqrt{2} w_0}{\gamma i\pi} \sin \frac{i\pi}{2} \sin \frac{i\pi a}{l} \qquad i = 1, 3, 5, \ldots \tag{8-133}$$

in which the undamped natural circular frequencies ω_i of the system are found to be

$$\omega_i = \sqrt{\left(\frac{i\pi}{l}\right)^4 \frac{EI}{\gamma} + \frac{k_0}{\gamma}} \tag{8-134}$$

The general solution of Eq. 8-133 is

$$q_i = C_i \cos \omega_i t + D_i \sin \omega_i t - \frac{2\sqrt{2} w_0}{\gamma i\pi \omega_i^2} \sin \frac{i\pi}{2} \sin \frac{i\pi a}{l} \qquad i = 1, 3, 5, \ldots \tag{8-135}$$

in which ω_i is as expressed in Eq. 8-134.

Substituting the initial conditions

$$(1) \quad y(x, 0) = 0 \qquad (q_i = 0)$$

$$(2) \quad \frac{\partial y}{\partial t}(x, 0) = 0 \qquad (\dot{q}_i = 0)$$

into Eq. 8-135 and its partial derivative with respect to t, we find that

$$D_i = 0$$

and

$$C_i = \frac{2\sqrt{2}w_0}{\gamma i \pi \omega_i^2}(-1)^{(i-1)/2} \sin \frac{i\pi a}{l}$$

in which $(-1)^{(i-1)/2} = \sin(i\pi/2)$ for $i = 1, 3, 5, \ldots$. Thus, Eq. 8-135 becomes

$$q_i = \frac{2\sqrt{2}w_0}{\gamma i \pi \omega_i^2}(-1)^{(i-1)/2} \sin \frac{i\pi a}{l}(\cos \omega_i t - 1) \qquad i = 1, 3, 5, \ldots$$

Recalling that $y = \Sigma\, \phi_i q_i$, the total response is obtained as

$$y = \frac{4w_0}{\gamma\pi} \sum_{i=1,3,\ldots}^{\infty} \frac{(-1)^{(i-1)/2}}{i\omega_i^2} \sin \frac{i\pi x}{l} \sin \frac{i\pi a}{l} (\cos \omega_i t - 1) \tag{8-136}$$

in which ω_i is as expressed in Eq. 8-134.

The reader should note that the series in Eq. 8-136 converges quite rapidly since ω_i^2 contains an i^4 as shown in Eq. 8-134. That is, the second term in the series in Eq. 8-136 contains $(3)^4$, the third term $(5)^4$, and so on. The response is thus due essentially to the contribution of the first mode ($i = 1$).

PROBLEMS

Problems 8-1 through 8-5 (Sections 8-1 through 8-4)

8-1. The ends of a tightly stretched string or cable of length l are fastened to rigid supports such that a tension T_0 exists. The damping is assumed to be viscous in nature (proportional to the velocity), and the damping coefficient is c per unit length. Determine (a) the partial differential equation of the string or cable and (b) the modal damping factor ζ_i in terms of the physical parameters of the system.

Partial ans: (b) $\zeta_i = \dfrac{cl}{2i\pi\sqrt{T_0\gamma}}$

8-2. Consider the torsional vibration of the circular bar of radius r shown in part a of the accompanying figure. The material of the bar has a mass density of ρ and behaves as a viscoelastic material such that

$$\tau = G\gamma + \alpha\dot{\gamma}$$

where τ = shear stress
γ = shear strain
$\dot{\gamma}$ = shear strain rate
G = shear modulus of elasticity
α = viscoelastic constant of material

From the elemental length dx of the bar shown in part b of the figure, it can be shown that the shear strain is given by

$$\gamma = r \frac{\partial \theta}{\partial x}$$

Recalling from elementary mechanics of materials that

$$\tau = \frac{Tr}{J}$$

in which T is the torque and J is the polar moment of inertia of the cross-sectional area of the bar, show that the partial differential equation of the circular bar is

$$\frac{G}{\rho} \frac{\partial^2 \theta}{\partial x^2} + \frac{\alpha}{\rho} \frac{\partial^3 \theta}{\partial x^2 \, \partial t^2} = \frac{\partial^2 \theta}{\partial t^2}$$

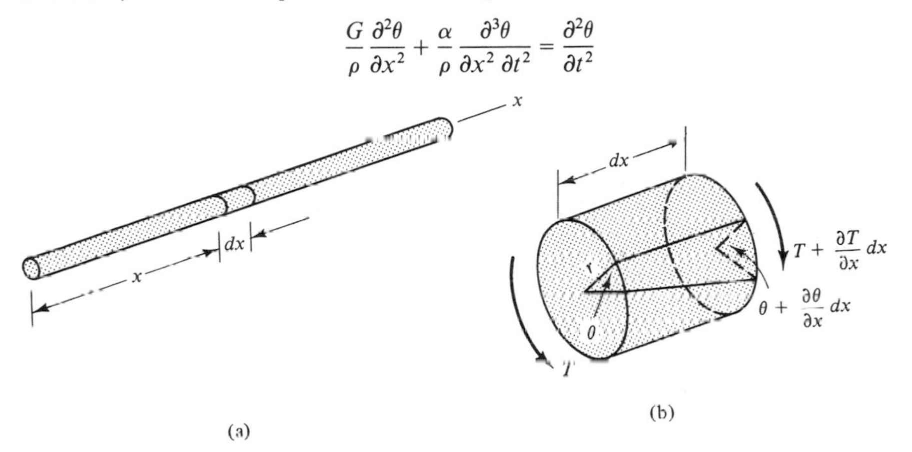

Prob. 8-2

8-3. A slender rod of length l is fixed at its left end and has a static force P applied at its right end as shown in the accompanying figure. Determine the motion of the rod at $x = l$ after the force P is suddenly removed.

Ans: $$u = \frac{8Pl}{A_e E \pi^2} \sum_{i=1,3,5,\ldots}^{\infty} \frac{(-1)^{(i-1)/2}}{i^2} \sin \frac{i\pi x}{2l} \cos \frac{i\pi ct}{2l}$$

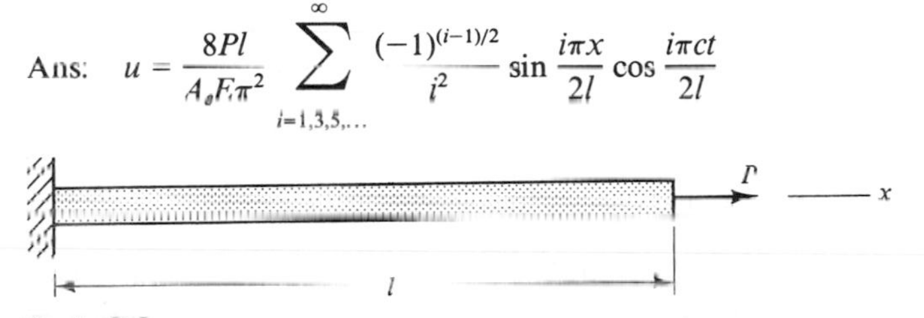

Prob. 8-3

8-4. A long flexible cable is attached to a mass M such that a tension T_0 exists in the cable, and the mass is attached to a rigid support through a spring of stiffness k as shown in the accompanying figure. Using the free body shown in part b of the figure, show that the frequency equation of the system is

$$\tan \frac{\omega_i l}{c} = \frac{(T_0/kl)(\omega_i l/c)}{(c^2 M/kl^2)(\omega_i l/c)^2 - 1}$$

and that the normal-mode functions are given by

$$\phi_i(x) = \sin \frac{\omega_i x}{c}$$

in which $c = \sqrt{T_0/\gamma}$.

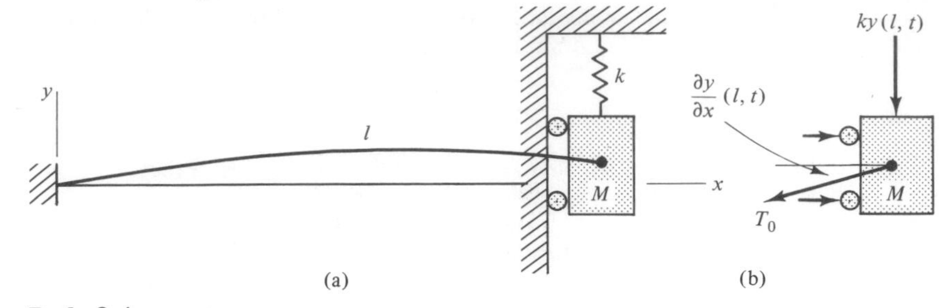

Prob. 8-4

8-5. Both the elastic rod and the helical spring shown in the accompanying figure are of length l and are fixed at both ends. Referring to Sec. 8-4 for a discussion of the equations defining c^2 shown in the figure, determine the frequency equation and the natural frequencies of the rod and of the spring.

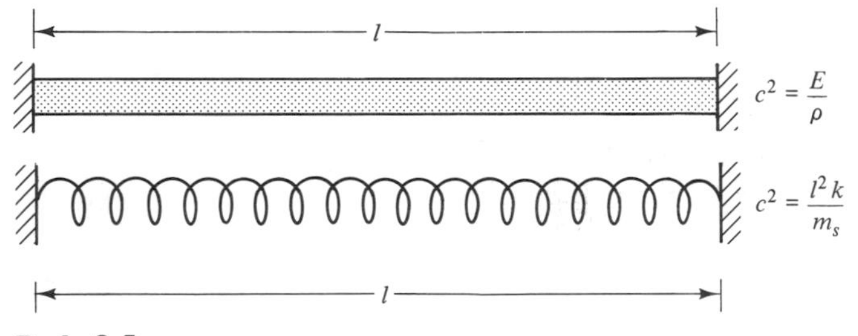

Prob. 8-5

Problems 8-6 through 8-9 (Sections 8-5 and 8-6)

8-6. Referring to Table 8-1 on page 608 verify the normal-mode functions shown there for a fixed-free (cantilever) beam that has the frequency equation

$$\cosh kl \cos kl = -1$$

It will be found helpful to note the following relationships:

$$\cosh^2 kl - \sinh^2 kl = 1$$
$$\cos^2 kl + \sin^2 kl = 1$$

8-7. Referring to Table 8-1 on page 608 verify the normal-mode functions shown there for a free-free beam that has the frequency equation

$$\cosh kl \cos kl = 1$$

It will be found helpful to note the relationships shown at the end of Prob. 8-6.

8-8. Referring to Table 8-1, on page 608, verify the normal-mode functions shown there for a fixed-fixed beam that has the frequency equation

$$\cosh kl \cos kl = 1$$

It will be found helpful to note the relationships shown at the end of Prob. 8-6.

8-9. A uniform beam of length l and mass per unit length γ is pinned at its upper end so that it hangs vertically as shown in the accompanying figure. Determine (a) the frequency equation of the system and its first three roots and (b) the normal-mode functions $\phi_i(x)$ so that

$$\int_0^l \phi_i^2(x)\, dx = l$$

Partial ans: (b) $\phi_i(x) = \sqrt{2}(\alpha_i \sinh k_i x + \sin k_i x)$ in which $\alpha_i = \dfrac{\sin k_i l}{\sinh k_i l}$

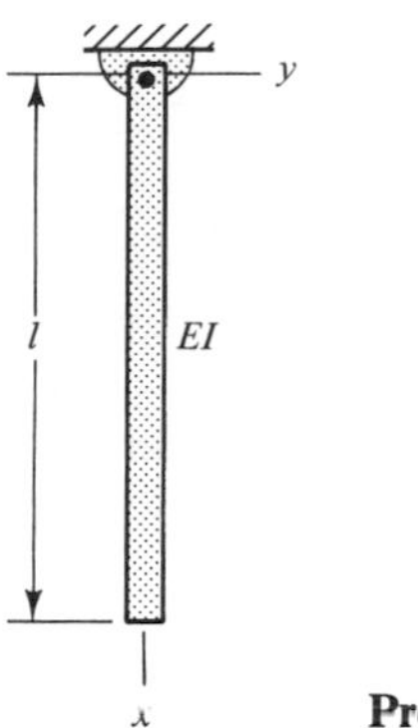

Prob. 8-9

Problems 8-10 through 8-18 (Sections 8-7)

8-10. A tightly stretched wire is subjected to a force $w(x, t) = g(x)f(t)$ as shown in the accompanying figure. Determine the differential equation of motion of the wire in terms of the generalized coordinate q_i, and show that the solution of the equation for $t \leq \tau$ is

$$q_i = \frac{2w_0 \cos i\pi}{i\pi\gamma\omega_i^2}\left(\cos \omega_i t - \frac{\sin \omega_i t}{\tau\omega_i} + \frac{t}{\tau} - 1\right)$$

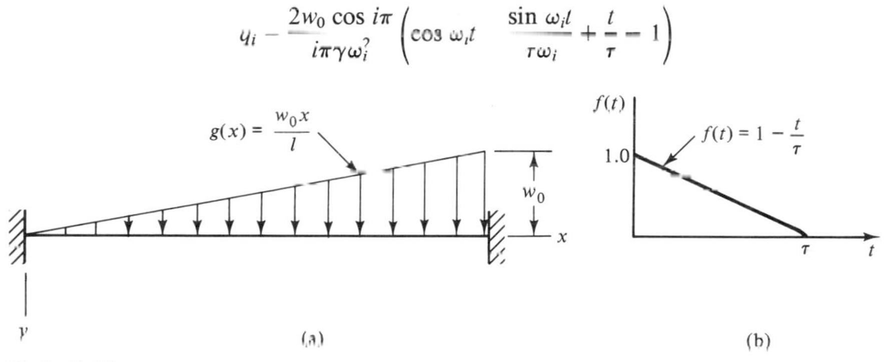

Prob. 8-10

8-11. A concentrated force P is moving along a tightly stretched wire with a constant velocity v as shown in the accompanying figure. Determine the motion of the wire at $x = l/2$ during the time interval that it takes for the force to move from $x = 0$ to $x = l$, assuming that the wire is initially at rest and that the damping is negligible.

Ans: $$y(l/2, t) = \sum_{i=1,3,\ldots}^{\infty} (-1)^{(i-1)/2} A_i\left(\sin \Omega_i t - \frac{\Omega_i}{\omega_i}\sin \omega_i t\right)$$

in which $$A_i = \left(\frac{2P}{\gamma l \omega_i^2}\right)\frac{1}{1 - (\Omega_i/\omega_i)^2}$$

$$\Omega_i = \frac{i\pi v}{l}$$

$$\omega_i = \frac{i\pi}{l}\sqrt{\frac{T_0}{\gamma}}$$

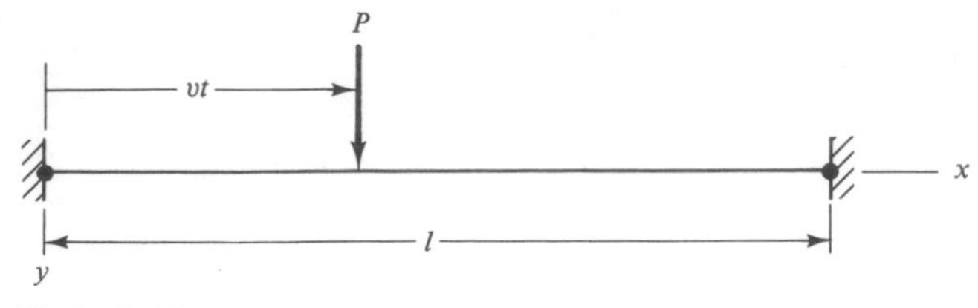

Prob. 8-11

8-12. A beam is simply supported by a rigid frame A as shown in the accompanying figure. When the frame is dropped from a height h onto the fixed foundation shown, it makes a plastic impact with the foundation (no rebound). Determine (a) the motion of the beam after impact and (b) the bending stress σ in the beam fibers that are a distance z from the neutral axis of a cross section of the beam located at $x = l/2$.

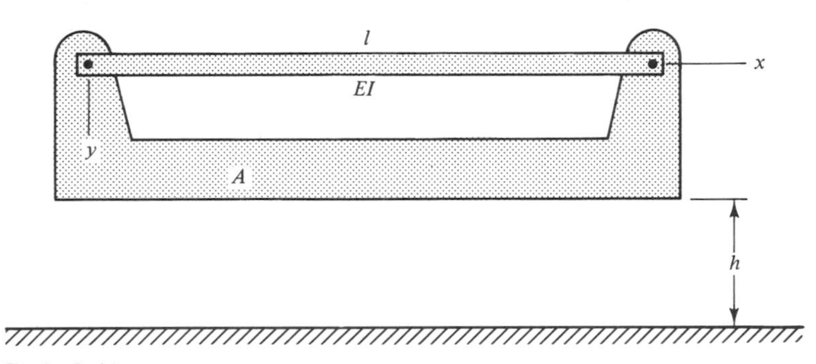

Prob. 8-12

8-13. Referring to Example 8-6, consider that the beam shown there on the elastic foundation is fixed-fixed, and determine (a) an expression for the lowest natural circular frequency ω_1 of the system formed by the beam and the elastic foundation and (b) the generalized nonpotential force Q_1 for the first mode due to the suddenly applied force intensity w_0 which is distributed uniformly over the entire length l of the beam.

Partial ans: (a) $\omega_1 = \sqrt{\dfrac{500.56EI}{\gamma l^4} + \dfrac{k_0}{\gamma}}$

8-14. A pinned-pinned beam is subjected to an impulsive force $P(t)$ at $x = d$ as shown in the accompanying figure. Considering that the damping in the beam is viscous, determine (a) the ordinary differential equation of motion for the generalized coordinate q_i and (b) the response of the beam at $x = l/2$ if the impulsive force is applied at $d = l/2$.

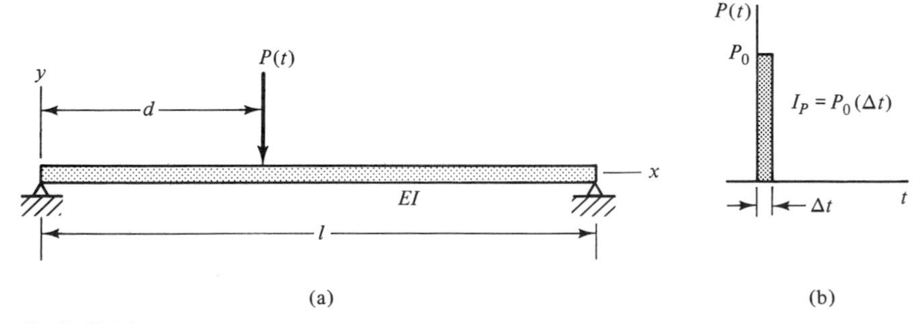

Prob. 8-14

8-15. A fixed-free beam is subjected to a suddenly applied moment $M(t)$ as shown in the accompanying figure. Determine the deflection y of the beam at $x = l$ as a function of time.

Ans: $y = \dfrac{M_0 l^3}{EI} \sum_{i=1,2,\ldots}^{\infty} \dfrac{\phi_i'(l)\phi_i(l)}{(k_i l)^4}(1 - \cos \omega_i t)$

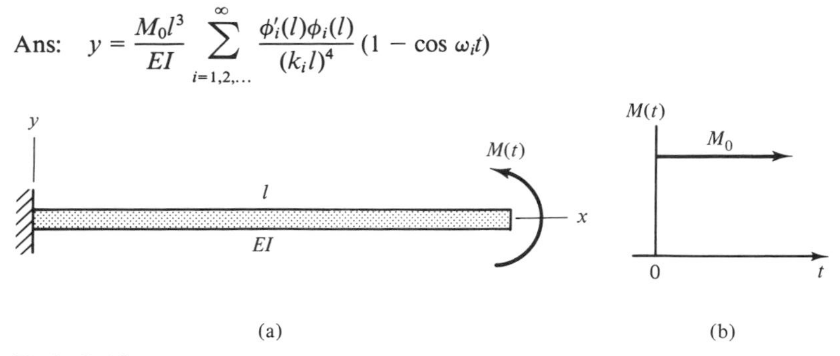

Prob. 8-15

8-16. In Example 8-6 the total response y of a pinned-pinned beam on an elastic foundation is given by the series in Eq. 8-136. Using this equation, determine an equation that will give the flexural stress σ in a beam fiber a distance z from the neutral axis of a beam cross section located at $x = l/2$. Discuss briefly the rate of convergence of the series obtained for the stress as compared with the rate of convergence of the series giving the deflection y in Eq. 8-136.

8-17. A beam 20 ft long is fabricated by gluing together four rough-dimension wooden planks to form the cross section shown in part c of the accompanying figure. The beam is simply supported and is subjected at its center to the step function shown in part b of the figure. The modulus of elasticity of the wood is $1.5(10)^6$ psi, and its specific weight is $w = 0.022$ lb/in.^{3} Considering the superposition of just the first three modes and neglecting damping, determine the absolute maximum response of the beam at $x = l/2$ in terms of F_0 if the rise time of the step function is 30 ms. [Hint: Refer to Eq. 4-26 for calculating $(q_i)_{\max}/(q_i)_{\text{static}}$ for the three normal modes.]

Ans: $(y_{\max})_{l/2} = 5.51(10)^{-4} F_0$ in.

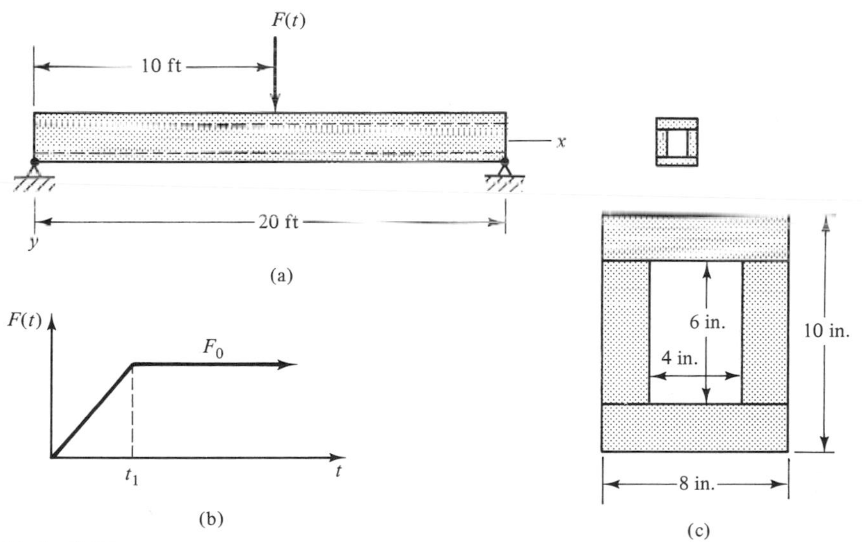

Prob. 8-17

8-18. The motion of a tightly stretched cable at $x = l$ is described by $\lambda e^{j\omega t}$ as shown in the accompanying figure. Determine the steady-state response of the cable.

Ans: $y = \lambda \dfrac{\sin(\omega x/c)}{\sin(\omega l/c)} \sin \omega t$

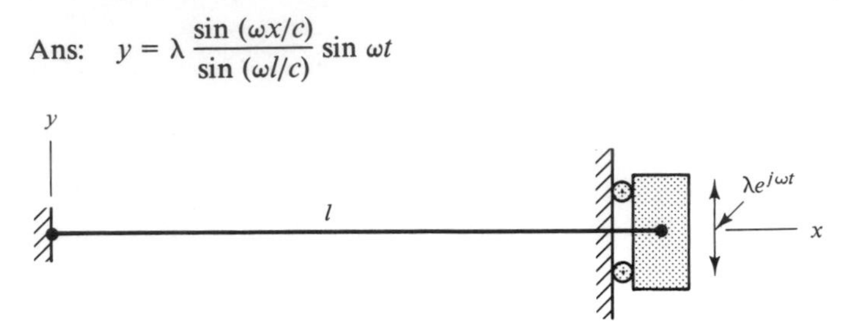

Prob. 8-18

Appendix A

Laplace Transforms

A-1 LAPLACE TRANSFORMATION

The Laplace transform of a function of time $f(t)$ for $t > 0$ was shown in Sec. 1-7 to be

$$L[f(t)] = g(s) = \int_0^\infty f(t)e^{-st}\,dt \tag{A-1}$$

in which L denotes "Laplace transform of" and s is a complex variable whose real part is greater than 0.

The inverse Laplace transform may be written in the form

$$f(t) = L^{-1}g(s) \tag{A-2}$$

A-2 LAPLACE TRANSFORMS OF DERIVATIVES

In the solution of differential equations by using Laplace transforms, it is necessary to take the Laplace transform of derivatives. Denoting the $L(x)$ as $\bar{x}$, the Laplace transform of the first derivative dx/dt is

$$L\left(\frac{dx}{dt}\right) = s\bar{x} - x_0 \tag{A-3}$$

in which x_0 is the initial value of the dependent variable x at $t = 0$. The Laplace transforms of higher-order derivatives such as the second and third derivatives are as follows:

$$L\left(\frac{d^2x}{dt^2}\right) = s^2\bar{x} - sx_0 - \dot{x}_0 \tag{A-4}$$

$$L\left(\frac{d^3x}{dt^3}\right) = s^3\bar{x} - s^2x_0 - s\dot{x}_0 - \ddot{x}_0 \tag{A-5}$$

in which x_0, $\dot{x}_0$, and $\ddot{x}_0$ are initial values of x, dx/dt, and d^2x/dt^2, respectively.

TABLE A-1 SHORT TABLE OF LAPLACE TRANSFORMS

	$g(s)$	$f(t)$
1	$\frac{1}{s}$	$u(t) = \begin{cases} 0 & t < 0 \\ 1 & t > 0 \end{cases}$ (unit step function)
2	1	$\delta(t) = \begin{cases} \infty & t = 0 \\ 0 & t \neq 0 \end{cases}$ (delta function)
3	e^{-sa}	$\delta(t - a) = \begin{cases} \infty & t = a \\ 0 & t \neq a \end{cases}$
4	$\frac{1}{s^n}$ $(n = 1, 2, \ldots)$	$\frac{t^{n-1}}{(n-1)!}$
5	$\frac{1}{s - a}$	e^{at}
6	$\frac{1}{(s + a)^2}$	$t\,e^{-at}$
7	$\frac{1}{(s - a)^2}$	$t\,e^{at}$
8	$\frac{1}{s^2 + a^2}$	$\frac{1}{a} \sin at$
9	$\frac{s}{s^2 + a^2}$	$\cos at$
10	$\frac{s}{(s^2 + a^2)^2}$	$\frac{t}{2a} \sin at$
11	$\frac{s^2 - a^2}{(s^2 + a^2)^2}$	$t \cos at$
12	$\frac{1}{s^2 - a^2}$	$\frac{1}{a} \sinh at$
13	$\frac{s}{s^2 - a^2}$	$\cosh at$
14	$\frac{1}{s(s^2 + a^2)}$	$\frac{1}{a^2}(1 - \cos at)$

TABLE A-1 (CONTINUED)

	$g(s)$	$f(t)$
15	$\dfrac{1}{s^2 + 2ns + p^2}$	$\dfrac{e^{-nt}}{\sqrt{p^2 - n^2}} \sin \sqrt{p^2 - n^2}\,t \qquad p^2 > n^2$
		$t\,e^{-nt} \qquad n^2 = p^2$
16	$\dfrac{s}{s^2 + 2ns + p^2}$	$\dfrac{-p}{\sqrt{p^2 - n^2}} e^{-nt} \sin(\sqrt{p^2 - n^2}\,t - \phi) \qquad p^2 > n^2$ $\tan \phi = \dfrac{\sqrt{p^2 - n^2}}{n}$
		$e^{-pt}(1 - pt) \qquad p^2 = n^2$
17	$\dfrac{s + a}{(s + a)^2 + k^2}$	$e^{-at} \cos kt$
18	$\dfrac{1}{s(s^2 + 2ns + p^2)}$	$\dfrac{1}{p^2} + \dfrac{e^{-nt}}{p\sqrt{p^2 - n^2}} \sin(\sqrt{p^2 - n^2}\,t - \phi)$ $\phi = \tan^{-1}\left[\dfrac{\sqrt{p^2 - n^2}}{-n}\right] \qquad (\phi > \pi/2)$ $p > n$

In general, the solution of differential equations by the use of Laplace transforms involves the following steps:

1. Obtain the Laplace transform of the various terms in the differential equation.
2. Algebraically obtain the Laplace transform for the dependent variable from the results of step 1.
3. Obtain the solution to the differential equation by finding the inverse of the Laplace transform determined for the dependent variable in step 2.

Steps 1 and 3 are usually performed by using a table of transforms. Table A-1 is a short table of Laplace transforms $g(s)$ and corresponding inverse Laplace transforms $f(t)$.

EXAMPLE A-1

The differential equation for a damped single-degree-of-freedom system subjected to a suddenly applied force (step function) of magnitude F_0 is

$$\ddot{x} + 2\zeta\omega_n\dot{x} + \omega_n^2 x = \frac{F_0}{m} \qquad (t > 0) \tag{A-6}$$

Consider that the system is initially at rest and determine the response $x(t)$ by the method of Laplace transforms.

Solution. Noting that the initial conditions are $x_0 = \dot{x}_0 = 0$, the Laplace transforms of the terms on the left-hand side of Eq. A-6 are readily written from Eqs. A-3 and A-4. From No. 1

of Table A-1, we see that $g(s)$ for a unit step function is $1/s$ so that the $L(F_0/m)$ is $(F_0/m)/s$ for the right-hand side of Eq. A-6. Thus, the Laplace transform of Eq. A-6 is found to be

$$s^2\bar{x} + 2\zeta\omega_n s\bar{x} + \omega_n^2\bar{x} = \frac{F_0/m}{s}$$

from which we obtain

$$\bar{x} = \frac{F_0/m}{s(s^2 + 2\zeta\omega_n s + \omega_n^2)} \quad \text{(A-7)}$$

in which $\bar{x}$ is the $L(x)$.

We now refer to No. 18 of Table A-1 to determine the inverse Laplace transform of Eq. A-7. Noting that $\zeta\omega_n$ and ω_n of Eq. A-7 correspond to n and p, respectively, of $g(s)$ shown in the table, we obtain

$$L^{-1}\bar{x} = x(t) = \frac{F_0}{m}\left[\frac{1}{\omega_n^2} + \frac{e^{-\zeta\omega_n t}}{\omega_n^2\sqrt{1-\zeta^2}}\sin(\omega_n\sqrt{1-\zeta^2}\,t - \phi)\right] \quad \text{(A-8)}$$

where $\phi = \tan^{-1}\left[\frac{\sqrt{1-\zeta^2}}{-\zeta]}\right]$

$\phi > \pi/2$

$\zeta < 1$

Since $k/m = \omega_n^2$, Eq. A-8 can be written in the familiar form

$$x(t) = \frac{F_0}{k}\left[1 + \frac{e^{-\zeta\omega_n t}}{\sqrt{1-\zeta^2}}\sin(\omega_n\sqrt{1-\zeta^2}\,t - \phi)\right] \quad \text{(A-9)}$$

REFERENCES

W. T. Thomson, *Laplace Transformation,* 2d ed., Prentice-Hall Inc., Englewood Cliffs, N.J., 1960.

Robert H. Cannon, Jr., *Dynamics of Physical Systems,* McGraw-Hill Book Co., New York, 1967.

Appendix B

Matrix Algebra

B-1 MULTIPLICATION

The product of two matrices $\mathbf{A} = [a_{ij}]$ and $\mathbf{B} = [b_{ij}]$ may be written as

$$[a_{ij}][b_{ij}] = [c_{ij}]$$

or, more simply, as

$$\mathbf{AB} = \mathbf{C}$$

The elements c_{ij} of the $\mathbf{C}$ matrix are obtained from

$$c_{ij} = \sum_{k=1} a_{ik}b_{kj} \tag{B-1}$$

where k identifies the kth element in the ith row of $\mathbf{A}$ and the kth element in the jth column of $\mathbf{B}$. To illustrate Eq. B-1, consider the equation

$$\begin{bmatrix} a_{11} & a_{12} & a_{13} \\ a_{21} & a_{22} & a_{23} \\ a_{31} & a_{32} & a_{33} \end{bmatrix} \begin{bmatrix} b_{11} & b_{12} & b_{13} \\ b_{21} & b_{22} & b_{23} \\ b_{31} & b_{32} & b_{33} \end{bmatrix} = \begin{bmatrix} c_{11} & c_{12} & c_{13} \\ c_{21} & c_{22} & c_{23} \\ c_{31} & c_{32} & c_{33} \end{bmatrix}$$

From Eq. B-1, the product of $\mathbf{A}$ and $\mathbf{B}$ gives

$$\begin{aligned} c_{11} &= a_{11}b_{11} + a_{12}b_{21} + a_{13}b_{31} \\ c_{21} &= a_{21}b_{11} + a_{22}b_{21} + a_{23}b_{31} \\ &\cdots\cdots\cdots\cdots\cdots\cdots \\ c_{33} &= a_{31}b_{13} + a_{32}b_{23} + a_{33}b_{33} \end{aligned}$$

It should now be apparent that matrix multiplication consists of a *row-on-column* multiplication sequence.

Matrix multiplication is not commutative. That is, $\mathbf{AB} \neq \mathbf{BA}$. In addition, the product of two matrices $\mathbf{AB}$ is defined only if the number of columns in $\mathbf{A}$ is equal to the number of rows in $\mathbf{B}$.

As shown in the following example, the multiplication of a matrix **A** by a *column* matrix **B** results in a column matrix:

$$\begin{bmatrix} 1 & 2 & 3 \\ 3 & 4 & 1 \\ 1 & 2 & 1 \end{bmatrix} \begin{Bmatrix} 1 \\ 2 \\ 3 \end{Bmatrix} = \begin{bmatrix} (1)(1) + (2)(2) + (3)(3) \\ (3)(1) + (4)(2) + (1)(3) \\ (1)(1) + (2)(2) + (1)(3) \end{bmatrix} = \begin{Bmatrix} 14 \\ 14 \\ 8 \end{Bmatrix}$$

The *identity* matrix **I**, or *unit* matrix, is an n-by-n matrix with *ones* on the main diagonal and zeros everywhere else. The identity matrix is to matrix algebra what the identity number (one) is to ordinary algebra. That is,

$$\mathbf{IA} = \mathbf{AI} = \mathbf{A} \tag{B-2}$$

Matrix multiplication is associative. That is,

$$\mathbf{ABC} = (\mathbf{AB})\mathbf{C} = \mathbf{A}(\mathbf{BC}) \tag{B-3}$$

Premultiplying the product **AB** by **C** means

$$\mathbf{CAB} = \mathbf{D}$$

If **AB** is to be *postmultiplied* by **C**, then

$$\mathbf{ABC} = \mathbf{E}$$

B-2 MATRIX INVERSION

The inverse of the matrix **A** is denoted as $\mathbf{A}^{-1}$. Also,

$$\mathbf{A}^{-1}\mathbf{A} = \mathbf{I}$$

in which matrix **A** is assumed to be nonsingular ($\det A \neq 0$), so that an inverse of **A** exists.

To invert the matrix

$$\mathbf{A} = \begin{bmatrix} 1 & 1 & 0 \\ 1 & 0 & 1 \\ 1 & 2 & 2 \end{bmatrix}$$

first write the *augmented* matrix

$$\begin{bmatrix} 1 & 1 & 0 & 1 & 0 & 0 \\ 1 & 0 & 1 & 0 & 1 & 0 \\ 1 & 2 & 2 & 0 & 0 & 1 \end{bmatrix}$$

The elements b_{ij} of the inverse matrix $\mathbf{A}^{-1}$ may be obtained by *successive* applications of the following equations on the augmented matrix:

$$b_{i-1,j-1} = a_{ij} - \frac{a_{1j}a_{i1}}{a_{11}} \qquad \begin{cases} 1 < i \leq n \\ 1 < j \leq m \\ a_{11} \neq 0 \end{cases} \tag{B-4}$$

$$b_{n,j-1} = \frac{a_{1j}}{a_{11}} \qquad \begin{cases} 1 < j \leq m \\ a_{11} \neq 0 \end{cases} \tag{B-5}$$

where m = maximum column number
n = maximum row number
i = row number of old matrix **A**
j = column number of old matrix **A**
a = an element of old matrix **A**
b = an element of new matrix **B**

Three successive applications of Eqs. B-4 and B-5 (starting with the augmented matrix) yield

$$\mathbf{A}^{-1} = \begin{bmatrix} \frac{2}{3} & \frac{2}{3} & -\frac{1}{3} \\ \frac{1}{3} & -\frac{2}{3} & \frac{1}{3} \\ -\frac{2}{3} & \frac{1}{3} & \frac{1}{3} \end{bmatrix}$$

B-3 TRANSPOSE OF A MATRIX

The transpose $\mathbf{A}^T$ of $\mathbf{A}$ is obtained by interchanging the rows and columns of $\mathbf{A}$. That is, if

$$\mathbf{A} = \begin{bmatrix} a_{11} & a_{12} & a_{13} \\ a_{21} & a_{22} & a_{23} \\ a_{31} & a_{32} & a_{33} \end{bmatrix}$$

then

$$\mathbf{A}^{\mathrm{T}} = \begin{bmatrix} a_{11} & a_{21} & a_{31} \\ a_{12} & a_{22} & a_{32} \\ a_{13} & a_{23} & a_{33} \end{bmatrix}$$

If $\mathbf{A}$ is a *symmetric* matrix ($a_{ij} = a_{ji}$), then

$$\mathbf{A} = \mathbf{A}^{\mathrm{T}}$$

From the definition of the transpose of a matrix, it follows that

$$[\mathbf{AB}]^{\mathrm{T}} = \mathbf{B}^{\mathrm{T}}\mathbf{A}^{\mathrm{T}} \tag{B-6}$$

B-4 ORTHOGONALITY PRINCIPLE OF SYMMETRIC MATRICES

The matrix equation for a set of n homogeneous algebraic equations may be written as

$$[\mathbf{A} - \lambda\mathbf{I}]\{\mathbf{X}\} = \mathbf{0}$$

or

$$\mathbf{AX} = \lambda\mathbf{X} \tag{B-7}$$

where $\mathbf{A}$ is a square symmetric matrix and the λ's are the eigenvalues corresponding to the eigenvectors $\mathbf{X}$. For the ith and jth eigenvalues and corresponding eigenvectors, we may write from Eq. B-7 the following:

$$\mathbf{AX}_i = \lambda_i\mathbf{X}_i \tag{B-8}$$

and

$$\mathbf{AX}_j = \lambda_j\mathbf{X}_j \tag{B-9}$$

Postmultiplying the transpose of Eq. B-8 by $\mathbf{X}_j$, gives

$$[\mathbf{AX}_i]^{\mathrm{T}}\mathbf{X}_j = \lambda_i\mathbf{X}_i^{\mathrm{T}}\mathbf{X}_j \tag{B-10}$$

From Eq. B-6, we note that $[\mathbf{AX}_i]^{\mathrm{T}} = \mathbf{X}_i^{\mathrm{T}}\mathbf{A}^{\mathrm{T}}$. Thus, Eq. B-10 may be written as

$$\mathbf{X}_i^{\mathrm{T}}\mathbf{A}^{\mathrm{T}}\mathbf{X}_j = \lambda_i\mathbf{X}_i^{\mathrm{T}}\mathbf{X}_j \tag{B-11}$$

Premultiplying Eq. B-9 by $\mathbf{X}_i^{\mathrm{T}}$ gives

$$\mathbf{X}_i^{\mathrm{T}}\mathbf{AX}_j = \lambda_j\mathbf{X}_i^{\mathrm{T}}\mathbf{X}_j \tag{B-12}$$

Noting that $\mathbf{A}^T = \mathbf{A}$, since $\mathbf{A}$ is a symmetric matrix, we obtain upon subtracting Eq. B-12 from Eq. B-11

$$(\lambda_i - \lambda_j)\mathbf{X}_i^T\mathbf{X}_j = 0$$

Since $\lambda_i \neq \lambda_j$, it follows that

$$\mathbf{X}_i^T\mathbf{X}_j = 0 \tag{B-13}$$

which states that the eigenvectors $\mathbf{X}_i$ and $\mathbf{X}_j$ are orthogonal. In summary,

$$\mathbf{X}_i^T\mathbf{X}_j = \begin{Bmatrix} 0 & i \neq j \\ G_{ii} & i = j \end{Bmatrix} \tag{B-14}$$

With $i = j$, G_{ii} is simply equal to the sum of the squares of the components of $\mathbf{X}_i$.

B-5 ORTHOGONALITY PRINCIPLE OF THE FORM AX = λBX

For many physical problems, the matrix form of a set of homogeneous equations does not occur in the standard form of Eq. B-7, but instead in the form

$$\mathbf{AX} = \lambda\mathbf{BX} \tag{B-15}$$

where $\mathbf{A}$ and $\mathbf{B}$ are square symmetric matrices. By following the procedure of B-4, it is easy to show that the orthogonality of the eigenvectors for matrices of the form of Eq. B-15 is

$$\mathbf{X}_i^T\mathbf{B}\mathbf{X}_j = \begin{Bmatrix} 0 & i \neq j \\ M_{ii} & i = j \end{Bmatrix} \tag{B-16}$$

Index

Absolute
  displacement, 412
  maximum relative displacement, 259
  maximum response, 36, 170, 441, 452
  maximum value, 36
  motion, 162, 164
  relative displacements, 412
Absorbers, 415, 421, 422
Accelerometer, 32, 190, 192
  compression type, 194
  high frequency, 192
  isoshear, 194
  piezoelectric, 32, 191, 193
  response, 193
  shear type, 194
Accelogram, 260
Airplane, 347, 447
Ai-Ting-Yua, 415
Algebra, matrix, 635
Algebraic equations, homogeneous, 313, 637
Algorithms
  Cooley-Tukey, 19
  Gauss-Jordan, 302
  Q-L, 343
Amplitude, 5, 69, 86, 97, 169
  curve, 86, 90, 93
  decay, 60, 72
  steady-state, 142
Angular
  deformation, 498
  displacements, 287, 490
  inertia effect, 51
  joint displacements, 498
  momentum, 50
  motion, 85, 145, 152, 287
Antenna, 159
ARBGEN.FOR program, 265
Arbitrary
  force, 228, 230
  function generator, 261, 452
  support acceleration, 243
Area moment of inertia, 78, 79
Assembling finite elements, 507, 512, 542
Augmented matrix, 636
Average damping factor, 68
Axes
  global, 507, 509
  local, 507, 509
Axial
  beam constraints, 109, 111
  deformation, 491, 535, 557
  extensions, 94, 109
  joint displacements, 492, 505
  loading, 74, 101
  rod deformations, 491
Axial vibration-slender rods, 379, 381, 591, 596

Bandwidth, 23
Beam elements. *See* Finite beam elements
Beams, 89
  axial extensions, 94

Beams (*Continued*)
cantilever, 7, 69, 76, 78, 98, 389, 421
composite, 103, 106
continuous, 532
deflection equations (Table 2-1), 78
fixed-fixed, 79, 80, 94
fixed-pinned, 80, 606
free-free, 528, 608
nonuniform, 98, 388, 518
of two materials, 103, 106
overhanging, 79, 300, 385
pinned-pinned, 78, 110, 604, 619, 622
simply supported, 63, 78, 107, 110, 238, 386, 437, 441
tapered, 389, 507
uniform, 91, 94, 111
Beam equations
nonuniform beams, 91
uniform beams, 90, 600
Beam shape functions, 90, 92, 94, 109, 501 (Table 8-1), 608
Beating, 160, 162
Bending deformations, 499, 505
Berns, H. D., 197
Best fit, 21
Bierman, R. L., 208
Blevins, R. D., 155, 157
Boat-and-trailer, 168
Boundary conditions
continuous systems, 594, 597, 598, 604, 606
finite element method, 500, 517
Boundary-value problems, 584
Buckling load, 101, 103

Cables, 358, 612, 614
Cams, 18, 174
Cannon, Robert H., Jr., 634
Cantilever beam, 7, 69, 76, 78, 98, 389, 421
Characteristic equations, 56
polynomial, 313, 315, 322
transcendental, 595, 606
Charge amplifier, 32, 189
Chen, P. S., 195
Choice of stiffness or flexibility matrix, 300
Cholesky decomposition, 367
Circular frequency, 6, 54, 62
damped, 62
undamped, 62
Coefficients
damping, 44, 305, 306, 434
flexibility, 283, 294, 324, 325, 327
influence, 283, 287, 288, 290, 292, 294, 324, 325, 327, 443
stiffness, 284, 287, 288, 290, 292, 327, 443
Column, 74, 101
Combined
bending/axial deformation, 505, 506
motion, 140
Complementary solutions, 71, 135
Complex
algebra, 8, 137
constants, 137, 603
elastic moduli, 207, 209, 590
exciting force, 139
Fourier series, 24, 174
notation, 8, 137
quantity, 8, 139
response, 139
shear moduli, 210
spring constants, 207
stiffness quantities, 207
vector, 8
Composite
beams, 103, 106
cross-sections, 105, 107
Compression accelerometer, 194
Computer function subprograms
F(T,I), 424, 428, 432, 433, 461
FUNCT(T), 261, 266, 461
FUNCT(T, R, DISP, VEL), 460
STEP(T,T1), 251
Computer programs
ARBGEN.FOR, 265
DEFLATEA.FOR, 352, 358, 360, 436
FINITEL.FOR, 541
INVERT.FOR, 302
JACOBI.FOR, 368, 436
MATMULT.FOR, 338
MODALEQ.FOR, 457, 461
SIM2.FOR, 425
SINEBASE.FOR, 257
STEP.FOR, 251
TRUSS.FOR, 561
Computer solutions-comments, 456
Computer subroutine subprograms
DECOMP, 374, 376, 544, 553, 563
JCBI, 544, 551, 563, 568
MATINV, 374, 544, 554, 563
MATMPY, 376, 544, 555, 563
POWERA, 352, 356
RKSFX, 423, 428
SEARCH, 376, 544, 556, 563
Computer subroutine file
LIBRARY.FOR, 544, 551
Conservative systems, 44
Constraint equations, 55, 484, 487, 490
Continuous frequency spectrum, 23, 28

Continuous systems, 3, 408, 584, 612
Conversion of units. *See* Inside cover
Convolution integral, 231, 235
Cooley, J. W., 19
Cooley-Tukey algorithm, 19
Coordinate coupling, 308
  dynamic, 308
  static, 308, 311
Coordinates
  generalized, 278, 287, 484, 500, 513, 529
  independent, 2, 278
  normal, 324, 334
  principal (normal), 324, 333, 334
  spatial, 584, 585
Coordinate transformations, 507
Coulomb damping, 70, 206
Coupled equations, 279, 310, 423, 516
Coupling
  dynamic, 308, 333
  static, 308, 311, 333
Crede, G. E., 247, 259, 420
Critical
  buckling load, 101, 103
  damping, 57
  speed, 147, 154
CSMP, 414
Curve fitting, 68

D'Alembert's principle, 51, 136, 585
Damped free vibration, 43, 60, 68, 72, 340
Damping
  average damping factor, 68
  coefficient, 44, 305, 306, 434
  Coulomb, 70
  critical, 57
  displacement, 207
  electrical, 192
  equivalent viscous, 201
  experimental measurement of, 66, 68
  factor, 44, 58, 62, 66, 69, 414, 434
  fluid, 203, 205
  forces, 44, 484
  frequency phase method, 197
  hysteretic, 207
  Kelvin model, 590
  logarithmic decrement, 66, 69
  matrix, 281, 306, 414, 431, 434, 517
  nonlinear, 203
  overdamped, 56
  proportional, 341
  quality factor, 199
  real, 59
  solid, 207
  structural, 207
  torsional, 145
  undamped, 59
  underdamped, 57
  velocity-squared, 203, 205
  viscoelastic, 590, 593
  viscous, 43, 44, 55, 60, 68, 169, 306, 614
Damping factor
  average, 68
  modal, 341
Decay, 60, 72
Decibel, 34
DECOMP subroutine subprogram, 374, 376, 544, 553, 563
Decoupling equations of motion, 329, 333
DEFLATEA.FOR program, 352, 358, 360, 436
Deflection curves, 102, 111
Deflection equations, table of, 78
Deformation
  angular, 498
  axial, 491, 505, 506, 535, 557
  bending, 499, 505, 506
  bending and axial, 505, 506
  shear, 455
  torsional, 498
Degrees-of-freedom
  definition of, 2
  infinite, 90
  multiple, 2, 3
Delta function, 29
Den Hartog, J. P., 157
Determinant, 313, 318
Diagonal matrix, 282, 312, 324, 329
Differential
  displacement, 484, 497, 504
  element, 84, 86, 601
  translation, 486
  work, 484, 486, 488, 497, 504, 615, 623
Digital frequency
  analyzer, 19
  counter, 196
Dirac delta function, 29
Discrete systems, 3, 408, 482
Displacement damping, 207
Displacement function, 17, 81, 83, 86, 93, 174, 497
Distributed parameters, 3
Domain
  frequency, 3
  time, 3, 253
Double-bar system, 290, 292
Double pendulum, 278
Drag coefficient, 156
Dry friction. *See* Coulomb damping
Duhamel's integral, 231, 235

Dynamic
  coupling, 308, 311, 333
  equilibrium, 51, 136
  matrix, 313

Earthquake
  response by modal analysis, 450, 452
  spectrums, 259, 260, 262, 452
  structural response to, 259, 450, 452
Effective
  mass, 87
  spring constant, 151
Eigenvalue problems-general procedures, 342
  deflation procedure, 363
  Hotelling's deflation method, 343, 349
  Jacobi's method, 343, 365
  polynomial method, 312
  power method, 343
Eigenvalues, 56, 311, 318, 324, 637
  ascending order, 350
  DEFLATEA.FOR program, 352, 358, 360, 436
  descending order, 363
  intermediate, 349
  JACOBI.FOR program, 368, 436
  largest, 344, 347
  lowest, 344, 345
Eigenvectors, 311, 322, 324, 637
  DEFLATEA.FOR program, 352, 358, 360, 436
  JACOBI.FOR program, 368, 436
  normalizing, 319, 323, 327, 352
  normalizing factor, 319
  orthogonality of, 637, 638
Elastic curve, 102
Elastic elements, 44, 74, 75, 78
Elastic moduli, 91, 98, 107, 207
Elastic forces, 47
Elastomers, 208, 210
El Centro earthquake, 453
Electrical damping, 192
Elements of physical models, 44
Energy
  dissipated by viscous damping, 202
  dissipation, 43
  kinetic, 81, 87, 93, 95, 100, 101, 153, 485, 488, 490, 493, 514, 613, 617
  potential, 82, 85, 87, 92, 95, 101, 153, 485, 488, 490, 494
  strain, 82, 87, 92, 95, 101, 153, 494, 514, 612, 617, 622
Equivalent
  cross sections, 105, 107
  damping coefficients, 177
  spring constants, 75, 144, 452
  springs, 74
  stiffness, 177
  viscous damping, 201, 206
  viscous damping coefficient, 206
Euclidean space, 365
Euler's
  beam equation, 601
  equation, 9
Examples
  1-1. Square wave-Fourier series, 16
  1-2. Sawtooth wave-Fourier series, 17
  1-3. Rectangular pulse-Fourier transform, 27
  1-4. Delta function-Fourier transform, 29
  1-5. Triangular wave (mean-square/root-mean-square value), 36
  2-1. Dynamic equilibrium, 54
  2-2. Damped/undamped natural frequencies, 61
  2-3. Response-initial conditions, 63
  2-4. Experimental damping factor, 68
  2-5. Springs in series, 76
  2-6. Rayleigh's energy method, 82 (rack-and-gear system)
  2-7. Rayleigh's energy method, 86 (rod-and-mass system)
  2-8. Rayleigh's energy method, 87 (inverted pendulum)
  2-9. Rayleigh's energy method, 94 (fixed-fixed beam)
  2-10. Rayleigh's energy method, 98 (nonuniform cantilever beam)
  2-11. Rayleigh's energy method, 101 (slender column)
  2-12. Beams of two materials, 106
  2-13. Effects of axial constraints, 110
  3-1. Harmonic force excitation, 143 (frame system)
  3-2. Harmonic force excitation, 145 (wind turbine system)
  3-3. Critical speed, 152 (disk-and-shaft system)
  3-4. Flow-induced vibration, 159
  3-5. Harmonic support excitation, 168 (boat-and-trailer system)
  3-6. Periodic support excitation, 171 (boat-and-trailer system)
  3-7. Periodic support excitation, 173 (cam-follower system)
  3-8. Vibration isolation, 183 (unbalanced rotating element)
  3-9. Vibration isolation, 183 (ship's rolling motion)
  3-10. Vibration isolation, 184 (changing mass of system)

3-11. Velocity-squared damping, 205 (equivalent viscous damping)
4-1. Transient response, 238 (step-function force on beam)
4-2. Transient response, 243 (sine-pulse force on system)
4-3. Sine-pulse force excitation, 253 (modification of STEP.FOR program)
4-4. Sine-pulse support excitation, 255 (modification of STEP.FOR program)
4-5. Arbitrary excitation, 263 (use of ARBGEN.FOR program)
5-1. Deriving differential equations, 279 (Newton's second law)
5-2. Differential equations in matrix form, 281
5-3. Deriving differential equations, 282 (Newton's second law)
5-4. Determining stiffness matrix, 284 (three-story building)
5-5. Determining stiffness matrix, 286 (disk-and-shaft system)
5-6. Determining stiffness matrix, 288 (five-story building)
5-7. Determining stiffness matrix, 290 (double-bar system)
5-8. Determining stiffness matrix, 292 (with gravity forces)
5-9. Determining flexibility matrix, 296 (three-story building)
5-10. Determining flexibility matrix, 297 (cantilever-beam-and-disk system)
5-11. Determining damping matrix, 306 (three-story building)
5-12. Determining eigenvalues/eigenvectors, 314 (shaft-and-disk system)
5-13. Determining natural frequencies/mode shapes, 317 (free–free shaft-and-disk system)
5-14. Determining roots of frequency equation, 320 (incremental search method)
5-15. Use of flexibility coefficients, 325 (shaft-and-disk system)
5-16. Generalized mass/stiffness elements, 331 (shaft-and-disk system)
5-17. Orthogonality relationships among modes, 331 (three-story building)
5-18. Time response by modal analysis, 337 (three-story building)
5-19. Determining eigenvalues by power method, 345 (spring-and-mass system)
5-20. Determining eigenvalues by power method, 346 (lumped-mass model of airplane)
5-21. Use of DEFLATEA.FOR program, 357 (lumped-mass models of tightly stretched cable)
5-22. Use of JACOBI.FOR program, 381 (lumped-mass model of rod)
5-23. Determining mass/stiffness elements, 383 (lumped-mass model, nonuniform rod)
5-24. Determining mass matrix, 385 (lumped-mass model, overhanging beam)
5-25. Determining natural frequencies/mode shapes, 386 (lumped-mass model, simply supported beam)
6-1. Nonpotential excitation vector, 410 (lumped-mass model of car)
6-2. Design of vibration absorber, 420 (motor-and-pump system)
6-3. Use of SIM2.FOR program, 430 (time response of four-story building)
6-4. Use of SIM2.FOR program, 432 (modification of function subprogram F(T,I))
6-5. Use of SIM2.FOR program, 433 (five-degree-of-freedom system)
6-6. Time response by modal analysis, 437 (lumped-mass model, simply supported beam)
6-7. Use of modal analysis/response spectrum, 440 (lumped-mass model, simply supported beam)
6-8. Steady-state response by modal analysis, 443 (damped shaft-and-disk system)
6-9. Steady-state response by modal analysis, 447 (lumped-mass model of airplane)
6-10. Earthquake excitation-modal analysis, 452 (five-story building)
6-11. Time response using MODALEQ.FOR program, 461 (tornadic wind on five-story building)
7-1. Differential equations by Lagrange's equations, 485 (damped pulley system)
7-2. Differential equations by Lagrange's equations, 487 (trolley-pendulum system)
7-3. Differential equations by La-

Examples (*Continued*)
grange's equations, 489 (shaft-and-gear system)
7-4. Global mass/stiffness matrices, 511 (for finite beam elements)
7-5. Differential equations by finite element method, 521 (uniform rod)
7-6. Forced response by modal analysis, 525 (uniform rod)
7-7. Finite element method/natural frequencies, 528 (lateral vibration of rocket)
7-8. Finite element method/natural frequencies, 532 (two-span continuous beam)
7-9. Finite element method/natural frequencies, 534 (frame)
7-10. FINITEL.FOR program/natural frequencies, 556 (two-bay frame system)
7-11. FINITEL.FOR program/natural frequencies, 559 (rocket)
7-12. FINITEL.FOR program/natural frequencies, 560 (axial vibration of a rod)
7-13. TRUSS.FOR program/natural frequencies, 569 (plane truss)
7-14. Response of truss by modal analysis, 570 (suddenly applied load on plane truss)
8-1. Motion of tightly stretched string, 589 (with initial configuration)
8-2. Determining natural frequencies/mode shapes, 593 (fixed-free viscoelastic rod)
8-3. Transcendental frequency equation, 596 (slender rod-and-mass system)
8-4. Steady-state response by modal superposition, 614 (tightly stretched cable)
8-5. Flexural stress due to impact, 618 (simply supported beam)
8-6. Forced response of beam on elastic foundation, 622 (simply supported beam)
A-1. Forced response using Laplace transforms (single-degree-of-freedom system)

Excitation
arbitrary function, 243
forces, 140, 410, 434, 436
general periodic, 170
harmonic force, 134, 414
harmonic support, 134, 414
nonharmonic force, 227
nonharmonic support, 227, 241
nonpotential forces/moments, 484
primary, 182
sawtooth, 174
secondary, 182
sine-pulse, 254, 430, 432
step-function, 228, 234, 246, 332, 441
support, 162, 227, 255, 264, 410, 412, 445, 447, 448
unbalanced rotating masses, 136, 140, 144, 145, 152, 183

Experimental damping measurement, 66, 68, 434
Exponential decay, 60

Faddeev, D. K., 349
Faddeeva, U. N., 349
Fast-Fourier transform, 19
Felgar, R. P., Jr., 609
FFT frequency analyzer, 19, 34
Fickes, J. D., 208
Finite beam elements, 499, 505, 528
bending/axial deformation, 505, 506
bending deformation, 499, 506
mass matrices of, 503, 506
stiffness matrices of, 504, 506
Finite element method, 482
mass matrices of, 503, 506
outline for use of, 518
stiffness matrices of, 504, 506
Finite elements
assembly of, 507, 512, 542
beam-axial/bending, 505, 506
beam-bending, 499, 506
rod-axial, 491, 521
rod-torsional, 498
truss, 562
FINITEL.FOR program, 541
Finite rod elements, 491, 521, 560
angular deformation, 499
axial deformation, 491
mass matrix of, 496, 506
stiffness matrix of, 496, 506
Fixed-axis rotation, 48, 50, 53, 88, 292
Fixed-fixed beam, 79, 80, 94, 608
Fixed-free beam, 608
Fixed-pinned beam, 80, 606, 608
Flat response, 186
Flexibility
coefficients, 283, 294, 324, 327, 328
matrix, 283, 295, 300, 345

Flexural vibration of beams, 617
Flow-induced vibration, 155, 158
Fluid damping, 203, 205
Forced
- response, 135, 414, 436, 520
- vibration, 1, 408, 414

Forces
- arbitrary, 228, 230
- axial, 74
- damping, 44, 484
- dynamic, 65, 134
- elastic, 47, 413
- excitation, 140, 410
- flow-induced, 155
- global, 510
- gravity, 47, 50, 87, 292
- harmonic, 134
- impulsive, 228, 229
- joint, 492, 497, 510, 516
- nonharmonic, 227
- nonpotential, 44, 484
- periodic, 136
- potential, 484
- pulse, 27, 243, 430, 432
- shear, 455
- sine pulse, 254, 430, 432
- springs, 47, 54, 74
- static, 65
- step function, 228, 232, 234, 246, 441
- suddenly applied, 228, 232, 234, 437, 622
- unbalanced, 136, 144, 145, 152, 183
- wind, 461, 611

Force transducer, 186, 188
Forray, M. O., 483, 514
FORTRAN programs. *See* Computer programs
Fourier integral, 22, 25
Fourier series, 11
- coefficients, 13
- complex form, 24
- even functions, 14
- full-range expansions, 11, 18
- half-range expansions, 14
- odd functions, 14, 17
- orthogonal relations, 10, 21, 588, 620
- truncated, 21

Fourier transform pair, 22, 26
Fourier transforms, 22
Frame, 144, 534, 541, 556
Free-free beam, 528, 608
Free-free system, 301, 317, 347, 364, 443, 529, 608
Freeman, W. H., 349
Free vibration
- damped, 1, 43, 60, 68, 72, 281, 340
- multiple-degree-of-freedom systems, 277
- record, 68
- undamped, 59, 282, 333, 433

Frequency, 5, 59
- analyzers, 19
- circular, 5, 6, 59
- damped, 59
- domain, 3
- equation, 313, 315, 321, 595, 606
- ratio, 187
- spectrum, 4, 17, 19, 20, 24, 28
- torsional, 157
- undamped, 59
- vortex shedding, 160

Frequency equations, 56
- polynomial, 313, 315, 322
- transcendental, 595, 606

Frequency-phase method, 197
Friction
- coefficient of, 70
- Coulomb, 70
- fluid, 203, 205
- internal, 207

Full matrix, 491
Full-range expansion/Fourier series, 11
F(T,I) function subprogram, 424, 428, 432, 433, 461
FUNCT(T), 261, 266, 461
FUNCT(T,R,DISP,VEL), 460
Function generator, arbitrary, 261
Function subprograms. *See* Computer function subprograms

Gammell, L. W., 120
Gauss-Jordan algorithm, 302
General comments-computer solutions, 456
General discussion/forced-vibration response, 414
Generalized
- coordinates, 278, 287, 484, 500, 513, 516, 529
- mass, 330, 434, 526
- nonpotential force, 484, 526, 615, 623
- nonpotential moment, 484
- stiffness, 330

General methods-eigenvalue problems, 342
General modal analysis procedure
- force excitation, 436
- support excitation, 445

General periodic excitation, 170
General plane motion, 53, 290

Global
  axes, 507
  forces, 510
  joint displacements, 508, 513, 529, 535
  mass matrices, 510, 511, 515, 523, 530, 536
  moments, 510
  stiffness matrices, 510, 511, 515, 523, 530, 536
Gourlay, A. R., 541
Gravity forces, 47, 50, 87, 292
Gretz, J. L., 120

Haack, D. C., 89
Half-power points, 189
Half-range expansions/Fourier series, 14
Hallowell, F. C., 415
Harmonic
  force excitations, 134, 414
  moment excitations, 134, 414
  motion, 3, 6, 7
  support excitations, 134, 414
Harris, C. M., 247, 259, 420
Helical spoilers, 159
Helical spring surge, 598
Hetenyi, M., 420
Higdon, A., 104
High-frequency transducer, 192
Hillberry, B. M., 208
Homogeneous
  algebraic equations, 313, 637, 638
  solutions, 71, 135, 137
Hotelling's deflation method, 343, 349
Housner, G. W., 259
Huebner, K. H., 483
Hysteretic damping, 207

Ideal impulse, 29
Ideal impulse function, 29
Identity matrix, 298, 324, 510
Ill-conditioned eigenvalue problem, 541
Imaginary part of vector, 9, 137
Impact, 63, 618
  hammer, 30, 32
  plastic, 618
Impulse, unit, 29, 31
Impulse function, 28
  ideal, 29
  unit, 28, 29
Impulsive forces, 228, 229
Incremental-search method, 321
Independent coordinates, 2, 278, 484
Inertia
  couple, 51, 52, 55
  effect, 45, 51, 55

Inertial frame of reference, 412
Influence coefficients, 283, 287, 290, 292, 294, 324, 327, 443
  damping, 305, 434
  flexibility, 283, 294, 324, 327
  stiffness, 284, 287, 290, 292, 327, 443
Initial
  conditions, 62, 72, 433, 439, 619, 624
  phase, 5
Instrumentation, 32
Intermediate eigenvalues, 349
Inverse of
  flexibility matrix, 295, 302
  mass matrix, 311, 312
  modal matrix, 338
  stiffness matrix, 295, 302
Inversion of a matrix, 302, 636
Inverted pendulum, 50, 87
INVERT.FOR program, 302
Isolation
  basic concepts, 179
  mounts, 177
Isoshear accelerometer, 194

JACOBI.FOR program, 368, 369
  used as subprogram JCBI, 544, 551, 563, 568
Jacobi's method, 364, 365, 378
JCBI subroutine subprogram, 544, 551, 563, 568
Joint, 482
  displacements, 492, 498, 505, 535
  forces, 492, 497, 510, 516
  moments, 484, 516, 533
  number matrix, 542
  torques, 499
Joint displacements
  angular, 498, 500
  axial, 492, 505
  global, 508, 513, 529, 535
  lateral, 500, 505
  local, 492, 493, 505, 535

Kelvin model, 590
Kimball, A. L., 207
Kinetic energy, 81, 87, 93, 95, 100, 153, 485, 488, 490, 493, 613, 617

Lagrange's equations, 483, 485, 488, 612
Laplace transform pair, 27
Laplace transforms, 26, 631, 632
Lateral joint displacements, 500
Lazan fatigue-testing machine, 213
Least-squares error fit, 21
LIBRARY.FOR subroutine file, 544, 551

Linear
  shape functions, 86, 492, 495
  system, 82
  transformation, 333, 434
  variable differential transformer, 144
Load cells, 186
Lobes (side), 32
Local
  axes, 507
  mass matrices, 491, 493, 496, 503, 506, 522, 530
  stiffness matrices, 491, 493, 496, 504, 506, 522, 530
Logarithmic decrement, 66, 69
Log-log graph, 182
Longitudinal vibration of slender rods, 379, 381, 591, 596, 616
Loss modulus, 209
Low-frequency transducer, 191
Lumped mass models of
  beams, 4, 379, 385, 386
  rods, 379, 380, 381, 383
  tightly stretched cable, 357, 612, 614
Lumped-mass systems definition, 277, 408
Lutes, L. D., 415
LVDT, 144

Magnification factor, 141, 186
Maheshwari, M., 259
Mass-effective, 87
Mass moment of inertia, 50, 83, 87, 620
Mass ratio, 417
Mathematical concepts
  complex algebra, 8
  complex Fourier series, 24
  complex notation, 8
  delta function, 29
  Duhamel's integral, 231, 235
  Euler's equation, 9
  Fourier integral, 22, 25
  Fourier series, 11
  Fourier transform, 22
  Fourier transform pair, 22, 26
  harmonic motion, 3, 6
  Laplace transform, 26
  Laplace transform pair, 27
  matrix algebra, 4
  orthogonal relations, 10, 21, 588, 620
  periodic functions, 12
  periodic motion, 11
  sinusoidal waves, 7
Mathematical models, 45, 137
MATINV subroutine subprogram, 374, 544, 554, 563
MATMPY subroutine subprogram, 376, 544, 555, 563
MATMULT.FOR program, 338
Matrices
  augmented, 636
  banded, 282
  damping, 281, 305, 414, 431, 434, 517
  diagonal, 282, 312, 324, 329
  flexibility, 283, 295, 300, 345
  full, 491
  generalized mass, 330, 434, 526
  generalized stiffness, 330
  global mass, 510, 515, 530, 537
  global stiffness, 510, 515, 530, 537
  identity, 298, 324, 510
  inverse of flexibility, 295, 302
  inverse of mass, 311, 312
  inverse of modal, 338
  joint number, 542
  local mass, 491, 493, 496, 503, 522, 530
  local stiffness, 491, 493, 496, 504, 522, 530
  mass, 281, 382, 491, 493, 496
  modal, 323, 334, 434, 438
  node-number, 542
  nondiagonal, 282, 328, 333
  null, 282
  orthogonal, 510
  plane rotation, 365
  positive definite, 366
  rectangular, 513, 536
  stiffness, 281, 284, 288, 290, 292, 300, 449, 491, 493, 496
  symmetric, 282, 296, 491, 637
  symmetric positive definite, 367
  system mass, 514, 523, 530, 537, 571
  system stiffness, 514, 524, 530, 538
  transformation, 507, 509, 511, 562
  transpose of, 637
  transpose of modal, 334
  transpose of rotation, 365
  transpose of transformation, 510
Matrix
  algebra, 4, 635
  inversion, 302, 636
  multiplication, 338, 351, 635
Mean-square value, 34, 35
Measurements
  frequency-phase method, 197
  quality factor, 199
  steady-state vibration, 194
  system damping, 66
Modal
  analysis, 333, 408, 414, 433, 450, 525
  damping factors, 341, 443

Modal (*Continued*)
  frequencies, 33
  matrix, 323, 334, 434, 438
  matrix transpose, 334
  superposition method, 414, 433, 453, 461, 612
MODALEQ.FOR program, 456, 461
Mode, 319, 440
  participation factor, 435, 446, 454
  rigid-body, 319, 348, 364, 443, 531, 560
  shapes, 90, 241, 311, 316, 322, 326, 349, 363, 388, 525, 532, 534, 540, 558, 594, 605, 607
Models
  lumped-mass, 3, 4, 357, 379, 380, 383, 385, 386
  mathematical, 45, 137
  physical, 44
Modes of vibration, 90, 198
Modulus
  complex elastic, 207, 210
  complex shear, 210
  elastic, 91, 98
  loss, 209
  shear, 210
  storage, 209
  Young's, 91, 98
Motion
  absolute, 162, 164
  angular, 85, 145, 152, 287
  combined, 140
  general plane, 53, 290
  harmonic, 3, 6, 7
  periodic, 11
  relative, 166, 167
  rotational, 148, 245
  simple harmonic, 7
  support, 162, 227, 255, 264

Natural
  circular frequencies, 6, 59, 77
  damped frequency, 61
  damped period, 61, 77
  frequencies, 5, 59, 61, 77, 80, 311, 381
  mode shapes. *See* Normal mode shapes
  motion, 43
  period, 59, 77
  response, 135
  undamped frequency, 59
  undamped period, 59, 61
Neutral
  axis, 91, 107
  equilibrium, 73
Newton's second law, 47, 241, 278, 279, 601
Nodal points, 90, 198, 440
Node-number matrix, 542
Nodes, 440, 482
Nondiagonal matrix, 282, 328, 333
Nonlinear
  damping, 203
  differential equations, 203, 489
Nonharmonic
  forces, 227, 229
  support excitation, 227, 241
Nonpotential
  damping forces, 484
  damping moments, 484
  excitation forces, 484
  excitation moments, 484
  forces, 484, 526, 615, 623
  joint forces, 497, 504
  joint moments, 490, 504
Nonuniform
  beam, 94, 98, 388, 518
  rods, 383
Normal
  coordinates, 324, 334
  mode shapes, 90, 241, 311, 316, 322, 326, 349, 363, 388, 525, 532, 534, 540, 558, 594, 605, 607
Normalizing
  eigenvector components, 319, 327
  factors, 319, 352
Normal mode functions, 595, 605, 608, 612, 617, 622
  orthogonal properties of, 609
  orthogonal relationships, 610, 611
  properties of, 609
  Table 8-1, 608
Normal modes, orthogonal properties of, 327, 328, 331, 609
Null matrix, 282
Numerical integration, 253

Ohlsen, E. H., 104
One-dimensional wave equation, 585, 590, 598
Ortega, James, 343
Orthogonal functions, 21, 588, 620
Orthogonality of eigenvectors, 638
Orthogonality principle of symmetric matrices, 637
Orthogonal matrices, 510
Orthogonal properties of
  normal mode functions, 609
  normal modes, 327, 328, 331, 334
  trigonometric functions, 21, 588, 620
Orthogonal relations, 10

Oscillograph record, 68, 70
Oscilloscope, 32
Overdamped systems, 56
Overhanging beam, 79, 300, 385

Pao, Y. C., 208, 483, 541
Parallel springs, 75
Partial differential equations, 584, 600
Particular solutions, 71, 135, 137, 175, 616
Pendulum
  double, 278
  inverted, 50, 87
  spring-restored, 50, 87
Period, 5, 59
Periodic
  excitation, 170
  forces, 136
  functions, 12
  motion, 11
Phase
  angle, 5, 139, 143, 147, 150, 165, 167, 175
  distortion, 189
  initial, 5
  shift, 189
Physical models, 44
Pian, T. H., 415
Piezoelectric accelerometer, 32, 191, 193
Pinned-pinned beam, 79, 110, 604, 608, 619, 622
Plane frames
  computer program, 541
  finite element analysis, 541, 556
Plane rotation matrices, 365
Plane trusses
  computer program, 561
  finite element analysis, 561, 569, 570
Plastic impact, 63, 618
Polynomials
  as characteristic equations, 313, 315, 322
  as shape functions, 501
Positive definite matrix, 366
Postmultiplication, 328, 351, 636, 637
Potential
  energy, 82, 85, 87, 92, 95, 101, 485, 488, 490, 494
  forces, 47, 484
  moments, 484
POWERA subroutine subprogram, 352, 356
Power method, 343
Premultiplication, 312, 318, 334, 351, 434, 511, 526, 636, 637
Principal
  coordinates, 324, 333, 334, 434, 441
  system, 415, 418
Programs. *See* Computer programs
Proportional damping, 341
Pulse
  rectangular, 27
  sine, 243, 430, 432
  train, 22
  unit, 29

Quality factor, 199
QL algorithm, 343

Rack-and-gear system, 82
Ralston, Anthony, 423
Ramp functions, 234, 236, 441
Rate of decay, 61
Rayleigh, J. W. S., 77
Rayleigh's energy method, 77, 83, 152
  applied to beams, 89
  obtaining differential equations, 83
  obtaining natural frequencies, 86, 89, 94, 98, 101, 103, 106, 110
Real part of vector, 137
Rectangular
  matrix, 513, 536
  pulse, 22, 27
  pulse train, 22
Reed, F. Everett, 420
Relative
  average maximum displacement, 228
  displacements, 413
  motion, 166
  velocity, 162, 191
Resonance, 2, 134, 142
Response
  flat, 34, 186
  forced, 135, 414, 436, 520
  free, 135, 137
  natural, 135
  relative steady-state, 168
  root-sum-square, 452
  spectrums, 246, 253, 254, 256, 259, 260, 262, 267, 440, 452
  steady-state, 136, 138, 166, 170, 176, 447, 616
  structural, 450
  total, 135
  transient, 136, 137, 245
Reynolds number, 155
Rigid-body modes, 319, 348, 364, 443, 531, 560
Riley, W. F., 104

Rise time, 234, 441
RKSFX subroutine subprogram, 423
Road contour, 172
Rod elements. *See* Finite rod elements
Rods
  nonuniform, 383
  uniform, 53, 61, 86, 379, 380, 381, 591, 596, 616
Root-mean-square value, 34, 35
Root-sum-square response, 452, 455
Rotating vectors, 4, 9, 139
Rotation, 56
Rubin, S., 247
Runge-Kutta method, 249, 414, 423

Sawtooth function, 17, 174
SEARCH subroutine subprogram, 376, 544, 556, 563
Seismometer, 190
Semidefinite systems, 301, 317, 347, 364, 443, 529
Semilog plot, 67, 70
Separation-of-variables method, 586, 602
Series springs, 75
Shaft-and-disk systems, 148, 152, 287, 298, 301, 314, 317, 443, 489
Shape functions, 81, 83, 87, 90, 92, 94, 97, 98, 109, 153, 491, 501
  beams, 90, 94, 109, 501
  linear, 86, 492, 493, 495
  polynomials, 501
  rods, 491, 492
  static deflection curves, 78, 90, 94
  trigonometric, 90, 94, 97, 98, 101, 493
Shear
  complex modulus, 210
  forces, 455
  type accelerometer, 194
Shock spectrum, 246
Side lobes, 32
Signal conditioner, 189
SIM2.FOR program, 423
Simple harmonic motion, 7
Simply supported beams, 63, 78, 107, 110, 238, 386, 437, 441
Simultaneous solutions, 423
SINEBASE.FOR program, 257
Sine pulse, 243, 430, 432
Sine-pulse force, 254, 430, 432
Single-degree-of-freedom systems, 2, 134
Single-valued function, 11
Sinusoidal waves, 4, 5
Solid damping, 207
Solutions
  complementary, 71, 135
  complete, 138, 139
  homogeneous, 71, 135, 137
  particular, 71, 135, 137, 139, 175
  simultaneous, 423
  steady-state, 140, 147, 149, 176, 447, 614
Solutions using Laplace transforms, 633
Solving differential equations of motion, 55
Spatial coordinates, 58, 584
Spectrum analyzers, 19
Spectrums, 17, 246, 254, 259, 260, 262, 267, 440, 452
Spring-and-mass systems, 44, 47, 81, 86, 135, 163, 165, 178, 187, 280, 301, 308, 345
Spring constants
  complex, 207
  effective, 151
  equivalent, 75, 144, 452
  table of, 78
  torsional, 286
Spring forces, 47, 54, 74
Spring-restored pendulum, 50, 87
Springs
  coil, 78
  combination series/parallel, 75
  equivalent, 74, 452
  helical, 598
  massless, 96, 101
  parallel, 75
  series, 75
  torsional, 78, 145, 286
Spring surge, 598
Square wave, 16
Static coupling, 311, 333
Static deflection curves, 78, 83, 86, 90
Static equilibrium position, 45, 47, 51
Steady-state
  relative response, 168
  response, 136, 170, 176, 447, 616
  vibration characteristics, 140
  vibration measurements, 194
STEP.FOR program, 251
Step function, 228, 232, 234, 246, 441
Stiffness
  coefficients, 284, 287, 288, 290, 292, 327, 443
  global matrices, 510, 511, 515, 523, 530, 536
  local matrices, 491, 493, 496, 504, 506, 522, 530
  matrix, 281, 300, 449, 496, 504
Stiles, W. B., 104
Stockbridge damper, 158
Storage modulus, 209

Strain
  distribution, 109
  energy, 82, 87, 91, 95, 101, 153, 494, 612, 617, 622
Stress
  beams, 104, 621
  normal, 104, 621
  rods, 591
Stress-strain diagram, 92
Striker tips, 32
Strouhal number, 156
Structural
  damping, 58, 207
  earthquake response, 259, 450, 452
Subroutine file, LIBRARY.FOR, 544, 551
Suddenly applied forces, 228, 232, 234, 437, 622
Support excitation, 162, 227, 241, 255, 264, 410, 412, 445, 447, 448
  arbitrary, 245
  nonharmonic, 241
Surge in helical springs, 598
Symmetric matrix, 282, 296, 491, 637
Synchronous whirl, 148, 151
System
  mass matrix, 523, 530, 537, 571
  stiffness matrix, 524, 530, 538
Systems
  absorber, 415
  conservative, 44
  continuous, 3, 408, 584
  discrete, 277, 408, 482
  elements of, 44
  free-free, 301, 317, 347, 364, 443, 528
  linear, 82
  lumped-mass, 4, 277, 408
  principal, 415, 418
  rotational, 245
  semidefinite, 301, 317, 347, 364, 443, 528
  unbalanced, 136, 140, 144, 152, 177, 183, 185
  unconstrained. *See* Free-free systems
  undamped, 59, 135
  unstable, 72, 587

Tables
  2-1, 78
  8-1, 608
  A-1, 632
Tacoma Narrows bridge, 157
Tapered cantilever beam, 389, 583
Taut strings, 585, 589, 612
Thomson, W. T., 634
Thornton, E. A., 483
Tightly stretched cables, 358, 612, 614
Time domain, 3, 253
Timoshenko, S., 593, 611
Torsional
  deformation, 498
  finite rod element, 499
  spring, 78, 145, 286
  vibration, 145, 157, 592, 617
Total response, 135, 137
Transcendental frequency equations, 595, 606
Transducers, 134, 186
  accelerometers, 32, 190, 193
  high-frequency, 192
  force, 186, 188
  load cells, 186
  low-frequency, 191
  LVDT, 144
  seismometer, 190
  vibration-measuring, 190
  vibrometer, 190, 191
Transformation
  linear, 333, 434
  matrix, 507, 509, 511, 562
Transformed cross section, 105, 107
Transform pairs
  Fourier, 22, 24, 26
  Laplace, 27
Transforms
  Fourier, 19
  Laplace, 26
Transient response, 136, 137, 245
Translation, 45, 56, 81
Transmissibility of forces, 177
Transpose of
  a matrix, 637
  modal matrix, 334
  transformation matrix, 510
  rotation matrix, 365
Trigonometric shape functions, 90, 94, 97, 98, 101, 493
Truncated Fourier series, 21
Trusses
  computer program, 561, 563
  finite element analysis of, 561, 569, 570
TRUSS.FOR program, 561, 563
Tukey, J. W., 19

Unbalanced systems, 136, 140, 144, 152, 177, 183, 185
Unconstrained systems. *See* Free-free systems

Uncoupled equations
  damped free vibration, 340
  force excitation, 434
  support excitation, 414, 445
  undamped free vibration, 333, 336, 337
Undamped natural frequency, 59
Undamped systems, 59, 135
Underdamped system, 57
Unit conversions. *See* Inside cover
Unit impulse, 29, 31
Unit impulse function, 29
Unstable system, 72
Use of initial conditions, 62

Velocity-squared damping, 203
Vibration
  absorbers, 415, 421, 422
  analyzers, 19, 34
  isolation, 177, 179
Vibration absorbers, 415
  design of, 418
  cantilever beam, 422
  spring-and-mass, 415
  temporary, 420
Vibrometer, 190, 191
Vigness, I., 247
Virtual displacement, 410, 415
Virtual work, 415
Viscoelastic damping, 590, 593
Viscous damping, 43, 44, 55, 60, 68, 169, 306, 614
  energy dissipated per cycle, 203
  equivalent, 201, 206
  model, 44, 69
Vortex shedding, 155, 158

Watson, Herbert S., 541
Wave equation, 585, 590, 598
  transcendental frequency equations, 595
  viscoelastic rods, 590
Weese, J. A., 104
Wheel rolling without slipping, 53
White noise, 28
Wilf, Herbert S., 423
Williams, Robert C., 349
Wind turbine, 146

Young, D. H., 593, 609, 611
Young's modulus, 91, 98

Zitek, S. J., 208
Zienkiewicz, O. C., 483